Student's Solutions Manual for

Physical Chemistry

Student's Solutions Manual for

Physical Chemistry

Sixth edition

P. W. Atkins
Professor of Chemistry, University of Oxford
and Fellow of Lincoln College

C. A. Trapp
Professor of Chemistry, University of Louisville,
Louisville, Kentucky, USA

M. P. Cady
Associate Professor of Chemistry,
Indiana University Southeast, New Albany, Indiana, USA

C. Giunta
Associate Professor of Chemistry,
Le Moyne College, Syracuse, NY, USA

W. H. FREEMAN AND COMPANY
NEW YORK

ISBN 0-7167-3167-3

Printed in the United States of America

Third Printing, 2000

Preface

This manual provides detailed solutions to all the end-of-chapter (**a**) Exercises and to a selection of slightly more than half the end-of-chapter Problems, including the new Additional Problems. Solutions to Exercises and Problems carried over from the fifth edition have been reworked, modified, or corrected when needed.

The solutions in this edition show a somewhat greater reliance on readily available plotting and statistical software than the previous edition, and this is particularly true of the Additional Problems, but almost all the end-of-chapter Exercises and Problems can still be solved with a modern hand-held scientific calculator. MathCad based solutions to the end-of-part Microprojects are provided in a separate manual.

In general, we have adhered rigorously to the rules for significant figures in displaying the final answers. However, when the results of intermediate answers are shown, they are often given with one more figure than would be justified by the data. These excess digits are indicated with an overline.

We have carefully cross-checked the solutions for errors and expect that most have been eliminated. This process was facilitated by our copy-editor, Claire Eisenhandler, who, in the course of her diligent review of the entire manuscript, found a few errors and ambiguities that we had missed. We would be grateful to any readers who bring any remaining errors to our attention.

We would also like to thank our publishers for their care and patience in guiding this complex, detailed project to completion.

P. W. A.
C. A. T.
M. P. C.
C. G.

Contents

Part 3: Change

Part 1: Equilibrium

1 The properties of gases

Solutions to exercises

E1.1 Boyle's law [6] provides the basis for the solution.

Since $pV = \text{constant}$, $p_f V_f = p_i V_i$

Solving for p_f, $p_f = \dfrac{V_i}{V_f} \times p_i$

$$V_i = 1.0\,\text{L} = 1\overline{000}\,\text{cm}^3, \qquad V_f = 100\,\text{cm}^3, \qquad p_i = 1.00\,\text{atm}$$

$$p_f = \frac{1\overline{000}\,\text{cm}^3}{100\,\text{cm}^3} \times 1.00\,\text{atm} = 10 \times 1.00\,\text{atm} = \boxed{10\,\text{atm}}$$

E1.2 **(a)** The perfect gas equation [12] is: $pV = nRT$

Solving for the pressure gives $p = \dfrac{nRT}{V}$

The amount of xenon is $n = \dfrac{131\,\text{g}}{131\,\text{g mol}^{-1}} = 1.00\,\text{mol}$

$$p = \frac{(1.00\,\text{mol}) \times (0.0821\,\text{L atm K}^{-1}\,\text{mol}^{-1}) \times (298.15\,\text{K})}{1.0\,\text{L}} = \boxed{24\,\text{atm}}$$

That is, the sample would exert a pressure of 24 atm if it were a perfect gas, not 20 atm.

(b) The van der Waals equation [39*a*] for the pressure of a gas is $p = \dfrac{nRT}{V - nb} - \dfrac{an^2}{V^2}$

For xenon, Table 1.6 gives $a = 4.250\,\text{L}^2\,\text{atm mol}^{-2}$ and $b = 5.105 \times 10^{-2}\,\text{L mol}^{-1}$. Inserting these constants, the terms in the equation for p become

$$\frac{nRT}{V - nb} = \frac{(1.00\,\text{mol}) \times (0.08206\,\text{L atm K}^{-1}\,\text{mol}^{-1}) \times (298.15\,\text{K})}{1.0\,\text{L} - \{(1.00\,\text{mol}) \times (5.105 \times 10^{-2}\,\text{L mol}^{-1})\}} = 25.\overline{8}\,\text{atm}$$

$$\frac{an^2}{V^2} = \frac{(4.250\,\text{L}^2\,\text{atm mol}^{-1}) \times (1.00\,\text{mol})^2}{(1.0\,\text{L})^2} = 4.2\overline{50}\,\text{atm}$$

Therefore, $p = 25.\overline{8}\,\text{atm} - 4.2\overline{50}\,\text{atm} = \boxed{22\,\text{atm}}$

E1.3 Boyle's law [6] in the form $p_f V_f = p_i V_i$ can be solved for either initial or final pressure, hence

$$p_i = \frac{V_f}{V_i} \times p_f$$

$$V_f = 4.65\,\text{L}, \qquad V_i = 4.65\,\text{L} + 2.20\,\text{L} = 6.85\,\text{L}, \qquad p_f = 3.78 \times 10^3\,\text{Torr}$$

Therefore,

(a) $$p_i = \left(\frac{4.65\,\text{L}}{6.85\,\text{L}}\right) \times (3.78 \times 10^3\,\text{Torr}) = \boxed{2.57 \times 10^3\,\text{Torr}}$$

(b) Since 1 atm = 760 Torr exactly, $p_i = (2.57 \times 10^3\,\text{Torr}) \times \left(\dfrac{1\,\text{atm}}{760\,\text{Torr}}\right) = \boxed{3.38\,\text{atm}}$

E1.4 Charles's law in the form $V = \text{constant} \times T$ [9] may be rewritten as $\frac{V}{T} = \text{constant}$ or $\frac{V_f}{T_f} = \frac{V_i}{T_i}$

Solving for T_f, $T_f = \frac{V_f}{V_i} \times T_i$, $V_i = 1.0\,\text{L}$, $V_f = 100\,\text{cm}^3$, $T_i = 298\,\text{K}$

$$T_f = \left(\frac{100\,\text{cm}^3}{1000\,\text{cm}^3}\right) \times (298\,\text{K}) = \boxed{30\,\text{K}}$$

E1.5 The perfect gas law, $pV = nRT$ [12], can be rearranged to $\frac{p}{T} = \frac{nR}{V} = \text{constant}$, if n and V are constant. Hence, $\frac{p_f}{T_f} = \frac{p_i}{T_i}$ or, solving for p_f, $p_f = \frac{T_f}{T_i} \times p_i$

Internal pressure = pump pressure + atmospheric pressure

$p_i = 24\,\text{lb in}^{-2} + 14.7\,\text{lb in}^{-2} = 38.\bar{7}\,\text{lb in}^{-2}$ $T_i = 268\,\text{K}\ (-5°\text{C})$, $T_f = 308\,\text{K}\ (35°\text{C})$

$$p_f = \frac{308\,\text{K}}{268\,\text{K}} \times 38.\bar{7}\,\text{lb in}^{-2} = 44.\bar{5}\,\text{lb in}^{-2}$$

Therefore, $p(\text{pump}) = 44.\bar{5}\,\text{lb in}^{-2} - 14.7\,\text{lb in}^{-2} = \boxed{30\,\text{lb in}^{-2}}$

Complications are those factors which destroy the constancy of V or n, such as the change in volume of the tyre, the change in rigidity of the material from which it is made, and loss of pressure by leaks and diffusion.

E1.6 The perfect gas law in the form $p = \frac{nRT}{V}$ [2] is appropriate. T and V are given; n needs to be calculated.

$$n = \frac{0.255\,\text{g}}{20.18\,\text{g mol}^{-1}} = 1.26 \times 10^{-2}\,\text{mol}, \qquad T = 122\,\text{K}, \qquad V = 3.00\,\text{L}$$

Therefore, upon substitution,

$$p = \frac{(1.26 \times 10^{-2}\,\text{mol}) \times (0.08206\,\text{L atm K}^{-1}\,\text{mol}^{-1}) \times (122\,\text{K})}{3.00\,\text{L}} = \boxed{4.20 \times 10^{-2}\,\text{atm}}$$

E1.7 The gas pressure is calculated as the force per unit area that a column of water of height 206.402 cm exerts on the gas due to its weight. The manometer is assumed to have uniform cross-sectional area, A.

Then force, $F = mg$, where m is the mass of the column of water and g is the acceleration of free fall. As in Example 1.2, $m = \rho \times V = \rho \times h \times A$ where $h = 206.402\,\text{cm}$ and A is the cross-sectional area.

$$p = \frac{F}{A} = \frac{\rho h A g}{A} = \rho h g$$

$$p = (0.99707\,\text{g cm}^{-3}) \times \left(\frac{1\,\text{kg}}{10^3\,\text{g}}\right) \times \left(\frac{10^6\,\text{cm}^3}{1\,\text{m}^3}\right) \times (206.402\,\text{cm}) \times \left(\frac{1\,\text{m}}{10^2\,\text{cm}}\right) \times (9.8067\,\text{m s}^{-2})$$

$$= 2.0182 \times 10^4\,\text{Pa}$$

$$V = (20.000\,\text{L}) \times \left(\frac{1\,\text{m}^3}{10^3\,\text{L}}\right) = 2.0000 \times 10^{-2}\,\text{m}^3$$

$$n = \frac{m}{M} = \frac{0.25132\,\text{g}}{4.00260\,\text{g mol}^{-1}} = 0.062789\,\text{mol}$$

The perfect gas equation [12] can be rearranged to give $R = \dfrac{pV}{nT}$

$$R = \frac{(2.0182 \times 10^4\,\text{Pa}) \times (2.0000 \times 10^{-2}\,\text{m}^3)}{(0.062789\,\text{mol}) \times (773.15\,\text{K})} = \boxed{8.3147\,\text{J K}^{-1}\,\text{mol}^{-1}}$$

The accepted value is $R = 8.3145\,\text{J K}^{-1}\,\text{mol}^{-1}$.

Although gas volume data should be extrapolated to $p = 0$ for the best value of R, helium is close to being a perfect gas under the conditions here, and thus a value of R close to the accepted value is obtained.

E1.8 Since $p < 1\,\text{atm}$, the approximation that the vapour is a perfect gas is adequate. Then (as in Exercise 1.7(b)),

$pV = nRT = \dfrac{m}{M}RT$. Upon rearrangement,

$$M = \rho\left(\frac{RT}{p}\right) = (3.71\,\text{g L}^{-1}) \times \frac{(0.0821\,\text{L atm mol}^{-1}\,\text{K}^{-1}) \times (773\,\text{K})}{(699\,\text{Torr}) \times \left(\frac{1\,\text{atm}}{760\,\text{Torr}}\right)} = 256\,\text{g mol}^{-1}$$

This molar mass must be an integral multiple of the molar mass of atomic sulfur; hence

$$\text{number of S atoms} = \frac{256\,\text{g mol}^{-1}}{32.0\,\text{g mol}^{-1}} = 8$$

The formula of the vapour is then $\boxed{S_8}$

E1.9 The partial pressure of the water vapour in the room is: $p_{H_2O} = (0.60) \times (26.74\,\text{Torr}) = 16\,\text{Torr}$

Assuming that the perfect gas equation [12] applies, with $n = \dfrac{m}{M}$, $pV = \dfrac{m}{M}RT$ or

$$m = \frac{pVM}{RT} = \frac{(16\,\text{Torr}) \times \left(\frac{1\,\text{atm}}{760\,\text{Torr}}\right) \times (400\,\text{m}^3) \times \left(\frac{10^3\,\text{L}}{\text{m}^3}\right) \times (18.02\,\text{g mol}^{-1})}{(0.0821\,\text{L atm K}^{-1}\,\text{mol}^{-1}) \times (300\,\text{K})}$$

$$= 6.2 \times 10^3\,\text{g} = \boxed{6.2\,\text{kg}}$$

E1.10 **(a)** For simplicity assume a container of volume 1 L. Then the total mass is

$$m_T = n_{N_2}M_{N_2} + n_{O_2}M_{O_2} = 1.146\,\text{g} \tag{1}$$

Assuming that air is a perfect gas, $p_T V = n_T RT$, where n_T is the total amount of gas

$$n_T = \frac{p_T V}{RT} = \frac{(740\,\text{Torr}) \times \left(\frac{1\,\text{atm}}{760\,\text{Torr}}\right) \times (1\,\text{L})}{(0.08206\,\text{L atm K}^{-1}\,\text{mol}^{-1}) \times (300\,\text{K})} = 0.0395\overline{5}\,\text{mol}$$

$$n_T = n_{N_2} + n_{O_2} = 0.0395\overline{5}\,\text{mol} \tag{2}$$

Equations (1) and (2) are simultaneous equations for the amounts of gas and may be solved for them. Inserting n_{O_2} from (2) into (1) we get

$$(n_{N_2}) \times (28.0136\,\text{g mol}^{-1}) + (0.0395\overline{5}\,\text{mol} - n_{N_2}) \times (31.9988\,\text{g mol}^{-1}) = 1.14\overline{6}\,\text{g}$$

$$(1.2655 - 1.14\overline{60})\,\text{g} = (3.98\overline{52}\,\text{g mol}^{-1}) \times (n_{N_2})$$

$$n_{N_2} = 0.0299\overline{9}\,\text{mol}$$

$$n_{O_2} = n_T - n_{N_2} = (0.0395\overline{5} - 0.0299\overline{9})\,\text{mol} = 9.5\overline{6} \times 10^{-3}\,\text{mol}$$

The mole fractions are $x_{N_2} = \dfrac{0.0299\bar{9}\,\text{mol}}{0.0395\bar{5}\,\text{mol}} = \boxed{0.758\bar{3}}$ $\quad x_{O_2} = \dfrac{9.56 \times 10^{-3}\,\text{mol}}{0.0395\bar{5}\,\text{mol}} = \boxed{0.241\bar{7}}$

The partial pressures are $p_{N_2} = (0.758\bar{3}) \times (740\,\text{Torr}) = \boxed{561\,\text{Torr}}$

$$p_{O_2} = (0.241\bar{7}) \times (740\,\text{Torr}) = \boxed{179\,\text{Torr}}$$

The sum checks, $(561 + 179)\,\text{Torr} = 740\,\text{Torr}$

(b) The simplest way to solve this part is to realize that n_T, p_T, and m_T remain the same as in part **(a)** as these are experimentally determined quantities. Thus the amounts, mole fractions, and partial pressures of N_2 and O_2 are reduced by 1.0% relative to part **(a)**.

$$x_{N_2} = (0.9900) \times (0.758\bar{3}) = \boxed{0.750\bar{7}}$$

$$x_{O_2} = (0.9900) \times (0.241\bar{7}) = \boxed{0.239\bar{3}}$$

$$x_{Ar} = 0.0100$$

$$\begin{aligned}
p_{N_2} &= x_{N_2} p_T = (0.750\bar{7}) \times (740\,\text{Torr}) = 555.\bar{5}\,\text{Torr} \\
p_{O_2} &= x_{O_2} p_T = (0.239\bar{3}) \times (740\,\text{Torr}) = 177.\bar{1}\,\text{Torr} \\
p_{Ar} &= x_{Ar} p_T = (0.0100) \times (740\,\text{Torr}) = \underline{\;\;7.4\,\text{Torr}\;\;} \\
& \qquad\qquad \text{Sum} = p_T = 740\,\text{Torr} \quad \text{(checks)}
\end{aligned}$$

We can also check this by seeing whether or not the total amount of gas, n_T, remains the same.

$$\begin{aligned}
n_{N_2} &= x_{N_2} \times n_T = (0.750\bar{7}) \times (0.0395\bar{5}\,\text{mol}) = 0.0296\bar{9}\,\text{mol} \\
n_{O_2} &= x_{O_2} \times n_T = (0.239\bar{3}) \times (0.0395\bar{5}\,\text{mol}) = 0.00946\,\text{mol} \\
n_{Ar} &= x_{Ar} \times n_T = (0.0100) \times (0.0395\bar{5}\,\text{mol}) = \underline{0.00040\,\text{mol}} \\
& \qquad\qquad \text{Sum} = n_T = 0.0395\bar{5} \quad \text{(checks)}
\end{aligned}$$

E1.11 This exercise uses the formula, $M = \rho\dfrac{RT}{p}$, which was developed and used in Exercises 1.7(b) and 1.8(a). Substituting the data, $M = \dfrac{(1.23\,\text{g L}^{-1}) \times (62.36\,\text{L Torr K}^{-1}\text{mol}^{-1}) \times (330\,\text{K})}{150\,\text{Torr}}$

$$= \boxed{169\,\text{g mol}^{-1}}$$

E1.12 The easiest way to solve this exercise is to assume a sample of mass 1.000 g, then calculate the volume at each temperature, plot the volume against the Celsius temperature, and extrapolate to $V = 0$.

Draw up the following table

$\theta/°\text{C}$	$\rho/(\text{g L}^{-1})$	$V/(\text{L g}^{-1})$
−85	1.877	0.5328
0	1.294	0.7728
100	0.946	$1.05\bar{7}$

V versus θ is plotted in Fig. 1.1. The extrapolation gives a value for absolute zero close to −273°C. Alternatively, one could use an equation for V as a linear function of θ, which is Charles's law, and solve for the value of absolute zero. $V = V_0 \times (1 + \alpha\theta)$

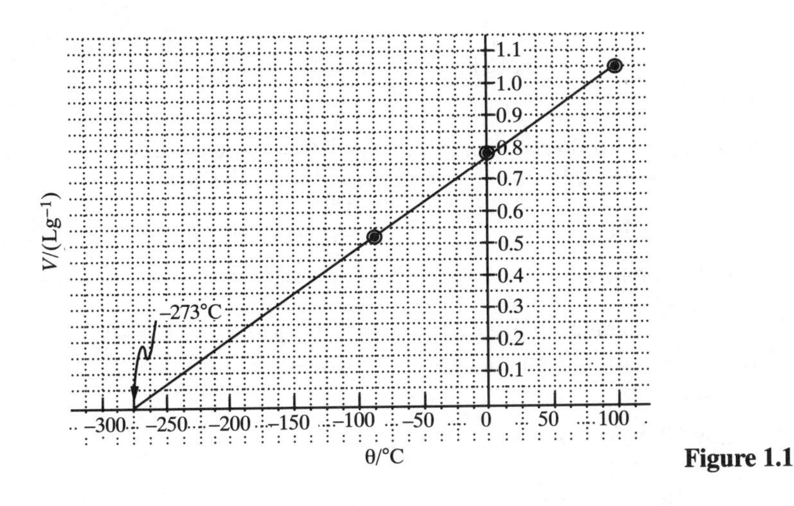

Figure 1.1

At absolute zero, $V = 0$, then $\theta(\text{abs.zero}) = -\dfrac{1}{\alpha}$. The value of α can be obtained from any one of the data points (except $\theta = 0$) as follows.

From $V = V_0 \times (1 + \alpha\theta)$,

$$\alpha = \frac{\left(\frac{V}{V_0} - 1\right)}{\theta} = \frac{\left(\frac{1.05\overline{7}}{0.7728}\right) - 1}{100°\text{C}} = 0.003678(°\text{C})^{-1}$$

$$-\frac{1}{\alpha} = -\frac{1}{0.003678(°\text{C})^{-1}} = \boxed{-272°\text{C}}$$

which is close to the value obtained graphically.

E1.13 **(a)** The formula for the mean speed is derived in Example 1.6 and is

$$\bar{c} = \left(\frac{8RT}{\pi M}\right)^{1/2}$$

Thus $\dfrac{\bar{c}(\text{H}_2)}{\bar{c}(\text{Hg})} = \left[\dfrac{M(\text{Hg})}{M(\text{H}_2)}\right]^{1/2} = \left(\dfrac{200.6\,\text{u}}{2.016\,\text{u}}\right)^{1/2} = \boxed{9.975}$

(b) The average kinetic energy involves the root mean square speed since the kinetic energy, ε, is given by $\bar{\varepsilon} = \dfrac{1}{2}m\langle v^2\rangle = \dfrac{1}{2}mc^2$ where $c = \sqrt{\langle v^2\rangle}$ and $c = \left(\dfrac{3RT}{M}\right)^{1/2}$ [21]

Thus, $\dfrac{\bar{\varepsilon}(\text{H}_2)}{\bar{\varepsilon}(\text{Hg})} = \dfrac{\frac{1}{2}m(\text{H}_2)\left[\frac{3RT}{M(\text{H}_2)}\right]}{\frac{1}{2}m(\text{Hg})\left[\frac{3RT}{M(\text{Hg})}\right]} = \boxed{1}$ since the masses, m, are proportional to the molar masses, M, $M = N_\text{A}m$.

Comment. Neither ratio is dependent on temperature and the ratio of energies is independent of both temperature and mass.

E1.14 **(a)** On the assumption that the gas is perfect, the temperature is easily calculated from eqn 2 after solving for T. $T = \dfrac{pV}{nR}$

$$n = \frac{1.0 \times 10^{23}\ \text{molecules}}{6.02 \times 10^{23}\ \text{molecules mol}^{-1}} = 0.16\overline{6}\,\text{mol}$$

$$T = \frac{(1.00 \times 10^5\ \text{Pa}) \times (1.0\,\text{L}) \times \left(\frac{1\,\text{m}^3}{10^3\,\text{L}}\right)}{(0.16\overline{6}\,\text{mol}) \times (8.314\,\text{J K}^{-1}\,\text{mol}^{-1})} = \boxed{72\,\text{K}}$$

(b) $c = \left(\frac{3RT}{M}\right)^{1/2}$ [21] $= \left(\frac{3 \times (8.314\,\mathrm{J\,K^{-1}\,mol^{-1}}) \times (72.\overline{46}\,\mathrm{K})}{2.016 \times 10^{-3}\,\mathrm{kg\,mol^{-1}}}\right)^{1/2} = \boxed{9.5 \times 10^{2}\,\mathrm{m\,s^{-1}}}$

(c) The temperature would not be different if they were O_2 molecules and exerted the same pressure in the same volume, but their root mean square speed would be different.

Comment. This exercise could have been solved by first obtaining the root mean square speed from $pV = \frac{1}{3}nMc^2$ [19] and then using eqn 21 to solve for the temperature. The results should be identical.

E1.15 The solution to this exercise is similar to that of Exercise 1.14(b)(b). Here p is calculated from the mean free path, rather than the mean free path from p, as in Exercise 1.14(b)(b).

$$\lambda = \frac{kT}{2^{1/2}\sigma p}\ [33] \quad \text{implies that} \quad p = \frac{kT}{2^{1/2}\sigma\lambda}$$

with $\lambda \approx 10\,\mathrm{cm} = \sqrt[3]{1000\,\mathrm{cm^3}}$

$$p = \frac{(1.381 \times 10^{-23}\,\mathrm{J\,K^{-1}}) \times (298.15\,\mathrm{K})}{(2^{1/2}) \times (0.36 \times 10^{-18}\,\mathrm{m^2}) \times (0.10\,\mathrm{m})} = \boxed{0.081\,\mathrm{Pa}}$$

This pressure corresponds to 8.0×10^{-7} atm and to 6.1×10^{-4} Torr, a pressure much larger than that of Exercise 1.14(b)(b).

E1.16 This exercise is similar to Exercise 1.14(b)(b) and the solution involves the same procedure.

$$\lambda = \frac{kT}{2^{1/2}\sigma p}[33] = \frac{(1.381 \times 10^{-23}\,\mathrm{J\,K^{-1}}) \times (217\,\mathrm{K})}{(2^{1/2}) \times (0.43 \times 10^{-18}\,\mathrm{m^2}) \times (0.050) \times (1.013 \times 10^{5}\,\mathrm{Pa})}$$
$$= \boxed{9.7 \times 10^{-7}\,\mathrm{m}}$$

E1.17 The collision frequency, z, is given by $z = \frac{2^{1/2}\sigma\bar{c}p}{kT}$ [31, 28], which becomes after substitution for

$$\bar{c} = \left(\frac{8RT}{\pi M}\right)^{1/2} \text{[Example 1.6]} = \left(\frac{8kT}{\pi m}\right)^{1/2}$$

$$z = \left(2^{1/2}\right) \times \sigma \times \left(\frac{8kT}{\pi m}\right)^{1/2} \times \left(\frac{p}{kT}\right) = \left(\frac{16}{\pi mkT}\right)^{1/2} \times \sigma p$$

$$= \left(\frac{16}{\pi \times (39.95) \times (1.6605 \times 10^{-27}\,\mathrm{kg}) \times (1.381 \times 10^{-23}\,\mathrm{J\,K^{-1}}) \times (298\,\mathrm{K})}\right)^{1/2}$$
$$\times (0.36 \times 10^{-18}\,\mathrm{m^2}) \times (p)$$
$$= (4.92 \times 10^{4}\,\mathrm{s^{-1}}) \times (p/\mathrm{Pa}) = (4.92 \times 10^{4}\,\mathrm{s^{-1}}) \times (1.0133 \times 10^{5}) \times (p/\mathrm{atm})$$
$$= (4.98 \times 10^{9}\,\mathrm{s^{-1}}) \times (p/\mathrm{atm})$$

Therefore

(a) $z = \boxed{5 \times 10^{10}\,\mathrm{s^{-1}}}$ when $p = 10$ atm,

(b) $z = \boxed{5 \times 10^{9}\,\mathrm{s^{-1}}}$ when $p = 1$ atm, and

(c) $z = \boxed{5 \times 10^{3}\,\mathrm{s^{-1}}}$ when $p = 10^{-6}$ atm. z is directly proportional to p at constant T.

E1.18 $$\lambda = \frac{kT}{2^{1/2}\sigma p}[33] = \frac{(1.381\times10^{-23}\,\mathrm{J\,K^{-1}})\times(298.15\,\mathrm{K})}{(2^{1/2})\times(0.43\times10^{-18}\,\mathrm{m^2})\times(p)} = \frac{6.8\times10^{-3}\,\mathrm{m}}{(p/\mathrm{Pa})} = \frac{6.7\times10^{-8}\,\mathrm{m}}{p/\mathrm{atm}}$$

(a) When $p = 10\,\mathrm{atm}$, $\lambda = 6.7\times10^{-9}\,\mathrm{m}$, or $\boxed{6.7\,\mathrm{nm}}$

(b) When $p = 1\,\mathrm{atm}$, $\lambda = \boxed{67\,\mathrm{nm}}$

(c) When $p = 10^{-6}\,\mathrm{atm}$, $\lambda = \boxed{6.7\,\mathrm{cm}}$

The mean free path is inversely proportional to p and to z (Exercise 1.17(a)).

E1.19 The Maxwell distribution of speeds is $f(v) = 4\pi\left(\frac{M}{2\pi RT}\right)^{3/2} v^2 e^{-Mv^2/2RT}$ [22]

The factor, $\frac{M}{2RT}$, can be evaluated as

$$\frac{M}{2RT} = \frac{28.02\times10^{-3}\,\mathrm{kg\,mol^{-1}}}{2\times(8.314\,\mathrm{J\,K^{-1}\,mol^{-1}})\times(500\,\mathrm{K})} = 3.37\times10^{-6}\,\mathrm{m^{-2}\,s^2}$$

Though $f(v)$ varies over the range 290 to 300 m s^{-1}, the variation is small over this small range and its value at the centre of the range can be used.

$$f(295\,\mathrm{m\,s^{-1}}) = (4\pi)\times\left(\frac{3.37\times10^{-6}\,\mathrm{m^{-2}\,s^2}}{\pi}\right)^{3/2}\times(295\,\mathrm{m\,s^{-1}})^2\times e^{(-3.37\times10^{-6})\times(295)^2}$$
$$= 9.06\times10^{-4}\,\mathrm{m^{-1}\,s}$$

Therefore, the fraction of molecules in the specified range is

$$f\times\Delta v = (9.06\times10^{-4}\,\mathrm{m^{-1}\,s})\times(10\,\mathrm{m\,s^{-1}}) = \boxed{9.06\times10^{-3}}$$

corresponding to 0.91 per cent.

Comment. This is a rather small percentage and suggests that the approximation of constancy of $f(v)$ over the range is adequate. To test the approximation $f(290\,\mathrm{m\,s^{-1}})$ and $f(300\,\mathrm{m\,s^{-1}})$ could be evaluated.

E1.20 **(a)** $p = \frac{nRT}{V}$ [2]

$n = 1.0\,\mathrm{mol}$, $T = 273.15\,\mathrm{K}$ (i) or $1000\,\mathrm{K}$ (ii)
$V = 22.414\,\mathrm{L}$ (i) or $100\,\mathrm{cm^3}$ (ii)

(i) $$p = \frac{(1.0\,\mathrm{mol})\times(8.206\times10^{-2}\,\mathrm{L\,atm\,K^{-1}\,mol^{-1}})\times(273.15\,\mathrm{K})}{22.414\,\mathrm{L}} = \boxed{1.0\,\mathrm{atm}}$$

(ii) $$p = \frac{(1.0\,\mathrm{mol})\times(8.206\times10^{-2}\,\mathrm{L\,atm\,K^{-1}\,mol^{-1}})\times(1000\,\mathrm{K})}{0.100\,\mathrm{L}} = \boxed{8.2\times10^2\,\mathrm{atm}}$$

(b) $p = \frac{nRT}{V-nb} - \frac{an^2}{V^2}$ [39*a*]

From Table 1.6, $a = 5.562\,\mathrm{L^2\,atm\,mol^{-2}}$ and $b = 6.380\times10^{-2}\,\mathrm{L\,mol^{-1}}$. Therefore,

$$\text{(i)}\quad \frac{nRT}{V-nb} = \frac{(1.0\,\text{mol}) \times (8.206 \times 10^{-2}\,\text{L atm K}^{-1}\,\text{mol}^{-1}) \times (273.15\,\text{K})}{[22.414 - (1.0) \times (6.380 \times 10^{-2})]\,\text{L}} = 1.0\overline{03}\,\text{atm}$$

$$\frac{an^2}{V^2} = \frac{(5.562\,\text{L}^2\,\text{atm mol}^{-2}) \times (1.0\,\text{mol})^2}{(22.414\,\text{L})^2} = 1.1\overline{1} \times 10^{-2}\,\text{atm}$$

and $p = 1.0\overline{03}\,\text{atm} - 1.1\overline{1} \times 10^{-2}\,\text{atm} = 0.992\,\text{atm} = \boxed{1.0\,\text{atm}}$

$$\text{(ii)}\quad \frac{nRT}{V-nb} = \frac{(1.0\,\text{mol}) \times (8.206 \times 10^{-2}\,\text{L atm K}^{-1}\,\text{mol}^{-1}) \times (1000\,\text{K})}{(0.100 - 0.06380)\,\text{L}}$$

$$= 2.2\overline{7} \times 10^3\,\text{atm}$$

$$\frac{an^2}{V^2} = \frac{(5.562\,\text{L}^2\,\text{atm mol}^{-1}) \times (1.0\,\text{mol})^2}{(0.100\,\text{L})^2} = 5.5\overline{6} \times 10^2\,\text{atm}$$

and $p = 2.2\overline{7} \times 10^3\,\text{atm} - 5.5\overline{6} \times 10^2\,\text{atm} = \boxed{1.7 \times 10^3\,\text{atm}}$

Comment. It is instructive to calculate the percentage deviation from perfect gas behaviour for (i) and (ii).

$$\text{(i)}\quad \frac{0.9\overline{92} - 1.0\overline{00}}{1.000} \times 100\% = \overline{0.8}\%$$

$$\text{(ii)}\quad \frac{(17 \times 10^2) - (8.2 \times 10^2)}{8.2 \times 10^2} \times 100\% = 10\overline{7}\%$$

Deviations from perfect gas behaviour are not observed at $p \approx 1\,\text{atm}$ except with very precise apparatus.

E1.21 The three equations in [40] are used. The van der Waals parameters a and b and the gas constant R are substituted into the equations.

$$V_c = 3b = 3 \times (0.0226\,\text{L mol}^{-1}) = \boxed{6.78 \times 10^{-2}\,\text{L mol}^{-1}}$$

$$p_c = \frac{a}{27b^2} = \frac{0.751\,\text{L}^2\,\text{atm mol}^{-2}}{27 \times (0.0226\,\text{L mol}^{-1})^2} = \boxed{54.5\,\text{atm}}$$

$$T_c = \frac{8a}{27Rb} = \frac{8 \times (0.751\,\text{L}^2\,\text{atm mol}^{-2})}{27 \times (8.206 \times 10^{-2}\,\text{L atm K}^{-1}\,\text{mol}^{-1}) \times (0.0226\,\text{L mol}^{-1})} = \boxed{120\,\text{K}}$$

E1.22 The definition of Z is used $Z = \dfrac{pV_m}{RT}[34] = \dfrac{V_m}{V_m^\circ}$

V_m is the actual molar volume, V_m° is the perfect gas molar volume. $V_m^\circ = \dfrac{RT}{p}$. Since V_m is 12 per cent smaller than that of a perfect gas, $V_m = 0.88V_m^\circ$, and

$$\textbf{(a)}\quad Z = \frac{0.88V_m^\circ}{V_m^\circ} = \boxed{0.88}$$

$$\textbf{(b)}\quad V_m = \frac{ZRT}{p} = \frac{(0.88) \times (8.206 \times 10^{-2}\,\text{L atm K}^{-1}\,\text{mol}^{-1}) \times (250\,\text{K})}{15\,\text{atm}} = \boxed{1.2\,\text{L}}$$

Since $V_m < V_m^\circ$ attractive forces dominate.

E1.23 The amount of gas is first determined from its mass; then the van der Waals equation is used to determine its pressure at the working temperature. The initial conditions of 300 K and 100 atm are in a sense superfluous information.

$$n = \frac{92.4\,\text{kg}}{28.02\times10^{-3}\,\text{kg mol}^{-1}} = 3.30\times10^{3}\,\text{mol}$$

$$V = 1.000\,\text{m}^3 = 1.000\times10^3\,\text{L}$$

$$p = \frac{nRT}{V-nb} - \frac{an^2}{V^2}[39a] = \frac{(3.30\times10^3\,\text{mol})\times(0.08206\,\text{L atm K}^{-1}\text{mol}^{-1})\times(500\,\text{K})}{(1.000\times10^3\,\text{L})-(3.30\times10^3\,\text{mol})\times(0.0391\,\text{L mol}^{-1})} - \frac{(1.408\,\text{L}^2\,\text{atm mol}^{-2})\times(3.30\times10^3\,\text{mol})^2}{(1.000\times10^3\,\text{L})^2}$$

$$= (155-15.3)\,\text{atm} = \boxed{140\,\text{atm}}$$

E1.24 **(a)** The molar volume is obtained from

$$\rho = \frac{M}{V_\text{m}} = \frac{\text{molar mass}}{\text{molar volume}} \quad \text{or} \quad V_\text{m} = \frac{M}{\rho} = \frac{18.02\,\text{g mol}^{-1}}{133.2\,\text{g L}^{-1}} = \boxed{0.1353\,\text{L mol}^{-1}}$$

$$Z = \frac{pV_\text{m}}{RT}[34] = \frac{(327.6\,\text{atm})\times(0.1353\,\text{L mol}^{-1})}{(0.08206\,\text{L atm K}^{-1}\,\text{mol}^{-1})\times(776.4\,\text{K})} = \boxed{0.6957}$$

(b) The van der Waals equation is

$$p = \frac{RT}{V_\text{m}-b} - \frac{a}{V_\text{m}^2}\ [39b]$$

Substituting this expression for p into Z [34] gives

$$Z = \frac{V_\text{m}}{V_\text{m}-b} - \frac{a}{V_\text{m}RT} = \frac{0.1353\,\text{L mol}^{-1}}{(0.1353\,\text{L mol}^{-1})-(0.03049\,\text{L mol}^{-1})} - \frac{5.536\,\text{L}^2\,\text{atm mol}^{-2}}{(0.1353\,\text{L mol}^{-1})\times(0.08206\,\text{L atm K}^{-1}\,\text{mol}^{-1})\times(776.4\,\text{K})}$$

$$= 1.291 - 0.642 = \boxed{0.649}$$

Comment. The difference is only about 5 per cent. Thus at this rather high pressure the van der Waals equation is still fairly accurate.

E1.25 **(a)** $p = \frac{nRT}{V}[2] = \frac{(10.0\,\text{mol})\times(0.08206\,\text{L atm K}^{-1}\,\text{mol}^{-1})\times(300\,\text{K})}{4.860\,\text{L}} = \boxed{50.7\,\text{atm}}$

(b)

$$p = \frac{nRT}{V-nb} - a\left(\frac{n}{V}\right)^2\ [39a]$$

$$= \frac{(10.0\,\text{mol})\times(0.08206\,\text{L atm K}^{-1}\,\text{mol}^{-1})\times(300\,\text{K})}{(4.860\,\text{L})-(10.0\,\text{mol})\times(0.06380\,\text{L mol}^{-1})} - (5.562\,\text{L}^2\text{atm mol}^{-2})\times\left(\frac{10.0\,\text{mol}}{4.860\,\text{L}}\right)^2$$

$$= 58.3\bar{1} - 23.5\bar{5} = \boxed{34.8\,\text{atm}}$$

The compression factor is calculated from its definition [34] after inserting $V_\text{m} = \frac{V}{n}$.

To complete the calculation of Z, a value for the pressure, p, is required. The implication in the definition [34] is that p is the actual pressure as determined experimentally. This pressure is neither the perfect gas pressure, nor the van der Waals pressure. However, on the assumption that the van der Waals equation provides a value for the pressure close to the experimental value, we can calculate the compression factor as follows

$$Z = \frac{pV}{nRT} = \frac{(34.8\,\text{atm}) \times (4.860\,\text{L})}{(10.0\,\text{mol}) \times (0.08206\,\text{L atm K}^{-1}\,\text{mol}^{-1}) \times (300\,\text{K})} = \boxed{0.687}$$

Comment. If the perfect gas pressure had been used Z would have been 1, the perfect gas value.

E1.26 $n = n(\text{H}_2) + n(\text{N}_2) = 2.0\,\text{mol} + 1.0\,\text{mol} = 3.0\,\text{mol} \qquad x_\text{J} = \frac{n_\text{J}}{n}$ [15]

(a) $x(\text{H}_2) = \frac{2.0\,\text{mol}}{3.0\,\text{mol}} = \boxed{0.67} \qquad x(\text{N}_2) = \frac{1.0\,\text{mol}}{3.0\,\text{mol}} = \boxed{0.33}$

(b) The perfect gas law is assumed to hold for each component individually as well as for the mixture as a whole. Hence, $p_\text{J} = n_\text{J}\frac{RT}{V}$ [14]

$$\frac{RT}{V} = \frac{(8.206 \times 10^{-2}\,\text{L atm K}^{-1}\,\text{mol}^{-1}) \times (273.15\,\text{K})}{22.4\,\text{L}} = 1.00\,\text{atm mol}^{-1}$$

$$p(\text{H}_2) = (2.0\,\text{mol}) \times (1.00\,\text{atm mol}^{-1}) = \boxed{2.0\,\text{atm}}$$

$$p(\text{N}_2) = (1.0\,\text{mol}) \times (1.00\,\text{atm mol}^{-1}) = \boxed{1.0\,\text{atm}}$$

(c) $p = p(\text{H}_2) + p(\text{N}_2)[13] = 2.0\,\text{atm} + 1.0\,\text{atm} = \boxed{3.0\,\text{atm}}$

Question. Does Dalton's law hold for a mixture of van der Waals gases?

E1.27 Equations [40] are solved for b and a, respectively, and yield $b = \frac{V_\text{c}}{3}$ and $a = 27b^2 p_\text{c} = 3V_\text{c}^2 p_\text{c}$

Substituting the critical constants

$$b = \frac{1}{3} \times (98.7\,\text{cm}^3\,\text{mol}^{-1}) = \boxed{32.9\,\text{cm}^3\,\text{mol}^{-1}}$$

$$a = 3 \times (98.7 \times 10^{-3}\,\text{L mol}^{-1})^2 \times (45.6\,\text{atm}) = \boxed{1.33\,\text{L}^2\,\text{atm mol}^{-2}}$$

Note that knowledge of the critical temperature, T_c, is not required.

As b is approximately the volume occupied per mole of particles

$$v_\text{mol} \approx \frac{b}{N_\text{A}} = \frac{32.9 \times 10^{-6}\,\text{m}^3\,\text{mol}^{-1}}{6.022 \times 10^{23}\,\text{mol}^{-1}} = 5.46 \times 10^{-29}\,\text{m}^3$$

Then, with $v_\text{mol} = \frac{4}{3}\pi r^3$, $r \approx \left(\frac{3}{4\pi} \times (5.46 \times 10^{-29}\,\text{m}^3)\right)^{1/3} = \boxed{0.24\,\text{nm}}$

E1.28 The Boyle temperature, T_B, is the temperature at which $B = 0$. In order to express T_B in terms of a and b, the van der Waals equation must be recast into the form of the virial equation.

$$p = \frac{RT}{V_\text{m} - b} - \frac{a}{V_\text{m}^2} \text{ [39}b\text{]}$$

Factoring out $\frac{RT}{V_m}$ yields $p = \frac{RT}{V_m}\left\{\frac{1}{1-b/V_m} - \frac{a}{RTV_m}\right\}$

So long as $b/V_m < 1$, the first term inside the brackets can be expanded using $(1-x)^{-1} = 1 + x + x^2 + \cdots$, which gives

$$p = \frac{RT}{V_m}\left\{1 + \left(b - \frac{a}{RT}\right) \times \left(\frac{1}{V_m}\right) + \cdots\right\}$$

We can now identify the second virial coefficient as $B = b - \frac{a}{RT}$

Since at the Boyle temperature $B = 0$, $T_B = \frac{a}{bR} = \frac{27T_c}{8}$

(a) From Table 1.6, $a = 6.579\,\mathrm{L^2\,atm\,mol^{-2}}$, $b = 5.622 \times 10^{-2}\,\mathrm{L\,mol^{-1}}$. Therefore,

$$T_B = \frac{6.579\,\mathrm{L^2\,atm\,mol^{-2}}}{(5.622 \times 10^{-2}\,\mathrm{L\,mol^{-1}}) \times (8.206 \times 10^{-2}\,\mathrm{L\,atm\,K^{-1}\,mol^{-1}})} = \boxed{1.4 \times 10^3\,\mathrm{K}}$$

(b) As in Exercise 1.27(a), $v_{mol} \approx \frac{b}{N_A} = \frac{5.622 \times 10^{-5}\,\mathrm{m^3\,mol^{-1}}}{6.022 \times 10^{23}\,\mathrm{mol^{-1}}} = 9.3 \times 10^{-29}\,\mathrm{m^3}$

$$r \approx \left(\frac{3}{4\pi} \times (9.3 \times 10^{-29}\,\mathrm{m^3})\right)^{1/3} = \boxed{0.28\,\mathrm{nm}}$$

E1.29 The reduced temperature and pressure of hydrogen are calculated from the relations

$$T_r = \frac{T}{T_c} \quad \text{and} \quad p_r = \frac{p}{p_c} \; [42]$$

$$T_r = \frac{298\,\mathrm{K}}{33.23\,\mathrm{K}} = 8.96\bar{8} \quad [T_c = 33.23\,\mathrm{K},\ \text{Table 1.5}]$$

$$p_r = \frac{1.0\,\mathrm{atm}}{12.8\,\mathrm{atm}} = 0.078\bar{1} \quad [p_c = 12.8\,\mathrm{atm},\ \text{Table 1.5}]$$

Hence, the gases named will be in corresponding states at $T = 8.96\bar{8} \times T_c$ and at $p = 0.078\bar{1} \times p_c$.

(a) For ammonia, $T_c = 405.5\,\mathrm{K}$ and $p_c = 111.3\,\mathrm{atm}$ (Table 1.5), so

$$T = (8.96\bar{8}) \times (405.5\,\mathrm{K}) = \boxed{3.64 \times 10^3\,\mathrm{K}}$$

$$p = (0.078\bar{1}) \times (111.3\,\mathrm{atm}) = \boxed{8.7\,\mathrm{atm}}$$

(b) For xenon, $T_c = 289.75\,\mathrm{K}$ and $p_c = 58.0\,\mathrm{atm}$, so

$$T = (8.96\bar{8}) \times (289.75\,\mathrm{K}) = \boxed{2.60 \times 10^3\,\mathrm{K}}$$

$$p = (0.078\bar{1}) \times (58.0\,\mathrm{atm}) = \boxed{4.5\,\mathrm{atm}}$$

(c) For helium, $T_c = 5.21\,\mathrm{K}$ and $p_c = 2.26\,\mathrm{atm}$, so

$$T = (8.96\bar{8}) \times (5.21\,\mathrm{K}) = \boxed{46.7\,\mathrm{K}}$$

$$p = (0.078\bar{1}) \times (2.26\,\mathrm{atm}) = \boxed{0.18\,\mathrm{atm}}$$

E1.30 The van der Waals equation [39b] is solved for b, which yields

$$b = V_m - \frac{RT}{\left(p + \frac{a}{V_m^2}\right)}$$

Substituting the data

$$b = 5.00 \times 10^{-4}\,\text{m}^3\,\text{mol}^{-1} - \frac{(8.314\,\text{J K}^{-1}\,\text{mol}^{-1}) \times (273\,\text{K})}{\left\{(3.0 \times 10^6\,\text{Pa}) + \left(\frac{0.50\,\text{m}^6\,\text{Pa mol}^{-2}}{(5.00\times10^{-4}\,\text{m}^3\,\text{mol}^{-1})^2}\right)\right\}}$$

$$= \boxed{0.46 \times 10^{-4}\,\text{m}^3\,\text{mol}^{-1}}$$

$$Z = \frac{pV_m}{RT}[34] = \frac{(3.0 \times 10^6\,\text{Pa}) \times (5.00 \times 10^{-4}\,\text{m}^3)}{(8.314\,\text{J K}^{-1}\,\text{mol}^{-1}) \times (273\,\text{K})} = \boxed{0.66}$$

Comment. The definition of Z involves the actual pressure, volume, and temperature and does not depend upon the equation of state used to relate these variables.

Solutions to problems

Solutions to numerical problems

P1.1 Boyle's law in the form $p_f V_f = p_i V_i$ is solved for V_f: $V_f = \frac{p_i}{p_f} \times V_i$

$$p_i = 1.0\,\text{atm}$$
$$p_f = p_{ex} + \rho g h[4] = p_i + \rho g h = 1.0\,\text{atm} + \rho g h$$
$$\rho g h = (1.025 \times 10^3\,\text{kg m}^{-3}) \times (9.81\,\text{m s}^{-2}) \times (50\,\text{m}) = 5.0\bar{3} \times 10^5\,\text{Pa}$$

Hence, $p_f = (1.0\bar{1} \times 10^5\,\text{Pa}) + (5.0\bar{3} \times 10^5\,\text{Pa}) = 6.0\bar{4} \times 10^5\,\text{Pa}$

$$V_f = \frac{1.0\bar{1} \times 10^5\,\text{Pa}}{6.0\bar{4} \times 10^5\,\text{Pa}} \times 3.0\,\text{m}^3 = \boxed{0.50\,\text{m}^3}$$

P1.3 Since the Neptunians know about perfect gas behaviour we may assume that they will write $pV = nRT$ at both temperatures. We may also assume that they will establish the size of their absolute unit to be the same as the °N, just as we write 1 K = 1°C. Thus

$$pV(T_1) = 28.0\,\text{L atm} = nRT_1 = nR \times (T_1 + 0°\text{N})$$
$$pV(T_2) = 40.0\,\text{L atm} = nRT_2 = nR \times (T_1 + 100°\text{N})$$

or $T_1 = \frac{28.0\,\text{L atm}}{nR}$ $\quad T_1 + 100°\text{N} = \frac{40.0\,\text{L atm}}{nR}$

Dividing, $\frac{T_1 + 100°\text{N}}{T_1} = \frac{40.0\,\text{L atm}}{28.0\,\text{L atm}} = 1.42\bar{9}$ or $T_1 + 100°\text{N} = 1.42\bar{9}T_1$, $T_1 = 233$ absolute units

As in the relationship between our Kelvin scale and Celsius scale $T = \theta -$ absolute zero (°N) so absolute zero (°N) = $\boxed{-233°\text{N}}$

Comment. To facilitate communication with Earth students we have converted the Neptunians' units of the pV product to units familiar to humans, that is L atm. However, we see from the solution that only the ratio of pV products is required, and that will be the same in any civilization.

Question. If the Neptunians' unit of volume is the lagoon (L), their unit of pressure is the poseidon (P), their unit of amount is the nereid (n), and their unit of absolute temperature is the titan (T), what is the value of the Neptunians' gas constant (R) in units of L, P, n, and T?

P1.5 Solving for n from the perfect gas equation [12] yields $n = \frac{pV}{RT}$ and $n = \frac{m}{M}$, hence $\rho = \frac{m}{V} = \frac{Mp}{RT}$

Rearrangement yields the desired relation, that is $\boxed{p = \rho\frac{RT}{M}}$, or $\frac{p}{\rho} = \frac{RT}{M}$, and $M = \frac{RT}{p/\rho}$

Draw up the following table and then plot $\frac{p}{\rho}$ versus p to find the zero pressure limit of $\frac{p}{\rho}$ where all gases behave ideally.

$$\rho/(\mathrm{g\,L^{-1}}) = \rho/(\mathrm{kg\,m^{-3}});$$

$$1\,\mathrm{Torr} = (1\,\mathrm{Torr}) \times \left(\frac{1\,\mathrm{atm}}{760\,\mathrm{Torr}}\right) \times \left(\frac{1.013 \times 10^5\,\mathrm{Pa}}{1\,\mathrm{atm}}\right) = 133.3\,\mathrm{Pa}$$

p/Torr	91.74	188.98	277.3	452.8	639.3	760.0
$p/(10^4\,\mathrm{Pa})$	1.223	2.519	3.696	6.036	8.522	10.132
$\rho/(\mathrm{kg\,m^{-3}})$	0.232	0.489	0.733	1.25	1.87	2.30
$\left(\frac{p}{\rho}\right)/(10^4\,\mathrm{m^2\,s^{-2}})$	5.27	5.15	5.04	4.83	4.56	4.41

$\frac{p}{\rho}$ is plotted in Fig. 1.2. A straight line fits the data rather well. The extrapolation to $p = 0$ yields an intercept of $5.39 \times 10^4\,\mathrm{m^2\,s^{-2}}$. Then

$$M = \frac{RT}{5.39 \times 10^4\,\mathrm{m^2\,s^{-2}}} = \frac{(8.314\,\mathrm{J\,K^{-1}\,mol^{-1}}) \times (298.15\,\mathrm{K})}{5.39 \times 10^4\,\mathrm{m^2\,s^{-2}}}$$

$$= 0.0460\,\mathrm{kg\,mol^{-1}} = \boxed{46.0\,\mathrm{g\,mol^{-1}}}$$

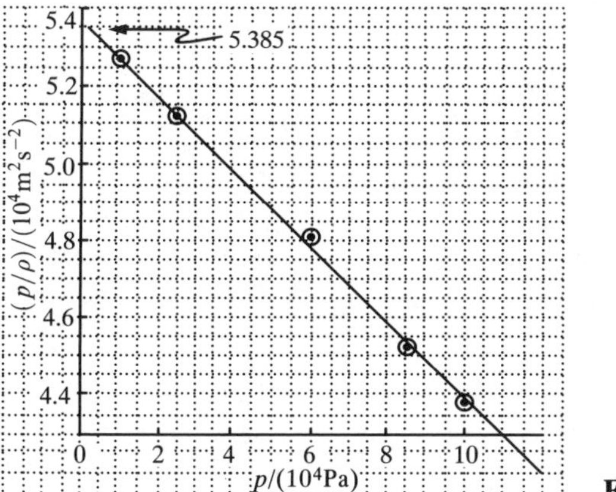

Figure 1.2

Comment. This method of the determination of the molar masses of gaseous compounds is due to Cannizarro who presented it at the Karlsruhe conference of 1860 which had been called to resolve the problem of the determination of the molar masses of atoms and molecules and the molecular formulas of compounds.

P1.7 $n = \frac{pV}{RT}$[2], $V = \frac{4\pi}{3}r^3 = \frac{4\pi}{3} \times (3.0\,\text{m})^3 = 11\bar{3}\,\text{m}^3$ = volume of balloon

$p = 1.0\,\text{atm}$, $T = 298\,\text{K}$

(a) $n = \frac{(1.0\,\text{atm}) \times (11\bar{3} \times 10^3\,\text{L})}{(8.206 \times 10^{-2}\,\text{L atm K}^{-1}\,\text{mol}^{-1}) \times (298\,\text{K})} = \boxed{4.6\bar{2} \times 10^3\,\text{mol}}$

(b) The mass that the balloon can lift is the difference between the mass of displaced air and the mass of the balloon. We assume that the mass of the balloon is essentially that of the gas it encloses.

Then $m(H_2) = nM(H_2) = (4.6\bar{2} \times 10^3\,\text{mol}) \times (2.02\,\text{g mol}^{-1}) = 9.3\bar{3} \times 10^3\,\text{g}$

Mass of displaced air $= (11\bar{3}\,\text{m}^3) \times (1.22\,\text{kg m}^{-3}) = 1.3\bar{8} \times 10^2\,\text{kg}$

Therefore, the payload is $(13\bar{8}\,\text{kg}) - (9.3\bar{3}\,\text{kg}) = \boxed{1.3 \times 10^2\,\text{kg}}$

(c) For helium, $m = nM(\text{He}) = (4.6\bar{2} \times 10^3\,\text{mol}) \times (4.00\,\text{g mol}^{-1}) = 18\,\text{kg}$

The payload is now $13\bar{8}\,\text{kg} - 18\,\text{kg} = \boxed{1.2 \times 10^2\,\text{kg}}$

P1.9 $\frac{p}{T} = \frac{nR}{V}$ = constant, if n and V are constant. Hence, $\frac{p}{T} = \frac{p_3}{T_3}$, where p is the measured pressure at temperature, T, and p_3 and T_3 are the triple point pressure and temperature, respectively. Rearranging, $p = \left(\frac{p_3}{T_3}\right)T$.

The ratio $\frac{p_3}{T_3}$ is a constant $= \frac{50.2\,\text{Torr}}{273.16\,\text{K}} = 0.183\bar{8}\,\text{Torr K}^{-1}$. Thus the change in p, Δp, is proportional to the change in temperature, ΔT. $\Delta p = (0.183\bar{8}\,\text{Torr K}^{-1}) \times (\Delta T)$

(a) $\Delta p = (0.183\bar{8}\,\text{Torr K}^{-1}) \times (1\,\text{K}) = \boxed{0.184\,\text{Torr}}$

(b) Rearranging, $p = \left(\frac{T}{T_3}\right)p_3 = \left(\frac{373.16\,\text{K}}{273.16\,\text{K}}\right) \times (50.2\,\text{Torr}) = \boxed{68.6\,\text{Torr}}$

(c) Since $\frac{p}{T}$ is a constant at constant n and V, it always has the value $0.183\bar{8}\,\text{Torr K}^{-1}$, hence

$$\Delta p = p_{374.15\,\text{K}} - p_{373.15\,\text{K}} = (0.183\bar{8}\,\text{Torr K}^{-1}) \times (1\,\text{K}) = \boxed{0.184\,\text{Torr}}$$

P1.11 The time in seconds for a disk to rotate 360° is the inverse of the frequency. The time for it to advance 2° is $\frac{\left(\frac{2°}{360°}\right)}{\nu}$. This is the time required for slots in neighbouring disks to coincide. For an atom to pass through all neighbouring slots it must have the speed $v_x = \frac{1.0\,\text{cm}}{\frac{\left(\frac{2}{360}\right)}{\nu}} = 180\,\nu\,\text{cm} = 180(\nu/\text{Hz})\,\text{cm s}^{-1}$

Hence, the distributions of the x-component of velocity are

ν/Hz	20	40	80	100	120
$v_x/(\text{cm s}^{-1})$	3600	7200	14400	18000	21600
$\mathcal{I}(40\,\text{K})$	0.846	0.513	0.069	0.015	0.002
$\mathcal{I}(100\,\text{K})$	0.592	0.485	0.217	0.119	0.057

Theoretically, the velocity distribution in the x-direction is

$$f(v_x) = \left(\frac{m}{2\pi kT}\right)^{1/2} e^{-mv_x^2/2kT} \quad [25, \text{with } M/R = m/k]$$

Therefore, as $\mathcal{I} \propto f$, $\mathcal{I} \propto \left(\frac{1}{T}\right)^{1/2} e^{-mv_x^2/2kT}$

$$\text{Since } \frac{mv_x^2}{2kT} = \frac{(83.8) \times (1.6605 \times 10^{-27}\,\text{kg}) \times \{1.80(\nu/\text{Hz})\,\text{m s}^{-1}\}^2}{(2) \times (1.381 \times 10^{-23}\,\text{J K}^{-1}) \times (T)} = \frac{1.63 \times 10^{-2}(\nu/\text{Hz})^2}{T/\text{K}}$$

We can write $\mathcal{I} \propto \left(\frac{1}{T/\text{K}}\right)^{1/2} e^{-1.63\times10^{-2}(\nu/\text{Hz})^2/(T/\text{K})}$ and draw up the following table, obtaining the constant of proportionality by fitting $\mathcal{I}$ to the value at $T = 40\,\text{K}$, $\nu = 80\,\text{Hz}$

ν/Hz	20	40	80	100	120
$\mathcal{I}$(40 K)	0.80	0.49	(0.069)	0.016	0.003
$\mathcal{I}$(100 K)	0.56	0.46	0.209	0.116	0.057

in fair agreement with the experimental data.

P1.12 For discrete rather than continuous variables the equation analogous to the equation for obtaining $\bar{c}$ (Example 1.6) is $\langle v_x \rangle = \sum_i v_{i,x}\left(\frac{N_i}{N}\right) = \frac{1}{N}\sum_i N_i v_{i,x}$ with $\left(\frac{N_i}{N}\right)$ the analogue of $f(v)$.

$$N = 40 + 62 + 53 + 12 + 2 + 38 + 59 + 60 + 2 = 328$$

(a)
$$\begin{aligned}\langle v_x \rangle &= \frac{1}{328}\{40 \times 80 + 62 \times 85 + \cdots + 2 \times 100 + 38 \times (-80) \\ &\quad + 59 \times (-85) + \cdots + 2 \times (-100)\}\,\text{km h}^{-1} \\ &= \boxed{2.8\,\text{km h}^{-1}}\ \text{East}\end{aligned}$$

(b)
$$\begin{aligned}\langle |v_x| \rangle &= \frac{1}{328}\{40 \times 80 + 62 \times 85 + \cdots + 2 \times 100 + 38 \times 80 \\ &\quad + 59 \times 85 + \cdots + 2 \times 100\}\,\text{km h}^{-1} \\ &= \boxed{86\,\text{km h}^{-1}}\end{aligned}$$

(c)
$$\langle v_x^2 \rangle = \frac{1}{328}\{40 \times 80^2 + 62 \times 85^2 + \cdots + 2 \times 100^2\}(\text{km h}^{-1})^2 = 7430\,(\text{km h}^{-1})^2$$

$$\sqrt{\langle v_x^2 \rangle} = \boxed{86\,\text{km h}^{-1}} \quad \left[\text{that } \sqrt{\langle v_x^2 \rangle} = \langle |v_x| \rangle \text{ in this case is coincidental.}\right]$$

P1.14 The work required for a mass, m, to go from a distance r from the centre of a planet of mass m' to infinity is

$$w = \int_r^{\infty} F\,\text{d}r$$

where F is the force of gravity and is given by Newton's law of universal gravitation, which is

$$F = \frac{Gmm'}{r^2}$$

G is the gravitational constant (not to be confused with g). Then

$$w' = \int_r^\infty \frac{Gmm'}{r^2}\,\mathrm{d}r = \frac{Gmm'}{r}$$

Since according to Newton's second law of motion, $F = mg$, we may make the identification

$$g = \frac{Gm'}{r^2}$$

Thus, $w = grm$. This is the kinetic energy that the particle must have in order to escape the planet's gravitational attraction at a distance r from the planet's centre; hence $w = \frac{1}{2}mv^2 = mgr$

$$v_\mathrm{e} = (2gR_\mathrm{p})^{1/2} \quad [R_\mathrm{p} = \text{radius of planet}]$$

which is the escape velocity.

(a) $v_\mathrm{e} = [(2) \times (9.81\,\mathrm{m\,s^{-2}}) \times (6.37 \times 10^6\,\mathrm{m})]^{1/2} = \boxed{11.2\,\mathrm{km\,s^{-1}}}$

(b) $g(\mathrm{Mars}) = \dfrac{m(\mathrm{Mars})}{m(\mathrm{Earth})} \times \dfrac{R(\mathrm{Earth})^2}{R(\mathrm{Mars})^2} \times g(\mathrm{Earth}) = (0.108) \times \left(\dfrac{6.37}{3.38}\right)^2 \times (9.81\,\mathrm{m\,s^{-2}})$

$$= 3.76\,\mathrm{m\,s^{-2}}$$

Hence, $v_\mathrm{e} = [(2) \times (3.76\,\mathrm{m\,s^{-2}}) \times (3.38 \times 10^6\,\mathrm{m})]^{1/2} = 5.0\,\mathrm{km\,s^{-1}}$

Since $\bar{c} = \left(\dfrac{8RT}{\pi M}\right)^{1/2}$, $T = \dfrac{\pi M\bar{c}^2}{8R}$

and we can draw up the following table

$10^{-3}\,T/\mathrm{K}$	H_2	He	O_2	
Earth	11.9	23.7	190	$[\bar{c} = 11.2\,\mathrm{km\,s^{-1}}]$
Mars	2.4	4.8	38	$[\bar{c} = 5.0\,\mathrm{km\,s^{-1}}]$

In order to calculate the proportion of molecules that have speeds exceeding the escape velocity, v_e, we must integrate the Maxwell distribution [22] from v_e to infinity.

$$P = \int_{v_\mathrm{e}}^\infty f(v)\,\mathrm{d}v = \int_{v_\mathrm{e}}^\infty 4\pi \left(\frac{m}{2\pi kT}\right)^{3/2} v^2 \mathrm{e}^{-mv^2/2kT}\,\mathrm{d}v \quad \left[\frac{M}{R} = \frac{m}{k}\right]$$

This integral cannot be evaluated analytically and must be expressed in terms of the error function. We proceed as follows.

Defining $\beta = \dfrac{m}{2kT}$ and $y^2 = \beta v^2$ gives $v = \beta^{-1/2}y$, $v^2 = \beta^{-1}y^2$, $v_\mathrm{e} = \beta^{-1/2}y_\mathrm{e}$,

$$y_\mathrm{e} = \beta^{1/2}v_\mathrm{e}, \quad \text{and} \quad \mathrm{d}v = \beta^{-1/2}\,\mathrm{d}y$$

$$P = 4\pi\left(\frac{\beta}{\pi}\right)^{3/2}\beta^{-1}\beta^{-1/2}\int_{\beta^{1/2}v_\mathrm{e}}^\infty y^2\mathrm{e}^{-y^2}\,\mathrm{d}y = \frac{4}{\pi^{1/2}}\int_{\beta^{1/2}v_\mathrm{e}}^\infty y^2\mathrm{e}^{-y^2}\,\mathrm{d}y$$

$$= \frac{4}{\pi^{1/2}}\left[\int_0^\infty y^2\mathrm{e}^{-y^2}\,\mathrm{d}y - \int_0^{\beta^{1/2}v_\mathrm{e}} y^2\mathrm{e}^{-y^2}\,\mathrm{d}y\right]$$

The first integral can be evaluated analytically; the second cannot.

$$\int_0^\infty y^2 e^{-y^2}\,dy = \frac{\pi^{1/2}}{4}, \text{ hence}$$

$$P = 1 - \frac{2}{\pi^{1/2}}\int_0^{\beta^{1/2}v_e} ye^{-y^2}(2y\,dy) = 1 - \frac{2}{\pi^{1/2}}\int_0^{\beta^{1/2}v_e} y\,d(-e^{-y^2})$$

This integral may be evaluated by parts

$$P = 1 - \frac{2}{\pi^{1/2}}\left[y(-e^{-y^2})\Big|_0^{\beta^{1/2}v_e} - \int_0^{\beta^{1/2}v_e}(-e^{-y^2})\,dy\right]$$

$$P = 1 + 2\left(\frac{\beta}{\pi}\right)^{1/2} v_e e^{-\beta v_e^2} - \frac{2}{\pi^{1/2}}\int_0^{\beta^{1/2}v_e} e^{-y^2}\,dy = 1 + 2\left(\frac{\beta}{\pi}\right)^{1/2} v_e e^{-\beta v_e^2} - \mathrm{erf}(\beta^{1/2}v_e)$$

$$= \mathrm{erfc}(\beta^{1/2}v_e) + 2\left(\frac{\beta}{\pi}\right)^{1/2} v_e e^{-\beta v_e^2} \quad [\mathrm{erfc}(z) = 1 - \mathrm{erf}(z)]$$

From $\beta = \frac{m}{2kT} = \frac{M}{2RT}$ and $v_e = (2gR_p)^{1/2}$

$$\beta^{1/2}v_e = \left(\frac{MgR_p}{RT}\right)^{1/2}$$

For H_2 on Earth at 240 K

$$\beta^{1/2}v_e = \left(\frac{(0.002016\,\mathrm{kg\,mol^{-1}}) \times (9.807\,\mathrm{m\,s^{-2}}) \times (6.37 \times 10^6\,\mathrm{m})}{(8.314\,\mathrm{J\,K^{-1}\,mol^{-1}}) \times (240\,\mathrm{K})}\right)^{1/2} = 7.94$$

$$P = \mathrm{erfc}(7.94) + 2\left(\frac{7.94}{\pi^{1/2}}\right)e^{-(7.94)^2} = (2.9 \times 10^{-29}) + (3.7 \times 10^{-27}) = \boxed{3.7 \times 10^{-27}}$$

at 1500 K

$$\beta^{1/2}v_e = \left(\frac{(0.002016\,\mathrm{kg\,mol^{-1}}) \times (9.807\mathrm{m\,s^{-2}}) \times (6.37 \times 10^6\,\mathrm{m})}{(8.314\,\mathrm{J\,K^{-1}\,mol^{-1}}) \times (1500\,\mathrm{K})}\right)^{1/2} = 3.18$$

$$P = \mathrm{erfc}(3.18) + 2\left(\frac{3.18}{\pi^{1/2}}\right)e^{-(3.18)^2} = (6.9 \times 10^{-6}) + (1.4\bar{6} \times 10^{-4}) = \boxed{1.5 \times 10^{-4}}$$

For H_2 on Mars at 240 K

$$\beta^{1/2}v_e = \left(\frac{(0.002016\,\mathrm{kg\,mol^{-1}}) \times (3.76\,\mathrm{m\,s^{-2}}) \times (3.38 \times 10^6\,\mathrm{m})}{(8.314\,\mathrm{J\,K^{-1}\,mol^{-1}}) \times (240\,\mathrm{K})}\right)^{1/2} = 3.58$$

$$P = \mathrm{erfc}(3.58) + 2\left(\frac{3.58}{\pi^{1/2}}\right)e^{-(3.58)^2} = (4.13 \times 10^{-7}) + (1.1\bar{0} \times 10^{-5}) = \boxed{1.1 \times 10^{-5}}$$

at 1500 K, $\beta^{1/2}v_e = 1.43$

$$P = \mathrm{erfc}(1.43) + (1.128) \times (1.43) \times e^{-(1.43)^2} = 0.0431 + 0.20\bar{9} = \boxed{0.25}$$

For He on Earth at 240 K

$$\beta^{1/2}v_e = \left(\frac{(0.004003\,\text{kg mol}^{-1}) \times (9.807\,\text{m s}^{-2}) \times (6.37 \times 10^6\,\text{m})}{(8.314\,\text{J K}^{-1}\,\text{mol}^{-1}) \times (240\,\text{K})}\right)^{1/2} = 11.1\overline{9}$$

$$P = \text{erfc}(11.2) + (1.128) \times (11.2) \times \text{e}^{-(11.2)^2} = 0 + (4 \times 10^{-54}) = \boxed{4 \times 10^{-54}}$$

at 1500 K, $\beta^{1/2}v_e = 4.48$

$$P = \text{erfc}(4.48) + (1.128) \times (4.48) \times \text{e}^{-(4.48)^2} = (2.36 \times 10^{-10}) + (9.7\overline{1} \times 10^{-9})$$
$$= \boxed{1.0 \times 10^{-8}}$$

For He on Mars at 240 K

$$\beta^{1/2}v_e = \left(\frac{(0.004003\,\text{kg mol}^{-1}) \times (3.76\,\text{m s}^{-2}) \times (3.38 \times 10^6\,\text{m})}{(8.314\,\text{J K}^{-1}\,\text{mol}^{-1}) \times (240\,\text{K})}\right)^{1/2} = 5.05$$

$$P = \text{erfc}(5.05) + (1.128) \times (5.05) \times \text{e}^{-(5.05)^2} = (9.21 \times 10^{-13}) + (4.7\overline{9} \times 10^{-11})$$
$$= \boxed{4.9 \times 10^{-11}}$$

at 1500 K, $\beta^{1/2}v_e = 2.02$

$$P = \text{erfc}(2.02) + (1.128) \times (2.02) \times \text{e}^{-(2.02)^2} = (4.28 \times 10^{-3}) + (0.040\overline{1}) = \boxed{0.044}$$

For O_2 on Earth it is clear that $P \approx 0$ at both temperatures.
For O_2 on Mars at 240 K, $\beta^{1/2}v_e = 14.3$

$$P = \text{erfc}(14.3) + (1.128) \times (14.3) \times \text{e}^{-(14.3)^2} = 0 + (2.5 \times 10^{-88}) = \boxed{2.5 \times 10^{-88}} \approx 0$$

at 1500 K, $\beta^{1/2}v_e = 5.71$

$$P = \text{erfc}(5.71) + (1.128) \times (5.71) \times \text{e}^{-(5.71)^2} = (6.7 \times 10^{-16}) + (4.46 \times 10^{-14})$$
$$= \boxed{4.5 \times 10^{-14}}$$

Based on these numbers alone, it would appear that H_2 and He would be depleted from the atmosphere of both Earth and Mars after many (millions?) years; that the rate on Mars, though still slow, would be many orders of magnitude larger than on Earth; that O_2 would be retained on Earth indefinitely; and that the rate of O_2 depletion on Mars would be very slow (billions of years?), though not totally negligible. The temperatures of both planets may have been higher in past times than they are now.

In the analysis of the data, we must remember that the proportions, P, are not rates of depletion, though the rates should be roughly proportional to P.

The results of the calculations are summarized in the following table.

	240 K			1500 K		
	H_2	He	O_2	H_2	He	O_2
P(Earth)	3.7×10^{-27}	4×10^{-54}	0	1.5×10^{-4}	1.0×10^{-8}	0
P(Mars)	1.1×10^{-5}	4.9×10^{-11}	0	0.25	0.044	4.5×10^{-14}

P1.16 From definition of Z [34] and the virial equation [36], Z may be expressed in virial form as

$$Z = 1 + B\left(\frac{1}{V_m}\right) + C\left(\frac{1}{V_m}\right)^2 + \cdots$$

Since $V_m = \frac{RT}{p}$ [assumption of perfect gas], $\frac{1}{V_m} = \frac{p}{RT}$; hence upon substitution, and dropping terms beyond the second power of $\left(\frac{1}{V_m}\right)$

$$Z = 1 + B\left(\frac{p}{RT}\right) + C\left(\frac{p}{RT}\right)^2$$

$$Z = 1 + (-21.7 \times 10^{-3}\,\mathrm{L\,mol^{-1}}) \times \left(\frac{100\,\mathrm{atm}}{(0.0821\,\mathrm{L\,atm\,K^{-1}mol^{-1}}) \times (273\,\mathrm{K})}\right)$$

$$+ (1.200 \times 10^{-3}\,\mathrm{L^2\,mol^{-2}}) \times \left(\frac{100\,\mathrm{atm}}{(0.0821\,\mathrm{L\,atm\,K^{-1}\,mol^{-1}}) \times (273\,\mathrm{K})}\right)^2$$

$$Z = 1 - (0.0968) + (0.0239) = \boxed{0.927}$$

$$V_m = (0.927) \times \left(\frac{RT}{p}\right) = (0.927) \times \left(\frac{(0.0821\,\mathrm{L\,atm\,K^{-1}\,mol^{-1}}) \times (273\,\mathrm{K})}{100\,\mathrm{atm}}\right) = \boxed{0.208\,\mathrm{L}}$$

Question. What is the value of Z obtained from the next approximation using the value of V_m just calculated? Which value of Z is likely to be more accurate?

P1.17 As indicated by eqns 35 and 36 the compression factor of a gas may be expressed as either a virial expansion in p or in $\left(\frac{1}{V_m}\right)$. The virial form of the van der Waals equation is derived in Exercise 1.28(a) and is $p = \frac{RT}{V_m}\left\{1 + \left(b - \frac{a}{RT}\right) \times \left(\frac{1}{V_m}\right) + \cdots\right\}$

Rearranging, $Z = \frac{pV_m}{RT} = 1 + \left(b - \frac{a}{RT}\right) \times \left(\frac{1}{V_m}\right) + \cdots$

On the assumption that the perfect gas expression for V_m is adequate for the second term in this expansion, we can readily obtain Z as a function of p.

$$Z = 1 + \left(\frac{1}{RT}\right) \times \left(b - \frac{a}{RT}\right) p + \cdots$$

(a) $T_c = 126.3\,\mathrm{K}$

$$V_m = \left(\frac{RT}{p}\right) \times Z = \frac{RT}{p} + \left(b - \frac{a}{RT}\right) + \cdots$$

$$= \frac{(0.08206\,\mathrm{L\,atm\,K^{-1}\,mol^{-1}}) \times (126.3\,\mathrm{K})}{10.0\,\mathrm{atm}}$$

$$+ \left\{(0.03913\,\mathrm{L\,mol^{-1}}) - \left(\frac{1.408\,\mathrm{L^2\,atm\,mol^{-2}}}{(0.08206\,\mathrm{L\,atm\,K^{-1}mol^{-1}}) \times (126.3\,\mathrm{K})}\right)\right\}$$

$$= (1.036 - 0.097)\,\mathrm{L\,mol^{-1}} = \boxed{0.939\,\mathrm{L\,mol^{-1}}}$$

$$Z = \left(\frac{p}{RT}\right) \times (V_m) = \frac{(10.0\,\mathrm{atm}) \times (0.939\,\mathrm{L\,mol^{-1}})}{(0.08206\,\mathrm{L\,atm\,K^{-1}\,mol^{-1}}) \times (126.3\,\mathrm{K})} = 0.906$$

(b) The Boyle temperature corresponds to the temperature at which the second virial coefficient is zero, hence correct to the first power in p, $Z = 1$, and the gas is close to perfect. However, if we assume the N_2 is a van der Waals gas, when the second virial coefficient is zero

$$\left(b - \frac{a}{RT_B}\right) = 0, \quad \text{or} \quad T_B = \frac{a}{bR}$$

$$T_B = \frac{1.408\,\text{L}^2\,\text{atm mol}^{-2}}{(0.03913\,\text{L mol}^{-1}) \times (0.08206\,\text{L atm K}^{-1}\,\text{mol}^{-1})} = 439\,\text{K}$$

The experimental value (Table 1.5) is 327.2 K. Insertion of this value of T in the expression for Z above would not yield a Z of unity. The discrepancy may be explained by two considerations.

1. Terms beyond the first power in p should not be dropped in the expansion for Z.
2. Nitrogen is only approximately a van der Waals gas.

P1.18 **(a)** $$V_m = \frac{\text{molar mass}}{\text{density}} = \frac{M}{\rho} = \frac{18.02\,\text{g mol}^{-1}}{1.332 \times 10^2\,\text{g L}^{-1}} = \boxed{0.1353\,\text{L mol}^{-1}}$$

(b) $$Z = \frac{pV_m}{RT}[34] = \frac{(327.6\,\text{atm}) \times (0.1353\,\text{L mol}^{-1})}{(0.08206\,\text{L atm K}^{-1}\,\text{mol}^{-1}) \times (776.4\,\text{K})} = \boxed{0.6957}$$

(c) Two expansions for Z based on the van der Waals equation are given in Problem 1.17. They are

$$Z = 1 + \left(b - \frac{a}{RT}\right) \times \left(\frac{1}{V_m}\right) + \cdots$$

$$= 1 + \left\{(0.03049\,\text{L mol}^{-1}) - \left(\frac{5.536\,\text{L}^2\,\text{atm mol}^{-2}}{(0.08206\,\text{L atm K}^{-1}\,\text{mol}^{-1}) \times (776.4\,\text{K})}\right)\right\}$$

$$\times \frac{1}{0.1353\,\text{L mol}^{-1}} = 1 - 0.4169 = 0.5831 \approx 0.58$$

$$Z = 1 + \left(\frac{1}{RT}\right) \times \left(b - \frac{a}{RT}\right) \times (p) + \cdots$$

$$= 1 + \frac{1}{(0.08206\,\text{L atm K}^{-1}\,\text{mol}^{-1}) \times (776.4\,\text{K})}$$

$$\times \left\{(0.03049\,\text{L mol}^{-1}) - \left(\frac{5.536\,\text{L}^2\,\text{atm mol}^{-2}}{(0.08206\,\text{L atm K}^{-1}\,\text{mol}^{-1}) \times (776.4\,\text{K})}\right)\right\} \times 327.6\,\text{atm}$$

$$= 1 - 0.2900 \approx \boxed{0.71}$$

In this case the expansion in p gives a value close to the experimental value; the expansion in $\frac{1}{V_m}$ is not as good. However, when terms beyond the second are included the results from the two expansions for Z converge.

Solutions to theoretical problems

P1.22 The most probable speed of a gas molecule corresponds to the condition that the Maxwell distribution be a maximum (it has no minimum); hence we find it by setting the first derivative of the function to zero and solve for the value of v for which this condition holds.

$$f(v) = 4\pi\left(\frac{m}{2\pi kT}\right)^{3/2} v^2 e^{-mv^2/2kT} = \text{const} \times v^2 e^{-mv^2/2kT} \qquad \left[\frac{M}{R} = \frac{m}{k}\right]$$

$$\frac{df(v)}{ds} = 0 \quad \text{when} \quad \left(2 - \frac{mv^2}{kT}\right) = 0$$

So, $\boxed{v(\text{most probable}) = c^* = \left(\frac{2kT}{m}\right)^{1/2} = \left(\frac{2RT}{M}\right)^{1/2}}$

The average kinetic energy corresponds to the average of $\frac{1}{2}mv^2$. The average is obtained by determining $\langle v^2\rangle = \int_0^\infty v^2 f(v)\,dv = 4\pi\left(\frac{m}{2\pi}\right)^{3/2} \times \left(\frac{1}{kT}\right)^{3/2} \int_0^\infty v^4\, e^{-mv^2/2kT}\,dv$

The integral evaluates to $\frac{3}{8}\pi^{1/2}\left(\frac{m}{2kT}\right)^{-5/2}$. Then

$$\langle v^2\rangle = 4\pi\left(\frac{m}{2\pi}\right)^{3/2} \times \left(\frac{1}{kT}\right)^{3/2} \times \left(\frac{3}{8}\pi^{1/2}\right) \times \left(\frac{2kT}{m}\right)^{5/2} = \frac{3kT}{m}$$

thus $\langle\varepsilon\rangle = \frac{1}{2}m\langle v^2\rangle = \frac{3}{2}kT$

P1.23 We proceed as in *Justification* 1.2 except that, instead of taking a product of three one-dimensional distributions in order to get the three-dimensional distribution, we make a product of two one-dimensional distributions.

$$f(v_x, v_y)\,dv_x\,dv_y = f(v_x^2)f(v_y^2)\,dv_x\,dv_y = \left(\frac{m}{2\pi kT}\right) e^{-mv^2/2kT}\,dv_x\,dv_y$$

where $v^2 = v_x^2 + v_y^2$. The probability $f(v)\,dv$ that the molecules have a two-dimensional speed, v, in the range $v,\ v + dv$ is the sum of the probabilities that it is in any of the area elements $dv_x\,dv_y$ in the circular shell of radius v. The sum of the area elements is the area of the circular shell of radius v and thickness dv which is $\pi(v + dv)^2 - \pi v^2 = 2\pi v\,dv$. Therefore

$$\boxed{f(v) = 2\pi\left(\frac{m}{2\pi kT}\right) ve^{-mv^2/2kT}} \quad \left[\frac{M}{R} = \frac{m}{k}\right]$$

The mean speed is determined as $\bar{c} = \int_0^\infty vf(v)\,dv = \int_0^\infty \frac{m}{kT}v^2 e^{-mv^2/2kT}\,dv$

Using standard integrals this evaluates to $\boxed{\bar{c} = \left(\frac{\pi kT}{2m}\right)^{1/2} = \left(\frac{\pi RT}{2M}\right)^{1/2}}$

Comment. The two-dimensional gas serves as a model of the motion of molecules on surfaces. See Chapter 28.

P1.25 Rewriting eqn 22 with $\left(\frac{M}{R}\right) = \left(\frac{m}{k}\right)$

$$f(v) = 4\pi\left(\frac{m}{2\pi kT}\right)^{3/2} v^2 e^{-mv^2/2kT}$$

The proportion of molecules with speeds less than c is

$$P = \int_0^c f(v)\,dv = 4\pi\left(\frac{m}{2\pi kT}\right)^{3/2} \int_0^c v^2 e^{-mv^2/2kT}\,dv$$

Defining $a \equiv \frac{m}{2kT}$

$$P = 4\pi\left(\frac{a}{\pi}\right)^{3/2} \int_0^c v^2 e^{-av^2}\,dv = -4\pi\left(\frac{a}{\pi}\right)^{3/2} \frac{d}{da} \int_0^c e^{-av^2}\,dv$$

Defining $x^2 \equiv av^2$, $\mathrm{d}v = a^{-1/2}\,\mathrm{d}x$

$$P = -4\pi\left(\frac{a}{\pi}\right)^{3/2}\frac{\mathrm{d}}{\mathrm{d}a}\left\{\frac{1}{a^{1/2}}\int_0^{ca^{1/2}} \mathrm{e}^{-x^2}\mathrm{d}x\right\}$$

$$= -4\pi\left(\frac{a}{\pi}\right)^{3/2}\left\{-\frac{1}{2}\left(\frac{1}{a}\right)^{3/2}\int_0^{ca^{1/2}} \mathrm{e}^{-x^2}\,\mathrm{d}x + \left(\frac{1}{a}\right)^{1/2}\frac{\mathrm{d}}{\mathrm{d}a}\int_0^{ca^{1/2}} \mathrm{e}^{-x^2}\,\mathrm{d}x\right\}$$

Then we use $\int_0^{ca^{1/2}} \mathrm{e}^{-x^2}\mathrm{d}x = \left(\frac{\pi^{1/2}}{2}\right)\mathrm{erf}(ca^{1/2})$

$$\frac{\mathrm{d}}{\mathrm{d}a}\int_0^{ca^{1/2}} \mathrm{e}^{-x^2}\,\mathrm{d}x = \left(\frac{\mathrm{d}ca^{1/2}}{\mathrm{d}a}\right)\times(\mathrm{e}^{-c^2a}) = \frac{1}{2}\left(\frac{c}{a^{1/2}}\right)\mathrm{e}^{-c^2a}$$

Where we have used $\frac{\mathrm{d}}{\mathrm{d}z}\int_0^z f(y)\,\mathrm{d}y = f(z)$

Substituting and cancelling we obtain $P = \mathrm{erf}(ca^{1/2}) - \frac{2ca^{1/2}}{\pi^{1/2}}\mathrm{e}^{-c^2a}$

Now, $c = \left(\frac{3kT}{m}\right)^{1/2}$, so $ca^{1/2} = \left(\frac{3kT}{m}\right)^{1/2}\times\left(\frac{m}{2kT}\right)^{1/2} = \left(\frac{3}{2}\right)^{1/2}$, and

$$P = \mathrm{erf}\left(\sqrt{\frac{3}{2}}\right) - \left(\frac{6}{\pi}\right)^{1/2}\mathrm{e}^{-3/2} = 0.92 - 0.31 = \boxed{0.61}$$

Therefore **(b)** $\boxed{\text{61 per cent}}$ of the molecules have a speed less than the root mean square speed and **(a)** $\boxed{\text{39 per cent}}$ have a speed greater than the root mean square speed.

(c) For the proportions in terms of the mean speed $\bar{c}$, replace c by $\bar{c} = \left(\frac{8kT}{\pi m}\right)^{1/2} = \left(\frac{8}{3\pi}\right)^{1/2} c$, so $\bar{c}a^{1/2} = \frac{2}{\pi^{1/2}}$.

Then $P = \mathrm{erf}(\bar{c}a^{1/2}) - \left(\frac{2\bar{c}a^{1/2}}{\pi^{1/2}}\right)\times(\mathrm{e}^{-\bar{c}^2a}) = \mathrm{erf}\left(\frac{2}{\pi^{1/2}}\right) - \frac{4}{\pi}\mathrm{e}^{-4/\pi} = 0.889 - 0.356 = \boxed{0.533}$

That is, $\boxed{\text{53 per cent}}$ of the molecules have a speed less than the mean, and $\boxed{\text{47 per cent}}$ have a speed greater than the mean.

P1.27 $Z = \frac{pV_\mathrm{m}}{RT} = \frac{1}{\left(1-\frac{b}{V_\mathrm{m}}\right)} - \frac{a}{RTV_\mathrm{m}}$ [see Exercise 1.28(a).]

which upon expansion of $\left(1-\frac{b}{V_\mathrm{m}}\right)^{-1} = 1 + \frac{b}{V_\mathrm{m}} + \left(\frac{b}{V_\mathrm{m}}\right)^2 + \cdots$ yields

$$Z = 1 + \left(b - \frac{a}{RT}\right)\times\left(\frac{1}{V_\mathrm{m}}\right) + b^2\left(\frac{1}{V_\mathrm{m}}\right)^2 + \cdots$$

We note that all terms beyond the second are necessarily positive, so only if

$$\frac{a}{RTV_\mathrm{m}} > \frac{b}{V_\mathrm{m}} + \left(\frac{b}{V_\mathrm{m}}\right)^2 + \cdots$$

can Z be less than one. If we ignore terms beyond $\dfrac{b}{V_m}$ the conditions are simply stated as

$$Z < 1 \quad \text{when } \frac{a}{RT} > b \qquad Z > 1 \quad \text{when } \frac{a}{RT} < b$$

Thus $Z < 1$ when attractive forces predominate, and $Z > 1$ when size effects (short-range repulsions) predominate.

P1.29 The critical point corresponds to a point of zero slope which is simultaneously a point of inflection in a plot of pressure versus molar volume. A critical point exists if there are values of p, V, and T which result in a point which satisfies these conditions.

$$p = \frac{RT}{V_m} - \frac{B}{V_m^2} + \frac{C}{V_m^3}$$

$$\left.\begin{aligned} \left(\frac{\partial p}{\partial V_m}\right)_T &= -\frac{RT}{V_m^2} + \frac{2B}{V_m^3} - \frac{3C}{V_m^4} = 0 \\ \left(\frac{\partial^2 p}{\partial V_m^2}\right)_T &= \frac{2RT}{V_m^3} - \frac{6B}{V_m^4} + \frac{12C}{V_m^5} = 0 \end{aligned}\right\} \text{at the critical point}$$

That is, $\left.\begin{aligned} -RT_cV_c^2 + 2BV_c - 3C = 0 \\ RT_cV_c^2 - 3BV_c + 6C = 0 \end{aligned}\right\}$

which solve to $V_c = \boxed{\dfrac{3C}{B}}$, $\boxed{T_c = \dfrac{B^2}{3RC}}$

Now use the equation of state to find p_c

$$p_c = \frac{RT_c}{V_c} - \frac{B}{V_c^2} + \frac{C}{V_c^3} = \left(\frac{RB^2}{3RC}\right) \times \left(\frac{B}{3C}\right) - B\left(\frac{B}{3C}\right)^2 + C\left(\frac{B}{3C}\right)^3 = \boxed{\frac{B^3}{27C^2}}$$

It follows that $Z_c = \dfrac{p_cV_c}{RT_c} = \left(\dfrac{B^3}{27C^2}\right) \times \left(\dfrac{3C}{B}\right) \times \left(\dfrac{1}{R}\right) \times \left(\dfrac{3RC}{B^2}\right) = \boxed{\dfrac{1}{3}}$

P1.30 $\dfrac{pV_m}{RT} = 1 + B'p + C'p^2 + \cdots \quad [35]$

$\dfrac{pV_m}{RT} = 1 + \dfrac{B}{V_m} + \dfrac{C}{V_m^2} + \cdots \quad [36]$

whence $B'p + C'p^2 + \cdots = \dfrac{B}{V_m} + \dfrac{C}{V_m^2} + \cdots$

Now multiply through by V_m, replace pV_m by $RT\{1 + (B/V_m) + \cdots\}$, and equate coefficients of powers of $\dfrac{1}{V_m}$: $B'RT + \dfrac{BB'RT + C'R^2T^2}{V_m} + \cdots = B + \dfrac{C}{V_m} + \cdots$

Hence, $B'RT = B$, implying that $\boxed{B' = \dfrac{B}{RT}}$

Also, $BB'RT + C'R^2T^2 = C$, or $B^2 + CR^2T^2 = C$, implying that $\boxed{C' = \dfrac{C - B^2}{R^2T^2}}$

P1.33 The critical temperature is that temperature above which the gas cannot be liquefied by the application of pressure alone. Below the critical temperature two phases, liquid and gas, may coexist at equilibrium; and in the two-phase region there is more than one molar volume corresponding to the same conditions of temperature and pressure. Therefore, any equation of state that can even approximately describe this situation must allow for more than one real root for the molar volume at some values of T and p, but as the temperature is increased above T_c, allows only one real root. Thus, appropriate equations of state must be equations of odd degree in V_m.

The equation of state for gas A may be rewritten $V_m^2 - \frac{RT}{p}V_m - \frac{RTb}{p} = 0$ which is a quadratic and never has just one real root. Thus, this equation can never model critical behaviour. It could possibly model in a very crude manner a two-phase situation, since there are some conditions under which a quadratic has two real positive roots, but not the process of liquefaction.

The equation of state of gas B is a first-degree equation in V_m and therefore can never model critical behaviour, the process of liquefaction, or the existence of a two-phase region.

A cubic equation is the equation of lowest degree which can show a cross-over from more than one real root to just one real root as the temperature increases. The van der Waals equation is a cubic equation in V_m.

P1.35 The pressure at the base of a column of height H is $p = \rho g H$ (Example 1.2). But the pressure at any altitude h within the atmospheric column of height H depends only on the air above it; therefore

$$p = \rho g(H - h) \quad \text{and} \quad \mathrm{d}p = -\rho g\,\mathrm{d}h$$

Since $\rho = \frac{pM}{RT}$ [Problem 1.5], $\mathrm{d}p = -\frac{pMg\,\mathrm{d}h}{RT}$, implying that $\frac{\mathrm{d}p}{p} = -\frac{Mg\,\mathrm{d}h}{RT}$

This relation integrates to $p = p_0 e^{-Mgh/RT}$

For air, $M \approx 29\,\mathrm{g\,mol^{-1}}$ and at 298 K

$$\frac{Mg}{RT} \approx \frac{(29 \times 10^{-3}\,\mathrm{kg\,mol^{-1}}) \times (9.81\,\mathrm{m s^{-2}})}{2.48 \times 10^3\,\mathrm{J\,mol^{-1}}} = 1.1\bar{5} \times 10^{-4}\,\mathrm{m^{-1}} \quad [1\,\mathrm{J} = 1\,\mathrm{kg\,m^2\,s^{-2}}]$$

(a) $h = 15\,\mathrm{cm}$

$$p = p_0 \times e^{(-0.15\,\mathrm{m})\times(1.1\bar{5}\times10^{-4}\,\mathrm{m^{-1}})} = 0.99\overline{998}\,p_0; \qquad \frac{p - p_0}{p_0} = \boxed{0.00}$$

(b) $h = 1350\,\mathrm{ft}$, which is equivalent to 412 m [1 inch = 2.54 cm]

$$p = p_0 \times e^{(-412\,\mathrm{m})\times(1.1\bar{5}\times10^{-4}\,\mathrm{m^{-1}})} = 0.95\,p_0; \qquad \frac{p - p_0}{p_0} = \boxed{-0.05}$$

Solutions to additional problems

P1.36 Avogadro's principle states that equal volumes of gas contain equal numbers of molecules. Consequently, the ratio of the masses of equal volumes of two gases is the ratio of their molecular masses and hence their molar masses. Thus, the density of a gas (mass per unit volume) is proportional to its molar mass (mass per mole), and ratios of densities are equal to ratios of molar masses. So the molar mass of water relative to that of hydrogen is

$$\frac{0.625}{0.0732} = \boxed{8.54}$$

For oxygen, we need the reaction stoichiometry: the mass of a unit volume of water vapour consists of the mass of a unit volume of hydrogen plus the mass of half a unit volume of oxygen. Thus, the density of oxygen would be expected to be 2 × (0.625 − 0.0732), and its molar mass relative to hydrogen is

$$\frac{2 \times (0.625 - 0.0732)}{0.0732} = \boxed{15.1}$$

Comment. The differences from the modern values are not unexpected.

P1.38 The virial equation is

$$pV_m = RT\left(1 + \frac{B}{V_m} + \frac{C}{V_m^2} + \cdots\right) \quad \text{or}$$

$$\frac{pV_m}{RT} = 1 + \frac{B}{V_m} + \frac{C}{V_m^2} + \cdots$$

(a) If we assume that the series may be truncated after the B term, then a plot of $\frac{pV_m}{RT}$ vs $\frac{1}{V_m}$ will have B as its slope and 1 as its y-intercept. Transforming the data gives

p/MPa	$V_m/(\text{L mol}^{-1})$	pV_m/RT	$(1/V_m)/(\text{mol L}^{-1})$
0.4000	6.2208	0.9976	0.1608
0.5000	4.9736	0.9970	0.2011
0.6000	4.1423	0.9964	0.2414
0.8000	3.1031	0.9952	0.3223
1.000	2.4795	0.9941	0.4033
1.500	1.6483	0.9912	0.6067
2.000	1.2328	0.9885	0.8112
2.500	0.98357	0.9858	1.017
3.000	0.81746	0.9832	1.223
4.000	0.60998	0.9782	1.639

A plot of the data in the third column against that of the fourth column is shown in Fig. 1.3. The data fit a straight line reasonably well, and the y-intercept is very close to 1. The regression yields $B = \boxed{-1.32 \times 10^{-2}\ \text{L mol}^{-1}}$

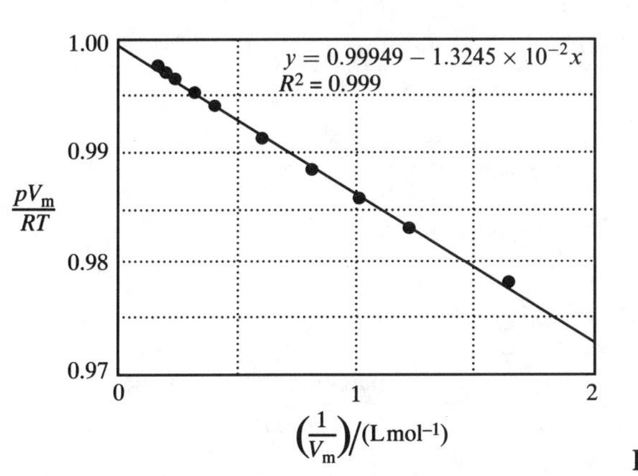

Figure 1.3

(b) A quadratic function fits the data somewhat better (Fig. 1.4) with a slightly better correlation coefficient and a y-intercept closer to 1. This fit implies that truncation of the viral series after the term with C is more accurate than after just the B term. The regression then yields

$$B = \boxed{-1.51 \times 10^{-2}\ \text{L mol}^{-1}} \quad \text{and} \quad C = \boxed{1.07 \times 10^{-3}\ \text{L}^2\ \text{mol}^{-2}}$$

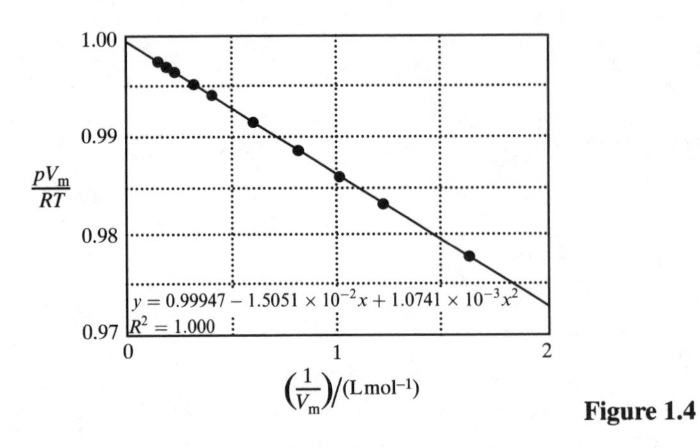

Figure 1.4

P1.40 Avogadro's principle states that equal volumes of gases represent equal amounts (moles) of the gases, so the volume mixing ratio is equal to the mole fraction. The definition of partial pressures is

$$p_J = x_J p$$

The perfect gas law is

$$pV = nRT \quad \text{so} \quad \frac{n_J}{V} = \frac{p_J}{RT} = \frac{x_J p}{RT}$$

(a) $$\frac{n(\text{CCl}_3\text{F})}{V} = \frac{(261 \times 10^{-12}) \times (1.0\,\text{atm})}{(0.08206\,\text{L atm K}^{-1}\text{mol}^{-1}) \times (10 + 273)\,\text{K}} = \boxed{1.1 \times 10^{-11}\,\text{mol L}^{-1}}$$

and $$\frac{n(\text{CCl}_2\text{F}_2)}{V} = \frac{(509 \times 10^{-12}) \times (1.0\,\text{atm})}{(0.08206\,\text{L atm K}^{-1}\text{mol}^{-1}) \times (10 + 273)\,\text{K}} = \boxed{2.2 \times 10^{-11}\,\text{mol L}^{-1}}$$

(b) $$\frac{n(\text{CCl}_3\text{F})}{V} = \frac{(261 \times 10^{-12}) \times (0.050\,\text{atm})}{(0.08206\,\text{L atm K}^{-1}\text{mol}^{-1}) \times (200\,\text{K})} = \boxed{8.0 \times 10^{-13}\,\text{mol L}^{-1}}$$

and $$\frac{n(\text{CCl}_2\text{F}_2)}{V} = \frac{(509 \times 10^{-12}) \times (0.050\,\text{atm})}{(0.08206\,\text{L atm K}^{-1}\text{mol}^{-1}) \times (200\,\text{K})} = \boxed{1.6 \times 10^{-12}\,\text{mol L}^{-1}}$$

P1.41 $$\text{Concentration of } ^1\text{H nuclei, } [^1\text{H}] = \frac{n_\text{H}}{V} = \frac{(\text{mass percentage}) \times (\text{density})}{100(\text{molar mass})}$$

$$= \frac{0.36(158\,\text{g cm}^{-3})}{1.0\,\text{g mol}^{-1}}$$

$$= 57\,\text{mol cm}^{-3}$$

$$\text{Concentration of } ^4\text{He nuclei, } [^4\text{He}] = \frac{n_\text{He}}{V} = \frac{(\text{mass percentage}) \times (\text{density})}{100(\text{molar mass})}$$

$$= \frac{0.64(158\,\text{g cm}^{-3})}{4.0\,\text{g mol}^{-1}}$$

$$= 25\,\text{mol cm}^{-3}$$

$$\text{Concentration of e}^- = [^1\text{H}] + 2[^4\text{He}] = (57 + 2 \times 25)\,\text{mol cm}^{-3} = 107\,\text{mol cm}^{-3}$$

Total concentration of gaseous particles $= (57 + 25 + 107)\,\text{mol cm}^{-3} = 189\,\text{mol cm}^{-3}$

$$r_{\text{H nucleus}} = (1.4 \times 10^{-13}\,\text{cm})(1)^{1/3} = 1.4 \times 10^{-13}\,\text{cm}$$

$$r_{\text{He nucleus}} = (1.4 \times 10^{-13}\,\text{cm})(4)^{1/3} = 2.2 \times 10^{-13}\,\text{cm}$$

(a) The excluded volume of a nuclear collisional *pair* is estimated to be equal to the volume of the dashed sphere in Fig. 1.5. The excluded volume of a single nucleus is 1/2 of this.

$$b \approx (N_{\text{A}}) \times \left(\frac{1}{2}\right) \times \left[\frac{4\pi}{3}(2r)^3\right] = \frac{16\pi}{3} N_{\text{A}} r^3$$

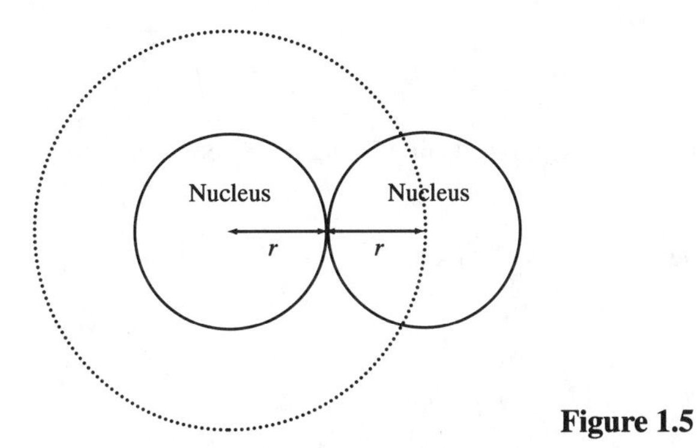

Figure 1.5

In this problem we have a mixture of hydrogen and helium nuclei so let us take 'r' to equal the weighted average of hydrogen and helium radii. This is, of course, a very simple estimate. Then

$$r \approx 0.36(1.4 \times 10^{-13}\,\text{cm}) + 0.64(2.2 \times 10^{-13}\,\text{cm})$$
$$r \approx 1.9 \times 10^{-13}\,\text{cm}$$
$$b \approx \frac{16\pi}{3}(6.022 \times 10^{23}\,\text{mol}^{-1}) \times (1.9 \times 10^{-13}\,\text{cm})^3$$
$$\boxed{b \approx 7.1 \times 10^{-14}\,\text{cm}^3\,\text{mol}^{-1}}$$
$$b(\text{per cm}^3) \approx 82\,\text{mol} \times 7.1 \times 10^{-14}\,\text{cm}^3\,\text{mol}^{-1}$$
$$\approx 5.8 \times 10^{-12}\,\text{cm}^3$$

This b is extraordinarily small compared to 1 cm^3 and the values of $[^1\text{H}]$ and $[^4\text{He}]$, so we may treat the nuclei as if they are points within any macroscopic volume. In the sense that the nuclei act as volumeless points, the perfect gas law would seem to be applicable. However, our analysis has not included details of the internuclear forces and these may be appreciably larger than the hard-sphere model estimate.

(b) $T_{\text{perfect}} = \dfrac{pV}{nR}$, where n = total number of moles of gaseous particles including the number of moles of electrons

$$= \frac{p}{\left(\frac{n}{V}\right) R}$$

$$= \frac{(2.5 \times 10^{11}\,\text{atm}) \times (1\,\text{L}/10^3\,\text{cm}^3)}{(189\,\text{mol cm}^{-3}) \times (0.0821\,\text{L atm K}^{-1}\,\text{mol}^{-1})} = \boxed{1.6 \times 10^7\,\text{K} = T_{\text{perfect}}}$$

(c)
$$T_{\text{van der Waals}} = \frac{V_m - b}{R}\left(p + \frac{a}{V_m^2}\right) \text{ [39]}$$
$$= \frac{p(V_m - b)}{R} \quad \text{assuming } a \approx 0$$
$$= \frac{p}{R}\left(\frac{V}{n} - b\right) = \frac{p}{R}\left\{\left(\frac{1}{\frac{n_{\text{total}}}{V}}\right) - b\right\}$$
$$= \left(\frac{2.5 \times 10^{11}\ \text{atm}}{0.0821\ \text{L atm K}^{-1}\ \text{mol}^{-1}}\right)$$
$$\times \left\{\frac{1}{189\ \text{mol cm}^{-3}} - 7.1 \times 10^{-14}\ \text{cm}^3\ \text{mol}^{-1}\right\}$$
$$= T_{\text{perfect}} - \frac{(7.1 \times 10^{-14}\ \text{cm}^3\ \text{mol}^{-1}) \times (2.5 \times 10^{11}\ \text{atm})}{(0.0821\ \text{L atm K}^{-1}\ \text{mol}^{-1})}$$

where the last term is negligible. Therefore,

$$T_{\text{van der Waals}} = T_{\text{perfect}}$$

P1.43 We want the height h at which $p(O_2) = \frac{1}{9}p(N_2)$. According to the barometric equation

$$p(O_2) = p_0(O_2)e^{-\left(\frac{M_{O_2}gh}{RT}\right)} \quad \text{where } p_0(O_2) = 0.20\ \text{bar}$$

and

$$p(N_2) = p_0(N_2)e^{-\left(\frac{M_{N_2}gh}{RT}\right)} \quad \text{where } p_0(N_2) = 0.80\ \text{bar}$$

$$p(O_2) = \frac{1}{9}p(N_2)$$

$$p_0(O_2)e^{-\left(\frac{M_{O_2}gh}{RT}\right)} = \frac{1}{9}p_0(N_2)e^{-\left(\frac{M_{N_2}gh}{RT}\right)}$$

$$e^{\left(\frac{(M_{N_2}-M_{O_2})gh}{RT}\right)} = \frac{p_0(N_2)}{9p_0(O_2)}$$

Solving for h,

$$h = \frac{RT}{(M_{N_2} - M_{O_2})g}\ln\left(\frac{p_0(N_2)}{9p_0(O_2)}\right)$$
$$= \frac{(8.31451\ \text{J K}^{-1}\ \text{mol}^{-1}) \times (298.15\ \text{K}) \times \ln\left(\frac{0.80\ \text{bar}}{9(0.20\ \text{bar})}\right)}{(0.02802\ \text{kg mol}^{-1} - 0.03200\ \text{kg mol}^{-1}) \times (9.80665\ \text{m s}^{-2})}$$
$$= 5.15 \times 10^4\ \text{m} = \boxed{51.5\ \text{km}}$$

The pressure at 51.5 km can be obtained by using the barometric equation twice, once for O_2 and again for N_2

$$p(O_2) = 0.20 \times e^{-\left(\frac{0.03200\ \text{kg mol}^{-1} \times 9.8067\ \text{m s}^{-2} \times 5.15 \times 10^4\ \text{m}}{8.314\ \text{J K}^{-1}\ \text{mol}^{-1} \times 298\ \text{K}}\right)}$$
$$= 2.9 \times 10^{-4}\ \text{bar}$$

$$p(N_2) = 0.80 \times e^{-\left(\frac{0.02802\ \text{kg mol}^{-1} \times 9.8067\ \text{m s}^{-2} \times 5.15 \times 10^4\ \text{m}}{8.314\ \text{J K}^{-1}\ \text{mol}^{-1} \times 298\ \text{K}}\right)}$$
$$= 2.66 \times 10^{-3}\ \text{bar}$$

$$p = p(O_2) + p(N_2) = \boxed{3.0 \times 10^{-3}\ \text{bar}}$$

2 The First Law: the concepts

Solutions to exercises

Assume all gases are perfect unless stated otherwise. Unless otherwise stated, thermochemical data are for 298 K.

E2.1 The physical definition of work is $dw = -F\,dz$ [6]

In a gravitational field the force is the weight of the object, which is $F = mg$

If g is constant over the distance the mass moves, dw may be intergrated to give the total work

$$w = -\int_{z_i}^{z_f} F\,dz = -\int_{z_i}^{z_f} mg\,dz = -mg(z_f - z_i) = -mgh \quad \text{where } h = (z_f - z_i)$$

(a) $w = (-1.0\,\text{kg}) \times (9.81\,\text{m s}^{-2}) \times (10\,\text{m}) = -98\,\text{J} = \boxed{98\,\text{J done}}$

(b) $w = (-1.0\,\text{kg}) \times (1.60\,\text{m s}^{-2}) \times (10\,\text{m}) = -16\,\text{J} = \boxed{16\,\text{J done}}$

E2.2 $w = -mgh$ [Exercise 2.1 **(a)**] $= (-65\,\text{kg}) \times (9.81\,\text{m s}^{-2}) \times (4.0\,\text{m})$

$= 2.6\,\text{kJ} = \boxed{2.6\,\text{kJ needed}}$

E2.3 This is an expansion against a constant external pressure; hence $w = -p_{ex}\Delta V$ [10]

$$p_{ex} = (1.0\,\text{atm}) \times (1.013 \times 10^5\,\text{Pa atm}^{-1}) = 1.0\bar{1} \times 10^5\,\text{Pa}$$
$$\Delta V = (100\,\text{cm}^2) \times (10\,\text{cm}) = 1.0 \times 10^3\,\text{cm}^3 = 1.0 \times 10^{-3}\,\text{m}^3$$
$$w = (-1.0\bar{1} \times 10^5\,\text{Pa}) \times (1.0 \times 10^{-3}\,\text{m}^3) = \boxed{-1.0 \times 10^2\,\text{J}} \text{ as } 1\,\text{Pa m}^3 = 1\,\text{J}$$

E2.4 For all cases $\Delta U = 0$, since the internal energy of a perfect gas depends only on temperature. (*Molecular interpretation* 2.2 and Section 3.1 for a more complete discussion.) From the definition of enthalpy, $H = U + pV$, $\Delta H = \Delta U + \Delta(pV) = \Delta U + \Delta(nRT)$ (perfect gas). Hence, $\Delta H = 0$ as well, at constant temperature for all processes in a perfect gas.

(a) $\boxed{\Delta U = \Delta H = 0}$

$$w = -nRT\ln\left(\frac{V_f}{V_i}\right) \text{ [13]}$$
$$= (-1.00\,\text{mol}) \times (8.314\,\text{J K}^{-1}\,\text{mol}^{-1}) \times (273\,\text{K}) \times \ln\left(\frac{44.8\,\text{L}}{22.4\,\text{L}}\right)$$
$$= -1.57 \times 10^3\,\text{J} = \boxed{-1.57\,\text{kJ}}$$
$$q = \Delta U - w \text{ (First Law)} = 0 + 1.57\,\text{kJ} = \boxed{+1.57\,\text{kJ}}$$

(b) $\boxed{\Delta U = \Delta H = 0}$

$$w = -p_{ex}\Delta V \text{ [10]} \quad \Delta V = (44.8 - 22.4)\,\text{L} = 22.4\,\text{L}$$
$$p_{ex} = p_f = \frac{nRT}{V_f} = \frac{(1.00\,\text{mol}) \times (0.08206\,\text{L atm K}^{-1}\,\text{mol}^{-1}) \times (273\,\text{K})}{44.8\,\text{L}} = 0.500\,\text{atm}$$
$$w = (-0.500\,\text{atm}) \times \left(\frac{1.013 \times 10^5\,\text{Pa}}{1\,\text{atm}}\right) \times (22.4\,\text{L}) \times \left(\frac{1\,\text{m}^3}{10^3\,\text{L}}\right)$$

$$= -1.13 \times 10^3\,\text{Pa m}^3 = -1.13 \times 10^3\,\text{J}$$
$$= \boxed{-1.13\,\text{kJ}}$$
$$q = \Delta U - w = 0 + 1.13\,\text{kJ} = \boxed{+1.13\,\text{kJ}}$$

(c) $\boxed{\Delta U = \Delta H = 0}$

$\boxed{w = 0}$ [free expansion] $q = \Delta U - w = 0 - 0 = \boxed{0}$

Comment. An isothermal free expansion of a perfect gas is also adiabatic.

E2.5 For a perfect gas at constant volume

$$\frac{p}{T} = \frac{nR}{V} = \text{constant}, \quad \text{hence,} \quad \frac{p_1}{T_1} = \frac{p_2}{T_2}$$
$$p_2 = \left(\frac{T_2}{T_1}\right) \times p_1 = \left(\frac{400\,\text{K}}{300\,\text{K}}\right) \times (1.00\,\text{atm}) = \boxed{1.33\,\text{atm}}$$
$$\Delta U = nC_{V,\text{m}}\Delta T[22b] = (n) \times \left(\tfrac{3}{2}R\right) \times (400\,\text{K} - 300\,\text{K})$$
$$= (1.00\,\text{mol}) \times \left(\tfrac{3}{2}\right) \times (8.314\,\text{J K}^{-1}\,\text{mol}^{-1}) \times (100\,\text{K})$$
$$= 1.25 \times 10^3\,\text{J} = \boxed{+1.25\,\text{kJ}}$$

$\boxed{w = 0}$ [constant volume] $\quad q = \Delta U - w$ [First Law] $= 1.25\,\text{kJ} - 0 = \boxed{+1.25\,\text{kJ}}$

E2.6 **(a)** $w = -p_{\text{ex}}\Delta V$ [10]

$p_{\text{ex}} = (200\,\text{Torr}) \times (133.3\,\text{Pa Torr}^{-1}) = 2.66\bar{6} \times 10^4\,\text{Pa}$

$\Delta V = 3.3\,\text{L} = 3.3 \times 10^{-3}\,\text{m}^3$

Therefore, $w = (-2.66\bar{6} \times 10^4\,\text{Pa}) \times (3.3 \times 10^{-3}\,\text{m}^3) = \boxed{-88\,\text{J}}$

(b) $w = -nRT \ln \dfrac{V_\text{f}}{V_\text{i}}$ [13]

$$n = \frac{4.50\,\text{g}}{16.04\,\text{g mol}^{-1}} = 0.280\bar{5}\,\text{mol}, \quad RT = 2.577\,\text{kJ mol}^{-1},\ V_\text{i} = 12.7\,\text{L},\ V_\text{f} = 16.0\,\text{L}$$
$$w = -(0.280\bar{5}\,\text{mol}) \times (2.577\,\text{kJ mol}^{-1}) \times \ln\left(\frac{16.0\,\text{L}}{12.7\,\text{L}}\right) = \boxed{-167\,\text{J}}$$

E2.7 $w = -nRT \ln \dfrac{V_\text{f}}{V_\text{i}}$ [13] $\quad V_\text{f} = \dfrac{1}{3}V_\text{i}$

$nRT = (5.20 \times 10^{-3}\,\text{mol}) \times (8.314\,\text{J K}^{-1}\,\text{mol}^{-1}) \times (260\,\text{K}) = 1.12\bar{4} \times 10^2\,\text{J}$

$w = -(1.12\bar{4} \times 10^2\,\text{J}) \times \ln\tfrac{1}{3} = \boxed{+123\,\text{J}}$

E2.8 $\Delta H = \Delta H_{\text{cond}} = -\Delta H_{\text{vap}} = (-1\,\text{mol}) \times (40.656\,\text{kJ mol}^{-1}) = \boxed{-40.656\,\text{kJ}}$

Since the condensation is done isothermally and reversibly, the external pressure is constant at 1.00 atm. Hence,

$$q = q_p = \Delta H = \boxed{-40.656\,\text{kJ}}$$
$$w = -p_{\text{ex}}\Delta V\ [10] \quad \Delta V = V_{\text{liq}} - V_{\text{vap}} \approx -V_{\text{vap}}\ [V_{\text{liq}} \ll V_{\text{vap}}]$$

On the assumption that $H_2O(g)$ is a perfect gas, $V_{vap} = \dfrac{nRT}{p}$ and $p = p_{ex}$, since the condensation is done reversibly. Hence,

$$w = nRT = (1.00\,\text{mol}) \times (8.314\,\text{J K}^{-1}\,\text{mol}^{-1}) \times (373\,\text{K}) = +3.10 \times 10^3\,\text{J} = \boxed{+3.10\,\text{kJ}}$$

From eqn 26 $\quad \Delta U = \Delta H - \Delta n_g RT \quad \Delta n_g = -1.00\,\text{mol}$

$$\Delta U = (-40.656\,\text{kJ}) + (1.00\,\text{mol}) \times (8.314\,\text{J K}^{-1}\,\text{mol}^{-1}) \times (373.15\,\text{K}) = \boxed{-37.55\,\text{kJ}}$$

E2.9 The chemical reaction that occurs is

$$\text{Mg(s)} + 2\text{HCl(aq)} \rightarrow \text{H}_2\text{(g)} + \text{MgCl}_2\text{(aq)}, \quad M(\text{Mg}) = 24.31\,\text{g mol}^{-1}$$

Work is done against the atmosphere by the expansion of the hydrogen gas produced in the reaction.

$$w = -p_{ex}\Delta V\ [10]$$

$$V_i = 0,\ V_f = \frac{nRT}{p_f},\ p_f = p_{ex} \qquad w = -p_{ex}(V_f - V_i) = (-p_{ex}) \times \frac{nRT}{p_{ex}} = -nRT$$

$$n = \frac{15\,\text{g}}{24.31\,\text{g mol}^{-1}} = 0.61\overline{7}\,\text{mol}, \quad RT = 2.479\,\text{kJ mol}^{-1}$$

Hence, $w = (-0.61\overline{7}\,\text{mol}) \times (2.479\,\text{kJ mol}^{-1}) = \boxed{-1.5\,\text{kJ}}$

E2.10

$$q = n\Delta H^{\ominus}_{fus} \quad \Delta H^{\ominus}_{fus} = 2.60\,\text{kJ mol}^{-1}\ [\text{Table 2.3}]$$

$$n = \frac{750 \times 10^3\,\text{g}}{22.99\,\text{g mol}^{-1}} = 3.26\overline{2} \times 10^4\,\text{mol}$$

$$q = (3.26\overline{2} \times 10^4\,\text{mol}) \times (2.60\,\text{kJ mol}^{-1}) = \boxed{+8.48 \times 10^4\,\text{kJ}}$$

E2.11 **(a)** $q = \Delta H$, since pressure is constant

$$\Delta H = \int_{T_i}^{T_f} dH, \quad dH = nC_{p,m}\,dT$$

$$d(H/\text{J}) = \{20.17 + 0.3665(T/\text{K})\}\,d(T/\text{K})$$

$$\Delta(H/\text{J}) = \int_{T_i}^{T_f} d(H/\text{J}) = \int_{298}^{473} \{20.17 + 0.3665(T/\text{K})\}\,d(T/\text{K})$$

$$= (20.17) \times (473 - 298) + \left(\frac{0.3665}{2}\right) \times \left(\frac{T}{\text{K}}\right)^2 \Bigg|_{298}^{473}$$

$$= (3.53\overline{0} \times 10^3) + (2.47\overline{25} \times 10^4)$$

$$q = \Delta H = \boxed{2.83 \times 10^4\,\text{J}} = \boxed{+28.3\,\text{kJ}}$$

$$w = -p_{ex}\Delta V\ [10], \quad p_{ex} = p$$

$$= -p\Delta V = -\Delta(pV)[\text{constant pressure}] = -\Delta(nRT)[\text{perfect gas}] = -nR\Delta T$$

$$= (-1.00\,\text{mol}) \times (8.314\,\text{J K}^{-1}\,\text{mol}^{-1}) \times (473\,\text{K} - 298\,\text{K}) = \boxed{-1.45 \times 10^3\,\text{J}}$$

$$= \boxed{-1.45\,\text{kJ}}$$

$$\Delta U = q + w = (28.3\,\text{kJ}) - (1.45\,\text{kJ}) = \boxed{+26.8\,\text{kJ}}$$

(b) The energy and enthalpy of a perfect gas depend on temperature alone (*Molecular interpretation* 2.2 and Exercise 2.4); hence it does not matter whether the temperature change is brought about at constant volume or constant pressure; ΔH and ΔU are the same.

$$\Delta H = \boxed{+28.3\,\text{kJ}}, \quad \Delta U = \boxed{+26.8\,\text{kJ}}, \quad w = \boxed{0} \quad \text{[constant volume]}$$

$$q = \Delta U - w = \boxed{+26.8\,\text{kJ}}$$

E2.12 For reversible adiabatic expansion

$$V_\text{f} T_\text{f}^c = V_\text{i} T_\text{i}^c \;[34] \quad \text{so} \quad T_\text{f} = T_\text{i}\left(\frac{V_\text{i}}{V_\text{f}}\right)^{1/c}$$

where $c = \dfrac{C_{V,\text{m}}}{R} = \dfrac{C_{p,\text{m}} - R}{R} = \dfrac{(20.786 - 8.3145)\,\text{J K}^{-1}\,\text{mol}^{-1}}{8.3145\,\text{J K}^{-1}\,\text{mol}^{-1}} = 1.500$

So the final temperature is

$$T_\text{f} = (273.15\,\text{K}) \times \left(\frac{1.0\,\text{L}}{3.0\,\text{L}}\right)^{1/1.500} = \boxed{13\underline{1}\,\text{K}}$$

E2.13 Reversible adiabatic work is

$$w = C_V \Delta T = C_V \Delta T = n(C_{p,\text{m}} - R) \times (T_\text{f} - T_\text{i})$$

where the temperatures are related by

$$T_\text{f} = T_\text{i}\left(\frac{V_\text{i}}{V_\text{f}}\right)^{1/c} \;[35] \quad \text{where } c = \frac{C_{V,\text{m}}}{R} = \frac{C_{p,\text{m}} - R}{R} = 3.463$$

So $T_\text{f} = [(27.0 + 273.15)\,\text{K}] \times \left(\dfrac{500 \times 10^{-3}\,\text{L}}{3.00\,\text{L}}\right)^{1/3.463} = 179\,\text{K}$

and $w = \left(\dfrac{2.45\,\text{g}}{44.0\,\text{g mol}^{-1}}\right) \times [(37.11 - 8.3145)\,\text{J K}^{-1}\text{mol}^{-1}] \times (300 - 179)\,\text{K} = \boxed{194\,\text{J}}$

E2.14 For reversible adiabatic expansion

$$p_\text{f} V_\text{f}^\gamma = p_\text{i} V_\text{i}^\gamma \;[36] \quad \text{so} \quad p_\text{f} = p_\text{i}\left(\frac{V_\text{i}}{V_\text{f}}\right)^\gamma = (57.4\,\text{kPa}) \times \left(\frac{1.0\,\text{L}}{2.0\,\text{L}}\right)^{1.4} = \boxed{22\,\text{kPa}}$$

E2.15 For reversible adiabatic expansion

$$p_\text{f} V_\text{f}^\gamma = p_\text{i} V_\text{i}^\gamma \;[36] \quad \text{so} \quad p_\text{f} = p_\text{i}\left(\frac{V_\text{i}}{V_\text{f}}\right)^\gamma$$

We need p_i, which we can obtain from the perfect gas law

$$pV = nRT \quad \text{so} \quad p = \frac{nRT}{V}$$

$$p_\text{i} = \frac{\left(\frac{2.4\,\text{g}}{44\,\text{g mol}^{-1}}\right) \times (0.08206\,\text{L atm K}^{-1}\text{mol}^{-1}) \times (278\,\text{K})}{1.0\,\text{L}} = \boxed{1.2\,\text{atm}}$$

$$p_\text{f} = (1.2\,\text{atm}) \times \left(\frac{1.0\,\text{L}}{2.0\,\text{L}}\right)^{1.4} = \boxed{0.45\,\text{atm}}$$

E2.16 The reaction for the combustion of butane is

$$C_4H_{10}(g) + \tfrac{13}{2}O_2(g) \rightarrow 4CO_2(g) + 5H_2O(l)$$
$$\Delta_c H^{\ominus}(C_4H_{10},\ g) = (4) \times (\Delta_f H^{\ominus}(CO_2,\ g)) + (5) \times (\Delta_f H^{\ominus}(H_2O, g)) - (1) \times (\Delta_f H^{\ominus}(C_4H_{10},\ g))$$

Solving for $\Delta_f H^{\ominus}(C_4H_{10},\ g)$ and looking up the other data in Tables 2.5 and 2.6, we obtain

$$\Delta_f H^{\ominus}(C_4H_{10},\ g) = (4) \times (-393.51\,\mathrm{kJ\,mol^{-1}}) + (5) \times (-285.83\,\mathrm{kJ\,mol^{-1}}) - (1) \times (-2878\,\mathrm{kJ\,mol^{-1}}) = \boxed{-125\,\mathrm{kJ\,mol^{-1}}}$$

Comment. This is very close to the value listed in Table 2.6. The small difference is undoubtedly the result of the error in the least precise value of the set of data, that for $\Delta_c H^{\ominus}(C_4H_{10},\ g)$.

E2.17 $C_p = \dfrac{q_p}{\Delta T}[29] = \dfrac{229\,\mathrm{J}}{2.55\,\mathrm{K}} = 89.8\,\mathrm{J\,K^{-1}}$ $\quad C_{p,\mathrm{m}} = \dfrac{C_p}{n} = \dfrac{89.8\,\mathrm{J\,K^{-1}}}{3.0\,\mathrm{mol}} = \boxed{30\,\mathrm{J\,K^{-1}\,mol^{-1}}}$

For a perfect gas

$$C_{p,\mathrm{m}} - C_{V,\mathrm{m}} = R\ [31]$$

$$C_{V,\mathrm{m}} = C_{p,\mathrm{m}} - R = (30 - 8.3)\,\mathrm{J\,K^{-1}\,mol^{-1}} = \boxed{22\,\mathrm{J\,K^{-1}\,mol^{-1}}}$$

E2.18 $q_p = \boxed{-1.2\,\mathrm{kJ}}$ [energy left the sample] $\quad \Delta H = q_p = \boxed{-1.2\,\mathrm{kJ}}$

$$C_p = \frac{q_p}{\Delta T} = \frac{-1.2\,\mathrm{kJ}}{-15\,\mathrm{K}} = \boxed{80\,\mathrm{J\,K^{-1}}}$$

E2.19 $q_p = C_p\Delta T[29] = nC_{p,\mathrm{m}}\Delta T = (3.0\,\mathrm{mol}) \times (29.4\,\mathrm{J\,K^{-1}\,mol^{-1}}) \times (25\,\mathrm{K}) = \boxed{+2.2\,\mathrm{kJ}}$

$$\Delta H = q_p\ [24b] = \boxed{+2.2\,\mathrm{kJ}}$$

$$\Delta U = \Delta H - \Delta(pV)(\text{From } H \equiv U + pV) = \Delta H - \Delta(nRT)[\text{perfect gas}] = \Delta H - nR\Delta T$$
$$= (2.2\,\mathrm{kJ}) - (3.0\,\mathrm{mol}) \times (8.314\,\mathrm{J\,K^{-1}\,mol^{-1}}) \times (25\,\mathrm{K}) = (2.2\,\mathrm{kJ}) - (0.62\,\mathrm{kJ})$$
$$= \boxed{+1.6\,\mathrm{kJ}}$$

E2.20 $q = \boxed{0}$ [adiabatic process]

$$w = -p_{\mathrm{ex}}\Delta V = (-600\,\mathrm{Torr}) \times \left(\frac{1.013 \times 10^5\,\mathrm{Pa}}{760\,\mathrm{Torr}}\right) \times (40 \times 10^{-3}\,\mathrm{m^3}) = \boxed{-3.2\,\mathrm{kJ}}$$

$\Delta U = w = \boxed{-3.2\,\mathrm{kJ}}\ [q = 0]$ or $C_V\Delta T = w\ [C_V = nC_{V,\mathrm{m}};\ C_{V,\mathrm{m}} = 21.1\,\mathrm{J\,K^{-1}\,mol^{-1}}]$

or

$$\Delta T = \frac{w}{C_V} = \frac{w}{nC_{V,\mathrm{m}}} = \frac{-3.2 \times 10^3\,\mathrm{J}}{(4.0\,\mathrm{mol}) \times (21.1\,\mathrm{J\,K^{-1}\,mol^{-1}})} = \boxed{-38\,\mathrm{K}}$$

$$\Delta H = \Delta U + \Delta(pV) = \Delta U + nR\Delta T$$
$$= (-3.2\,\mathrm{kJ}) + (4.0\,\mathrm{mol}) \times (8.314\,\mathrm{J\,K^{-1}\,mol^{-1}}) \times (-38\,\mathrm{K}) = \boxed{-4.5\,\mathrm{kJ}}$$

Question. Calculate the final pressure of the gas.

E2.21 $q = \boxed{0}$ [adiabatic process]

$$\Delta U = nC_{V,m}\Delta T \text{ [perfect gas]} = (3.0\,\text{mol}) \times (27.5\,\text{J K}^{-1}\,\text{mol}^{-1}) \times (50\,\text{K}) = \boxed{+4.1\,\text{kJ}}$$

$$w = \Delta U - q = 4.1\,\text{kJ} - 0 = \boxed{+4.1\,\text{kJ}}$$

$$\Delta H = \Delta U + nR\Delta T \quad [\Delta(pV) = \Delta(nRT) = nR\Delta T]$$

$$= (4.1\,\text{kJ}) + (3.0\,\text{mol}) \times (8.314\,\text{J K}^{-1}\,\text{mol}^{-1}) \times (50\,\text{K}) = \boxed{+5.4\,\text{kJ}}$$

$$V_i = \frac{nRT_i}{p_i} = \frac{(3.0\,\text{mol}) \times (8.206 \times 10^{-2}\,\text{L atm K}^{-1}\,\text{mol}^{-1}) \times (200\,\text{K})}{(2.0\,\text{atm})} = 24.6\,\text{L}$$

$$V_f = V_i\left(\frac{T_i}{T_f}\right)^c \text{ [34]}, \qquad c = \frac{C_V}{R} = \frac{27.5\,\text{J K}^{-1}\,\text{mol}^{-1}}{8.314\,\text{J K}^{-1}\,\text{mol}^{-1}} = 3.31$$

$$V_f = (24.6\,\text{L}) \times \left(\frac{200\,\text{K}}{250\,\text{K}}\right)^{3.31} = \boxed{11.8\,\text{L}}$$

$$p_f = \frac{nRT_f}{V_f} = \frac{(3.0\,\text{mol}) \times (8.206 \times 10^{-2}\,\text{L atm K}^{-1}\,\text{mol}^{-1}) \times (250\,\text{K})}{11.8\,\text{L}} = \boxed{5.2\,\text{atm}}$$

E2.22

$$V_i = \frac{nRT_i}{p_i} = \frac{(1.0\,\text{mol}) \times (8.206 \times 10^{-2}\,\text{L atm K}^{-1}\,\text{mol}^{-1}) \times (310\,\text{K})}{3.25\,\text{atm}} = \boxed{7.8\bar{3}\,\text{L}}$$

$$\gamma = \frac{C_p}{C_V} = \frac{C_V + R}{C_V} = \frac{(20.8 + 8.31)\text{J K}^{-1}\,\text{mol}^{-1}}{20.8\,\text{J K}^{-1}\,\text{mol}^{-1}} = 1.40 \qquad \frac{1}{\gamma} = 0.714$$

$$V_f = V_i\left(\frac{p_i}{p_f}\right)^{1/\gamma} \text{[36]} = (7.8\bar{3}\,\text{L}) \times \left(\frac{3.25\,\text{atm}}{2.50\,\text{atm}}\right)^{0.714} = \boxed{9.4\bar{4}\,\text{L}}$$

$$T_f = \frac{p_f V_f}{nR} = \frac{(2.50\,\text{atm}) \times (9.4\bar{4}\,\text{L})}{(1.0\,\text{mol}) \times (8.206 \times 10^{-2}\,\text{L atm K}^{-1}\,\text{mol}^{-1})} = \boxed{28\bar{8}\,\text{K}}$$

$$w = C_V(T_f - T_i) \text{ [33]} = (20.8\,\text{J K}^{-1}\,\text{mol}^{-1}) \times (1.0\,\text{mol}) \times (288\,\text{K} - 310\,\text{K})$$

$$= \boxed{-0.46\,\text{kJ}}$$

E2.23 For this small temperature range α may be assumed to be constant; hence

$$dV = \left(\frac{\partial V}{\partial T}\right)_p dT \text{[pressure constant]} = \alpha V\,dT$$

$$\Delta V \approx \alpha V \Delta T \text{ [the change in } V \text{ is small; hence } V \approx \text{constant]}$$

Mercury, $\alpha = 1.82 \times 10^{-4}\,\text{K}^{-1}$,

$$\Delta V \approx (1.82 \times 10^{-4}\,\text{K}^{-1}) \times (1.0\,\text{cm}^3) \times (5\,\text{K}) \approx 9.\bar{1} \times 10^{-4}\,\text{cm}^3 = \boxed{+0.9\,\text{mm}^3}$$

E2.24 In an adiabatic process, $q = \boxed{0}$. Work against a constant external pressure is

$$w = -p_{ex}\Delta V = -(1.0\,\text{atm}) \times (1.01 \times 10^5\,\text{Pa atm}^{-1}) \times \frac{(20\,\text{cm}) \times (10\,\text{cm}^2)}{(100\,\text{cm m}^{-1})^3} = \boxed{-20\,\text{J}}$$

$$\Delta U = q + w = \boxed{-20\,\text{J}}$$

$$w = C_V\Delta T = n(C_{p,m} - R)\Delta T \quad \text{so} \quad \Delta T = \frac{w}{n(C_{p,m} - R)},$$

$$\Delta T = \frac{-20\,\text{J}}{(2.0\,\text{mol}) \times (37.11 - 8.3145\,\text{J K}^{-1}\,\text{mol}^{-1})} = \boxed{-0.35\,\text{K}}$$

$$\Delta H = \Delta U + \Delta(pV) = \Delta U + nR\Delta T$$

$$= -20\,\text{J} + (2.0\,\text{mol}) \times (8.3145\,\text{J K}^{-1}\,\text{mol}^{-1}) \times (-0.35\,\text{K}) = \boxed{-26\,\text{J}}$$

E2.25 The amount of Xe in the sample is

$$n = \frac{65.0\,\text{g}}{131.3\,\text{g mol}^{-1}} = 0.495\,\text{mol}$$

(a) For reversible adiabatic expansion

$$p_f V_f^{\gamma} = p_i V_i^{\gamma} \quad \text{so} \quad V_f = V_i \left(\frac{p_i}{p_f}\right)^{1/\gamma},$$

$$\text{where} \quad \gamma = \frac{C_{p,m}}{C_{V,m}} \quad \text{where} \quad C_{V,m} = (20.79 - 8.3145)\,\text{J K}^{-1}\text{mol}^{-1} = 12.48\,\text{J K}^{-1}\,\text{mol}^{-1},$$

$$\text{so } \gamma = \frac{20.79\,\text{J K}^{-1}\,\text{mol}^{-1}}{12.48\,\text{J K}^{-1}\,\text{mol}^{-1}} = 1.666,$$

$$\text{and } V_i = \frac{nRT_i}{p_i} = \frac{(0.495\,\text{mol}) \times (0.08206\,\text{L atm K}^{-1}\text{mol}^{-1}) \times (298\,\text{K})}{2.00\,\text{atm}} = 6.05\,\text{L},$$

$$\text{so} \quad V_f = V_i \left(\frac{p_i}{p_f}\right)^{1/\gamma} = (6.05\,\text{L}) \times \left(\frac{2.00\,\text{atm}}{1.00\,\text{atm}}\right)^{(1/1.666)} = 9.17\,\text{L}$$

$$T_f = \frac{p_f V_f}{nR} = \frac{(1.00\,\text{atm}) \times (9.17\,\text{L})}{(0.495\,\text{mol}) \times (0.08206\,\text{L atm K}^{-1}\,\text{mol}^{-1})} = \boxed{226\,\text{K}}$$

(b) For adiabatic expansion against a constant external pressure

$$w = -p_{ex}\Delta V = C_V \Delta T \quad \text{so} \quad -p_{ex}(V_f - V_i) = C_V(T_f - T_i)$$

In addition, the perfect gas law holds

$$p_f V_f = nRT_f$$

Solve the latter for T_f in terms of V_f, and insert into the previous relationship to solve for V_f

$$T_f = \frac{p_f V_f}{nR} \quad \text{so} \quad -p_{ex}(V_f - V_i) = C_V \left(\frac{p_f V_f}{nR} - T_i\right)$$

Collecting terms gives

$$C_V T_i + p_{ex} V_i = V_f \left(p_{ex} + \frac{C_V p_f}{nR}\right) \quad \text{so} \quad V_f = \frac{C_V T_i + p_{ex} V_i}{p_{ex} + \frac{C_{V,m} p_f}{R}},$$

$$V_f = \frac{(12.48\,\text{J K}^{-1}\,\text{mol}^{-1}) \times (0.495\,\text{mol}) \times (298\,\text{K}) + (1.00\,\text{atm}) \times (1.01 \times 10^5\,\text{Pa atm}^{-1}) \times \left(\frac{6.05\,\text{L}}{1000\,\text{L m}^{-3}}\right)}{\left(1.00\,\text{atm} + \frac{(12.48\,\text{J K}^{-1}\,\text{mol}^{-1}) \times (1.00\,\text{atm})}{8.3145\,\text{J K}^{-1}\,\text{mol}^{-1}}\right) \times (1.01 \times 10^5\,\text{Pa atm}^{-1})}$$

$$V_f = 9.71 \times 10^{-3}\,\text{m}^3$$

Finally, the temperature is

$$T_f = \frac{p_f V_f}{nR} = \frac{(1.00\,\text{atm}) \times (1.01 \times 10^5\,\text{Pa atm}^{-1}) \times (9.71 \times 10^{-3}\,\text{m}^3)}{(0.495\,\text{mol}) \times (8.3145\,\text{J K}^{-1}\,\text{mol}^{-1})} = \boxed{238\,\text{K}}$$

E2.26 $q_p = n\Delta_{vap}H^{\ominus}$ [constant pressure] $= (0.50\,\text{mol}) \times (26.0\,\text{kJ mol}^{-1}) = \boxed{+13\,\text{kJ}}$

$$w = -p_{ex}\Delta V\ [10] \approx -p_{ex}V(\text{g})[V(\text{g}) \gg V(\text{l})] \approx -(p_{ex}) \times \left(\frac{nRT}{p_{ex}}\right) = -nRT$$

Therefore, $w \approx (-0.50\,\text{mol}) \times (8.314\,\text{J K}^{-1}\,\text{mol}^{-1}) \times (250\,\text{K}) = \boxed{-1.0\,\text{kJ}}$

$$\Delta H = q_p\,[24b] = \boxed{+13\,\text{kJ}} \qquad \Delta U = q + w = (13\,\text{kJ}) - (1.0\,\text{kJ}) = \boxed{+12\,\text{kJ}}$$

E2.27 $C_6H_5C_2H_5(l) + \frac{21}{2}O_2(g) \rightarrow 8CO_2(g) + 5H_2O(l)$

$$\begin{aligned}\Delta_c H^\ominus &= 8\Delta_f H^\ominus(CO_2, g) + 5\Delta_f H^\ominus(H_2O, l) - \Delta_f H^\ominus(C_6H_5C_2H_5, l)\\ &= \{(8)\times(-393.51) + (5)\times(-285.83) - (-12.5)\}\,\mathrm{kJ\,mol^{-1}}\\ &= \boxed{-4564.7\,\mathrm{kJ\,mol^{-1}}}\end{aligned}$$

E2.28 The reaction is $C_6H_{12}(l) + H_2(g) \rightarrow C_6H_{14}(l) \quad \Delta_r H^\ominus = ?$

From Table 2.5 and the information in the exercise

$$C_6H_{12}(l) + 9O_2(g) \rightarrow 6CO_2(g) + 6H_2O(l) \quad \Delta_c H^\ominus = -4003\,\mathrm{kJ\,mol^{-1}}$$
$$C_6H_{14}(l) + \tfrac{19}{2}O_2(g) \rightarrow 6CO_2(g) + 7H_2O(l) \quad \Delta_c H^\ominus = -4163\,\mathrm{kJ\,mol^{-1}}$$

The difference of these reactions is

$$C_6H_{12}(l) + H_2(l) \rightarrow C_6H_{14}(l) + \tfrac{1}{2}O_2(g) \quad \Delta_r H^\ominus = +160\,\mathrm{kJ\,mol^{-1}}$$

This reaction may be converted to the desired reaction by subtracting from it

$$H_2O(l) \rightarrow H_2(g) + \tfrac{1}{2}O_2(g) \quad \Delta_r H^\ominus = -\Delta_f H^\ominus(H_2O, l) = 285.83\,\mathrm{kJ\,mol^{-1}}$$

This gives $C_6H_{12}(l) + H_2(g) \rightarrow C_6H_{14}(l) \quad \Delta_r H^\ominus = \boxed{-126\,\mathrm{kJ\,mol^{-1}}}$

E2.29 First $\Delta_f H^\ominus[(CH_2)_3,\ g]$ is calculated, and then that result is used to calculate $\Delta_r H^\ominus$ for the isomerization.

$$(CH_2)_3(g) + \tfrac{9}{2}O_2(g) \rightarrow 3CO_2(g) + 3H_2O(l) \quad \Delta_c H^\ominus = -2091\,\mathrm{kJ\,mol^{-1}}$$

$$\begin{aligned}\Delta_f H^\ominus\{(CH_2)_3,\ g\} &= -\Delta_c H^\ominus + 3\Delta_f H^\ominus(CO_2,\ g) + 3\Delta_f H^\ominus(H_2O,\ g)\\ &= \{+2091 + (3)\times(-393.51) + (3)\times(-285.83)\}\,\mathrm{kJ\,mol^{-1}}\\ &= \boxed{+53\,\mathrm{kJ\,mol^{-1}}}\end{aligned}$$

$$(CH_2)_3(g) \rightarrow C_3H_6(g) \quad \Delta_r H^\ominus = ?$$

$$\begin{aligned}\Delta_r H^\ominus &= \Delta_f H^\ominus(C_3H_6, g) - \Delta_f H^\ominus\{(CH_2)_3, g\}\\ &= (20.42 - 53)\,\mathrm{kJ\,mol^{-1}} = \boxed{-33\,\mathrm{kJ\,mol^{-1}}}\end{aligned}$$

E2.30 The formation reaction of liquid methylacetate is

$$3C(s) + 3H_2(g) + O_2(g) \rightarrow CH_3COOOCH_3(l) \quad \Delta_f H^\ominus = -442\,\mathrm{kJ\,mol^{-1}}$$
$$\Delta U = \Delta H - \Delta n_g RT\ [26], \quad \Delta n_g = -4\,\mathrm{mol},$$
$$\Delta n_g RT = (-4\,\mathrm{mol}) \times (2.479\,\mathrm{kJ\,mol^{-1}}) = -9.916\,\mathrm{kJ}$$

Therefore $\Delta_f U^\ominus = (-442\,\mathrm{kJ\,mol^{-1}}) + (9.9\,\mathrm{kJ\,mol^{-1}}) = \boxed{-432\,\mathrm{kJ\,mol^{-1}}}$

E2.31 $C = \dfrac{q}{\Delta T}$ [17] and $q = I\mathcal{V}t$ [18]

Thus $C = \dfrac{I\mathcal{V}t}{\Delta T} = \dfrac{(3.20\,\mathrm{A}) \times (12.0\,\mathrm{V}) \times (27.0\,\mathrm{s})}{1.617\,\mathrm{K}} = \boxed{641\,\mathrm{J\,K^{-1}}} \quad (1\,\mathrm{J} = 1\,\mathrm{A\,V\,s})$

E2.32 For naphthalene the reaction is $C_{10}H_8(s) + 12O_2(g) \rightarrow 10CO_2(g) + 4H_2O(l)$

A bomb calorimeter gives $q_V = n\Delta_c U^\ominus$ rather than $q_p = n\Delta_c H^\ominus$; thus we need

$$\Delta_c U^\ominus = \Delta_c H^\ominus - \Delta n_g RT \text{ [26]}, \quad \Delta n_g = -2\,\text{mol}$$

$$\Delta_c H^\ominus = -5157\,\text{kJ mol}^{-1} \text{ [Table 2.5]} \quad \text{assume } T \approx 298\,\text{K}$$

$$\Delta_c U^\ominus = (-5157\,\text{kJ mol}^{-1}) - (-2) \times (8.3 \times 10^{-3}\,\text{kJ K}^{-1}\,\text{mol}^{-1}) \times (298\,\text{K})$$

$$= -5152\,\text{kJ mol}^{-1}$$

$$|q| = |q_V| = |n\Delta_c U^\ominus| = \left(\frac{120 \times 10^{-3}\,\text{g}}{128.18\,\text{g mol}^{-1}}\right) \times (5152\,\text{kJ mol}^{-1}) = 4.82\bar{3}\,\text{kJ}$$

$$C = \frac{|q|}{\Delta T} = \frac{4.82\bar{3}\,\text{kJ}}{3.05\,\text{K}} = \boxed{1.58\,\text{kJ K}^{-1}}$$

When phenol is used the reaction is

$$C_6H_5OH(s) + \tfrac{15}{2}O_2(g) \rightarrow 6CO_2(g) + 3H_2O(l)$$

$$\Delta_c H^\ominus = -3054\,\text{kJ mol}^{-1} \text{ [Table 2.5]}$$

$$\Delta_c U^\ominus = \Delta_c H^\ominus - \Delta n_g RT, \quad \Delta n_g = -\tfrac{3}{2}\,\text{mol}$$

$$= (-3054\,\text{kJ mol}^{-1}) + \left(\tfrac{3}{2}\right) \times (8.314 \times 10^{-3}\,\text{kJ K}^{-1}\,\text{mol}^{-1}) \times (298\,\text{K})$$

$$= -3050\,\text{kJ mol}^{-1}$$

$$|q| = \left(\frac{100 \times 10^{-3}\,\text{g}}{94.12\,\text{g mol}^{-1}}\right) \times (3050\,\text{kJ mol}^{-1}) = 3.24\bar{1}\,\text{kJ}$$

$$\Delta T = \frac{|q|}{C} = \frac{3.24\bar{1}\,\text{kJ}}{1.58\,\text{kJ K}^{-1}} = \boxed{+2.05\,\text{K}}$$

Comment. In this case $\Delta_c U^\ominus$ and $\Delta_c H^\ominus$ differed by ≈ 0.1 per cent. Thus, to within 3 significant figures, it would not have mattered if we had used $\Delta_c H^\ominus$ instead of $\Delta_c U^\ominus$, but for very precise work it would.

E2.33 **(a)** $q_V = n\Delta_c U^\ominus$; hence

$$|\Delta_c U^\ominus| = \frac{q_V}{n} = \frac{C\Delta T}{n}[q_V = C \times \Delta T] = \frac{MC\Delta T}{m} \quad \left[n = \frac{m}{M},\ m = \text{mass}\right]$$

Therefore, since $M = 180.16\,\text{g mol}^{-1}$,

$$|\Delta_c U^\ominus| = \frac{(180.16\,\text{g mol}^{-1}) \times (641\,\text{J K}^{-1}) \times (7.793\,\text{K})}{0.3212\,\text{g}} = 280\bar{2}\,\text{kJ mol}^{-1}$$

Since the combustion is exothermic, $\Delta_c U^\ominus = \boxed{-2.80\,\text{MJ mol}^{-1}}$

(b) The combustion reaction is

$$C_6H_{12}O_6(s) + 6O_2(g) \rightarrow 6CO_2(g) + 6H_2O(l) \quad \Delta n_g = 0$$

Hence, $\Delta_c U^\ominus = \Delta_c H^\ominus$ [26]; $\quad \Delta_c H^\ominus = \boxed{-2.80\,\text{MJ mol}^{-1}}$

(c) For the enthalpy of formation we combine

$$6CO_2(g) + 6H_2O(l) \rightarrow C_6H_{12}O_6(s) + 6O_2(s) \quad \Delta H^\ominus = +2.80\,\text{MJ mol}^{-1}$$

$$6C(s) + 6O_2(g) \rightarrow 6CO_2(g) \quad \Delta H^\ominus = 6 \times \Delta_f H^\ominus(CO_2,\ g)$$

$$6H_2(g) + 3O_2(g) \rightarrow 6H_2O(l) \quad \Delta H^\ominus = 6 \times \Delta_f H^\ominus(H_2O, l)$$

The sum of these three reactions is

$$6C(s) + 6H_2(g) + 3O_2(g) \rightarrow C_6H_{12}O_6(s)$$

$$\Delta_f H^\ominus = \{(2.80) + (6) \times (-0.3935) + (6) \times (-0.2858)\}\ \text{MJ mol}^{-1}$$

$$= \boxed{-1.28\ \text{MJ mol}^{-1}}$$

E2.34 $AgCl(s) \rightarrow Ag^+(aq) + Cl^-(aq)$

$$\Delta_{sol} H^\ominus = \Delta_f H^\ominus(Ag^+, aq) + \Delta_f H^\ominus(Cl^-, aq) - \Delta_f H^\ominus(AgCl, s)$$

$$= (105.58) + (167.16) - (-127.07)\ \text{kJ mol}^{-1}$$

$$= \boxed{+65.49\ \text{kJ mol}^{-1}}$$

E2.35 $NH_3(g) + SO_2(g) \rightarrow NH_3SO_2(s) \quad \Delta_r H^\ominus = -40\ \text{kJ mol}^{-1}$

$$\Delta_r H^\ominus = \Delta_f H^\ominus(NH_3SO_2, s) - \Delta_f H^\ominus(NH_3, g) - \Delta_f H^\ominus(SO_2, g)$$

Solving for $\Delta_f H^\ominus(NH_3SO_2, s)$ yields

$$\Delta_f H^\ominus(NH_3SO_2, s) = \Delta_f H^\ominus(NH_3, g) + \Delta_f H^\ominus(SO_2, g) + \Delta_r H^\ominus$$

$$= (-46.11 - 296.83 - 40)\ \text{kJ mol}^{-1} = \boxed{-383\ \text{kJ mol}^{-1}}$$

E2.36 $q_p = n\Delta_c H^\ominus$ [constant-pressure process] $= \left(\dfrac{1.5\ \text{g}}{342.3\ \text{g mol}^{-1}}\right) \times (-5645\ \text{kJ mol}^{-1})$

$$= -25\ \text{kJ} = \boxed{25\ \text{kJ released}}$$

Effective work available $\approx (25\ \text{kJ}) \times (0.25) = 6.2\bar{5}\ \text{kJ}$

Since $w = mgh$, with $m \approx 65$ kg [mass of average human]

$$h \approx \frac{6.2\bar{5} \times 10^3\ \text{J}}{(65\ \text{kg}) \times (9.81\ \text{m s}^{-2})} = \boxed{9.8\ \text{m}}$$

E2.37 **(a)** $\Delta_c H^\ominus(l) = \Delta_{vap} H^\ominus + \Delta_c H(g) = (15\ \text{kJ mol}^{-1}) - (2220\ \text{kJ mol}^{-1}) = \boxed{-2205\ \text{kJ mol}^{-1}}$

(b) $\Delta_c U^\ominus(l) = \Delta_c H^\ominus(l) - \Delta n_g RT, \quad \Delta n_g = -2$

$$= (-2205\ \text{kJ mol}^{-1}) + (2) \times (2.479\ \text{kJ mol}^{-1}) = \boxed{-2200\ \text{kJ mol}^{-1}}$$

E2.38 **(a)** $CH_4(g) + 2O_2(g) \rightarrow CO_2(g) + 2H_2O(l) \quad \Delta_r H^\ominus = -890\ \text{kJ mol}^{-1}$

(b) $2C(s) + H_2(g) \rightarrow C_2H_2(g) \quad \Delta_r H^\ominus = +227\ \text{kJ mol}^{-1}$

(c) $NaCl(s) \rightarrow NaCl(aq) \quad \Delta_r H^\ominus = +3.9\ \text{kJ mol}^{-1}$

$\Delta_r H^\ominus > 0$ indicates an endothermic reaction and $\Delta_r H^\ominus < 0$ an exothermic reaction. Therefore, **(a)** is $\boxed{\text{exothermic}}$; **(b)** and **(c)** are $\boxed{\text{endothermic}}$.

Stoichiometric coefficients of products are positive, those of reactants are negative; hence

(a) $0 = CO_2 + 2H_2O - CH_4 - 2O_2;$

$\nu(CO_2) = +1,\ \nu(H_2O) = +2,\ \nu(CH_4) = -1,\ \nu(O_2) = -2$

(b) $0 = C_2H_2 - 2C - H_2; \quad \nu(C_2H_2) = +1,\ \nu(C) = -2,\ \nu(H_2) = -1$

(c) $0 = Na^+(aq) + Cl^-(aq) - NaCl(s); \quad \nu(Na^+) = +1,\ \nu(Cl^-) = +1,\ \nu(NaCl) = -1$

E2.39 In each case $\Delta_r H^\ominus = \sum_J \nu_J \Delta_f H^\ominus(J)$ [43]

(a) $\Delta_r H^\ominus = \Delta_f H^\ominus(N_2O_4, g) - 2\Delta_f H^\ominus(NO_2, g) = (9.16) - (2) \times (33.18)\,\text{kJ mol}^{-1}$

$= \boxed{-57.20\,\text{kJ mol}^{-1}}$

(b) $\Delta_r H^\ominus = \Delta_f H^\ominus(NH_4Cl, s) - \Delta_f H^\ominus(NH_3, g) - \Delta_f H^\ominus(HCl, g)$

$= \{(-314.43) - (-46.11) - (-92.31)\}\,\text{kJ mol}^{-1} = \boxed{-176.01\,\text{kJ mol}^{-1}}$

E2.40 **(a)** reaction(3) $= (-2) \times$ reaction (1) + reaction (2) $\quad \Delta n_g = -2$

The enthalpies of reactions are combined in the same manner as the equations (Hess's law).

$$\Delta_r H^\ominus(3) = (-2) \times \Delta_r H^\ominus(1) + \Delta_r H^\ominus(2)$$

$$= \{(-2) \times (-184.62) + (-483.64)\}\,\text{kJ mol}^{-1}$$

$$= \boxed{-114.40\,\text{kJ mol}^{-1}}$$

$$\Delta_r U^\ominus = \Delta_r H^\ominus - \Delta n_g RT\ [26] = (-114.40\,\text{kJ mol}^{-1}) - (-2) \times (2.48\,\text{kJ mol}^{-1})$$

$$= \boxed{-109.44\,\text{kJ mol}^{-1}}$$

(b) $\Delta_f H^\ominus$ refers to the formation of one mole of the compound, hence

$$\Delta_f H^\ominus(J) = \frac{\Delta_r H^\ominus(J)}{\nu_J}$$

$$\Delta_f H^\ominus(HCl, g) = \frac{-184.62}{2}\,\text{kJ mol}^{-1} = \boxed{-92.31\,\text{kJ mol}^{-1}}$$

$$\Delta_f H^\ominus(H_2O, g) = \frac{-483.64}{2}\,\text{kJ mol}^{-1} = \boxed{-241.82\,\text{kJ mol}^{-1}}$$

E2.41 $\Delta_r H^\ominus = \Delta_r U^\ominus + \Delta n_g RT\ [26]; \quad \Delta n_g = +2$

$= (-1373\,\text{kJ mol}^{-1}) + 2 \times (2.48\,\text{kJ mol}^{-1}) = \boxed{-1368\,\text{kJ mol}^{-1}}$

Comment. As a number of these exercises have shown, the use of $\Delta_r H^\ominus$ as an approximation for $\Delta_r U^\ominus$ is often valid.

E2.42 In each case, the strategy is to combine reactions in such a way that the combination corresponds to the formation reaction desired. The enthalpies of the reactions are then combined in the same manner as the equations to yield the enthalpies of formation.

(a)

	$\Delta_r H^\ominus/(\text{kJ mol}^{-1})$
$K(s) + \frac{1}{2}Cl_2(g) \rightarrow KCl(s)$	-436.75
$KCl(s) + \frac{3}{2}O_2(g) \rightarrow KClO_3(s)$	$\frac{1}{2} \times (89.4)$
$K(s) + \frac{1}{2}Cl_2(g) + \frac{3}{2}O_2(g) \rightarrow KClO_3(s)$	-392.1

Hence, $\Delta_f H^\ominus(KClO_3, s) = \boxed{-392.1\,\text{kJ mol}^{-1}}$

(b)

	$\Delta_r H^\ominus/(\text{kJ mol}^{-1})$
$Na(s) + \frac{1}{2}O_2(g) + \frac{1}{2}H_2(g) \rightarrow NaOH(s)$	-425.61
$NaOH(s) + CO_2(g) \rightarrow NaHCO_3(s)$	-127.5
$C(s) + O_2(g) \rightarrow CO_2(g)$	-393.51
$Na(s) + C(s) + \frac{1}{2}H_2(g) + \frac{3}{2}O_2(g) \rightarrow NaHCO_3(s)$	-946.6

Hence, $\Delta_f H^\ominus(NaHCO_3, s) = \boxed{-946.6\,\text{kJ mol}^{-1}}$

(c)

	$\Delta_r H^\ominus/(\text{kJ mol}^{-1})$
$\frac{1}{2}N_2(g) + \frac{1}{2}O_2(g) \rightarrow NO(g)$	$+90.25$
$NO(g) + \frac{1}{2}Cl_2(g) \rightarrow NOCl(g)$	$-\frac{1}{2}(75.5)$
$\frac{1}{2}N_2(g) + \frac{1}{2}O_2(g) + \frac{1}{2}Cl_2(g) \rightarrow NOCl(g)$	$+52.5$

Hence, $\Delta_f H^\ominus(NOCl, g) = \boxed{+52.5\,\text{kJ mol}^{-1}}$

E2.43 When the heat capacities of all substances participating in a chemical reaction are assumed to be constant over the range of temperatures involved Kirchoff's law [45] integrates to

$$\Delta_r H^\ominus(T_2) = \Delta_r H^\ominus(T_1) + \Delta_r C_p(T_2 - T_1) \quad \text{[Example 2.7]}$$

$$\Delta_r C_p = \sum_J \nu_J C_{p,m}(J) \ [47]$$

$$\Delta_r C_p = C_p(N_2O_4, g) - 2\,C_p(NO_2, g) = (77.28) - (2) \times (37.20\,\text{J K}^{-1}\,\text{mol}^{-1})$$
$$= +2.88\,\text{J K}^{-1}\,\text{mol}^{-1}$$

$$\Delta_r H^\ominus(373\,\text{K}) = \Delta_r H^\ominus(298\,\text{K}) + \Delta_r C_p \Delta T$$
$$= (-57.20\,\text{kJ mol}^{-1}) + (2.88\,\text{J K}^{-1}) \times (75\,\text{K})$$
$$= \{(-57.20) + (0.22)\}\,\text{kJ mol}^{-1}$$
$$= \boxed{-56.98\,\text{kJ mol}^{-1}}$$

E2.44 **(a)** $\Delta_r H^\ominus = \sum_J \nu_J \Delta_f H^\ominus(J)$ [43]

$$\Delta_r H^\ominus(298\,\text{K}) = [(-110.53) - (-241.82)]\,\text{kJ mol}^{-1} = \boxed{+131.29\,\text{kJ mol}^{-1}}$$

$$\Delta_r U^\ominus(298\,\text{K}) = \Delta_r H^\ominus(298\,\text{K}) - \Delta n_g RT \ [26]$$
$$= (131.29\,\text{kJ mol}^{-1}) - (1) \times (2.48\,\text{kJ mol}^{-1})$$
$$= \boxed{+128.81\,\text{kJ mol}^{-1}}$$

(b) $\Delta_r H^\ominus(378\,\text{K}) = \Delta_r H^\ominus(298\,\text{K}) + \Delta_r C_p(T_2 - T_1)$ [Example 2.7]

$$\Delta_r C_p = C_{p,m}(CO, g) + C_{p,m}(H_2, g) - C_{p,m}(C, gr) - C_{p,m}(H_2O, g)$$
$$= (29.14 + 28.82 - 8.53 - 33.58) \times 10^{-3}\,\text{kJ K}^{-1}\,\text{mol}^{-1}$$
$$= 15.85 \times 10^{-3}\,\text{kJ K}^{-1}\,\text{mol}^{-1}$$

$$\Delta_r H^\ominus(378\,\text{K}) = (131.29\,\text{kJ mol}^{-1}) + (15.85 \times 10^{-3}\,\text{kJ K}^{-1}\,\text{mol}^{-1}) \times (80\,\text{K})$$
$$= (131.29 + 1.27)\,\text{kJ mol}^{-1} = \boxed{+132.56\,\text{kJ mol}^{-1}}$$

$$\Delta_r U^\ominus(378\,\text{K}) = \Delta_r H^\ominus(378\,\text{K}) - (1) \times (3.14\,\text{kJ mol}^{-1}) = (132.56 - 3.14)\,\text{kJ mol}^{-1}$$
$$= \boxed{+129.42\,\text{kJ mol}^{-1}}$$

Comment. The differences between both $\Delta_r H^\ominus$ and $\Delta_r U^\ominus$ at the two temperatures are small and justify the use of the approximation the $\Delta_r C_p$ is a constant.

E2.45 Since enthalpy is a state function, $\Delta_r H$ for the process (see Fig. 2.1)

$$Mg^{2+}(g) + 2Cl(g) + 2e^- \rightarrow MgCl_2(aq)$$

is independent of path; therefore the change in enthalpy for the path on the left is equal to the change in enthalpy for the path on the right. All numerical values are in kJ mol^{-1}.

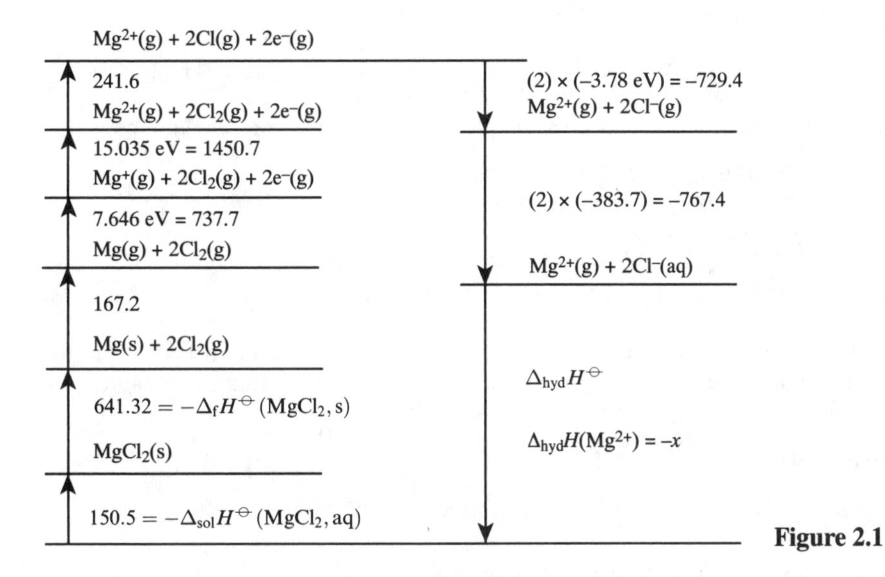

Figure 2.1

The cycle is the distance traversed upward along the left plus the distance traversed downward on the right. The sum of these distances is zero. Note that $E_{ea} = -\Delta_{eg} H^\ominus$. Therefore,

$$(150.5) + (641.32) + (167.2) + (737.7) + (1450.7)$$
$$+(241.6) + (-729.4) + (-767.4) + (-x) = 0$$

Solving to $x = 1892.2$, which yields

$$\Delta_{hyd} H^\ominus(Mg^{2+}) = \boxed{-1892.2\,\text{kJ mol}^{-1}}$$

E2.46 **(a)** Cyclohexane is composed of 6 $C(H)_2(C)_2$ groups, so

$$\Delta_f H^\ominus = 6 \times \Delta_f H^\ominus[C(H)_2(C)_2] = 6 \times (-20.7\,\text{kJ mol}^{-1})$$
$$= \boxed{-124.2\,\text{kJ mol}^{-1}}$$

(b) 2,4-dimethylhexane is composed of four $C(H)_3(C)$ groups, two $C(H)_2(C_2)$ groups, and two $C(H)(C)_3$ groups, so

$$\Delta_f H^\ominus = 4 \times (-42.17\,\text{kJ mol}^{-1}) + 2 \times (-20.7\,\text{kJ mol}^{-1}) + 2 \times (-6.19\,\text{kJ mol}^{-1})$$
$$= \boxed{-222.46\,\text{kJ mol}^{-1}}$$

Solutions to problems

Assume all gases are perfect unless stated otherwise. Unless otherwise stated, thermochemical data are for 298 K.

Solutions to numerical problems

P2.1 Since houses are not air-tight, some of the air originally in the house escapes to the outside and in the process does work against the atmosphere. Air may be assumed to have an effective molar mass of $29\,\mathrm{g\,mol^{-1}}$.

mass of air (20°C) $= (1.21\,\mathrm{kg\,m^{-3}}) \times 600\,\mathrm{m^3} = 726\,\mathrm{kg}$

amount of air (20°C) $= n(20°\mathrm{C}) = \dfrac{m}{M} = \dfrac{726\,\mathrm{kg}}{0.029\,\mathrm{kg\,mol^{-1}}} = 2.5\overline{03} \times 10^4\,\mathrm{mol}$

The heating of a house is a constant-pressure process, hence for a constant volume

$$n \propto \frac{1}{T} \quad \text{and} \quad \frac{n(25°\mathrm{C})}{n(20°\mathrm{C})} = \frac{T(20°\mathrm{C})}{T(25°\mathrm{C})} = \frac{293\,\mathrm{K}}{298\,\mathrm{K}}$$

$$n(25°\mathrm{C}) = \left(\frac{293\,\mathrm{K}}{298\,\mathrm{K}}\right) \times (2.5\overline{03} \times 10^4\,\mathrm{mol}) = 2.4\overline{61} \times 10^4\,\mathrm{mol}$$

Thus, $2.4\overline{61} \times 10^4$ mol of air has been heated from 20°C to 25°C. In addition 4.2×10^2 mol of air which has escaped into the outsifde air at 20°C has done work against the atmosphere ($p_{\mathrm{ext}} = 1.00\,\mathrm{atm}$).

$$V(\text{escaped air}) = \left(\frac{n(\text{escaped})}{n(\text{total})}\right) \times (600\,\mathrm{m^3}) = \left(\frac{4.2 \times 10^2\,\mathrm{mol}}{2.5 \times 10^4\,\mathrm{mol}}\right) \times (600\,\mathrm{m^3}) = 10\,\mathrm{m^3}$$

$$q_p = \Delta H = nC_{p,\mathrm{m}}\Delta T\ [28b]; \qquad \Delta U = nC_{V,\mathrm{m}}\Delta T\ [22b]$$

Since the gas is assumed to be a perfect diatomic

$$C_{p,\mathrm{m}} = \frac{7}{2}R, \qquad C_{V,\mathrm{m}} = \frac{5}{2}R$$

Hence for the air which remained in the house

$$\Delta H(\text{internal}) = (2.4\overline{61} \times 10^4\,\mathrm{mol}) \times \left(\tfrac{7}{2}\right) \times (8.314\,\mathrm{J\,K^{-1}mol^{-1}}) \times (5\,\mathrm{K}) = 3.6 \times 10^6\,\mathrm{J}$$
$$= +3.6 \times 10^3\,\mathrm{kJ}$$

$$\Delta U(\text{internal}) = (2.4\overline{61} \times 10^4\,\mathrm{mol}) \times \left(\tfrac{5}{2}\right) \times (8.314\,\mathrm{J\,K^{-1}mol^{-1}}) \times (5\,\mathrm{K})$$
$$= +2.6 \times 10^6\,\mathrm{J} = \boxed{+2.6 \times 10^3\,\mathrm{kJ}}$$

The complete answer for ΔH(total) must take into account the energy expended as work against the atmosphere by the expanding air

$$w(\text{by air}) = -p_{\mathrm{ext}}\Delta V[10] = (-1.013 \times 10^5\,\mathrm{Pa}) \times (10\,\mathrm{m^3}) = -1.0 \times 10^6\,\mathrm{J}$$

Thus the total heat which had to have been supplied to the air is

$$\Delta H(\text{total}) = \Delta H(\text{internal}) + w = (3.6 \times 10^3\,\mathrm{kJ}) - (1.0 \times 10^3\,\mathrm{kJ}) = \boxed{+2.6 \times 10^3\,\mathrm{kJ}}$$

Hence, $\Delta H(\text{total}) = \Delta U(\text{internal})$

P2.3 **(a)** The work done on the gas in section B is

$$w_B = -nRT \ln\left(\frac{V_f}{V_i}\right) [13] = (-2.00\,\text{mol}) \times (8.314\,\text{J K}^{-1}\text{mol}^{-1}) \times (300\,\text{K}) \times \ln\left(\frac{1.00\,\text{L}}{2.00\,\text{L}}\right) = 3.46 \times 10^3\,\text{J}$$

Therefore, the work done by the gas in section A is $w_A = \boxed{-3.46 \times 10^3\,\text{J}}$

(b) $\boxed{\Delta U_B = 0}$ [constant temperature]

(c) $q_B = \Delta U_B - w_B = 0 - (3.46 \times 10^3\,\text{J}) = \boxed{-3.46 \times 10^3\,\text{J}}$

(d) Since the volume in section B is decreased by a factor of $\frac{1}{2}$, the pressure in B is doubled, and, since $p_A = p_B$, $p_{f,A} = 2p_{i,A}$. From the perfect gas law

$$\frac{T_{f,A}}{T_{i,A}} = \frac{p_{f,A}V_{f,A}}{p_{i,A}V_{i,A}} = \frac{(2p_{i,A}) \times (3.00\,\text{L})}{(p_{i,A}) \times (2.00\,\text{L})} = 3.00$$

Hence, $T_{f,A} = 3.00\,T_{i,A} = (3.00) \times (300\,\text{K}) = 900\,\text{K}$

$$\Delta U_A = nC_{V,m}\Delta T = (2.00\,\text{mol}) \times (20.0\,\text{J K}^{-1}\text{mol}^{-1}) \times (600\,\text{K}) = \boxed{+2.40 \times 10^4\,\text{J}}$$

(e) $q_A = \Delta U_A - w_A = (2.40 \times 10^4\,\text{J}) - (-3.46 \times 10^3\,\text{J}) = \boxed{+2.75 \times 10^4\,\text{J}}$

P2.4 The temperatures are readily obtained from the perfect gas equation, $T = \dfrac{pV}{nR}$

$$T_1 = \frac{(1.00\,\text{atm}) \times (22.4\,\text{L})}{(1.00\,\text{mol}) \times (0.0821\,\text{L atm mol}^{-1}\,\text{K}^{-1})} = \boxed{273\,\text{K}}$$

Similarly, $T_2 = \boxed{546\,\text{K}}$, $T_3 = \boxed{273\,\text{K}}$

Step $1 \to 2$

$$w = -p_{ex}\Delta V = -p\Delta V = -nR\Delta T \quad (\Delta(pV) = \Delta(nRT))$$

$$w = -(1.00\,\text{mol}) \times (8.314\,\text{J K}^{-1}\,\text{mol}^{-1}) \times (546 - 273)\,\text{K} = \boxed{-2.27\,\text{kJ}}$$

$$\Delta U = nC_{V,m}\Delta T = (1.00\,\text{mol}) \times \tfrac{3}{2} \times (8.314\,\text{J K}^{-1}\,\text{mol}^{-1}) \times (273\,\text{K}) = \boxed{+3.40 \times 10^3\,\text{J}}$$

$$q = \Delta U - w = (3.40 \times 10^3 + 2.27 \times 10^3)\,\text{J} = \boxed{+5.67 \times 10^3\,\text{J}}$$

$$\Delta H = q = q_p = \boxed{+5.67 \times 10^3\,\text{J}}$$

Step $2 \to 3$

$$w = \boxed{0} \quad \text{[constant volume]}$$

$$q = \Delta U = nC_{V,m}\Delta T = (1.00\,\text{mol}) \times \left(\tfrac{3}{2}\right) \times (8.314\,\text{J K}^{-1}\,\text{mol}^{-1}) \times (-273\,\text{K}) = \boxed{-3.40\,\text{kJ}}$$

From $H \equiv U + pV$

$$\Delta H = \Delta U + \Delta(pV) = \Delta U + \Delta(nRT) = \Delta U + nR\Delta T$$
$$= (-3.40 \times 10^3\,\text{J}) + (1.00\,\text{mol}) \times (8.314\,\text{J K}^{-1}\,\text{mol}^{-1}) \times (-273\,\text{K}) = \boxed{-5.67\,\text{kJ}}$$

Step $3 \to 1$

ΔU and ΔH are $\boxed{\text{zero}}$ for an isothermal process in a perfect gas; hence

$$-q = w = -nRT \ln \frac{V_1}{V_3} = (-1.00\,\text{mol}) \times (8.314\,\text{J K}^{-1}\,\text{mol}^{-1}) \times (273\,\text{K}) \times \ln\left(\frac{22.4\,\text{L}}{44.8\,\text{L}}\right)$$

$$= \boxed{+1.57 \times 10^3\,\text{J}} \qquad q = \boxed{-1.57 \times 10^3\,\text{J}}$$

Total cycle

State	p/atm	V/L	T/K
1	1.00	22.44	273
2	1.00	44.8	546
3	0.50	44.8	273

Step	Process	q/kJ	w/kJ	$\Delta U/\text{kJ}$	$\Delta H/\text{kJ}$
$1 \to 2$	p constant at p_{ex}	+5.67	−2.27	+3.40	+5.67
$2 \to 3$	V constant	−3.40	0	−3.40	−5.67
$3 \to 1$	Isothermal, reversible	−1.57	+1.57	0	0
Cycle		+0.70	−0.70	0	0

Comment. All values can be determined unambiguously. The net result of the overall process is that 700 J of heat has been converted to work.

P2.7 The heat supplied by the electric heater is

$$q_p = I\mathcal{V}t\ [18] = (0.232\,\text{A}) \times (12.0\,\text{V}) \times (650\,\text{s}) = 1.81 \times 10^3\,\text{J} = 1.81\,\text{kJ}$$
$$\Delta H = q_p = 1.81\,\text{kJ} \quad [\text{constant pressure}]$$
$$\Delta_{\text{vap}} H = \frac{\Delta H}{n} = \left(\frac{102\,\text{g mol}^{-1}}{1.871\,\text{g}}\right) \times (1.81\,\text{kJ}) = \boxed{+98.7\,\text{kJ mol}^{-1}}$$
$$\Delta_{\text{vap}} U = \Delta_{\text{vap}} H - \Delta n_{\text{g}} RT\ [20], \quad \Delta n_{\text{g}} = +1$$
$$= (98.7\,\text{kJ mol}^{-1}) - (8.314\,\text{J K}^{-1}\,\text{mol}^{-1}) \times (351\,\text{K}) = \boxed{+95.8\,\text{kJ mol}^{-1}}$$

P2.9 The formation reaction is

$$2\text{C(s)} + 3\text{H}_2\text{(g)} \to \text{C}_2\text{H}_6\text{(g)} \quad \Delta_{\text{f}} H^{\ominus}(\text{T}) = -84.68\,\text{kJ mol}^{-1}$$

In order to determine $\Delta_{\text{f}} H^{\ominus}(350\,\text{K})$ we employ Kirchhoff's law [45]; $T_2 = 350\,\text{K}$, $T_1 = 298\,\text{K}$

$$\Delta_{\text{f}} H^{\ominus}(T_2) = \Delta_{\text{f}} H^{\ominus}(T_1) + \int_{T_1}^{T_2} \Delta_{\text{r}} C_p\,\text{d}T$$
$$\Delta_{\text{r}} C_p = \sum_{\text{J}} \nu_{\text{J}} C_{p,\text{m}}(\text{J}) = C_{p,\text{m}}(\text{C}_2\text{H}_6) - 2C_{p,\text{m}}(\text{C}) - 3C_{p,\text{m}}(\text{H}_2)$$

From Table 2.2

$$C_{p,\text{m}}(\text{C}_2\text{H}_6)/(\text{J K}^{-1}\,\text{mol}^{-1}) = (14.73) + \left(\frac{0.1272}{\text{K}}\right) T$$

$$C_{p,\mathrm{m}}(\mathrm{C, s})/(\mathrm{J\,K^{-1}\,mol^{-1}}) = (16.86) + \left(\frac{4.77\times10^{-3}}{\mathrm{K}}\right)T - \left(\frac{8.54\times10^{5}\,\mathrm{K}^2}{T^2}\right)$$

$$C_{p,\mathrm{m}}(\mathrm{H_2, g})/(\mathrm{J\,K^{-1}\,mol^{-1}}) = (27.28) + \left(\frac{3.26\times10^{-3}}{\mathrm{K}}\right)T + \left(\frac{0.50\times10^{5}\,\mathrm{K}^2}{T^2}\right)$$

$$\Delta_\mathrm{r}C_p/(\mathrm{J\,K^{-1}\,mol^{-1}}) = (-100.83) + \left(\frac{0.1079\,T}{\mathrm{K}}\right) + \left(\frac{1.56\times10^{6}\,\mathrm{K}^2}{T^2}\right)$$

$$\int_{T_1}^{T_2}\frac{\Delta_\mathrm{r}C_p\,\mathrm{d}T}{\mathrm{J\,K^{-1}\,mol^{-1}}} = (-100.83)\times(T_2-T_1) + \left(\tfrac{1}{2}\right)\times(0.1079\,\mathrm{K}^{-1})\times({T_2}^2-{T_1}^2)$$
$$-(1.56\times10^{6}\,\mathrm{K}^2)\times\left(\frac{1}{T_2}-\frac{1}{T_1}\right)$$
$$= (-100.83)\times(52\,\mathrm{K}) + \left(\tfrac{1}{2}\right)\times(0.1079)\times(350^2-298^2)\,\mathrm{K}$$
$$-(1.56\times10^{6})\times\left(\frac{1}{350}-\frac{1}{298}\right)\mathrm{K}$$
$$= -2.65\times10^{3}\,\mathrm{K}$$

Multiplying by the units $\mathrm{J\,K^{-1}\,mol^{-1}}$, we obtain

$$\int_{T_1}^{T_2}\Delta_\mathrm{r}C_p\,\mathrm{d}T = (-2.65\times10^{3}\,\mathrm{K})\times(\mathrm{J\,K^{-1}\,mol^{-1}}) = -2.65\times10^{3}\,\mathrm{J\,mol^{-1}}$$
$$= -2.65\,\mathrm{kJ\,mol^{-1}}$$

Hence
$$\Delta_\mathrm{f}H^\ominus(350\,\mathrm{K}) = \Delta_\mathrm{f}H^\ominus(298\,\mathrm{K}) - 2.65\,\mathrm{kJ\,mol^{-1}}$$
$$= (-84.68\,\mathrm{kJ\,mol^{-1}}) - (2.65\,\mathrm{kJ\,mol^{-1}})$$
$$= \boxed{-87.33\,\mathrm{kJ\,mol^{-1}}}$$

P2.10 The calorimeter is a constant-volume calorimeter as described in the text (Section 2.4); therefore

$$\Delta U = q_V$$

The calorimeter constant is determined from the data for the combustion of benzoic acid

$$\Delta U = \left(\frac{0.825\,\mathrm{g}}{122.12\,\mathrm{g\,mol^{-1}}}\right)\times(-3251\,\mathrm{kJ\,mol^{-1}}) = -21.9\bar{6}\,\mathrm{kJ}$$

Since $\Delta T = 1.940\,\mathrm{K}$, $C = \dfrac{|q|}{\Delta T} = \dfrac{21.9\bar{6}\,\mathrm{kJ}}{1.940\,\mathrm{K}} = 11.3\bar{2}\,\mathrm{kJ\,K^{-1}}$

For D-ribose, $\Delta U = -C\Delta T = (-11.3\bar{2}\,\mathrm{kJ\,K^{-1}})\times(0.910\,\mathrm{K})$

Therefore, $\Delta_\mathrm{r}U = \dfrac{\Delta U}{n} = (-11.3\bar{2}\,\mathrm{kJ\,K^{-1}})\times(0.910\,\mathrm{K})\times\left(\dfrac{150.13\,\mathrm{g\,mol^{-1}}}{0.727\,\mathrm{g}}\right) = -212\bar{7}\,\mathrm{kJ\,mol^{-1}}$

The combustion reaction for D-ribose is

$$C_5H_{10}O_5(s) + 5O_2(g) \rightarrow 5CO_2(g) + 5H_2O(l), \quad \Delta n_\mathrm{g} = 0$$

$$\Delta_\mathrm{c}H = \Delta_\mathrm{c}U = \boxed{-2130\,\mathrm{kJ\,mol^{-1}}}$$

The enthalpy of formation is obtained from the sum

	$\Delta H/(\text{kJ mol}^{-1})$
$5CO_2(g) + 5H_2O(l) \rightarrow C_5H_{10}O_5(s) + 5O_2(g)$	2130
$5C(s) + 5O_2(g) \rightarrow 5CO_2(g)$	$5 \times (-393.51)$
$5H_2(g) + \frac{5}{2}O_2(g) \rightarrow 5H_2O(l)$	$5 \times (-285.83)$
$5C(s) + 5H_2(g) + \frac{5}{2}O_2(g) \rightarrow C_5H_{10}O_5(s)$	-1267

Hence, $\Delta_f H = \boxed{-1267\,\text{kJ mol}^{-1}}$

P2.12 The complete aerobic oxidation is

$$C_{12}H_{22}O_{11} + 12O_2 \rightarrow 12CO_2 + 11H_2 \quad \Delta_c H^{\ominus} = -5645\,\text{kJ mol}^{-1}$$

The anaerobic hydrolysis to lactic acid is

$$C_{12}H_{22}O_{11} + H_2O \rightarrow 4CH_3CH(OH)COOH$$
$$\begin{aligned}\Delta H^{\ominus} &= 4\Delta_f H^{\ominus}(\text{lactic acid}) - \Delta_f H^{\ominus}(\text{sucrose}) - \Delta_f H^{\ominus}(H_2O,\ l)\\ &= \{(4) \times (-694.0) - (-2222) - (-285.8)\}\,\text{kJ mol}^{-1}\\ &= -268\,\text{kJ mol}^{-1}\end{aligned}$$

Therefore, $\Delta_c H^{\ominus}$ is $\boxed{\text{more exothermic by } 5376\,\text{kJ mol}^{-1}}$ than the hydrolysis reaction.

Solutions to theoretical problems

P2.14 Since ΔU is independent of path $\Delta U(A \rightarrow B) = q(ACB) + w(ACB) = 80\,J - 30\,J = 50\,J$

(a) $\Delta U = 50\,J = q(ADB) + w(ADB)$

$q(ADB) = 50\,J - (-10\,J) = \boxed{+60\,J}$

(b) $q(B \rightarrow A) = \Delta U(B \rightarrow A) - w(B \rightarrow A) = -50\,J - (+20\,J) = \boxed{-70\,J}$
The system liberates heat.

(c) $\Delta U(ADB) = \Delta U(A \rightarrow D) + \Delta U(D \rightarrow B); \quad 50\,J = 40\,J + \Delta U(D \rightarrow B)$
$\Delta U(D \rightarrow B) = 10\,J = q(D \rightarrow B) + w(D \rightarrow B); \quad w(D \rightarrow B) = 0,$

hence $q(D \rightarrow B) = \boxed{+10\,J}$

$q(ADB) = 60\,J[\text{part a}] = q(A \rightarrow D) + q(D \rightarrow B)$
$60\,J = q(A \rightarrow D) + 10\,J; \quad q(A \rightarrow D) = \boxed{+50\,J}$

P2.15 The enthalpy of a perfect gas depends only on temperature; hence

$$\Delta H = nC_{p,m}(T_f - T_i)$$

This applies for any temperature change in a perfect gas including a reversible adiabatic one. The strategy is then to show that

$$\int_i^f dH = \int_i^f V\,dp = nC_{p,m}(T_f - T_i)$$

For a reversible, adiabatic change, $pV^{\gamma} = \text{const}(A)$, so $V = \dfrac{A}{p^{1/\gamma}}$

$$\Delta H = A\int_{\mathrm{i}}^{\mathrm{f}} \frac{\mathrm{d}p}{p^{1/\gamma}} = \left\{\frac{A}{1-\frac{1}{\gamma}}\right\} \times \left(\frac{1}{p^{1/\gamma-1}}\right)\Bigg|_{p_\mathrm{i}}^{p_\mathrm{f}} = \left(\frac{\gamma A}{\gamma-1}\right) \times \left(\frac{1}{p_\mathrm{f}^{1/\gamma-1}} - \frac{1}{p_\mathrm{i}^{1/\gamma-1}}\right)$$

$$= \left(\frac{\gamma A}{\gamma-1}\right) \times \left(\frac{p_\mathrm{f}}{p_\mathrm{f}^{1/\gamma}} - \frac{p_\mathrm{i}}{p_\mathrm{i}^{1/\gamma}}\right)$$

$$= \left(\frac{\gamma}{\gamma-1}\right) \times (p_\mathrm{f}V_\mathrm{f} - p_\mathrm{i}V_\mathrm{i}) = \left(\frac{nR\gamma}{\gamma-1}\right) \times (T_\mathrm{f} - T_\mathrm{i})$$

$$\frac{\gamma}{\gamma-1} = \frac{1}{\left(1-\frac{1}{\gamma}\right)} = \frac{1}{\left(1-\frac{C_{V,\mathrm{m}}}{C_{p,\mathrm{m}}}\right)} = \frac{C_{p,\mathrm{m}}}{C_{p,\mathrm{m}} - C_{V,\mathrm{m}}} = \frac{C_{p,\mathrm{m}}}{R}$$

Hence, $\boxed{\Delta H = nC_{p,\mathrm{m}}(T_\mathrm{f} - T_\mathrm{i})}$, and the supposition is proven.

Solutions to additional problems

P2.17 The reaction is

$$C_{60}\,(s) + 60\,O_2(g) \rightarrow 60\,CO_2(g)$$

Because the reaction does not change the number of moles of gas, $\Delta_\mathrm{c}H = \Delta_\mathrm{c}U$.

Then

$$\Delta_\mathrm{c}H^{\ominus} = (-36.0334\,\mathrm{kJ\,g^{-1}}) \times (60 \times 12.011\,\mathrm{g\,mol^{-1}}) = \boxed{-25\,968\,\mathrm{kJ\,mol^{-1}}}$$

$$\Delta_\mathrm{c}H^{\ominus} = 60\,\Delta_\mathrm{f}H^{\ominus}(CO_2) - 60\,\Delta_\mathrm{f}H^{\ominus}(O_2) - \Delta_\mathrm{f}H^{\ominus}(C_{60}),$$

so

$$\Delta_\mathrm{f}H^{\ominus}(C_{60}) = 60\,\Delta_\mathrm{f}H^{\ominus}(CO_2) - 60\,\Delta_\mathrm{f}H^{\ominus}(O_2) - \Delta_\mathrm{c}H^{\ominus}$$

$$= [60(-393.51) - 60(0) - (-25968)]\,\mathrm{kJ\,mol^{-1}} = \boxed{2357\,\mathrm{kJ\,mol^{-1}}}$$

P2.19 The three possible fates of the radical are

(a) *tert*-$C_4H_9 \rightarrow$ *sec*-C_4H_9,

(b) *tert*-$C_4H_9 \rightarrow C_3H_6 + CH_3$,

(c) *tert*-$C_4H_9 \rightarrow C_2H_4 + C_2H_5$.

The three corresponding enthalpy changes are

(a) $\Delta_\mathrm{r}H^{\ominus} = \Delta_\mathrm{f}H^{\ominus}(sec\text{-}C_4H_9) - \Delta_\mathrm{f}H^{\ominus}(tert\text{-}C_4H_9) = (67.5 - 51.3)\mathrm{kJ\,mol^{-1}}$
$= \boxed{16.2\,\mathrm{kJ\,mol^{-1}}}$

(b) $\Delta_\mathrm{r}H^{\ominus} = \Delta_\mathrm{f}H^{\ominus}(C_3H_6) + \Delta_\mathrm{f}H^{\ominus}(CH_3) - \Delta_\mathrm{f}H^{\ominus}(tert\text{-}C_4H_9)$
$= (20.42 + 145.49 - 51.3)\,\mathrm{kJ\,mol^{-1}} = \boxed{114.6\,\mathrm{kJ\,mol^{-1}}}$

(c) $\Delta_\mathrm{r}H^{\ominus} = \Delta_\mathrm{f}H^{\ominus}(C_2H_4) + \Delta_\mathrm{f}H^{\ominus}(C_2H_5) - \Delta_\mathrm{f}H^{\ominus}(tert\text{-}C_4H_9)$

$= (52.26 + 121.0 - 51.3)\,\mathrm{kJ\,mol^{-1}} = \boxed{122.0\,\mathrm{kJ\,mol^{-1}}}$

P2.21 **(a)** $\Delta_r H^{\ominus} = \Delta_f H^{\ominus}(SiH_2) + \Delta_f H^{\ominus}(H_2) - \Delta_f H^{\ominus}(SiH_4),$

$$= (274 + 0 - 34.3)\ \text{kJ mol}^{-1} = \boxed{240\ \text{kJ mol}^{-1}}$$

(b) $\Delta_r H^{\ominus} = \Delta_f H^{\ominus}(SiH_2) + \Delta_f H^{\ominus}(SiH_4) - \Delta_f H^{\ominus}(Si_2H_6),$

$$= (274 + 34.3 - 80.3)\ \text{kJ mol}^{-1} = \boxed{228\ \text{kJ mol}^{-1}}$$

P2.23 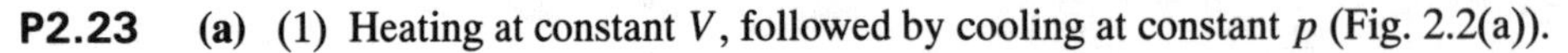
(a) (1) Heating at constant V, followed by cooling at constant p (Fig. 2.2(a)).

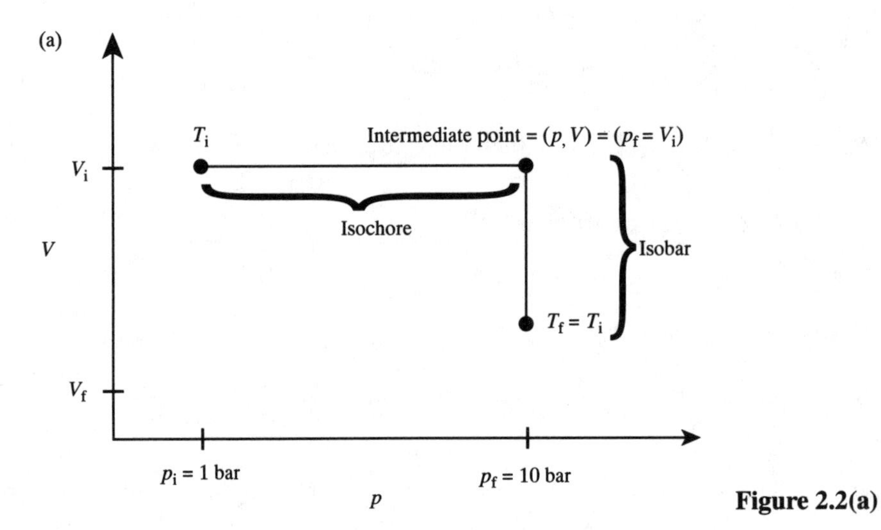

Figure 2.2(a)

Isochoric step

$$pV^n = C \quad \text{or} \quad V = \left(\frac{C}{p}\right)^{1/n}$$

This says that V = constant when $\boxed{n = \infty}$

Isobaric step

$$pV^n = C \quad \text{or} \quad p = \frac{C}{V^n}$$

This says that p = constant when $\boxed{n = 0}$

(2) Adiabatic compression, followed by cooling at constant V (Fig. 2.2(b)).

Adiabat

pV^{γ} = constant where $\gamma = C_p/C_V$ [36] so pV^n = constant provided that $\boxed{n = \gamma}$

Isochore

(See part (1) above.) $\boxed{n = \infty}$

(b) (1) *Initial state*

$$p_i = 1.00\ \text{bar}, \qquad T_i = 298.15\ \text{K}$$

$$V_i = \frac{nRT_i}{p_i} = \frac{(1\ \text{mol}) \times (0.08315\ \text{L bar K}^{-1}\ \text{mol}^{-1}) \times (298.15\ \text{K})}{1.00\ \text{bar}} = 24.79\ \text{L}$$

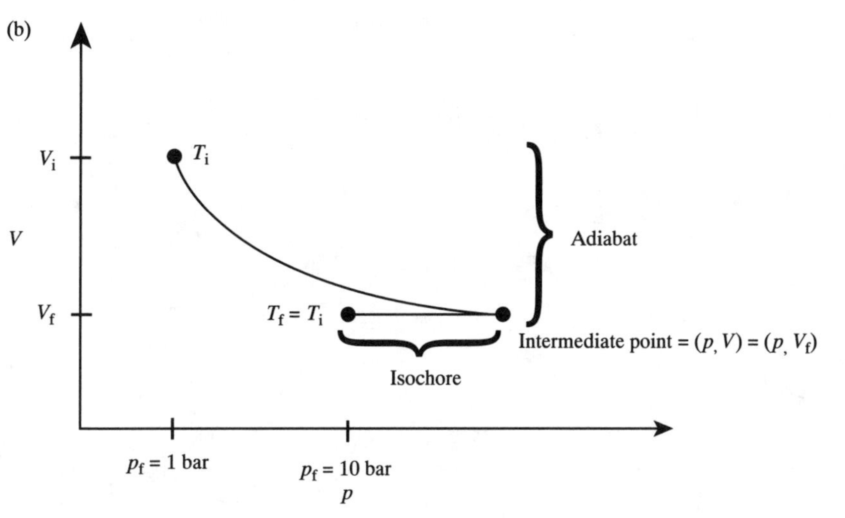

Figure 2.2(b)

Intermediate state

$$p = 10\,\text{bar}, \qquad V = 24.79\,\text{L}$$

$$T = \frac{pV}{nR} = \frac{(10.0\,\text{bar}) \times (24.79\,\text{L})}{(1\,\text{mol}) \times (0.08315\,\text{L bar K}^{-1}\,\text{mol}^{-1})} = 2981\,\text{K}$$

Final state

$$p_f = 10.0\,\text{bar}, \qquad T_f = 298.15\,\text{K}$$

$$V_f = \frac{1}{10} V_i = 2.479\,\text{L}$$

Isochoric step

$$w = -\int p\,\text{d}V = \boxed{0}$$

$$\Delta U = C_V \Delta T = (C_p - nR)\Delta T$$

$$= \tfrac{5}{2} nR\Delta T = \tfrac{5}{2} n(8.3145\,\text{J K}^{-1}\,\text{mol}^{-1}) \times (2981 - 298)\,\text{K}$$

$$\Delta U = \boxed{55.8\,\text{kJ}}$$

$$q = \Delta U - w = \boxed{55.8\,\text{kJ}}$$

$$\Delta H = \Delta U + \Delta(pV) = \Delta U + nR\Delta T$$

$$= 55.8\,\text{kJ} + (1\,\text{mol}) \times (8.3145\,\text{J K}^{-1}\,\text{mol}^{-1}) \times (2981 - 298)\,\text{K}$$

$$\Delta H = \boxed{78.1\,\text{kJ}}$$

Isobaric step

$$w = -\int p\,dV = -p_f \Delta V = -10\,\text{bar}(2.479\text{ L} - 24.79\text{ L})$$

$$= 223\text{ L bar}\left(\frac{8.315\text{ J}}{0.08315\text{ L bar}}\right) = \boxed{22.3\text{ kJ}}$$

$$\Delta U = \tfrac{5}{2}nR\Delta T = \boxed{-55.8\text{ kJ}}$$

$$q = \Delta U - w = -55.8\text{ kJ} - 22.3\text{ kJ} = \boxed{-78.1\text{ kJ}}$$

$$\Delta H = \Delta U + \Delta(pV) = \Delta U + nR\Delta T = -78.1\text{ kJ}$$

Overall

$$w = w_{\text{isochoric}} + w_{\text{isobaric}} = \boxed{22.3\text{ kJ}}$$

$$\Delta U = \Delta U_{\text{isochoric}} + \Delta U_{\text{isobaric}} = \boxed{0}$$

$$q = q_{\text{isochoric}} + q_{\text{isobaric}} = 55.8\text{ kJ} - 78.1\text{ kJ}$$

$$q = \boxed{-22.3\text{ kJ}}$$

$$\Delta H = \Delta H_{\text{isochoric}} + \Delta H_{\text{isobaric}} = \boxed{0}$$

(2) *Intermediate state*

$$V = V_f; \qquad \gamma = C_p/C_V = \left(\tfrac{7}{2}\right)/\left(\tfrac{5}{2}\right) = 7/5$$

$$p_i V_i^{\gamma} = p V_f^{\gamma}$$

$$p = \left(\frac{V_i}{V_f}\right)^{\gamma} p_i = \left(\frac{24.79\text{ L}}{2.479\text{ L}}\right)^{7/5}(1.00\text{ bar}) = 25.12\text{ bar}$$

$$T = \frac{pV}{nR} = \frac{(25.12\text{ bar}) \times (2.479\text{ L})}{(1\text{ mol}) \times (0.08315\text{ L bar K}^{-1}\text{ mol}^{-1})} = 749\text{ K}$$

Adiabatic step

$$q = \boxed{0}$$

$$\Delta U = C_V \Delta T = (1\text{ mol}) \times \left(\tfrac{5}{2}\right) \times (8.3145\text{ J K}^{-1}\text{ mol}^{-1}) \times (749 - 298)\text{ K}$$

$$\Delta U = \boxed{9.37\text{ kJ}}$$

$$w = \Delta U - q = \boxed{9.37\text{ kJ}}$$

$$\Delta H = \Delta U + \Delta(pV) = \Delta U + nR\Delta T$$

$$= 9.37\text{ kJ} + (1\text{ mol}) \times (8.3145\text{ J K}^{-1}\text{ mol}^{-1}) \times (749 - 298)\text{ K}$$

$$\Delta H = \boxed{13.1\text{ kJ}}$$

Isochoric step

$$w = -\int p\,dV = \boxed{0}$$

$$\Delta U = C_V\Delta T = \tfrac{5}{2}nR\Delta T$$

$$= \tfrac{5}{2}(1\text{ mol}) \times (8.3145\text{ J K}^{-1}\text{ mol}^{-1}) \times (298 - 749)\text{K}$$

$$\Delta U = \boxed{-9.37\text{ kJ}}$$

$$q = \Delta U - w = \boxed{-9.37\text{ kJ}}$$

$$\Delta H = \Delta U + nR\Delta T$$

$$= -9.37\text{ kJ} + (1\text{ mol}) \times (8.3145\text{ J K}^{-1}\text{ mol}^{-1}) \times (298 - 749)\text{K}$$

$$\Delta H = \boxed{-13.1\text{ kJ}}$$

Overall

$$w = w_{\text{adiabat}} + w_{\text{isochore}} = \boxed{9.37\text{ kJ}}$$

$$q = q_{\text{adiabat}} + q_{\text{isochore}} = \boxed{-9.37\text{ kJ}}$$

$$\Delta U = \Delta U_{\text{adiabat}} + \Delta U_{\text{isochore}} = \boxed{0}$$

$$\Delta H = \Delta H_{\text{adiabat}} + \Delta H_{\text{isochore}} = \boxed{0}$$

P2.25 Data: methane–octane normal alkanes

$\Delta_c H/(\text{kJ mol}^{-1})$	−890	−1560	−2220	−2878	−3537	−4163	−5471
$M/(\text{g mol}^{-1})$	16.04	30.07	44.10	58.13	72.15	86.18	114.23

Suppose that $\Delta_c H = kM^n$. There are two methods by which a regression analysis can be used to determine the values of k and n. If you have a software package that can perform a 'power fit' of the type $Y = aX^b$, the analysis is direct using $Y = \Delta_c H$ and $X = M$. Then, $k = a$ and $n = b$, alternatively, taking the logarithm yields another equation—one of linear form

$$\ln|\Delta_c H| = \ln|k| + n\ln M \quad \text{where } k < 0$$

This equation suggests a linear regression fit of $\ln(\Delta_c H)$ against $\ln M$ (Fig. 2.3). The intercept is $\ln k$ and the slope is n. Linear regression fit

$$\ln|k| = 4.2112, \quad \text{standard deviation} = 0.0480; \quad k = -e^{4.2112} = \boxed{-67.44}$$

$$\boxed{n = 0.9253}, \quad \text{standard deviation} = 0.0121$$

$$R = 1.000$$

This is a good regression fit; essentially all of the variation is explained by the regression.

For decane the experimental value of $\Delta_c H$ equals $-6772.5\text{ kJ mol}^{-1}$ (*CRC Handbook of Chemistry and Physics*). The predicted value is

$$\Delta_c H = kM^n = -67.44(142.28)^{(0.9253)}\text{ kJ mol}^{-1}$$

$$\boxed{\Delta_c H = -6625.5\text{ kJ mol}^{-1}}$$

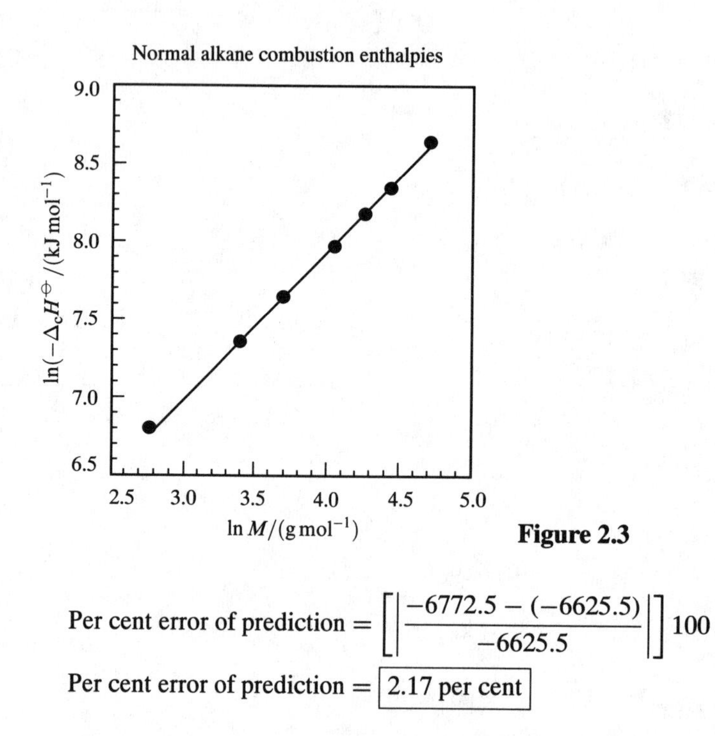

Figure 2.3

$$\text{Per cent error of prediction} = \left[\left|\frac{-6772.5-(-6625.5)}{-6625.5}\right|\right]100$$

$$\text{Per cent error of prediction} = \boxed{2.17 \text{ per cent}}$$

P2.26 (a) The process is polytropic provided that $pV^n = \text{constant} = C$. Taking the logarithm, $\ln p = \ln C - n\ln V$. Thus, if the plot of $\ln p$ against $\ln V$ is linear, the process is polytropic and the plot slope equals $-n$. The plot is shown in Fig. 2.4. A linear regression analysis of the plot provides the following

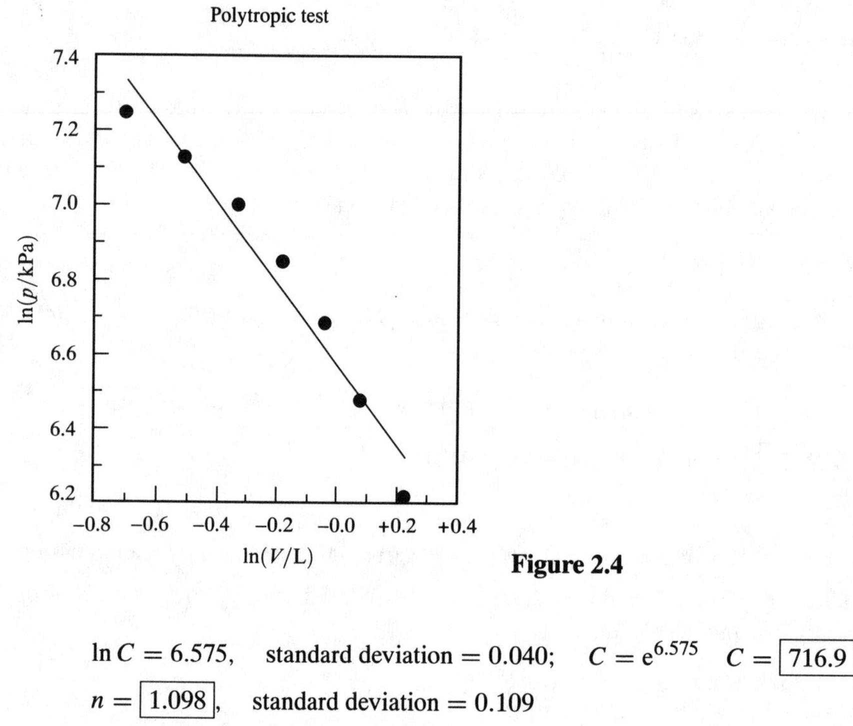

Figure 2.4

$\ln C = 6.575$, standard deviation $= 0.040$; $C = e^{6.575}$ $C = \boxed{716.9}$

$n = \boxed{1.098}$, standard deviation $= 0.109$

$R = 0.9761$

Thus 97.61 per cent of the variation is apparently explained by the regression. However, the plot appears to vary noticeably from the linear in a systematic manner, so we conclude that the process is *not polytropic*. An R value of 0.9761, although apparently close to 1.000, does not satisfy the criterion of a good fit to a linear plot; an R value > 0.99 would be desirable.
The polytropic test plot suggests that a polynomial fit of the form

$$\ln(p/\text{kPa}) = a + b\ln(V/\text{L}) + C[\ln(V/\text{L})]^2$$

would provide a good empirical fit of the data.
Polynomial fit

$a = 6.6071$, standard deviation $= 0.0083$

$b = -1.5217$, standard deviation $= 0.0432$

$c = -0.08976$, standard deviation $= 0.0795$

$R = 0.99928$

99.93 per cent of the variation is explained by the regression.

(b) Since the data have proven not to be polytropic, $p(V)$ is not a simple function that may be used in the work integral. In the absence of the simple integrand that can yield an analytic function for w, a numerical integration is appropriate.

$$w = -\int p_{\text{ext}}\,\text{d}V = -\int_{V_i=1.25\,\text{L}}^{V_f=0.50\,\text{L}} p\,\text{d}V \text{ [assuming mechanical equilibrium]}$$

$$= -(-685)\,\text{kPa L}\left(\frac{10^{-3}\,\text{m}^3}{\text{L}}\right) = \boxed{685\,\text{J}}$$

(c) We know the values of p, T, and V for the initial state. This data can be used to calculate n. p and T can be used in the cubic polynomial form of the van der Waals equation. The root is V_m

$$V_m^3 - \left(b + \frac{RT}{p}\right)V_m^2 + \left(\frac{a}{p}\right)V_m - \frac{ab}{p} = 0 \quad \text{[Example 1.7]}$$

$$p = 5.00 \times 10^5\,\text{Pa}\left(\frac{1\,\text{atm}}{1.01325 \times 10^5\,\text{Pa}}\right) = 4.93\,\text{atm}$$

$$T = (273.15 + 30)\,\text{K} = 303.15\,\text{K}$$

$$a = 4.225\,\text{L}^2\,\text{atm mol}^{-2}$$

$$b = 3.707 \times 10^{-2}\,\text{L mol}^{-1}$$

$$b + \frac{RT}{p} = 3.707 \times 10^{-2}\,\text{L mol}^{-1} + \frac{(0.0820578\,\text{L atm K}^{-1}\,\text{mol}^{-1}) \times (303.15\,\text{K})}{4.93\,\text{atm}}$$

$$= 5.08\,\text{L mol}^{-1}$$

$$\frac{a}{p} = \frac{4.225\,\text{L}^2\,\text{atm mol}^{-2}}{4.93\,\text{atm}} = 0.857\,\text{L}^2\,\text{mol}^{-2}$$

$$\frac{ab}{p} = (0.857\,\text{L}^2\,\text{mol}^{-2}) \times (3.707 \times 10^{-2}\,\text{L mol}^{-1}) = 0.0318\,\text{L}^3\,\text{mol}^{-3}$$

Therefore, $V_m^3 - (5.08\,\text{L mol}^{-1})V_m^2 + (0.857\,\text{L}^2\,\text{mol}^{-2})V_m - 0.0318\,\text{L}^3\,\text{mol}^{-3} = 0$
The real root of this equation is

$$V_m = 4.91\,\text{L mol}^{-1}$$

$$n = V/V_m = 1.25\,\text{L}/(4.91\,\text{L mol}^{-1})$$

$$n = \boxed{0.255\,\text{mol}}$$

Now we can use the van der Waals equation and the values of n, p, and V for the final state to calculate the final temperature. For the final state

$$V_m = \frac{0.50\ \text{L}}{0.255\ \text{mol}} = 1.96\ \text{L}\,\text{mol}^{-1}$$

$$p = 1400\ \text{kPa}\left(\frac{1\ \text{atm}}{101.325\ \text{kPa}}\right) = 13.82\ \text{atm}$$

$$T = \frac{(V_m - b) \times \left(p + \frac{a}{V_m^2}\right)}{R} \quad [1.39]$$

$$= \frac{\left(1.96\ \text{L}\,\text{mol}^{-1} - 0.03707\ \text{L}\,\text{mol}^{-1}\right) \times \left(13.82\ \text{atm} + \frac{4.225\ \text{L}^2\ \text{atm}\ \text{mol}^{-2}}{(1.96\ \text{L}\,\text{mol}^{-1})^2}\right)}{0.082058\ \text{L}\,\text{atm}\,\text{K}^{-1}\,\text{mol}^{-1}}$$

$$T = \boxed{350\ \text{K or } 76.5°\text{C}}$$

3 The First Law: the machinery

Solutions to exercises

Assume that all gases are perfect and that all data refer to 298 K unless stated otherwise.

E3.1 **(a)** $\dfrac{\partial^2 f}{\partial y \partial x} = \dfrac{\partial}{\partial y}(2xy) = 2x \qquad \dfrac{\partial^2 f}{\partial x \partial y} = \dfrac{\partial}{\partial x}(x^2 + 6y) = 2x$

(b) $\dfrac{\partial^2 f}{\partial y \partial x} = \dfrac{\partial}{\partial y}(\cos xy - xy \sin xy)$

$$= -x \sin xy - x \sin xy - x^2 y \cos xy = -2x \sin xy - x^2 y \cos xy$$

$$\frac{\partial^2 f}{\partial x \partial y} = \frac{\partial}{\partial x}(-x^2 \sin xy) = -2x \sin xy - x^2 y \cos xy$$

E3.2 $\mathrm{d}z = \left(\dfrac{\partial z}{\partial x}\right)_y \mathrm{d}x + \left(\dfrac{\partial z}{\partial y}\right)_x \mathrm{d}y$ [*Further information* 1]

$$\left(\frac{\partial z}{\partial x}\right)_y = \left(\frac{\partial(ax^2y^3)}{\partial x}\right)_y = 2axy^3 \qquad \left(\frac{\partial z}{\partial y}\right)_x = \left(\frac{\partial(ax^2y^3)}{\partial y}\right)_x = 3ax^2y^2$$

$$\mathrm{d}z = \boxed{2axy^3\,\mathrm{d}x + 3ax^2y^2\,\mathrm{d}y}$$

E3.3 **(a)** $\mathrm{d}z = \left(\dfrac{\partial z}{\partial x}\right)_y \mathrm{d}x + \left(\dfrac{\partial z}{\partial y}\right)_x \mathrm{d}y$ [*Further information* 1]

$$\left(\frac{\partial z}{\partial x}\right)_y = (2x - 2y + 2) \qquad \left(\frac{\partial z}{\partial y}\right)_x = (4y - 2x - 4)$$

$$\mathrm{d}z = \boxed{(2x - 2y + 2)\,\mathrm{d}x + (4y - 2x - 4)\,\mathrm{d}y}$$

(b) $\dfrac{\partial^2 z}{\partial y \partial x} = \dfrac{\partial}{\partial y}(2x - 2y + 2) = -2 \qquad \dfrac{\partial^2 z}{\partial x \partial y} = \dfrac{\partial}{\partial x}(4y - 2x - 4) = -2$

Comment. The total differential of a function is necessarily an exact differential.

E3.4 $\mathrm{d}z = \left(\dfrac{\partial z}{\partial x}\right)_y \mathrm{d}x + \left(\dfrac{\partial z}{\partial y}\right)_x \mathrm{d}y$ [*Further information* 1]

$$\left(\frac{\partial z}{\partial x}\right)_y = \left(y + \frac{1}{x}\right) \qquad \left(\frac{\partial z}{\partial y}\right)_x = (x - 1)$$

$$\mathrm{d}z = \boxed{\left(y + \frac{1}{x}\right)\mathrm{d}x + (x - 1)\,\mathrm{d}y}$$

A differential is exact if it satisfies the condition

$$\frac{\partial^2 z}{\partial y \partial x} = \frac{\partial^2 z}{\partial x \partial y} \quad \text{[\textit{Further information} 1]}$$

$$\left(\frac{\partial}{\partial y}\right)\left(\frac{\partial z}{\partial x}\right)_y = \left(\frac{\partial}{\partial y}\right)\left(y + \frac{1}{x}\right) = 1 \qquad \left(\frac{\partial}{\partial x}\right)\left(\frac{\partial z}{\partial y}\right)_x = \left(\frac{\partial}{\partial x}\right)(x - 1) = 1$$

Comment. The total differential of a function is necessarily exact as is demonstrated here by the reciprocity test.

E3.5 $C_V = \left(\frac{\partial U}{\partial T}\right)_V$

$$\boxed{\left(\frac{\partial C_V}{\partial V}\right)_T = \left(\frac{\partial}{\partial V}\left(\frac{\partial U}{\partial T}\right)_V\right)_T = \left(\frac{\partial}{\partial T}\left(\frac{\partial U}{\partial V}\right)_T\right)_V}$$ [derivatives may be taken in any order]

$\left(\frac{\partial U}{\partial V}\right)_T = 0$ for a perfect gas [Section 3.1]

Hence, $\boxed{\left(\frac{\partial C_V}{\partial V}\right)_T = 0}$

E3.6 $H = U + pV$

$$\left(\frac{\partial H}{\partial U}\right)_p = \boxed{1 + p\left(\frac{\partial V}{\partial U}\right)_p}$$

E3.7 $V = V(p, T)$; hence, $\mathrm{d}V = \boxed{\left(\frac{\partial V}{\partial p}\right)_T \mathrm{d}p + \left(\frac{\partial V}{\partial T}\right)_p \mathrm{d}T}$

We use $\alpha = \left(\frac{1}{V}\right)\left(\frac{\partial V}{\partial T}\right)_p$ [7] and $\kappa_T = -\left(\frac{1}{V}\right)\left(\frac{\partial V}{\partial p}\right)_T$ [13] and obtain

$$\mathrm{d}\ln V = \frac{1}{V}\mathrm{d}V = \left(\frac{1}{V}\right)\left(\frac{\partial V}{\partial p}\right)_T \mathrm{d}p + \left(\frac{1}{V}\right)\left(\frac{\partial V}{\partial T}\right)_p \mathrm{d}T = \boxed{-\kappa_T\,\mathrm{d}p + \alpha\,\mathrm{d}T}$$

E3.8 $\left(\frac{\partial U}{\partial V}\right)_T = \left(\frac{\partial}{\partial V}\left(\frac{3}{2}nRT\right)\right)_T = \boxed{0}$

$H = U + pV = U + nRT$ [$pV = nRT$]

$$\left(\frac{\partial H}{\partial V}\right)_T = \left(\frac{\partial U}{\partial V}\right)_T + \left(\frac{\partial nRT}{\partial V}\right)_T = 0 + 0 = \boxed{0}$$

E3.9 $V = V(T, p)$; hence, $\mathrm{d}V = \left(\frac{\partial V}{\partial T}\right)_p \mathrm{d}T + \left(\frac{\partial V}{\partial p}\right)_T \mathrm{d}p$

Dividing each term by $(\mathrm{d}T)_V$ we obtain

$$\left(\frac{\partial V}{\partial T}\right)_V = 0 = \left(\frac{\partial V}{\partial T}\right)_p\left(\frac{\partial T}{\partial T}\right)_V + \left(\frac{\partial V}{\partial p}\right)_T\left(\frac{\partial p}{\partial T}\right)_V$$

or $0 = \left(\frac{\partial V}{\partial T}\right)_p + \left(\frac{\partial V}{\partial p}\right)_T\left(\frac{\partial p}{\partial T}\right)_V$

or $\left(\frac{\partial p}{\partial T}\right)_V = -\frac{\left(\frac{\partial V}{\partial T}\right)_p}{\left(\frac{\partial V}{\partial P}\right)_T} = \frac{\left(\frac{1}{V}\right)\left(\frac{\partial V}{\partial T}\right)_p}{-\left(\frac{1}{V}\right)\left(\frac{\partial V}{\partial p}\right)_T} = \frac{\alpha}{\kappa_T}$

E3.10 $$\mu = \left(\frac{\partial T}{\partial p}\right)_H [14] = \lim_{\Delta p \to 0}\left(\frac{\Delta T}{\Delta p}\right)_H \approx \frac{\Delta T}{\Delta p} \text{ [for } \mu \text{ constant over this temperature range]}$$

$$\mu = \frac{-22\,\text{K}}{-31\,\text{atm}} = \boxed{0.71\,\text{K atm}^{-1}}$$

E3.11 $$U_\text{m} = U_\text{m}(T, V_\text{m}); \qquad \text{d}U_\text{m} = \left(\frac{\partial U_\text{m}}{\partial T}\right)_{V_\text{m}} \text{d}T + \left(\frac{\partial U_\text{m}}{\partial V_\text{m}}\right)_T \text{d}V_{\text{m}T} \qquad [\pi_T = (\partial U_\text{m}/\partial V)_T]$$

For an isothermal expansion $\text{d}T = 0$; hence

$$\text{d}U_\text{m} = \left(\frac{\partial U_\text{m}}{\partial V_\text{m}}\right)_T \text{d}V_\text{m} = \frac{a}{V_\text{m}^2}\,\text{d}V_\text{m}$$

$$\Delta U_\text{m} = \int_{V_\text{m,1}}^{V_\text{m,2}} \text{d}U_\text{m} = \int_{V_\text{m,1}}^{V_\text{m,2}} \frac{a}{V_\text{m}^2}\,\text{d}V_\text{m} = a\int_{1.00\,\text{L mol}^{-1}}^{24.8\,\text{L mol}^{-1}} \frac{\text{d}V_\text{m}}{V_\text{m}^2} = -\frac{a}{V_\text{m}}\Big|_{1.00\,\text{L mol}^{-1}}^{24.8\,\text{L mol}^{-1}}$$

$$= -\frac{a}{24.8\,\text{L mol}^{-1}} + \frac{a}{1.00\,\text{L mol}^{-1}} = \frac{23.8a}{24.8\,\text{L mol}^{-1}} = 0.959\overline{7}\,\text{mol L}^{-1}a;$$

$a = 1.408\,\text{L}^2\,\text{atm mol}^{-2}$ [Table 1.6]

$$\Delta U_\text{m} = (0.959\overline{7}\,\text{mol L}^{-1}) \times (1.408\,\text{L}^2\,\text{atm mol}^{-2})$$

$$= (1.351\,\text{L atm mol}^{-1}) \times \left(\frac{10^{-3}\,\text{m}^3}{\text{L}}\right) \times \left(\frac{1.013 \times 10^5\,\text{Pa}}{\text{atm}}\right)$$

$$= \boxed{+137\,\text{J mol}^{-1}}$$

$$w = -\int p\,\text{d}V_\text{m}$$

For a van der Waals gas

$$p = \frac{RT}{V_\text{m} - b} - \frac{a}{V_\text{m}^2}$$

Hence,

$$w = -\int \left(\frac{RT}{V_\text{m} - b}\right)\text{d}V_\text{m} + \int \frac{a}{V_\text{m}^2}\,\text{d}V_\text{m} = -q + \Delta U_\text{m}$$

Therefore,

$$q = \int_{1.0\,\text{L mol}^{-1}}^{24.8\,\text{L mol}^{-1}} \left(\frac{RT}{V_\text{m} - b}\right)\text{d}V_\text{m} = -RT\ln(V_\text{m} - b)\Big|_{1.00\,\text{L mol}^{-1}}^{24.8\,\text{L mol}^{-1}}$$

$$= RT\ln\left(\frac{(24.8) - (3.9 \times 10^{-2})}{(1.00) - (3.9 \times 10^{-2})}\right)$$

$$= (8.314\,\text{J K}^{-1}\,\text{mol}^{-1}) \times (298\,\text{K}) \times (3.25) = \boxed{+8.05 \times 10^3\,\text{J mol}^{-1}}$$

$$w = -q + \Delta U_\text{m} = -(8.05 \times 10^3\,\text{J mol}^{-1}) + (137\,\text{J mol}^{-1}) = \boxed{-7.91 \times 10^3\,\text{J mol}^{-1}}$$

E3.12 $$\alpha = \left(\frac{1}{V}\right)\left(\frac{\partial V}{\partial T}\right)_p [7]; \qquad \alpha_{320} = \left(\frac{1}{V_{320}}\right)\left(\frac{\partial V}{\partial T}\right)_{p,320}$$

$$V_{320} = V_{300}\{(0.75) + (3.9 \times 10^{-4}) \times (320) + (1.48 \times 10^{-6}) \times (320)^2\} = (V_{300}) \times (1.026)$$

$$\frac{1}{V_{320}} = \left(\frac{1}{1.02\bar{6}}\right) \times \left(\frac{1}{V_{300}}\right) = \frac{0.974}{V_{300}}$$

$$\left(\frac{\partial V}{\partial T}\right)_p = V_{300}(3.9 \times 10^{-4}/\text{K} + 2.96 \times 10^{-6} T/\text{K}^2)$$

$$\left(\frac{\partial V}{\partial T}\right)_{p,320} = V_{300}(3.9 \times 10^{-4}/\text{K} + 2.96 \times 10^{-6} \times 320/\text{K}) = 1.34 \times 10^{-3}\,\text{K}^{-1} V_{300}$$

$$\alpha_{320} = \left(\frac{1}{V_{320}}\right)\left(\frac{\partial V}{\partial T}\right)_{p,320} = \left(\frac{0.974}{V_{300}}\right) \times (1.3 \times 10^{-3}\,\text{K}^{-1}\,V_{300})$$

$$= (0.974) \times (1.34 \times 10^{-3}\,\text{K}^{-1})$$

$$= \boxed{1.31 \times 10^{-3}\,\text{K}^{-1}}$$

Comment. Knowledge of the density at 300 K is not required to solve this exercise, though it would be required to obtain V_{300} and V_{320} in absolute rather than relative form.

E3.13 $$\kappa_T = -\left(\frac{1}{V}\right)\left(\frac{\partial V}{\partial p}\right)_T \text{ [13]; thus } \left(\frac{\partial V}{\partial p}\right)_T = -\kappa_T V$$

$$\text{d}V = \left(\frac{\partial V}{\partial p}\right)_T \text{d}p \text{ [at constant } T\text{]; then } \text{d}V = -\kappa_T V\,\text{d}p \text{ or } \frac{\text{d}V}{V} = -\kappa_T\,\text{d}p$$

Substituting $V = \frac{m}{\rho}$ and $\text{d}V = -\frac{m}{\rho^2}\text{d}\rho$; $\frac{\text{d}V}{V} = -\frac{\text{d}\rho}{\rho} = -\kappa_T\,\text{d}p$

Therefore, $\frac{\delta\rho}{\rho} \approx \kappa_T \delta p$

For $\frac{\delta\rho}{\rho} = 0.08 \times 10^{-2} = 8 \times 10^{-4}$, $\delta p \approx \frac{8 \times 10^{-4}}{\kappa} = \frac{8 \times 10^{-4}}{7.35 \times 10^{-7}\,\text{atm}^{-1}} = \boxed{1.1\bar{1} \times 10^3\,\text{atm}}$

E3.14 $$\left(\frac{\partial H_\text{m}}{\partial p}\right)_T = -\mu C_{p,\text{m}}[15] = (-0.25\,\text{K atm}^{-1}) \times (29\,\text{J K}^{-1}\,\text{mol}^{-1}) = \boxed{-7.2\,\text{J atm}^{-1}\,\text{mol}^{-1}}$$

$$\text{d}H = n\left(\frac{\partial H_\text{m}}{\partial p}\right)_T \text{d}p = -n\mu C_{p,\text{m}}\,\text{d}p$$

$$\Delta H = \int_{p_1}^{p_2} -n\mu C_{p,\text{m}}\,\text{d}p = -n\mu C_{p,\text{m}}(p_2 - p_1) \quad [\mu \text{ and } C_p \text{ are constants}]$$

$$\Delta H = -n\mu C_{p,\text{m}}(-75\,\text{atm}) = (-15\,\text{mol}) \times (+7.2\,\text{J atm}^{-1}\,\text{mol}^{-1}) \times (-75\,\text{atm}) = +8.1\,\text{kJ}$$

$$q(\text{supplied}) = +\Delta H = \boxed{+8.1\,\text{kJ}}$$

E3.15 $$\mu = \left(\frac{\partial T}{\partial p}\right)_H = \lim_{\Delta p \to 0}\left(\frac{\Delta T}{\Delta p}\right)$$

If Δp is not so large as to produce a ΔT which is a large fraction of T we may write approximately

$$\mu \approx \frac{\Delta T}{\Delta p} \quad \text{or} \quad \Delta p \approx \frac{\Delta T}{\mu}$$

For $\Delta T = -5.0\,\text{K}$,

$$\Delta p \approx \frac{-5.0\,\text{K}}{1.2\,\text{K atm}^{-1}} = \boxed{-4.2\,\text{atm}}$$

Solutions to problems

Assume that all gases are perfect and that all data refer to 298 K unless stated otherwise.

Solutions to numerical problems

P3.2 **(a)** $\mathrm{d}U_\mathrm{m} = \left(\frac{\partial U_\mathrm{m}}{\partial T}\right)_p \mathrm{d}T + \left(\frac{\partial U_\mathrm{m}}{\partial p}\right)_T \mathrm{d}p$

If we assume that $\left(\frac{\partial U_\mathrm{m}}{\partial p}\right)_T$ is small, and if the change in temperature is not large (10 K probably qualifies as small), then we may write

$$\Delta U_\mathrm{m} \approx \left(\frac{\partial U_\mathrm{m}}{\partial T}\right)_p \Delta T \qquad \left(\frac{\partial U_\mathrm{m}}{\partial T}\right)_p = C_{V,\mathrm{m}} + \alpha V_\mathrm{m}\left(\frac{\partial U}{\partial V}\right)_T = C_{V,\mathrm{m}} + \alpha\pi_T V_\mathrm{m} \ [8]$$

Since $C_{p,\mathrm{m}} - C_{V,\mathrm{m}} = \alpha V_\mathrm{m}(p + \pi_T)$ [*Justification* 3.2]

$$\pi_T = \frac{C_{p,\mathrm{m}} - C_{V,\mathrm{m}}}{\alpha V_\mathrm{m}} - p$$

and hence

$$\left(\frac{\partial U_\mathrm{m}}{\partial T}\right)_p = C_{V,\mathrm{m}} + \alpha V_\mathrm{m}\left(\frac{C_{p,\mathrm{m}} - C_{V,\mathrm{m}}}{\alpha V} - p\right) = C_{p,\mathrm{m}} - \alpha p V_\mathrm{m}$$

$$C_{p,\mathrm{m}} = 75.29\,\mathrm{J\,K^{-1}\,mol^{-1}}\ \text{[Table 2.6]}, \qquad \alpha = 2.1 \times 10^{-4}\,\mathrm{K^{-1}}\ \text{[Table 3.1]}$$

$$V_\mathrm{m} = 18.02\,\mathrm{g\,mol^{-1}}/\rho[\rho = 0.997\,\mathrm{g\,cm^{-3}}\ \text{at } 25^\circ\mathrm{C}] = 18.07 \times \mathrm{cm^3\,mol^{-1}}$$
$$= 18.07 \times 10^{-6}\,\mathrm{m^3\,mol^{-1}}$$

Therefore, $\left(\frac{\partial U_\mathrm{m}}{\partial T}\right)_p = (75.29\,\mathrm{J\,K^{-1}\,mol^{-1}}) - (2.1 \times 10^{-4}\,\mathrm{K^{-1}}) \times (1.013 \times 10^5\,\mathrm{Pa})$

$$\times (18.07 \times 10^{-6}\,\mathrm{m^3\,mol^{-1}})$$
$$= (75.29\,\mathrm{J\,K^{-1}\,mol^{-1}}) - (3.8 \times 10^{-4}\,\mathrm{J\,K^{-1}\,mol^{-1}})$$
$$= 75.29\,\mathrm{J\,K^{-1}\,mol^{-1}}$$

Therefore, $\Delta U_\mathrm{m} \approx (75.29\,\mathrm{J\,K^{-1}\,mol^{-1}}) \times (10\,\mathrm{K}) = \boxed{+0.75\,\mathrm{kJ\,mol^{-1}}}$

(b) $\mathrm{d}H_\mathrm{m} = \left(\frac{\partial H_\mathrm{m}}{\partial T}\right)_p \mathrm{d}T + \left(\frac{\partial H_\mathrm{m}}{\partial p}\right)_T \mathrm{d}p$

Assuming that $\left(\frac{\partial H_\mathrm{m}}{\partial p}\right)_T$ is small and that ΔT is not large, we may write

$$\Delta H_\mathrm{m} \approx \left(\frac{\partial H_\mathrm{m}}{\partial T}\right)_p \Delta T = C_{p,\mathrm{m}}\Delta T = (75.29\,\mathrm{J\,K^{-1}\,mol^{-1}}) \times (10\,\mathrm{K})$$
$$= 7.5 \times 10^2\,\mathrm{J\,mol^{-1}} = \boxed{+0.75\,\mathrm{kJ\,mol^{-1}}}$$

The difference is

$$\Delta H_\mathrm{m} - \Delta U_\mathrm{m} = \alpha p V_\mathrm{m} \Delta T = +3.8\,\mathrm{mJ\,mol^{-1}}$$

which is the change in energy as a result of doing expansion work.

P3.3 $T_\mathrm{f} = \left(\dfrac{p_\mathrm{f}}{p_\mathrm{i}}\right)^{1/c\gamma} \times (T_\mathrm{i})$ [Exercise 2.25 (a)]; hence, $c\gamma \ln \dfrac{T_\mathrm{f}}{T_\mathrm{i}} = \ln \dfrac{p_\mathrm{f}}{p_\mathrm{i}}$

Since $c\gamma = \left(\dfrac{C_{V,\mathrm{m}}}{R}\right) \times \left(\dfrac{C_{p,\mathrm{m}}}{C_{V,\mathrm{m}}}\right) = \dfrac{C_{p,\mathrm{m}}}{R}$

$$C_{p,\mathrm{m}} = R\frac{\ln\left(\frac{p_\mathrm{f}}{p_\mathrm{i}}\right)}{\ln\left(\frac{T_\mathrm{f}}{T_\mathrm{i}}\right)} = (8.314\,\mathrm{J\,K^{-1}\,mol^{-1}}) \times \left(\frac{\ln\left(\frac{613.85}{1522.2}\right)}{\ln\left(\frac{248.44}{298.15}\right)}\right) = \boxed{41.40\,\mathrm{J\,K^{-1}\,mol^{-1}}}$$

P3.4 $\mathrm{d}H = \left(\dfrac{\partial H}{\partial T}\right)_p \mathrm{d}T + \left(\dfrac{\partial H}{\partial p}\right)_T \mathrm{d}p$ or $\mathrm{d}H = \left(\dfrac{\partial H}{\partial p}\right)_T \mathrm{d}p$ [constant temperature]

$$\left(\frac{\partial H}{\partial p}\right)_T = -\mu C_{p,\mathrm{m}}[15] = -\left(\frac{2a}{RT} - b\right)$$
$$= -\left(\frac{(2) \times (3.60\,\mathrm{L^2\,atm\,mol^{-2}})}{(0.0821\,\mathrm{L\,atm\,K^{-1}\,mol^{-1}}) \times (300\,\mathrm{K})} - (0.044\,\mathrm{L\,mol^{-1}})\right)$$
$$= -0.248\bar{3}\,\mathrm{L\,mol^{-1}}$$

$$\Delta H = \int_{p_\mathrm{i}}^{p_\mathrm{f}} \mathrm{d}H = \int_{p_\mathrm{i}}^{p_\mathrm{f}} (-0.248\bar{3}\,\mathrm{L\,mol^{-1}})\,\mathrm{d}p = -0.248\bar{3}(p_\mathrm{f} - p_\mathrm{i})\,\mathrm{L\,mol^{-1}}$$

$$p = \frac{RT}{V_\mathrm{m} - b} - \frac{a}{V_\mathrm{m}^2} \quad [1.39b]$$

$$p_\mathrm{i} = \left(\frac{(0.0821\,\mathrm{L\,atm\,K^{-1}\,mol^{-1}}) \times (300\,\mathrm{K})}{(20.0\,\mathrm{L\,mol^{-1}}) - (0.044\,\mathrm{L\,mol^{-1}})}\right) - \left(\frac{3.60\,\mathrm{L^2\,atm\,mol^{-2}}}{(20.0\,\mathrm{L\,mol^{-1}})^2}\right) = 1.22\bar{5}\,\mathrm{atm}$$

$$p_\mathrm{f} = \left(\frac{(0.0821\,\mathrm{L\,atm\,K^{-1}\,mol^{-1}}) \times (300\,\mathrm{K})}{(10.0\,\mathrm{L\,mol^{-1}}) - (0.044\,\mathrm{L\,mol^{-1}})}\right) - \left(\frac{3.60\,\mathrm{L^2\,atm\,mol^{-2}}}{(10.0\,\mathrm{L\,mol^{-1}})^2}\right) = 2.43\bar{8}\,\mathrm{atm}$$

$$\Delta H = (-0.248\bar{3}\,\mathrm{L\,mol^{-1}}) \times (2.43\bar{8}\,\mathrm{atm} - 1.225\,\mathrm{atm})$$
$$= (-0.301\,\mathrm{L\,atm\,mol^{-1}}) \times \left(\frac{10^{-3}\,\mathrm{m^3}}{\mathrm{L}}\right) \times \left(\frac{1.013 \times 10^5\,\mathrm{Pa}}{\mathrm{atm}}\right) = \boxed{-30.5\,\mathrm{J\,mol^{-1}}}$$

Solutions to theoretical problems

P3.6 $\oint \mathrm{d}z = \displaystyle\int_{(0,0)}^{(1,1)} \mathrm{d}z + \int_{(1,1)}^{(0,0)} \mathrm{d}z$

along $y = x$

$$\int_{(0,0)}^{(1,1)} \mathrm{d}z = \int_{(0,0)}^{(1,1)} (xy\,\mathrm{d}x + xy\,\mathrm{d}y) = \int_{(0,0)}^{(1,1)} (x^2\,\mathrm{d}x + y^2\,\mathrm{d}y) = \frac{1}{3} + \frac{1}{3} = \frac{2}{3}$$

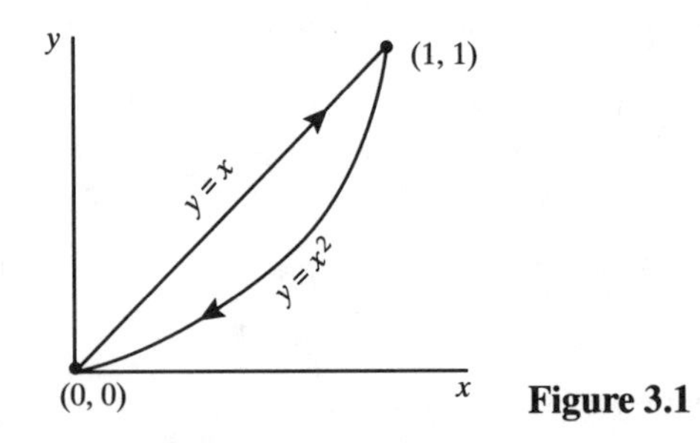

Figure 3.1

along $y = x^2$

$$\int_{(1,1)}^{(0,0)} \mathrm{d}z = \int_{(1,1)}^{(0,0)} (x^3\,\mathrm{d}x + y^{3/2}\,\mathrm{d}y) = -\frac{1}{4} - \frac{2}{5} = -\frac{13}{20}$$

The sum is $\left(\frac{2}{3} - \frac{13}{20}\right) = \frac{1}{60} \neq 0$. Therefore, this differential is $\boxed{\text{not exact}}$, since $\oint \mathrm{d}z = 0$ if $\mathrm{d}z$ = exact.

P3.7 A differential $\mathrm{d}f = g\,\mathrm{d}x + h\,\mathrm{d}y$ is exact if

$$\left(\frac{\partial g}{\partial y}\right)_x = \left(\frac{\partial h}{\partial x}\right)_y \quad [\textit{Further information } 1]$$

$$\left[\frac{\partial\left(\frac{RT}{p}\right)}{\partial T}\right]_p = \frac{R}{p} \quad \left[\frac{\partial(-R)}{\partial p}\right]_T = 0$$

Since $\frac{R}{p} \neq 0$, this differential is $\boxed{\text{not exact}}$, and hence cannot be the differential of a property of the system.

$$\frac{1}{T}\,\mathrm{d}q = \frac{R}{p}\,\mathrm{d}p - \frac{R}{T}\,\mathrm{d}T \qquad \left[\frac{\partial\left(\frac{R}{p}\right)}{\partial T}\right]_p = 0 \qquad \left[\frac{\partial\left(-\frac{R}{T}\right)}{\partial p}\right]_T = 0$$

Therefore, $\frac{\mathrm{d}q}{T}$ $\boxed{\text{is an exact differential}}$ and is the differential of a property of the system. This property will be identified with the entropy in Chapter 4.

P3.11 Using the permuter (Relation 3, *Further information* 1)

$$\left(\frac{\partial p}{\partial T}\right)_V = -\left(\frac{\partial p}{\partial V}\right)_T \left(\frac{\partial V}{\partial T}\right)_p$$

Substituting into the given expression for $C_p - C_V$

$$C_p - C_V = -T\left(\frac{\partial p}{\partial V}\right)_T \left(\frac{\partial V}{\partial T}\right)_p^2$$

Using the inverter [Relation 2, *Further information* 1]

$$C_p - C_V = -\frac{T\left(\frac{\partial V}{\partial T}\right)_p^2}{\left(\frac{\partial V}{\partial p}\right)_T}$$

With $pV = nRT$

$$\left(\frac{\partial V}{\partial T}\right)_p^2 = \left(\frac{nR}{p}\right)^2; \qquad \left(\frac{\partial V}{\partial p}\right)_T = -\frac{nRT}{p^2}$$

$$C_p - C_V = \frac{-T\left(\frac{nR}{p}\right)^2}{-\frac{nRT}{p^2}} = \boxed{nR}$$

P3.12 $U = U(T, V)$

Hence, $\mathrm{d}U = \left(\frac{\partial U}{\partial T}\right)_V \mathrm{d}T + \left(\frac{\partial U}{\partial V}\right)_T \mathrm{d}V = C_V\,\mathrm{d}T + \pi_T\,\mathrm{d}V$

Thus, if $\pi_T = 0$

$$\Delta U = \int_{\mathrm{i}}^{\mathrm{f}} \mathrm{d}U = \int_{T_\mathrm{i}}^{T_\mathrm{f}} C_V\,\mathrm{d}T$$

Therefore, if C_V and $\Delta T = T_\mathrm{f} - T_\mathrm{i}$ are known, ΔU may be calculated.

In the Joule experiment, a gas is expanded freely in a water bath (Figure 3.3 of text). Hence, $w = 0$. Heat transferred to the water bath may be determined from the change in temperature, ΔT, of the bath. Those gases for which $\Delta T = 0$ are defined as perfect gases. Since $\Delta T = 0$, $q = 0$, and $\Delta U = 0$.

$$\pi_T = \left(\frac{\partial U}{\partial V}\right)_T = \lim_{\Delta V \to 0}\left(\frac{\Delta U}{\Delta V}\right)_T = 0$$

Therefore, whether or not a process in a perfect gas is one of constant volume, $\Delta U = \int_{T_\mathrm{i}}^{T_\mathrm{f}} C_V\,\mathrm{d}T$ applies.

P3.14 $\mu \equiv \left(\frac{\partial T}{\partial p}\right)_H$

Use of the permuter (Relation 3, *Further information* 1) yields

$$\mu = -\frac{\left(\frac{\partial H}{\partial p}\right)_T}{C_{p,\mathrm{m}}}$$

$$\left(\frac{\partial H}{\partial p}\right)_T = \left(\frac{\partial U}{\partial p}\right)_T + \left[\frac{\partial (pV_\mathrm{m})}{\partial p}\right]_T = \left(\frac{\partial U}{\partial V_\mathrm{m}}\right)_T\left(\frac{\partial V_\mathrm{m}}{\partial p}\right)_T + \left[\frac{\partial (pV_\mathrm{m})}{\partial p}\right]_T$$

Use the virial expansion of the van der Waals equation in terms of p. (See the solution to Problem 1.17 and Section 1.4 of the text.)

$$pV_\mathrm{m} = RT\left[1 + \frac{1}{RT}\left(b - \frac{a}{RT}\right)p + \cdots\right]$$

$$\left[\frac{\partial (pV_\mathrm{m})}{\partial p}\right]_T \approx b - \frac{a}{RT}, \quad \left(\frac{\partial V_\mathrm{m}}{\partial p}\right)_T \approx -\frac{RT}{p^2}$$

Substituting $\left(\frac{\partial H}{\partial p}\right)_T \approx \left(\frac{a}{V_\mathrm{m}^2}\right) \times \left(-\frac{RT}{p^2}\right) + \left(b - \frac{a}{RT}\right) \approx \frac{-aRT}{(pV_\mathrm{m})^2} + \left(b - \frac{a}{RT}\right)$

Since $\left(\frac{\partial H}{\partial p}\right)_T$ is in a sense a correction term, that is, it approaches zero for a perfect gas, little error will be introduced by the approximation, $(pV_\mathrm{m})^2 = (RT)^2$.

Thus $\left(\frac{\partial H}{\partial p}\right)_T \approx \left(-\frac{a}{RT}\right) + \left(b - \frac{a}{RT}\right) = \left(b - \frac{2a}{RT}\right)$ and $\mu = \frac{\left(\frac{2a}{RT} - b\right)}{C_{p,\mathrm{m}}}$

P3.16 $H_m = H_m(T, p)$

$$dH_m = \left(\frac{\partial H_m}{\partial T}\right)_p dT + \left(\frac{\partial H_m}{\partial p}\right)_T dp$$

Since $dT = 0$

$$dH_m = \left(\frac{\partial H_m}{\partial p}\right)_T dp \qquad \left(\frac{\partial H_m}{\partial p}\right)_T = -\mu C_{p,m}[15] = -\left(\frac{2a}{RT} - b\right)$$

$$\Delta H_m = \int_{p_i}^{p_f} dH_m = -\int_{p_i}^{p_f}\left(\frac{2a}{RT} - b\right)dp = -\left(\frac{2a}{RT} - b\right)(p_f - p_i)$$

$$= -\left(\frac{(2)\times(1.408\,L^2\,atm\,mol^{-2})}{(0.08206\,L\,atm\,K^{-1}\,mol^{-1})\times(300\,K)} - (0.03913\,L\,mol^{-1})\right)$$

$$\times(1.00\,atm - 500\,atm)$$

$$= (37.5\,atm)\times\left(\frac{10^{-3}\,m^3}{1\,L}\right)\times\left(\frac{1.013\times10^5\,Pa}{1\,atm}\right) = 3.80\times10^3\,J = \boxed{+3.80\,kJ}$$

Comment. Note that it is not necessary to know the value of $C_{p,m}$.

P3.19 $\alpha = \frac{1}{V}\left(\frac{\partial V}{\partial T}\right)_p = \frac{1}{V\left(\frac{\partial T}{\partial V}\right)_p}$ [Relation 2, *Further information* 1]

$$= \frac{1}{V}\times\frac{1}{\left(\frac{T}{V-nb}\right) - \left(\frac{2na}{RV^3}\right)\times(V-nb)} \text{ [Problem 3.18]}$$

$$= \frac{(RV^2)\times(V-nb)}{(RTV^3) - (2na)\times(V-nb)^2}$$

$$\kappa_T = -\frac{1}{V}\left(\frac{\partial V}{\partial p}\right)_T = \frac{-1}{V\left(\frac{\partial p}{\partial V}\right)_T} \text{ [Relation 2]}$$

$$= -\frac{1}{V}\times\frac{1}{\left(\frac{-nRT}{(V-nb)^2}\right) + \left(\frac{2n^2a}{V^3}\right)} \text{ [Problem 3.17]}$$

$$= \boxed{\frac{V^2(V-nb)^2}{nRTV^3 - 2n^2a(V-nb)^2}}$$

Then $\frac{\kappa_T}{\alpha} = \frac{V-nb}{nR}$, implying that $\kappa_T R = \alpha(V_m - b)$

From the definitions of α and κ_T above

$$\frac{\kappa_T}{\alpha} = \frac{-\left(\frac{\partial V}{\partial p}\right)_T}{\left(\frac{\partial V}{\partial T}\right)_p} = \frac{-1}{\left(\frac{\partial p}{\partial V}\right)_T\left(\frac{\partial V}{\partial T}\right)_p} \text{ [Relation 2]}$$

$$= \left(\frac{\partial T}{\partial p}\right)_V \text{ [Chain relation, Further information 1]}$$

$$= \frac{V-nb}{nR} \text{ [Problem 3.18]},$$

$$\kappa_T R = \frac{\alpha(V-nb)}{n}$$

Hence, $\kappa_T R = \alpha(V_m - b)$

P3.21 Work with the left-hand side of the relation to be proved and show that after manipulation using the general relations between partial derivatives and the given equation for $\left(\frac{\partial U}{\partial V}\right)_T$, the right-hand side is produced.

$$\begin{aligned}
\left(\frac{\partial H}{\partial p}\right)_T &= \left(\frac{\partial H}{\partial V}\right)_T\left(\frac{\partial V}{\partial p}\right)_T \text{ [change of variable]}\\
&= \left(\frac{\partial(U+pV)}{\partial V}\right)_T\left(\frac{\partial V}{\partial p}\right)_T \text{ [definition of } H\text{]}\\
&= \left(\frac{\partial U}{\partial V}\right)_T\left(\frac{\partial V}{\partial p}\right)_T + \left(\frac{\partial pV}{\partial V}\right)_T\left(\frac{\partial V}{\partial p}\right)_T\\
&= \left\{T\left(\frac{\partial p}{\partial T}\right)_V - p\right\}\left(\frac{\partial V}{\partial p}\right)_T + \left(\frac{\partial pV}{\partial p}\right)_T \left[\text{equation for } \left(\frac{\partial U}{\partial V}\right)_T\right]\\
&= T\left(\frac{\partial p}{\partial T}\right)_V\left(\frac{\partial V}{\partial p}\right)_T - p\left(\frac{\partial V}{\partial p}\right)_T + V + p\left(\frac{\partial V}{\partial p}\right)_T\\
&= T\left(\frac{\partial p}{\partial T}\right)_V\left(\frac{\partial V}{\partial p}\right)_T + V = \frac{-T}{\left(\frac{\partial T}{\partial V}\right)_p} + V \text{ [chain relation]}\\
&= \boxed{-T\left(\frac{\partial V}{\partial T}\right)_p + V} \text{ [Relation 2, } \textit{Further information } 1\text{]}
\end{aligned}$$

P3.23 $$c = \left(\frac{RT\gamma}{M}\right)^{1/2}, \quad p = \rho\frac{RT}{M}, \quad \text{so} \quad \frac{RT}{M} = \frac{p}{\rho}; \quad \text{hence} \quad \boxed{c = \left(\frac{\gamma p}{\rho}\right)^{1/2}}$$

For argon, $\gamma = \frac{5}{3}$, so $c = \left(\frac{(8.314\,\mathrm{J\,K^{-1}\,mol^{-1}}) \times (298\,\mathrm{K}) \times \frac{5}{3}}{39.95 \times 10^{-3}\,\mathrm{kg\,mol^{-1}}}\right)^{1/2} = \boxed{322\,\mathrm{m\,s^{-1}}}$

Solutions to additional problems

P3.24 The coefficient of thermal expansion is

$$\alpha = \frac{1}{V}\left(\frac{\partial V}{\partial T}\right)_p \approx \frac{\Delta V}{V\Delta T} \quad \text{so} \quad \Delta V = \alpha V \Delta T$$

This change in volume is equal to the change in height (sea level rise, Δh) times the area of the ocean (assuming that area remains constant). We will use α of pure water, although the oceans are complex solutions. For a 2°C rise

$$\Delta V = (2.1 \times 10^{-4}\,\mathrm{K^{-1}}) \times (1.37 \times 10^{9}\,\mathrm{km^3}) \times (2.0\,\mathrm{K}) = 5.8 \times 10^{5}\,\mathrm{km^3}$$

so $\Delta h = \frac{\Delta V}{A} = 1.6 \times 10^{-3}\,\mathrm{km} = \boxed{1.6\,\mathrm{m}}$

Since the rise in sea level is directly proportional to the rise in temperature, $\Delta T = 1°\mathrm{C}$ would lead to $\Delta h = \boxed{0.80\,\mathrm{m}}$ and $\Delta T = 3.5°\mathrm{C}$ would lead to $\Delta h = \boxed{2.8\,\mathrm{m}}$

Comment. More detailed models of climate change predict somewhat smaller rises, but the same order of magnitude.

P3.26 We compute μ from

$$\mu = -\frac{1}{C_p}\left(\frac{\partial H}{\partial p}\right)_T$$

and we estimate $\left(\dfrac{\partial H}{\partial p}\right)_T$ from the enthalpy and pressure data. We are given both enthalpy and heat capacity data on a mass basis rather than a molar basis; however, the masses will cancel, so we need not convert to a molar basis.

(a) At 300 K

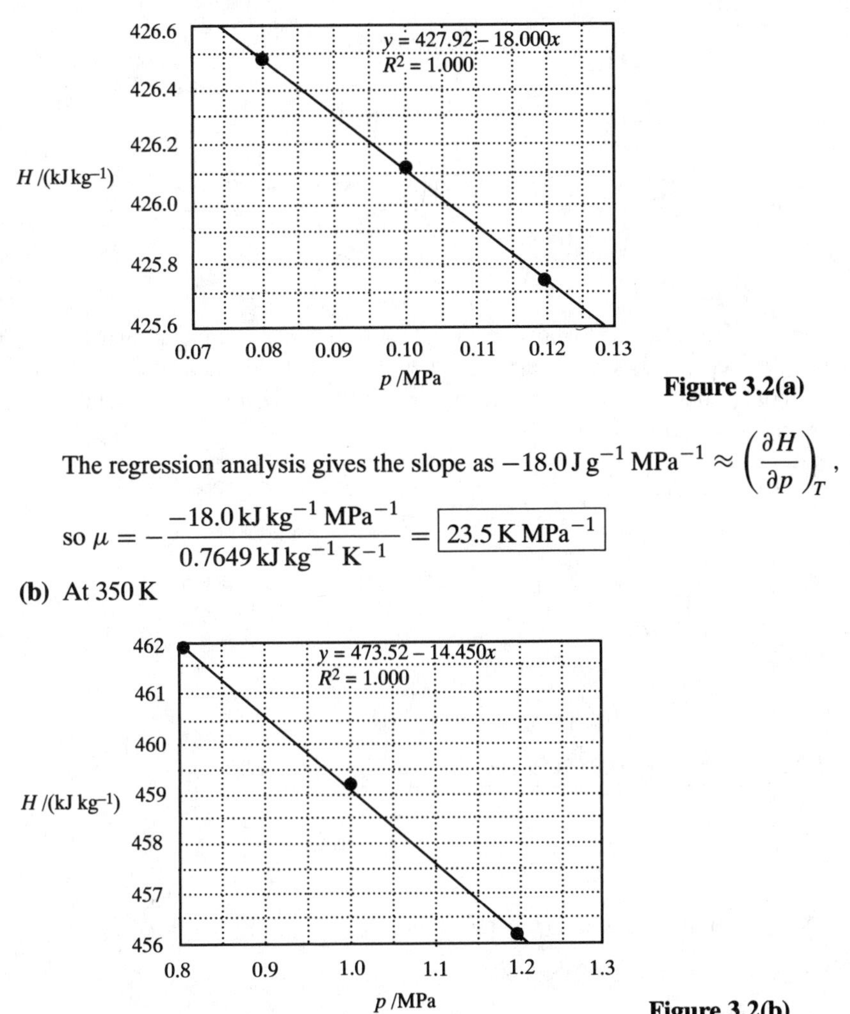

Figure 3.2(a)

The regression analysis gives the slope as $-18.0\,\mathrm{J\,g^{-1}\,MPa^{-1}} \approx \left(\dfrac{\partial H}{\partial p}\right)_T$,

so $\mu = -\dfrac{-18.0\,\mathrm{kJ\,kg^{-1}\,MPa^{-1}}}{0.7649\,\mathrm{kJ\,kg^{-1}\,K^{-1}}} = \boxed{23.5\,\mathrm{K\,MPa^{-1}}}$

(b) At 350 K

Figure 3.2(b)

The regression analysis gives the slope as $-14.5\,\mathrm{J\,g^{-1}\,MPa^{-1}} \approx \left(\dfrac{\partial H}{\partial p}\right)_T$,

so $\mu = -\dfrac{-14.5\,\mathrm{kJ\,kg^{-1}\,MPa^{-1}}}{1.0392\,\mathrm{kJ\,kg^{-1}\,K^{-1}}} = \boxed{14.0\,\mathrm{K\,MPa^{-1}}}$

P3.27 The system is shown in Fig. 3.3.

Initial equilibrium state

$m = 1.00\,\mathrm{mol}$ diatomic gas in each section

$p_{\mathrm{i}} = 1.00\,\mathrm{bar}$

$T_{\mathrm{i}} = 298\,\mathrm{K}$

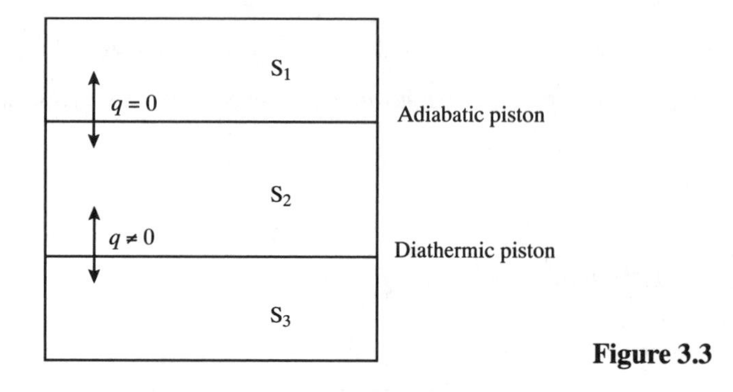

Figure 3.3

For each section $V_i = \dfrac{nRT_i}{p_i} = \dfrac{(1\,\text{mol}) \times (0.083145\,\text{L bar K}^{-1}\,\text{mol}^{-1}) \times (298\,\text{K})}{1.00\,\text{bar}} = 24.8\,\text{L}$

$V_{\text{total}} = 3V_i = 74.3\,\text{L} = \text{constant}$

Final equilibrium state

S_2 and S_3 have experienced an adiabatic process that has changed the temperature of S_3 to $T_3 = 348\,\text{K}$. Since S_2 and S_3 are separated by a diathermic wall, we can conclude that the final temperature of S_2 is identical to that of S_3 so that $T_2 = T_3$. For the same reason $V_2 = V_3$ and at equilibrium $p_1 = p_2 = p_3$ (so-called mechanical equilibrium).

$$T_3 = T_i\left(\frac{V_i}{V_3}\right)^{1/c} \quad [2.35] \quad \text{where } c = \frac{C_{V,\text{m}}}{R} = \frac{5}{2}$$

$$\left(\frac{T_3}{T_i}\right)^c = \frac{V_i}{V_3} \quad \text{or} \quad V_3 = V_i\left(\frac{T_i}{T_3}\right)^c$$

$$V_3 = 24.8\,\text{L}\left(\frac{298\,\text{K}}{348\,\text{K}}\right)^{5/2} = \boxed{16.8\,\text{L}} = V_2$$

$$V_1 = V - 2V_2 = 74.3\,\text{L} - 2(16.8\,\text{L}) = \boxed{40.7\,\text{L}}$$

$$p_3 = \frac{nRT_3}{V_3} = \frac{(1\,\text{mol}) \times (0.08315\,\text{L bar K}^{-1}\text{mol}^{-1}) \times (348\,\text{K})}{16.8\,\text{L}}$$

$$p_3 = \boxed{1.72\,\text{bar}} = p_1 = p_2$$

$$T_1 = \frac{p_1 V_1}{nR} = \frac{(1.72\,\text{bar}) \times (40.7\,\text{L})}{(1\,\text{mol}) \times (0.08315\,\text{L bar K}^{-1}\,\text{mol}^{-1})}$$

$$T_1 = \boxed{842\,\text{K}} \quad T_2 = T_3 = \boxed{348\,\text{K}}$$

$$\Delta U_1 = n_1 C_V \Delta T_1 = \frac{5}{2}(8.315\,\text{J K}^{-1}\,\text{mol}^{-1}) \times (1\,\text{mol}) \times (842\,\text{K} - 298\,\text{K})$$

$$\Delta U_1 = \boxed{11.3\,\text{kJ}}$$

$$\Delta U_3 = n_3 C_V \Delta T_3 = \frac{5}{2}(8.315\,\text{J K}^{-1}\,\text{mol}^{-1}) \times (1\,\text{mol}) \times (348\,\text{K} - 298\,\text{K})$$

$$\Delta U_3 = \boxed{1.04\,\text{kJ}} = \Delta U_2$$

$$\Delta U_{\text{total}} = \Delta U_1 + \Delta U_2 + \Delta U_3 = 11.3\,\text{kJ} + 2(1.04\,\text{kJ}) = \boxed{13.4\,\text{kJ}}$$

Notice that it does not matter whether the piston between chambers 2 and 3 is diathermic or adiabatic. The result is identical in these two cases because chamber 2 is receiving no heat from chamber 1 that can be distributed to chamber 3. That is, even with a diathermic piston between chambers 2 and 3 the heat flow from chamber 2 to chamber 3, q_{23}, equals zero.

The proof is as follows.

$$\Delta U_2 = w_2 + q_2 = C_V \Delta T_2$$
$$= -\int p\,\mathrm{d}V_2 + \underbrace{q_{12}}_{0} + \underbrace{q_{32}}_{-q_{23}} = C_V \Delta T_2$$

Therefore

$$\Delta U_2 = -\int p\,\mathrm{d}V_2 - q_{23} = C_V \Delta T_2 \quad (1)$$

Likewise,

$$\Delta U_3 = -\int p\,\mathrm{d}V_3 + q_{23} = C_V \Delta T_3 \quad (2)$$

Subtracting (2) from (1):

$$-\int p\,\mathrm{d}\underbrace{(V_2 - V_3)}_{0} - 2q_{23} = C_V \underbrace{(T_2 - T_3)}_{0}$$

Therefore, $q_{23} = 0$ for the presence of the diathermic wall with the particular heat flow of this problem.

P3.29 $\mu = \left(\frac{\partial T}{\partial p}\right)_H = -\frac{1}{C_p}\left(\frac{\partial H}{\partial p}\right)_T$ [14 and 15]

$$\mu = \frac{1}{C_p}\left\{T\left(\frac{\partial V}{\partial T}\right)_p - V\right\} \quad \text{[Problem 3.21]}$$

But $V = \frac{RT}{p} + b$ or $\left(\frac{\partial V}{\partial T}\right)_p = \frac{R}{p}$

Therefore,

$$\mu = \frac{1}{C_p}\left\{\frac{RT}{p} - V\right\} = \frac{1}{C_p}\left\{\frac{RT}{p} - \frac{RT}{p} - b\right\}$$
$$= \frac{-b}{C_p}$$

Since $b > 0$ and $C_p > 0$, we conclude that for this gas $\mu < 0$ or $\left(\frac{\partial T}{\partial p}\right)_H < 0$. This says that when the pressure drops during a Joule–Thomson expansion the temperature must **increase**

4 The Second Law: the concepts

Solutions to exercises

Assume that all gases are perfect and that data refer to 298 K unless otherwise stated.

E4.1 Assume that the block is so large that its temperature does not change significantly as a result of the heat transfer. Then

$$\Delta S = \int_i^f \frac{dq_{rev}}{T}[3] = \frac{1}{T}\int_i^f dq_{rev}\ [\text{constant } T] = \frac{q_{rev}}{T}$$

(a) $\Delta S = \dfrac{25\times10^3\,J}{273.15\,K} = \boxed{92\,J\,K^{-1}}$ **(b)** $\Delta S = \dfrac{25\times10^3\,J}{373.15\,K} = \boxed{67\,J\,K^{-1}}$

E4.2 $$S_m(T_f) = S_m(T_i) + \int_{T_i}^{T_f}\frac{C_{V,m}}{T}\,dT\ [20, \text{with } C_{V,m} \text{ in place of } C_{p,m}]$$

If we assume that neon is a perfect gas then $C_{V,m}$ may be taken to be constant and given by

$$C_{V,m} = C_{p,m} - R; \qquad C_{p,m} = 20.786\,J\,K^{-1}\,mol^{-1}\ [\text{Table 2.6}]$$
$$= (20.786 - 8.314)\,J\,K^{-1}\,mol^{-1}$$
$$= 12.472\,J\,K^{-1}\,mol^{-1}$$

Integrating, we obtain

$$S_m(500\,K) = S_m(298\,K) + C_{V,m}\ln\frac{T_f}{T_i}$$
$$= (146.22\,J\,K^{-1}\,mol^{-1}) + (12.472\,J\,K^{-1}\,mol^{-1})\ln\left(\frac{500\,K}{298\,K}\right)$$
$$= (146.22 + 6.45)\,J\,K^{-1}\,mol^{-1} = \boxed{152.67\,J\,K^{-1}\,mol^{-1}}$$

E4.3 $$\Delta S(\text{system}) = nC_{p,m}\ln\left(\frac{T_f}{T_i}\right)\ [20]$$

$$C_{p,m} = C_{V,m} + R[2.31] = \tfrac{3}{2}R + R = \tfrac{5}{2}R$$

$$\Delta S = (1.00\,mol)\times\left(\frac{5}{2}\right)\times(8.314\,J\,K^{-1}\,mol^{-1})\times\ln\left(\frac{573\,K}{373\,K}\right) = \boxed{8.92\,J\,K^{-1}}$$

E4.4 Since entropy is a state function, ΔS may be calculated from the most convenient path, which in this case corresponds to constant-pressure heating followed by constant-temperature compression.

$$\Delta S = nC_{p,m}\ln\left(\frac{T_f}{T_i}\right)[20, \text{at } p_i] + nR\ln\left(\frac{V_f}{V_i}\right)[17, \text{at } T_f]$$

Since pressure and volume are inversely related (Boyle's law), $\dfrac{V_f}{V_i} = \dfrac{p_i}{p_f}$. Hence,

$$\Delta S = nC_{p,m}\ln\left(\frac{T_f}{T_i}\right) - nR\ln\left(\frac{p_f}{p_i}\right) = (3.00\,mol)\times\left(\tfrac{5}{2}\right)\times(8.314\,J\,K^{-1}\,mol^{-1})\times\ln\left(\tfrac{398\,K}{298\,K}\right)$$
$$-(3.00\,mol)\times(8.314\,J\,K^{-1}\,mol^{-1})\times\ln\left(\frac{5.00\,atm}{1.00\,atm}\right)$$
$$= (18.0\bar{4} - 40.1\bar{4})\,J\,K^{-1} = \boxed{-22.1\,J\,K^{-1}}$$

Though ΔS(system) is negative, the process can still occur spontaneously if ΔS(total) is positive.

E4.5 $q = q_{\text{rev}} = \boxed{0}$ [adiabatic reversible process]

$$\Delta S = \int_{\text{i}}^{\text{f}} \frac{\text{d}q_{\text{rev}}}{T} = \boxed{0}$$

$$\Delta U = nC_{V,\text{m}}\Delta T\ [2.22b] = (3.00\,\text{mol}) \times (27.5\,\text{J K}^{-1}\,\text{mol}^{-1}) \times (50\,\text{K}) = 4.1 \times 10^3\,\text{J}$$

$$= \boxed{+4.1\,\text{kJ}}$$

$w = \Delta U$ [First Law with $q = 0$]

$$\Delta H = nC_{p,\text{m}}\Delta T\ [2.28b]$$

$$C_{p,\text{m}} = C_{V,\text{m}} + R[2.31] = (27.5 + 8.3)\,\text{J K}^{-1}\,\text{mol}^{-1} = 35.8\,\text{J K}^{-1}\,\text{mol}^{-1}$$

$$\Delta H = (3.00\,\text{mol}) \times (35.8\,\text{J K}^{-1}\,\text{mol}^{-1}) \times (50\,\text{K}) = 5.4 \times 10^3\,\text{J} = \boxed{+5.4\,\text{kJ}}$$

Comment. Neither initial nor final pressures and volumes are needed for the solution to this exercise.

E4.6

$$\Delta S = nC_{V,\text{m}} \ln\left(\frac{T_\text{f}}{T_\text{i}}\right) + nR\ln\left(\frac{V_\text{f}}{V_\text{i}}\right) \quad \text{[Example 4.3]}$$

$$C_{V,\text{m}} = C_{p,\text{m}} - R = \tfrac{5}{2}R - R = \tfrac{3}{2}R$$

$$\Delta S = (1.00\,\text{mol}) \times \left(\frac{3}{2}\right) \times (8.314\,\text{J K}^{-1}\,\text{mol}^{-1}) \ln\left(\frac{600\,\text{K}}{300\,\text{K}}\right)$$

$$+ (1.00\,\text{mol}) \times (8.314\,\text{J K}^{-1}\,\text{mol}^{-1}) \times \ln\left(\frac{50.0\,\text{L}}{30.0\,\text{L}}\right) = \boxed{+12.9\,\text{J K}^{-1}}$$

E4.7 $\Delta S = \dfrac{q_{\text{rev}}}{T}$ [constant temperature]

If reversible $q = q_{\text{rev}}$.

$$q_{\text{rev}} = T\Delta S = (500\,\text{K}) \times (2.41\,\text{J K}^{-1}) = 1.21\,\text{kJ}$$

$$1.21\,\text{kJ} \neq 1.00\,\text{kJ} = q$$

Therefore, the process is $\boxed{\text{not reversible}}$

E4.8 (a) $\Delta H = \displaystyle\int_{T_1}^{T_2} nC_{p,\text{m}}\,\text{d}T$ [constant pressure]

$C_{p,\text{m}}/(\text{J K}^{-1}\,\text{mol}^{-1}) = (a + bT)$ [Table 2.2], $a = 20.67$, $b = 12.38 \times 10^{-3}\,\text{K}^{-1}$

$$\Delta H = \int_{T_1}^{T_2} n(a + bT)\,\text{J K}^{-1}\,\text{mol}^{-1}\,\text{d}T$$

$$= na(T_2 - T_1)\,\text{J K}^{-1}\,\text{mol}^{-1} + \tfrac{1}{2}nb(T_2^2 - T_1^2)\,\text{J K}^{-1}\,\text{mol}^{-1}$$

$$n = \frac{1.75 \times 10^3\,\text{g}}{26.98\,\text{g mol}^{-1}} = 64.8\bar{6}\,\text{mol}$$

$$\Delta H = \big[(64.8\bar{6}\,\text{mol}) \times (20.67) \times (265 - 300)\,\text{K}$$

$$+ \left(\tfrac{1}{2}\right) \times (64.8\bar{6}\,\text{mol}) \times (12.38 \times 10^{-3}\,\text{K}^{-1}) \times (265^2 - 300^2)\,\text{K}^2\big]\,\text{J K}^{-1}\,\text{mol}^{-1}$$

$$\Delta H = -54.9 \times 10^3\,\text{J} = \boxed{-54.9\,\text{kJ}}$$

(b) $$\Delta S = \int_{T_1}^{T_2} \frac{nC_{p,\text{m}}}{T}\,\text{d}T[19] = \int_{T_1}^{T_2} \frac{n(a+bT)}{T}\,\text{J K}^{-1}\,\text{mol}^{-1}\,\text{d}T$$

$$= na\ln\left(\frac{T_2}{T_1}\right)\text{J K}^{-1}\,\text{mol}^{-1} + nb(T_2 - T_1)\,\text{J K}^{-1}\,\text{mol}^{-1}$$

$$= (64.86\,\text{mol}) \times (20.67) \times \ln\left(\frac{265}{300}\right)\text{J K}^{-1}\,\text{mol}^{-1}$$

$$+ (64.86\,\text{mol}) \times (12.38 \times 10^{-3}\,\text{K}) \times (265\,\text{K} - 300\,\text{K})\,\text{J K}^{-1}\,\text{mol}^{-1}$$

$$\Delta S = \boxed{-195\,\text{J K}^{-1}}$$

E4.9 $$\Delta S = nR\ln\left(\frac{V_\text{f}}{V_\text{i}}\right)[17]; \qquad \frac{p_\text{i}}{p_\text{f}} = \frac{V_\text{f}}{V_\text{i}} \quad \text{[Boyle's law]}$$

$$\Delta S = nR\ln\left(\frac{p_\text{i}}{p_\text{f}}\right) = \left(\frac{25\,\text{g}}{16.04\,\text{g mol}^{-1}}\right) \times (8.314\,\text{J K}^{-1}\,\text{mol}^{-1}) \times \ln\left(\frac{18.5\,\text{atm}}{2.5\,\text{atm}}\right)$$

$$= \boxed{+26\,\text{J K}^{-1}}$$

E4.10 $$\Delta S = nR\ln\left(\frac{V_\text{f}}{V_\text{i}}\right)[17]$$

The number of moles (or nR) and then V_f need to be determined

$$nR = \frac{p_\text{i}V_\text{i}}{T_\text{i}} = \frac{(1.00\,\text{atm}) \times (15.0\,\text{L})}{250\,\text{K}} = \frac{(1.01\bar{3} \times 10^5\,\text{Pa}) \times (15.0 \times 10^{-3}\,\text{m}^3)}{250\,\text{K}} = 6.08\,\text{J K}^{-1}$$

$$\ln\frac{V_\text{f}}{V_\text{i}} = \frac{\Delta S}{nR} = \frac{-5.0\,\text{J K}^{-1}}{6.08\,\text{J K}^{-1}} = -0.82\bar{3}$$

Hence, $V_\text{f} = V_\text{i}\text{e}^{-0.82\bar{3}} = (15.0\,\text{L}) \times (0.43\bar{9}) = \boxed{6.6\,\text{L}}$

E4.11 Find the common final temperature T_f by noting that the heat lost by the hot sample is gained by the cold sample

$$-n_1C_{p,\text{m}}(T_\text{f} - T_{\text{i}1}) = n_2C_{p,\text{m}}(T_\text{f} - T_{\text{i}2})$$

Hence, $T_\text{f} = \dfrac{n_1T_{\text{i}1} + n_2T_{\text{i}2}}{n_1 + n_2}$

Since $\dfrac{n_1}{n_2} = \dfrac{1}{2}$, $T_\text{f} = \dfrac{1}{3}(353\,\text{K} + 2 \times 283\,\text{K}) = 306\,\text{K}$

The total change in entropy is that of the 50 g sample (ΔS_1) plus that of the 100 g sample (ΔS_2).

$$\Delta S = \Delta S_1 + \Delta S_2 = n_1C_{p,\text{m}}\ln\frac{T_\text{f}}{T_{\text{i}1}} + n_2C_{p,\text{m}}\ln\frac{T_\text{f}}{T_{\text{i}2}} \quad \text{[constant pressure, 20]}$$

$$= \left(\frac{50\,\text{g}}{18.02\,\text{g mol}^{-1}}\right) \times (75.5\,\text{J K}^{-1}\,\text{mol}^{-1}) \times \left(\ln\frac{306}{353} + 2\ln\frac{306}{283}\right) = \boxed{+2.8\,\text{J K}^{-1}}$$

E4.12 Since the container is isolated, the heat flow is zero and therefore $\boxed{\Delta H = 0}$ since the masses of the bricks are equal, the final temperature must be their mean temperature, 50°C.

Specific heat capacities are heat capacities per gram and are related to the molar heat capacities by

$$C_\text{s} = \frac{C_\text{m}}{M} \quad [C_{p,\text{m}} \approx C_{V,\text{m}} = C_\text{m}]$$

So $nC_{\rm m} = mC_{\rm s}$ $(nM = m)$

$$\Delta H(\text{individual}) = mC_{\rm s}\Delta T = 1.00 \times 10^4\,\text{g} \times 0.385\,\text{J K}^{-1}\,\text{g}^{-1} \times (\pm 50\,\text{K})$$

$$= \boxed{\pm 1.9 \times 10^2\,\text{kJ}}$$

$$\Delta S = mC_{\rm s}\ln\left(\frac{T_{\rm f}}{T_{\rm i}}\right)\ [20]$$

$$\Delta S_1 = (10.0 \times 10^3\,\text{g}) \times (0.385\,\text{J K}^{-1}\,\text{g}^{-1}) \times \ln\left(\frac{323\,\text{K}}{273\,\text{K}}\right) = -5.541 \times 10^2\,\text{J K}^{-1}$$

$$\Delta S_2 = (10.0 \times 10^3\,\text{g}) \times (0.385\,\text{J K}^{-1}\,\text{g}^{-1}) \times \ln\left(\frac{323}{273}\right) = 6.475 \times 10^2\,\text{J K}^{-1}$$

$$\Delta S_{\rm tot} = \Delta S_1 + \Delta S_2 = \boxed{+93.4\,\text{J K}^{-1}}$$

Comment. The positive value of $\Delta S_{\rm tot}$ corresponds to a spontaneous process.

E4.13 **(a)** $q = \boxed{0}$ [adiabatic]

(b) $$w = -p_{\rm ex}\Delta V[2.10] = -(1.01 \times 10^5\,\text{Pa}) \times (20\,\text{cm}) \times (10\,\text{cm}^2) \times \left(\frac{10^{-6}\,\text{m}^3}{\text{cm}^3}\right) = \boxed{-20\,\text{J}}$$

(c) $\Delta U = q + w = 0 - 20\,\text{J} = \boxed{-20\,\text{J}}$

(d) $\Delta U = nC_{V,\rm m}\Delta T$ [2.22*b*]

$$\Delta T = \frac{-20\,\text{J}}{(2.0\,\text{mol}) \times (28.8\,\text{J K}^{-1}\,\text{mol}^{-1})} = \boxed{-0.34\bar{7}\,\text{K}}$$

(e) $$\Delta S = nC_{V,\rm m}\ln\left(\frac{T_{\rm f}}{T_{\rm i}}\right) + nR\ln\left(\frac{V_{\rm f}}{V_{\rm i}}\right)$$ [Example 4.3 and Exercise 4.6(**a**)]

$$T_{\rm f} = T_{\rm i} - 0.34\bar{7}\,\text{K} = (298.15\,\text{K}) - (0.34\bar{7}\,\text{K}) = 297.80\bar{3}\,\text{K}$$

$$V_{\rm i} = \frac{nRT}{p_{\rm i}} = \frac{(2.0\,\text{mol}) \times (0.08206\,\text{L atm K}^{-1}\,\text{mol}^{-1}) \times (298.15\,\text{K})}{10\,\text{atm}} = 4.\overline{893}\,\text{L}$$

$$V_{\rm f} = V_{\rm i} + \Delta V = 4.\overline{893} + 0.20\,\text{L} = 5.\overline{093}\,\text{L}$$

Substituting these values into the expression for ΔS above gives

$$\Delta S = (2.0\,\text{mol}) \times (28.8\,\text{J K}^{-1}\,\text{mol}^{-1}) \times \ln\left(\frac{297.80\bar{3}\,\text{K}}{298.15\,\text{K}}\right)$$

$$+(2.0\,\text{mol}) \times (8.314\,\text{J K}^{-1}\,\text{mol}^{-1})\ln\left(\frac{5.\overline{093}\,\text{L}}{4.893}\right)$$

$$= (-0.067\bar{1} + 0.66\bar{6})\,\text{J K}^{-1} = \boxed{+0.60\,\text{J K}^{-1}}$$

E4.14 **(a)** $$\Delta_{\rm vap}S = \frac{\Delta_{\rm vap}H}{T_{\rm b}} = \frac{29.4 \times 10^3\,\text{J mol}^{-1}}{334.88\,\text{K}} = \boxed{+87.8\,\text{J K}^{-1}\,\text{mol}^{-1}}$$

(b) If the vaporization occurs reversibly, $\Delta S_{\rm tot} = 0$, so $\Delta S_{\rm surr} = \boxed{-87.8\,\text{J K}^{-1}\,\text{mol}^{-1}}$

E4.15 In each case

$$\Delta_{\rm r}S^{\ominus} = \sum_{\rm J} \nu_{\rm J}S_{\rm m}^{\ominus}(\text{J})\ [22b]$$

with $S_{\rm m}^{\ominus}$ values obtained from Tables 2.5 and 2.6.

(a) $$\Delta_r S^{\ominus} = 2S_m^{\ominus}(CH_3COOH, l) - 2S_m^{\ominus}(CH_3CHO, g) - S_m^{\ominus}(O_2, g)$$
$$= [(2 \times 159.8) - (2 \times 250.3) - 205.14]\,J\,K^{-1}\,mol^{-1} = \boxed{-386.1\,J\,K^{-1}\,mol^{-1}}$$

(b) $$\Delta_r S^{\ominus} = 2S_m^{\ominus}(AgBr, s) + S_m^{\ominus}(Cl_2, g) - 2S_m^{\ominus}(AgCl, s) - S_m^{\ominus}(Br_2, l)$$
$$= [(2 \times 107.1) + (223.07) - (2 \times 96.2) - (152.23)]\,J\,K^{-1}\,mol^{-1}$$
$$= \boxed{+92.6\,J\,K^{-1}\,mol^{-1}}$$

(c) $$\Delta_r S^{\ominus} = S_m^{\ominus}(HgCl_2, s) - S_m^{\ominus}(Hg, l) - S_m^{\ominus}(Cl_2, g)$$
$$= [146.0 - 76.02 - 223.07]\,J\,K^{-1}\,mol^{-1} = \boxed{-153.1\,J\,K^{-1}\,mol^{-1}}$$

E4.16 In each case we use

$$\Delta_r G^{\ominus} = \Delta_r H^{\ominus} - T\Delta_r S^{\ominus}\ [39]$$

along with

$$\Delta_r H^{\ominus} = \sum_J \nu_J \Delta_f H^{\ominus}(J)\ [2.43]$$

(a) $$\Delta_r H^{\ominus} = 2\Delta_f H^{\ominus}(CH_3COOH, l) - 2\Delta_f H^{\ominus}(CH_3CHO, g)$$
$$= [2 \times (-484.5) - 2 \times (-166.19)]\,kJ\,mol^{-1} = -636.6\bar{2}\,kJ\,mol^{-1}$$
$$\Delta_r G^{\ominus} = -636.6\bar{2}\,kJ\,mol^{-1} - (298.15\,K) \times (-386.1\,J\,K^{-1}\,mol^{-1}) = \boxed{-521.5\,kJ\,mol^{-1}}$$

(b) $$\Delta_r H^{\ominus} = 2\Delta_f H^{\ominus}(AgBr, s) - 2\Delta_f H^{\ominus}(AgCl, s)$$
$$= [2 \times (-100.37) - 2 \times (-127.07)]\,kJ\,mol^{-1} = +53.40\,kJ\,mol^{-1}$$
$$\Delta_r G^{\ominus} = +53.40\,kJ\,mol^{-1} - (298.15\,K) \times (+92.6)\,J\,K^{-1}\,mol^{-1} = \boxed{+25.8\,kJ\,mol^{-1}}$$

(c) $$\Delta_r H^{\ominus} = \Delta_f H^{\ominus}(HgCl_2, s) = -224.3\,kJ\,mol^{-1}$$
$$\Delta_r G^{\ominus} = -224.3\,kJ\,mol^{-1} - (298.15\,K) \times (-153.1\,J\,K^{-1}\,mol^{-1}) = \boxed{-178.7\,kJ\,mol^{-1}}$$

E4.17 In each case $\Delta_r G^{\ominus} = \sum_J \nu_J \Delta_f G^{\ominus}(J)$

with $\Delta_f G^{\ominus}(J)$ values from Table 2.6.

(a) $$\Delta_r G^{\ominus} = 2\Delta_f G^{\ominus}(CH_3COOH, l) - 2\Delta_f G^{\ominus}(CH_3CHO, g)$$
$$= [2 \times (-389.9) - 2 \times (-128.86)]\,kJ\,mol^{-1}$$
$$= \boxed{-522.1\,kJ\,mol^{-1}}$$

(b) $$\Delta_r G^{\ominus} = 2\Delta_f G^{\ominus}(AgBr, s) - 2\Delta_f G^{\ominus}(AgCl, s) = [2 \times (-96.90) - 2 \times (-109.79)]\,kJ\,mol^{-1}$$
$$= \boxed{+25.78\,kJ\,mol^{-1}}$$

(c) $$\Delta_r G^{\ominus} = \Delta_f G^{\ominus}(HgCl_2, s)$$
$$= \boxed{-178.6\,kJ\,mol^{-1}}$$

Comment. In each case these values of $\Delta_r G^{\ominus}$ agree closely with the calculated values in Exercise 4.16(**a**).

E4.18 $$\Delta_r G^{\ominus} = \Delta_r H^{\ominus} - T\Delta_r S\ [39] \quad \Delta_r H^{\ominus} = \sum_J \nu_J \Delta_f H^{\ominus}(J)\ [2.43]$$

$$\Delta_r S^{\ominus} = \sum_J \nu_J S_m^{\ominus}(J)\ [22b]$$

$$\Delta_r H^\ominus = 2\Delta_f H^\ominus(H_2O, l) - 4\Delta_f H^\ominus(HCl, g) = \{2 \times (-285.83) - 4 \times (-92.31)\}\,kJ\,mol^{-1}$$
$$= -202.42\,kJ\,mol^{-1}$$
$$\Delta_r S^\ominus = 2S_m^\ominus(Cl_2, g) + 2S_m^\ominus(H_2O, l) - 4S_m^\ominus(HCl, g) - S_m^\ominus(O_2, g)$$
$$= [(2 \times 69.91) + (2 \times 223.07) - (4 \times 186.91) - (205.14)]\,J\,K^{-1}\,mol^{-1}$$
$$= -366.82\,J\,K^{-1}\,mol^{-1} = -0.36682\,kJ\,K^{-1}\,mol^{-1}$$
$$\Delta_r G^\ominus = -202.42\,kJ\,mol^{-1} - (298.15\,K) \times (-0.36682\,kJ\,K^{-1}\,mol^{-1}) = \boxed{-93.05\,kJ\,mol^{-1}}$$

Question. Repeat the calculation based on $\Delta_f G^\ominus$ data of Table 2.6. What difference, if any, is there from the value above?

E4.19 The formation reaction for phenol is

$$6C(s) + 3H_2(g) + \tfrac{1}{2}O_2(g) \rightarrow C_6H_5OH(s)$$
$$\Delta_f G^\ominus = \Delta_f H^\ominus - T\Delta_f S^\ominus \quad [39]$$

$\Delta_f H^\ominus$ is to be obtained from $\Delta_c H^\ominus$ for phenol and data from Tables 2.5 and 2.6. Thus

$$C_6H_5OH(s) + 7O_2(g) \rightarrow 6CO_2(g) + 3H_2O(l)$$
$$\Delta_c H^\ominus = 6\Delta_f H^\ominus(CO_2, g) + 3\Delta_f H^\ominus(H_2O, l) - \Delta_f H^\ominus(C_6H_5OH, s)$$

Hence $\Delta_f H^\ominus(C_6H_5OH, s) = 6\Delta_f H^\ominus(CO_2, g) + 3\Delta_f H^\ominus(H_2O, l) - \Delta_c H^\ominus$

$$= [6 \times (-393.51) + 3 \times (-285.83) - (-3054)]\,kJ\,mol^{-1}$$
$$= -164.\overline{55}\,kJ\,mol^{-1}$$

$$\Delta_f S^\ominus = \sum_J \nu_J S_m^\ominus(J) \quad [22b]$$

$$\Delta_f S^\ominus = S_m^\ominus(C_6H_5OH, s) - 6S_m^\ominus(C, s) - 3S_m^\ominus(H_2, g) - \tfrac{1}{2}S_m^\ominus(O_2, g)$$
$$= \left[144.0 - (6 \times 5.740) - (3 \times 130.68) - \left(\tfrac{1}{2} \times 205.14\right)\right]\,J\,K^{-1}\,mol^{-1}$$
$$= -385.0\bar{5}\,J\,K^{-1}\,mol^{-1}$$

Hence $\Delta_f G^\ominus = -164.5\bar{5}\,kJ\,mol^{-1} - (298.15\,K) \times (-385.0\bar{5}\,J\,K^{-1}\,mol^{-1}) = \boxed{-50\,kJ\,mol^{-1}}$

E4.20 **(a)** $\Delta S(\text{gas}) = nR\ln\frac{V_f}{V_i}[17] = \left(\frac{14\,g}{28.02\,g\,mol^{-1}}\right) \times (8.314\,J\,K^{-1}\,mol^{-1}) \times (\ln 2)$

$$= \boxed{+2.9\,J\,K^{-1}}$$

$\Delta S(\text{surroundings}) = \boxed{-2.9\,J\,K^{-1}}$ [overall zero entropy production]

$\Delta S(\text{total}) = \boxed{0}$ [reversible process]

(b) $\Delta S(\text{gas}) = \boxed{+2.9\,J\,K^{-1}}$ [S a state function]

$\Delta S(\text{surroundings}) = \boxed{0}$ [the surroundings do not change]

$\Delta S(\text{total}) = \boxed{+2.9\,J\,K^{-1}}$

(c) $\Delta S(\text{gas}) = \boxed{0}$ [$q_{rev} = 0$]

$\Delta S(\text{surroundings}) = \boxed{0}$ [no heat is transferred to the surroundings]

$\Delta S(\text{total}) = \boxed{0}$

E4.21 The same final state is attained if the change takes place in two stages, one being isothermal compression

$$\Delta S_1 = nR\ln\frac{V_f}{V_i}[17] = nR\ln\frac{1}{2} = -nR\ln 2$$

and the second, heating at constant volume

$$\Delta S_2 = nC_{V,m}\ln\frac{T_f}{T_i}[20] = nC_{V,m}\ln 2$$

The overall entropy change is therefore

$$\Delta S = -nR\ln 2 + nC_{V,m}\ln 2 = \boxed{n(C_{V,m} - R)\ln 2}$$

E4.22 $CH_4(g) + 2O_2(g) \rightarrow CO_2(g) + 2H_2O(l)$

$$\Delta_r G^\ominus = \sum_J \nu_J \Delta_f G^\ominus(J)\ [40b]$$

$$\Delta_r G^\ominus = \Delta_f G^\ominus(CO_2, g) + 2\Delta_f G^\ominus(H_2O, l) - \Delta_f G^\ominus(CH_4, g)$$
$$= \{-394.36 + (2\times -237.13) - (-50.72)\}\ \text{kJ mol}^{-1} = -817.90\ \text{kJ mol}^{-1}$$

Therefore, the maximum non-expansion work is $\boxed{817.90\ \text{kJ mol}^{-1}}$ [since $|w_e| = |\Delta G|$].

E4.23 $\varepsilon = 1 - \dfrac{T_c}{T_h}$ [11]

(a) $\varepsilon = 1 - \dfrac{333\ \text{K}}{373\ \text{K}} = \boxed{0.11}$ [11 per cent efficiency]

(b) $\varepsilon = 1 - \dfrac{353\ \text{K}}{573\ \text{K}} = \boxed{0.38}$ [38 per cent efficiency]

Solutions to problems

Assume that all gases are perfect and that data refer to 298 K unless otherwise stated.

Solutions to numerical problems

P4.3 **(a)** $q(\text{total}) = q(H_2O) + q(Cu) = 0$, hence $-q(H_2O) = q(Cu)$

$q(H_2O) = n(-\Delta_{vap}H) + nC_{p,m}(H_2O, l)\Delta T(H_2O)$

$q(Cu) = mC_s\Delta T(Cu) \quad C_s = 0.385\ \text{J K}^{-1}\text{g}^{-1}$

$(1.00\ \text{mol}) \times (40.656\times 10^3\ \text{J mol}^{-1}) - (1.00\ \text{mol}) \times (75.3\ \text{J K}^{-1}\text{mol}^{-1}) \times (\theta - 100°\text{C})$
$= (2.00\times 10^3\ \text{g}) \times (0.385\ \text{J K}^{-1}\text{g}^{-1}) \times \theta$

Solving for θ, $\theta = 57.0°\text{C} = \boxed{330.2\ \text{K}}$

$q(Cu) = (2.00\times 10^3\ \text{g}) \times (0.385\ \text{J K}^{-1}\text{g}^{-1}) \times (57.0\ \text{K}) = 4.39\times 10^4\ \text{J} = \boxed{43.9\ \text{kJ}}$

$q(H_2O) = \boxed{-43.9\ \text{kJ}}$

$\Delta S(\text{total}) = \Delta S(H_2O) + \Delta S(Cu)$

$$\Delta S(H_2O) = \frac{-n\Delta_{vap}H}{T_b} + nC_{p,m}\ln\left(\frac{T_f}{T_i}\right)\ [20]$$
$$= -\frac{(1.00\ \text{mol}) \times (40.656\times 10^3\ \text{J mol}^{-1})}{373.2\ \text{K}}$$

$$+ (1.00\,\text{mol}) \times (75.3\,\text{J K}^{-1}\,\text{mol}^{-1}) \times \ln\left(\frac{330.2\,\text{K}}{373.2\,\text{K}}\right)$$

$$= -108.9\,\text{J K}^{-1} - 9.22\,\text{J K}^{-1} = \boxed{-118.\bar{1}\,\text{J K}^{-1}}$$

$$\Delta S(\text{Cu}) = mC_\text{s} \ln \frac{T_\text{f}}{T_\text{i}} = (2.00 \times 10^3\,\text{g}) \times (0.385\,\text{J K}^{-1}\,\text{g}^{-1}) \times \ln\left(\frac{330.2\,\text{K}}{273.2\,\text{K}}\right)$$

$$= \boxed{145.\bar{9}\,\text{J K}^{-1}}$$

$$\Delta S(\text{total}) = -118.\bar{1}\,\text{J K}^{-1} + 145.\bar{9}\,\text{J K}^{-1} = \boxed{28\,\text{J K}^{-1}}$$

This process is spontaneous since ΔS (surroundings) is zero and, hence, $\Delta S(\text{universe}) = \Delta S(\text{total}) = \text{positive}$.

(b) The volume of the container may be calculated from the perfect gas law.

$$V = \frac{nRT}{p} = \frac{(1.00\,\text{mol}) \times (0.08206\,\text{L atm K}^{-1}\,\text{mol}^{-1}) \times (373.2\,\text{K})}{1.00\,\text{atm}} = 30.6\,\text{L}$$

At 57.0°C the vapour pressure of water is 130 Torr (HCP). The amount of water vapour present at equilibrium is then

$$n = \frac{pV}{RT} = \frac{(130\,\text{Torr}) \times \left(\frac{1\,\text{atm}}{760\,\text{Torr}}\right) \times (30.6\,\text{L})}{(0.08206\,\text{L atm K}^{-1}\,\text{mol}^{-1}) \times (330.2\,\text{K})} = 0.193\,\text{mol}$$

This is a substantial fraction of the original amount of water and cannot be ignored. Consequently the calculation needs to be redone taking into account the fact that only a part, n_1, of the vapour condenses into a liquid while the remainder $(1.00\,\text{mol} - n_1)$ remains gaseous. The heat flow involving water, then, becomes

$$\begin{aligned} q(\text{H}_2\text{O}) &= -n_1\Delta_\text{vap}H + n_1C_{p,\text{m}}(\text{H}_2\text{O, l})\Delta T(\text{H}_2\text{O}) \\ &\quad +(1.00\,\text{mol} - n_1)C_{p,\text{m}}(\text{H}_2\text{O, g})\Delta T(\text{H}_2\text{O}) \end{aligned}$$

Because n_1 depends on the equilibrium temperature through

$$n_1 = 1.00\,\text{mol} - \frac{pV}{RT}$$

where p is the vapour pressure of water, we will have two unknowns (p and T) in the equation $-q(\text{H}_2\text{O}) = q(\text{Cu})$. There are two ways out of this dilemma: (1) p may be expressed as a function of T by use of the Clapeyron equation (Chapter 6), or (2) by use of successive approximations. Redoing the calculation with

$$n_1 = (1.00\,\text{mol}) - (0.193\,\text{mol}) = 0.80\bar{7}\,\text{mol}$$

(noting that $C_{p,\text{m}}(\text{H}_2\text{O, g}) = (75.3{-}41.9)\,\text{J mol}^{-1}\,\text{K}^{-1}$ (Problem 4.1)) yields a final temperature of 47.2°C. At this temperature, the vapour pressure of water is 80.41 Torr, corresponding to

$$n_1 = (1.00\,\text{mol}) - (0.123\,\text{mol}) = 0.87\bar{7}\,\text{mol}$$

The recalculated final temperature is 50.8°C. The successive approximations eventually converge to yield a value of $\boxed{49.9°\text{C} = 323.2\,\text{K}}$ for the final temperature. Using this value of the final temperature, the heat transferred and the various entropies are calculated as in part **(a)**.

P4.4 This problem concerns the same system and the same changes of state as Problem 2.3. The final temperature of section A was there calculated to be 900 K.

(a) $$\Delta S_A = nC_{V,m}\ln\left(\frac{T_{A,f}}{V_{A,i}}\right) + nR\ln\left(\frac{V_{A,f}}{V_{A,i}}\right) \text{ [Example 4.3]}$$

$$= (2.0\,\text{mol}) \times (20\,\text{J K}^{-1}\,\text{mol}^{-1}) \times \ln\left(\frac{900\,\text{K}}{300\,\text{K}}\right)$$

$$+ (2.00\,\text{mol}) \times (8.314\,\text{J K}^{-1}\,\text{mol}^{-1}) \times \ln\left(\frac{3.00\,\text{L}}{2.00\,\text{L}}\right)$$

$$= \boxed{50.7\,\text{J K}^{-1}}$$

$$\Delta S_B = nR\ln\left(\frac{V_{B,f}}{V_{B,i}}\right) = (2.00\,\text{mol}) \times (8.314\,\text{J K}^{-1}\,\text{mol}^{-1}) \times \ln\left(\frac{1.00\,\text{L}}{2.00\,\text{L}}\right)$$

$$= \boxed{-11.5\,\text{J K}^{-1}}$$

(b) In the solution to Problem 2.3 the reversible work in sections A and B was calculated:

$$w_A = -3.46 \times 10^3\,\text{J}, \qquad w_B = 3.46 \times 10^3\,\text{J}, \qquad w_{max} = w_{rev} = \Delta A \text{ [35]}$$

But this relation holds only at constant temperature; hence

$$\Delta A_B = w_B = +3.46 \times 10^3\,\text{J} = \boxed{+3.46\,\text{kJ}} \text{ [constant temperature]}$$
$$\Delta A_A \neq w_A \text{ [temperature not constant]}$$

We might expect that ΔA_A is negative, since w_A is negative; but based on the information provided we can only state that it is $\boxed{\text{indeterminate}}$

(c) Under constant-temperature conditions

$$\Delta G = \Delta H - T\Delta S$$

In section B, $\Delta H_B = 0$ [constant temperature, perfect gas]

$$\Delta S_B = -11.5\,\text{J K}^{-1}$$
$$\Delta G_B = -T_B\Delta S_B = -(300\,\text{K}) \times (-11.5\,\text{J K}^{-1}) = \boxed{3.46 \times 10^3\,\text{J}}$$

ΔG_A is $\boxed{\text{indeterminate}}$ in both magnitude and sign. A resolution of this problem is only possible based on additional relations developed in Chapters 5 and 19.

(d) $$\Delta S(\text{total system}) = \Delta S_A + \Delta S_B = (50.7 - 11.5)\,\text{J K}^{-1} = \boxed{+39.2\,\text{J K}^{-1}}$$

If the process has been carried out reversibly as assumed in the statement of the problem we can say

$$\Delta S(\text{system}) + \Delta S(\text{surroundings}) = 0$$

Hence, $\Delta S(\text{surroundings}) = \boxed{-39.2\,\text{J K}^{-1}}$

Question. Can you design this process such that heat is added to section A reversibly?

P4.5

	Step 1	*Step 2*	*Step 3*	*Step 4*	*Cycle*
q	+11.5 kJ	0	−5.74 kJ	0	5.8 kJ
w	−11.5 kJ	−3.74 kJ	+5.74 kJ	3.74 kJ	−5.8 kJ
ΔU	0	−3.74 kJ	0	+3.74 kJ	0
ΔH	0	−6.23 kJ	0	+6.23 kJ	0
ΔS	$+19.1\,\mathrm{J\,K^{-1}}$	0	$-19.1\,\mathrm{J\,K^{-1}}$	0	0
$\Delta S_{\rm sur}$	$-19.1\,\mathrm{J\,K^{-1}}$	0	$+19.1\,\mathrm{J\,K^{-1}}$	0	0
$\Delta S_{\rm tot}$	0	0	0	0	0

Step 1

$$\Delta U = \Delta H = \boxed{0} \text{ [isothermal]}$$

$$w = -nRT\ln\left(\frac{V_{\rm f}}{V_{\rm i}}\right) = nRT\ln\left(\frac{p_{\rm f}}{p_{\rm i}}\right) \text{ [2.13, and Boyle's law]}$$

$$= (1.00\,\mathrm{mol}) \times (8.314\,\mathrm{J\,K^{-1}\,mol^{-1}}) \times (600\,\mathrm{K}) \times \ln\left(\frac{1.00\,\mathrm{atm}}{10.0\,\mathrm{atm}}\right) = \boxed{-11.5\,\mathrm{kJ}}$$

$$q = -w = \boxed{11.5\,\mathrm{kJ}}$$

$$\Delta S = nR\ln\left(\frac{V_{\rm f}}{V_{\rm i}}\right) [17] = -nR\ln\left(\frac{p_{\rm f}}{p_{\rm i}}\right) \text{ [Boyle's law]}$$

$$= -(1.00\,\mathrm{mol}) \times (8.314\,\mathrm{J\,K^{-1}\,mol^{-1}}) \times \ln\left(\frac{1.00\,\mathrm{atm}}{10.0\,\mathrm{atm}}\right) = \boxed{+19.1\,\mathrm{J\,K^{-1}}}$$

$$\Delta S(\text{sur}) = -\Delta S(\text{system}) \text{ [reversible process]} = \boxed{-19.1\,\mathrm{J\,K^{-1}}}$$

$$\Delta S_{\rm tot} = \Delta S(\text{system}) + \Delta S(\text{sur}) = \boxed{0}$$

Step 2

$$q = \boxed{0} \text{ [adiabatic]}$$

$$\Delta U = nC_{V,\rm m}\Delta T \text{ [2.22}b\text{]}$$

$$= (1.00\,\mathrm{mol}) \times \left(\tfrac{3}{2}\right) \times (8.314\,\mathrm{J\,K^{-1}\,mol^{-1}}) \times (300\,\mathrm{K} - 600\,\mathrm{K}) = \boxed{-3.74\,\mathrm{kJ}}$$

$$w = \Delta U = \boxed{-3.74\,\mathrm{kJ}}$$

$$\Delta H = \Delta U + \Delta(pV) = \Delta U + nR\Delta T$$

$$= (-3.74\,\mathrm{kJ}) + (1.00\,\mathrm{mol}) \times (8.314\,\mathrm{J\,K^{-1}\,mol^{-1}}) \times (-300\,\mathrm{K})$$

$$= \boxed{-6.23\,\mathrm{kJ}}$$

$$\Delta S = \Delta S(\text{sur}) = \boxed{0} \text{ [reversible adiabatic process]}$$

$$\Delta S_{\rm tot} = \boxed{0}$$

Step 3
These quantities may be calculated in the same manner as for *Step 1* or more easily as follows

$$\Delta U = \Delta H = \boxed{0} \text{ [isothermal]}$$

$$\varepsilon = 1 - \frac{T_{\rm c}}{T_{\rm h}}[11] = 1 - \frac{300\,\mathrm{K}}{600\,\mathrm{K}} = 0.500 = 1 + \frac{q_{\rm c}}{q_{\rm h}}[10]$$

$$q_{\rm c} = -0.500\,q_{\rm h} = -(0.500) \times (11.5\,\mathrm{kJ}) = -5.74\,\mathrm{kJ}$$

$$q_{\rm c} = \boxed{-5.74\,\mathrm{kJ}} \qquad w = -q_{\rm c} = \boxed{5.74\,\mathrm{kJ}}$$

$$\Delta S = -\Delta S\ (Step\ 1)\ [\text{initial and final temperature reversed}] = \boxed{-19.1\,\mathrm{J\,K^{-1}}}$$

$$\Delta S(\text{sur}) = -\Delta S(\text{system}) = \boxed{+19.1\,\mathrm{J\,K^{-1}}}$$

$$\Delta S_{\text{tot}} = \boxed{0}$$

Step 4

ΔU and ΔH are the negative of their values in *Step 2*. (Initial and final temperatures reversed.)

$$\Delta U = \boxed{+3.74\,\mathrm{kJ}}, \qquad \Delta H = \boxed{+6.23\,\mathrm{kJ}}, \qquad q = \boxed{0}\ [\text{adiabatic}]$$

$$w = \Delta U = \boxed{+3.74\,\mathrm{kJ}}$$

$$\Delta S = \Delta S(\text{sur}) = \boxed{0}\ [\text{reversible adiabatic process}]$$

$$\Delta S_{\text{tot}} = \boxed{0}$$

Cycle

$$\Delta U = \Delta H = \Delta S = \boxed{0} \quad [\Delta(\text{state function}) = 0 \text{ for any cycle}]$$

$$\Delta S(\text{sur}) = 0\ [\text{all reversible processes}]$$

$$\Delta S_{\text{tot}} = \boxed{0}$$

$$q(\text{cycle}) = (11.5 - 5.74)\,\mathrm{kJ} = \boxed{5.8\,\mathrm{kJ}} \qquad w(\text{cycle}) = -q(\text{cycle}) = \boxed{-5.8\,\mathrm{kJ}}$$

P4.6

	q	w	$\Delta U = \Delta H$	ΔS	ΔS_{sur}	ΔS_{tot}
Path(a)	2.74 kJ	−2.74 kJ	0	$9.13\,\mathrm{J\,K^{-1}}$	$-9.13\,\mathrm{J\,K^{-1}}$	0
Path(b)	1.66 kJ	−1.66 kJ	0	$9.13\,\mathrm{J\,K^{-1}}$	$-5.53\,\mathrm{J\,K^{-1}}$	$3.60\,\mathrm{J\,K^{-1}}$

Path (a)

$$w = -nRT\ln\left(\frac{V_\mathrm{f}}{V_\mathrm{i}}\right) = -nRT\ln\left(\frac{p_\mathrm{i}}{p_\mathrm{f}}\right)\ [\text{Boyle's law}]$$

$$= -(1.00\,\mathrm{mol}) \times (8.314\,\mathrm{J\,K^{-1}\,mol^{-1}}) \times (300\,\mathrm{K}) \times \ln\left(\frac{3.00\,\mathrm{atm}}{1.00\,\mathrm{atm}}\right) = -2.74 \times 10^3\,\mathrm{J}$$

$$= \boxed{-2.74\,\mathrm{kJ}}$$

$$\Delta H = \Delta U = \boxed{0}\ [\text{isothermal process in perfect gas}]$$

$$q = \Delta U - w = 0 - (-2.74\,\mathrm{kJ}) = \boxed{+2.74\,\mathrm{kJ}}$$

$$\Delta S = \frac{q_{\text{rev}}}{T}\ [\text{Example 4.1}] = \frac{2.74 \times 10^3\,\mathrm{J}}{300\,\mathrm{K}} = \boxed{+9.13\,\mathrm{J\,K^{-1}}}$$

$$\Delta S_{\text{tot}} = \boxed{0}\ [\text{reversible process}]$$

$$\Delta S_{\text{sur}} = \Delta S_{\text{tot}} - \Delta S = 0 - 9.13\,\mathrm{J\,K^{-1}} = \boxed{-9.13\,\mathrm{J\,K^{-1}}}$$

Path (b)

$$w = -p_{\text{ex}}(V_\mathrm{f} - V_\mathrm{i}) = -p_{\text{ex}}\left(\frac{nRT}{p_\mathrm{f}} - \frac{nRT}{p_\mathrm{i}}\right)\ [\text{perfect gas}] = -nRT\left(\frac{p_{\text{ex}}}{p_\mathrm{f}} - \frac{p_{\text{ex}}}{p_\mathrm{i}}\right)$$

$$= -(1.00\,\mathrm{mol}) \times (8.314\,\mathrm{J\,K^{-1}\,mol^{-1}}) \times (300\,\mathrm{K}) \times \left(\frac{1.00\,\mathrm{atm}}{1.00\,\mathrm{atm}} - \frac{1.00\,\mathrm{atm}}{3.00\,\mathrm{atm}}\right)$$

$$= -1.66 \times 10^3\,\text{J} = \boxed{-1.66\,\text{kJ}}$$

$$\Delta H = \Delta U = \boxed{0} \text{ [isothermal process in perfect gas]}$$

$$q = \Delta U - w = 0 - (-1.66\,\text{kJ}) = \boxed{+1.66\,\text{kJ}}$$

$$\Delta S = \frac{q_{\text{rev}}}{T} = \boxed{+9.13\,\text{J K}^{-1}} \quad [\Delta S \text{ is independent of path}]$$

$$\Delta S_{\text{sur}} = \frac{q_{\text{sur}}}{T_{\text{sur}}} = \frac{-q}{T_{\text{sur}}} = \frac{-1.66 \times 10^3\,\text{J}}{300\,\text{K}} = \boxed{-5.53\,\text{J K}^{-1}}$$

$$\Delta S_{\text{tot}} = \Delta S + \Delta S_{\text{sur}} = (9.13 - 5.53)\,\text{J K}^{-1} = \boxed{+3.60\,\text{J K}^{-1}}$$

P4.7

	q	$w = \Delta U$	ΔH	ΔS	ΔS_{sur}	ΔS_{tot}
Path(a)	0	-9.1×10^2 J	-1.5×10^3 J	0	0	0
Path(b)	0	-7.5×10^2 J	-1.2×10^3J	$+1.12\,\text{J K}^{-1}$	0	$+1.12\,\text{J K}^{-1}$

$$C_{p,\text{m}} = C_{V,\text{m}} + R = \frac{3}{2}R + R = \frac{5}{2}R, \qquad \gamma = \frac{C_{p,\text{m}}}{C_{V,\text{m}}} = \frac{5}{3}, \qquad c = \frac{C_{V,\text{m}}}{R} = \frac{\frac{3}{2}R}{R} = \frac{3}{2}$$

(a) $$T_{\text{f}} = \left(\frac{V_{\text{i}}}{V_{\text{f}}}\right)^{1/c} T_{\text{i}}\ [2.35] = \left(\frac{V_{\text{i}}}{V_{\text{f}}}\right)^{\gamma-1} T_{\text{i}} \quad \left[\frac{1}{c} = \frac{R}{C_{V,\text{m}}} = \frac{C_{p,\text{m}} - C_{V,\text{m}}}{C_{V,\text{m}}} = \gamma - 1\right]$$

$$p_{\text{i}}V_{\text{i}}^{\gamma} = p_{\text{f}}V_{\text{f}}^{\gamma}\ [2.36] \quad \text{or} \quad \frac{V_{\text{i}}}{V_{\text{f}}} = \left(\frac{p_{\text{f}}}{p_{\text{i}}}\right)^{1/\gamma}$$

Substituting into the expression for T_{f} above

$$T_{\text{f}} = \left(\frac{p_{\text{f}}}{p_{\text{i}}}\right)^{(\gamma-1)/\gamma} T_{\text{i}} = \left(\frac{p_{\text{i}}}{p_{\text{f}}}\right)^{(1-\gamma)/\gamma} T_{\text{i}}$$

$$= \left(\frac{1.00\,\text{atm}}{0.50\,\text{atm}}\right)^{[1-(5/3)]/(5/3)} \times (300\,\text{K}) = \boxed{227\,\text{K}}$$

$$w = \Delta U = nC_{V,\text{m}}\Delta T = (1.00\,\text{mol}) \times \left(\tfrac{3}{2}\right) \times (8.314\,\text{J K}^{-1}\,\text{mol}^{-1}) \times (227.\bar{4} - 300\,\text{K})$$

$$= \boxed{-9.1 \times 10^2\,\text{J}}$$

$$\Delta H = nC_{p,\text{m}}\Delta T = (1.00\,\text{mol}) \times \left(\tfrac{5}{2}\right) \times (8.314\,\text{J K}^{-1}\,\text{mol}^{-1}) \times (-72.\bar{6}\,\text{K})$$

$$= \boxed{-1.5 \times 10^3\,\text{J}}$$

$$\Delta S_{\text{tot}} = \boxed{0} \text{ [reversible process]} = \Delta S + \Delta S_{\text{sur}}$$

$$\Delta S_{\text{sur}} = \boxed{0} \text{ [adiabatic process]}; \quad \text{hence,} \quad \Delta S = \boxed{0}$$

(b) $$\Delta U = w \text{ [adiabatic process]}$$

$$\Delta U = nC_{V,\text{m}}(T_{\text{f}} - T_{\text{i}})$$

$$w = -p_{\text{ex}}(V_{\text{f}} - V_{\text{i}}) = -p_{\text{ex}}\left(\frac{nRT_{\text{f}}}{p_{\text{f}}} - \frac{nRT_{\text{i}}}{p_{\text{i}}}\right)$$

Solving for T_{f}, with $p_{\text{ex}} = p_{\text{f}} = 0.50\,\text{atm}$, $p_{\text{i}} = 1.00\,\text{atm}$

$$T_{\text{f}} = T_{\text{i}} \times \left\{\frac{C_{V,m} + \left(\frac{p_{\text{ex}}R}{p_{\text{i}}}\right)}{C_{V,m} + \left(\frac{p_{\text{ex}}R}{p_{\text{f}}}\right)}\right\} = (300\,\text{K}) \times \left(\frac{\frac{3}{2}R + \frac{1}{2}R}{\frac{3}{2}R + R}\right) = (300\,\text{K}) \times \frac{4}{5} = \boxed{240\,\text{K}}$$

$$w = \Delta U = (1.00\,\text{mol}) \times \left(\tfrac{3}{2}\right) \times (8.314\,\text{J K}^{-1}\,\text{mol}^{-1}) \times (240\,\text{K} - 300\,\text{K})$$

$$= \boxed{-7.5 \times 10^2\,\text{J}}$$

$$\Delta H = nC_{p,\text{m}}\Delta T = (1.00\,\text{mol}) \times \left(\tfrac{5}{2}\right) \times (8.314\,\text{J K}^{-1}\,\text{mol}^{-1}) \times (-60\,\text{K})$$

$$= \boxed{-1.2 \times 10^3\,\text{J}}$$

$$\Delta S = nC_{p,\text{m}} \ln\left(\frac{T_\text{f}}{T_\text{i}}\right) - nR\ln\left(\frac{p_\text{f}}{p_\text{i}}\right) \text{ [Exercise 4.4]}$$

$$= (1.00\,\text{mol}) \times \left(\frac{5}{2}\right) \times (8.314\,\text{J K}^{-1}\,\text{mol}^{-1}) \times \ln\left(\frac{240\,\text{K}}{300\,\text{K}}\right)$$

$$- (1.00\,\text{mol}) \times (8.314\,\text{J K}^{-1}\,\text{mol}^{-1}) \times \ln\left(\frac{0.50\,\text{atm}}{1.00\,\text{atm}}\right) = \boxed{+1.12\,\text{J K}^{-1}}$$

$$\Delta S_\text{sur} = \boxed{0} \text{ [adiabatic process]}$$

$$\Delta S_\text{tot} = \Delta S + \Delta S_\text{sur} = 1.12\,\text{J K}^{-1} + 0 = \boxed{+1.12\,\text{J K}^{-1}}$$

P4.11 $C(s) + \tfrac{1}{2}O_2(g) + 2H_2(g) \rightarrow CH_3OH(l), \quad \Delta n_\text{g} = -2.5\,\text{mol}$

$\Delta G = \Delta H - T\Delta S$ [constant temperature] $\quad \Delta H = \Delta U + \Delta(pV)$

Therefore, $\Delta G = \Delta U - T\Delta S + \Delta(pV) = \Delta A + \Delta(pV)$ and

$$\Delta_\text{f}A^\ominus = \Delta_\text{f}G^\ominus - \Delta(pV) = \Delta_\text{f}G^\ominus - \Delta n_\text{g}(RT) \text{ [perfect gases]} = \Delta_\text{f}G^\ominus + 2.5RT$$

$$= [(-166.27) + (2.5) \times (2.479)]\,\text{kJ mol}^{-1} = \boxed{-160.07\,\text{kJ mol}^{-1}}$$

P4.13 **(a)** Under constant-temperature conditions

$$\Delta A = w_\text{max} \text{ [35]}$$

Since $\Delta A = \Delta G - \Delta(pV)$, it is convenient to first work part **(b)**.

(b) Under constant-temperature and pressure conditions

$$\Delta G = w_\text{e,max} \text{ [38]}$$

Using the same cycle as in Problem 4.1, with

$$\Delta C_{p,\text{m}} \equiv C_{p,\text{m}}(\text{liq}) - C_{p,\text{m}}(\text{gas})\cdots$$

$$\Delta G(T) = \Delta H(T) - T\Delta S(T)$$

$$= \Delta H(T_\text{f}) - \Delta C_{p,\text{m}}(T - T_\text{f}) - T\left(\Delta S(T_\text{f}) - \Delta C_{p,\text{m}} \ln\frac{T}{T_\text{f}}\right)$$

$$= \Delta H(T_\text{f}) - \frac{T}{T_\text{f}}\Delta H(T_\text{f}) - \Delta C_{p,\text{m}}\left(T - T_\text{f} - T\ln\frac{T}{T_\text{f}}\right);$$

$$\Delta H(T_\text{f}) = -\Delta_\text{fus}H(T_\text{f})$$

$$= \left(\frac{T}{T_\text{f}} - 1\right)\Delta_\text{fus}H(T_\text{f}) - \Delta C_{p,\text{m}}\left(T - T_\text{f} - T\ln\frac{T}{T_\text{f}}\right)$$

$T = 268\,\text{K}, \;\; T_\text{f} = 273\,\text{K}, \;\; \Delta_\text{fus}H = 6.01\,\text{kJ mol}^{-1}, \;\; \Delta C_{p,\text{m}} = +37.3\,\text{J K}^{-1}\,\text{mol}^{-1}$:

$$\Delta G(268\,\text{K}) = \left(\frac{268}{273} - 1\right) \times (6.01\,\text{kJ mol}^{-1}) - (37.3\,\text{J mol}^{-1})$$

$$\times \left(268 - 273 - 268\ln\frac{268}{273}\right) = \boxed{-0.11\,\text{kJ mol}^{-1}}$$

Returning to part **(a)** we use

$$\Delta A = \Delta G - \Delta(pV) = \Delta G - p\Delta V\,[\text{constant pressure}] = \Delta G - pM\Delta\left(\frac{1}{\rho}\right)$$
$$= (-0.11\,\text{kJ mol}^{-1}) - (1.013\times10^{5}\,\text{Pa})\times(18.02\times10^{-3}\,\text{kg mol}^{-1})$$
$$\times\left(\frac{1}{917\,\text{kg m}^{-3}} - \frac{1}{999\,\text{kg m}^{-3}}\right)$$
$$= (-0.11\,\text{kJ mol}^{-1}) - (1.6\times10^{-4}\,\text{kJ mol}^{-1}) = -0.11\,\text{kJ mol}^{-1}$$

Therefore

(a) Maximum work is $\boxed{0.11\,\text{kJ mol}^{-1}}$

(b) Maximum non-expansion work is also $\boxed{0.11\,\text{kJ mol}^{-1}}$

However, there is a slight difference of $1.6\times10^{-4}\,\text{kJ mol}^{-1}$ between the two values.

P4.14 $S_{\text{m}}(T) = S_{\text{m}}(0) + \displaystyle\int_0^T \frac{C_{p,\text{m}}\,\text{d}T}{T}$ [19]

From the data, draw up the following table

T/K	10	15	20	25	30	50
$\dfrac{C_{p,\text{m}}}{T}/(\text{J K}^{-2}\,\text{mol}^{-1})$	0.28	0.47	0.540	0.564	0.550	0.428

T/K	70	100	150	200	250	298
$\dfrac{C_{p,\text{m}}}{T}/(\text{J K}^{-2}\,\text{mol}^{-1})$	0.333	0.245	0.169	0.129	0.105	0.089

Plot $C_{p,\text{m}}/T$ against T (Fig. 4.1). This has been done on two scales. The region 0 to 10 K has been constructed using $C_{p,\text{m}} = aT^3$, fitted to the point at $T = 10\,\text{K}$, at which $C_{p,\text{m}} = 2.8\,\text{J K}^{-1}\,\text{mol}^{-1}$, so $a = 2.8\times10^{-3}\,\text{J K}^{-4}\,\text{mol}^{-1}$. The area can be determined (primitively) by counting squares, which gives area A $= 38.28\,\text{J K}^{-1}\,\text{mol}^{-1}$, area B (up to 0°C) $= 25.60\,\text{J K}^{-1}\,\text{mol}^{-1}$, area B (up to 25°C) $= 27.80\,\text{J K}^{-1}\,\text{mol}^{-1}$. Hence

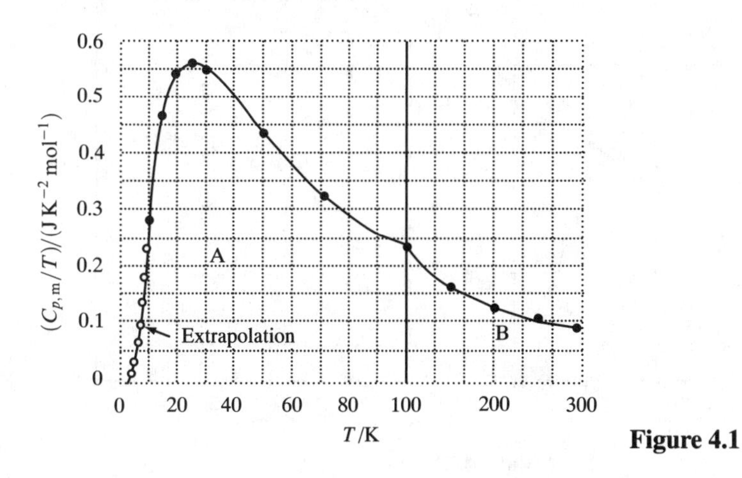

Figure 4.1

(a) $S_m(273\,K) = S_m(0) + \boxed{63.88\,J\,K^{-1}\,mol^{-1}}$

(b) $S_m(298\,K) = S_m(0) + \boxed{66.08\,J\,K^{-1}\,mol^{-1}}$

P4.17 $\Delta_r G^{\ominus} = \Delta_r H^{\ominus} - T\Delta_r S^{\ominus} = 26.120\,kJ\,mol^{-1}$

$\Delta_r H^{\ominus} = +55.000\,kJ\,mol^{-1}$

It is convenient to first work part **(b)**.

(b) Hence $\Delta_r S^{\ominus} = \dfrac{(55.000 - 26.120)\,kJ\,mol^{-1}}{298.15\,K} = \boxed{+96.864\,J\,K^{-1}\,mol^{-1}}$

$$\Delta_r S^{\ominus} = 4S_m^{\ominus}(K^+, aq) + S_m^{\ominus}([Fe(CN)_6]^{4-}, aq) + 3S_m^{\ominus}(H_2O, l) - S_m^{\ominus}(K_4[Fe(CN)_6]\cdot 3H_2O, s)$$

(a) Therefore,

$$\begin{aligned} S_m^{\ominus}([Fe(CN)_6]^{4-}, aq) &= \Delta_r S^{\ominus} - 4S_m^{\ominus}(K^+, aq) - 3S_m^{\ominus}(H_2O, l) \\ &\quad + S_m^{\ominus}(K_4[Fe(CN)_6]\cdot 3\,H_2O, s) \\ &= [96.864 - (4 \times 102.5) - (3 \times 69.9) + (599.7)]\,J\,K^{-1}\,mol^{-1} \\ &= \boxed{+76.9\,J\,K^{-1}\,mol^{-1}} \end{aligned}$$

P4.19 Draw up the following table and proceed as in Problem 4.14.

T/K	14.14	16.33	20.03	31.15	44.08	64.81
$(C_{p,m}/T)/(J\,K^{-2}\,mol^{-1})$	0.671	0.778	0.908	1.045	1.063	1.024

T/K	100.90	140.86	183.59	225.10	262.99	298.06
$(C_{p,m}/T)/(J\,K^{-2}\,mol^{-1})$	0.942	0.861	0.787	0.727	0.685	0.659

Plot $C_{p,m}$ against T (Fig. 4.2(a)) and $C_{p,m}/T$ against T (Fig. 4.2(b)), extrapolating to $T = 0$ with $C_{p,m} = aT^3$ fitted at T=14.14 K, which gives $a = 3.36\,mJ\,K^{-4}\,mol^{-1}$. Integration by determining the area under the curve then gives

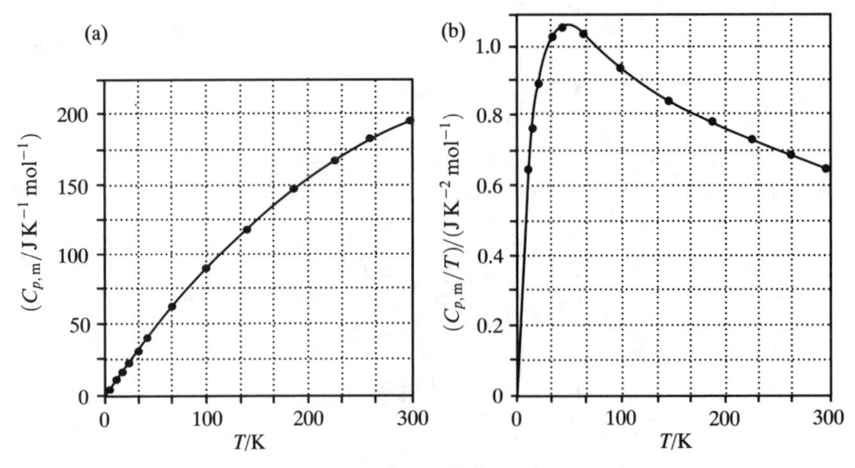

Figure 4.2

$$\int_0^{298\,\mathrm{K}} C_{p,\mathrm{m}}\,\mathrm{d}T = 34.4\,\mathrm{kJ\,mol^{-1}}, \quad \text{so} \quad H_\mathrm{m}(298\,\mathrm{K}) = H_\mathrm{m}(0) + \boxed{34.4\,\mathrm{kJ\,mol^{-1}}}$$

$$\int_0^{298\,\mathrm{K}} \frac{C_{p,\mathrm{m}}\,\mathrm{d}T}{T} = 243\,\mathrm{J\,K^{-1}mol^{-1}}, \quad \text{so} \quad S_\mathrm{m}(298\,\mathrm{K}) = S_\mathrm{m}(0) + \boxed{243\,\mathrm{J\,K^{-1}\,mol^{-1}}}$$

Solutions to theoretical problems

P4.21 Paths A and B in Fig. 4.3 are the reversible adiabatic paths which are assumed to cross at state 1. Path C (dashed) is an isothermal path which connects the adiabatic paths at states 2 and 3. Now go round the cycle (1→ 2, step 1; 2→ 3, step 2; 3→ 1, step 3).

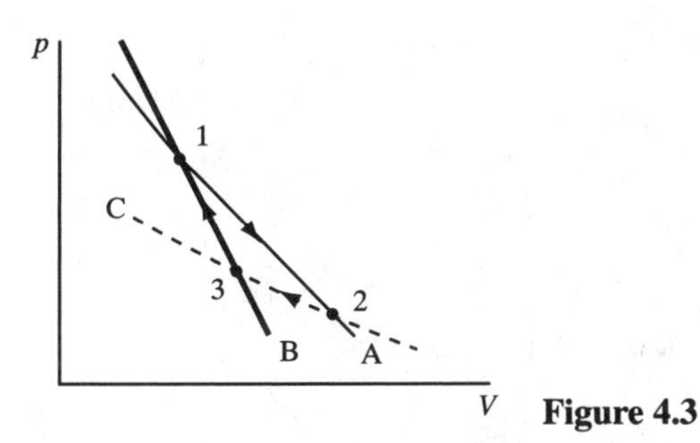

Figure 4.3

Step 1 $\Delta U_1 = q_1 + w_1 = w_1$ $[q_1 = 0,$ adiabatic$]$

Step 2 $\Delta U_2 = q_2 + w_2 = 0$ [isothermal step, energy depends on temperature only]

Step 3 $\Delta U_3 = q_3 + w_3 = w_3$ $[q_3 = 0,$ adiabatic$]$

For the cycle $\Delta U = 0 = w_1 + q_2 + w_2 + w_3$ or $w(\text{net}) = w_1 + w_2 + w_3 = -q_2$

But, $\Delta U_1 = -\Delta U_3[\Delta T_1 = -\Delta T_2]$; hence $w_1 = -w_3$, and $w(\text{net}) = w_2 = -q_2$, or $-w(\text{net}) = q_2$.

Thus, a net amount of work has been done by the system from heat obtained from a heat reservoir at the temperature of step 2, without at the same time transferring heat from a hot to a cold reservoir. This violates the Kelvin statement of the Second Law. Therefore, the assumption that the two adiabatic reversible paths may intersect is disproven.

Question. May any adiabatic paths intersect, reversible or not?

P4.24 $T = T(p, H)$

$$\mathrm{d}T = \left(\frac{\partial T}{\partial p}\right)_H \mathrm{d}p + \left(\frac{\partial T}{\partial H}\right)_p \mathrm{d}H$$

The Joule–Thomson expansion is a constant-enthalpy process (Section 3.2). Hence,

$$\mathrm{d}T = \left(\frac{\partial T}{\partial p}\right)_H \mathrm{d}p = \mu\,\mathrm{d}p$$

$$\Delta T = \int_{p_\mathrm{i}}^{p_\mathrm{f}} \mu\,\mathrm{d}p = \mu\,\Delta p \quad [\mu \text{ is constant}]$$

$$= (0.21\,\mathrm{K\,atm^{-1}}) \times (1.00\,\mathrm{atm} - 100\,\mathrm{atm}) = \boxed{-21\,\mathrm{K}}$$

$T_\mathrm{f} = T_\mathrm{i} + \Delta T = (373 - 21)\,\mathrm{K} = 352\,\mathrm{K}$ [Mean $T = 363\,\mathrm{K}$]

$S = S(T, p)$

Therefore,

$$\mathrm{d}S = \left(\frac{\partial S}{\partial T}\right)_p \mathrm{d}T + \left(\frac{\partial S}{\partial p}\right)_T \mathrm{d}p$$

$$\left(\frac{\partial S}{\partial T}\right)_p = \frac{C_p}{T} \qquad \left(\frac{\partial S}{\partial p}\right)_T = -\left(\frac{\partial V}{\partial T}\right)_p \qquad \text{[Table 5.1]}$$

For $V_\mathrm{m} = \dfrac{RT}{p}(1+Bp)$

$$\left(\frac{\partial V_\mathrm{m}}{\partial T}\right)_p = \frac{R}{p}(1+Bp)$$

Then

$$\mathrm{d}S_\mathrm{m} = \frac{C_{p,\mathrm{m}}}{T}\,\mathrm{d}T - \frac{R}{p}(1+Bp)\,\mathrm{d}p$$

or

$$\mathrm{d}S_\mathrm{m} = \frac{C_{p,\mathrm{m}}}{T}\,\mathrm{d}T - \frac{R}{p}\mathrm{d}p - RB\,\mathrm{d}p$$

Upon integration

$$\begin{aligned}\Delta S_\mathrm{m} &= \int_{T_1,p_1}^{T_2,p_2} \mathrm{d}S_\mathrm{m} = C_{p,\mathrm{m}} \ln\left(\frac{T_2}{T_1}\right) - R\ln\left(\frac{p_2}{p_1}\right) - RB(p_2-p_1) \\ &= \frac{5}{2}R\ln\left(\frac{352}{373}\right) - R\ln\left(\frac{1}{100}\right) - R\left(-\frac{0.525\ \mathrm{atm}^{-1}}{363}\right)\times(-99\ \mathrm{atm}) \\ &= \boxed{+35.9\ \mathrm{J\,K^{-1}\,mol^{-1}}}\end{aligned}$$

P4.25 The Otto cycle is represented in Fig. 4.4. Assume one mole of air.

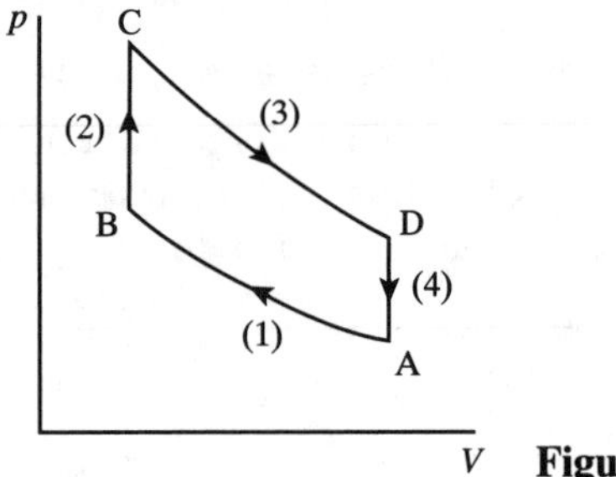

Figure 4.4

$$\varepsilon = \frac{|w|_\text{cycle}}{|q_2|} \quad [9]$$

$$w_\text{cycle} = w_1 + w_3 = \Delta U_1 + \Delta U_3[q_1 = q_3 = 0] = C_V(T_\mathrm{B} - T_\mathrm{A}) + C_V(T_\mathrm{D} - T_\mathrm{C}) \quad [2.32]$$

$$q_2 = \Delta U_2 = C_V(T_\mathrm{C} - T_\mathrm{B})$$

$$\varepsilon = \frac{|T_\mathrm{B} - T_\mathrm{A} + T_\mathrm{D} - T_\mathrm{C}|}{|T_\mathrm{C} - T_\mathrm{B}|} = 1 - \left(\frac{T_\mathrm{D} - T_\mathrm{A}}{T_\mathrm{C} - T_\mathrm{B}}\right)$$

We know that

$$\frac{T_\mathrm{A}}{T_\mathrm{B}} = \left(\frac{V_\mathrm{B}}{V_\mathrm{A}}\right)^{1/c} \quad \text{and} \quad \frac{T_\mathrm{D}}{T_\mathrm{C}} = \left(\frac{V_\mathrm{C}}{V_\mathrm{D}}\right)^{1/c} \quad [2.34]$$

and since $V_B = V_C$ and $V_A = V_D$, $\frac{T_A}{T_B} = \frac{T_D}{T_C}$, or $T_D = \frac{T_A T_C}{T_B}$

Then $\varepsilon = 1 - \frac{\frac{T_A T_C}{T_B} - T_A}{T_C - T_B} = 1 - \frac{T_A}{T_B}$ or $\boxed{\varepsilon = 1 - \left(\frac{V_B}{V_A}\right)^{1/c}}$

Assume $C_{V,m}(\text{air}) = \frac{5}{2}R$, then $c = \frac{2}{5}$.

For $\frac{V_A}{V_B} = 10$, $\varepsilon = 1 - \left(\frac{1}{10}\right)^{2/5} = \boxed{0.47}$

$$\Delta S_1 = \Delta S_3 = \Delta S_{\text{sur},1} = \Delta S_{\text{sur},3} = \boxed{0} \quad \text{[adiabatic reversible steps]}$$

$$\Delta S_2 = C_{V,m} \ln\left(\frac{T_C}{T_B}\right)$$

At constant volume $\left(\frac{T_C}{T_B}\right) = \left(\frac{p_C}{p_B}\right) = 5.0$

$$\Delta S_2 = \left(\tfrac{5}{2}\right) \times (8.314\,\text{J K}^{-1}\,\text{mol}^{-1}) \times (\ln 5.0) = \boxed{+33\,\text{J K}^{-1}}$$

$$\Delta S_{\text{sur},2} = -\Delta S_2 = \boxed{-33\,\text{J K}^{-1}}$$

$$\Delta S_4 = -\Delta S_2\left[\frac{T_C}{T_D} = \frac{T_B}{T_A}\right] = \boxed{-33\,\text{J K}^{-1}}$$

$$\Delta S_{\text{sur},4} = -\Delta S_4 = \boxed{+33\,\text{J K}^{-1}}$$

P4.26 The efficiency of any reversible engine in which the working substance is a perfect gas is given by

$$\varepsilon_{\text{rev}} = \frac{|w|}{q_h}[9] = 1 - \frac{T_c}{T_h} = 1 + \frac{q_{c,\min}}{q_h} \text{ [8, with } q_c = q_{c,\min}]$$

Therefore, for a perfect gas [*Justification* 4.1]

$$-\frac{q_{c,\min}}{q_h} = \frac{T_c}{T_h} \quad \text{or} \quad -\frac{q_{T,\min}}{q_h} = \frac{T}{T_h} \quad \text{and} \quad T = -\frac{q_{T,\min}}{q_h} \times T_h$$

But for a reversible engine employing any working substance (including a perfect gas)

$$-\frac{q_{c,\min}}{q_h} = \frac{T_c^a}{T_h^a} \quad \text{[8, Section 4.2(b)]},$$

where the symbol T^a is used to indicate the absolute temperature based on the Second Law. Thus,

$$T^a = -\frac{q_{T,\min}}{q_h} \times T_h^a$$

Since $q_{T,\min}$ and q_h are experimentally measured heats, and are the same no matter what temperature scale is employed

$$\frac{T}{T_h} = \frac{T^a}{T_h^a}$$

Thus T and T^a differ from each other by at most a constant numerical factor which becomes 1 if T_h and T_h^a are both assigned the same value, say 273.16 at the triple point of water.

Solutions to additional problems

P4.28 The groups in C_2H_5 are (C–C)(H)$_2$ and C–(C)(H)$_3$, so

$$S^{\ominus}_{\text{int}} = (135.9 + 126.8)\ \text{J K}^{-1}\ \text{mol}^{-1} = 262.7\ \text{J K}^{-1}\ \text{mol}^{-1},$$

and

$$S^{\ominus} = S^{\ominus}_{\text{int}} - R \ln \sigma = (262.7 - 8.3145 \ln 6)\ \text{J K}^{-1}\ \text{mol}^{-1}$$
$$= \boxed{247.8\ \text{J K}^{-1}\ \text{mol}^{-1}}$$

The groups in *sec*-C_4H_9 are C–(C)(H)$_3$, C–(C)(C)(H)$_2$, C–(C)$_2$(H) and C–(C)(H)$_3$, so

$$S^{\ominus}_{\text{int}} = (126.8 + 42.0 + 59.3 + 126.8)\ \text{J K}^{-1}\ \text{mol}^{-1} = 354.9\ \text{J K}^{-1}\ \text{mol}^{-1},$$

and

$$S^{\ominus} = S^{\ominus}_{\text{int}} - R \ln \sigma = (354.9 - 8.3145 \ln 9)\ \text{J K}^{-1}\ \text{mol}^{-1}$$
$$= \boxed{336.6\ \text{J K}^{-1}\ \text{mol}^{-1}}$$

The groups in *tert*-C_4H_9 are (C)–(C)$_3$ and 3C–(C)(H)$_3$, so

$$S^{\ominus}_{\text{int}} = [-29.2 + 3(126.8)]\ \text{J K}^{-1}\ \text{mol}^{-1} = 351.2\ \text{J K}^{-1}\ \text{mol}^{-1},$$

and

$$S^{\ominus} = S^{\ominus}_{\text{int}} - R \ln \sigma = (351.2 - 8.3145 \ln 81)\ \text{J K}^{-1}\ \text{mol}^{-1}$$
$$= \boxed{314.7\ \text{J K}^{-1}\ \text{mol}^{-1}}$$

P4.30 The entropy at 200 K is calculated from

$$S^{\ominus}_{\text{m},200} = S^{\ominus}_{\text{m},100} + \int_{100\ \text{K}}^{200\ \text{K}} \frac{C_{p,\text{m}}\, \text{d}T}{T}$$

The integrand may be evaluated for each of the data points, and the numerical integration carried out by a standard procedure such as the trapezoid rule (taking the integral within any interval as the mean value of the integrand times the length of the interval). Programs for performing this integration are readily available for personal computers. Many graphing calculators will also perform this numerical integration.

The transformed data appear below.

T/K	100	120	140	150	160	180	200
$C_{p,\text{m}}/(\text{J K}^{-1}\ \text{mol}^{-1})$	23.00	23.74	24.25	24.44	24.61	24.89	25.11
$\dfrac{C_{p,\text{m}}}{T}/(\text{J K}^{-2}\ \text{mol}^{-1})$	0.230	0.1978	0.1732	0.1629	0.1538	0.1383	0.1256

Integration by the trapezoid rule yields

$$S^{\ominus}_{\text{m},200} = (29.79 + 16.81)\ \text{J K}^{-1}\ \text{mol}^{-1} = \boxed{46.60\ \text{J K}^{-1}\ \text{mol}^{-1}}$$

Taking $C_{p,\text{m}}$ constant yields

$$S^{\ominus}_{\text{m},200} = S^{\ominus}_{\text{m},100} + C_{p,\text{m}} \ln(200\ \text{K}/100\ \text{K})$$
$$= [29.79 + 24.44 \ln(200\ \text{K}/100\ \text{K})]\ \text{J K}^{-1}\ \text{mol}^{-1} = \boxed{46.73\ \text{J K}^{-1}\ \text{mol}^{-1}}$$

The difference is slight.

P4.32 Polytropic process: $pV^n = C = \text{constant}$.

If $n = 0$, then $pV^n = pV^0 = p = \text{constant}$ and the process is isobaric.

If $n = 1$, then $pV^n = pV = \text{constant}$. But the perfect gas equation of state says that $pV = nRT$ so we note that both of these equations are correct provided that $T = \text{constant}$. The process is isothermal (Fig. 4.5(a)).

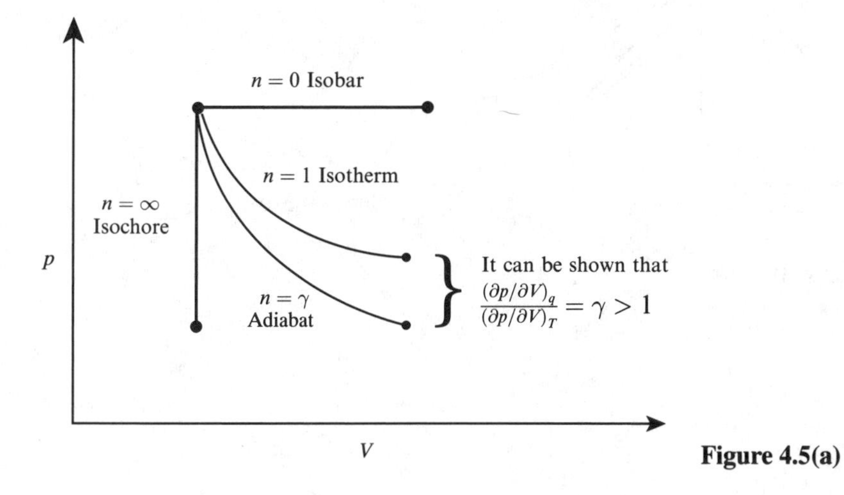

Figure 4.5(a)

If $n = \gamma$, then $pV^n = pV^\gamma = C$ and the process is adiabatic.

In the limit as $n \to \infty$ we need to examine the process equation in the form $V = \left(\frac{C}{p}\right)^{1/n}$. Then, $\lim_{n\to\infty} V = \left(\frac{C}{p}\right)^0 = \text{constant}$ and we find that the process is isochoric (Fig. 4.5(a)).

Consider the functional behaviour of $\partial V/\partial p$

$$\frac{\partial V}{\partial p} = \frac{\partial}{\partial p}\left(\frac{C}{p}\right)^{1/n} = -\frac{C^{1/n}p^{-\left(\frac{1}{n}+1\right)}}{n}$$

In all cases $p > 0$, $V > 0$, $C > 0$ and $p^{-\left(\frac{1}{n}+1\right)} > 0$.

Therefore we may write

$$\frac{\partial V}{\partial p} = -\frac{1}{n}\left|C^{1/n}p^{-\left(\frac{1}{n}+1\right)}\right|$$

and, in the special case for which $n < 0$, this becomes

$$\frac{\partial V}{\partial p} = \left|\frac{1}{n}C^{1/n}p^{-\left(\frac{1}{n}+1\right)}\right| \qquad n < 0$$

This equation *cannot* be a description of a real substance because it says that, as pressure increases, volume increases.

Nevertheless, we can construct plots of p against V and T against S for these physically unrealizable cases (Fig. 4.5(b)).

To determine the qualitative behaviour of $S(T)$ for each process, consider the following observations.

Adiabatic, reversible process [2]

$$\mathrm{d}S = \mathrm{d}q/T = 0 \quad \text{or} \quad S = \text{constant}$$

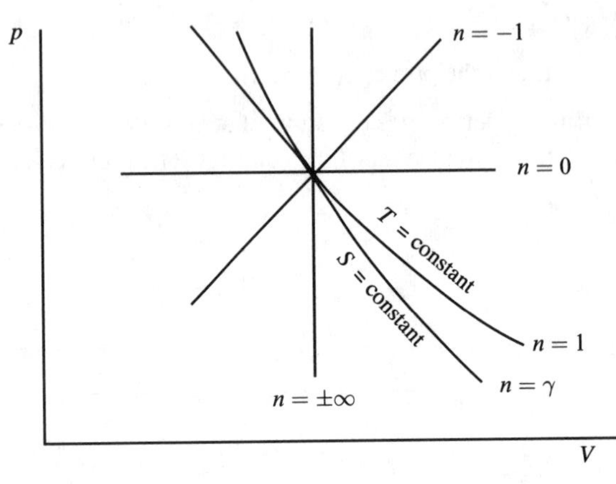

Figure 4.5(b)

Isobaric process [19]

$$\left(\frac{\partial S}{\partial T}\right)_p = \frac{C_p}{T} > 0 \quad \text{so} \quad \left(\frac{\partial S}{\partial T}\right)_p \sim \frac{1}{T}$$

Isochoric process

$$\mathrm{d}q_V = C_V\,\mathrm{d}T$$

$$\mathrm{d}S = \frac{\mathrm{d}q_V}{T} = \frac{C_V\,\mathrm{d}T}{T}$$

$$\left(\frac{\partial S}{\partial T}\right)_V = \frac{C_V}{T} > 0 \quad \text{so} \quad \left(\frac{\partial S}{\partial T}\right)_V \sim \frac{1}{T}$$

Because $C_p > C_V$ [3.22], $\left(\frac{\partial S}{\partial T}\right)_p > \left(\frac{\partial S}{\partial T}\right)_V$

Plots of T against S for all values of n, even the physically unrealizable cases are shown in Fig. 4.5(c).

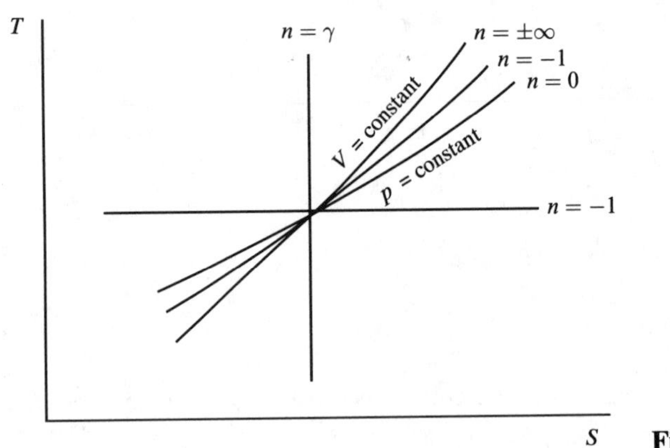

Figure 4.5(c)

5 The Second Law: the machinery

Solutions to exercises

Assume all gases are perfect and that the temperature is 298 K unless stated otherwise.

E5.1

$$\alpha = \left(\frac{1}{V}\right)\left(\frac{\partial V}{\partial T}\right)_p \; [3.7]; \qquad \kappa_T = -\left(\frac{1}{V}\right) \times \left(\frac{\partial V}{\partial p}\right)_T \; [3.13]$$

$$\left(\frac{\partial S}{\partial V}\right)_T = \left(\frac{\partial p}{\partial T}\right)_V = -\left(\frac{\partial V}{\partial T}\right)_p\left(\frac{\partial p}{\partial V}\right)_T \text{ [Relation no. 3, \textit{Further information} 1]}$$

$$= -\frac{\left(\frac{\partial V}{\partial T}\right)_p}{\left(\frac{\partial V}{\partial p}\right)_T} \text{ [Relation no. 2, \textit{Further information} 1]}$$

$$= -\frac{\left(\frac{1}{V}\right)\left(\frac{\partial V}{\partial T}\right)_p}{\left(\frac{1}{V}\right)\left(\frac{\partial V}{\partial p}\right)_T} = \boxed{+\frac{\alpha}{\kappa_T}}$$

E5.2

$$\Delta G = nRT\ln\left(\frac{p_f}{p_i}\right) \text{ [Example 5.2]} = nRT\ln\left(\frac{V_i}{V_f}\right) \text{ [Boyle's law]}$$

$$= (3.0\times10^{-3}\,\text{mol})\times(8.314\,\text{J K}^{-1}\,\text{mol}^{-1})\times(300\,\text{K})\times\ln\left(\frac{36}{60}\right) = \boxed{-3.8\,\text{J}}$$

E5.3

$$\Delta G = G_f - G_i$$

$$\left(\frac{\partial G}{\partial T}\right)_p = -S \text{ [10]; \quad hence} \quad \left(\frac{\partial G_f}{\partial T}\right)_p = -S_f, \quad \text{and} \quad \left(\frac{\partial G_i}{\partial T}\right)_p = -S_i$$

$$\Delta S = S_f - S_i = -\left(\frac{\partial G_f}{\partial T}\right)_p + \left(\frac{\partial G_i}{\partial T}\right)_p = -\left(\frac{\partial(G_f - G_i)}{\partial T}\right)_p$$

$$= -\left(\frac{\partial \Delta G}{\partial T}\right)_p = -\frac{\partial}{\partial T}\left(-85.40\,\text{J} + 36.5\,\text{J}\times\frac{T}{\text{K}}\right)$$

$$= \boxed{-36.5\,\text{J K}^{-1}}$$

E5.4 $dG = -S\,dT + V\,dp$ [9] at constant T, $dG = V\,dp$; therefore

$$\Delta G = \int_{p_i}^{p_f} V\,dp$$

$$V = V_1(1-\kappa_T p) \quad V_1 = V(1\,\text{atm}) = \frac{m}{\rho} \quad \kappa_T = 76.8\times10^{-6}\,\text{atm}^{-1} \quad \text{[Table 3.1]}$$

Then,

$$V = \frac{m}{p}(1 - 7.68\times10^{-5}\,\text{atm}^{-1}p)$$

$$\Delta G = \int_{1\,\text{atm}}^{3000\,\text{atm}} \frac{m}{\rho}(1 - 7.68\times10^{-5}\,\text{atm}^{-1}p)\,dp$$

$$= \int_{1\,\text{atm}}^{3000\,\text{atm}} \frac{m}{\rho}\,dp - \int_{1\,\text{atm}}^{3000\,\text{atm}} \frac{m}{\rho}\times(7.68\times10^{-5}\,\text{atm}^{-1})p\,dp$$

$$= \frac{m}{\rho} \times 2999\,\text{atm} - (7.68 \times 10^{-5}\,\text{atm}^{-1}) \times \frac{m}{\rho} \times (9.00 \times 10^6\,\text{atm}^3)$$

$$= \frac{m}{\rho}(2999\,\text{atm} - 691\,\text{atm}) = \frac{35\,\text{g}}{0.789\,\text{g cm}^{-3}} \times 2308\,\text{atm}$$

$$= 44.\bar{4}\,\text{cm}^3 \times \frac{10^{-6}\,\text{m}^3}{\text{cm}^3} \times 2308\text{atm} \times (1.013 \times 10^{-5}\,\text{Pa atm}^{-1})$$

$$= 10.\bar{4}\,\text{kJ} = \boxed{10\,\text{kJ}}$$

E5.5 **(a)** $\Delta S = nR\ln\left(\frac{V_f}{V_i}\right)$ [4.17] $= nR\ln\left(\frac{p_i}{p_f}\right)$ [Boyle's law]

Taking inverse logarithms

$$p_f = p_i e^{-\Delta S/nR} = (3.50\,\text{atm}) \times e^{-(25.0\,\text{J K}^{-1})/(2.00\times 8.314\,\text{J K}^{-1}\,\text{mol}^{-1})}$$

$$= (3.50\,\text{atm}) \times e^{1.50} = \boxed{15.7\,\text{atm}}$$

(b) $\Delta G = nRT\ln\left(\frac{p_f}{p_i}\right)$ [Example 5.2]

$$= -T\Delta S\ [\Delta H = 0,\ \text{constant temperature, perfect gas}]$$

$$= (-330\,\text{K}) \times (-25.0\,\text{J K}^{-1}) = \boxed{+8.25\,\text{kJ}}$$

E5.6 $\Delta\mu = \mu_f - \mu_i = RT\ln\left(\frac{p_f}{p_i}\right)$ [20] $= (8.314\,\text{J K}^{-1}\,\text{mol}^{-1}) \times (313\,\text{K}) \times \ln\left(\frac{29.5}{1.8}\right)$

$$= \boxed{+7.3\,\text{kJ mol}^{-1}}$$

E5.7 $\mu^0 = \mu^\ominus + RT\ln\left(\frac{p}{p^\ominus}\right)$ [20 with $\mu = \mu^0$]

$$\mu = \mu^\ominus + RT\ln\left(\frac{f}{p^\ominus}\right)\ [21]$$

$$\mu - \mu^0 = RT\ln\frac{f}{p}\,[21\text{ minus }20]; \quad \frac{f}{p} = \phi$$

$$= RT\ln\phi = (8.314\,\text{J K}^{-1}\,\text{mol}^{-1}) \times (200\,\text{K}) \times (\ln 0.72) = \boxed{-0.55\,\text{kJ mol}^{-1}}$$

E5.8 $B' = \frac{B}{RT}$ [Problem 1.30] $= \frac{-81.7 \times 10^{-6}\,\text{m}^3\,\text{mol}^{-1}}{(8.314\,\text{J K}^{-1}\,\text{mol}^{-1}) \times (373\,\text{K})} = \boxed{-2.63 \times 10^{-8}\,\text{Pa}^{-1}}$

$$\phi = e^{B'p+\cdots}\ [25]$$

$$= e^{(-2.63\times10^{-8}\,\text{Pa}^{-1})\times(50)\times(1.013\times10^5\,\text{Pa})}\quad [\text{truncating series after term in } B']$$

$$= e^{-0.13\bar{3}} = \boxed{0.88}$$

E5.9 $\Delta G = nV_m\Delta p[15] = V\Delta p = (1.0 \times 10^{-3}\,\text{m}^3) \times (99) \times (1.013 \times 10^5\,\text{Pa})$

$$= 10\,\text{kPa m}^3 = \boxed{+10\,\text{kJ}}$$

E5.10 $\Delta G_m = RT\ln\frac{p_f}{p_i}[16] = (8.314\,\text{J K}^{-1}\,\text{mol}^{-1}) \times (298\,\text{K}) \times \ln\left(\frac{100.0}{1.0}\right) = \boxed{+11\,\text{kJ mol}^{-1}}$

E5.11 An equation of state is a functional relationship between the state properties, p, V_m, and T. From the definition

$$A \equiv U - TS \qquad dA = dU - T\,dS - S\,dT$$

Using [2], $dA = -S\,dT - p\,dV_m$; hence

$$p = -\left(\frac{\partial A}{\partial V_m}\right)_T = -\frac{a}{V_m^2} + RT \times \left(\frac{1}{V_m - b}\right) = \boxed{\frac{RT}{V_m - b} - \frac{a}{V_m^2}}$$

which is the van der Waals equation.

E5.12 $\left(\frac{\partial S}{\partial V}\right)_T = \left(\frac{\partial p}{\partial T}\right)_V$ [Table 5.1]

For a van der Waals gas

$$p = \frac{nRT}{V - nb} - \frac{n^2 a}{V^2}$$

Hence, $\left(\frac{\partial S}{\partial V}\right)_T = \left(\frac{\partial p}{\partial T}\right)_V = \boxed{\frac{nR}{V - nb}}$

$$dS = \left(\frac{\partial S}{\partial V}\right)_T dV[\text{constant temperature}] = \left(\frac{\partial p}{\partial T}\right)_V dV = \frac{nR}{V - nb} dV$$

$$\Delta S = \int_{V_i}^{V_f} dS = \int_{V_i}^{V_f} \frac{nR}{V - nb} dV = \boxed{nR \ln\left(\frac{V_f - nb}{V_i - nb}\right)}$$

For a perfect gas $\Delta S = nR \ln\left(\frac{V_f}{V_i}\right)$

$$\frac{V_f - nb}{V_i - nb} > \frac{V_f}{V_i},$$

Therefore, ΔS will be greater for a van der Waals gas.

Solutions to problems

Solutions to numerical problems

P5.1 The Gibbs–Helmholtz equation [11] may be recast into an analogous equation involving ΔG and ΔH, since

$$\left(\frac{\partial \Delta G}{\partial T}\right)_p = \left(\frac{\partial G_f}{\partial T}\right)_p - \left(\frac{\partial G_i}{\partial T}\right)_p$$

and $\Delta H = H_f - H_i$

Thus, $\left(\frac{\partial}{\partial T} \frac{\Delta_r G^\ominus}{T}\right)_p = -\frac{\Delta_r H^\ominus}{T^2}$

$$d\left(\frac{\Delta_r G^\ominus}{T}\right) = \left(\frac{\partial}{\partial T} \frac{\Delta_r G^\ominus}{T}\right)_p dT[\text{constant pressure}] = -\frac{\Delta_r H^\ominus}{T^2} dT$$

$$\Delta\left(\frac{\Delta_r G^\ominus}{T}\right) = -\int_{T_c}^{T} \frac{\Delta_r H^\ominus\, dT}{T^2}$$

$$\approx -\Delta_r H^\ominus \int_{T_c}^{T} \frac{dT}{T^2} = \Delta_r H^\ominus \left(\frac{1}{T} - \frac{1}{T_c}\right) \quad [\Delta_r H^\ominus \text{ assumed constant}]$$

Therefore, $\dfrac{\Delta_r G^{\ominus}(T)}{T} - \dfrac{\Delta_r G^{\ominus}(T_c)}{T_c} \approx \Delta_r H^{\ominus}\left(\dfrac{1}{T} - \dfrac{1}{T_c}\right)$

and so

$$\Delta_r G^{\ominus}(T) = \frac{T}{T_c}\Delta_r G^{\ominus}(T_c) + \left(1 - \frac{T}{T_c}\right)\Delta_r H^{\ominus}(T_c)$$

$$= \tau\Delta_r G^{\ominus}(T_c) + (1-\tau)\Delta_r H^{\ominus}(T_c) \quad \tau = \frac{T}{T_c}$$

For the reaction

$$2CO(g) + O_2(g) \rightarrow 2CO_2(g)$$

$$\Delta_r G^{\ominus}(T_c) = 2\Delta_f G^{\ominus}(CO_2,\ g) - 2\Delta_f G^{\ominus}(CO,\ g)$$

$$= [2 \times (-394.36) - 2 \times (-137.17)]\,\mathrm{kJ\,mol^{-1}} = -514.38\,\mathrm{kJ\,mol^{-1}}$$

$$\Delta_r H^{\ominus}(T_c) = 2\Delta_f H^{\ominus}(CO_2,\ g) - 2\Delta_f H^{\ominus}(CO,\ g)$$

$$= [2 \times (-393.51) - 2 \times (-110.53)]\,\mathrm{kJ\,mol^{-1}} = -565.96\,\mathrm{kJ\,mol^{-1}}$$

Therefore, since $\tau = \dfrac{375}{298.15} = 1.25\bar{8}$

$$\Delta_r G^{\ominus}(375\,\mathrm{K}) = \{(1.25\bar{8}) \times (-514.38) + (1 - 1.25\bar{8}) \times (-565.96)\}\,\mathrm{kJ\,mol^{-1}}$$

$$= \boxed{-501\,\mathrm{kJ\,mol^{-1}}}$$

P5.4 A graphical integration of $\ln\phi = \displaystyle\int_0^p \left(\frac{Z-1}{p}\right)\mathrm{d}p$ [24] is performed. We draw up the following table

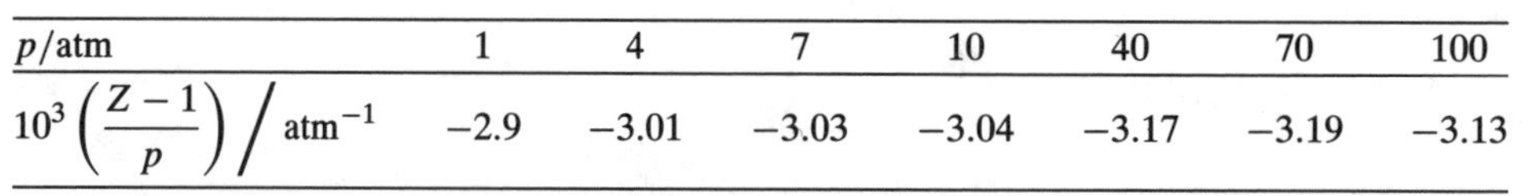

p/atm	1	4	7	10	40	70	100
$10^3\left(\dfrac{Z-1}{p}\right)\Big/\mathrm{atm^{-1}}$	−2.9	−3.01	−3.03	−3.04	−3.17	−3.19	−3.13

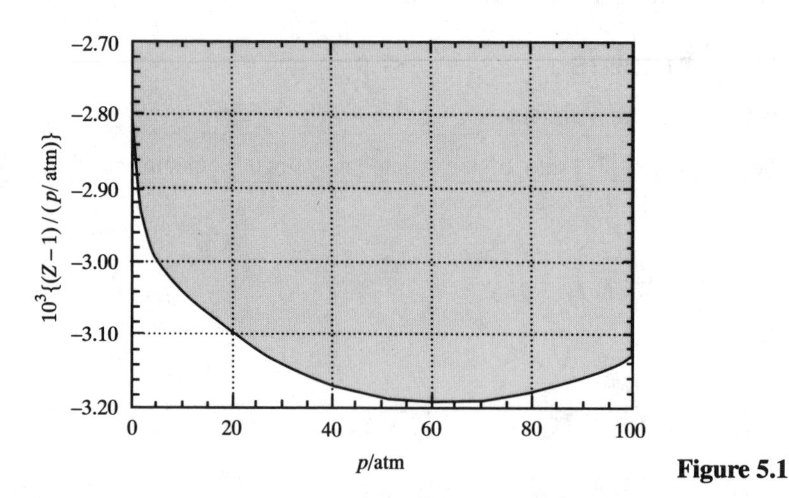

Figure 5.1

The points are plotted in Fig. 5.1. The integral is the shaded area which has the value −0.313, so at 100 atm

$$\phi = e^{-0.313} = 0.73$$

and the fugacity of oxygen is 100 atm × 0.73 = $\boxed{73\ \mathrm{atm}}$

Solutions to theoretical problems

P5.6 $H \equiv U + pV$

$$\mathrm{d}H = \mathrm{d}U + p\,\mathrm{d}V + V\,\mathrm{d}p = T\,\mathrm{d}S - p\,\mathrm{d}V\ [2] + p\,\mathrm{d}V + V\,\mathrm{d}p = T\,\mathrm{d}S + V\,\mathrm{d}p$$

Since H is a state function, $\mathrm{d}H$ is exact, and it follows that

$$\left(\frac{\partial H}{\partial S}\right)_p = T \quad \text{and} \quad \boxed{\left(\frac{\partial V}{\partial S}\right)_p = \left(\frac{\partial T}{\partial p}\right)_S}$$

Similarly, $A \equiv U - TS$

$$\mathrm{d}A = \mathrm{d}U - T\,\mathrm{d}S - S\mathrm{d}T = T\,\mathrm{d}S - p\,\mathrm{d}V\ [2] - T\,\mathrm{d}S - S\mathrm{d}T = -p\,\mathrm{d}V - S\,\mathrm{d}T$$

Since $\mathrm{d}A$ is exact,

$$\boxed{\left(\frac{\partial S}{\partial V}\right)_T = \left(\frac{\partial p}{\partial T}\right)_V}$$

P5.7 $\left(\frac{\partial S}{\partial V}\right)_T = \left(\frac{\partial p}{\partial T}\right)_V = \boxed{\frac{\alpha}{\kappa_T}}$ [Exercise 5.1(**a**)]

$$\left(\frac{\partial V}{\partial S}\right)_p = \left(\frac{\partial T}{\partial p}\right)_S$$

$$\left(\frac{\partial T}{\partial p}\right)_S = -\left(\frac{\partial T}{\partial S}\right)_p\left(\frac{\partial S}{\partial p}\right)_T \text{ [permuter]} = -\frac{\left(\frac{\partial S}{\partial p}\right)_T}{\left(\frac{\partial S}{\partial T}\right)_p} \text{ [inversion]}$$

$$\left(\frac{\partial S}{\partial p}\right)_T = -\left(\frac{\partial V}{\partial T}\right)_p \quad \text{[Maxwell relation]}$$
$$= -\alpha V$$

at constant p

$$\mathrm{d}S = \left(\frac{\partial S}{\partial T}\right)_p \mathrm{d}T$$

and

$$\mathrm{d}S = \frac{\mathrm{d}q_{\text{rev}}}{T} = \frac{\mathrm{d}H}{T} = \frac{C_p\,\mathrm{d}T}{T} \quad [\mathrm{d}q_p = \mathrm{d}H]$$

Therefore, $\left(\frac{\partial S}{\partial T}\right)_p = \frac{C_p}{T}$

and $\left(\frac{\partial V}{\partial S}\right)_p = \boxed{\frac{\alpha T V}{C_p}}$

P5.9 $\left(\frac{\partial S}{\partial V}\right)_T = \left(\frac{\partial p}{\partial T}\right)_V$ [Maxwell relation]; $\quad \left(\frac{\partial p}{\partial T}\right)_V = \left(\frac{\partial}{\partial T}\left(\frac{nRT}{V}\right)\right)_V = \frac{nR}{V}$

$$\mathrm{d}S = \left(\frac{\partial S}{\partial V}\right)_T \mathrm{d}V \text{ [constant temperature]} = nR\frac{\mathrm{d}V}{V} = nR\,\mathrm{d}\ln V$$

$$S = \int \mathrm{d}S = \int nR\,\mathrm{d}\ln V$$

$$S = nR\ln V + \text{constant} \quad \text{or} \quad S \propto R\ln V$$

P5.11 Start from the relation

$$dH = T\,dS + V\,dp \text{ [Problem 5.6]}$$

Divide by dV at constant T, which gives

$$\left(\frac{\partial H}{\partial V}\right)_T = T\left(\frac{\partial S}{\partial V}\right)_T + V\left(\frac{\partial p}{\partial V}\right)_T = T\left(\frac{\partial p}{\partial T}\right)_V \text{ [Maxwell relation]} + V\left(\frac{\partial p}{\partial V}\right)_T$$

Inserting $\left(\frac{\partial p}{\partial V}\right)_T = -\frac{\left(\frac{\partial T}{\partial V}\right)_p}{\left(\frac{\partial T}{\partial p}\right)_V}$ [permuter followed by inversion]

yields $\left(\frac{\partial H}{\partial V}\right)_V = \left[T - V\left(\frac{\partial T}{\partial V}\right)_p\right]\left(\frac{\partial p}{\partial T}\right)_V$

Now note that $-(V^2)\left(\frac{\partial p}{\partial T}\right)_V\left(\frac{\partial(T/V)}{\partial V}\right)_p = -(V^2)\left(\frac{\partial p}{\partial T}\right)_V$

$$\times\left[\left(\frac{1}{V}\right)\times\left(\frac{\partial T}{\partial V}\right)_p - \left(\frac{T}{V^2}\right)\right]$$

$$= \left[T - V\left(\frac{\partial T}{\partial V}\right)_p\right]\left(\frac{\partial p}{\partial T}\right)_V = \left(\frac{\partial H}{\partial V}\right)_T$$

which is the relation to be proved.

P5.14 $\mu_J = \left(\frac{\partial T}{\partial V}\right)_U \qquad C_V = \left(\frac{\partial U}{\partial T}\right)_V$

$$\mu_J C_V = \left(\frac{\partial T}{\partial V}\right)_U\left(\frac{\partial U}{\partial T}\right)_V = \frac{-1}{\left(\frac{\partial V}{\partial U}\right)_T} \text{ [chain relation]}$$

$$= -\left(\frac{\partial U}{\partial V}\right)_T \text{ [inversion]} = p - T\left(\frac{\partial p}{\partial T}\right)_V \text{ [8]}$$

$$\left(\frac{\partial p}{\partial T}\right)_V = \frac{-1}{\left(\frac{\partial T}{\partial V}\right)_p\left(\frac{\partial V}{\partial p}\right)_T} \text{ [chain relation]} = \frac{-\left(\frac{\partial V}{\partial T}\right)_p}{\left(\frac{\partial V}{\partial p}\right)_T} = \frac{\alpha}{\kappa_T}$$

Therefore, $\boxed{\mu_J C_V = p - \frac{\alpha T}{\kappa_T}}$

P5.17 $\left(\frac{\partial}{\partial T}\left(\frac{\Delta_r G}{T}\right)\right)_p = \frac{-\Delta_r H}{T^2}$ [11 and Problem 5.1]

(a) $\int d\left(\frac{\Delta_r G}{T}\right) = -\int \frac{\Delta_r H\,dT}{T^2} \approx -\Delta_r H\int\frac{dT}{T^2}$ [$\Delta_r H$ constant]

$$\frac{\Delta_r G'}{T'} - \frac{\Delta_r G}{T} = \Delta_r H\left(\frac{1}{T'} - \frac{1}{T}\right)$$

$$\Delta_r G' = \frac{T'}{T}\Delta_r G + \left(1 - \frac{T'}{T}\right)\Delta_r H$$

$$= \boxed{\tau\Delta_r G + (1-\tau)\Delta_r H} \text{ with } \tau = \frac{T'}{T} \text{ [Problem 5.1]}$$

(b) $\Delta_r H(T'') = \Delta_r H(T) + (T'' - T)\Delta_r C_p$ [given, T'' is the variable]

$$\frac{\Delta_r G'}{T'} - \frac{\Delta_r G}{T} = -\Delta_r H \int_T^{T'} \frac{dT''}{T''^2} - \Delta_r C_p \int_T^{T'} \frac{(T'' - T)\,dT''}{T''^2}$$
$$= \left(\frac{1}{T'} - \frac{1}{T}\right)\Delta_r H - \Delta_r C_p \ln\frac{T'}{T} - T\Delta_r C_p\left(\frac{1}{T'} - \frac{1}{T}\right)$$

Therefore, with $\tau = \dfrac{T'}{T}$

$$\Delta_r G' = \tau\Delta_r G + (1-\tau)\Delta_r H - T'\Delta_r C_p \ln\tau - T\Delta_r C_p(1-\tau)$$
$$= \boxed{\tau\Delta_r G + (1-\tau)(\Delta_r H - T\Delta_r C_p) - T'\Delta_r C_p \ln\tau}$$

P5.19 $S = S(T, V)$

$$dS = \left(\frac{\partial S}{\partial T}\right)_V dT + \left(\frac{\partial S}{\partial V}\right)_T dV$$

$$T\,dS = T\left(\frac{\partial S}{\partial T}\right)_V dT + T\left(\frac{\partial S}{\partial V}\right)_T dV$$

Now, $\left(\frac{\partial S}{\partial T}\right)_V = \left(\frac{\partial S}{\partial U}\right)_V \left(\frac{\partial U}{\partial T}\right)_V = \frac{1}{T} \times C_V$ [4]

$$\left(\frac{\partial S}{\partial V}\right)_T = \left(\frac{\partial p}{\partial T}\right)_V \quad \text{[Maxwell relation]}$$

Hence, $\boxed{T\,dS = C_V\,dT + T\left(\frac{\partial p}{\partial T}\right)_V dV}$

For a reversible, isothermal expansion, $T\,dS = dq_{rev}$; therefore

$$dq_{rev} = T\left(\frac{\partial p}{\partial T}\right)_V dV = \frac{nRT}{V - nb}\,dV$$

$$q_{rev} = nRT \int_{V_i}^{V_f} \frac{dV}{V - nb} = \boxed{nRT \ln\left(\frac{V_f - nb}{V_i - nb}\right)}$$

P5.21 $G' = G + \int_0^p V\,dp = G + V_0 \int_0^p e^{-p/p^*}\,dp$ [V_0 is a constant] $= \boxed{G + p^* V_0(1 - e^{-p/p^*})}$

$$\Delta G = p^* V_0(1 - e^{-p/p^*})$$

Since $e^{-p/p^*} < 1$ if $p > 0$, ΔG is positive. When the pressure is reduced to zero from a positive value, ΔG decreases to zero. Under constant-temperature conditions (p and V not constant) it is ΔA that determines the direction of natural change.

$$dA = -S\,dT - p\,dV \text{ [Exercise 5.11(a)]} = -p\,dV \quad \text{[constant temperature]}$$

Since $dV = -\dfrac{V_0}{p^*} e^{-p/p^*}\,dp$, it is clear that $\boxed{V \text{ increases}}$ when pressure is relaxed [$dV > 0$ when $dp < 0$]. Substituting for dV, $dA = \dfrac{p}{p^*} V_0 e^{-p/p^*}\,dp$, so a decrease in p is spontaneous, as ΔA will be negative.

P5.22 $\ln\phi = \int_0^p \left(\frac{Z-1}{p}\right) \mathrm{d}p$ [24]

$$Z = 1 + \frac{B}{V_\mathrm{m}} + \frac{C}{{V_\mathrm{m}}^2} = 1 + B'p + C'p^2 + \cdots$$

with $B' = \frac{B}{RT}$, $C' = \frac{C - B^2}{R^2T^2}$ [Problem 1.30]

$$\frac{Z-1}{p} = B' + C'p + \cdots$$

Therefore, $\ln\phi = \int_0^p B'\,\mathrm{d}p + \int_0^p C'p\,\mathrm{d}p + \cdots = B'p + \frac{1}{2}C'p^2 + \cdots = \boxed{\frac{Bp}{RT} + \frac{(C-B^2)p^2}{2R^2T^2} + \cdots}$

For argon, $\frac{Bp}{RT} = \frac{(-21.13 \times 10^{-3}\,\mathrm{L\,mol^{-1}}) \times (1.00\,\mathrm{atm})}{(8.206 \times 10^{-2}\,\mathrm{L\,atm\,K^{-1}\,mol^{-1}}) \times (273\,\mathrm{K})} = -9.43 \times 10^{-4}$

$$\frac{(C-B^2)p^2}{2R^2T^2} = \frac{\{(1.054 \times 10^{-3}\,\mathrm{L^2\,mol^{-2}}) - (-21.13 \times 10^{-3}\,\mathrm{L\,mol^{-1}})^2\} \times (1.00\,\mathrm{atm})^2}{(2) \times \{(8.206 \times 10^{-2}\,\mathrm{L\,atm\,K^{-1}\,mol^{-1}}) \times (273\,\mathrm{K})\}^2}$$
$$= 6.05 \times 10^{-7}$$

Therefore, $\ln\phi = (-9.43 \times 10^{-4}) + (6.05 \times 10^{-7}) = -9.42 \times 10^{-4}$; $\phi = 0.9991$

Hence, $f = (1.00\,\mathrm{atm}) \times (0.9991) = \boxed{0.999\overline{1}\,\mathrm{atm}}$

Solutions to additional problems

P5.24 The relative increase in water vapour in the atmosphere at constant relative humidity is the same as the relative increase in the equilibrium vapour pressure of water. Examination of the chemical potential will help us estimate this increase. At equilibrium, the vapour and liquid have the same chemical potential. So at the current temperature

$$\mu_\mathrm{liq}(T_0) = \mu_\mathrm{vap}(T_0) \quad \text{so} \quad \mu^\ominus_\mathrm{liq}(T_0) = \mu^\ominus_\mathrm{vap}(T_0) + RT_0 \ln p_0,$$

where the subscript 0 refers to the current equilibrium and p is the pressure divided by the standard pressure. The Gibbs function changes with temperature as follows

$$(\partial G/\partial T) = -S \quad \text{so} \quad \mu^\ominus_\mathrm{liq}(T_1) = \mu^\ominus_\mathrm{liq}(T_0) - (\Delta T)S^\ominus_\mathrm{liq}$$

and similarly for the vapour. Thus, at the higher temperature

$$\mu^\ominus_\mathrm{liq}(T_0) - (\Delta T)S^\ominus_\mathrm{liq} = \mu^\ominus_\mathrm{vap}(T_0) - (\Delta T)S^\ominus_\mathrm{vap} + R(T_0 + \Delta T)\ln p$$

Solving both of these expressions for $\mu^\ominus_\mathrm{liq}(T_0) - \mu^\ominus_\mathrm{vap}(T_0)$ and equating them leads to

$$(\Delta T)(S^\ominus_\mathrm{liq} - S^\ominus_\mathrm{vap}) + R(T_0 + \Delta T)\ln p = RT_0 \ln p_0$$

Isolating p leads to

$$\ln p = \frac{(\Delta T)(S^\ominus_\mathrm{vap} - S^\ominus_\mathrm{liq})}{R(T_0 + \Delta T)} + \frac{T_0 \ln p_0}{T_0 + \Delta T}$$

$$p = \exp\left(\frac{(\Delta T)(S^\ominus_\mathrm{vap} - S^\ominus_\mathrm{liq})}{R(T_0 + \Delta T)}\right) p_0^{(T_0/(T_0+\Delta T))}$$

So $$p = \exp\left(\frac{(2.0\,\mathrm{K}) \times (188.83 - 69.91)\,\mathrm{J\,mol^{-1}\,K^{-1}}}{(8.3145\,\mathrm{J\,mol^{-1}\,K^{-1}}) \times (290 + 2.0\,\mathrm{K})}\right) \times (0.0189)^{(290\,\mathrm{K}/(290+2.0)\mathrm{K})},$$

$$p = 0.0214$$

which represents a $\boxed{\text{13 per cent}}$ increase.

P5.26 The logarithm of the fugacity coefficient is given by

$$\ln \phi = \int_0^p \left(\frac{Z-1}{p}\right) \mathrm{d}p$$

The integrand may be evaluated for each of the experimental points, and the numerical integration carried out by a standard procedure such as the trapezoid rule (taking the integral within any interval as the mean value of the intergrand times the length of the interval). The transformed data as computed using the trapezoid rule appear below, along with a plot of the integrand (Fig. 5.2).

p/bar	Z	$[(Z-1)/p]$/bar^{-1}	$\ln \phi$	ϕ
0.500	0.99412	−0.0118	−0.00294	0.997
1.013	0.98896	−0.0109	−0.00875	0.991
2.00	0.97942	−0.0103	−0.0192	0.981
3.00	0.96995	−0.0100	−0.0294	0.971
5.00	0.95133	−0.00973	−0.0491	0.952
10.00	0.90569	−0.00943	−0.0970	0.908
20.0	0.81227	−0.00939	−0.191	0.826
30.0	0.70177	−0.00994	−0.288	$\boxed{0.750}$
42.4	0.47198	−0.0125	−0.427	0.653
50.0	0.22376	−0.0155	−0.533	0.587
70.0	0.26520	−0.0105	−0.793	0.452
100.0	0.34920	−0.00651	−1.05	0.351
200	0.62362	−0.00188	−1.47	0.230
300	0.88288	−0.000390	−1.58	0.206
500	1.37109	0.000742	−1.55	0.213
1000	2.48836	0.00149	−0.989	$\boxed{0.372}$

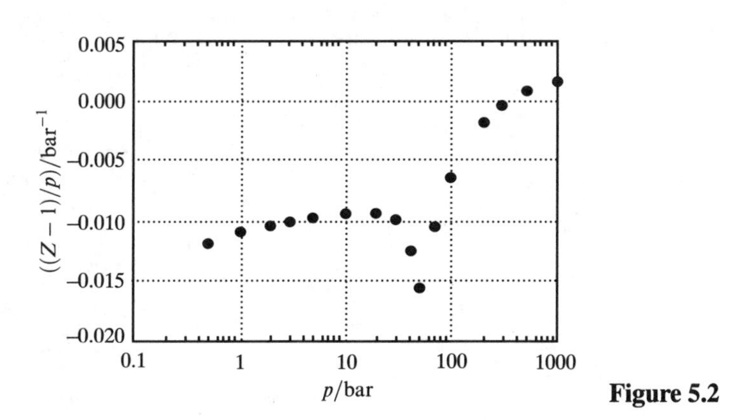

Figure 5.2

P5.28 (A) $\mathrm{d}p = [2(V-b)/RT]\,\mathrm{d}V + [(V-b)^2/RT^2]\,\mathrm{d}T$

$$\left(\frac{\partial p}{\partial V}\right)_T = \frac{2(V-b)}{RT}; \qquad \left(\frac{\partial p}{\partial T}\right)_V = \frac{(V-b)^2}{RT^2}$$

(B) $\mathrm{d}p = -[RT/(V-b)^2]\,\mathrm{d}V + [R/(V-b)]\,\mathrm{d}T$

$$\left(\frac{\partial p}{\partial V}\right)_T = -\frac{RT}{(V-b)^2}; \qquad \left(\frac{\partial p}{\partial T}\right)_V = \frac{R}{V-b}$$

It is expected that for a real substance $(\partial p/\partial V)_T < 0$. Option (B) has this property, but option (A) has the unreal $(\partial p/\partial V)_T > 0$ for $V > b$. Additionally, it is expected that a real substance will have the property that $(\partial p/\partial T)_V > 0$ and $\lim_{V\to\infty}\left(\frac{\partial p}{\partial T}\right)_V = 0$. Both option (A) and option (B) have $(\partial p/\partial T) > 0$. Option (B) has the desired properties as $V \to \infty$. However, Option (A) has a $(\partial p/\partial T)_V$ that 'explodes' to infinity as $V \to \infty$ and this is definitely not good—it is physically unreal.

We conclude that Option (B) is the description of choice. Taking the guess that

$$\boxed{p = \frac{RT}{V-b}} + \text{constant}$$

we see that this equation has the desired properties that $(\partial p/\partial V)_T = -RT/(V-b)^2$ and $(\partial p/\partial_T)_V = R/(V-b)$. It is the equation of state.

P5.29 $\left(\frac{\partial S_\mathrm{m}}{\partial p}\right)_T = -\left(\frac{\partial V_\mathrm{m}}{\partial T}\right)_p$ [Maxwell relation; Table 5.1]

Using $V_\mathrm{m} = M/\rho$ where M is molar mass and ρ is density,

$$\left(\frac{\partial S_\mathrm{m}}{\partial p}\right)_T = -\left(\frac{\partial (M/\rho)}{\partial T}\right)_p = \boxed{\frac{M}{\rho^2}\left(\frac{\partial \rho}{\partial T}\right)_p = \left(\frac{\partial S_\mathrm{m}}{\partial p}\right)_T}$$

When $\theta < 4°\mathrm{C}$, $(\partial \rho/\partial T)_p > 0$. So we conclude that for $0°\mathrm{C} \le \theta \le 4°\mathrm{C}$ entropy $\boxed{\text{increases}}$ with an increase in pressure at constant T (i.e. $(\partial S_\mathrm{m}/\partial p)_T > 0$).

When $\theta = 4°\mathrm{C}$, $(\partial \rho/\partial T)_p = 0$, because this is the temperature of maximum density. Consequently, $(\partial S_\mathrm{m}/\partial p)_T = 0$ and entropy $\boxed{\text{remains constant}}$ with pressure.

When $\theta > 4°\mathrm{C}$, $(\partial \rho/\partial T) < 0$ so $(\partial S_\mathrm{m}/\partial p)_T < 0$ and entropy $\boxed{\text{decreases}}$ with increasing pressure at constant T.

6 Physical transformations of pure substances

Solutions to exercises

E6.1 On the assumption that the vapour is a perfect gas and that $\Delta_{vap}H$ is independent of temperature, we may write

$$p = p^*e^{-\chi}, \quad \chi = \left(\frac{\Delta_{vap}H}{R}\right) \times \left(\frac{1}{T} - \frac{1}{T^*}\right) \text{ [12]}, \quad \ln\frac{p^*}{p} = \chi$$

$$\frac{1}{T} = \frac{1}{T^*} + \frac{R}{\Delta_{vap}H}\ln\frac{p^*}{p}$$

$$= \frac{1}{297.25\,\text{K}} + \frac{8.314\,\text{J K}^{-1}\,\text{mol}^{-1}}{28.7\times10^3\,\text{J mol}^{-1}}\ln\frac{400\,\text{Torr}}{500\,\text{Torr}} = 3.30\bar{0}\times10^{-3}\,\text{K}^{-1}$$

Hence, $T = \boxed{303\,\text{K}} = \boxed{30^\circ\text{C}}$

E6.2 $$\frac{dp}{dT} = \frac{\Delta_{trs}S}{\Delta_{trs}V} \text{ [6]}$$

$$\Delta_{fus}S = \Delta_{fus}V \times \left(\frac{dp}{dT}\right) \approx \Delta_{fus}V \times \frac{\Delta p}{\Delta T}$$

[$\Delta_{fus}S$ and $\Delta_{fus}V$ assumed independent of temperature.]

$$\Delta_{fus}S = [(163.3 - 161.0)\times10^{-6}\,\text{m}^3\,\text{mol}^{-1}] \times \left(\frac{(100-1)\times(1.013\times10^5\,\text{Pa})}{(351.26-350.75)\,\text{K}}\right)$$

$$= \boxed{+45.2\bar{3}\,\text{J K}^{-1}\,\text{mol}^{-1}}$$

$$\Delta_{fus}H = T_f\Delta S = (350.75\,\text{K})\times(45.23\,\text{J K}^{-1}\,\text{mol}^{-1}) = \boxed{+16\,\text{kJ mol}^{-1}}$$

E6.3 The expression for $\ln p$ is the indefinite integral of eqn 11

$$\int d\ln p = \int \frac{\Delta_{vap}H}{RT^2}\,dT; \quad \ln p = \text{constant} - \frac{\Delta_{vap}H}{RT}$$

Therefore, $\Delta_{vap}H = (2501.8\,\text{K})\times R = (2501.8\,\text{K})\times(8.314\,\text{J K}^{-1}\,\text{mol}^{-1}) = \boxed{+20.80\,\text{kJ mol}^{-1}}$

E6.4 **(a)** The indefinitely integrated form of eqn 11 is used as in Exercise 6.3.

$$\ln p = \text{constant} - \frac{\Delta_{vap}H}{RT}, \quad \text{or} \quad \log p = \text{constant} - \frac{\Delta_{vap}H}{2.303RT}$$

Therefore,

$$\Delta_{vap}H = (2.303)\times(1780\,\text{K})\times R = (2.303)\times(1780\,\text{K})\times(8.314\,\text{J K}^{-1}\,\text{mol}^{-1})$$

$$= \boxed{+34.08\,\text{kJ mol}^{-1}}$$

(b) The boiling point corresponds to $p = 1.000\,\text{atm} = 760\,\text{Torr}$.

$$\log 760 = 7.960 - \frac{1780\,\text{K}}{T_b}$$

$$T_b = \boxed{350.5\,\text{K}}$$

E6.5

$$\Delta T \approx \frac{\Delta_{\text{fus}} V}{\Delta_{\text{fus}} S} \times \Delta p\text{[6, and Exercise 6.2]}$$

$$\approx \frac{T_{\text{f}} \Delta_{\text{fus}} V}{\Delta_{\text{fus}} H} \times \Delta p = \frac{T_{\text{f}} \Delta p M}{\Delta_{\text{fus}} H} = \Delta\left(\frac{1}{\rho}\right) \quad [V_{\text{m}} = M/\rho]$$

$$\approx \left(\frac{(278.6\,\text{K}) \times (999) \times (1.013 \times 10^{5}\,\text{Pa}) \times (78.12 \times 10^{-3}\,\text{kg mol}^{-1})}{10.59 \times 10^{3}\,\text{J mol}^{-1}}\right) \times \left(\frac{1}{879\,\text{kg m}^{-3}} - \frac{1}{891\,\text{kg m}^{-3}}\right) \approx 3.18\,\text{K}$$

Therefore, at 1000 atm, $T_{\text{f}} \approx 278.6 + 3.18 = \boxed{281.8\,\text{K}}$ [8.7°C]

E6.6 The rate of loss of mass of water may be expressed as

$$\frac{\text{d}m}{\text{d}t} = \frac{\text{d}n}{\text{d}t} \times M_{\text{H}_2\text{O}}; \qquad n = \frac{q}{\Delta_{\text{vap}} H}$$

$$\frac{\text{d}n}{\text{d}t} = \frac{\left(\frac{\text{d}q}{\text{d}t}\right)}{\Delta_{\text{vap}} H} = \frac{(1.2 \times 10^{3}\,\text{W m}^{-2}) \times (50\,\text{m}^{2})}{44.0 \times 10^{3}\,\text{J mol}^{-1}} = 1.4\,\text{mol s}^{-1}$$

$$\frac{\text{d}m}{\text{d}t} = (1.4\,\text{mol s}^{-1}) \times (18.02\,\text{g mol}^{-1}) = \boxed{25\,\text{g s}^{-1}}$$

E6.7 Assume perfect gas behaviour.

$$n = \frac{pV}{RT} \qquad n = \frac{m}{M} \qquad V = 75\,\text{m}^{3}$$

$$m = \frac{pVM}{RT}$$

(a) $$m = \frac{(24\,\text{Torr}) \times (75 \times 10^{3}\,\text{L}^{3}) \times (18.02\,\text{g mol}^{-1})}{(62.364\,\text{L Torr K}^{-1}\,\text{mol}^{-1}) \times (298.15\,\text{K})} = \boxed{1.7\,\text{kg}}$$

(b) $$m = \frac{(98\,\text{Torr}) \times (75 \times 10^{3}\,\text{L}^{3}) \times (78.11\,\text{g mol}^{-1})}{(62.364\,\text{L Torr K}^{-1}\,\text{mol}^{-1}) \times (298.15\,\text{K})} = \boxed{31\,\text{kg}}$$

(c) $$m = \frac{(1.7 \times 10^{-3}\,\text{Torr}) \times (75 \times 10^{3}\,\text{L}^{3}) \times (200.59\,\text{g mol}^{-1})}{(62.364\,\text{L Torr K}^{-1}\,\text{mol}^{-1}) \times (298.15\,\text{K})} = \boxed{1.4\,\text{g}}$$

Question. Assuming all the mercury vapour breathed remains in the body, how long would it take to accumulate 1.4 g? Make reasonable assumptions about the volume and frequency of a breath.

E6.8 The volume decreases as the vapour is cooled from 400 K to 373 K. At the latter temperature, the vapour condenses to a liquid and there is a large decrease in volume. The liquid cools with only a small decrease in volume until the temperature reaches 273 K, when it freezes. The negative slope of the solid/liquid curve shows that the volume of the sample will then increase slightly if the pressure is maintained. Ice remains at 260 K. There will be a pause in the rate of cooling at 373 K (about 40 kJ mol^{-1} of energy is released as heat) and a pause at 273 K (when about 6 kJ mol^{-1} is released).

E6.9 The Clausius–Clapeyron equation [11] integrates to the form [12] which may be rewritten as

$$\ln\left(\frac{p_2}{p_1}\right) = \frac{\Delta_{\text{vap}} H}{R} \times \left(\frac{1}{T_1} - \frac{1}{T_2}\right)$$

(a) $$\ln\left(\frac{40\,\text{Torr}}{10\,\text{Torr}}\right)=\left(\frac{\Delta_{\text{vap}}H}{8.314\,\text{J K}^{-1}\,\text{mol}^{-1}}\right)\times\left(\frac{1}{359.0\,\text{K}}-\frac{1}{392.5\,\text{K}}\right)$$

$$1.38\overline{6}=\Delta_{\text{vap}}H\times(2.8\overline{6}\times10^{-5}\,\text{J}^{-1}\,\text{mol})$$

$$\Delta_{\text{vap}}H=\boxed{48.\overline{5}\,\text{kJ mol}^{-1}}$$

(b) The normal boiling point corresponds to a vapour pressure of 760 Torr. Using the data at 119.3°C

$$\ln\left(\frac{760\,\text{Torr}}{40\,\text{Torr}}\right)=\left(\frac{48.\overline{5}\times10^{3}\,\text{J mol}^{-1}}{8.314\,\text{J K}^{-1}\,\text{mol}^{-1}}\right)\times\left(\frac{1}{392.5\,\text{K}}-\frac{1}{T_{\text{b}}}\right)$$

$$2.94\overline{4}=14.\overline{86}-\frac{5\overline{831}\,\text{K}}{T_{\text{b}}};\quad T_{\text{b}}=48\overline{9}\,\text{K}=\boxed{21\overline{6}\,^{\circ}\text{C}}$$

[The accepted value is 218°C.]

(c) $$\Delta_{\text{vap}}S(T_{\text{b}})=\frac{\Delta_{\text{vap}}H(T_{\text{b}})}{T_{\text{b}}}\approx\frac{48.5\times10^{3}\,\text{J mol}^{-1}}{489\,\text{K}}=\boxed{99\,\text{J K}^{-1}\,\text{mol}^{-1}}$$

E6.10 $$\Delta T=T_{\text{f}}(50\,\text{bar})-T_{\text{f}}(1\,\text{bar})\approx\frac{T_{\text{f}}\Delta pM}{\Delta_{\text{fus}}H}\Delta\left(\frac{1}{\rho}\right)\quad\text{[Exercise 6.5]}$$

$$\Delta_{\text{fus}}H=6.01\,\text{kJ mol}^{-1}\quad\text{[Table 2.3]}$$

$$\Delta T=\left(\frac{(273.15\,\text{K})\times(49\times10^{5}\,\text{Pa})\times(18\times10^{-3}\,\text{kg mol}^{-1})}{6.01\times10^{3}\,\text{J mol}^{-1}}\right)\times\left(\frac{1}{1.00\times10^{3}\,\text{kg m}^{-3}}-\frac{1}{9.2\times10^{2}\,\text{kg m}^{3}}\right)=-0.35\,\text{K}$$

$$T_{\text{f}}(50\,\text{bar})=(273.15\,\text{K})-(0.35\,\text{K})=\boxed{272.80\,\text{K}}$$

E6.11 $$\Delta_{\text{vap}}H=\Delta_{\text{vap}}U+\Delta_{\text{vap}}(pV)=40.656\,\text{kJ mol}^{-1}\quad\text{[Table 2.3]}$$

$$\Delta_{\text{vap}}(pV)=p\Delta_{\text{vap}}V=p(V_{\text{gas}}-V_{\text{liq}})\approx pV_{\text{gas}}$$

$$=RT\ \text{[per mole of a perfect gas]}$$

$$=(8.314\,\text{J K}^{-1}\,\text{mol}^{-1})\times(373.2\,\text{K})=3.102\times10^{3}\,\text{kJ mol}^{-1}$$

$$\text{Fraction}=\frac{\Delta_{\text{vap}}(pV)}{\Delta_{\text{vap}}H}=\frac{3.102\times10^{3}\,\text{kJ mol}^{-1}}{40.656\,\text{kJ mol}^{-1}}=\boxed{0.07630}\approx7.6\text{ per cent}$$

E6.12 $$p=p^{*}\text{e}^{2\gamma V_{\text{m}}/rRT}\ [17]$$

$$V_{\text{m}}=\frac{M}{\rho}=\frac{18.02\,\text{g mol}^{-1}}{0.9982\,\text{g cm}^{-3}}=18.05\,\text{cm}^{3}\,\text{mol}^{-1}=1.805\times10^{-5}\,\text{m}^{3}\,\text{mol}^{-1}$$

$$\frac{2\gamma V_{\text{m}}}{rRT}=\frac{(2)\times(7.275\times10^{-2}\,\text{N m}^{-1})\times(1.805\times10^{-5}\,\text{m}^{3}\,\text{mol}^{-1})}{(1.0\times10^{-8}\,\text{m})\times(8.314\,\text{J K}^{-1}\,\text{mol}^{-1})\times(293\,\text{K})}=0.10\overline{78}$$

$$p=(2.3\,\text{kPa})\times\text{e}^{0.10\overline{78}}=\boxed{2.6\,\text{kPa}}$$

E6.13 $$\gamma=\tfrac{1}{2}\rho ghr\ [19]=\left(\tfrac{1}{2}\right)\times(998.2\,\text{kg m}^{-3})\times(9.807\,\text{m s}^{-2})\times(4.96\times10^{-2}\,\text{m})\times(3.00\times10^{-4}\,\text{m})$$

$$=7.28\times10^{-2}\,\text{kg s}^{-2}=\boxed{7.28\times10^{-2}\,\text{N m}^{-1}}$$

This value is in agreement with Table 6.1.

E6.14 $p_{\text{in}} - p_{\text{out}} = \frac{2\gamma}{r}[16] = \frac{(2) \times (7.275 \times 10^{-2}\,\text{N m}^{-1})}{2.00 \times 10^{-7}\,\text{m}}$ [Table 6.1] $= \boxed{7.28 \times 10^5\,\text{Pa}}$

Comment. Pressure differentials for small droplets are quite large.

Solutions to problems

Solutions to numerical problems

P6.1 At the triple point, T_3, the vapour pressures of liquid and solid are equal, hence

$$10.5916 - \frac{1871.2\,\text{K}}{T_3} = 8.3186 - \frac{1425.7\,\text{K}}{T_3}; \quad T_3 = \boxed{196.0\,\text{K}}$$

$$\log(p_3/\text{Torr}) = \frac{-1871.2\,\text{K}}{196.0\,\text{K}} + 10.5916 = 1.044\bar{7}; \quad p_3 = \boxed{11.1\,\text{Torr}}$$

P6.2 Use the definite integral form of the Clausius–Clapeyron equation (Exercise 6.9(**a**)).

$$\ln\left(\frac{p_2}{p_1}\right) = \frac{\Delta_{\text{vap}}H}{R} \times \left(\frac{1}{T_1} - \frac{1}{T_2}\right); \quad T_1 = \text{normal boiling point}; \quad p_1 = 1.000\,\text{atm}$$

$$\ln(p_2/\text{atm}) = \left(\frac{20.25 \times 10^3\,\text{J mol}^{-1}}{8.314\,\text{J K}^{-1}\,\text{mol}^{-1}}\right) \times \left(\frac{1}{244.0\,\text{K}} - \frac{1}{313.2\,\text{K}}\right) = 2.206$$

$$p_2 = \boxed{9.07\,\text{atm}} \approx 9\,\text{atm}$$

Comment. Three significant figures are not really warranted in this answer because of the approximations employed.

P6.4 (**a**) $\left(\frac{\partial \mu(\text{l})}{\partial T}\right)_p - \left(\frac{\partial \mu(\text{s})}{\partial T}\right)_p = -S_{\text{m}}(\text{l}) + S_{\text{m}}(\text{s})$ [Section 6.7, eqn 13]

$$= -\Delta_{\text{fus}}S = \frac{-\Delta_{\text{fus}}H}{T_{\text{f}}}; \quad \Delta_{\text{fus}}H = 6.01\,\text{kJ mol}^{-1} \text{ [Table 2.3]}$$

$$= \frac{-6.01\,\text{kJ mol}^{-1}}{273.15\,\text{K}} = \boxed{-22.0\,\text{J K}^{-1}\,\text{mol}^{-1}}$$

(**b**) $\left(\frac{\partial \mu(\text{g})}{\partial T}\right)_p - \left(\frac{\partial \mu(\text{l})}{\partial T}\right)_p = -S_{\text{m}}(\text{g}) + S_{\text{m}}(\text{l}) = -\Delta_{\text{vap}}S$

$$= \frac{-\Delta_{\text{vap}}H}{T_{\text{b}}} = \frac{-40.6\,\text{kJ mol}^{-1}}{373.15\,\text{K}} = \boxed{-109.0\,\text{J K}^{-1}\,\text{mol}^{-1}}$$

(**c**) $\Delta\mu \approx \left(\frac{\partial \mu}{\partial T}\right)_p \Delta T = -S_{\text{m}}\Delta T$ [1]

$$\begin{aligned}\Delta\mu(\text{l}) - \Delta\mu(\text{s}) &= \mu(\text{l}, -5°\text{C}) - \mu(\text{l}, 0°\text{C}) - \mu(\text{s}, -5°\text{C}) + \mu(\text{s}, 0°\text{C}) \\ &= \mu(\text{l}, -5°\text{C}) - \mu(\text{s}, -5°\text{C})[\mu(\text{l}, 0°\text{C}) = \mu(\text{s}, 0°\text{C})] \\ &\approx -\{S_{\text{m}}(\text{l}) - S_{\text{m}}(\text{s})\}\Delta T \approx -\Delta_{\text{fus}}S\Delta T \\ &= -(5\,\text{K}) \times (-22.0\,\text{J K}^{-1}\,\text{mol}^{-1}) = \boxed{+11\bar{0}\,\text{J mol}^{-1}}\end{aligned}$$

Since $\mu(\text{l}, -5°\text{C}) > \mu(\text{s}, -5°\text{C})$, there is a thermodynamic tendency to freeze.

P6.6

$$\frac{dp}{dT} = \frac{\Delta_{fus}S}{\Delta_{fus}V}[6] = \frac{\Delta_{fus}H}{T\Delta_{fus}V}$$

$$\Delta T = \int_{T_{m,1}}^{T_{m,2}} dT = \int_{p_{top}}^{p_{bot}} \frac{T_m \Delta_{fus}V}{\Delta_{fus}H}\,dp$$

$$\Delta T \approx \frac{T_m \Delta_{fus}V}{\Delta_{fus}H} \times \Delta p \quad [T_m,\ \Delta_{fus}H,\text{ and } \Delta_{fus}V \text{ assumed constant}]$$

$$\Delta p = p_{bot} - p_{top} = \rho g h$$

Therefore

$$\Delta T = \frac{T_m \rho g h \Delta_{fus}V}{\Delta_{fus}H}$$

$$= \frac{(234.3\,\text{K}) \times (13.6 \times 10^3\,\text{kg m}^{-3}) \times (9.81\,\text{m s}^{-2}) \times (10\,\text{m}) \times (0.517 \times 10^{-6}\,\text{m}^3\,\text{mol}^{-1})}{2.292 \times 10^3\,\text{J mol}^{-1}}$$

$$= 0.070\,\text{K}$$

Therefore, the freezing point changes to $\boxed{234.4\,\text{K}}$

P6.8 $\frac{d\ln p}{dT} = \frac{\Delta_{vap}H}{RT^2}$ [11], yields upon indefinite integration

$$\ln p = \text{constant} - \frac{\Delta_{vap}H}{RT}$$

Therefore, plot $\ln p$ against $\frac{1}{T}$ and identify $\frac{-\Delta_{vap}H}{R}$ as its slope. Construct the following table

$\theta/°\text{C}$	0	20	40	50	70	80	90	100
T/K	273	293	313	323	343	353	363	373
$1000\,\text{K}/T$	3.66	3.41	3.19	3.10	2.92	2.83	2.75	2.68
$\ln p/\text{Torr}$	2.67	3.87	4.89	5.34	6.15	6.51	6.84	7.16

The points are plotted in Fig. 6.1. The slope is −4569 K, so

$$\frac{-\Delta_{vap}H}{R} = -4569\,\text{K}, \quad \text{or} \quad \Delta_{vap}H = \boxed{+38.0\,\text{kJ mol}^{-1}}$$

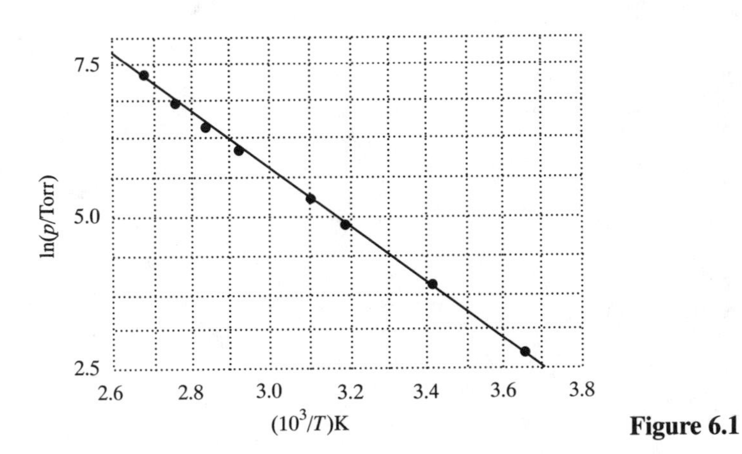

Figure 6.1

The normal boiling point occurs at $p = 760\,\text{Torr}$, or at $\ln(p/\text{Torr}) = 6.633$, which from the figure corresponds to $1000\,\text{K}/T \approx 2.80$. Therefore, $T_b = \boxed{357\,\text{K}(84°\text{C})}$. The accepted value is 83°C.

P6.10 The equations describing the coexistence curves for the three states are

(a) Solid–liquid boundary

$$p = p^* + \frac{\Delta_{\text{fus}}H}{\Delta_{\text{fus}}V}\ln\frac{T}{T^*}\ [8]$$

(b) Liquid–vapour boundary

$$p = p^*\text{e}^{-\chi}, \quad \chi = \frac{\Delta_{\text{vap}}H}{R}\times\left(\frac{1}{T}-\frac{1}{T^*}\right)\ [12]$$

(c) Solid–vapour boundary

$$p = p^*\text{e}^{-\chi}, \quad \chi = \frac{\Delta_{\text{sub}}H}{R}\times\left(\frac{1}{T}-\frac{1}{T^*}\right)\ [\text{similar to } 12]$$

We need $\Delta_{\text{sub}}H = \Delta_{\text{fus}}H + \Delta_{\text{vap}}H = 41.4\,\text{kJ mol}^{-1}$

$$\Delta_{\text{fus}}V = M\times\left(\frac{1}{\rho(\text{l})}-\frac{1}{\rho(\text{s})}\right) = \left(\frac{78.11\,\text{g mol}^{-1}}{\text{g cm}^{-3}}\right)\times\left(\frac{1}{0.879}-\frac{1}{0.891}\right)$$
$$= +1.19\bar{7}\,\text{cm}^3\,\text{mol}^{-1}$$

After insertion of these numerical values into the above equations, we obtain

(a)
$$p = p^* + \left(\frac{10.6\times10^3\,\text{J mol}^{-1}}{1.197\times10^{-6}\,\text{m}^3\,\text{mol}^{-1}}\right)\ln\frac{T}{T^*}$$
$$= p^* + 8.85\bar{5}\times10^9\,\text{Pa}\ln\frac{T}{T^*} = p^* + (6.64\times10^7\,\text{Torr})\ln\frac{T}{T^*}\quad(1\,\text{Torr} = 133.322\,\text{Pa})$$

This line is plotted as a in Fig. 6.2, starting at $(p^*, T^*) = (36\,\text{Torr}, 5.50°\text{C}(278.65\,\text{K}))$.

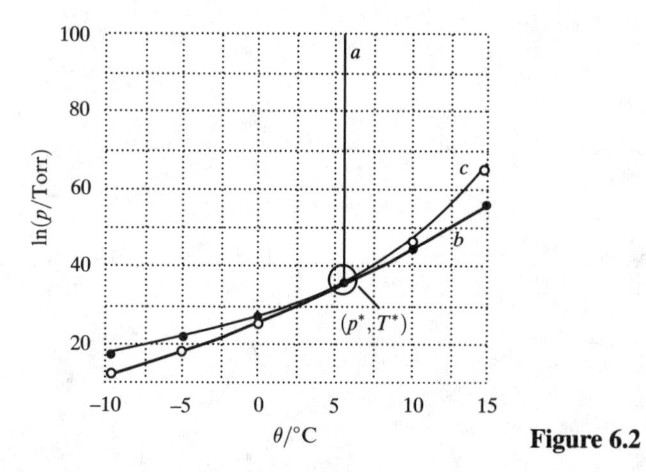

Figure 6.2

(b) $$\chi = \left(\frac{30.8 \times 10^3\,\mathrm{J\,mol^{-1}}}{8.314\,\mathrm{J\,K^{-1}\,mol^{-1}}}\right) \times \left(\frac{1}{T} - \frac{1}{T^*}\right) = (370\overline{5}\,\mathrm{K}) \times \left(\frac{1}{T} - \frac{1}{T^*}\right)$$

$$p = p^* \mathrm{e}^{-370\overline{5}\,\mathrm{K} \times (1/T - 1/T^*)}$$

This equation is plotted as line b in Fig. 6.2, starting from $(p^*, T^*) = (36\,\mathrm{Torr}, 5.50°\mathrm{C}$ $(278.65\,\mathrm{K}))$.

(c) $$\chi = \left(\frac{41.4 \times 10^3\,\mathrm{J\,mol^{-1}}}{8.314\,\mathrm{J\,K^{-1}\,mol^{-1}}}\right) \times \left(\frac{1}{T} - \frac{1}{T^*}\right) = (49\overline{8}0\,\mathrm{K}) \times \left(\frac{1}{T} - \frac{1}{T^*}\right)$$

$$p = p^* \mathrm{e}^{-498\overline{0}\,\mathrm{K} \times (1/T - 1/T^*)}$$

These points are plotted as line c in Fig. 6.2, starting at (36 Torr, 5.50°C).

The lighter lines in Fig. 6.2 represent extensions of lines b and c into regions where the liquid and solid states respectively are not stable.

Solutions to theoretical problems

P6.12 $\mathrm{d}H = C_p\,\mathrm{d}T + V\,\mathrm{d}p$, implying that $\mathrm{d}\Delta H = \Delta C_p\,\mathrm{d}T + \Delta V\,\mathrm{d}p$

However, along a phase boundary $\mathrm{d}p$ and $\mathrm{d}T$ are related by

$$\frac{\mathrm{d}p}{\mathrm{d}T} = \frac{\Delta H}{T\Delta V} \text{ [Clapeyron equation, e.g. 6,7, or 10]}$$

Therefore,

$$\mathrm{d}\Delta H = \left(\Delta C_p + \Delta V \times \frac{\Delta H}{T\Delta V}\right)\mathrm{d}T = \left(\Delta C_p + \frac{\Delta H}{T}\right)\mathrm{d}T \quad \text{and} \quad \frac{\mathrm{d}H}{\mathrm{d}T} = \Delta C_p + \frac{\Delta H}{T}$$

Then, since

$$\frac{\mathrm{d}}{\mathrm{d}T}\left(\frac{\Delta H}{T}\right) = \frac{1}{T}\frac{\mathrm{d}\Delta H}{\mathrm{d}T} - \frac{\Delta H}{T^2} = \frac{1}{T}\left(\frac{\mathrm{d}\Delta H}{\mathrm{d}T} - \frac{\Delta H}{T}\right)$$

substituting the first result gives

$$\frac{\mathrm{d}}{\mathrm{d}T}\left(\frac{\Delta H}{T}\right) = \frac{\Delta C_p}{T}$$

Therefore, $$\mathrm{d}\left(\frac{\Delta H}{T}\right) = \frac{\Delta C_p\,\mathrm{d}T}{T} = \boxed{\Delta C_p\,\mathrm{d}\ln T}$$

P6.15 In each phase the slopes are given by

$$\left(\frac{\partial \mu}{\partial T}\right)_p = -S_\mathrm{m} \text{ [1]}$$

The curvatures of the graphs of μ against T are given by

$$\left(\frac{\partial^2 \mu}{\partial T^2}\right)_p = -\left(\frac{\partial S_\mathrm{m}}{\partial T}\right)_p = \boxed{-\frac{1}{T} \times C_{p,\mathrm{m}}} \text{ [Problem 5.7]}$$

Since $C_{p,\mathrm{m}}$ is necessarily positive, the curvatures in all states of matter are necessarily negative. $C_{p,\mathrm{m}}$ is often largest for the liquid state, though not always; but it is the ratio $C_{p,\mathrm{m}}/T$ that determines the magnitude of the curvature, so no precise answer can be given for the state with greatest curvature. It depends upon the substance.

P6.16 (1) $V = V(T, p)$

$$\mathrm{d}V = \left(\frac{\partial V}{\partial T}\right)_p \mathrm{d}T + \left(\frac{\partial V}{\partial p}\right)_T \mathrm{d}p$$

$$\left(\frac{\partial V}{\partial T}\right)_p = \alpha V, \qquad \left(\frac{\partial V}{\partial p}\right)_T = -\kappa_T V;$$

hence, $\mathrm{d}V = \alpha V\,\mathrm{d}T - \kappa_T V\,\mathrm{d}p$

This equation applies to both phases 1 and 2, and since V is continuous through a second-order transition

$$\alpha_1\,\mathrm{d}T - \kappa_{T,1}\,\mathrm{d}p = \alpha_2\,\mathrm{d}T - \kappa_{T,2}\,\mathrm{d}p$$

Solving for $\frac{\mathrm{d}p}{\mathrm{d}T}$ yields $\boxed{\frac{\mathrm{d}p}{\mathrm{d}T} = \frac{\alpha_2 - \alpha_1}{\kappa_{T,2} - \kappa_{T,1}}}$

(2) $S_\mathrm{m} = S_\mathrm{m}(T, p)$

$$\mathrm{d}S_\mathrm{m} = \left(\frac{\partial S_\mathrm{m}}{\partial T}\right)_p \mathrm{d}T + \left(\frac{\partial S_\mathrm{m}}{\partial p}\right)_T \mathrm{d}p$$

$$\left(\frac{\partial S_\mathrm{m}}{\partial T}\right)_p = \frac{C_{p,\mathrm{m}}}{T}\ \text{[Problem 5.7]} \qquad \left(\frac{\partial S_\mathrm{m}}{\partial p}\right)_T = -\left(\frac{\partial V_\mathrm{m}}{\partial T}\right)_p\ \text{[Maxwell relation]}$$

$$= -\alpha V_\mathrm{m}$$

Thus, $\mathrm{d}S_\mathrm{m} = \frac{C_{p,\mathrm{m}}}{T}\,\mathrm{d}T - \alpha V_\mathrm{m}\,\mathrm{d}p.$

This relation applies to both phases. For second-order transitions both S_m and V_m are continuous through the transition, $S_{\mathrm{m},1} = S_{\mathrm{m},2}$, $V_{\mathrm{m},1} = V_{\mathrm{m},2} = V_\mathrm{m}$, so that

$$\frac{C_{p,\mathrm{m},1}}{T}\,\mathrm{d}T - \alpha_1 V_\mathrm{m}\,\mathrm{d}p = \frac{C_{p,\mathrm{m},2}}{T}\,\mathrm{d}T - \alpha_2 V_\mathrm{m}\,\mathrm{d}p$$

Solving for $\frac{\mathrm{d}p}{\mathrm{d}T}$ yields $\boxed{\frac{\mathrm{d}p}{\mathrm{d}T} = \frac{C_{p,\mathrm{m},2} - C_{p,\mathrm{m},1}}{T V_\mathrm{m}(\alpha_2 - \alpha_1)}}$

The Clapeyron equation cannot apply because both ΔV and ΔS are zero through a second-order transition, resulting in an indeterminate form $\frac{0}{0}$.

Solutions to additional problems

P6.18 **(a)** The phase boundary is plotted in Fig. 6.3

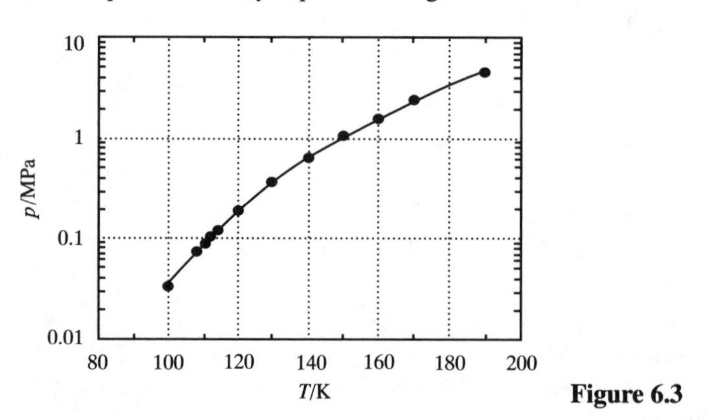

Figure 6.3

(b) The standard boiling point is the temperature at which the liquid is in equilibrium with the standard pressure of 1 bar (0.1 MPa). Interpolation of the plotted points gives $T_b = \boxed{112\,\text{K}}$

(c) The slope of the liquid–vapour coexistence curve is given by

$$\frac{\mathrm{d}p}{\mathrm{d}T} = \frac{\Delta_{\text{vap}}H}{T\Delta_{\text{vap}}V} \quad \text{so} \quad \Delta_{\text{vap}}H = (T\Delta_{\text{vap}}V)\frac{\mathrm{d}p}{\mathrm{d}T}$$

The slope can be obtained graphically or by fitting the points nearest the boiling point. Then $\mathrm{d}p/\mathrm{d}T = 8.14 \times 10^{-3}\,\text{MPa K}^{-1}$, so

$$\Delta_{\text{vap}}H = (112\,\text{K}) \times \left(\frac{(8.89 - 0.0380)\,\text{L mol}^{-1}}{1000\,\text{L m}^3}\right) \times (8.14\,\text{kPa K}^{-1}) = \boxed{8.07\,\text{kJ mol}^{-1}}$$

P6.21 C(graphite) $\rightleftharpoons$ C(diamond) $\Delta_r G^{\ominus} = 2.8678\,\text{kJ mol}^{-1}$ at $\mathfrak{T}$.

We want the pressure at which $\Delta_r G = 0$; above that pressure the reaction will be spontaneous. Equation 5.10 determines the rate of change of $\Delta_r G$ with p at constant T.

$$(1) \quad \left(\frac{\partial \Delta_r G}{\partial p}\right)_T = \Delta_r V = (V_D - V_G)M$$

where M is the molar mass of carbon; V_D and V_G are the specific volumes of diamond and graphite, respectively.

$\Delta_r G(\mathfrak{T}, p)$ may be expanded in a Taylor series around the pressure $p^{\ominus} = 100\,\text{kPa}$ at $\mathfrak{T}$.

$$(2) \quad \Delta_r G(\mathfrak{T}, p) = \Delta_r G^{\ominus}(\mathfrak{T}, p^{\ominus}) + \left(\frac{\partial \Delta_r G^{\ominus}(\mathfrak{T}, p^{\ominus})}{\partial p}\right)_T (p - p^{\ominus}) + \frac{1}{2}\left(\frac{\partial^2 \Delta_r G^{\ominus}(\mathfrak{T}, p^{\ominus})}{\partial p^2}\right)_T (p - p^{\ominus})^2 + \theta(p - p^{\ominus})^3$$

We will neglect the third and higher-order terms; the derivative of the first-order term can be calculated with eqn 1. An expression for the derivative of the second-order term can be derived with eqn 1

$$(3) \quad \left(\frac{\partial^2 \Delta_r G}{\partial p^2}\right)_T = \left\{\left(\frac{\partial V_D}{\partial p}\right)_T - \left(\frac{\partial V_G}{\partial p}\right)_T\right\} M = \{V_G \kappa_T(\text{G}) - V_D \kappa_T(\text{D})\}M \; [3.13]$$

Calculating the derivatives of eqns 1 and 2 at $\mathfrak{T}$ and $p^{\ominus}$

$$(4) \quad \left(\frac{\partial \Delta_r G(\mathfrak{T}, p^{\ominus})}{\partial p}\right)_T = (0.284 - 0.444) \times \left(\frac{\text{cm}^3}{\text{g}}\right) \times \left(\frac{12.01\,\text{g}}{\text{mol}}\right) = -1.92\,\text{cm}^3\,\text{mol}^{-1}$$

$$(5) \quad \left(\frac{\partial^2 \Delta_r G(\mathfrak{T}, p^{\ominus})}{\partial p^2}\right)_T = \{0.444(3.04 \times 10^{-8}) - 0.284(0.187 \times 10^{-8})\} \times \left(\frac{\text{cm}^3\,\text{kPa}^{-1}}{\text{g}}\right) \times \left(\frac{12.01\,\text{g}}{\text{mol}}\right)$$

$$= 1.56 \times 10^{-7}\,\text{cm}^3\,(\text{kPa})^{-1}\,\text{mol}^{-1}$$

It is convenient to convert the value of $\Delta_r G^{\ominus}$ to the units $\text{cm}^3\,\text{kPa mol}^{-1}$

$$\Delta_r G^{\ominus} = 2.8678\,\text{kJ mol}^{-1}\left(\frac{8.315 \times 10^{-2}\,\text{L bar K}^{-1}\,\text{mol}^{-1}}{8.315\,\text{J K}^{-1}\,\text{mol}^{-1}}\right) \times \left(\frac{10^3\,\text{cm}^3}{\text{L}}\right) \times \left(\frac{10^5\,\text{Pa}}{\text{bar}}\right)$$

(6) $\Delta_r G^\ominus = 2.8678 \times 10^6\ \text{cm}^3\ \text{kPa mol}^{-1}$

Setting $\chi = p - p^\ominus$, eqns 2 and 3–6 give

$$2.8678 \times 10^6\ \text{cm}^3\ \text{kPa mol}^{-1} - (1.92\ \text{cm}^3\ \text{mol}^{-1})\chi + (7.80 \times 10^{-8}\ \text{cm}^3\ \text{kPa}^{-1}\ \text{mol}^{-1})\chi^2 = 0$$

when $\Delta_r G(T, p) = 0$. One real root of this equation is

$$\chi = 1.60 \times 10^6\ \text{kPa} = p - p^\ominus \text{ or}$$

$$p = 1.60 \times 10^6\ \text{kPa} - 10^2\ \text{kPa}$$

$$= 1.60 \times 10^6\ \text{kPa} = \boxed{1.60 \times 10^4\ \text{bar}}$$

Above this pressure the reaction is spontaneous. The other real root is much higher: 2.3×10^7 kPa.

Question. What interpretation might you give to the other real root?

7 Simple mixtures

Solutions to exercises

E7.1 Let A denote acetone and C chloroform. The total volume of the solution is

$$V = n_A V_A + n_C V_C$$

V_A and V_C are given; hence we need to determine n_A and n_C in 1.000 kg of the solution with the stated mole fraction. The total mass of the sample is $m = n_A M_A + n_C M_C$ (a). We also know that

$$x_A = \frac{n_A}{n_A + n_C}, \quad \text{implies that} \quad (x_A - 1)n_A + x_A n_C = 0$$

and hence that

$$-x_C n_A + x_A n_C = 0 \quad \text{(b)}$$

On solving (a) and (b), we find

$$n_A = \left(\frac{x_A}{x_C}\right) \times n_C, \quad n_C = \frac{m x_C}{x_A M_A + x_C M_C}$$

Since $x_C = 0.4693$, $x_A = 1 - x_C = 0.5307$,

$$n_C = \frac{(0.4693) \times (1000\,\text{g})}{[(0.5307) \times (58.08) + (0.4693) \times (119.37)]\,\text{g mol}^{-1}} = 5.404\,\text{mol}$$

$$n_A = \left(\frac{0.5307}{0.4693}\right) \times (5.404)\,\text{mol} = 6.111\,\text{mol}$$

The total volume, $V = n_A V_A + n_B V_B$, is therefore

$$\begin{aligned} V &= (6.111\,\text{mol}) \times (74.166\,\text{cm}^3\,\text{mol}^{-1}) + (5.404\,\text{mol}) \times (80.235\,\text{cm}^3\,\text{mol}^{-1}) \\ &= \boxed{886.8\,\text{cm}^3} \end{aligned}$$

E7.2 Let A denote water and B ethanol. The total volume of the solution is

$$V = n_A V_A + n_B V_B$$

We are given V_A, we need to determine n_A and n_B in order to solve for V_B. Assume we have 100 cm^3 of solution, then the mass of solution is

$$m = d \times V = (0.914\,\text{g cm}^{-3}) \times (100\,\text{cm}^3) = 91.4\,\text{g}$$

of which 45.7 g is water and 45.7 g ethanol.

$$\begin{aligned} 100\,\text{cm}^3 &= \left(\frac{45.7\,\text{g}}{18.02\,\text{g mol}^{-1}}\right) \times (17.4\,\text{cm}^3\,\text{mol}^{-1}) + \left(\frac{45.7\text{g}}{46.07\,\text{g mol}^{-1}}\right) \times V_B \\ &= 44.1\overline{3}\,\text{cm}^3 + 0.99\overline{20}\,\text{mol} \times V_B \end{aligned}$$

$$V_B = \frac{55.8\overline{7}\,\text{cm}^3}{0.9920\,\text{mol}} = \boxed{56.3\,\text{cm}^3\,\text{mol}^{-1}}$$

E7.3 Check whether $\frac{p_B}{x_B}$ is equal to a constant (K_B)

x	0.005	0.012	0.019
p/x	6.4×10^3	6.4×10^3	6.4×10^3 kPa

Hence, $K_B \approx \boxed{6.4 \times 10^3\,\text{kPa}}$

E7.4 In Exercise 7.3(**a**), the Henry's law constant was determined for concentrations expressed in mole fractions. Thus the concentration in molality must be converted to mole fraction.

$$m(\text{GeCl}_4) = 1000\,\text{g, corresponding to}$$

$$n(\text{GeCl}_4) = \frac{1000\,\text{g}}{214.39\,\text{g mol}^{-1}} = 4.664\,\text{mol}, \qquad n(\text{HCl}) = 0.10\,\text{mol}$$

Therefore, $x = \dfrac{0.10\,\text{mol}}{(0.10\,\text{mol}) + (4.664\,\text{mol})} = 0.021\bar{0}$

From $K_B = 6.4 \times 10^3$ kPa (Exercise 7.3(**a**)), $p = (0.021\bar{0} \times 6.4 \times 10^3\,\text{kPa}) = \boxed{1.3 \times 10^2\,\text{kPa}}$

E7.5 Because the mole fraction of B is small,

$$x_B = \frac{n_B}{n_A + n_B} \approx \frac{n_B}{n_A}$$

The amount of solvent molecules in 1 kg of solvent of molar mass M is

$$n_A = \frac{1\,\text{kg}}{M}$$

Therefore,

$$x_B = \frac{n_B}{n_A} = n_B \times \frac{M}{1\,\text{kg}} = b_B \times M$$

where b_B is the molality of B. Hence, from eqn 30,

$$\Delta T = \left(\frac{RT^{*2}M}{\Delta_{\text{vap}}H}\right) b_B$$

and we can identify the ebullioscopic constant as

$$K_b = \frac{RT^{*2}M}{\Delta_{\text{vap}}H}$$

$$= \frac{(8.314\,\text{J K}^{-1}\,\text{mol}^{-1}) \times (349.9\,\text{K})^2 \times (153.81 \times 10^{-3}\,\text{kg mol}^{-1})}{30.0 \times 10^3\,\text{J mol}^{-1}}$$

$$= \boxed{5.22\,\text{K}/(\text{mol kg}^{-1})}$$

$$K_f = \frac{RT^{*2}M}{\Delta_{\text{fus}}H} \text{ [by analogy with the above]}$$

$$= \frac{(8.314\,\text{J K}^{-1}\,\text{mol}^{-1}) \times (250.3\,\text{K})^2 \times (153.81 \times 10^{-3}\,\text{kg mol}^{-1})}{2.47 \times 10^3\,\text{J mol}^{-1}}$$

$$= \boxed{32\,\text{K}/(\text{mol kg}^{-1})}$$

E7.6 We assume that the solvent, benzene, is ideal and obeys Raoult's law.

Let B denote benzene and A the solute; then

$$p_B = x_B p_B^* \quad \text{and} \quad x_B = \frac{n_B}{n_A + n_B}$$

Hence, $p_B = \dfrac{n_B p_B^*}{n_A + n_B}$; which solves to

$$n_A = \frac{n_B(p_B^* - p_B)}{p_B}$$

Then, since $n_A = \dfrac{m_A}{M_A}$, where m_A is the mass of A present,

$$M_A = \frac{m_A p_B}{n_B(p_B^* - p_B)} = \frac{m_A M_B p_B}{m_B(p_B^* - p_B)}$$

From the data

$$M_A = \frac{(19.0\,\text{g}) \times (78.11\,\text{g mol}^{-1}) \times (386\,\text{Torr})}{(500\,\text{g}) \times (400 - 386)\,\text{Torr}} = \boxed{82\,\text{g mol}^{-1}}$$

E7.7 $M_B = \dfrac{\text{mass of B}}{n_B}$ [B = compound]

$n_B = \text{mass of } CCl_4 \times b_B$ [b_B = molality of B]

$b_B = \dfrac{\Delta T}{K_f}$ [34]; thus

$$M_B = \frac{\text{mass of B} \times K_f}{\text{mass of } CCl_4 \times \Delta T} \qquad K_f = 30\,\text{K}/(\text{mol kg}^{-1})\ [\text{Table 7.2}]$$

$$M_B = \frac{(100\,\text{g}) \times (30\,\text{K kg mol}^{-1})}{(0.750\,\text{kg}) \times (10.5\,\text{K})} = \boxed{381\,\text{g mol}^{-1}}$$

E7.8 $\Delta T = K_f b_B$ [34] $\quad b_B = \dfrac{n_B}{\text{mass of water}} \approx \dfrac{n_B}{V\rho}$ [dilute solution]

$\rho \approx 10^3\,\text{kg m}^{-3}$ [density of solution ≈ density of water]

$n_B \approx \dfrac{\Pi V}{RT}$ [37] $\qquad \Delta T \approx K_f \times \dfrac{\Pi}{RT\rho}$

with $K_f = 1.86\,\text{K}/(\text{mol kg}^{-1})$ [Table 7.2]

$$\Delta T \approx \frac{(1.86\,\text{K kg mol}^{-1}) \times (120 \times 10^3\,\text{Pa})}{(8.314\,\text{J K}^{-1}\,\text{mol}^{-1}) \times (300\,\text{K}) \times (10^3\,\text{kg m}^{-3})} = 0.089\,\text{K}$$

Therefore, the solution will freeze at about $\boxed{-0.09°\text{C}}$

Comment. Osmotic pressures are inherently large. Even dilute solutions with small freezing point depressions have large osmotic pressures.

E7.9 $\Delta_{\text{mix}}G = nRT\{x_A \ln x_A + x_B \ln x_B\}$ [17] $x_A = x_B = 0.5, \quad n = \frac{pV}{RT}$

Therefore,

$$\Delta_{\text{mix}}G = (pV) \times \left(\frac{1}{2}\ln\frac{1}{2} + \frac{1}{2}\ln\frac{1}{2}\right) = -pV\ln 2$$

$$= (-1.0) \times (1.013 \times 10^5\ \text{Pa}) \times (5.0 \times 10^{-3}\ \text{m}^3) \times (\ln 2)$$

$$= -3.5 \times 10^2\ \text{J} = \boxed{-0.35\ \text{kJ}}$$

$$\Delta_{\text{mix}}S = -nR\{x_A \ln x_A + x_B \ln x_B\} = \frac{-\Delta_{\text{mix}}G}{T}\ [18] = \frac{-0.35\ \text{kJ}}{298\ \text{K}} = \boxed{+1.2\ \text{J K}^{-1}}$$

E7.10 $\Delta_{\text{mix}}S = -nR\sum_J x_J \ln x_J$ [18]

Therefore, for molar amounts,

$$\Delta_{\text{mix}}S = -R\sum_J x_J \ln x_J$$

$$= -R[(0.782\ln 0.782) + (0.209\ln 0.209) + (0.009\ln 0.009) + (0.0003\ln 0.0003)]$$

$$= 0.564R = \boxed{+4.7\ \text{J K}^{-1}\ \text{mol}^{-1}}$$

E7.11 Hexane and heptane form nearly ideal solutions, therefore eqn 18 applies.

$\Delta_{\text{mix}}S = -nR(x_A \ln x_A + x_B \ln x_B)$ [18]

We need to differentiate eqn 18 with respect to x_A and look for the value of x_A at which the derivative is zero. Since $x_B = 1 - x_A$, we need to differentiate

$$\Delta_{\text{mix}}S = -nR\{x_A \ln x_A + (1 - x_A)\ln(1 - x_A)\}$$

This gives $\left(\text{using}\ \frac{\text{d}\ln x}{\text{d}x} = \frac{1}{x}\right)$

$$\frac{\text{d}\Delta_{\text{mix}}S}{\text{d}x_A} = -nR\{\ln x_A + 1 - \ln(1 - x_A) - 1\} = -nR\ln\frac{x_A}{1 - x_A}$$

which is zero when $x_A = \frac{1}{2}$. Hence, the maximum entropy of mixing occurs for the preparation of a mixture that contains equal mole fractions of the two components.

(a) $$\frac{n(\text{Hex})}{n(\text{Hep})} = 1 = \frac{\left(\frac{m(\text{Hex})}{M(\text{Hex})}\right)}{\left(\frac{m(\text{Hep})}{M(\text{Hep})}\right)}$$

(b) $$\frac{m(\text{Hex})}{m(\text{Hep})} = \frac{M(\text{Hex})}{M(\text{Hep})} = \frac{86.17\ \text{g mol}^{-1}}{100.20\ \text{g mol}^{-1}} = \boxed{0.8600}$$

E7.12 $p = xK$ [25], $K = 1.25 \times 10^6$ Torr $\quad x = \frac{n(\text{CO}_2)}{n(\text{CO}_2) + n(\text{H}_2\text{O})} \approx \frac{n(\text{CO}_2)}{n(\text{H}_2\text{O})}$

Therefore, with 1.00 kg H_2O

$$n(\text{CO}_2) \approx xn(\text{H}_2\text{O}) \quad \text{with} \quad n(\text{H}_2\text{O}) = \frac{1.00 \times 10^3\ \text{g}}{18.02\ \text{g mol}^{-1}} \quad \text{and} \quad x = \frac{p}{K}$$

Hence $n(\text{CO}_2) \approx \left(\frac{10^3\ \text{g}}{18.02\ \text{g mol}^{-1}}\right) \times \left(\frac{p}{1.25 \times 10^6\ \text{Torr}}\right) \approx (4.44 \times 10^{-5}\ \text{mol}) \times (p/\text{Torr})$

(a) $p = 0.10 = 76\,\text{Torr}$

Hence, $n(CO_2) = (4.44 \times 10^{-5}\,\text{mol}) \times (76) = 3.4 \times 10^{-3}\,\text{mol}$. The solution is therefore $\boxed{3.4\,\text{mmol kg}^{-1}}$ in CO_2.

(b) $p = 1.0\,\text{atm}$; since $n \propto p$, the solution is $\boxed{34\,\text{mmol kg}^{-1}}$ in CO_2.

E7.13 Use the result established in Exercise 7.12(**a**) that the amount of CO_2 in 1 kg of water is given by

$$n(CO_2) = (4.4 \times 10^{-5}\,\text{mol}) \times (p/\text{Torr})$$

and substitute $p \approx (5.0) \times (760\,\text{Torr}) = 3.8 \times 10^3\,\text{Torr}$, to give

$$n(CO_2) = (4.4 \times 10^{-5}\,\text{mol}) \times (3.8 \times 10^3) = 0.17\,\text{mol}$$

Hence, the molality of the solution is about $\boxed{0.17\,\text{mol kg}^{-1}}$ and, since molalities and molar concentrations for dilute aqueous solutions are approximately equal, the molar concentration is about $0.17\,\text{mol L}^{-1}$.

E7.14 $\Delta T = K_f b_B$ [34]; $K_f = 1.86\,\text{K kg mol}^{-1}$ [Table 7.2]

$$\Delta T = (1.86\,\text{K kg mol}^{-1}) \times \frac{\left(\frac{7.5\,\text{g}}{342.3\,\text{g mol}^{-1}}\right)}{0.25\,\text{kg}} = 0.16\,\text{K}$$

Hence, the freezing point will be approximately $\boxed{-0.16°\text{C}}$.

E7.15 The solubility in grams of anthracene per kg of benzene can be obtained from its mole fraction with use of the equation

$$\ln x_B = \frac{\Delta_{\text{fus}}H}{R} \times \left(\frac{1}{T^*} - \frac{1}{T}\right) \quad [36; \text{B, the solute, is anthracene}]$$

$$= \left(\frac{28.8 \times 10^3\,\text{J mol}^{-1}}{8.314\,\text{J K}^{-1}\,\text{mol}^{-1}}\right) \times \left(\frac{1}{490.15\,\text{K}} - \frac{1}{298.15\,\text{K}}\right) = -4.55$$

Therefore, $x_B = e^{-4.55} = 0.0106$

Since $x_B \ll 1$, $x(\text{anthracene}) \approx \dfrac{n(\text{anthracene})}{n(\text{benzene})}$

Therefore, in 1 kg of benzene,

$$n(\text{anthr.}) \approx x(\text{anthr.}) \times \left(\frac{1000\,\text{g}}{78.11\,\text{g mol}^{-1}}\right) \approx (0.0106) \times (12.80\,\text{mol}) = 0.136\,\text{mol}$$

The molality of the solution is therefore $0.136\,\text{mol kg}^{-1}$. Since $M = 178\,\text{g mol}^{-1}$, 0.136 mol corresponds to $\boxed{24\,\text{g anthracene}}$ in 1 kg of benzene.

E7.16 The best value of the molar mass is obtained from values of the data extrapolated to zero concentration, since it is under this condition that eqn 37 applies.

$$\Pi V = n_B RT \; [37], \quad \text{so} \quad \Pi = \frac{mRT}{MV} = \frac{cRT}{M}, \quad c = \frac{m}{V}$$

$$\Pi = \rho g h \; [\text{hydrostatic pressure}], \quad \text{so} \quad h = \left(\frac{RT}{\rho g M}\right) c$$

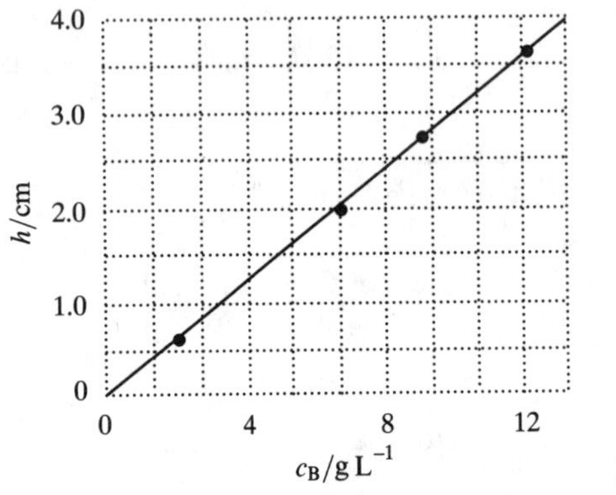

Figure 7.1

Hence, plot h against c and identify the slope as $\dfrac{RT}{\rho g M}$. Figure 7.1 shows the plot of the data.

The slope of the line is $0.29\,\text{cm}/(\text{g L}^{-1})$, so

$$\frac{RT}{\rho g M} = \frac{0.29\,\text{cm}}{\text{g L}^{-1}} = 0.29\,\text{cm L g}^{-1} = 0.29 \times 10^{-2}\,\text{m}^4\,\text{kg}^{-1}$$

Therefore,

$$M = \frac{RT}{(\rho g) \times (0.29 \times 10^{-2}\,\text{m}^4\,\text{kg}^{-1})}$$
$$= \frac{(8.314\,\text{J K}^{-1}\,\text{mol}^{-1}) \times (298.15\,\text{K})}{(1.004 \times 10^3\,\text{kg m}^{-3}) \times (9.81\,\text{m s}^{-2}) \times (0.29 \times 10^{-2}\,\text{m}^4\,\text{kg}^{-1})} = \boxed{87\,\text{kg mol}^{-1}}$$

E7.17 *For* A (Raoult's law basis; concentration in mole fraction)

$$a_A = \frac{p_A}{p_A^*}\,[42] = \frac{250\,\text{Torr}}{300\,\text{Torr}} = \boxed{0.833}; \qquad \gamma_A = \frac{a_A}{x_A} = \frac{0.833}{0.90} = \boxed{0.93}$$

For B (Henry's law basis; concentration in mole fraction)

$$a_B = \frac{p_B}{K_B}\,[49] = \frac{25\,\text{Torr}}{200\,\text{Torr}} = \boxed{0.125}; \qquad \gamma_B = \frac{a_B}{x_B} = \frac{0.125}{0.10} = \boxed{1.25}$$

For B (Henry's law basis; concentration in molality)

An equation analogous to eqn 49 is used $a_B = \dfrac{p_B}{K'_B}$ with a modified Henry's law constant K'_B which corresponds to the pressure of B in the limit of very low molalities.

$$p_B = \frac{b_B}{b^\ominus} \times K'_B$$

is analogous to $p_B = x_B K_B$. Since x_B and b_B are related as $b_B = \dfrac{x_B}{M_A x_A}$

K'_B and K_B are related as $K'_B = x_A M_A b^\ominus K_B$

We also need M_A

$$M_A = \frac{x_B}{x_A b_B} = \frac{0.10}{(0.90) \times (2.22\,\text{mol kg}^{-1})} = 0.050\,\text{kg mol}^{-1}$$

Then, $K'_B = (0.90) \times (0.050\,\text{kg mol}^{-1}) \times (1\,\text{mol kg}^{-1}) \times (200\,\text{Torr}) = 9.0\,\text{Torr}$

and $a_B = \dfrac{25\,\text{Torr}}{9.0\,\text{Torr}} = \boxed{2.8} \qquad \gamma_B = \dfrac{a_B}{\left(\frac{b_B}{b^\ominus}\right)} = \dfrac{2.8}{2.22} = \boxed{1.25}$

Comment. The two methods for the 'solute' B give different values for the activities. This is reasonable since the chemical potentials in the reference states $\mu^\dagger$ and $\mu^\ominus$ are different.

Question. What are the activity and activity coefficient of B in the Raoult's law basis?

E7.18 In an ideal dilute solution the solvent (CCl_4) obeys Raoult's law and the solute (Br_2) obeys Henry's law; hence

$$p(\text{CCl}_4) = x(\text{CCl}_4)p^*(\text{CCl}_4)\ [23] = (0.950) \times (33.85\,\text{Torr}) = \boxed{32.2\,\text{Torr}}$$

$$p(\text{Br}_2) = x(\text{Br}_2)K(\text{Br}_2)\ [25] = (0.050) \times (122.36\,\text{Torr}) = \boxed{6.1\,\text{Torr}}$$

$$p(\text{Total}) = (32.2 + 6.1)\text{Torr} = \boxed{38.3\,\text{Torr}}$$

The composition of the vapour in equilibrium with the liquid is

$$y(\text{CCl}_4) = \frac{p(\text{CCl}_4)}{p(\text{Total})} = \frac{32.2\,\text{Torr}}{38.3\,\text{Torr}} = \boxed{0.841}$$

$$y(\text{Br}_2) = \frac{p(\text{Br}_2)}{p(\text{Total})} = \frac{6.1\,\text{Torr}}{38.3\,\text{Torr}} = \boxed{0.16}$$

E7.19 Let A = acetone and M = methanol

$$y_A = \frac{p_A}{p_A + p_M}[\text{Dalton's law}] = \frac{p_A}{760\,\text{Torr}} = 0.516$$

$$p_A = 392\,\text{Torr}, \quad p_M = 368\,\text{Torr}$$

$$a_A = \frac{p_A}{p_A^*}\,[42] = \frac{392\,\text{Torr}}{786\,\text{Torr}} = \boxed{0.499} \qquad a_M = \frac{p_M}{p_M^*} = \frac{368\,\text{Torr}}{551\,\text{Torr}} = \boxed{0.668}$$

$$\gamma_A = \frac{a_A}{x_A} = \frac{0.499}{0.400} = \boxed{1.25} \qquad \gamma_M = \frac{a_M}{x_M} = \frac{0.668}{0.600} = \boxed{1.11}$$

Solutions to problems

Solutions to numerical problems

P7.1 $p_A = y_A p$ and $p_B = y_B p$ (Dalton's law). Hence, draw up the following table

p_A/kPa	0	1.399	3.566	5.044	6.996	7.940	9.211	10.105	11.287	12.295
x_A	0	0.0898	0.2476	0.3577	0.5194	0.6036	0.7188	0.8019	0.9105	1
y_A	0	0.0410	0.1154	0.1762	0.2772	0.3393	0.4450	0.5435	0.7284	1

p_B/kPa	0	4.209	8.487	11.487	15.462	18.243	23.582	27.334	32.722	36.066
x_B	0	0.0895	0.1981	0.2812	0.3964	0.4806	0.6423	0.7524	0.9102	1
y_B	0	0.2716	0.4565	0.5550	0.6607	0.7228	0.8238	0.8846	0.9590	1

The data are plotted in Fig. 7.2.

We can assume, at the lowest concentrations of both A and B, that Henry's law will hold. The Henry's law constants are then given by

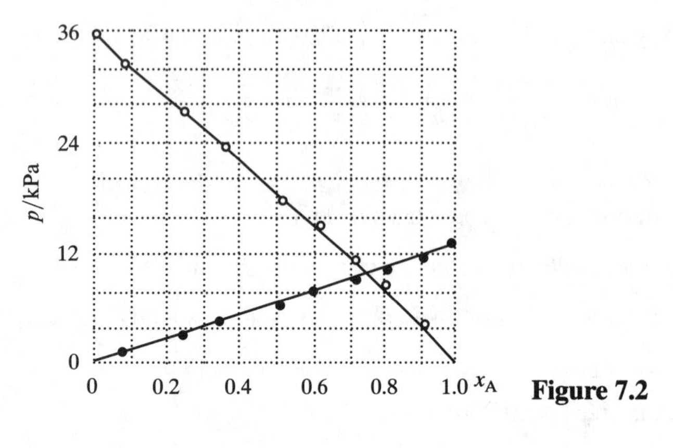

Figure 7.2

$$K_A = \frac{p_A}{x_A} = \boxed{15.58\,\text{kPa}} \text{ from the point at } x_A = 0.0898$$

$$K_B = \frac{p_B}{x_B} = \boxed{47.03\,\text{kPa}} \text{ from the point at } x_B = 0.0895$$

P7.2
$$V_A = \left(\frac{\partial V}{\partial n_A}\right)_{n_B} [1, \text{A} = \text{NaCl(aq)}, \text{B} = \text{water}] = \left(\frac{\partial V}{\partial b}\right)_{n(\text{H}_2\text{O})} \text{mol}^{-1} \; [\text{with } b \equiv b/(\text{mol kg}^{-1})]$$

$$= \left((16.62) + \tfrac{3}{2} \times (1.77) \times (b)^{1/2} + (2) \times (0.12b)\right) \text{cm}^3\,\text{mol}^{-1}$$

$$= \boxed{17.5\,\text{cm}^3\,\text{mol}^{-1}} \quad \text{when} \quad b = 0.100$$

For a solution consisting of 0.100 mol NaCl and 1.000 kg of water, corresponding to 55.49 mol H_2O, the total volume is given both by

$$V = [(1003) + (16.62) + (0.100) \times (1.77) \times (0.100)^{3/2} + (0.12) \times (0.100)^2]\,\text{cm}^3$$
$$= 1004.7\,\text{cm}^3$$

and by $V = n(\text{NaCl})V_{\text{NaCl}} + n(\text{H}_2\text{O})V_{\text{H}_2\text{O}}$ [3] $= (0.100\,\text{mol}) \times (17.5\,\text{cm}^3\,\text{mol}^{-1}) + (55.49\,\text{mol}) \times V_{\text{H}_2\text{O}}$

Therefore, $V_{\text{H}_2\text{O}} = \dfrac{1004.7\,\text{cm}^3 - 1.75\,\text{cm}^3}{55.49\,\text{mol}} = \boxed{18.07\,\text{cm}^3\,\text{mol}^{-1}}$

Comment. Within four significant figures, this result is the same as the molar volume of pure water at 25°C.

Question. How does the partial molar volume of NaCl(aq) in this solution compare to molar volume of pure solid NaCl?

P7.4 Let $m(\text{CuSO}_4)$, which is the mass of $CuSO_4$ dissolved in 100 g of solution, be represented by

$$w = \frac{100m_B}{m_A + m_B} = \text{mass per cent of CuSO}_4$$

where m_B is the mass of $CuSO_4$ and m_A is the mass of water. Then using

$$\rho = \frac{m_A + m_B}{V} \qquad n_A = \frac{m_A}{M_A}$$

the procedure runs as follows

$$V_A = \left(\frac{\partial V}{\partial n_A}\right)_{n_B} = \left(\frac{\partial V}{\partial m_A}\right)_B M_A$$

$$= \frac{\partial}{\partial m_A}\left(\frac{m_A + m_B}{\rho}\right) \times M_A$$

$$= \frac{M_A}{\rho} + (m_A + m_B) M_A \frac{\partial}{\partial m_A}\frac{1}{\rho}$$

$$\frac{\partial}{\partial m_A}\frac{1}{\rho} = \left(\frac{\partial w}{\partial m_A}\right)\frac{\partial}{\partial w}\frac{1}{\rho} = \frac{-w}{m_A + m_B}\frac{\partial}{\partial w}\frac{1}{\rho}$$

Therefore,

$$V_A = \frac{M_A}{\rho} - w M_A \frac{\partial}{\partial w}\frac{1}{\rho}$$

and hence

$$\frac{1}{\rho} = \frac{V_A}{M_A} + w\frac{d}{dw}\left(\frac{1}{\rho}\right)$$

Therefore, plot $1/\rho$ against w and extrapolate the tangent to $w = 100$ to obtain V_B/M_B. For the actual procedure, draw up the following table

w	5	10	15	20
$\rho/(\mathrm{g\,cm^{-3}})$	1.051	1.107	1.167	1.230
$1/(\rho/\mathrm{g\,cm^{-3}})$	0.951	0.903	0.857	0.813

The values of $1/\rho$ are plotted against w in Fig. 7.3.

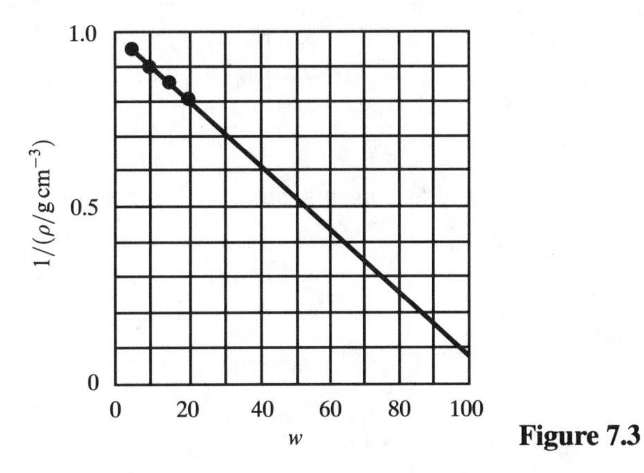

Figure 7.3

Four tangents are drawn to the curve at the four values of w. As the curve is a straight line to within the precision of the data, all four tangents are coincident and all four intercepts are equal at $0.075\,\mathrm{g^{-1}\,cm^3}$. Thus

$$V(\mathrm{CuSO_4}) = 0.075\,\mathrm{g^{-1}\,cm^3} \times 159.6\,\mathrm{g\,mol^{-1}} = \boxed{12.0\,\mathrm{cm^3\,mol^{-1}}}$$

P7.6 **(a)** On a Raoult's law basis, $a = \frac{p}{p^*}$, $a = \gamma x$, and $\gamma = \frac{p}{xp^*}$. On a Henry's law basis, $a = \frac{p}{K}$, and $\gamma = \frac{p}{xK}$. The vapour pressures of the pure components are given in the table of data and are: $p_I^* = 353.4\,\mathrm{Torr}$, $p_A^* = 280.4\,\mathrm{Torr}$.

(b) The Henry's law constants are determined by plotting the data and extrapolating the low concentration data to $x = 1$. The data are plotted in Fig. 7.4. K_A and K_I are estimated as graphical tangents at $x_I = 1$ and $x_I = 0$, respectively. The values obtained are: $K_A = \boxed{450\text{ Torr}}$ and $K_I = \boxed{465\text{ Torr}}$

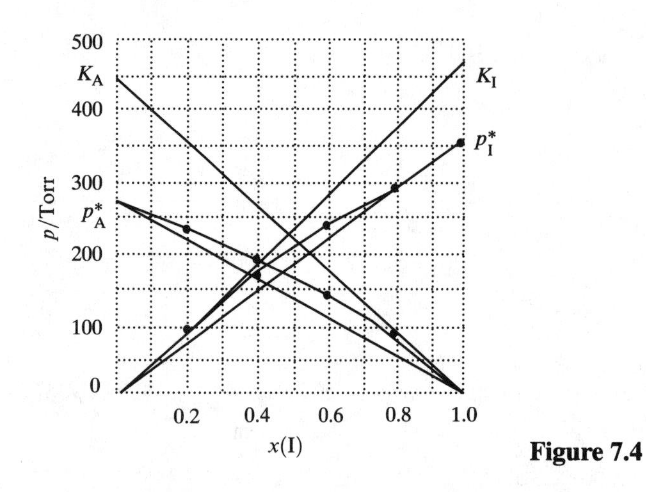

Figure 7.4

Then draw up the following table based on the values of the partial pressures obtained from the plots at the values of x(I) given in the figure.

x_I	0	0.2	0.4	0.6	0.8	1.0
p_I/Torr	0	92	165	230	290	353.4‡
p_A/Torr	280.4†	230	185	135	80	0
γ_I(R)	—	1.30	1.17	1.09	1.03	1.000 $[p_I/x_I p^*_I]$
γ_A(R)	1.000	1.03	1.10	1.20	1.43	— $[p_A/x_A p^*_A]$
γ_I(H)	1.000	0.990	0.887	0.824	0.780	0.760 $[p_I/x_I K_I]$

†The value of p^*_A; ‡the value of p^*_I.

Question. In this problem both I and A were treated as solvents, but only I as a solute. Extend the table by including a row for γ_A(H).

Solutions to theoretical problems

P7.8
$$\mu_A = \left(\frac{\partial G}{\partial n_A}\right)_{n_B} [4] = \mu^\circ_A + \left(\frac{\partial}{\partial n_A}(nG^E)\right)_{n_B} \quad [\mu^\circ_A \text{ is ideal value} = \mu^*_A + RT\ln x_A]$$

$$\left(\frac{\partial nG^E}{\partial n_A}\right)_{n_B} = G^E + n\left(\frac{\partial G^E}{\partial n_A}\right)_{n_B} = G^E + n\left(\frac{\partial x_A}{\partial n_A}\right)_B\left(\frac{\partial G^E}{\partial x_A}\right)_B$$

$$= G^E + n \times \frac{x_B}{n} \times \left(\frac{\partial G^E}{\partial x_A}\right)_B \quad [\partial x_A/\partial n_A = x_B/n]$$

$$= gRTx_A(1-x_A) + (1-x_A)gRT(1-2x_A)$$

$$= gRT(1-x_A)^2 = gRTx_B^2$$

Therefore, $\mu_A = \boxed{\mu^*_A + RT\ln x_A + gRTx_B^2}$

P7.10 $n_A dV_A + n_B dV_B = 0$ [Example 7.1]

Hence $\dfrac{n_A}{n_B}\,dV_A = -dV_B$

Therefore, by integration,

$$V_B(x_A) - V_B(0) = -\int_{V_A(0)}^{V_A(x_A)} \frac{n_A}{n_B}\,dV_A = -\int_{V_A(0)}^{V_A(x_A)} \frac{x_A dV_A}{1-x_A} \quad [n_A = x_A n,\ n_B = x_B n]$$

Therefore, $V_B(x_A, x_B) = V_B(0,1) - \displaystyle\int_{V_A(0)}^{V_A(x_A)} \frac{x_A\,dV_A}{1-x_A}$

P7.12 $$\phi = -\frac{\ln a_A}{r} \qquad \text{(a)}$$

Therefore, $d\phi = -\dfrac{1}{r}\,d\ln a_A + \dfrac{1}{r^2}\ln a_A\,dr$

$$d\ln a_A = \frac{1}{r}\ln a_A\,dr - r\,d\phi \qquad \text{(b)}$$

From the Gibbs–Duhem equation, $x_A\,d\mu_A + x_B\,d\mu_B = 0$, which implies that (since $\mu = \mu^{\ominus} + RT\ln a$, $d\mu_A = RT\,d\ln a_A$, $d\mu_B = RT\,d\ln a_B$)

$$d\ln a_B = -\frac{x_A}{x_B}\,d\ln a_A = -\frac{d\ln a_A}{r} = -\frac{1}{r^2}\ln a_A\,dr + d\phi\ \text{[from (b)]} = \frac{1}{r}\phi\,dr + d\phi\ \text{[from (a)]}$$

$$= \phi\,d\ln r + d\phi$$

Subtract $d\ln r$ from both sides, to obtain

$$d\ln\frac{a_B}{r} = (\phi - 1)\,d\ln r + d\phi = \frac{(\phi-1)}{r}\,dr + d\phi$$

Then, by integration and noting that $\ln\left(\dfrac{a_B}{r}\right)_{r=0} = \ln\left(\dfrac{\gamma_B x_B}{r}\right)_{r=0} = \ln(\gamma_B)_{r=0} = \ln 1 = 0$

$$\ln\frac{a_B}{r} = \boxed{\phi - \phi(0) + \int_0^r \left(\frac{\phi-1}{r}\right) dr}$$

Solutions to additional problems

P7.14 The partial molar volume of cyclohexane is

$$V_c = \left(\frac{\partial V}{\partial n_c}\right)_{p,T,n_2}$$

A similar expression holds for V_p. V_c can be evaluated graphically by plotting V against n_c and finding the slope at the desired point. In a similar manner, V_p can be evaluated by plotting V against n_p. To find V_c, V is needed at a variety of n_c while holding n_p constant, say at 1.0000 mol; likewise to find V_p, V is needed at a variety of n_p while holding n_c constant.The mole fraction in this system is

$$x_c = \frac{n_c}{n_c + n_p} \quad \text{so} \quad n_c = \frac{x_c n_p}{1 - x_c}$$

From n_c and n_p, the mass of the sample can be calculated, and the volume can be calculated from

$$V = \frac{m}{\rho} = \frac{n_c M_c + n_p M_p}{\rho}$$

The following table is drawn up

$n_c/\text{mol}(n_p = 1)$	V/cm^3	x_c	$\rho/\text{g cm}^{-3}$	$n_p/\text{mol}(n_c = 1)$	V/cm^3
2.295	529.4	0.6965	0.7661	0.4358	230.7
3.970	712.2	0.7988	0.7674	0.2519	179.4
9.040	1264	0.9004	0.7697	0.1106	139.9

These values are plotted in Fig. 7.5(a) and (b).

These plots show no curvature, so in this case, perhaps due to the limited number of data points, the molar volumes are independent of the mole numbers and are

$$V_c = \boxed{109.0\,\text{cm}^3\,\text{mol}^{-1}} \quad \text{and} \quad V_p = \boxed{279.3\,\text{cm}^3\,\text{mol}^{-1}}$$

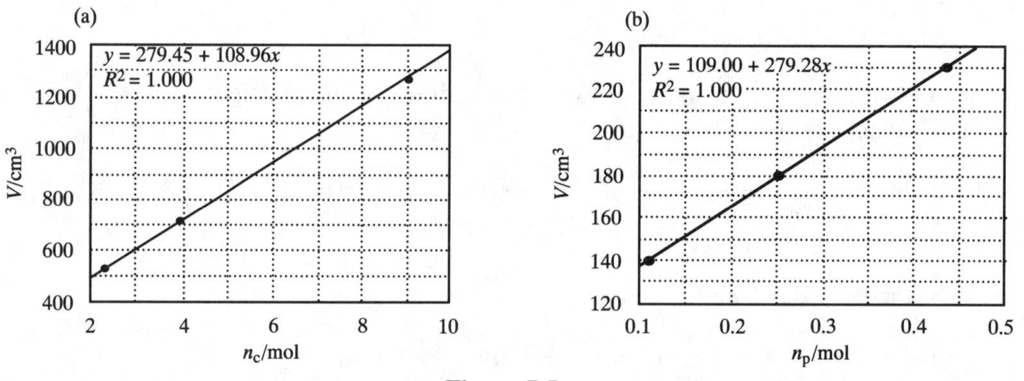

Figure 7.5

P7.16 The activity of a solvent is

$$a_A = \frac{p_A}{p_A^*} = x_A \gamma_A$$

so the activity coefficient is

$$\gamma_A = \frac{p_A}{x_A p_A^*} = \frac{y_A p}{x_A p_A^*}$$

where the last equality applies Dalton's law of partial pressures to the vapour phase.

Substituting the data, the following table of results is obtained.

p/kPa	x_T	y_T	γ_T	γ_E
23.40	0.000	0.000		
21.75	0.129	0.065	0.418	0.998
20.25	0.228	0.145	0.490	1.031
18.75	0.353	0.285	0.576	1.023
18.15	0.511	0.535	0.723	0.920
20.25	0.700	0.805	0.885	0.725
22.50	0.810	0.915	0.966	0.497
26.30	1.000	1.000		

P7.18 $S = S_0 e^{\tau/T}$ may be written in the form $\ln S = \ln S_0 + \dfrac{\tau}{T}$ which indicates that a plot of $\ln S$ against $1/T$ should be linear with slope τ and intercept $\ln S_0$. Linear regression analysis gives $\boxed{\tau = 165\,\text{K}}$, standard deviation $= 2\,\text{K}$

$\ln(S_0/\text{mol L}^{-1}) = 2.990$, standard deviation $= 0.007$; $S_0 = e^{2.990}\,\text{mol L}^{-1} = \boxed{19.89\,\text{mol L}^{-1}}$

$R = \boxed{0.99978}$

The linear regression explains 99.98 per cent of the variation.

Equation 36 is

$$x_B = e^{-\left(\frac{\Delta_{\text{fus}}H}{R}\left(\frac{1}{T}-\frac{1}{T^*}\right)\right)} = e^{-\Delta_{\text{fus}}H/RT}e^{\Delta_{\text{fus}}H/RT^*}$$

Comparing to $S = S_0 e^{\tau/T}$, we see that

$$\boxed{S_0 = e^{-\Delta_{\text{fus}}H/RT^*}}$$

where T^* is the normal melting point of the solute and $\Delta_{\text{fus}}H$ is its heat of fusion $\boxed{\tau = \Delta_{\text{fus}}H/R}$

8 Phase diagrams

Solutions to exercises

E8.1 An expression for composition of the solution in terms of its vapour pressure is required. This is obtained from Dalton's law and Raoult's law as follows

$$p = p_A + p_B[\text{Dalton's law}] = x_A p_A^* + (1 - x_A) p_B^*$$

Solving for x_A, $x_A = \dfrac{p - p_B^*}{p_A^* - p_B^*}$

For boiling under 0.50 atm (380 Torr) pressure, the combined vapour pressure, p, must be 380 Torr; hence $x_A = \dfrac{380 - 150}{400 - 150} = \boxed{0.920}$, $x_B = \boxed{0.080}$

The composition of the vapour is given by eqn 5

$$y_A = \frac{x_A p_A^*}{p_B^* + (p_A^* - p_B^*)x_A} = \frac{0.920 \times 400}{150 + (400 - 150) \times 0.920} = \boxed{0.968}$$

and $y_B = 1 - 0.968 = \boxed{0.032}$

E8.2 The vapour pressures of components A and B may be expressed in terms of both their composition in the vapour and in the liquid. The pressures are the same whatever the expression; hence the expressions can be set equal to each other and solved for the composition.

$$p_A = y_A p = 0.350p = x_A p_A^* = x_A \times (575\,\text{Torr})$$
$$p_B = y_B p = (1 - y_A)p = 0.650p = x_B p_B^* = (1 - x_A) \times (390\,\text{Torr})$$

Therefore, $\dfrac{y_A p}{y_B p} = \dfrac{x_A p_A^*}{x_B p_B^*}$

Hence $\dfrac{0.350}{0.650} = \dfrac{575 x_A}{390(1 - x_A)}$

which solves to $x_A = \boxed{0.268}$, $x_B = 1 - x_A = \boxed{0.732}$

and, since $0.350p = x_A p_A^*$

$$p = \frac{x_A p_A^*}{0.350} = \frac{(0.268) \times (575\,\text{Torr})}{0.350} = \boxed{440\,\text{Torr}}$$

E8.3 **(a)** Check to see if Raoult's law holds; if it does the solution is ideal.

$$p_A = x_A p_A^* = (0.6589) \times (957\,\text{Torr}) = 630.6\,\text{Torr}$$
$$p_B = x_A p_B^* = (0.3411) \times (379.5\,\text{Torr}) = 129.\bar{4}\,\text{Torr}$$
$$p = p_A + p_B = 760\,\text{Torr} = 1\,\text{atm}$$

Since this is the pressure at which boiling occurs, Raoult's law holds and $\boxed{\text{the solution is ideal}}$

(b) $y_A = \dfrac{p_A}{p}[4] = \dfrac{630.6\,\text{Torr}}{760\,\text{Torr}} = \boxed{0.830}$ $\quad y_B = 1 - y_A = 1.000 - 0.830 = \boxed{0.170}$

E8.4 **(a)** $p(\text{total}) = p_{DE} + p_{DP}$ [Dalton's law] $= x_{DE} p_{DE}^* + x_{DP} p_{DP}^*$ [Raoult's law, 3]

$x_{DE} = z_{DE}$, $\quad x_{DP} = 1 - z_{DE}$ [system all liquid]

$p(\text{total}) = (0.60) \times (172\,\text{Torr}) + (0.40) \times (128\,\text{Torr}) = 10\bar{3} + 51 = \boxed{15\bar{4}\,\text{Torr}}$

(b) $y_{DE} = \frac{p_{DE}}{p}\,[4] = \frac{10\bar{3}\text{ Torr}}{154\text{ Torr}} = \boxed{0.67}$ $\quad y_{DP} = 1 - y_{DE} = \boxed{0.33}$

E8.5 The data are plotted in Fig. 8.1. From the graph, the vapour in equilibrium with a liquid of composition **(a)** $x_M = 0.25$ is determined from the tie line labelled a in the figure extending from $x_M = 0.25$ to $\boxed{y_M = 0.36}$, **(b)** $x_O = 0.25$ is determined from the tie line labelled b in the figure extending from $x_M = 0.75$ to $\boxed{y_M = 0.82}$

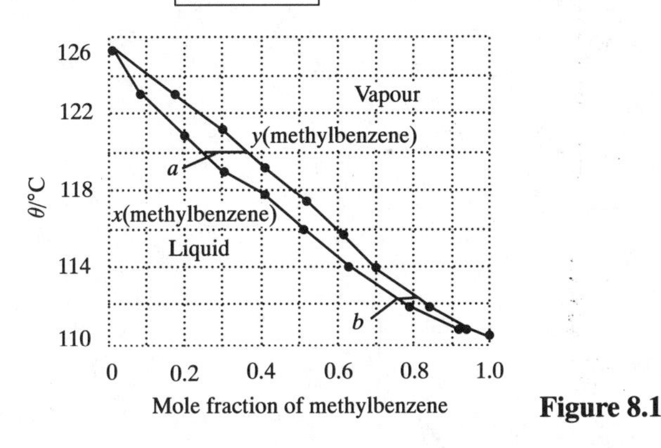

Figure 8.1

E8.6 **(a)** Though there are three constituents, salt, water, and water vapour, there is an equilibrium condition between liquid water and its vapour. Hence, $\boxed{C = 2}$

(b) Disregarding the water vapour for the reasons in **(a)** there are seven species: Na^+, H^+, $H_2PO_4^-$, HPO_4^{2-}, PO_4^{3-}, H_2O, OH^-. There are also three equilibria, namely

$$H_2PO_4^- \rightleftharpoons H^+ + HPO_4^{2-}$$
$$HPO_4^{2-} \rightleftharpoons H^+ + PO_4^{3-}$$
$$H^+ + OH^- \rightleftharpoons H_2O$$

(These could all be written as Brønsted equilibria without changing the conclusions.) There are also two conditions of electrical neutrality, namely

$$[Na^+] = [\text{phosphates}], \qquad [H^+] = [OH^-] + [\text{phosphates}]$$

where $[\text{phosphates}] = [H_2PO_4^-] + 2[HPO_4^{2-}] + 3[PO_4^{3-}]$
Hence, the number of independent components is

$$C = 7 - (3 + 2) = \boxed{2}$$

E8.7 $CuSO_4 \cdot 5H_2O(s) \rightleftharpoons CuSO_4(s) + 5H_2O(g)$

There are two solids, but one solid phase, as well as a gaseous phase; hence $\boxed{P = 2}$ Assuming all the water and $CuSO_4$ are formed by the dehydration, their amounts are then fixed by the equilibrium; hence $\boxed{C = 2}$

E8.8 **(a)** The two components are Na_2SO_4 and H_2O (proton transfer equilibria to give HSO_4^- etc. do not change the number of independent components) so $\boxed{C = 2}$ There are three phases present (solid salt, liquid solution, vapour), so $\boxed{P = 3}$

(b) The variance is $F = C - P + 2 = 2 - 3 + 2 = \boxed{1}$

Either pressure or temperature may be considered the independent variable, but not both as long as the equilibrium is maintained. If the pressure is changed, the temperature must be changed to maintain the equilibrium.

E8.9 See Figs 8.2(a) and (b).

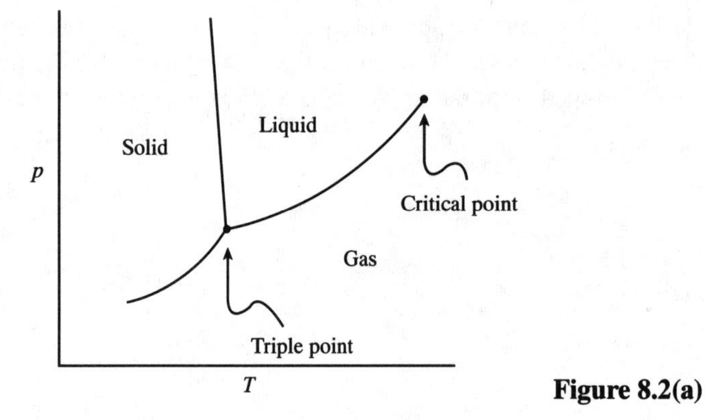

Figure 8.2(a)

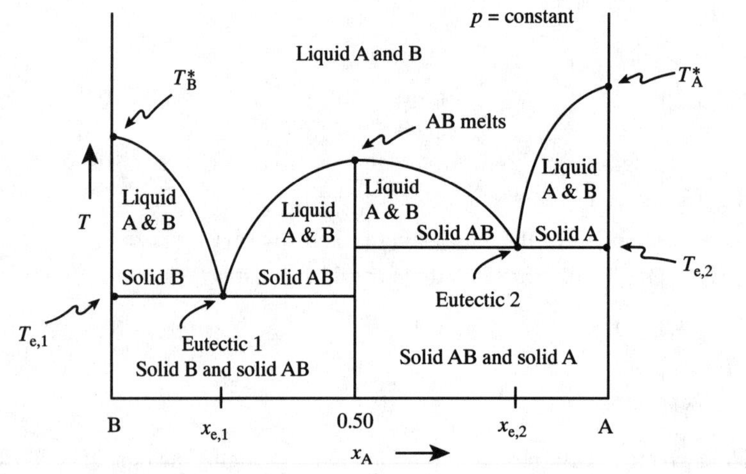

Figure 8.2(b)

E8.10 See Fig. 8.3.

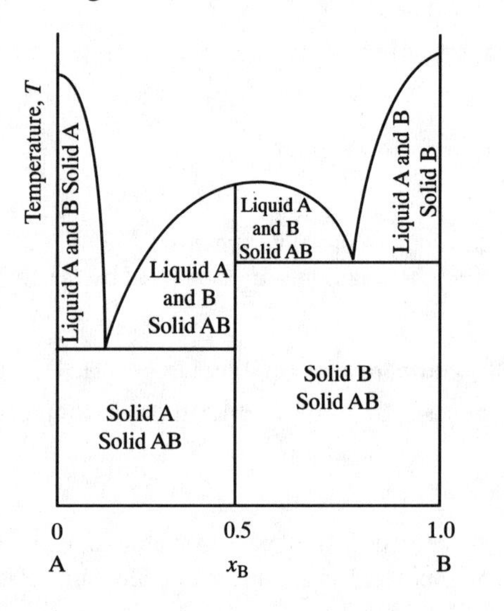

Figure 8.3

E8.11 See Fig. 8.4.

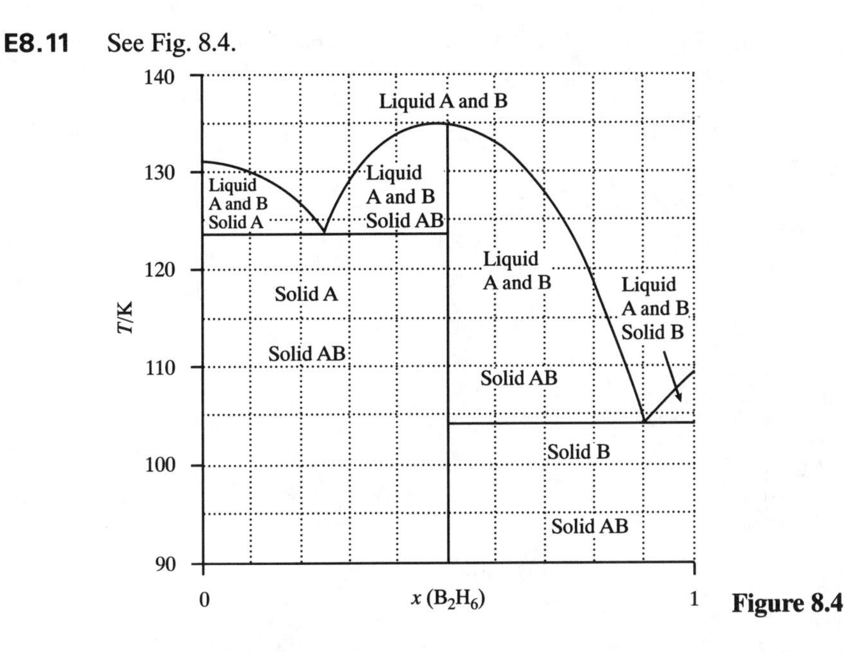

Figure 8.4

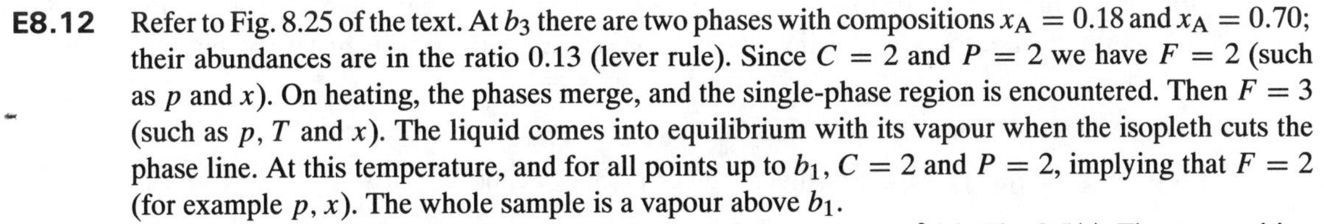

E8.12 Refer to Fig. 8.25 of the text. At b_3 there are two phases with compositions $x_A = 0.18$ and $x_A = 0.70$; their abundances are in the ratio 0.13 (lever rule). Since $C = 2$ and $P = 2$ we have $F = 2$ (such as p and x). On heating, the phases merge, and the single-phase region is encountered. Then $F = 3$ (such as p, T and x). The liquid comes into equilibrium with its vapour when the isopleth cuts the phase line. At this temperature, and for all points up to b_1, $C = 2$ and $P = 2$, implying that $F = 2$ (for example p, x). The whole sample is a vapour above b_1.

E8.13 The incongruent melting point (Section 8.6) is marked as $T_1 = 400°\mathrm{C}$ in Fig. 8.5(a). The composition of the eutectic is marked as $x_e (\approx 0.30)$ in the figure. Its melting point is T_2 ($\approx 200°\mathrm{C}$)

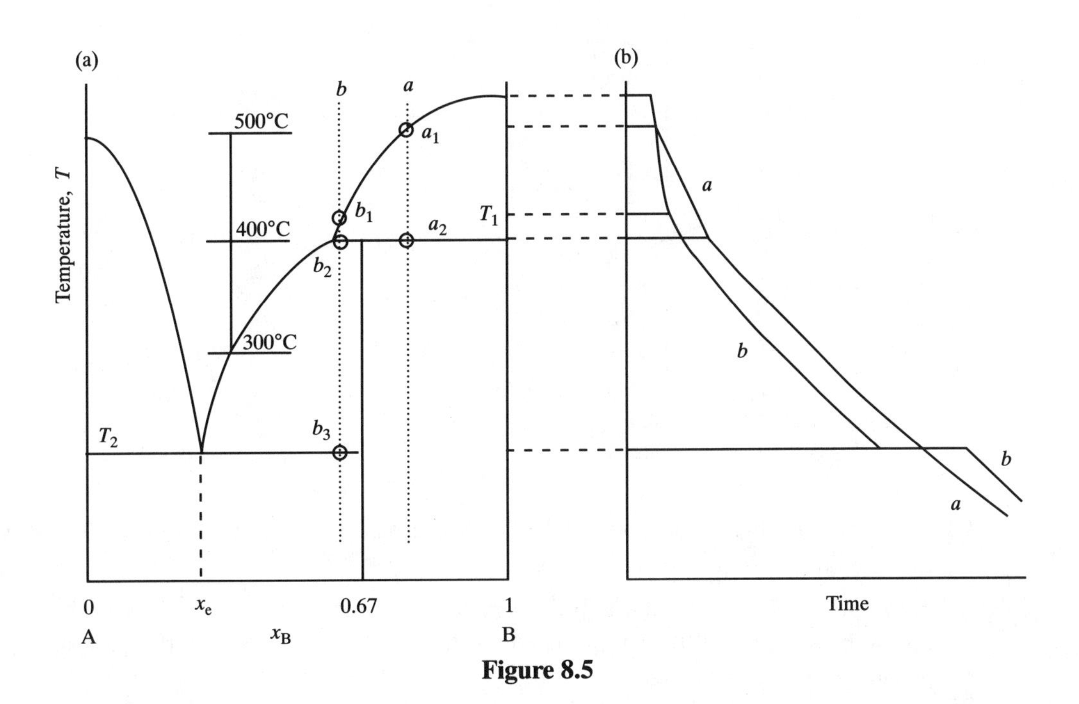

Figure 8.5

E8.14 The cooling curves are shown in Fig. 8.5(b). Note the breaks (abrupt change in slope) at temperatures corresponding to points a_1, a_2, b_1, b_2. Also note the eutectic halt at b_3.

E8.15 Refer to Fig. 8.6.

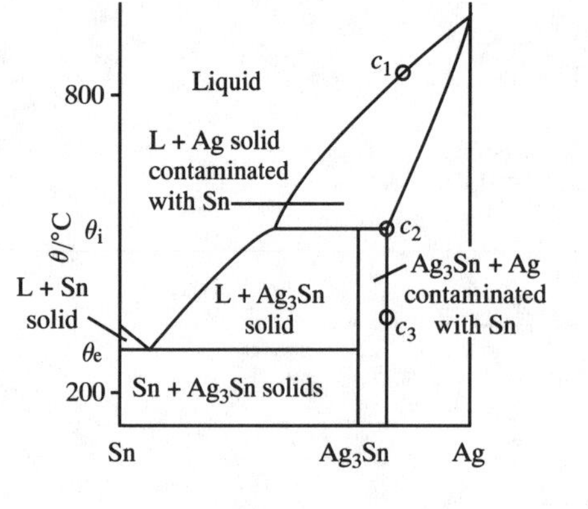

Figure 8.6

(a) The solubility of silver in tin at 800°C is determined by the point c_1 (at higher proportions of silver, the system separates into two phases). The point c_1 corresponds to $\boxed{\text{80 per cent}}$ silver by mass.

(b) See point c_2. The compound Ag_3Sn decomposes at this temperature.

(c) The solubility of Ag_3Sn in silver is given point c_3 at 300°C.

E8.16 **(a)** See Figs. 8.7(a) and (b).

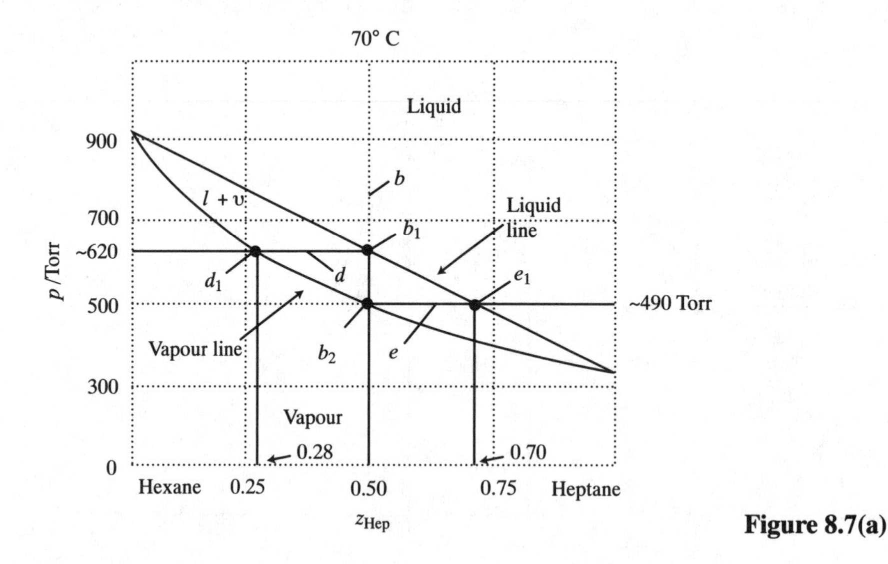

Figure 8.7(a)

(b) Follow line b in Fig. 8.7(a) down to the liquid line which intersects at point b_1. The vapour pressure at b_1 is $\approx \boxed{620\ \text{Torr}}$

(c) Follow line b in Fig. 8.7(a) down to the vapour line which intersects at point b_2. The vapour pressure at b_2 is $\approx \boxed{490\ \text{Torr}}$ From points b_1 to b_2, the system changes from essentially all liquid to essentially all vapour.

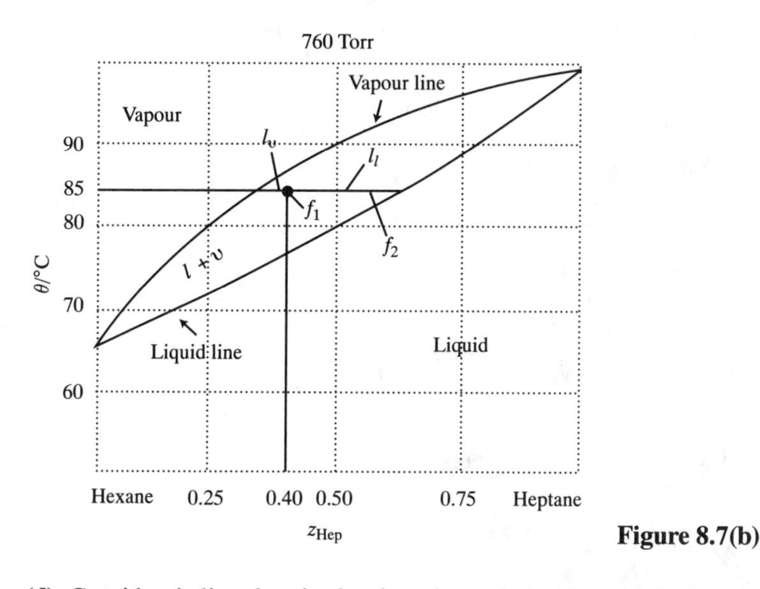

Figure 8.7(b)

(d) Consider tie line d; point b_1 gives the mole fractions of the liquid, which are

$$x(\text{Hep}) = 0.50 = 1 - x(\text{Hex}) \qquad x(\text{Hex}) = \boxed{0.50}$$

Point d_1 gives the mole fractions in the vapour which are

$$y(\text{Hep}) \approx 0.28 = 1 - y(\text{Hex}) \qquad y(\text{Hex}) \approx \boxed{0.72}$$

The initial vapour is richer in the more volatile component, hexane.

(e) Consider tie line e; point b_2 gives the mole fractions in the vapour, which are

$$y(\text{Hep}) = 0.50 = 1 - y(\text{Hex}) \qquad y(\text{Hex}) = \boxed{0.50}$$

Point e_1, gives the mole fractions in the liquid, which are

$$x(\text{Hep}) = 0.70 = 1 - x(\text{Hex}) \qquad x(\text{Hex}) = \boxed{0.30}$$

(f) Consider tie line f. The section, l_l, from point f_1 to the liquid line gives the relative amount of vapour; the section, l_v, from point f_1 to the liquid line gives the relative amount of liquid. That is

$$n_v l_v = n_l l_l \ [7] \quad \text{or} \quad \frac{n_v}{n_l} = \frac{l_l}{l_v} \approx \frac{6}{1}$$

Since the total amount is 2 mol, $n_v \approx \boxed{1.7}$ and $n_l \approx \boxed{0.3\,\text{mol}}$

E8.17 The phase diagram is drawn in Fig. 8.8.

E8.18 The cooling curves are sketched in Fig. 8.9. Note the breaks and halts. The breaks correspond to changes in the rate of cooloing due to the freezing out of a solid which releases its heat of fusion and thus slows down the cooling process. The halts correspond to the existence of three phases and hence no variance until one of the phases disappears.

E8.19 The phase diagram is sketched in Fig. 8.10.

(a) The mixture has a single liquid phase at all compositions.

(b) When the composition reaches $x(C_6F_{14}) = 0.24$ the mixture separates into two liquid phases of compositions $x = 0.24$ and 0.48. The relative amounts of the two phases change until the composition reaches $x = 0.48$. At all mole fractions greater than 0.48 in C_6F_{14} the mixture forms a single liquid phase.

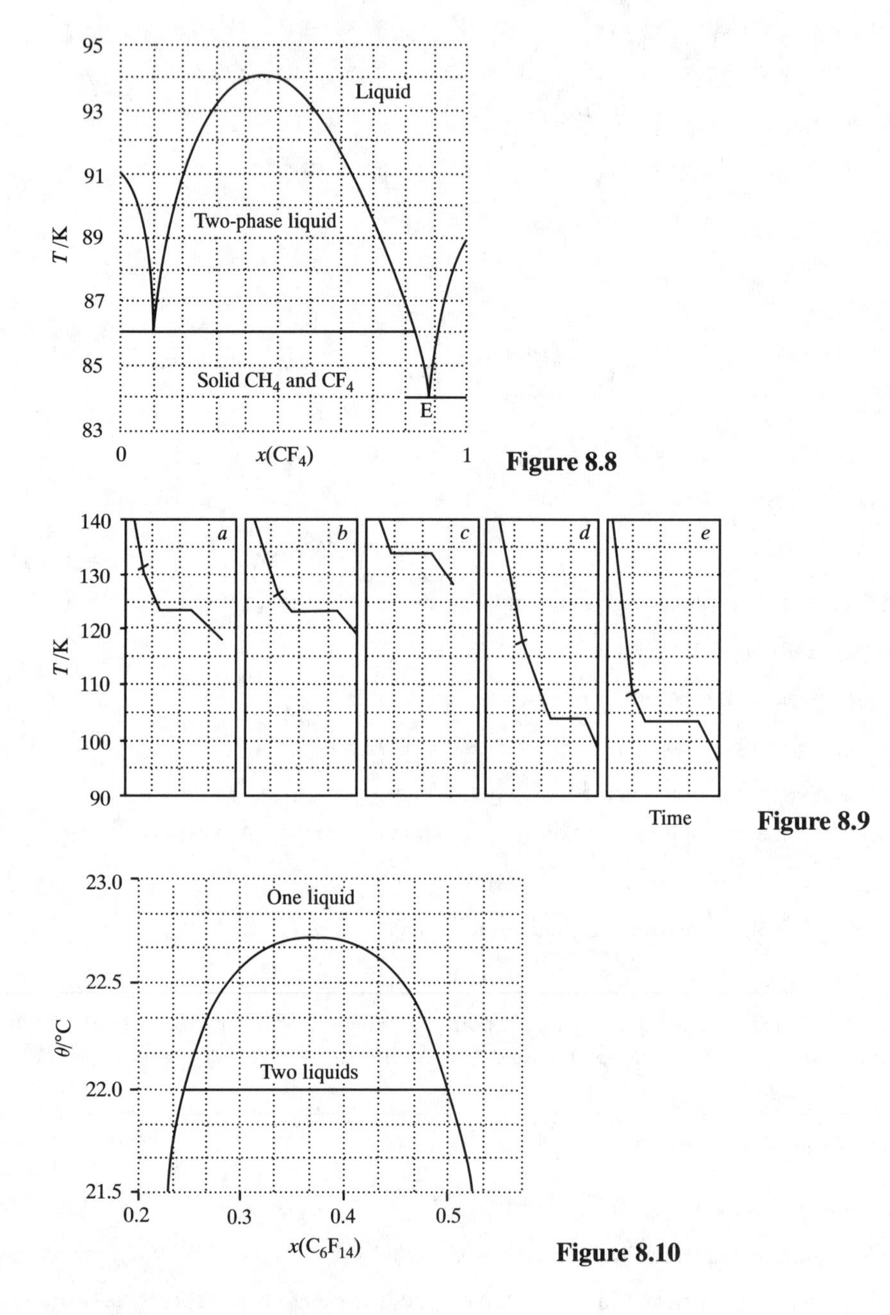

Figure 8.8

Figure 8.9

Figure 8.10

Solutions to problems

Solutions to numerical problems

P8.2 The data are plotted in Fig. 8.11.

(a) As the solid composition $x(\text{MgO}) = 0.3$ is heated, liquid begins to form when the solid (lower) line is reached $\boxed{\text{at } 2150°\text{C}}$

(b) From the tie line at 2200°C, the liquid composition is $y(\text{MgO}) = \boxed{0.18}$ and the solid $x(\text{MgO}) = \boxed{0.35}$

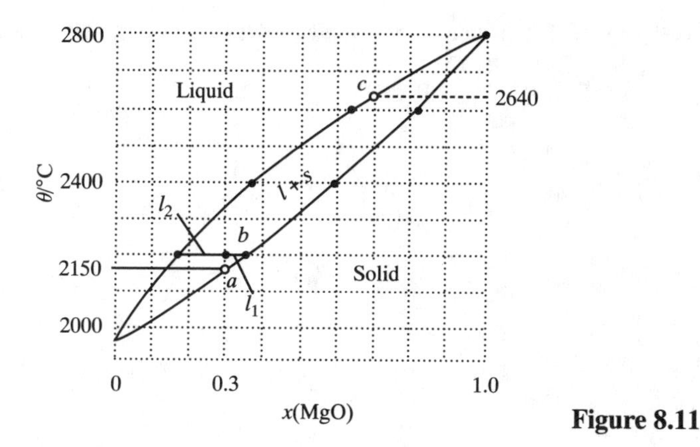

Figure 8.11

The proportions of the two phases are given by the lever rule,

$$\frac{l_1}{l_2} = \frac{n(\text{liq})}{n(\text{sol})} = \frac{0.05}{0.12} = \boxed{0.4}$$

(c) Solidification begins at point c, corresponding to $\boxed{2640°\text{C}}$

P8.4 The phase diagram is shown in Fig. 8.12(a). The values of x_S corresponding to the three compounds are: (1) P_4S_3, 0.43; (2) P_4S_7, 0.64. (3) P_4S_{10}, 0.71.

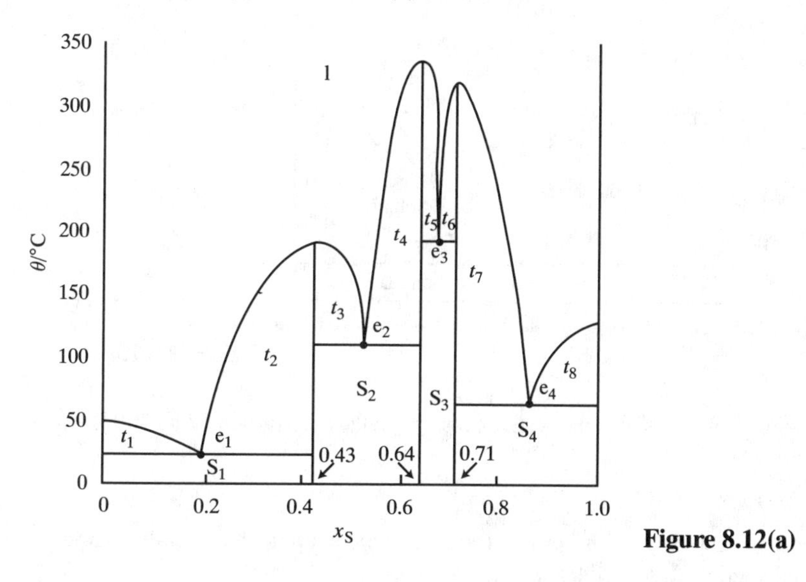

Figure 8.12(a)

The diagram has four eutectics labelled e_1, e_2, e_3, and e_4; eight two-phase liquid–solid regions, t_1 through t_8; and four two-phase solid regions, S_1, S_2, S_3, and S_4. The composition and physical state of the regions are as follows:

l: liquid S and P; S_1: solid P and solid P_4S_3; S_2: solid P_4S_3 and solid P_4S_7;
S_3: solid P_4S_7 and P_4S_{10}; S_4: solid P_4S_{10} and solid S

t_1: liquid P and S and solid P
t_2: liquid P and S and solid P_4S_3
t_3: liquid P and S and solid P_4S_3
t_4: liquid P and S and solid P_4S_7
t_5: liquid P and S and solid P_4S_7
t_6: liquid P and S and solid P_4S_{10}
t_7: liquid P and S and solid P_4S_{10}
t_8: liquid P and S and solid S

A break in the cooling curve (Fig. 18.12(b)) occurs at point $b_1 \approx 125°C$ as a result of solid P_4S_3 forming; a eutectic halt occurs at point $e_1 \approx 20°C$.

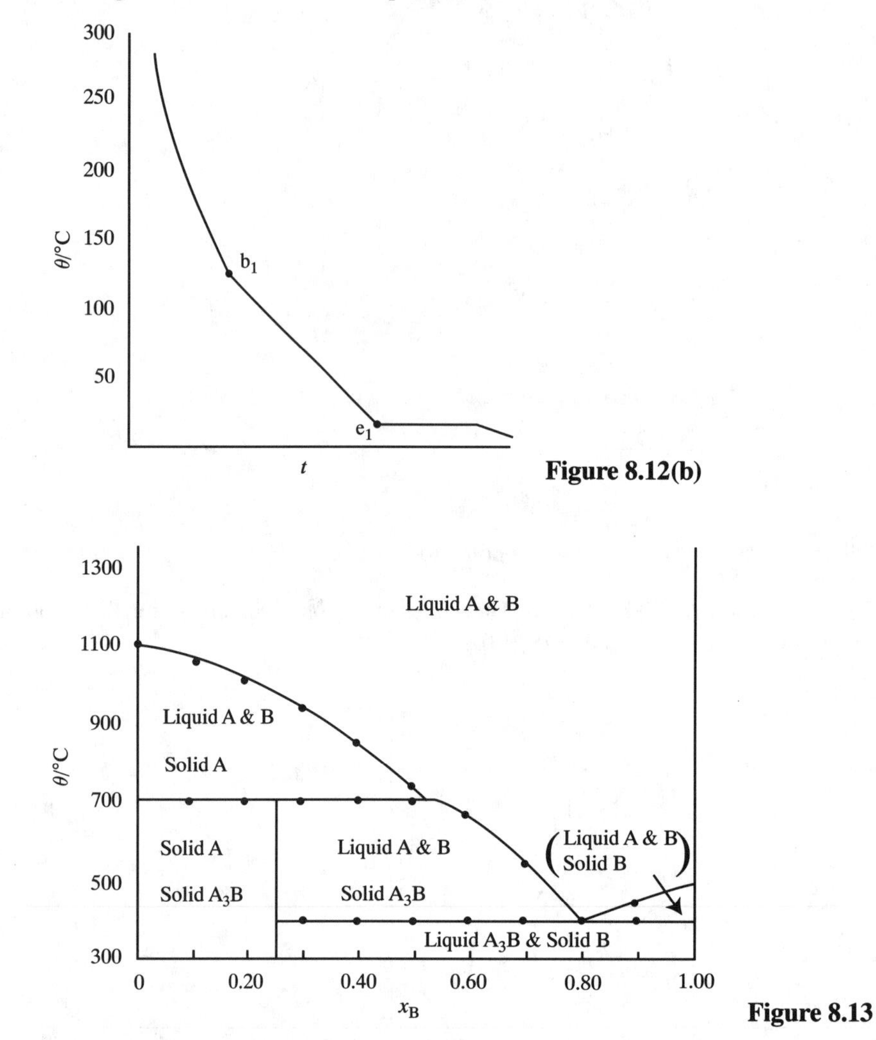

Figure 8.12(b)

P8.5

Figure 8.13

A compound with probable formula A_3B exists. It melts incongruently at 700 K, undergoing the peritectic reaction

$$A_3B(s) \rightarrow A(s) + (A + B, l)$$

The proportions of A and B in the product are dependent upon the overall composition and the temperature. A eutectic exists at 400 K and $x_B \approx 0.83$.

P8.7 The data are plotted in Fig. 8.14.

From the upper and lower extremes of the two-phase region we find $T_{uc} = \boxed{122°C}$ and $T_{lc} = \boxed{8°C}$. According to the phase diagram, miscibility is complete up to point a. Therefore, before that point is reached, $P = 1$, $C = 2$, implying that $F = 3(p, T, \text{ and } x)$. Two phases occur at a corresponding to $w(\text{MP}) = 0.18$ and 0.84. At that point, $P = 2$, $C = 2$, and $F = 2(p, \text{ and } x \text{ or } T)$. At the point a' there are two phases of composition $w = 0.18$ and 0.84. They are present in the ratio $\dfrac{a'' - a'}{a' - a} = 2$ with the former dominant. At a'' there are still two phases with those compositions, but the former ($w = 0.18$) is present only as a trace. One more drop takes the system into the one-phase region.

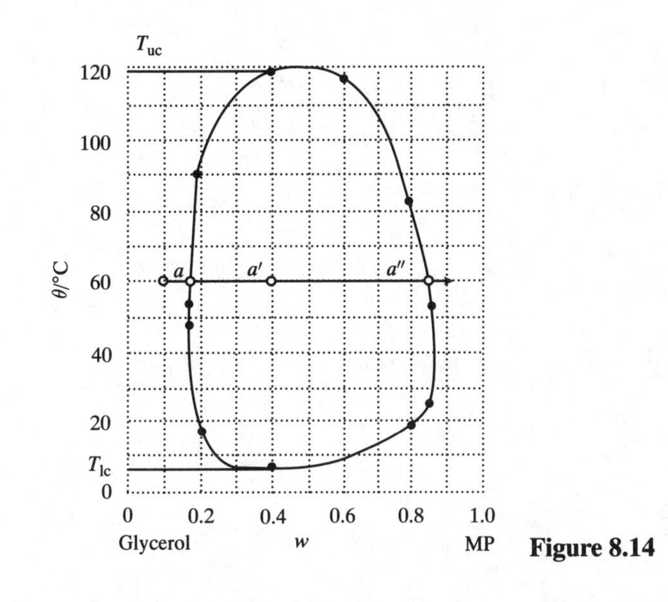

Figure 8.14

P8.9 The data are plotted in Fig. 8.15. At 360°C, $K_2FeCl_4(s)$ appears. The solution becomes richer in $FeCl_2$ until the temperature reaches 351°C, at which point $KFeCl_3(s)$ also appears. Below 351°C the system is a mixture of $K_2FeCl_4(s)$ and $KFeCl_3(s)$.

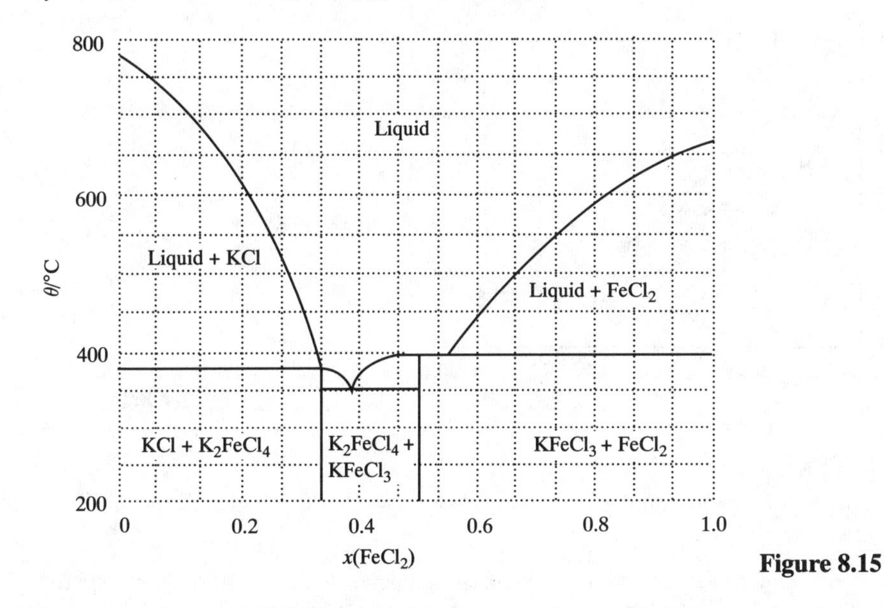

Figure 8.15

Solutions to theoretical problems

P8.11 The implication of this problem is that energy in the form of heat may be transferred between phases and that the volumes of the phases may also change. However, $U_\alpha + U_\beta =$ constant and $V_\alpha + V_\beta =$ constant. Hence,

$$\mathrm{d}U_\beta = -\mathrm{d}\,U_\alpha \text{ (b)} \quad \text{and} \quad \mathrm{d}V_\beta = -\mathrm{d}V_\alpha \text{ (c)}$$

The general condition of equilibrium in an isolated system is $\mathrm{d}S = 0$; hence

$$\mathrm{d}S = \mathrm{d}S_\alpha + \mathrm{d}S_\beta = 0 \text{ (a)}$$

$$S = S(U, V)$$

$$dS = \left(\frac{\partial S_\alpha}{\partial U_\alpha}\right)_{V_\alpha} dU_\alpha + \left(\frac{\partial S_\alpha}{\partial V_\alpha}\right)_{U_\alpha} dV_\alpha + \left(\frac{\partial S_\beta}{\partial U_\beta}\right)_{V_\beta} dU_\beta + \left(\frac{\partial S_\beta}{\partial V_\beta}\right)_{U_\beta} dV_\beta$$

Using conditions (b) and (c), and eqn 5.4

$$dS = \left(\frac{1}{T_\alpha} - \frac{1}{T_\beta}\right) dU_\alpha + \left(\frac{p_\alpha}{T_\alpha} - \frac{p_\beta}{T_\beta}\right) dV_\alpha = 0$$

The only way in which this expression may, in general, equal zero is for

$$\frac{1}{T_\alpha} - \frac{1}{T_\beta} = 0 \quad \text{and} \quad \frac{p_\alpha}{T_\alpha} - \frac{p_\beta}{T_\beta} = 0$$

Therefore, $\boxed{T_\alpha = T_\beta \text{ and } p_\alpha = p_\beta}$

Solutions to additional problems

P8.12 **(a)** The data, including that for pure chlorobenzene, are plotted in Fig. 8.16.

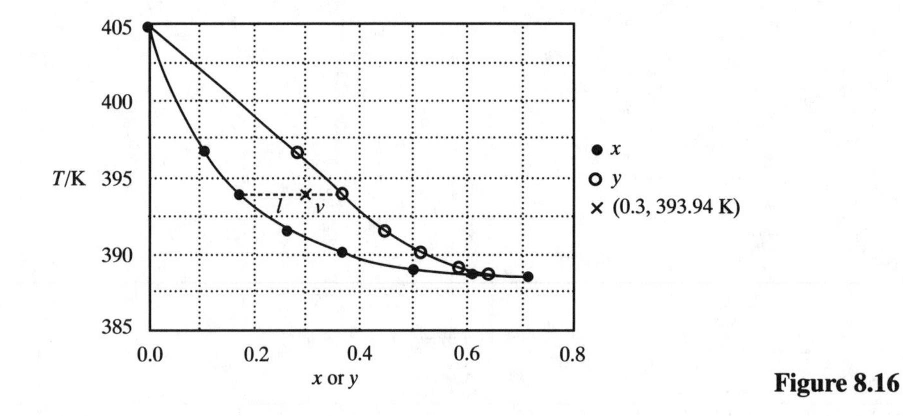

Figure 8.16

(b) The smooth curve through the x, T data crosses $x = 0.300$ at 391.0 K, the boiling point of the mixture.

(c) We need not interpolate data, for 393.94 K is a temperature for which we have experimental data. The mole fraction of 1-butanol in the liquid phase is 0.1700 and in the vapour phase 0.3691. According to the lever rule, the proportions of the two phases are in an inverse ratio of the distances their mole fractions are from the composition point in question. That is

$$\frac{n_{\text{liq}}}{n_{\text{vap}}} = \frac{v}{l} = \frac{0.3691 - 0.300}{0.300 - 0.1700} = \boxed{0.532}$$

P8.15 $p_A = a_A p_A^* = \gamma_A x_A p_A^*$ [7.42]

$$\gamma_A = \frac{p_A}{x_A p_A^*} = \frac{y_A p}{x_A p_A^*}$$

Sample calculation at 80 K

$$\gamma_{O_2}(80\,\text{K}) = \frac{0.11(100\,\text{kPa})}{0.34(225\,\text{Torr})}\left(\frac{760\,\text{Torr}}{101.325\,\text{kPa}}\right)$$

$$\gamma_{O_2}(80\,\text{K}) = 1.079$$

Summary

T/K	77.3	78	80	82	84	86	88	90.2
γ_{O_2}	—	0.877	1.079	1.039	0.995	0.993	0.990	0.987

To within the experimental uncertainties the solution appears to be ideal ($\gamma = 1$). The low value at 78 K may be caused by nonideality; however, the larger relative uncertainty in $y(O_2)$ is probably the origin of the low value.

A temperature–composition diagram is shown in Fig. 8.17(a). The near ideality of this solution is, however, best shown in the pressure–composition diagram of Fig. 8.17(b). The liquid line is essentially a straight line as predicted for an ideal solution.

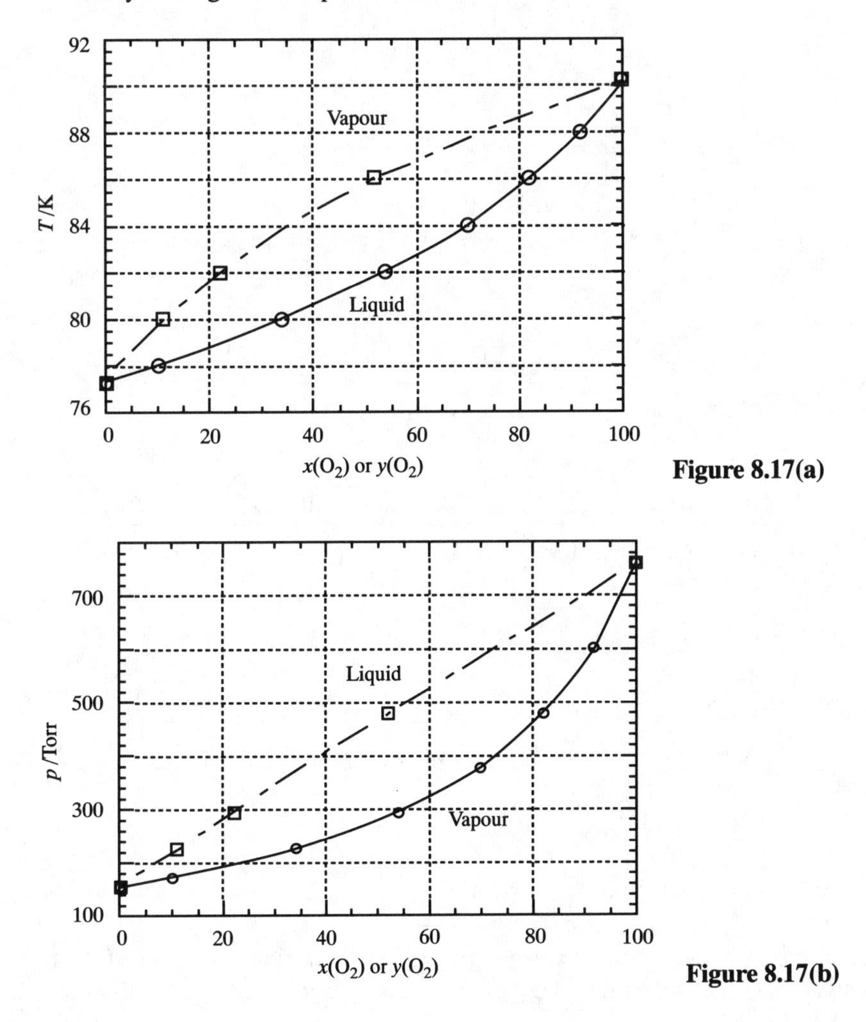

Figure 8.17(a)

Figure 8.17(b)

9 Chemical equilibrium

Solutions to exercises

E9.1 $\Delta_r G^\ominus = -RT \ln K[8] = (-8.314\,\mathrm{J\,K^{-1}\,mol^{-1}}) \times (400\,\mathrm{K}) \times (\ln 2.07) = \boxed{-2.42\,\mathrm{kJ\,mol^{-1}}}$

E9.2 $\Delta_r G^\ominus = -RT \ln K\ [8]$

Taking inverse logarithms of both sides of this equation yields

$$K = \mathrm{e}^{-\Delta_r G^\ominus/RT} = \mathrm{e}^{+3.67\times 10^3\,\mathrm{J\,mol^{-1}}/(8.314\,\mathrm{J\,K^{-1}\,mol^{-1}}\times 400\,\mathrm{K})} = \boxed{3.01}$$

E9.3 We draw up the following equilibrium table (Example 9.2). α is the equilibrium extent of dissociation.

	H_2O	H_2	O_2
Amount at equilibrium	$(1-\alpha)n$	αn	$\frac{1}{2}\alpha n$
Mole fraction	$\dfrac{1-\alpha}{1+\frac{1}{2}\alpha}$	$\dfrac{\alpha}{1+\frac{1}{2}\alpha}$	$\dfrac{\frac{1}{2}\alpha}{1+\frac{1}{2}\alpha}$
Partial pressure	$\dfrac{(1-\alpha)p}{1+\frac{1}{2}\alpha}$	$\dfrac{\alpha p}{1+\frac{1}{2}\alpha}$	$\dfrac{\frac{1}{2}\alpha p}{1+\frac{1}{2}\alpha}$

(a) $K = \prod_{\mathrm{J}} a_{\mathrm{J}}^{\nu_{\mathrm{J}}}$ [17]; $\quad a_{\mathrm{J}} = \dfrac{p_{\mathrm{J}}}{p^\ominus}$ [assume gases are perfect]

$$K = \frac{\left(\frac{p_{H_2}}{p^\ominus}\right)^2 \times \left(\frac{p_{O_2}}{p^\ominus}\right)}{\left(\frac{p_{H_2O}}{p^\ominus}\right)^2}[17] = \frac{\left(\frac{\alpha p}{(1+\frac{1}{2}\alpha)p^\ominus}\right)^2 \times \left(\frac{\frac{1}{2}\alpha p}{(1+\frac{1}{2}\alpha)p^\ominus}\right)}{\left(\frac{(1-\alpha)p}{(1+\frac{1}{2}\alpha)p^\ominus}\right)^2}$$

$$= \frac{\alpha^3 p}{2(1-\alpha)^2 \times \left(1+\frac{1}{2}\alpha\right)p^\ominus} = \frac{(0.0177)^3}{2(1-0.0177)^2 \times \left(1+\frac{1}{2}\times 0.0177\right)}$$

$$= 2.84\overline{8} \times 10^{-6} = \boxed{2.85 \times 10^{-6}}$$

(b) $\Delta_r G^\ominus = -RT \ln K\ [8]$

$$= -(8.314\,\mathrm{J\,K^{-1}\,mol^{-1}}) \times (2257\,\mathrm{K}) \times \ln(2.84\overline{8} \times 10^{-6}) = 2.40 \times 10^5\,\mathrm{J\,mol^{-1}}$$

$$= \boxed{+240\,\mathrm{kJ\,mol^{-1}}}$$

(c) $\Delta_r G = \boxed{0}$ [the system is at equilibrium]

Comment. The equilibrium constant always applies to the reaction as written. If the reaction had been written as $H_2O(g) \rightleftharpoons H_2(g) + \frac{1}{2}O_2(g)$ as in Example 9.2 the value of K would have been 1.69×10^{-3}, which compares favourably to the approximate value 2.08×10^{-3} calculated there, at a slightly different temperature.

E9.4 We draw up the following equilibrium table

	N_2O_4	NO_2
Amount at equilibrium	$(1-\alpha)n$	$2\alpha n$
Mole fraction	$\dfrac{1-\alpha}{1+\alpha}$	$\dfrac{2\alpha}{1+\alpha}$
Partial pressure	$\dfrac{(1-\alpha)p}{1+\alpha}$	$\dfrac{2\alpha p}{1+\alpha}$

(a) Assuming the gases are perfect $a_J = \left(\dfrac{p_J}{p^\ominus}\right)$; hence

$$K = \frac{\left(\frac{p_{NO_2}}{p^\ominus}\right)^2}{\left(\frac{p_{N_2O_4}}{p^\ominus}\right)} [17] = \frac{4\alpha^2 p}{(1-\alpha^2)p^\ominus} = \frac{4\alpha^2}{(1-\alpha^2)} \ [p = p^\ominus]$$

$$K = \frac{(4) \times (0.1846)^2}{1-(0.1846)^2} = \boxed{0.1411}$$

(b) $$\Delta_r G^\ominus = -RT \ln K [8] = -(8.314\,\mathrm{J\,K^{-1}\,mol^{-1}}) \times (298.2\,\mathrm{K}) \times \ln(0.1411)$$

$$= 4.855 \times 10^3\,\mathrm{J\,mol^{-1}} = \boxed{+4.855\,\mathrm{kJ\,mol^{-1}}}$$

(c) $$\ln K(100°\mathrm{C}) = \ln K(25°\mathrm{C}) - \frac{\Delta_r H^\ominus}{R}\left(\frac{1}{373.2\,\mathrm{K}} - \frac{1}{298.2\,\mathrm{K}}\right) [26]$$

$$\ln K(100°\mathrm{C}) = \ln(0.1411) - \left(\frac{57.2 \times 10^3\,\mathrm{J\,mol^{-1}}}{8.314\,\mathrm{J\,K^{-1}\,mol^{-1}}}\right) \times (-6.739 \times 10^{-4}\,\mathrm{K^{-1}}) = 2.678$$

$$K(100°\mathrm{C}) = \boxed{14.556}$$

Comment. In this case the increase in temperature results in a considerable shift in the equilibrium amounts of the substances in the reaction. The value of K changes from less than 1 to greater that 1. The value $\Delta_r G^\ominus$ calculated in (b) compares favourably to the value 4.73 kJ mol^{-1} determined from the data of Table 2.6.

E9.5 **(a)** $$\Delta_r G^\ominus = \sum_J \nu_J \Delta_f G^\ominus(J) \ [12]$$

$$\nu(\mathrm{Pb}) = 1, \quad \nu(\mathrm{CO_2}) = 1, \quad \nu(\mathrm{PbO}) = -1, \quad \nu(\mathrm{CO}) = -1$$

The equation is

$$0 = \mathrm{Pb(s)} + \mathrm{CO_2(g)} - \mathrm{PbO(s)} - \mathrm{CO(g)}$$

$$\Delta_r G^\ominus = \Delta_f G^\ominus(\mathrm{Pb, s}) + \Delta_f G^\ominus(\mathrm{CO_2, g}) - \Delta_f G^\ominus(\mathrm{PbO, s, red}) - \Delta_f G^\ominus(\mathrm{CO, g})$$

$$= (-394.36\,\mathrm{kJ\,mol^{-1}}) - (-188.93\,\mathrm{kJ\,mol^{-1}}) - (-137.17\,\mathrm{kJ\,mol^{-1}})$$

$$= \boxed{-68.26\,\mathrm{kJ\,mol^{-1}}}$$

$$\ln K = \frac{-\Delta_r G^\ominus}{RT} [8] = \frac{+68.26 \times 10^3\,\mathrm{J\,mol^{-1}}}{(8.314\,\mathrm{J\,K^{-1}\,mol^{-1}}) \times (298\,\mathrm{K})} = 27.55; \quad K = \boxed{9.2 \times 10^{11}}$$

(b) $\Delta_r H^\ominus = \Delta_f H^\ominus(\text{Pb, s}) + \Delta_f H^\ominus(\text{CO}_2, \text{g}) - \Delta_f H^\ominus(\text{PbO, s, red}) - \Delta_f H^\ominus(\text{CO, g})$
$= (-393.51\,\text{kJ mol}^{-1}) - (-218.99\,\text{kJ mol}^{-1}) - (-110.53\,\text{kJ mol}^{-1})$
$= -63.99\,\text{kJ mol}^{-1}$

$$\ln K(400\,\text{K}) = \ln K(298) - \frac{\Delta_r H^\ominus}{R}\left(\frac{1}{400\,\text{K}} - \frac{1}{298\,\text{K}}\right) \text{ [26]}$$

$$= 27.55 - \left(\frac{-63.99 \times 10^3\,\text{J mol}^{-1}}{8.314\,\text{J K}^{-1}\,\text{mol}^{-1}}\right) \times (-8.55\bar{7} \times 10^{-4}\,\text{K}^{-1}) = 20.9\bar{6}$$

$$K(400\,\text{K}) = \boxed{1.3 \times 10^9}$$

$$\Delta_r G^\ominus(400\,\text{K}) = -RT \ln K(400\,\text{K})[8] = -(8.314\,\text{J K}^{-1}\,\text{mol}^{-1}) \times (400\,\text{K}) \times (20.9\bar{6})$$
$$= -6.97 \times 10^4\,\text{J mol}^{-1} = \boxed{-69.7\,\text{kJ mol}^{-1}}$$

Comment. $\Delta_r G^\ominus(400\,\text{K})$ could have been determined directly from its value at 298 K by using the integrated form of the Gibbs–Helmholtz equation, (third equation in *Justification* 9.3), rather than by first calculating $K(400\,\text{K})$ as in the solution above.

Question. What is the value of $\Delta_r G^\ominus(400\,\text{K})$ for this reaction obtained by the method suggested in the Comment?

E9.6 Draw up the following equilibrium table

	A	B	C	D	Total
Initial amounts/mol	1.00	2.00	0	1.00	4.00
Stated change/mol			+0.90		
Implied change/mol	−0.60	−0.30	+0.90	+0.60	
Equilibrium amounts/mol	0.40	1.70	0.90	1.60	4.60
Mole fractions	0.087	0.370	0.196	0.348	1.001

(a) The mole fractions are given in the table.

(b) $K_x = \prod_J x_J^{\nu_J}$ [analogous to eqn 17 and *Illustration*, p. 224]

$$K_x = \frac{(0.196)^3 \times (0.348)^2}{(0.087)^2 \times (0.370)} = 0.32\bar{6} = \boxed{0.33}$$

(c) $p_J = x_J p, \quad p = 1\,\text{bar}, \quad p^\ominus = 1\,\text{bar}$

Assuming that the gases are perfect, $a_J = \dfrac{p_J}{p^\ominus}$, hence

$$K = \frac{(p_C/p^\ominus)^3 \times (p_D/p^\ominus)^2}{(p_A/p^\ominus)^2 \times (p_B/p^\ominus)}$$
$$= \frac{x_C^3 x_D^2}{x_A^2 x_B} \times \left(\frac{p}{p^\ominus}\right)^2 = K_x \quad \text{when } p = 1.00\,\text{bar} = \boxed{0.33}$$

(d) $\Delta_r G^\ominus = -RT \ln K = -(8.314\,\text{J K}^{-1}\,\text{mol}^{-1}) \times (298\,\text{K}) \times (\ln 0.32\bar{6})$
$= \boxed{+2.8 \times 10^3\,\text{J mol}^{-1}}$

E9.7 At 1280 K, $\Delta_r G^{\ominus} = +33 \times 10^3\,\mathrm{J\,mol^{-1}}$; thus

$$\ln K_1(1280\,\mathrm{K}) = -\frac{\Delta_r G^{\ominus}}{RT} = -\frac{33 \times 10^3\,\mathrm{J\,mol^{-1}}}{(8.314\,\mathrm{J\,K^{-1}\,mol^{-1}}) \times (1280\,\mathrm{K})} = -3.1\overline{0}$$

$$K_1 = \boxed{0.045}$$

$$\ln K_2 = \ln K_1 - \frac{\Delta_r H^{\ominus}}{R}\left(\frac{1}{T_2} - \frac{1}{T_1}\right) \;[26]$$

We look for the temperature T_2 that corresponds to $\ln K_2 = \ln(1) = 0$. This is the crossover temperature. Solving for T_2 from eqn 26 with $\ln K_2 = 0$, we obtain

$$\frac{1}{T_2} = \frac{R \ln K_1}{\Delta_r H^{\ominus}} + \frac{1}{T_1} = \left(\frac{(8.314\,\mathrm{J\,K^{-1}\,mol^{-1}}) \times (-3.1\overline{0})}{224 \times 10^3\,\mathrm{J\,mol^{-1}}}\right) + \left(\frac{1}{1280\,\mathrm{K}}\right)$$

$$= 6.6\overline{6} \times 10^{-4}\,\mathrm{K^{-1}}$$

$$T_2 = \boxed{15\overline{00}\,\mathrm{K}}$$

E9.8 Given $\ln K = -1.04 - \dfrac{1088\,\mathrm{K}}{T} + \dfrac{1.51 \times 10^5\,\mathrm{K^2}}{T^2}$

and since $\dfrac{\mathrm{d}\ln K}{\mathrm{d}(1/T)} = \dfrac{-\Delta_r H^{\ominus}}{R}$ [24*b*]

$$\frac{-\Delta_r H^{\ominus}}{R} = -1088\,\mathrm{K} + \frac{(2) \times (1.51 \times 10^5\,\mathrm{K^2})}{T}$$

Then, at 400 K

$$\Delta_r H^{\ominus} = \left(1088\,\mathrm{K} - \frac{3.02 \times 10^5\,\mathrm{K^2}}{400\,\mathrm{K}}\right) \times (8.314\,\mathrm{J\,K^{-1}\,mol^{-1}}) = \boxed{+2.77\,\mathrm{kJ\,mol^{-1}}}$$

$$\Delta_r G^{\ominus} = -RT \ln K[8] = RT \times \left(1.04 + \frac{1088\,\mathrm{K}}{T} - \frac{1.51 \times 10^5\,\mathrm{K^2}}{T^2}\right)$$

$$= RT \times \left(1.04 + \frac{1088\,\mathrm{K}}{400\,\mathrm{K}} - \frac{1.51 \times 10^5\,\mathrm{K^2}}{(400\,\mathrm{K})^2}\right) = +9.37\,\mathrm{kJ\,mol^{-1}}$$

$$= \Delta_r H^{\ominus} - T\Delta_r S^{\ominus} \;[4.39]$$

Therefore, $\Delta_r S^{\ominus} = \dfrac{\Delta_r H^{\ominus} - \Delta_r G^{\ominus}}{T} = \dfrac{2.77\,\mathrm{kJ\,mol^{-1}} - 9.37\,\mathrm{kJ\,mol^{-1}}}{400\,\mathrm{K}} = \boxed{-16.5\,\mathrm{J\,K^{-1}\,mol^{-1}}}$

E9.9 Let B = borneol and I = isoborneol

$$\Delta_r G = \Delta G^{\ominus} + RT \ln Q \;[10], \quad Q = \frac{p_\mathrm{I}}{p_\mathrm{B}} \;[14]$$

$$p_\mathrm{B} = x_\mathrm{B} p = \frac{0.15\,\mathrm{mol}}{0.15\,\mathrm{mol} + 0.30\,\mathrm{mol}} \times 600\,\mathrm{Torr} = 200\,\mathrm{Torr}; \qquad p_\mathrm{I} = p - p_\mathrm{B} = 400\,\mathrm{Torr}$$

$$Q = \frac{400\,\mathrm{Torr}}{200\,\mathrm{Torr}} = 2.00$$

$$\Delta_r G = (+9.4\,\mathrm{kJ\,mol^{-1}}) + (8.314\,\mathrm{J\,K^{-1}\,mol^{-1}}) \times (503\,\mathrm{K}) \times (\ln 2.00) = \boxed{+12.3\,\mathrm{kJ\,mol^{-1}}}$$

E9.10 $K_x = \prod_J x_J^{\nu_J}$ [analogous to eqn 17, *Justification* 9.2]

The relation of K_x to K is established in the *Illustration* on p. 224

$$K = \prod_J \left(\frac{p_J}{p^\ominus}\right)^{\nu_J} \left[17 \text{ with } a_J = \frac{p_J}{p^\ominus}\right]$$

$$= \prod_J x_J^{\nu_J} \times \left(\frac{p}{p^\ominus}\right)^{\sum_J \nu_J} [p_J = x_J p] = K_x \times \left(\frac{p}{p^\ominus}\right)^{\nu} \left[\nu \equiv \sum_J \nu_J\right]$$

Therefore, $K_x = K\left(\frac{p}{p^\ominus}\right)^{-\nu}$ $K_x \propto p^{-\nu}$ [K and $p^\ominus$ are constants]

$\nu = 1 + 1 - 1 = 1$, thus $K_x(2\,\text{bar}) = \frac{1}{2}K_x(1\,\text{bar})$; percentage change is 50 per cent

E9.11 Let B = borneol and I = isoborneol

$$K = K_x \times \left(\frac{p}{p^\ominus}\right)^{\nu} \text{[Exercise 9.10]} = \frac{x_I}{x_B}[\nu = 1 - 1 = 0] = \frac{1 - x_B}{x_B}$$

Hence, $x_B = \frac{1}{1 + K} = \frac{1}{1 + 0.106} = 0.904$

$x_I = 0.096$

The initial amounts of the isomers are

$$n_B = \frac{7.50\,\text{g}}{M}, \quad n_I = \frac{14.0\,\text{g}}{M}, \quad n = \frac{21.5\bar{0}\,\text{g}}{M}$$

The total amount remains the same, but at equilibrium

$$\frac{n_B}{n} = x_B = \boxed{0.904}, \quad x_I = \boxed{0.096}, \quad n_B = (0.904) \times \left(\frac{21.5\bar{0}\,\text{g}}{M}\right)$$

The mass of borneol at equilibrium is therefore

$$m_B = n_B \times M = (0.904) \times (21.5\bar{0}\,\text{g}) = 19.4\,\text{g}$$

and the mass of isoborneol is

$$m_I = n_I \times M = (0.096) \times (21.5\bar{0}\,\text{g}) = 2.1\,\text{g}$$

E9.12 $\Delta_r G^\ominus = -RT \ln K$ [8]

Hence, a value of $\Delta_r G^\ominus < 0$ at 298 K corresponds to $K > 1$.

(a) $\Delta_r G^\ominus/(\text{kJ mol}^{-1}) = (-202.87) - (-95.30 - 16.45) = -91.12$, $\boxed{K > 1}$

(b) $\Delta_r G^\ominus/(\text{kJ mol}^{-1}) = (3) \times (-856.64) - (2) \times (-1582.3) = +594.7$, $\boxed{K < 1}$

(c) $\Delta_r G^\ominus/(\text{kJ mol}^{-1}) = (-100.4) - (-33.56) = -66.8$, $\boxed{K > 1}$

E9.13 Le Chatelier's principle in the form of the rules in the first paragraph of Section 9.3 is employed. Thus we determine whether $\Delta_r H^\ominus$ is positive or negative by using the $\Delta_f H^\ominus$ values of Table 2.6.

(a) $\Delta_r H^\ominus/(\text{kJ mol}^{-1}) = (-314.43) - (-46.11 - 92.31) = -176.01$

(b) $\Delta_r H^\ominus/(\text{kJ mol}^{-1}) = (3) \times (-910.94) - (2) \times (-1675.7) = +618.6$

(c) $\Delta_r H^\ominus/(\text{kJ mol}^{-1}) = (-100.0) - (-20.63) = -79.4$

Since **(a)** and **(c)** are exothermic, an increase in temperature favours the reactants; $\boxed{\textbf{(b)}}$ is endothermic, and an increase in temperature favours the products.

E9.14 $$\ln \frac{K'}{K} = \frac{\Delta_r H^\ominus}{R}\left(\frac{1}{T} - \frac{1}{T'}\right) \; [26]$$

Therefore, $\Delta_r H^\ominus = \dfrac{R \ln \frac{K'}{K}}{\left(\frac{1}{T} - \frac{1}{T'}\right)}$

$T' = 308\,\text{K}$; hence, with $\dfrac{K'}{K} = \kappa$

$$\Delta_r H^\ominus = \frac{(8.314\,\text{J K}^{-1}\,\text{mol}^{-1}) \times (\ln \kappa)}{\left(\frac{1}{298\,\text{K}} - \frac{1}{308\,\text{K}}\right)} = 76\,\text{kJ mol}^{-1} \times \ln \kappa$$

Therefore

(a) $\kappa = 2, \quad \Delta_r H^\ominus = (76\,\text{kJ mol}^{-1}) \times (\ln 2) = \boxed{+53\,\text{kJ mol}^{-1}}$

(b) $\kappa = \frac{1}{2}, \quad \Delta_r H^\ominus = (76\,\text{kJ mol}^{-1}) \times \left(\ln \frac{1}{2}\right) = \boxed{-53\,\text{kJ mol}^{-1}}$

E9.15 $$\Delta_r G = \Delta G^\ominus + RT \ln Q \; [10]; \quad Q = \prod_J a_J^{\nu_J} \; [14]$$

for $\frac{1}{2}N_2(g) + \frac{3}{2}H_2(g) \rightarrow NH_3(g)$

$$Q = \frac{\left(\frac{p(NH_3)}{p^\ominus}\right)}{\left(\frac{p(N_2)}{p^\ominus}\right)^{1/2}\left(\frac{p(H_2)}{p^\ominus}\right)^{3/2}} \quad \left[a_J = \frac{p_J}{p^\ominus} \text{ for perfect gases}\right]$$

$$= \frac{p(NH_3)p^\ominus}{p(N_2)^{1/2}p(H_2)^{3/2}} = \frac{4.0}{(3.0)^{1/2} \times (1.0)^{3/2}} = \frac{4.0}{\sqrt{3.0}}$$

Therefore, $\Delta_r G = (-16.45\,\text{kJ mol}^{-1}) + RT \ln \dfrac{4.0}{\sqrt{3.0}} = (-16.45\,\text{kJ mol}^{-1}) + (2.07\,\text{kJ mol}^{-1})$

$$= \boxed{-14.38\,\text{kJ mol}^{-1}}$$

Since $\Delta_r G < 0$, the spontaneous direction of reaction is $\boxed{\text{toward products}}$

E9.16 The reaction is

$$CaCO_3(s) \rightleftharpoons CaO(s) + CO_2(g)$$

For the purposes of this exercise we may assume that the required temperature is that temperature at which the $K = 1$ which corresponds to a pressure of 1 bar for the gaseous product. For $K = 1$, $\ln K = 0$ and $\Delta_r G^\ominus = 0$.

$$\Delta_r G^\ominus = \Delta_r H^\ominus - T\Delta_r S^\ominus = 0 \quad \text{when} \quad \Delta_r H^\ominus = T\Delta_r S^\ominus$$

Therefore, the decomposition temperature (when $K = 1$) is

$$T = \frac{\Delta_r H^{\ominus}}{\Delta_r S^{\ominus}}$$

$CaCO_3(s) \rightarrow CaO(s) + CO_2(g)$

$\Delta_r H^{\ominus} = (-635.09) - (393.51) - (-1206.9)\ \text{kJ mol}^{-1} = +178.3\ \text{kJ mol}^{-1}$

$\Delta_r S^{\ominus} = (39.75) + (213.74) - (92.9)\ \text{J K}^{-1}\ \text{mol}^{-1} = +160.6\ \text{J K}^{-1}\ \text{mol}^{-1}$

$$T = \frac{178.3 \times 10^3\ \text{J mol}^{-1}}{160.6\ \text{J K}^{-1}\ \text{mol}^{-1}} = \boxed{1110\ \text{K}}\ (840°\text{C})$$

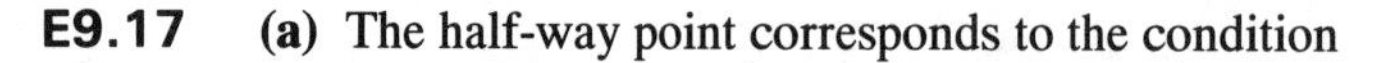

E9.17 **(a)** The half-way point corresponds to the condition

$$[\text{Acid}] = [\text{Salt}]$$

for which $pK_a = \text{pH}$ [45]

Hence, $\boxed{pK_a = 5.40}$ and $K_a = 10^{-5.40} = \boxed{4.0 \times 10^{-6}}$

(b) When the solution is $[\text{Acid}] = 0.015\ \text{M}$

$$\text{pH} = \tfrac{1}{2}pK_a - \tfrac{1}{2}\log[\text{Acid}]\ [40] = \tfrac{1}{2} \times (5.40) - \tfrac{1}{2} \times (-1.82) = \boxed{3.61}$$

E9.18 **(a)** NH_4Cl

In water, the NH_4^+ acts as an acid in the Brønsted equilibrium

$$NH_4^+(aq) + H_2O(l) \rightleftharpoons NH_3(aq) + H_3O^+(aq) \quad K_a = \frac{[H_3O^+][NH_3]}{[NH_4^+]}$$

$[NH_3] \approx [H_3O^+]$, because the water autoprotolysis can be ignored in the presence of a weak acid (NH_4^+); therefore,

$$K_a \approx \frac{[H_3O^+]^2}{[NH_4{}^+]} \approx \frac{[H_3O^+]^2}{S}$$

where S is the nominal concentration of the salt. Therefore,

$$[H_3O^+] \approx (SK_a)^{1/2}$$

and $\text{pH} \approx \tfrac{1}{2}pK_a - \tfrac{1}{2}\log S[48] \approx \tfrac{1}{2} \times (9.25) - \tfrac{1}{2} \times (\log 0.10) = \boxed{5.13}$

(b) $NaCH_3CO_2$

The $CH_3CO_2^-$ ion acts as a weak base

$$CH_3CO_2^-(aq) + H_2O(l) \rightleftharpoons CH_3COOH(aq) + OH^-(aq) \quad K_b = \frac{[CH_3COOH][OH^-]}{[CH_3CO_2^-]}$$

Then, since $[CH_3COOH] \approx [OH^-]$ and $[CH_3CO_2^-] \approx S$, the nominal concentration of the salt,

$$K_b \approx \frac{[OH^-]^2}{S}, \quad \text{implying that} \quad [OH^-] \approx (SK_b)^{1/2}$$

Therefore, $\text{pOH} = \tfrac{1}{2}pK_b - \tfrac{1}{2}\log S$

However, $\mathrm{pH} + \mathrm{pOH} = \mathrm{p}K_\mathrm{w}$, so $\mathrm{pH} = \mathrm{p}K_\mathrm{w} - \mathrm{pOH}$

$$\mathrm{p}K_\mathrm{a} + \mathrm{p}K_\mathrm{b} = \mathrm{p}K_\mathrm{w}, \quad \text{so} \quad \mathrm{p}K_\mathrm{b} = \mathrm{p}K_\mathrm{w} - \mathrm{p}K_\mathrm{a}$$

Therefore, $\mathrm{pH} = \mathrm{p}K_\mathrm{w} - \frac{1}{2}(\mathrm{p}K_\mathrm{w} - \mathrm{p}K_\mathrm{a}) + \frac{1}{2}\log S = \frac{1}{2}\mathrm{p}K_\mathrm{w} + \frac{1}{2}\mathrm{p}K_\mathrm{a} + \frac{1}{2}\log S$

$$= \tfrac{1}{2} \times (14.00) + \tfrac{1}{2} \times (4.75) + \tfrac{1}{2} \times (\log 0.10) = \boxed{8.88}$$

(c) $\mathrm{CH_3COOH(aq) + H_2O(l) \rightleftharpoons H_3O^+(aq) + CH_3CO_2^-(aq)} \quad K_\mathrm{a} = \dfrac{[\mathrm{H_3O^+}][\mathrm{CH_3CO_2^-}]}{[\mathrm{CH_3COOH}]}$

Since we can ignore the water autoprotolysis, $[\mathrm{H_3O^+}] \approx [\mathrm{CH_3CO_2^-}]$, so

$$K_\mathrm{a} \approx \frac{[\mathrm{H_3O^+}]^2}{A}$$

where $A = [\mathrm{CH_3COOH}]$, the nominal acid concentration (the ionization is small). Therefore,

$$[\mathrm{H_3O^+}] \approx (AK_\mathrm{a})^{1/2}, \quad \text{implying that} \quad \mathrm{pH} \approx \tfrac{1}{2}\mathrm{p}K_\mathrm{a} - \tfrac{1}{2}\log A \ [40]$$

Hence, $\mathrm{pH} \approx \frac{1}{2} \times (4.75) - \frac{1}{2} \times (\log 0.100) = \boxed{2.88}$

E9.19 The pH of a solution in which the nominal salt concentration is S is

$$\mathrm{pH} = \tfrac{1}{2}\mathrm{p}K_\mathrm{w} + \tfrac{1}{2}\mathrm{p}K_\mathrm{a} + \tfrac{1}{2}\log S \quad [47, \text{Exercise 9.18(b)}]$$

The volume of the solution at the stoichiometric point is

$$V = (25.00\,\mathrm{mL}) + (25.00\,\mathrm{mL}) \times \left(\frac{0.100\,\mathrm{M}}{0.150\,\mathrm{M}}\right) = 41.67\,\mathrm{mL}$$

and the concentration of salt is

$$S = (0.100\,\mathrm{M}) \times \left(\frac{25.00\,\mathrm{mL}}{41.67\,\mathrm{mL}}\right) = 0.0600\,\mathrm{M}$$

Hence, with $\mathrm{p}K_\mathrm{a} = 3.86$,

$$\mathrm{pH} = \tfrac{1}{2} \times (14.00) + \tfrac{1}{2} \times (3.86) + \tfrac{1}{2} \times (\log 0.0600) = \boxed{8.3}$$

E9.20 One procedure is to plot eqn 43. An alternative procedure is to estimate some of the points using the expressions given in Fig. 9.11 of the text. Initially only the salt is present, and we use eqn 47 (as in Exercise 9.19)

$$\mathrm{pH} = \tfrac{1}{2}\mathrm{p}K_\mathrm{a} + \tfrac{1}{2}\mathrm{p}K_\mathrm{w} + \tfrac{1}{2}\log S, \quad \log S = -1.00$$
$$= \tfrac{1}{2}(4.75 + 14.00 - 1.00) = 8.88 \qquad \text{(a)}$$

When $A' \approx S$, use the Henderson–Hasselbalch equation [43]

$$\mathrm{pH} = \mathrm{p}K_\mathrm{a} - \log\frac{A'}{S} = 4.75 - \log\frac{A'}{0.10} = 3.75 - \log A' \qquad \text{(b)}$$

When so much acid has been added that $A' \gg S$, use the 'weak acid alone' formula [40]

$$\mathrm{pH} = \tfrac{1}{2}\mathrm{p}K_\mathrm{a} - \tfrac{1}{2}\log A \qquad \text{(c)}$$

We can draw up the following table

$A(\text{or}A')$	0	0.06	0.08	0.10	0.12	0.14	0.6	0.8	1.0
pH	8.88	4.97	4.85	4.75	4.67	4.60	2.49	2.43	2.38
Formula	(a)			(b)				(c)	

The results are plotted in Fig. 9.1.

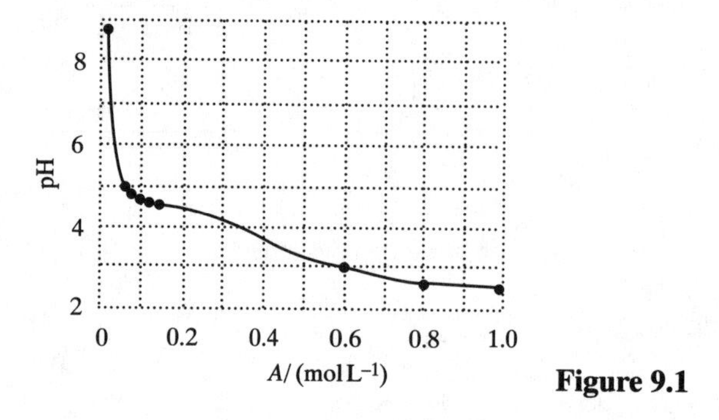

Figure 9.1

E9.21 According to the Henderson–Hasselbalch equation [44] the pH of a buffer varies about a central value given by $\text{p}K_\text{a}$. For the $\frac{[\text{acid}]}{[\text{salt}]}$ ratio to be neither very large nor very small we require $\text{p}K_\text{a} \approx \text{pH}$ (buffer).

(a) For pH ≈ 2.2 use $\boxed{Na_2HPO_4 + H_3PO_4}$ since

$$H_3PO_4 + H_2O \rightleftharpoons H_3O^+ + H_2PO_4^- \qquad \text{p}K_\text{a} = 2.12$$

(b) For pH ≈ 7 use $\boxed{NaH_2PO_4 + Na_2HPO_4}$ since

$$H_2PO_4^- + H_2O \rightleftharpoons H_3O^+ + HPO_4^{2-} \qquad \text{p}K_\text{a} = 7.2$$

Solutions to problems

Solutions to numerical problems

P9.1 **(a)** $\Delta_\text{r}G^\ominus = -RT\ln K = -(8.314\,\text{J K}^{-1}\,\text{mol}^{-1}) \times (298\,\text{K}) \times (\ln 0.164) = 4.48 \times 10^3\,\text{J mol}^{-1}$

$$= \boxed{+4.48\,\text{kJ mol}^{-1}}$$

(b) Draw up the following equilibrium table

	I_2	Br_2	IBr
Amounts	—	$(1-\alpha)n$	$2\alpha n$
Mole fractions	—	$\frac{(1-\alpha)}{(1+\alpha)}$	$\frac{2\alpha}{(1+\alpha)}$
Partial pressure	—	$\frac{(1-\alpha)p}{(1+\alpha)}$	$\frac{2\alpha p}{(1+\alpha)}$

$$K = \prod_\text{J} a_\text{J}^{\nu_\text{J}}\ [17] = \frac{\left(\frac{p_{\text{IBr}}}{p^\ominus}\right)^2}{\frac{p_{\text{Br}_2}}{p^\ominus}}\text{[perfect gases]} = \frac{\left\{(2\alpha)^2\frac{p}{p^\ominus}\right\}}{(1-\alpha)\times(1+\alpha)} = \frac{\left(4\alpha^2\frac{p}{p^\ominus}\right)}{1-\alpha^2} = 0.164$$

With $p = 0.164$ atm,

$$4\alpha^2 = 1 - \alpha^2 \qquad \alpha^2 = \tfrac{1}{5} \qquad \alpha = 0.447$$

$$p_{\mathrm{IBr}} = \frac{2\alpha}{1+\alpha} \times p = \frac{(2)\times(0.447)}{1+0.447} \times (0.164\,\mathrm{atm}) = \boxed{0.101\,\mathrm{atm}}$$

(c) The equilibrium table needs to be modified as follows

$$p = p_{\mathrm{I_2}} + p_{\mathrm{Br_2}} + p_{\mathrm{IBr}}$$

$$p_{\mathrm{Br_2}} = x_{\mathrm{Br_2}}p, \qquad p_{\mathrm{IBr}} = x_{\mathrm{IBr}}p, \qquad p_{\mathrm{I_2}} = x_{\mathrm{I_2}}p$$

with $x_{\mathrm{Br_2}} = \dfrac{(1-\alpha)n}{(1+\alpha)n + n_{\mathrm{I_2}}}$ [n = amount of Br_2 introduced into container]

and $x_{\mathrm{IBr}} = \dfrac{2\alpha n}{(1+\alpha)n + n_{\mathrm{I_2}}}$

K is constructed as above [17], but with these modified partial pressures. In order to complete the calculation additional data are required, namely, the amount of Br_2 introduced, n, and the equilibrium vapour pressure of $I_2(s)$. $n_{\mathrm{I_2}}$ can be calculated from a knowledge of the volume of the container at equilibrium which is most easily determined by successive approximations since $p_{\mathrm{I_2}}$ is small.

Question. What is the partial pressure of IBr(g) if 0.0100 mol of Br_2(g) is introduced into the container? The partial pressure of I_2(s) at 25°C is 0.305 Torr.

P9.4 $\quad CO_2(g) \rightleftharpoons CO(g) + \tfrac{1}{2}O_2(g)$

Draw up the following equilibrium table

	CO_2	CO	O_2
Amounts	$(1-\alpha)n$	αn	$\frac{1}{2}\alpha n$
Mole fractions	$\dfrac{(1-\alpha)}{\left(1+\frac{\alpha}{2}\right)}$	$\dfrac{\alpha}{\left(1+\frac{\alpha}{2}\right)}$	$\dfrac{\frac{1}{2}\alpha}{\left(1+\frac{\alpha}{2}\right)}$
Partial pressures	$\dfrac{(1-\alpha)p}{\left(1+\frac{\alpha}{2}\right)}$	$\dfrac{\alpha p}{\left(1+\frac{\alpha}{2}\right)}$	$\dfrac{\alpha p}{2\left(1+\frac{\alpha}{2}\right)}$

$$K = \prod_{\mathrm{J}} a_{\mathrm{J}}^{\nu_{\mathrm{J}}}\,[17] = \frac{\left(\frac{p_{\mathrm{CO}}}{p^{\ominus}}\right)\times\left(\frac{p_{\mathrm{O_2}}}{p^{\ominus}}\right)^{1/2}}{\left(\frac{p_{\mathrm{CO_2}}}{p^{\ominus}}\right)} = \frac{\left(\frac{\alpha}{1+(\alpha/2)}\right)\times\left(\frac{\alpha/2}{1+(\alpha/2)}\right)^{1/2}\times\left(\frac{p}{p^{\ominus}}\right)^{1/2}}{\left(\frac{1-\alpha}{1+(\alpha/2)}\right)}$$

$$K \approx \frac{\alpha^{3/2}}{\sqrt{2}} \quad [\alpha \ll 1 \text{ at all the specified temperatures}]$$

$$\Delta_r G^{\ominus} = -RT\ln K\ [8]$$

The calculated values of K and $\Delta_r G$ are given in the table below. From any two pairs of K and T, $\Delta_r H^{\ominus}$ may be calculated.

$$\ln K_2 = \ln K_1 - \frac{\Delta_r H^{\ominus}}{R}\left(\frac{1}{T_2} - \frac{1}{T_1}\right)\ [26]$$

Solving for $\Delta_r H^\ominus$

$$\Delta_r H^\ominus = \frac{R\ln\left(\frac{K_2}{K_1}\right)}{\left(\frac{1}{T_1}-\frac{1}{T_2}\right)}\text{[Exercise 9.14]} = \frac{(8.314\,\mathrm{J\,K^{-1}\,mol^{-1}})\times\ln\left(\frac{7.23\times10^{-6}}{1.22\times10^{-6}}\right)}{\left(\frac{1}{1395\,\mathrm{K}}-\frac{1}{1498\,\mathrm{K}}\right)}$$

$$= \boxed{3.00\times10^{5}\,\mathrm{J\,mol^{-1}}}$$

$$\Delta_r S^\ominus = \frac{\Delta_r H^\ominus - \Delta_r G^\ominus}{T}$$

The calculated values of $\Delta_r S^\ominus$ are also given in the table.

T/K	1395	1443	1498
$\alpha/10^{-4}$	1.44	2.50	4.71
$K/10^{-6}$	1.22	2.80	7.23
$\Delta_r G^\ominus/(\mathrm{kJ\,mol^{-1}})$	158	153	147
$\Delta_r S^\ominus/(\mathrm{J\,K^{-1}\,mol^{-1}})$	102	102	102

Comment. $\Delta_r S^\ominus$ is essentially constant over this temperature range but it is much different from its value at 25°C. $\Delta_r H^\ominus$, however, is only slightly different.

Question. What are the values of $\Delta_r H^\ominus$ and $\Delta_r S^\ominus$ at 25°C for this reaction?

P9.6 $\Delta_r G^\ominus(\mathrm{H_2CO, g}) = \Delta_r G^\ominus(\mathrm{H_2CO, l}) + \Delta_{vap} G^\ominus(\mathrm{H_2CO, l})$

For $\mathrm{H_2CO(l)} \rightleftharpoons \mathrm{H_2CO(g)}$, $K(\mathrm{vap}) = \frac{p}{p^\ominus}$

$$\Delta_{vap} G^\ominus = -RT\ln K(\mathrm{vap}) = -RT\ln\frac{p}{p^\ominus}$$

$$= -(8.314\,\mathrm{J\,K^{-1}\,mol^{-1}})\times(298\,\mathrm{K})\times\ln\left(\frac{1500\,\mathrm{Torr}}{750\,\mathrm{Torr}}\right) = -1.72\,\mathrm{kJ\,mol^{-1}}$$

Therefore, for the reaction

$$\mathrm{CO(g)} + \mathrm{H_2(g)} \rightleftharpoons \mathrm{H_2CO(g)},$$

$$\Delta_r G^\ominus = (+28.95) + (-1.72)\,\mathrm{kJ\,mol^{-1}} = +27.23\,\mathrm{kJ\,mol^{-1}}$$

Hence, $K = e^{(-27.23\times10^{3}\,\mathrm{J\,mol^{-1}})/(8.314\,\mathrm{J\,K^{-1}\,mol^{-1}})\times(298\,\mathrm{K})} = e^{-10.99} = \boxed{1.69\times10^{-5}}$

P9.8 The equilibrium to be considered is (A = gas)

$$\mathrm{A(g, 1\,bar)} \rightleftharpoons \mathrm{A(soln)} \quad K = \frac{(c/c^\ominus)}{(p/p^\ominus)} = \frac{s}{s^\ominus}$$

$$\Delta_r H^\ominus = -R\times\frac{\mathrm{d}\ln K}{\mathrm{d}\left(\frac{1}{T}\right)}\ [24b]$$

$$\ln K = \ln\left(\frac{s}{s^\ominus}\right) = 2.303\log\left(\frac{s}{s^\ominus}\right)$$

$$\Delta_r H^\ominus(\mathrm{H_2}) = -(2.303)\times(R)\times\frac{\mathrm{d}}{\mathrm{d}\left(\frac{1}{T}\right)}\left(-5.39-\frac{768\,\mathrm{K}}{T}\right) = 2.303R\times768\,\mathrm{K}$$

$$= \boxed{+14.7\,\mathrm{kJ\,mol^{-1}}}$$

$$\Delta_r H^{\ominus}(\mathrm{CO}) = -(2.303) \times (R) \times \frac{\mathrm{d}}{\mathrm{d}(\frac{1}{T})}\left(-5.98 - \frac{980\,\mathrm{K}}{T}\right) = 2.303R \times 980\,\mathrm{K}$$

$$= \boxed{+18.8\,\mathrm{kJ\,mol^{-1}}}$$

Solutions to theoretical problems

P9.10 $K = K_\phi K_p$, but $\left(\frac{\partial K}{\partial p}\right)_T = 0$ [21]

Therefore, $\left(\frac{\partial K}{\partial p}\right)_T = K_\phi \left(\frac{\partial K_p}{\partial p}\right)_T + K_p \left(\frac{\partial K_\phi}{\partial p}\right)_T = 0$

which implies that $\left(\frac{\partial K_\phi}{\partial p}\right)_T = -\left(\frac{\partial K_p}{\partial p}\right)_T \left(\frac{K_\phi}{K_p}\right)$

and therefore if K_p increases with pressure, K_ϕ must decrease (because K_ϕ/K_p is positive).

P9.11 We draw up the following table using the stoichiometry $\mathrm{A} + 3\mathrm{B} \rightarrow 2\mathrm{C}$ and $\Delta_{n_J} = \nu_J \xi$

	A	B	C	Total
Initial amount/mol	1	3	0	4
Change, Δn_J/mol	$-\xi$	-3ξ	$+2\xi$	
Equilibrium amount/mol	$1-\xi$	$3(1-\xi)$	2ξ	$2(2-\xi)$
Mole fraction	$\frac{1-\xi}{2(2-\xi)}$	$\frac{3(1-\xi)}{2(2-\xi)}$	$\frac{\xi}{2-\xi}$	1

$$K = \frac{\left(\frac{p_C}{p^{\ominus}}\right)^2}{\left(\frac{p_A}{p^{\ominus}}\right)\left(\frac{p_B}{p^{\ominus}}\right)^3} = \frac{x_C^2}{x_A x_B^3} \times \left(\frac{p^{\ominus}}{p}\right)^2 = \frac{\xi^2}{(2-\xi)^2} \times \frac{2(2-\xi)}{1-\xi} \times \frac{2^3(2-\xi)^3}{3^3(1-\xi)^3} \times \left(\frac{p^{\ominus}}{p}\right)^2$$

$$= \frac{16(2-\xi)^2\xi^2}{27(1-\xi)^4} \times \left(\frac{p^{\ominus}}{p}\right)^2$$

Since K is independent of the pressure

$$\frac{(2-\xi)^2\xi^2}{(1-\xi)^4} = a^2\left(\frac{p}{p^{\ominus}}\right)^2 \quad a^2 = \frac{27}{16}K,\ \text{a constant}$$

Therefore $(2-\xi)\xi = a\left(\frac{p}{p^{\ominus}}\right) \times (1-\xi)^2$

$$\left(1 + \frac{ap}{p^{\ominus}}\right)\xi^2 - 2\left(1 + \frac{ap}{p^{\ominus}}\right)\xi + \frac{ap}{p^{\ominus}} = 0$$

which solves to $\boxed{\xi = 1 - \left(\frac{1}{1 + ap/p^{\ominus}}\right)^{1/2}}$

We choose the root with the negative sign because ξ lies between 0 and 1. The variation of ξ with p is shown in Fig. 9.2.

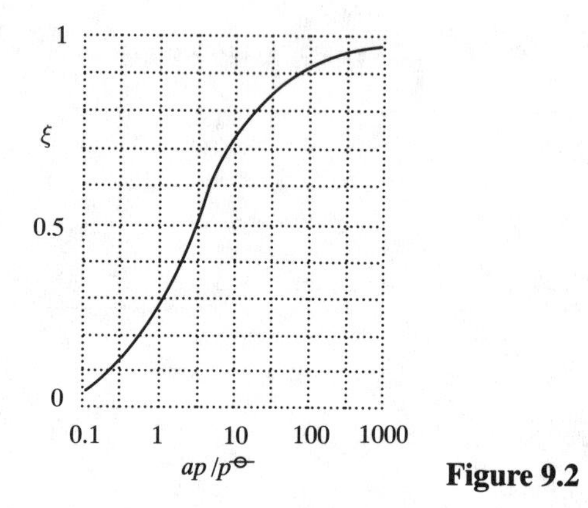

Figure 9.2

P9.13 $\Delta_r G = \Delta_r H - T\Delta_r S$

$$\Delta_r H' = \Delta_r H + \int_T^{T'} \Delta_r C_p \, dT \ [2.45]$$

$$\Delta_r S' = \Delta_r S + \int_T^{T'} \frac{\Delta_r C_p}{T} \, dT \ [4.19]$$

$$\Delta_r G' = \Delta_r G + \int_T^{T'} \Delta_r C_p \, dT + (T - T')\Delta_r S - T' \int_T^{T'} \frac{\Delta_r C_p}{T} \, dT$$

$$= \Delta_r G + (T - T')\Delta_r S + \int_T^{T'} \left(1 - \frac{T'}{T}\right) \Delta_r C_p \, dT$$

$$\Delta_r C_p = \Delta a + T\Delta b + \frac{\Delta c}{T^2}$$

$$\left(1 - \frac{T'}{T}\right) \Delta_r C_p = \Delta a + T\Delta b + \frac{\Delta c}{T^2} - \frac{T'\Delta a}{T} - T'\Delta b - \frac{T'\Delta c}{T^3}$$

$$= \Delta a - T'\Delta b + T\Delta b - \frac{T'\Delta a}{T} + \frac{\Delta c}{T^2} - \frac{T'\Delta c}{T^3}$$

$$\int_T^{T'} \left(1 - \frac{T'}{T}\right) \Delta_r C_p \, dT = (\Delta a - T'\Delta b)(T' - T) + \frac{1}{2}(T'^2 - T^2)\Delta b - T'\Delta a \ln \frac{T'}{T}$$

$$+ \Delta c \left(\frac{1}{T} - \frac{1}{T'}\right) - \frac{1}{2}T'\Delta c \left(\frac{1}{T^2} - \frac{1}{T'^2}\right)$$

Therefore, $\boxed{\Delta_r G' = \Delta_r G + (T - T')\Delta_r S + \alpha\Delta a + \beta\Delta b + \gamma\Delta c}$

where $\alpha = T' - T - T' \ln \dfrac{T'}{T}$

$$\beta = \frac{1}{2}(T'^2 - T^2) - T'(T' - T)$$

$$\gamma = \frac{1}{T} - \frac{1}{T'} + \frac{1}{2}T'\left(\frac{1}{T'^2} - \frac{1}{T^2}\right)$$

For water,

$$H_2(g) + \tfrac{1}{2}O_2(g) \rightarrow H_2O(l) \quad \Delta_f G^\ominus(T) = -237.13\ \text{kJ mol}^{-1}$$

$$\Delta_r S^\ominus(T) = -163.34\ \text{J K}^{-1}\ \text{mol}^{-1}$$

$$\Delta a = a(H_2O) - a(H_2) - \tfrac{1}{2}a(O_2) = (75.29 - 27.88 - 14.98)\,J\,K^{-1}\,mol^{-1}$$
$$= +33.03\,J\,K^{-1}\,mol^{-1}$$
$$\Delta b = [(0) - (3.26 \times 10^{-3}) - (2.09 \times 10^{-3})]\,J\,K^{-2}\,mol^{-1} = -5.35 \times 10^{-3}\,J\,K^{-2}\,mol^{-1}$$
$$\Delta c = [(0) - (0.50 \times 10^5) + (0.83 \times 10^5)]\,J\,K\,mol^{-1} = +0.33 \times 10^5\,J\,K\,mol^{-1}$$
$$T = 298\,K, \qquad T' = 372\,K, \text{ so}$$
$$\alpha = -8.5\,K, \qquad \beta = -2738\,K^2, \qquad \gamma = -8.288 \times 10^{-5}\,K^{-1}$$

and so

$$\begin{aligned}\Delta_f G^{\ominus}(372\,K) &= (-237.13\,kJ\,mol^{-1}) + (-74\,K) \times (-163.34\,J\,K^{-1}\,mol^{-1}) \\ &\quad + (-8.5\,K) \times (33.03 \times 10^{-3}\,kJ\,K^{-1}\,mol^{-1}) \\ &\quad + (-2738\,K^2) \times (-5.35 \times 10^{-6}\,kJ\,K^{-2}\,mol^{-1}) \\ &\quad + (-8.288 \times 10^{-5}\,K^{-1}) \times (0.33 \times 10^2\,kJ\,K\,mol^{-1}) \\ &= [(-237.13) + (12.09) - (0.28) + (0.015) - (0.003)]\,kJ\,mol^{-1} \\ &= \boxed{-225.31\,kJ\,mol^{-1}}\end{aligned}$$

Note that the β and γ terms are not significant (for this reaction and temperature range).

Solutions to additional problems

P9.14 If we knew $\Delta_r H^{\ominus}$ for this reaction, we could calculate $\Delta_f H^{\ominus}(HClO)$ from

$$\Delta_r H^{\ominus} = 2\Delta_f H^{\ominus}(HClO) - \Delta_f H^{\ominus}(Cl_2O) - \Delta_f H^{\ominus}(H_2O)$$

We can find $\Delta_r H^{\ominus}$ if we know $\Delta_r G^{\ominus}$ and $\Delta_r S^{\ominus}$, since

$$\Delta_r G^{\ominus} = \Delta_r H^{\ominus} - T\Delta_r S^{\ominus}$$

And we can find $\Delta_r G^{\ominus}$ from the equilibrium constant.

$$K = \exp(-\Delta_r G^{\ominus}/RT) \quad \text{so} \quad \Delta_r G^{\ominus} = -RT\ln K,$$
$$\Delta_r G^{\ominus} = -(8.3145 \times 10^{-3}\,kJ\,K^{-1}\,mol^{-1}) \times (298\,K)\ln 8.2 \times 10^{-2}$$
$$= 6.2\,kJ\,mol^{-1}$$
$$\Delta_r H^{\ominus} = \Delta_r G^{\ominus} + T\Delta_r S^{\ominus}$$
$$= 6.2\,kJ\,mol^{-1} + (298\,K) \times (16.38 \times 10^{-3}\,kJ\,K^{-1}\,mol^{-1}),$$
$$\Delta_r H^{\ominus} = 11.1\,kJ\,mol^{-1}$$

Finally

$$\Delta_f H^{\ominus}(HClO) = 1/2[\Delta_r H^{\ominus} + \Delta_f H^{\ominus}(Cl_2O) + \Delta_f H^{\ominus}(H_2O)],$$
$$\Delta_f H^{\ominus}(HClO) = 1/2[11.1 + 77.2 + (-241.82)]\,kJ\,mol^{-1}$$
$$= \boxed{76.8\,kJ\,mol^{-1}}$$

P9.16 According to Henry's law

$$p_{CO_2} = x_{CO_2}K_H,$$

$$\text{so } x_{CO_2} = \frac{p_{CO_2}}{K_H} = \frac{(3.6 \times 10^{-4}\,atm) \times (760\,Torr\,atm)}{1.25 \times 10^6\,Torr} = 2.2 \times 10^{-7}$$

This mole fraction of aqueous CO_2 in natural rainwater corresponds to the following molality

$$b_{CO_2} = \frac{n_{CO_2}}{1\,\text{kg}\,H_2O} \approx \frac{n_{H_2O}x_{CO_2}}{1\,\text{kg}\,H_2O} = \frac{x_{CO_2}}{M_{H_2O}} = \frac{2.2 \times 10^{-7}}{18.015 \times 10^{-3}\,\text{kg mol}^{-1}}$$
$$= 1.2 \times 10^{-5}\,\text{mol kg}^{-1}$$

Assuming that aqueous CO_2 forms H_2CO_3 quantitatively and that H_2CO_3 is an ideally dilute solute, the reaction which determines the pH is

$$H_2CO_3 \rightleftharpoons H^+ + HCO_3^- \quad \text{with} \quad K = \frac{a_{H^+}a_{HCO_3^-}}{b_{H_2CO_3}} \approx \frac{a_{H^+}^2}{b_{H_2CO_3}}$$

The last approximation assumes that any other sources of H^+ or HCO_3^- are negligible. Assuming further that ionization is so small that $b_{H_2CO_3}$ is virtually unchanged by it, we have

$$a_{H^+} = \sqrt{Kb_{H_2CO_3}} = \sqrt{(4.3 \times 10^{-7}) \times (1.2 \times 10^{-5})} = 2.3 \times 10^{-6}$$

So $\text{pH} = -\log a_{H^+} = \boxed{5.64}$

The pre-industrial atmosphere had $x_{CO_2} = 1.7 \times 10^{-7}$, $b_{CO_2} = 9.4 \times 10^{-6}\,\text{mol kg}^{-1}$, $a_{H^+} = 2.0 \times 10^{-6}$, and $\text{pH} = \boxed{5.70}$

P9.19 Let a be the mole fraction of perylene, b that of benzo(e)pyrene, and c that of benzo(a)pyrene. The mole fractions add up to unity

$$a + b + c = 1$$

And the mole ratios of these isomers are equal to equilibrium constants

$$\text{perylene} \rightleftharpoons \text{benzo(e)pyrene} \quad b/a = K_1 \qquad (1)$$
$$\text{benzo(e)pyrene} \rightleftharpoons \text{benzo(a)pyrene} \quad c/b = K_2 \qquad (2)$$

(The third such equilibrium would not yield any additional information.) So we have three equations in three unknowns

$$b = aK_1 = c/K_2 \quad \text{so} \quad c = aK_1K_2,$$

and $a + aK_1 + aK_1K_2 = 1 = a(1 + K_1 + K_1K_2)$ so $a = (1 + K_1 + K_1K_2)^{-1}$

We need the equilibrium constants, which are given by

$$K = \exp\left(\frac{-\Delta_r G^\ominus}{RT}\right) = \exp\left(\frac{-\Delta_r H^\ominus}{RT}\right)\exp\left(\frac{\Delta_r S^\ominus}{R}\right)$$

For reaction (1)

$$\Delta_r H^\ominus = (253.2 - 253.2)\,\text{kJ mol}^{-1} = 0.0$$

and $\Delta_r S^\ominus = (993.7 - 987.9)\,\text{J mol}^{-1}\,\text{K}^{-1} = 5.8\,\text{J mol}^{-1}\,\text{K}^{-1}$

$$\text{so } K = \exp\left(\frac{5.8\,\text{J mol}^{-1}\,\text{K}^{-1}}{8.3145\,\text{J mol}^{-1}\,\text{K}^{-1}}\right) = 2.0$$

For reaction (2)

$$\Delta_r H^{\ominus} = (262.4 - 253.2)\,\text{kJ mol}^{-1} = 9.2\,\text{kJ mol}^{-1}$$

and $\Delta_r S^{\ominus} = (999.4 - 993.7)\,\text{J mol}^{-1}\,\text{K}^{-1} = 5.7\,\text{J mol}^{-1}\,\text{K}^{-1}$

so $K = \exp\left(\dfrac{-9.2\,\text{kJ mol}^{-1}}{(8.3145\,\text{J mol}^{-1}\,\text{K}^{-1})(1000\,\text{K})}\right)\exp\left(\dfrac{5.7\,\text{J mol}^{-1}\,\text{K}^{-1}}{8.3145\,\text{J mol}^{-1}\,\text{K}^{-1}}\right) = 0.66$

Now we can go back and evaluate the terms

$$a = (1 + K_1 + K_1K_2)^{-1} = (1 + 2.0 + 2.0 \times 0.66)^{-1} = \boxed{0.23}$$

$$b = aK_1 = (0.23) \times (2.0) = \boxed{0.46}$$

and $c = aK_1K_2 = (0.23) \times (2.0) \times (0.66) = \boxed{0.30}$

10 Equilibrium electrochemistry

Solutions to exercises

E10.1 $CuSO_4(aq)$ and $ZnSO_4(aq)$ are strong electrolytes; therefore the net ionic equation is

$$Zn(s) + Cu^{2+}(aq) \rightarrow Zn^{2+}(aq) + Cu(s)$$

$$\Delta_r H^{\ominus} = \Delta_f H^{\ominus}(Zn^{2+}, aq) - \Delta_f H^{\ominus}(Cu^{2+}, aq)$$

$$= (-153.89\,\text{kJ mol}^{-1}) - (64.77\,\text{kJ mol}^{-1}) = \boxed{-218.66\,\text{kJ mol}^{-1}}$$

Comment. $SO_4^{2-}(aq)$ is a spectator ion and was ignored in the determination of $\Delta_r H^{\ominus}$ above. This is justifiable because $\Delta_r H^{\ominus}$ refers to the standard state of all species participating in the reaction.

E10.2 $HgCl_2(s) \rightleftharpoons Hg^{2+}(aq) + 2Cl^-(aq)$

$$K = \prod_J a_J^{\nu_J}$$

Since the solubility is expected to be low, we may (initially) ignore activity coefficients. Hence

$$K = \frac{b(Hg^{2+})}{b^{\ominus}} \times \frac{b^2(Cl^-)}{(b^{\ominus})^2} \qquad b(Cl^-) = 2b(Hg^{2+}) = 2s$$

$$K = \frac{s(2s)^2}{(b^{\ominus})^3} = \frac{4s^3}{(b^{\ominus})^3} \qquad s = b(Hg^{2+}) = \left(\frac{1}{4}K\right)^{1/3} b^{\ominus}$$

K may be determined from

$$\ln K = \boxed{\frac{-\Delta_r G^{\ominus}}{RT}} \quad \text{[Chapter 9]}$$

$$\Delta_r G^{\ominus} = \Delta_r G^{\ominus}(Hg^{2+}, aq) + 2\Delta_f G^{\ominus}(Cl^-, aq) - \Delta_r G^{\ominus}(HgCl_2, s)$$

$$= [(+164.40) + (2) \times (-131.23) - (-178.6)]\,\text{kJ mol}^{-1} = +80.54\,\text{kJ mol}^{-1}$$

$$\ln K = \frac{-80.54 \times 10^3\,\text{J mol}^{-1}}{(8.314\,\text{J K}^{-1}\,\text{mol}^{-1}) \times (298.15\,\text{K})} = -32.49$$

Hence $K = 7.75 \times 10^{-15}$ and $s = \boxed{1.25 \times 10^{-5}\,\text{mol L}^{-1}}$

E10.3 A procedure similar to that outlined in Section 10.1 and Fig. 10.1 of the text is followed.

		$\Delta G^{\ominus}/(\text{kJ mol}^{-1})$		
		Cl^-	F^-	
Dissociation of H_2	$\frac{1}{2}H_2 \rightarrow H$	+218	+218	(Table 14.3)
Ionization of H	$H \rightarrow H^+ + e^-$	+1312	+1312	(Table 13.4)
Hydration of H^+	$H^+(g) \rightarrow H^+(aq)$	x	x	
Dissociation of X_2	$\frac{1}{2}X_2 \rightarrow X$	+121	78	(Table 14.3)
Electron gain by X	$X + e^- \rightarrow X^-$	−348.7	−322	(Table 13.5)
Hydration of X^-	$X^-(g) \rightarrow X^-(aq)$	y	y'	
Overall		$\Delta_f G^{\ominus}(Cl^-)$	$\Delta_f G^{\ominus}(F^-)$	

Hence, $\Delta_f G^{\ominus}(Cl^-) = x + y + 1302\,\text{kJ mol}^{-1}$

$\Delta_f G^{\ominus}(F^-) = x + y' + 1286\,\text{kJ mol}^{-1}$

and $\Delta_f G^{\ominus}(Cl^-) - \Delta_f G^{\ominus}(F^-) = y - y' + 16\,kJ\,mol^{-1}$

The ratio of hydration Gibbs energies is

$$\frac{\Delta_{solv}G^{\ominus}(F^-)}{\Delta_{solv}G^{\ominus}(Cl^-)} = \frac{r(Cl^-)}{r(F^-)}[5] = \frac{181\,pm}{131\,pm}\,[\text{Table 21.3}] = 1.38$$

$$\Delta_{solv}G^{\ominus} = -\frac{z_i^2}{r_i/pm} \times (6.86 \times 10^4\,kJ\,mol^{-1})$$

$$z^2 = 1, \qquad r(Cl^-) = 181\,pm\,[\text{Table 21.3}]$$

$$\Delta_{solv}G^{\ominus}(Cl^-) = -\frac{6.86 \times 10^4\,kJ\,mol^{-1}}{181} = -379\,kJ\,mol^{-1}$$

$$\Delta_{solv}G^{\ominus}(F^-) = (1.38) \times (-379\,kJ\,mol^{-1}) = -523\,kJ\,mol^{-1}$$

and $\Delta_f G^{\ominus}(Cl^-) - \Delta_f G^{\ominus}(F^-) = [(-379) - (-523) + (16)]\,kJ\,mol^{-1} = +160\,kJ\,mol^{-1}$

Hence (Table 2.6)

$$\Delta_f G^{\ominus}(F^-) = [(-131.23) - (160)]\,kJ\,mol^{-1} = \boxed{-291\,kJ\,mol^{-1}}$$

(The 'experimental' value, Table 2.6, is $-278.79\,kJ\,mol^{-1}$.)

E10.4 $I = \frac{1}{2}\sum_i (b_i/b^{\ominus})z_i^2$ [20]

and for an M_pX_q salt, $(b_+/b^{\ominus}) = p(b/b^{\ominus})$, $(b_-/b^{\ominus}) = q(b/b^{\ominus})$, so

$$I = \frac{1}{2}(pz_+^2 + qz_-^2)\left(\frac{b}{b^{\ominus}}\right)$$

(a) $I(KCl) = \frac{1}{2}(1 \times 1 + 1 \times 1)\left(\frac{b}{b^{\ominus}}\right) = \left(\frac{b}{b^{\ominus}}\right)$

(b) $I(FeCl_3) = \frac{1}{2}(1 \times 3^2 + 3 \times 1)\left(\frac{b}{b^{\ominus}}\right) = 6\left(\frac{b}{b^{\ominus}}\right)$

(c) $I(CuSO_4) = \frac{1}{2}(1 \times 2^2 + 1 \times 2^2)\left(\frac{b}{b^{\ominus}}\right) = 4\left(\frac{b}{b^{\ominus}}\right)$

E10.5 $I = I(KCl) + I(CuSO_4) = \left(\frac{b}{b^{\ominus}}\right)(KCl) + 4\left(\frac{b}{b^{\ominus}}\right)(CuSO_4)$ [Exercise 10.4(a)]

$$= (0.10) + (4) \times (0.20) = \boxed{0.90}$$

Comment. Note that the ionic strength of a solution of more than one electrolyte may be calculated by summing the ionic strengths of each electrolyte considered as a separate solution, as in the solution to this exercise, or by summing the product $\frac{1}{2}\left(\frac{b_i}{b^{\ominus}}\right)z_i^2$ for each individual ion, as in the definition of I [20].

E10.6 $I = I(KNO_3) = \left(\frac{b}{b^{\ominus}}\right)(KNO_3) = 0.150$

Therefore, the ionic strengths of the added salts must be 0.100.

(a) $I(\mathrm{Ca(NO_3)_2}) = \frac{1}{2}(2^2+2)\left(\frac{b}{b^{\ominus}}\right) = 3\left(\frac{b}{b^{\ominus}}\right)$

Therefore, the solution should be made $\frac{1}{3} \times 0.100\,\mathrm{mol\,kg^{-1}} = 0.0333\,\mathrm{mol\,kg^{-1}}$ in $Ca(NO_3)_2$. The mass that should be added to 500 g of the solution is therefore

$$(0.500\,\mathrm{kg}) \times (0.0333\,\mathrm{mol\,kg^{-1}}) \times (164\,\mathrm{g\,mol^{-1}}) = \boxed{2.73\,\mathrm{g}}$$

(b) $I(\mathrm{NaCl}) = \left(\frac{b}{b^{\ominus}}\right)$; therefore, with $b = 0.100\,\mathrm{mol\,kg^{-1}}$

$$(0.500\,\mathrm{kg}) \times (0.100\,\mathrm{mol\,kg^{-1}}) \times (58.4\,\mathrm{g\,mol^{-1}}) = \boxed{2.92\,\mathrm{g}}$$

(We are neglecting the fact that the mass of solution is slightly different from the mass of solvent.)

E10.7 $I(\mathrm{KCl}) = (b/b^{\ominus})$, $\quad I(\mathrm{CuSO_4}) = 4(b/b^{\ominus})$ [Exercise 10.4(a)]

For $I(\mathrm{KCl}) = I(\mathrm{CuSO_4})$, $(b/b^{\ominus})(\mathrm{KCl}) = 4(b/b^{\ominus})(\mathrm{CuSO_4})$

Therefore, if $b(\mathrm{KCl}) = 1.00\,\mathrm{mol\,kg^{-1}}$, we require $b(\mathrm{CuSO_4}) = \boxed{0.25\,\mathrm{mol\,kg^{-1}}}$

E10.8 $\gamma_{\pm} = (\gamma_+^p \gamma_-^q)^{1/s} \quad s = p + q$ [16]

For $CaCl_2$, $p = 1,\ q = 2,\ s = 3,\ \boxed{\gamma_{\pm} = (\gamma_+\gamma_-^2)^{1/3}}$

E10.9 These concentrations are sufficiently dilute for the Debye–Hückel limiting law to give a good approximate value for the mean ionic activity coefficient. Hence

$$\log \gamma_{\pm} = -|z_+z_-|AI^{1/2} \text{ [19]}$$

$$I = \frac{1}{2}\sum_i z_i^2\left(\frac{b_i}{b^{\ominus}}\right) \text{ [20]} = \frac{1}{2}[(4\times 0.010) + (1\times 0.020) + (1\times 0.030) + (1\times 0.030)]$$

$$= \boxed{0.060}$$

$$\log \gamma_{\pm} = -2\times 1\times 0.509\times (0.060)^{1/2} = -0.249\overline{4}; \qquad \gamma_{\pm} = 0.56\overline{3} = \boxed{0.56}$$

E10.10 $I(\mathrm{LaCl_3}) = \frac{1}{2}(3^2+3)\left(\frac{b}{b^{\ominus}}\right) = 6\left(\frac{b}{b^{\ominus}}\right) = 3.000$

From the limiting law [19]

$$\log \gamma_{\pm} = -0.509|z_+z_-|I^{1/2} = (-0.509)\times(3)\times(3.000)^{1/2} = -2.64\overline{5}$$

Hence $\gamma_{\pm} = 2.3\times 10^{-3}$

and the error is $\boxed{1\times 10^4 \text{ per cent}}$

Comment. It is not surprising that the limiting law provides such a poor prediction of $\gamma_{\pm}$ for this (3, 1) electrolyte at this high concentration.

E10.11 $\log \gamma_{\pm} = -\frac{A|z_+z_-|I^{1/2}}{1+BI^{1/2}}$ [39]

Solving for B,

$$B = -\left(\frac{1}{I^{1/2}} + \frac{A|z_+z_-|}{\log \gamma_{\pm}}\right)$$

For HBr, $I = \left(\frac{b}{b^\ominus}\right)$ and $|z_+z_-| = 1$; so

$$B = -\left(\frac{1}{(b/b^\ominus)^{1/2}} + \frac{0.509}{\log \gamma_\pm}\right)$$

Hence, draw up the following table

$(b/b^\ominus)$	5.0×10^{-3}	10.0×10^{-3}	20.0×10^{-3}
$\gamma_\pm$	0.930	0.907	0.879
B	2.01	2.01	2.02

The constancy of B indicates that the mean ionic activity coefficient of HBr obeys the extended Debye–Hückel law very well.

E10.12 $CaF_2(s) \rightleftharpoons Ca^{2+}(aq) + 2F^-(aq) \quad K_s = 3.9 \times 10^{-11}$

$$\begin{aligned}\Delta_r G^\ominus &= -RT \ln K_s \\ &= -(8.314\,\mathrm{J\,K^{-1}\,mol^{-1}}) \times (298.15\,\mathrm{K}) \times (\ln 3.9 \times 10^{-11}) = +59.4\,\mathrm{kJ\,mol^{-1}} \\ &= \Delta_f G^\ominus(CaF_2, aq) - \Delta_f G^\ominus(CaF_2, s)\end{aligned}$$

Hence, $\Delta_f G^\ominus(CaF_2, aq) = \Delta G^\ominus + \Delta_f G^\ominus(CaF_2, s)$

$$= [59.4 - 1167]\,\mathrm{kJ\,mol^{-1}} = \boxed{-1108\,\mathrm{kJ\,mol^{-1}}}$$

E10.13 The Nernst equation may be applied to individual reduction potentials as well as to overall cell potentials (Section 10.5). Hence

$$E(H^+/H_2) = \frac{RT}{F}\ln\frac{a(H^+)}{\left(\frac{f_{H_2}}{p^\ominus}\right)^{1/2}} \quad [56]$$

$$\text{and } \Delta E = E_1 - E_2 = \frac{RT}{F}\ln\frac{a_1(H^+)}{a_2(H^+)}\ [f_{H_2}\text{is constant}] = \frac{RT}{F}\ln\frac{\gamma_\pm b_1}{\gamma_\pm b_2} \quad [\gamma_+ \approx \gamma_\pm]$$

$$= (25.7\,\mathrm{mV}) \times \ln\left(\frac{(20.0) \times (0.879)}{(5.0) \times (0.930)}\right) = \boxed{34.2\,\mathrm{mV}}$$

Comment. Strictly $a(H^+) = \gamma(H^+)b(H^+)$, but $\gamma(H^+)$ cannot be determined from the data provided. However, since the solution is dilute, it is a valid approximation to replace $\gamma(H^+)$ with $\gamma_\pm$.

E10.14 We begin by choosing, based on an educated guess, the right and left electrodes.

R: $Cl_2(g) + 2e^- \rightarrow 2Cl^-(aq) \quad E_R^\ominus = +1.36\,\mathrm{V}$ [Table 10.7]

L: $Mn^{2+}(aq) + 2e^- \rightarrow Mn(s) \quad E_L^\ominus = ?$

The cell corresponding to these half-reaction is

$$Mn|MnCl_2(aq)|Cl_2(g)|Pt; \quad E_{cell}^\ominus = E_R^\ominus - E_L^\ominus = 1.36\,\mathrm{V} - E^\ominus(Mn, Mn^{2+})$$

Hence, $E^\ominus(Mn, Mn^{2+}) = 1.36\,\mathrm{V} - 2.54\,\mathrm{V} = \boxed{-1.18\,\mathrm{V}}$

Comment. With this choice of the right and left electrodes $E_{cell}^\ominus > 0$; the opposite choice would have resulted in $E_{cell}^\ominus < 0$ and could not have corresponded to the thermodynamically spontaneous reaction given.

E10.15 The cell notation specifies the right and left electrodes. Note that for proper cancellation we must equalize the number of electrons in half-reactions being combined.

		$E^{\ominus}$
(a)	R: $2Ag^+(aq) + 2e^- \rightarrow 2Ag(s)$	+0.80 V
	L: $Zn^+(aq) + 2e^- \rightarrow Zn(s)$	−0.76 V
	Overall (R − L): $2Ag^+(aq) + Zn(s) \rightarrow 2Ag(s) + Zn^{2+}(aq)$	+1.56 V
(b)	R: $2H^+(aq) + 2e^- \rightarrow H_2(g)$	0
	L: $Cd^{2+}(aq) + 2e^- \rightarrow Cd(s)$	−0.40 V
	Overall (R − L): $Cd(s) + 2H^+(aq) \rightarrow Cd^{2+}(aq) + H_2(g)$	+0.40 V
(c)	R: $Cr^{3+}(aq) + 3e^- \rightarrow Cr(s)$	−0.74 V
	L: $3[Fe(CN)_6]^{3-}(aq) + 3e^- \rightarrow 3[Fe(CN)_6]^{4-}(aq)$	+0.36 V
	Overall (R − L): $Cr^{3+}(aq) + 3[Fe(CN)_6]^{4-}(aq) \rightarrow Cr(s) + 3[Fe(CN)_6]^{3-}(aq)$	−1.10 V

Comment. Those cells for which $E^{\ominus} > 0$ may operate as spontaneous galvanic cells under standard conditions. Those for which $E^{\ominus} < 0$ may operate as non-spontaneous electrolytic cells. Recall that $E^{\ominus}$ informs us of the spontaneity of a cell under standard conditions only. For other conditions we require E.

E10.16 The conditions (concentrations, etc.) under which these reactions occur are not given. For the purposes of this exercise we assume standard conditions. The specification of the right and left electrodes is determined by the direction of the reaction as written. As always, in combining half-reactions to form an overall cell reaction we must write the half-reactions with equal number of electrons to ensure proper cancellation. We first identify the half-reactions, and then set up the corresponding cell.

		$E^{\ominus}$
(a)	R: $Cu^{2+}(aq) + 2e^- \rightarrow Cu(s)$	+0.34 V
	L: $Zn^{2+}(aq) + 2e^- \rightarrow Zn(s)$	−0.76 V
	Hence the cell is	
	$Zn(s)|ZnSO_4(aq)||CuSO_4(aq)|Cu(s)$	+1.10 V
(b)	R: $AgCl(s) + e^- \rightarrow Ag(s) + Cl^-(aq)$	+0.22 V
	L: $H^+(aq) + e^- \rightarrow \frac{1}{2}H_2(g)$	0
	and the cell is	
	$Pt|H_2(g)|H^+(aq)|AgCl(s)|Ag(s)$	
	or $Pt|H_2(g)|HCl(aq)|AgCl(s)|Ag(s)$	+0.22 V
(c)	R: $O_2(g) + 4H^+(aq) + 4e^- \rightarrow 2H_2O(l)$	+1.23 V
	L: $4H^+(aq) + 4e^- \rightarrow 2H_2(g)$	0
	and the cell is	
	$Pt|H_2(g)|H^+(aq), H_2O(l)|O_2(g)|Pt$	+1.23 V

Comment. All of these cells have $E^{\ominus} > 0$, corresponding to a spontaneous cell reaction under standard conditions. If $E^{\ominus}$ had turned out to be negative, the spontaneous reaction would have been the reverse of the one given, with the right and left electrodes of the cell also reversed.

E10.17 See the solutions to Exercise 10.15**(a)**, where we have used $E^{\ominus} = E^{\ominus}_R - E^{\ominus}_L$, with standard electrode potentials from Table 10.7.

E10.18 See the solutions to Exercise 10.16(**a**), where we have used $E^{\ominus} = E_R^{\ominus} - E_L^{\ominus}$.

E10.19 In each case find $E^{\ominus} = E_R^{\ominus} - E_L^{\ominus}$ from the data in Table 10.7, then use

$$\Delta_r G^{\ominus} = -\nu F E^{\ominus} \text{ [43]}$$

(a) $2Na(s) + 2H_2O(l) \rightarrow 2NaOH(aq) + H_2(g)$ $\quad E^{\ominus} = +1.88\,V$ [Exercise 10.16(**b**)(**a**)]
Therefore, with $\nu = 2$

$$\Delta_r G^{\ominus} = (-2) \times (96.485\,kC\,mol^{-1}) \times (1.88\,V) = \boxed{-363\,kJ\,mol^{-1}}$$

(b) $2K(s) + 2H_2O(l) \rightarrow 2KOH(aq) + H_2(g)$

$$E^{\ominus} = E^{\ominus}(H_2O, OH^-, H_2) - E^{\ominus}(K, K^+)$$
$$= -0.83\,V - (-2.93\,V) = +2.10\,V \quad \text{with } \nu = 2$$

Therefore,

$$\Delta_r G^{\ominus} = (-2) \times (96.485\,kC\,mol^{-1}) \times (2.10\,V) = \boxed{-405\,kJ\,mol^{-1}}$$

E10.20 **(a)** $$E^{\ominus} = \frac{-\Delta G^{\ominus}}{\nu F}\text{[43]} = \frac{+62.5\,kJ\,mol^{-1}}{(2) \times (96.485\,kC\,mol^{-1})} = \boxed{+0.324\,V}$$

(b) $E^{\ominus} = E_R^{\ominus} - E_L^{\ominus} = E^{\ominus}(Fe^{3+}, Fe^{2+}) - E^{\ominus}(Ag, Ag_2CrO_4, CrO_4^{2-})$
Therefore, $E^{\ominus}(Ag, Ag_2CrO_4, CrO_4^{2-}) = E^{\ominus}(Fe^{3+}, Fe^{2+}) - E^{\ominus}$

$$= [+0.77 - 0.324]\,V = \boxed{+0.45\,V}$$

E10.21 When combining two half-reactions to correspond to an overall cell reaction which is spontaneous, the combination must be such that the electrons in the half-reactions cancel and that $E_{cell} > 0$. Thus

R: $O_2(g) + 4H^+(aq) + 4e^- \rightarrow 2H_2O(l)$ $\quad E^{\ominus} = +1.23\,V$
L: $2Ag_2S(s) + 4e^- \rightarrow 4Ag(s) + 2S^{2-}(aq)$ $\quad E^{\ominus} = -0.69\,V$
R − L: $4Ag(s) + 2S^{2-}(aq) + O_2(g) + 4H^+(aq) \rightarrow 2Ag_2S(s) + 2H_2O(l)$

$$E^{\ominus} = E_R^{\ominus} - E_L^{\ominus} = (1.23\,V) - (-0.69\,V) = \boxed{+1.92\,V}$$

Comment. Under standard conditions $E = E^{\ominus} > 0$ and the reaction is spontaneous. Because of the large positive $E^{\ominus}$, the reaction is likely to remain spontaneous unless the conditions are changed drastically to make $E < 0$.

Question. Can you devise conditions such that $E < 0$?

E10.22

	$E^{\ominus}$
R: $Cd^{2+}(aq) + 2e^- \rightarrow Cd(s)$	$-0.40\,V$
L: $2AgBr(s) + 2e^- \rightarrow 2Ag(s) + 2Br^-(aq)$	$+0.07\,V$

Hence, overall (R − L)

$$Cd^{2+}(aq) + 2Ag(s) + 2Br^-(aq) \rightarrow Cd(s) + 2AgBr(s) \qquad -0.47\,V$$

$$Q = \frac{1}{a(Cd^{2+})a^2(Br^-)} \qquad E = E^{\ominus} + \frac{RT}{2F}\ln a(Cd^{2+})a^2(Br^-)$$

$$a(Cd^{2+}) = \gamma_+ b_+; \qquad a(Br^-) = \gamma_- b_- \quad \left[b \equiv \frac{b}{b^{\ominus}}\right]$$

$$b_+ = 0.010\,mol\,kg^{-1}, \qquad b_- = 0.050\,mol\,kg^{-1}$$

We assume that $\gamma_+(Cd^{2+}) \approx \gamma_\pm\{Cd(NO_3)_2\}$ and that $\gamma_-(Br^-) \approx \gamma_\pm(KBr)$; hence

$$E = E^\ominus + \frac{RT}{2F}\ln b(Cd^{2+})b^2(Br^-) + \frac{2.303RT}{2F}\log \gamma_\pm\{Cd(NO_3)_2\}\gamma_\pm^2(KBr)$$

$$\log \gamma_\pm\{Cd(NO_3)_2\} \approx -A|z_+z_-| \times I^{1/2}, \quad I = 3b = 0.030\,\text{mol kg}^{-1}$$

$$\approx -(0.509) \times (2) \times (0.030)^{1/2} = -0.18$$

$$\log \gamma_\pm(KBr) \approx -A|z_+z_-| \times I^{1/2}, \quad I = b$$

$$\approx -(0.509) \times (1) \times (0.050)^{1/2} = -0.11$$

Hence, $E = (-0.47\,\text{V}) + \left(\frac{25.693\,\text{mV}}{2}\right) \times \ln(0.010 \times 0.050^2)$

$$+\left(\frac{(2.303) \times (25.693\,\text{mV})}{2}\right) \times (-0.18 + 2 \times (-0.11)) = \boxed{-0.62\,\text{V}}$$

E10.23 In each case $\ln K = \frac{\nu F E^\ominus}{RT}$ [47]

(a) $Sn(s) + Sn^{4+}(aq) \rightleftharpoons 2Sn^{2+}(aq)$

R: $Sn^{4+} + 2e^- \rightarrow Sn^{2+}(aq)$ $\quad +0.15\,\text{V}$

L: $Sn^{2+}(aq) + 2e^- \rightarrow Sn(s)$ $\quad -0.14\,\text{V}$

$\qquad E^\ominus = +0.29\,\text{V}$

$$\ln K = \frac{(2) \times (0.29\,\text{V})}{25.693\,\text{mV}} = 22.\bar{6}, \qquad K = \boxed{6.5 \times 10^9}$$

(b) $Sn(s) + 2AgCl(s) \rightleftharpoons SnCl_2(aq) + 2Ag(s)$

R: $2AgCl(s) + 2e^- \rightarrow 2Ag(s) + 2Cl^-(aq)$ $\quad +0.22\,\text{V}$

L: $Sn^{2+}(aq) + 2e^- \rightarrow Sn(s)$ $\quad -0.14\,\text{V}$

$\qquad +0.36\,\text{V}$

$$\ln K = \frac{(2) \times (0.36\,\text{V})}{25.693\,\text{mV}} = +28.\bar{0}, \qquad K = \boxed{1.5 \times 10^{12}}$$

E10.24 We need to obtain $E^\ominus$ for the couple

(3) $Au^{3+}(aq) + 2e^- \rightarrow Au^+(aq)$

from the values of $E^\ominus$ for the couples

(1) $Au^+(aq) + e^- \rightarrow Au(s)$ $\qquad E_1^\ominus = 1.69\,\text{V}$

(2) $Au^{3+}(aq) + 3e^- \rightarrow Au(s)$ $\qquad E_2^\ominus = 1.40\,\text{V}$

We see that (3) = (2) – (1), therefore

$$\Delta_r G_3^\ominus = \Delta_r G_1^\ominus - \Delta_r G_2^\ominus$$

$$-\nu_3 F E_3^\ominus = -\nu_1 F E_1^\ominus - \nu_2 F E_2^\ominus$$

Solving for $E_3^\ominus$ we obtain

$$E_3^\ominus = \frac{\nu_2 E_2^\ominus - \nu_1 E_1^\ominus}{\nu_3} = \frac{(3) \times (1.40\,\text{V}) - (1) \times (1.69\,\text{V})}{2} = 1.26\,\text{V}$$

Then,

$$\text{R:}\quad Au^{3+}(aq) + 2e^- \rightarrow Au^+(aq) \qquad E_R^\ominus = 1.26\,V$$
$$\text{L:}\quad 2Fe^{3+}(aq) + 2e^- \rightarrow 2Fe^{2+}(aq) \qquad E_L^\ominus = 0.77\,V$$
$$\text{R} - \text{L:}\quad 2Fe^{2+}(aq) + Au^{3+}(aq) \rightarrow 2Fe^{3+}(aq) + Au^+(aq)$$
$$E^\ominus = E_R^\ominus - E_L^\ominus = (1.26\,V) - (0.77\,V) = \boxed{+0.49\,V}$$
$$\ln K = \frac{\nu F E^\ominus}{RT}[47] = \frac{(2) \times (0.49\,V)}{25.7 \times 10^{-3}\,V} = 38.\bar{1}, \qquad K = \boxed{\bar{4} \times 10^{16}}$$

E10.25 First assume all activity coefficients are 1 and calculate K_s°, the ideal solubility product constant.

(1) $AgCl(s) \rightleftharpoons Ag^+(aq) + Cl^-(aq)$

Since all stoichiometric coefficients are 1

$$S(AgCl) = b(Ag^+) = b(Cl^-)$$

Hence, $K_s^\circ = \dfrac{b(Ag^+)b(Cl^-)}{b^{\ominus 2}} = \dfrac{b^2(Ag^+)}{b^{\ominus 2}} = \dfrac{S^2}{b^{\ominus 2}} = (1.34 \times 10^{-5})^2 = \boxed{1.80 \times 10^{-10}}$

(2) $BaSO_4(s) \rightleftharpoons Ba^{2+}(aq) + SO_4^{2-}(aq)$

$$S(BaSO_4) = b(Ba^{2+}) = b(SO_4^{2-})$$

As above, $K_s^\circ = \dfrac{S^2}{b^{\ominus 2}} = (9.51 \times 10^{-4})^2 = \boxed{9.04 \times 10^{-7}}$

Now redo the calculation taking into account the deviation of the activity coefficients from 1 in order to obtain K_s, the time thermodynamic solubility product constant. We assume that the activity coefficients can be estimated from the Debye–Hückel limiting law since the concentrations of the ions are low.

For both $AgCl(s)$ and $BaSO_4(s)$

$$a_+ = \frac{\gamma_+ b_+}{b^\ominus}, \qquad a_- = \frac{\gamma_- b_-}{b^\ominus}$$
$$K_s = a_+ a_- = \gamma_+\gamma_- \left(\frac{b_+}{b^\ominus}\right) \times \left(\frac{b_-}{b^\ominus}\right) = \gamma_+\gamma_- K_s^\circ, \qquad \gamma_+\gamma_- = \gamma_\pm^2$$

Thus, $K_s = \gamma_\pm^2 K_s^\circ$

$$\log \gamma_\pm = -|z_+ z_-| A I^{1/2}, \quad A = 0.509$$

For AgCl, $I = S$, $|z_+ z_-| = 1$, and so

$$\log \gamma_\pm = -(0.509) \times (1.34 \times 10^{-5})^{1/2} = -1.86 \times 10^{-3}, \qquad \gamma_\pm \approx 0.9957$$

Hence, $K_s = \gamma_\pm^2 \times K_s^\circ \approx \boxed{0.991\,K_s^\circ}$

For $BaSO_4$, $I = 4S$, $|z_+ z_-| = 4$, and so

$$\log \gamma_\pm = -(0.509) \times (4) \times \left[(4) \times (9.51 \times 10^{-4})\right]^{1/2} = -0.126, \qquad \gamma_\pm \approx 0.75$$

Hence, $K_s = \gamma_\pm^2 K_s^\circ \approx (0.75)^2 K_s^\circ \approx \boxed{0.56\,K_s^\circ}$

Thus, the neglect of activity coefficients is significant for $BaSO_4$.

E10.26 A Nernst equation can be written for a half-reaction as well as for a whole-cell reaction.

The half-reaction is (Table 10.7)

$$Cr_2O_7^{2-}(aq) + 14\,H^+(aq) + 6e^- \rightarrow 2Cr^{3+}(aq) + 7H_2O(l)$$

The reaction quotient is

$$Q = \frac{a^2(Cr^{3+})}{a(Cr_2O_7^{2-})a^{14}(H^+)} \qquad \nu = 6$$

Hence, $\boxed{E = E^\ominus - \frac{RT}{6F}\ln\frac{a^2(Cr^{3+})}{a(Cr_2O_7^{2-})a^{14}(H^+)}}$

E10.27
R: $2AgCl(s) + 2e^- \rightarrow 2Ag(s) + 2Cl^-(aq) \quad +0.22\,V$
L: $2H^+(aq) + 2e^- \rightarrow H_2(g) \quad 0$

Overall, R − L: $2AgCl(s) + H_2(g) \rightarrow 2Ag(s) + 2H^+(aq) + 2Cl^-(aq)$

$$Q = a^2(H^+)a^2(Cl^-)[\nu = 2] = a^4(H^+) \quad [\text{Assume } a(H^+) \approx a(Cl^-)]$$

Therefore, from the Nernst equation [45],

$$E = E^\ominus - \frac{RT}{2F}\ln a^4(H^+)$$
$$= E^\ominus - \frac{2RT}{F}\ln a(H^+) = E^\ominus + (2)\times(2.303)\times\left(\frac{RT}{F}\right)\times \text{pH}$$

Hence,

$$\text{pH} = \left(\frac{F}{(2)\times(2.303RT)}\right)\times(E - E^\ominus)$$
$$= \frac{E - 0.22\,V}{0.1183\,V} = \frac{(0.322\,V) - (0.22\,V)}{0.1183\,V} = \boxed{0.86}$$

Comment. This value of the pH corresponds roughly to a concentration of $H^+(aq)$ of about $0.1\,mol\,kg^{-1}$. At this rather high concentration the assumption that the activities of $H^+(aq)$ and $Cl^-(aq)$ are equal may not be justified.

E10.28 The left electrode contains no AgBr(s); hence the electrode reactions are

R: $AgBr(s) + e^- \rightarrow Ag(s) + Br^-(aq)$
L: $Ag^+(aq) + e^- \rightarrow Ag(s)$
Overall: $AgBr(s) \rightarrow Ag^+(aq) + Br^-(aq)$

Therefore, since the cell reaction is the solubility equilibrium, for a saturated solution there is no further tendency to dissolve and so $\boxed{E = 0}$

E10.29
R: $Ag^+(aq) + e^- \rightarrow Ag(s) \quad +0.80\,V$
L: $AgI(s) + e^- \rightarrow Ag(s) + I^-(aq) \quad -0.15\,V$
$E^\ominus = E_R^\ominus - E_L^\ominus = 0.95\,V$

Overall(R − L): $Ag^+(aq) + I^-(aq) \rightarrow AgI(s) \quad \nu = 1$

$$\ln K = \frac{\nu F E^\ominus}{RT}[47] = \frac{0.95\,V}{25.693\times10^{-3}\,V} = 36.\overline{975}$$

$$K = \boxed{\overline{1}\times10^{16}}$$

However, $K_s = K^{-1}$ since the solubility equilibrium is written as the reverse of the cell reaction. Therefore, (b) $K_s = \boxed{\bar{1} \times 10^{-16}}$. The solubility is obtained from $b(Ag^+) \approx b(I^-)$ and $S = b(Ag^+)$, so $K_s \approx b^2(Ag^+)$ implying that (a) $S = (K_s)^{1/2} = \boxed{\bar{1} \times 10^{-8}\,\text{mol kg}^{-1}}$

Solutions to problems

Solutions to numerical problems

P10.2 (a) $$I = \frac{1}{2}\left\{\left(\frac{b}{b^{\ominus}}\right)_+ z_+^2 + \left(\frac{b}{b^{\ominus}}\right)_- z_-^2\right\} [21] = 4\left(\frac{b}{b^{\ominus}}\right)$$

For $CuSO_4$, $I = (4) \times (1.0 \times 10^{-3}) = \boxed{4.0 \times 10^{-3}}$

For $ZnSO_4$, $I = (4) \times (3.0 \times 10^{-3}) = \boxed{1.2 \times 10^{-2}}$

(b) $\log \gamma_\pm = -|z_+z_-|AI^{1/2}$

$$\log \gamma_\pm(CuSO_4) = -(4) \times (0.509) \times (4.0 \times 10^{-3})^{1/2} = -0.12\overline{88}$$
$$\gamma_\pm(CuSO_4) = \boxed{0.74}$$
$$\log \gamma_\pm(ZnSO_4) = -(4) \times (0.509) \times (1.2 \times 10^{-2})^{1/2} = -0.22\overline{30}$$
$$\gamma_\pm(ZnSO_4) = \boxed{0.60}$$

(c) The reaction in the Daniell cell is

$$Cu^{2+}(aq) + SO_4^{2-}(aq) + Zn(s) \rightarrow Cu(s) + Zn^{2+}(aq) + SO_4^{2-}(aq)$$

Hence, $$Q = \frac{a(Zn^{2+})a(SO_4^{2-}, R)}{a(Cu^{2+})a(SO_4^{2-}, L)}$$
$$= \frac{\gamma_+ b_+(Zn^{2+})\gamma_- b_-(SO_4^{2-}, R)}{\gamma_+ b_+(Cu^{2+})\gamma_- b_-(SO_4^{2-}, L)} \quad \left[b \equiv \frac{b}{b^{\ominus}} \text{ here and below}\right]$$

where the designations R and L refer to the right and left sides of the equation for the cell reaction and all b are assumed to be unitless, that is, $\frac{b}{b^{\ominus}}$.

$$b_+(Zn^{2+}) = b_-(SO_4^{2-}, R) = b(ZnSO_4)$$
$$b_+(Cu^{2+}) = b_-(SO_4^{2-}, L) = b(CuSO_4)$$

Therefore,
$$Q = \frac{\gamma_\pm^2(ZnSO_4)b^2(ZnSO_4)}{\gamma_\pm^2(CuSO_4)b^2(CuSO_4)} = \frac{(0.60)^2 \times (3.0 \times 10^{-3})^2}{(0.74)^2 \times (1.0 \times 10^{-3})^2} = 5.9\bar{2} = \boxed{5.9}$$

(d) $$E^{\ominus} = -\frac{\Delta_r G^{\ominus}}{\nu F}[44] = \frac{-(-212.7 \times 10^3\,\text{J mol}^{-1})}{(2) \times (9.6485 \times 10^4\,\text{C mol}^{-1})} = \boxed{+1.102\,\text{V}}$$

(e) $$E = E^{\ominus} = -\frac{25.693 \times 10^{-3}\,\text{V}}{\nu}\ln Q = (1.102\,\text{V}) - \left(\frac{25.693 \times 10^{-3}\,\text{V}}{2}\right)\ln(5.9\bar{2})$$
$$= (1.102\,\text{V}) - (0.023\,\text{V}) = \boxed{+1.079\,\text{V}}$$

P10.3 The electrode half-reactions and their potentials are

	$E^{\ominus}$
R: $Q(aq) + 2H^+(aq) + 2e^- \rightarrow QH_2(aq)$	0.6994 V
L: $Hg_2Cl_2(s) + 2e^- \rightarrow 2Hg(l) + 2Cl^-(aq)$	0.2676 V
Overall, (R − L): $Q(aq) + 2H^+(aq) \rightarrow QH_2(aq) + Hg_2Cl_2(s)$	0.4318 V

$$Q(\text{reaction quotient}) = \frac{a(QH_2)}{a(Q)a^2(H^+)a^2(Cl^-)}$$

Since quinhydrone is an equimolecular complex of Q and QH_2, $m(Q) = m(QH_2)$, and since their activity coefficients are assumed to be 1 or to be equal, we have $a(QH_2) \approx a(Q)$. Thus

$$Q = \frac{1}{a^2(H^+)a^2(Cl^-)} \qquad E = E^{\ominus} - \frac{25.7\,\text{mV}}{\nu}\ln Q \quad [\textit{Illustration}, \text{p. } 260]$$

$$\ln Q = \frac{\nu(E^{\ominus} - E)}{25.7\,\text{mV}} = \frac{(2) \times (0.4318 - 0.190)\,\text{V}}{25.7 \times 10^{-3}\,\text{V}} = 18.8\overline{2} \qquad Q = 1.\overline{49} \times 10^8$$

$$a^2(H^+) = (\gamma_+ b_+)^2; \qquad a^2(Cl^-) = (\gamma_- b_-)^2 \quad \left[b \equiv \frac{b}{b^{\ominus}}\right]$$

For HCl(aq), $b_+ = b_- = b$, and if the activity coefficients are assumed equal, $a^2(H^+) = a^2(Cl^-)$; hence

$$Q = \frac{1}{a^2(H^+)a^2(Cl^-)} = \frac{1}{a^4(H^+)}$$

Thus, $a(H^+) = \left(\frac{1}{Q}\right)^{1/4} = \left(\frac{1}{1.49 \times 10^8}\right)^{1/4} = 9 \times 10^{-3}$

$$\text{pH} = -\log a(H^+) = \boxed{2.0}$$

P10.5 **(a)** $E = E^{\ominus} - \frac{25.693\,\text{mV}}{\nu}\ln Q \quad [45, 25°\text{C}]$

$$Q = a(Zn^{2+})a^2(Cl^-)$$

$$= \gamma_+\left(\frac{b}{b^{\ominus}}\right)(Zn^{2+})\gamma_-^2\left(\frac{b}{b^{\ominus}}\right)^2(Cl^-) \qquad b(Zn^{2+}) = b,\ b(Cl^-) = 2b,\ \gamma_+\gamma_-^2 = \gamma_\pm^3$$

Therefore, $Q = \gamma_\pm^3 \times 4b^3 \quad \left[b \equiv \frac{b}{b^{\ominus}} \text{ here and below}\right]$

and $E = E^{\ominus} - \frac{25.693\,\text{mV}}{2}\ln(4b^3\gamma_\pm^3) = E^{\ominus} - \left(\frac{3}{2}\right) \times (25.693\,\text{mV}) \times \ln(4^{1/3}b\gamma_\pm)$

$$= \boxed{E^{\ominus} - (38.54\,\text{mV}) \times \ln(4^{1/3}b) - (38.54\,\text{mV})\ln(\gamma_\pm)}$$

(b) $E^{\ominus}(\text{Cell}) = E_R^{\ominus} - E_L^{\ominus} = E^{\ominus}(Hg_2Cl_2, Hg) - E^{\ominus}(Zn^{2+}, Zn)$

$$= (0.2676\,\text{V}) - (-0.7628\,\text{V}) = \boxed{+1.0304\,\text{V}}$$

(c) $\Delta_r G = -\nu FE = -(2) \times (9.6485 \times 10^4\,\text{C mol}^{-1}) \times (1.2272\,\text{V}) = \boxed{-236.81\,\text{kJ mol}^{-1}}$

$$\Delta_r G^{\ominus} = -\nu FE^{\ominus} = -(2) \times (9.6485 \times 10^4\,\text{C mol}^{-1}) \times (1.0304\,\text{V}) = \boxed{-198.84\,\text{kJ mol}^{-1}}$$

$$\ln K = -\frac{\Delta_r G^{\ominus}}{RT} = \frac{1.9884 \times 10^5\,\text{J mol}^{-1}}{(8.3145\,\text{J K}^{-1}\,\text{mol}^{-1}) \times (298.15\,\text{K})} = 80.211 \quad K = \boxed{6.84 \times 10^{34}}$$

(d) From part **(a)**

$$1.2272\,\text{V} = 1.0304\,\text{V} - (38.54\,\text{mV}) \times \ln(4^{1/3} \times 0.0050) - (38.54\,\text{mV}) \times \ln \gamma_{\pm}$$

$$\ln \gamma_{\pm} = -\frac{(1.2272\,\text{V}) - (1.0304\,\text{V}) - (0.186\bar{4}\,\text{V})}{0.03854\,\text{V}} = -0.269\bar{8}; \qquad \gamma_{\pm} = \boxed{0.763}$$

(e) $\log \gamma_{\pm} = -|z_{-}z_{+}|AI^{1/2}$ [19]

$$I = \frac{1}{2}\sum_{i} z_i^2 \left(\frac{b_i}{b^{\ominus}}\right) \text{ [20]}$$

$b(\text{Zn}^{2+}) = b = 0.0050\,\text{mol kg}^{-1} \qquad b(\text{Cl}^{-}) = 2b = 0.010\,\text{mol kg}^{-1}$

$I = \frac{1}{2}[(4) \times (0.0050) + (0.010)] = 0.015$

$\log \gamma_{\pm} = -(2) \times (0.509) \times (0.015)^{1/2} = -0.12\bar{5}; \qquad \gamma_{\pm} = \boxed{0.75}$

This compares remarkably well to the value obtained from experimetal data in part **(d)**.

(f)

$$\Delta_r S = -\left(\frac{\partial \Delta_r G}{\partial T}\right)_p$$

$$= \nu F \left(\frac{\partial E}{\partial T}\right)_p \text{ [60]} = (2) \times (9.6485 \times 10^4\,\text{C mol}^{-1}) \times (-4.52 \times 10^{-4}\,\text{V K}^{-1})$$

$$= \boxed{-87.2\,\text{J K}^{-1}\,\text{mol}^{-1}}$$

$$\Delta_r H = \Delta_r G + T\Delta_r S = (-236.81\,\text{kJ mol}^{-1}) + (298.15\,\text{K}) \times (-87.2\,\text{J K}^{-1}\,\text{mol}^{-1})$$

$$= \boxed{-262.4\,\text{kJ mol}^{-1}}$$

P10.6 $H_2(g)|HCl(aq)|Hg_2Cl_2(s)|Hg(l)$

$$E = E^{\ominus} - \frac{RT}{F}\ln a(\text{H}^{+})a(\text{Cl}^{-}) \text{ [45]}$$

$$a(\text{H}^{+}) = \gamma_{+}b_{+} = \gamma_{+}b; \qquad a(\text{Cl}^{-}) = \gamma_{-}b_{-} = \gamma_{-}b \qquad \left[b = \frac{b}{b^{\ominus}}\text{ here and below}\right]$$

$$a(\text{H}^{+})a(\text{Cl}^{-}) = \gamma_{+}\gamma_{-}b^2 = \gamma_{\pm}^2 b^2$$

$$E = E^{\ominus} - \frac{2RT}{F}\ln b - \frac{2RT}{F}\ln \gamma_{\pm} \qquad \text{(a)}$$

Converting from natural logarithms to common logarithms (base 10) in order to introduce the Debye–Hückel expression, we obtain

$$E = E^{\ominus} - \frac{(2.303) \times 2RT}{F}\log b - \frac{(2.303) \times 2RT}{F}\log \gamma_{\pm}$$

$$= E^{\ominus} - (0.1183\,\text{V})\log b - (0.1183\,\text{V})\log \gamma_{\pm}$$

$$= E^{\ominus} - (0.1183\,\text{V})\log b - (0.1183\,\text{V})\left[-|z_{+}z_{-}|AI^{1/2}\right]$$

$$= E^{\ominus} - (0.1183\,\text{V})\log b + (0.1183\,\text{V}) \times A \times b^{1/2} \quad [I = b]$$

Rearranging,

$$E + (0.1183\,\text{V})\log b = E^{\ominus} + \text{constant} \times b^{1/2}$$

Therefore, plot $E + (0.1183\,\text{V})\log b$ against $b^{1/2}$, and the intercept at $b = 0$ is $E^{\ominus}/\text{V}$. Draw up the following table

$b/(\text{mmol kg}^{-1})$	1.6077	3.0769	5.0403	7.6938	10.9474
$\left(\frac{b}{b^{\ominus}}\right)^{1/2}$	0.04010	0.05547	0.07100	0.08771	0.1046
$E/\text{V} + (0.1183)\log b$	0.27029	0.27109	0.27186	0.27260	0.27337

The points are plotted in Fig. 10.1. The intercept is at 0.26840, so $E^{\ominus} = +0.26840$ V. A least–squares best fit gives $E^{\ominus} = \boxed{+0.26843 \text{ V}}$ and a coefficient of determination equal to 0.99895.

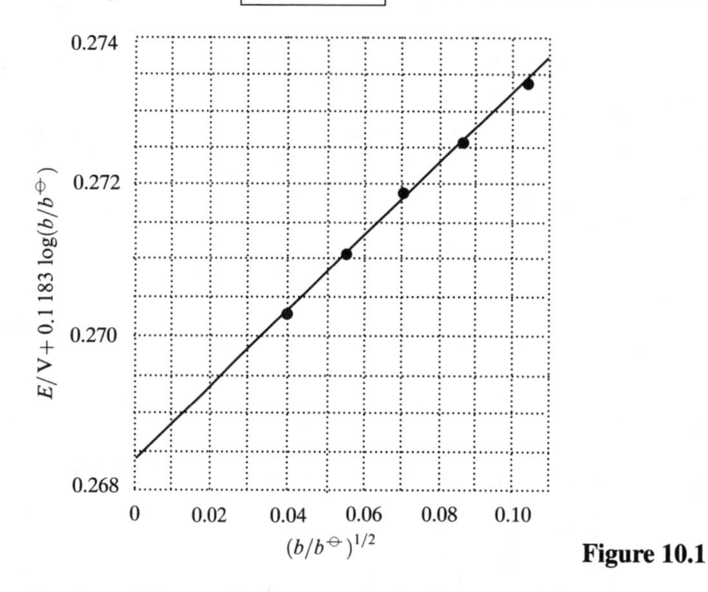

Figure 10.1

For the activity coefficients we obtain from equation (a)

$$\ln \gamma_{\pm} = \frac{E^{\ominus} - E}{2RT/F} - \ln \frac{b}{b^{\ominus}} = \frac{0.26843 - E/\text{V}}{0.05139} - \ln \frac{b}{b^{\ominus}}$$

and we draw up the following table

$b/(\text{mmol kg}^{-1})$	1.6077	3.0769	5.0403	7.6938	10.9474
$\ln \gamma_{\pm}$	−0.03465	−0.05038	−0.06542	−0.07993	−0.09500
$\gamma_{\pm}$	0.9659	0.9509	0.9367	0.9232	0.9094

P10.9 $H_2(g)|HCl(b)|AgCl(s)|Ag(s)$

$\frac{1}{2}H_2(g) + AgCl(s) \rightarrow HCl(aq) + Ag(s)$

$$E = E^{\ominus} - \frac{RT}{F}\ln a(\text{H}^+)a(\text{Cl}^-) = E^{\ominus} - \frac{2RT}{F}\ln b - \frac{2RT}{F}\ln \gamma_{\pm}$$

$\left[b \equiv \frac{b}{b^{\ominus}}, \text{ here and below}\right]$

$$= E^{\ominus} - \frac{2RT}{F}\ln b - (2) \times (2.303)\frac{RT}{F}\log \gamma_{\pm}$$

$$= E^{\ominus} - \frac{2RT}{F}\ln b - (2) \times (2.303)\frac{RT}{F}\left[-0.509b^{1/2} + kb\right] \quad [I = b]$$

Therefore, with $\frac{2RT}{F} \times 2.303 = 0.1183$ V,

$$E/\text{V} + 0.1183\log b - 0.0602b^{1/2} = E^{\ominus}/\text{V} - 0.1183kb$$

hence, with $y = E/\mathrm{V} + 0.1183 \log b - 0.0602b^{1/2}$,

$$\boxed{y = E^{\ominus}/\mathrm{V} - 0.1183kb}$$

We now draw up the following table

$b/(\mathrm{mmol\,kg^{-1}})$	123.8	25.63	9.138	5.619	3.215
y	0.2135	0.2204	0.2216	0.2218	0.2221

(a) The data plotted in Fig. 10.2, and extrapolate to 0.2223 V; hence $E^{\ominus} = \boxed{+0.2223\ \mathrm{V}}$

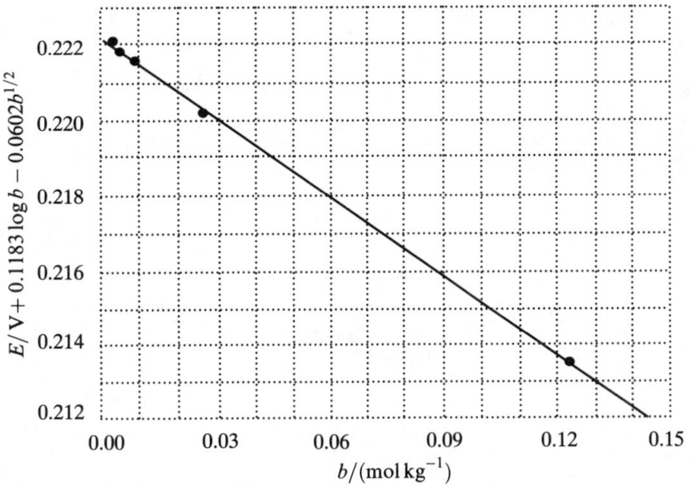

Figure 10.2

(b) $$E = E^{\ominus} - \frac{2RT}{F} \ln b - \frac{2RT}{F} \ln \gamma_{\pm}$$

and so $$\ln \gamma_{\pm} = \frac{E^{\ominus} - E - 0.0514\ \mathrm{V} \ln b}{0.0514\ \mathrm{V}} = \frac{(0.2223) - (0.3524) - (0.0514) \ln(0.100)}{0.0514}$$

$$= -0.228\overline{5}, \qquad \text{implying that } \gamma_{\pm} = \boxed{0.796}$$

Since $a(\mathrm{H^+}) = \frac{\gamma_{\pm} b}{b^{\ominus}}$, $a(\mathrm{H^+}) = (0.796) \times (0.100) = 0.0796$, and hence

$$\mathrm{pH} = -\log a(\mathrm{H^+}) = -\log(0.0796) = \boxed{1.10}$$

P10.10 According to the Debye–Hückel limiting law

$$\log \gamma_{\pm} = -0.509|z_+ z_-| I^{1/2} = -0.509 \left(\frac{b}{b^{\ominus}}\right)^{1/2} \quad [19]$$

We draw up the following table

$b/(\mathrm{mmol\,kg^{-1}})$	1.0	2.0	5.0	10.0	20.0
$I^{1/2}$	0.032	0.045	0.071	0.100	0.141
$\gamma_{\pm}$(calc)	0.964	0.949	0.920	0.889	0.847
$\gamma_{\pm}$(exp)	0.9649	0.9519	0.9275	0.9024	0.8712
$\log \gamma_{\pm}$(calc)	−0.0161	−0.0228	−0.0360	−0.0509	−0.0720
$\log \gamma_{\pm}$(exp)	−0.0155	−0.0214	−0.0327	−0.0446	−0.0599

The points are plotted against $I^{1/2}$ in Fig. 10.3. Note that the limiting slopes of the calculated and experimental curves coincide. A sufficiently good value of B in the extended Debye–Hückel law may be obtained by assuming that the constant A in the extended law is the same as A in the limiting law. Using the data at $20.0\,\text{mmol kg}^{-1}$ we may solve for B.

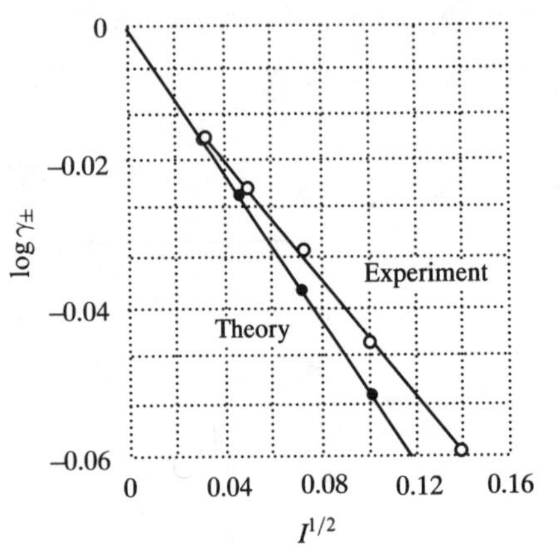

Figure 10.3

$$B = -\frac{A}{\log\gamma_\pm} - \frac{1}{I^{1/2}} = -\frac{0.509}{(-0.0599)} - \frac{1}{0.141} = 1.40\overline{5}$$

Thus,

$$\log\gamma_\pm = -\frac{0.509 I^{1/2}}{1 + 1.40\overline{5} I^{1/2}}$$

In order to determine whether or not the fit is improved, we use the data at $10.0\,\text{mmol kg}^{-1}$

$$\log\gamma_\pm = \frac{-(0.509)\times(0.100)}{(1)+(1.405)\times(0.100)} = -0.0446$$

which fits the data almost exactly. The fits to the other data points will also be almost exact.

P10.12

$$\text{HA(aq)} \rightarrow \text{H}^+\text{(aq)} + \text{A}^-\text{(aq)}$$

Molalities $\quad (1-\alpha)b \qquad \alpha b \qquad \alpha b$

$$K_a = \frac{a(\text{H}^+)a(\text{A}^-)}{a(\text{HA})} = \frac{\gamma_\pm^2 b(\text{H}^+)b(\text{A}^-)}{b(\text{HA})} = \gamma_\pm^2 K_a' \quad \left[b \equiv \frac{b}{b^\ominus}\right]$$

$$K_a' = \frac{b(\text{H}^+)b(\text{A}^-)}{b(\text{HA})} = \frac{\alpha^2 b}{1-\alpha}$$

Hence,

$$\log K_a' = \log K_a - 2\log\gamma_\pm = \log K_a + 2AI^{1/2} \text{ [Debye–Hückel limiting law]}$$

$$= \log K_a + 2A(\alpha b)^{1/2} \quad [I = \alpha b]$$

We therefore construct the following table

$\frac{1000b}{b^{\ominus}}$	0.0280	0.1114	0.2184	1.0283	2.414	5.9115
$1000\left(\frac{\alpha b}{b^{\ominus}}\right)^{1/2}$	3.89	6.04	7.36	11.3	14.1	17.9
$10^5 \times K'_a$	1.768	1.779	1.781	1.799	1.809	1.822
$\log K'_a$	−4.753	−4.750	−4.749	−4.745	−4.743	−4.739

$\log K'_a$ is plotted against $\left(\frac{\alpha b}{b^{\ominus}}\right)^{1/2}$ in Fig. 10.4, and we see that a good straight line is obtained.

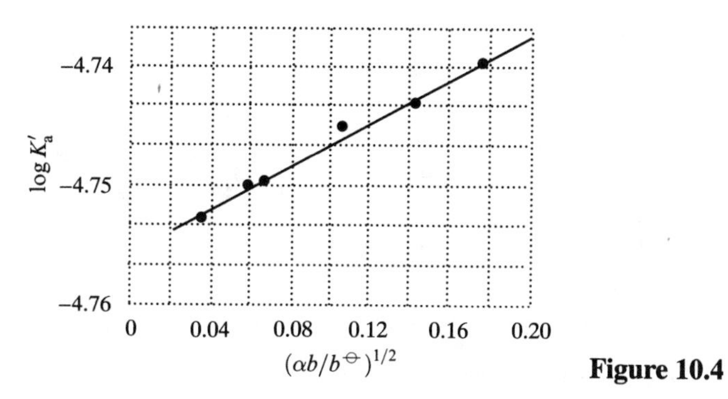

Figure 10.4

Solutions to theoretical problems

P10.15 $MX(s) \rightleftharpoons M^+(aq) + X^-(aq)$

$$K_s = a(M^+)a(X^-) = b(M^+)b(X^-)\gamma_\pm^2 \quad \left[b \equiv \frac{b}{b^{\ominus}}\right]$$

$$b(M^+) = b(X^-) = S = I$$

$$\ln \gamma_\pm = 2.303 \log \gamma_\pm = (-2.303) \times (0.509) \times S^{1/2} = -1.172S^{1/2}$$

$$\gamma_\pm = e^{-(1.172S^{1/2})},$$

Hence, $\frac{K_s}{\gamma_\pm^2} = S^2$ implying that

$$S = \frac{K_s^{1/2}}{\gamma_\pm} = \boxed{K_s e^{(1.172S^{1/2})}}$$

P10.18 $A(s) \rightleftharpoons A(l)$

$$\mu_A^*(s) = \mu_A^*(l) + RT \ln a_A$$

and $\Delta_{fus}G = \mu_A^*(l) - \mu_A^*(s) = -RT \ln a_A$

Hence, $\ln a_A = \frac{-\Delta_{fus}G}{RT}$

$$\frac{d \ln a_A}{dT} = -\frac{1}{R}\frac{d}{dT}\left(\frac{\Delta_{fus}G}{T}\right) = \frac{\Delta_{fus}H}{RT^2} \text{ [Gibbs–Helmholtz]}$$

For $\Delta T = T_f^* - T$, $\mathrm{d}\Delta T = -\mathrm{d}T$ and

$$\frac{\mathrm{d}\ln a_A}{\mathrm{d}\Delta T} = \frac{-\Delta_{\text{fus}}H}{RT^2} \approx \frac{-\Delta_{\text{fus}}H}{RT_f^2}$$

But $K_f = \dfrac{RT_f^2 M_A}{\Delta_{\text{fus}}H}$ [Chapter 7]

Therefore,

$$\frac{\mathrm{d}\ln a_A}{\mathrm{d}\Delta T} = \frac{-M_A}{K_f} \quad \text{and} \quad \mathrm{d}\ln a_A = \frac{-M_A\,\mathrm{d}\Delta T}{K_f}$$

According to the Gibbs–Duhem equation (Chapter 7)

$$n_A\,\mathrm{d}\mu_A + n_B\,\mathrm{d}\mu_B = 0$$

which implies that

$$n_A\,\mathrm{d}\ln a_A + n_B\,\mathrm{d}\ln a_B = 0 \quad [\mu = \mu^{\ominus} + RT\ln a]$$

and hence that $\mathrm{d}\ln a_A = -\dfrac{n_B}{n_A}\,\mathrm{d}\ln a_B$

Hence, $\dfrac{\mathrm{d}\ln a_B}{\mathrm{d}\Delta T} = \dfrac{n_A M_A}{n_B K_f} = \dfrac{1}{b_B K_f}$ [for $n_A M_A = 1\,\mathrm{kg}$]

We know from the Gibbs–Duhem equation that

$$x_A\,\mathrm{d}\ln a_A + x_B\,\mathrm{d}\ln a_B = 0$$

and hence that $\displaystyle\int \mathrm{d}\ln a_A = -\int \frac{x_B}{x_A}\,\mathrm{d}\ln a_B$

Therefore $\ln a_A = -\displaystyle\int \frac{x_B}{x_A}\,\mathrm{d}\ln a_B$

The osmotic coefficient was defined in Problem 7.12 as

$$\phi = -\frac{1}{r}\ln a_A = -\frac{x_A}{x_B}\ln a_A$$

Therefore,

$$\phi = \frac{x_A}{x_B}\int \frac{x_B}{x_A}\,\mathrm{d}\ln a_B = \frac{1}{b}\int_0^b b\,\mathrm{d}\ln a_B = \frac{1}{b}\int_0^b b\,\mathrm{d}\ln \gamma b = \frac{1}{b}\int_0^b b\,\mathrm{d}\ln b + \frac{1}{b}\int_0^b b\,\mathrm{d}\ln\gamma$$

$$= 1 + \frac{1}{b}\int_0^b b\,\mathrm{d}\ln\gamma$$

From the Debye–Hückel limiting law,

$$\ln\gamma = -A'b^{1/2} \quad [A' = 2.303A]$$

Hence, $\mathrm{d}\ln\gamma = -\frac{1}{2}A'b^{-1/2}\,\mathrm{d}b$ and so

$$\phi = 1 + \frac{1}{b}\left(-\frac{1}{2}A'\right)\int_0^b b^{1/2}\,\mathrm{d}b = 1 - \frac{1}{2}\left(\frac{A'}{b}\right) \times \frac{2}{3}b^{3/2} = \boxed{1 - \frac{1}{3}A'b^{1/2}}$$

Comment. For the depression of the freezing point in a 1,1–electrolyte

$$\ln a_A = \frac{-\Delta_{fus}G}{RT} + \frac{\Delta_{fus}G}{RT^*}$$

and hence $-r\phi = \dfrac{-\Delta_{fus}H}{R}\left(\dfrac{1}{T} - \dfrac{1}{T^*}\right)$

Therefore, $\phi = \dfrac{\Delta_{fus}Hx_A}{Rx_B}\left(\dfrac{1}{T} - \dfrac{1}{T^*}\right) = \dfrac{\Delta_{fus}Hx_A}{Rx_B}\left(\dfrac{T^* - T}{TT^*}\right) \approx \dfrac{\Delta_{fus}Hx_A\Delta T}{Rx_BT^{*2}}$

$$\approx \frac{\Delta_{fus}H\Delta T}{\nu Rb_BT^{*2}M_A}$$

where $\nu = 2$. Therefore, since $K_f = \dfrac{MRT^{*2}}{\Delta_{fus}H}$

$$\boxed{\phi = \frac{\Delta T}{2b_BK_f}}$$

Solutions to additional problems

P10.19 **(a)** We seek a redox couple with a reduction potential that is more negative than that of Eu^{3+}/Eu but not as negative as that of Yb^{3+}/Yb, so that the standard potential of the reduction of Eu^{3+} is positive while that of Yb^{3+} (and all the other lanthanides) is negative. The couple, then, must have a reduction potential $\boxed{\text{between } -1.991 \text{ and } -2.19\text{ V}}$. In addition, there must not be any more favourable reaction for the reducing agent (e.g. further oxidation with a potential of greater than 2.19 V). $\boxed{\text{Scandium}}$ fits these criteria; the Sc^{3+}/Sc couple has $E^{\ominus} = -2.09$ V.

(b) For Eu to deposit spontaneously, we must have $E > 0$ for the reaction

$$Eu^{3+} + Sc \rightarrow Eu + Sc^{3+}$$

The cell potential is given by the Nernst equation

$$E_{Eu} = E^{\ominus}_{Eu} - \frac{RT}{3F}\ln Q = E^{\ominus}_{Eu} - \frac{RT}{3F}\ln\frac{a(Sc^{3+})}{a(Eu^{3+})} > 0$$

$$\frac{a(Sc^{3+})}{a(Eu^{3+})} < \exp\left(\frac{3FE^{\ominus}_{Eu}}{RT}\right)$$

where $E^{\ominus}_{Eu} = -1.991\text{ V} - (-2.09\text{ V}) = 0.10\text{ V}$

We must have $E < 0$ for the reaction

$$Yb^{3+} + Sc \rightarrow Yb + Sc^{3+},\ \text{so}$$

$$E_{Yb} = E^{\ominus}_{Yb} - \frac{RT}{3F}\ln Q = E^{\ominus}_{Yb} - \frac{RT}{3F}\ln\frac{a(Sc^{3+})}{a(Yb^{3+})} < 0$$

$$\frac{a(Sc^{3+})}{a(Yb^{3+})} > \exp\left(\frac{3FE^{\ominus}_{Yb}}{RT}\right) \quad \text{or} \quad \frac{a(Yb^{3+})}{a(Sc^{3+})} < \exp\left(\frac{-3FE^{\ominus}_{Yb}}{RT}\right)$$

where $E^{\ominus}_{Yb} = -2.19\text{ V} - (-2.09\text{ V}) = -0.10\text{ V}$

Clearly both criteria require the proper amount of scandium to be used, so there are two criteria which must be satisfied separately. The question, however, asked about the implications for the ratio of lanthanides. So, bearing in mind the requirement that the proper amount of scandium be present, the constraint on the lanthanides is

$$\frac{a(\mathrm{Yb^{3+}})}{a(\mathrm{Eu^{3+}})} < \exp\left(\frac{3(96485\,\mathrm{C\,mol^{-1}}) \times (0.20\,\mathrm{V})}{(8.3145\,\mathrm{J\,mol^{-1}\,K^{-1}}) \times (298\,\mathrm{K})}\right) = \boxed{1.4 \times 10^{10}}$$

Thus, the separation could work over a large range of concentrations.

P10.22

$$\mathrm{Hg_2Cl_2(s) + 2e^- \rightleftharpoons 2Hg(s) + 2Cl^-(aq)} \qquad E^{\ominus}_{\text{cathode}} = 0.27\,\mathrm{V}$$

$$\mathrm{H_2(g) \rightleftharpoons H^+(aq) + 2e^-} \qquad E^{\ominus}_{\text{anode}} = 0.00\,\mathrm{V}$$

$$\mathrm{Hg_2Cl_2(s) + H_2(g) \rightleftharpoons 2H^+(aq) + 2Hg(s) + 2Cl^-(aq)}$$

$$E^{\ominus} = E^{\ominus}_{\text{cathode}} - E^{\ominus}_{\text{anode}} = 0.27\,\mathrm{V}$$

$\nu = 2$

(a) $$E = E^{\ominus} - \frac{RT}{\nu F}\ln Q = E^{\ominus} - \frac{RT}{\nu F}\ln\left[\frac{a^2_{\mathrm{Cl^-}}a^2_{\mathrm{H^+}}}{f_{\mathrm{H_2}}/p^{\ominus}}\right]$$

assuming that the ion activity coefficients are not affected by pressure and that H_2 behaves as a perfect gas

$$\Delta E = E(p_2) - E(p_1) = \frac{RT}{\nu F}\ln\frac{f_2}{f_1} = \frac{RT}{\nu F}\ln\frac{p_2}{p_1}$$

Alternatively, with $p_{\text{ref}} = 1.00\,\mathrm{atm}$ this can be written as

$$\Delta E(p) = \frac{RT}{\nu F}\ln\frac{p}{p_{\text{ref}}} \qquad (1)$$

$$\left(\frac{\partial \Delta E}{\partial p}\right)_T = \frac{RT}{\nu F p} \qquad (2)$$

Equation 1 indicates that a plot of ΔE against $\ln p$ should be linear if the perfect gas assumption is valid. The accompanying plot is indeed linear below about 100 atm but is considerably nonlinear at higher pressures.

(b) An empirical equation for $\Delta E(p)$ is suggested by the substitution of $f = \phi p$ [5.22] into the equation for ΔE,

$$\Delta E = \frac{RT}{\nu F}\ln\frac{f_2}{f_1} = \frac{RT}{\nu F}\left\{\ln\frac{p_2}{p_1} + \ln\frac{\phi_2}{\phi_1}\right\}$$

Equation 5.25 relates ϕ to p and the virial coefficients

$\ln\phi = B'p + \frac{1}{2}C'p^2 + \cdots$

which suggests the empirical form

$$\Delta E = \frac{RT}{\nu F}\{\ln p + c_1 + c_2 p + c_3 p^2 + \cdots\}$$

where the values of c_1, c_2 and c_3 are regression parameters. Regression fit with $\frac{RT}{\nu F} = \frac{25.693\,\mathrm{mV}}{2} = 12.847\,\mathrm{mV}$ yields

$c_1 = -0.01685$, standard deviation $= 0.02471$

$$c_2 = 6.288 \times 10^{-4}\ \text{atm}^{-1}, \quad \text{standard deviation} = 1.362 \times 10^{-4}\text{atm}^{-1}$$
$$c_3 = 7.663 \times 10^{-8}\text{atm}^{-1}, \quad \text{standard deviation} = 1.355 \times 10^{-7}\text{atm}^{-1}$$
$$R = 0.999\,760$$

The standard deviation of c_1 allows for the conclusion that $c_1 = 0$. The large standard deviation of c_3 also indicates that its inclusion is superfluous. Redoing the regression analysis with the simpler form

$$\boxed{\Delta E = \frac{RT}{\nu F}\{\ln p + Cp\}} \qquad \boxed{\left(\frac{\partial E}{\partial p}\right) = \frac{RT}{\nu F}\left\{\frac{1}{p} + C\right\}}$$

We find that

$$\boxed{C = 6.665 \times 10^{-4}\text{atm}^{-1}}, \quad \text{standard deviation} = 2.4 \times 10^{-5}\text{atm}^{-1}$$
$$R = 0.999\,400$$

The correlation coefficient is modestly smaller for the simpler regression fit, the standard deviation of the regression coefficient is much smaller. The simpler equation is the form of choice to describe this data (Fig. 10.5).

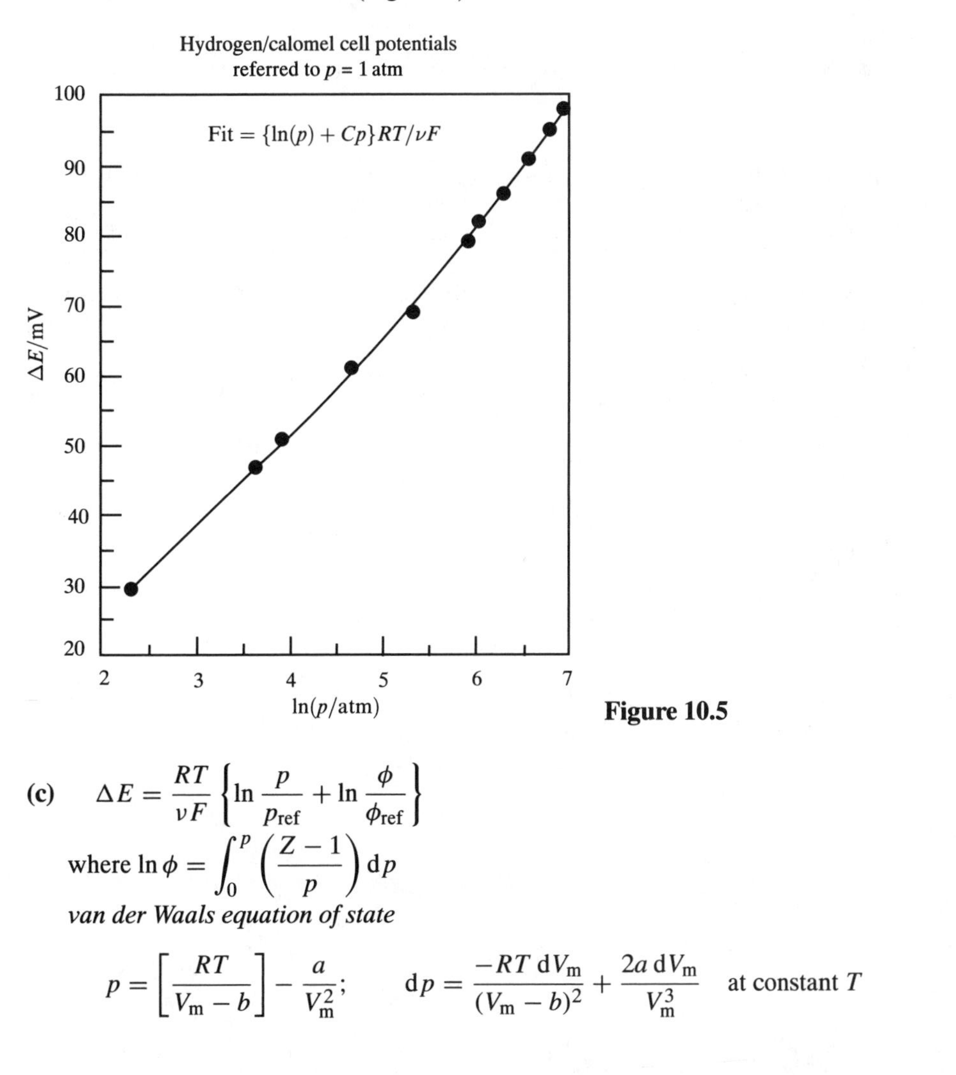

Figure 10.5

(c) $$\Delta E = \frac{RT}{\nu F}\left\{\ln\frac{p}{p_{\text{ref}}} + \ln\frac{\phi}{\phi_{\text{ref}}}\right\} \qquad (3)$$

where $\ln\phi = \displaystyle\int_0^p \left(\frac{Z-1}{p}\right) \mathrm{d}p$

van der Waals equation of state

$$p = \left[\frac{RT}{V_{\text{m}} - b}\right] - \frac{a}{V_{\text{m}}^2}; \qquad \mathrm{d}p = \frac{-RT\,\mathrm{d}V_{\text{m}}}{(V_{\text{m}} - b)^2} + \frac{2a\,\mathrm{d}V_{\text{m}}}{V_{\text{m}}^3} \quad \text{at constant } T$$

$$\ln\phi = \int_0^p \left(\frac{\frac{pV_m}{RT}-1}{p}\right)dp = \int_\infty^{V_m}\left(\frac{V_m}{RT}-\frac{1}{p}\right)\times\left\{\frac{-RT}{(V_m-b)^2}+\frac{2a}{V_m^3}\right\}dV_m$$

The integral may be numerically performed and substituted into eqn (1) for the evaluation of ΔE.

Empirical virial equation

$$Z = 1 + Ap + Bp^2 \text{ where } A = 5.37\times10^{-4}\,\text{atm}^{-1} \quad \text{and} \quad B = 3.5\times10^{-8}\,\text{atm}^{-2}$$

$$\ln\phi = \int_0^p \left[\frac{(1+Ap+Bp^2)-1}{p}\right]dp = \int_0^p [A+Bp]\,dp$$

$$\ln\phi = Ap + \frac{Bp^2}{2}; \qquad \ln\phi_{\text{ref}} = A\,\text{atm} + B\,\text{atm}^2/2$$

$$\boxed{\Delta E = \frac{RT}{\nu F}\left\{\ln\left(\frac{p}{\text{atm}}\right) + A(p-1\,\text{atm}) + \frac{B}{2}(p-1\,\text{atm})^2\right\}}$$

(d) From eqn 1

$$\phi/\phi_{\text{ref}} = \left(\frac{p_{\text{ref}}}{p}\right)e^{\nu F\Delta E/RT} = \left(\frac{1\,\text{atm}}{p}\right)e^{\nu F\Delta E/RT}$$

p/atm	10	38	51	108	210	380	430	560	720	900	1020
ΔE/mV	29.5	47	51	61	69	79	82	86	91	95	98
ϕ/ϕ_{ref}	0.9937	1.0211	1.0387	1.0683	1.0241	1.2326	1.3758	1.4423	1.6555	1.8082	2.0151

As pressure increases, the values of ϕ are greater than 1 and increase. This means the repulsive forces dominate.

Part 2: Structure

11 Quantum theory: introduction and principles

Solutions to exercises

E11.1 The power radiated divided by the area is the excitance, M

$$M = \sigma T^4 \ [2b]$$

Hence, the power, P is

$$P = \sigma T^4 \times A = (5.67 \times 10^{-8}\,\mathrm{W\,m^{-2}\,K^{-4}}) \times (1500\,\mathrm{K})^4 \times (6.0\,\mathrm{m^2}) = \boxed{1.7\,\mathrm{MW}}$$

E11.2 The energy of each photon is hc/λ; hence the total energy is

$$E = Nhc/\lambda = P \times t$$

Solving for the power, P

$$\begin{aligned} P = Nhc/\lambda t &= (8.0 \times 10^7) \times (6.62 \times 10^{-34}\,\mathrm{J\,s}) \times (2.998 \times 10^8\,\mathrm{m\,s^{-1}})/\{(325 \times 10^{-9}\,\mathrm{m}) \\ &\quad \times (3.8 \times 10^{-3}\,\mathrm{s})\} \\ &= \boxed{1.3 \times 10^{-8}\,\mathrm{W}} \end{aligned}$$

E11.3 The Wien displacement law [1] is used to obtain the wavelength corresponding to the maximum (greatest intensity) in Fig. 11.1 of the text.

$$\begin{aligned} \lambda_{\max} &= \frac{c_2}{5T}[c_2 = 1.44\,\mathrm{cm\,K}] = \frac{1.44\,\mathrm{cm\,K}}{(5) \times (11000\,\mathrm{K})} = 2.62 \times 10^{-5}\,\mathrm{cm} \\ &= 2.62 \times 10^{-7}\,\mathrm{m} = \boxed{262\,\mathrm{nm}} \end{aligned}$$

Comment. This wavelength is in the ultraviolet region of the electromagnetic spectrum. Compare to the Sun which is the subject of Exercise 11.13.

E11.4 $\lambda = \dfrac{h}{p}[13] = \dfrac{h}{mv}$

Hence, $v = \dfrac{h}{m\lambda} = \dfrac{6.63 \times 10^{-34}\,\mathrm{J\,s}}{(9.11 \times 10^{-31}\,\mathrm{kg}) \times (0.030\,\mathrm{m})} = \boxed{0.0242\,\mathrm{m\,s^{-1}}}$ very slow!

E11.5 $\lambda = \dfrac{h}{p}[13] = \dfrac{h}{mv} = \dfrac{6.626 \times 10^{-34}\,\mathrm{J\,s}}{(9.109 \times 10^{-31}\,\mathrm{kg}) \times \left(\frac{1}{137}\right) \times (2.998 \times 10^8\,\mathrm{m\,s^{-1}})}$

$$= 3.32\overline{4} \times 10^{-10}\,\mathrm{m} = \boxed{332\,\mathrm{pm}}$$

Comment. One wavelength of the matter wave of an electron with this velocity just fits in the first Bohr orbit. The velocity of the electron in the first Bohr classical orbit is thus $\dfrac{1}{137}c$.

Question. What is the wavelength of an electron with velocity approaching the speed of light? Such velocities can be achieved with particle accelerators.

E11.6 If we assume that photons obey a relation analogous to the deBroglie relation we may write

$$p = \frac{h}{\lambda} = \frac{6.626 \times 10^{-34}\,\mathrm{J\,s}}{750 \times 10^{-9}\,\mathrm{m}} = \boxed{8.83 \times 10^{-28}\,\mathrm{kg\,m\,s^{-1}}}$$

For an electron with the same momentum

$$v = \frac{p}{m} = \frac{8.83 \times 10^{-28}\,\mathrm{kg\,m\,s^{-1}}}{9.11 \times 10^{-31}\,\mathrm{kg}} = \boxed{9.69 \times 10^{2}\,\mathrm{m\,s^{-1}}}$$

E11.7 This is essentially the photoelectric effect with the work function Φ being the ionization energy I. Hence,

$$\frac{1}{2}m_{\mathrm{e}}v^2 = h\nu - I = \frac{hc}{\lambda} - I$$

Solving for λ

$$\lambda = \frac{hc}{I + \frac{1}{2}mv^2} = \frac{(6.626 \times 10^{-34}\,\mathrm{J\,s}) \times (2.998 \times 10^{8}\,\mathrm{m\,s^{-1}})}{(3.44 \times 10^{-18}\,\mathrm{J}) + \left(\frac{1}{2}\right) \times (9.109 \times 10^{-31}\,\mathrm{kg}) \times (1.03 \times 10^{6}\,\mathrm{m\,s^{-1}})^2}$$

$$= 5.06 \times 10^{-8}\,\mathrm{m} = \boxed{50.6\,\mathrm{nm}}$$

Question. What is the energy of the photon?

E11.8 $\Delta p \approx 0.0100$ per cent of $p_0 = p_0 \times (1.00 \times 10^{-4}) = m_{\mathrm{p}}v \times (1.00 \times 10^{-4})\quad (p_0 = m_{\mathrm{p}}v)$

$$\Delta q \approx \frac{\hbar}{2\Delta p}[46] \approx \frac{1.055 \times 10^{-34}\,\mathrm{J\,s}}{(2) \times (1.673 \times 10^{-27}\,\mathrm{kg}) \times (4.5 \times 10^{5}\,\mathrm{m\,s^{-1}}) \times (1.00 \times 10^{-4})}$$

$$\approx 7.0\bar{1} \times 10^{-10}\,\mathrm{m}, \quad \text{or} \quad \boxed{0.70\,\mathrm{nm}}$$

E11.9 $E = h\nu = \dfrac{hc}{\lambda}, \qquad E(\text{per mole}) = N_{\mathrm{A}}E = \dfrac{N_{\mathrm{A}}hc}{\lambda}$

$hc = (6.62608 \times 10^{-34}\,\mathrm{J\,s}) \times (2.99792 \times 10^{8}\,\mathrm{m\,s^{-1}}) = 1.986 \times 10^{-25}\,\mathrm{J\,m}$

$N_{\mathrm{A}}hc = (6.02214 \times 10^{23}\,\mathrm{mol^{-1}}) \times (1.986 \times 10^{-25}\,\mathrm{J\,m}) = 0.1196\,\mathrm{J\,m\,mol^{-1}}$

Thus, $E = \dfrac{1.986 \times 10^{-25}\,\mathrm{J\,m}}{\lambda}; \qquad E(\text{per mole}) = \dfrac{0.1196\,\mathrm{J\,m\,mol^{-1}}}{\lambda}$

We can therefore draw up the following table

λ/nm	E/J	E/(kJ mol^{-1})
(a) 600	3.31×10^{-19}	199
(b) 550	3.61×10^{-19}	218
(c) 400	4.97×10^{-19}	299

E11.10 Assuming that the H atom is free and stationary, if a photon is absorbed, the atom acquires its momentum p. It therefore reaches a speed v such that $p = mv$. Thus,

$$v = \frac{p}{m_{\mathrm{H}}} = \frac{p}{1.674 \times 10^{-27}\,\mathrm{kg}}$$

$$[m_{\mathrm{H}} = 1.008\,\mathrm{u} = (1.008) \times (1.6605 \times 10^{-27}\,\mathrm{kg}) = 1.674 \times 10^{-27}\,\mathrm{kg}]$$

We draw up the following table using the information in the table above and $p = \dfrac{h}{\lambda}$.

λ/nm	$p/(\mathrm{kg\,m\,s^{-1}})$	$v/(\mathrm{m\,s^{-1}})$
600	1.10×10^{-27}	0.66
550	1.20×10^{-27}	0.72
400	1.66×10^{-27}	0.99

E11.11 The total energy emitted in a period τ is $P\tau$. The energy of a photon of 650 nm light is $E = \dfrac{hc}{\lambda}$ with $\lambda = 650\,\mathrm{nm}$. The total number of photons emitted in an interval τ is then the total energy divided by the energy per photon.

$$N = \frac{P\tau}{E} = \frac{P\tau\lambda}{hc}$$

DeBroglie's relation applies to each photon and thus the total momentum imparted to the glow-worm is

$$p = \frac{Nh}{\lambda} = \frac{P\tau\lambda}{hc} \times \frac{h}{\lambda} = \frac{P\tau}{c}$$

$$P = 0.10\,\mathrm{W} = 0.10\,\mathrm{J\,s^{-1}}, \qquad \tau = 10\,\mathrm{y}, \qquad p = mv$$

Hence the final speed is

$$v = \frac{P\tau}{cm} = \frac{(0.10\,\mathrm{J\,s^{-1}}) \times (3.16 \times 10^{8}\,\mathrm{s})}{(2.998 \times 10^{8}\,\mathrm{m\,s^{-1}}) \times (5.0 \times 10^{-3}\,\mathrm{kg})} = \boxed{21\,\mathrm{m\,s^{-1}}}$$

Comment. Note that the answer is independent of the wavelength of the radiation emitted: the greater the wavelength the smaller the photon momentum, but the greater the number of photons emitted.

Question. If this glow-worm eventually turns into a firefly which glows for 1 s intervals while flying with a speed of $0.1\,\mathrm{m\,s^{-1}}$, what additional speed does the 1 s glowing impart to the firefly? Ignore any frictional effects of air.

E11.12 Power is energy per unit time; hence

$$N = \frac{P}{h\nu}[P = \text{power in J s}^{-1}] = \frac{P\lambda}{hc}$$

$$= \frac{P\lambda}{(6.626 \times 10^{-34}\,\mathrm{J\,s}) \times (2.998 \times 10^{8}\,\mathrm{m\,s^{-1}})} = \frac{(P/\mathrm{W}) \times (\lambda/\mathrm{nm})\,\mathrm{s^{-1}}}{1.99 \times 10^{-16}}$$

$$= 5.03 \times 10^{15}(P/\mathrm{W}) \times (\lambda/\mathrm{nm})\,\mathrm{s^{-1}}$$

(a) $N = (5.03 \times 10^{15}) \times (1.0) \times (550\,\mathrm{s^{-1}}) = \boxed{2.8 \times 10^{18}\,\mathrm{s^{-1}}}$

(b) $N = (5.03 \times 10^{15}) \times (100) \times (550\,\mathrm{s^{-1}}) = \boxed{2.8 \times 10^{20}\,\mathrm{s^{-1}}}$

E11.13 From Wien's law,

$$T\lambda_{\max} = \frac{1}{5}c_2, \quad c_2 = 1.44\,\mathrm{cm\,K}\,[1]$$

Therefore, $T = \dfrac{1.44\,\mathrm{cm\,K}}{(5) \times (480 \times 10^{-7}\,\mathrm{cm})} = \boxed{6000\,\mathrm{K}}$

E11.14 $E_K = \frac{1}{2}mv^2 = h\nu - \Phi = \frac{hc}{\lambda} - \Phi$ [12]

$\Phi = 2.14\,\text{eV} = (2.14) \times (1.602 \times 10^{-19}\,\text{J}) = 3.43 \times 10^{-19}\,\text{J}$

(a) $\frac{hc}{\lambda} = \frac{(6.626 \times 10^{-34}\,\text{J s}) \times (2.998 \times 10^{8}\,\text{m s}^{-1})}{700 \times 10^{-9}\,\text{m}} = 2.84 \times 10^{-19}\,\text{J} < \Phi$, so $\boxed{\text{no ejection}}$ occurs

(b) $\frac{hc}{\lambda} = 6.62 \times 10^{-19}\,\text{J}$

$$E_K = \frac{1}{2}mv^2 = (6.62 - 3.43) \times 10^{-19}\,\text{J} = \boxed{3.19 \times 10^{-19}\,\text{J}}$$

$$v = \left(\frac{2E_K}{m}\right)^{1/2} = \left(\frac{(2) \times (3.19 \times 10^{-19}\,\text{J})}{9.109 \times 10^{-31}\,\text{kg}}\right)^{1/2} = \boxed{837\,\text{km s}^{-1}}$$

E11.15 $\Delta E = \hbar\omega = h\nu = \frac{h}{T}$ $\left[T = \text{period} = \frac{1}{\nu} = \frac{2\pi}{\omega}\right]$

(a) $\Delta E = \frac{6.626 \times 10^{-34}\,\text{J s}}{10^{-15}\,\text{s}} = \boxed{7 \times 10^{-19}\,\text{J}}$,

corresponding to $N_A \times (7 \times 10^{-19}\,\text{J}) = \boxed{400\,\text{kJ mol}^{-1}}$

(b) $\Delta E = \frac{6.626 \times 10^{-34}\,\text{J s}}{10^{-14}\,\text{s}} = \boxed{7 \times 10^{-20}\,\text{J}}$, $\boxed{40\,\text{kJ mol}^{-1}}$

(c) $\Delta E = \frac{6.626 \times 10^{-34}\,\text{J s}}{1\,\text{s}} = \boxed{7 \times 10^{-34}\,\text{J}}$, $\boxed{4 \times 10^{-13}\,\text{kJ mol}^{-1}}$

E11.16 $\lambda = \frac{h}{p} = \frac{h}{mv}$ [13]

(a) $\lambda = \frac{6.626 \times 10^{-34}\,\text{J s}}{(1.0 \times 10^{-3}\,\text{kg}) \times (1.0 \times 10^{-2}\,\text{m s}^{-1})} = \boxed{6.6 \times 10^{-29}\,\text{m}}$

(b) $\lambda = \frac{6.626 \times 10^{-34}\,\text{J s}}{(1.0 \times 10^{-3}\,\text{kg}) \times (1.00 \times 10^{5}\,\text{m s}^{-1})} = \boxed{6.6 \times 10^{-36}\,\text{m}}$

(c) $\lambda = \frac{6.626 \times 10^{-34}\,\text{J s}}{(4.003) \times (1.6605 \times 10^{-27}\,\text{kg}) \times (1000\,\text{m s}^{-1})} = \boxed{99.7\,\text{pm}}$

Comment. The wavelengths in **(a)** and **(b)** are smaller than the dimensions of any known particle, whereas that in **(c)** is comparable to atomic dimensions.

Question. For stationary particles, $v = 0$, corresponding to an infinite wavelength. What meaning can be ascribed to this result?

E11.17 The minimum uncertainty in position and momentum is given by the uncertainly principle in the form

$\Delta p \Delta q \geq \frac{1}{2}\hbar$ [46], with the choice of the equality

$\Delta p = m\Delta v$

$$\Delta v_{\text{min}} = \frac{\hbar}{2m\Delta q} = \frac{1.055 \times 10^{-34}\,\text{J s}}{(2) \times (0.500\,\text{kg}) \times (1.0 \times 10^{-6}\,\text{m})} = \boxed{1.1 \times 10^{-28}\,\text{m s}^{-1}}$$

$$\Delta q_{\text{min}} = \frac{\hbar}{2m\Delta v} = \frac{1.055 \times 10^{-34}\,\text{J s}}{(2) \times (5.0 \times 10^{-3}\,\text{kg}) \times (1 \times 10^{-5}\,\text{m s}^{-1})} = \boxed{1 \times 10^{-27}\,\text{m}}$$

Comment. These uncertainties are extremely small; thus, the ball and bullet are effectively classical particles.

Question. If the ball were stationary (no uncertainty in position) the uncertainty in speed would be infinite. Thus, the ball could have a very high speed, contradicting the fact that it is stationary. What is the resolution of this apparent paradox?

E11.18 In this case the work function is the ionization energy of the electron

$$\frac{1}{2}mv^2 = h\nu - I\ [12], \quad \nu = \frac{c}{\lambda}$$

$$I = \frac{hc}{\lambda} - \frac{1}{2}mv^2$$

$$= \frac{(6.626\times10^{-34}\,\mathrm{J\,s})\times(2.998\times10^{8}\,\mathrm{m\,s^{-1}})}{150\times10^{-12}\,\mathrm{m}} - \left(\frac{1}{2}\right)\times(9.109\times10^{-31}\,\mathrm{kg})$$

$$\times(2.14\times10^{7}\,\mathrm{m\,s^{-1}})^2$$

$$= \boxed{1.12\times10^{-15}\,\mathrm{J}}$$

Solutions to problems

Solutions to numerical problems

P11.1 A cavity approximates an ideal black body; hence the Planck distribution applies

$$\rho = \frac{8\pi hc}{\lambda^5}\left(\frac{1}{e^{hc/\lambda kT}-1}\right)\ [5]$$

Since the wavelength range is small (5 nm) we may write as a good approximation

$$\Delta E = \rho\Delta\lambda, \quad \lambda \approx 652.5\,\mathrm{nm}$$

$$\frac{hc}{\lambda k} = \frac{(6.626\times10^{-34}\,\mathrm{J\,s})\times(2.998\times10^{8}\,\mathrm{m\,s^{-1}})}{(6.525\times10^{-7}\,\mathrm{m})\times(1.381\times10^{-23}\,\mathrm{J\,K^{-1}})} = 2.205\times10^{4}\,\mathrm{K}$$

$$\frac{8\pi hc}{\lambda^5} = \frac{(8\pi)\times(6.626\times10^{-34}\,\mathrm{J\,s})\times(2.998\times10^{8}\,\mathrm{m\,s^{-1}})}{(652.5\times10^{-9}\,\mathrm{m})^5} = 4.221\times10^{7}\,\mathrm{J\,m^{-4}}$$

$$\Delta E = (4.221\times10^{7}\,\mathrm{J\,m^{-4}})\times\left(\frac{1}{e^{(2.205\times10^4\,\mathrm{K})/T}-1}\right)\times(5\times10^{-9}\,\mathrm{m})$$

(a) $T = 298\,\mathrm{K}, \quad \Delta E = \frac{0.211\,\mathrm{J\,m^{-3}}}{e^{(2.205\times10^4)/298}-1} = \boxed{1.6\times10^{-33}\,\mathrm{J\,m^{-3}}}$

(b) $T = 3273\,\mathrm{K}, \quad \Delta E = \frac{0.211\,\mathrm{J\,m^{-3}}}{e^{(2.205\times10^4)/3273}-1} = \boxed{2.5\times10^{-4}\,\mathrm{J\,m^{-3}}}$

Comment. The energy density in the cavity does not depend on the volume of the cavity, but the total energy in any given wavelength range does, as well as the total energy over all wavelength ranges.

Question. What is the total energy in this cavity within the range 650–655 nm at the stated temperatures?

P11.2 $\lambda_{\max}T = \dfrac{hc}{5k}, \quad \left[1, \text{ and } c_2 = \dfrac{hc}{k}\right]$

Therefore, $\lambda_{\max} = \dfrac{hc}{5k} \times \dfrac{1}{T}$ and if we plot $\lambda_{\max}$ against $\dfrac{1}{T}$ we can obtain h from the slope. We draw up the following table

$\theta/°\mathrm{C}$	1000	1500	2000	2500	3000	3500
T/K	1273	1773	2273	2773	3273	3773
$10^4/(T/\mathrm{K})$	7.86	5.64	4.40	3.61	3.06	2.65
$\lambda_{\max}/\mathrm{nm}$	2181	1600	1240	1035	878	763

The points are plotted in Fig. 11.1. From the graph, the slope is $2.73 \times 10^6\,\mathrm{nm}/(1/\mathrm{K})$, that is,

$$\frac{hc}{5k} = 2.73 \times 10^6 \frac{\mathrm{nm}}{1/\mathrm{K}} = 2.73 \times 10^{-3}\,\mathrm{m\,K}$$

$$\text{and } h = \frac{(5) \times (1.38066 \times 10^{-23}\,\mathrm{J\,K^{-1}}) \times (2.73 \times 10^{-3}\,\mathrm{m\,K})}{2.99792 \times 10^8\,\mathrm{m\,s^{-1}}} = \boxed{6.29 \times 10^{-34}\,\mathrm{J\,s}}$$

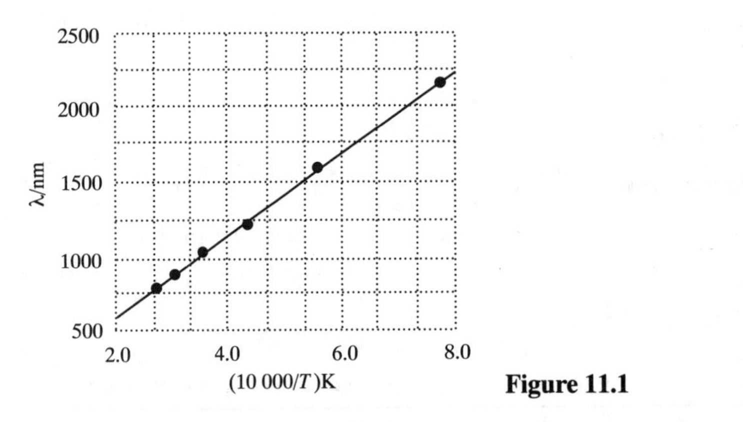

Figure 11.1

Comment. Planck's estimate of the constant h in his first paper of 1900 on black body radiation was 6.55×10^{-27} erg sec(1 erg $= 10^{-7}$ J) which is remarkably close to the current value of 6.626×10^{-34} J s. Also from his analysis of the experimental data he obtained values of k (the Boltzmann constant), N_A (the Avogadro constant), and e (the fundamental charge). His values of these constants remained the most accurate for almost 20 years.

P11.5 The full solution of the Schrödinger equation for the problem of a particle in a one-dimensional box is given in Chapter 12. Here we need only the wavefunction which is provided. It is the square of the wavefunction that is related to the probability. Here $\psi^2 = \dfrac{2}{L}\sin^2\dfrac{\pi x}{L}$ and the probability that the particle will be found between a and b is

$$P(a,b) = \int_a^b \psi^2\,\mathrm{d}x \text{ [Section 11.4]}$$

$$= \frac{2}{L}\int_a^b \sin^2\frac{\pi x}{L}\,\mathrm{d}x = \left(\frac{x}{L} - \frac{1}{2\pi}\sin\frac{2\pi x}{L}\right)\Bigg|_a^b$$

$$= \frac{b-a}{L} - \frac{1}{2\pi}\left(\sin\frac{2\pi b}{L} - \sin\frac{2\pi a}{L}\right)$$

$L = 10.0\,\mathrm{nm}$

(a) $$P(4.95, 5.05) = \frac{0.10}{10.0} - \frac{1}{2\pi}\left(\sin\frac{(2\pi)\times(5.05)}{10.0} - \sin\frac{(2\pi)\times(4.95)}{10.0}\right)$$
$$= 0.010 + 0.010 = \boxed{0.020}$$

(b) $$P(1.95, 2.05) = \frac{0.10}{10.0} - \frac{1}{2\pi}\left(\sin\frac{(2\pi)\times(2.05)}{10.0} - \sin\frac{(2\pi)\times(1.95)}{10.0}\right)$$
$$= 0.010 - 0.0031 = \boxed{0.007}$$

(c) $$P(9.90, 10.0) = \frac{0.10}{10.0} - \frac{1}{2\pi}\left(\sin\frac{(2\pi)\times(10.0)}{10.0} - \sin\frac{(2\pi)\times(9.90)}{10.0}\right)$$
$$= 0.010 - 0.009993 = \boxed{7\times10^{-6}}$$

(d) $P(5.0, 10.0) = \boxed{0.5}$ [by symmetry]

(e) $$P\left(\frac{1}{3}L, \frac{2}{3}L\right) = \frac{1}{3} - \frac{1}{2\pi}\left(\sin\frac{4\pi}{3} - \sin\frac{2\pi}{3}\right) = \boxed{0.61}$$

Solutions to theoretical problems

P11.8 We require $\int \psi^*\psi\,\mathrm{d}\tau = 1$, and so write $\psi = Nf$ and find N for the given f.

(a) $$N^2\int_0^L \sin^2\frac{n\pi x}{L}\,\mathrm{d}x = \frac{1}{2}N^2\int_0^L\left(1-\cos\frac{2n\pi x}{L}\right)\mathrm{d}x \text{ [trigonometric identity]}$$
$$= \frac{1}{2}N^2\left(x - \frac{L}{2n\pi}\sin\frac{2n\pi x}{L}\right)\Bigg|_0^L$$
$$= \frac{L}{2}N^2 = 1 \quad \text{if } \boxed{N = \left(\frac{2}{L}\right)^{1/2}}$$

(b) $$N^2\int_{-L}^{L} c^2\,\mathrm{d}x = 2N^2c^2L = 1 \quad \text{if } \boxed{N = \frac{1}{c(2L)^{1/2}}}$$

(c) $$N^2\int_0^\infty \mathrm{e}^{-2r/a}r^2\,\mathrm{d}r\int_0^\pi \sin\theta\,\mathrm{d}\theta\int_0^{2\pi}\mathrm{d}\phi \qquad [\mathrm{d}\tau = r^2\sin\theta\,\mathrm{d}r\,\mathrm{d}\theta\,\mathrm{d}\phi]$$
$$= N^2\left(\frac{a^3}{4}\right)\times(2)\times(2\pi) = 1 \quad \text{if } \boxed{N = \frac{1}{(\pi a^3)^{1/2}}}$$

(d) $$N^2\int_0^\infty r^2\times r^2\mathrm{e}^{-r/a}\,\mathrm{d}r\int_0^\pi \sin^3\theta\,\mathrm{d}\theta\int_0^{2\pi}\cos^2\phi\,\mathrm{d}\phi \qquad [x = r\cos\phi\sin\theta]$$
$$= N^2 4!a^5\times\frac{4}{3}\times\pi = 32\pi a^5N^2 = 1 \quad \text{if } \boxed{N = \frac{1}{(32\pi a^5)^{1/2}}}$$

We have used $\int \sin^3\theta\,\mathrm{d}\theta = -\frac{1}{3}(\cos\theta)(\sin^2\theta + 2)$, as found in tables of integrals and

$$\int_0^{2\pi}\cos^2\phi\,\mathrm{d}\phi = \int_0^{2\pi}\sin^2\phi\,\mathrm{d}\phi$$

by symmetry with $\int_0^{2\pi}(\cos^2\phi + \sin^2\phi)\,\mathrm{d}\phi = \int_0^{2\pi}\mathrm{d}\phi = 2\pi$

P11.10 In each case form $\hat{\Omega} f$. If the result is ωf where ω is a constant, then f is an eigenfunction of the operator $\hat{\Omega}$ and ω is the eigenvalue [31].

(a) $\frac{d}{dx} e^{ikx} = ik\, e^{ikx}$; $\boxed{\text{yes; eigenvalue} = ik}$

(b) $\frac{d}{dx} \cos kx = -k \sin kx$; no.

(c) $\frac{d}{dx} k = 0$; $\boxed{\text{yes; eigenvalue} = 0}$

(d) $\frac{d}{dx} kx = k = \frac{1}{x} kx$; no [$1/x$ is not a constant].

(e) $\frac{d}{dx} e^{-\alpha x^2} = -2\alpha x e^{-\alpha x^2}$; no [$-2\alpha x$ is not a constant].

P11.12 Follow the procedure of Problem 11.10

(a) $\frac{d^2}{dx^2} e^{ikx} = -k^2\, e^{ikx}$; yes; eigenvalue $= \boxed{-k^2}$

(b) $\frac{d^2}{dx^2} \cos kx = -k^2 \cos kx$; yes; eigenvalue $= \boxed{-k^2}$

(c) $\frac{d^2}{dx^2} k = 0$; yes; eigenvalue $= \boxed{0}$

(d) $\frac{d^2}{dx^2} kx = 0$; yes; eigenvalue $= \boxed{0}$

(e) $\frac{d^2}{dx^2} e^{-\alpha x^2} = (-2\alpha + 4\alpha^2 x^2) e^{-\alpha x^2}$; no.

Hence, $\boxed{\text{(a, b, c, d) are eigenfunctions of } \frac{d^2}{dx^2}\text{; (b, d) are eigenfunctions of } \frac{d^2}{dx^2}\text{, but not of } \frac{d}{dx}}$

P11.13 $\psi = (\cos \chi) e^{ikx} + (\sin \chi) e^{-ikx} = c_1 e^{ikx} + c_2 e^{-ikx}$ The linear momentum operator is

$$\hat{p}_x = \frac{\hbar}{i} \frac{d}{dx} \quad [32]$$

As demonstrated in the text (Section 11.5, eqn 35), e^{ikx} is an eigenfunction of $\hat{p}_x$ with eigenvalue $+k\hbar$; likewise e^{-ikx} is an eigenfunction of $\hat{p}_x$ with eigenvalue $-k\hbar$. Therefore, by the principle of linear superposition (Section 11.5(d), *Justification* 11.2),

(a) $P = c_1^2 = \boxed{\cos^2 \chi}$

(b) $P = c_2^2 = \boxed{\sin^2 \chi}$

(c) $c_1^2 = 0.90 = \cos^2 \chi$, so $\cos \chi = 0.95$

$c_2^2 = 0.10 = \sin^2 \chi$, so $\sin \chi = \pm 0.32$; hence

$$\boxed{\psi = 0.95 e^{ikx} \pm 0.32 e^{-ikx}}$$

P11.16 $\langle r\rangle = N^2\int \psi^* r\psi\,\mathrm{d}\tau, \qquad \langle r^2\rangle = N^2\int \psi^* r^2\psi\,\mathrm{d}\tau$

(a) $\psi = \left(2-\dfrac{r}{a_0}\right)\mathrm{e}^{-r/2a_0}, \qquad N = \left(\dfrac{1}{32\pi a_0{}^3}\right)^{1/2}$ [Problem 11.9]

$$\langle r\rangle = \frac{1}{32\pi a_0{}^3}\int_0^\infty r\left(2-\frac{r}{a_0}\right)^2 r^2\mathrm{e}^{-r/a_0}\,\mathrm{d}r\times 4\pi \quad \left[\int_0^\pi \sin\theta\,\mathrm{d}\theta\int_0^{2\pi}\mathrm{d}\phi = 4\pi\right]$$

$$= \frac{1}{8a_0{}^3}\int_0^\infty\left(4r^3-\frac{4r^4}{a_0}+\frac{r^5}{a_0{}^2}\right)\mathrm{e}^{-r/a_0}\,\mathrm{d}r$$

$$= \frac{1}{8a_0{}^3}(4\times 3!a_0{}^4 - 4\times 4!a_0{}^4 + 5!a_0{}^4) = \boxed{6a_0} \quad \left[\int_0^\infty x^n\mathrm{e}^{-ax}\,\mathrm{d}x = \frac{n!}{a^{n+1}}\right]$$

$$\langle r^2\rangle = \frac{1}{8a_0{}^3}\int_0^\infty\left(4r^4-\frac{4r^5}{a_0}+\frac{r^6}{a_0{}^2}\right)\mathrm{e}^{-r/a_0}\,\mathrm{d}r = \frac{1}{8a_0{}^3}(4\times 4! - 4\times 5! + 6!)a_0{}^5$$

$$= \boxed{42a_0{}^2}$$

(b) $\psi = Nr\sin\theta\cos\phi\,\mathrm{e}^{-r/2a_0}, \qquad N = \left(\dfrac{1}{32\pi a_0{}^5}\right)^{1/2}$ [Problem 11.9]

$$\langle r\rangle = \frac{1}{32\pi a_0{}^5}\int_0^\infty r^5\mathrm{e}^{-r/a_0}\,\mathrm{d}r\times\frac{4\pi}{3} = \frac{1}{24a_0{}^5}\times 5!a_0{}^6 = \boxed{5a_0}$$

$$\langle r^2\rangle = \frac{1}{24a_0{}^5}\int_0^\infty r^6\mathrm{e}^{-r/a_0}\,\mathrm{d}r = \frac{1}{24a_0{}^5}\times 6!a_0{}^7 = \boxed{30a_0{}^2}$$

P11.17 $\psi = \left(\dfrac{1}{\pi a_0{}^3}\right)^{1/2}\mathrm{e}^{-r/a_0}$ [Example 11.4]

(a) $\langle V\rangle = \int\psi^*\hat{V}\psi\,\mathrm{d}\tau \quad \left[V = -\dfrac{e^2}{4\pi\varepsilon_0 r}, \text{ Section 13.1}\right]$

$$\langle V\rangle = \int\psi^*\left(\frac{-e^2}{4\pi\varepsilon_0}\cdot\frac{1}{r}\right)\psi\,\mathrm{d}\tau = \frac{1}{\pi a_0{}^3}\left(\frac{-e^2}{4\pi\varepsilon_0}\right)\int_0^\infty r\mathrm{e}^{-2r/a_0}\,\mathrm{d}r\times 4\pi$$

$$= \frac{1}{\pi a_0{}^3}\left(\frac{-e^2}{4\pi\varepsilon_0}\right)\times\left(\frac{a_0}{2}\right)^2\times 4\pi = \boxed{\frac{-e^2}{4\pi\varepsilon_0 a_0}}$$

(b) For three-dimensional systems such as the hydrogen atom the kinetic energy operator is

$$\hat{T} = -\frac{\hbar^2}{2m_\mathrm{e}}\nabla^2 \text{ [Table 11.1, } m_\mathrm{e}\approx\mu \text{ for the hydrogen atom]}$$

$$\nabla^2 = \frac{\partial^2}{\partial r^2}+\frac{2}{r}\frac{\partial}{\partial r}+\frac{1}{r^2}\Lambda^2 = \left(\frac{1}{r}\right)\times\left(\frac{\partial^2}{\partial r^2}\right)r+\frac{1}{r^2}\Lambda^2$$

$\Lambda^2\psi = 0 \quad$ [ψ has no angular coordinates]

$$\nabla^2\psi = \left(\frac{1}{\pi a_0^3}\right)^{1/2} \times \left(\frac{1}{r}\right) \times \left(\frac{d^2}{dr^2}\right) r e^{-r/a_0}$$
$$= \left(\frac{1}{\pi a_0^3}\right)^{1/2} \times \left[-\left(\frac{2}{a_0 r}\right) + \frac{1}{a_0^2}\right] e^{-r/a_0}$$

Then, $\langle T\rangle = -\left(\frac{\hbar^2}{2m_e}\right) \times \left(\frac{1}{\pi a_0^3}\right) \int_0^{2\pi} d\phi \int_0^{\pi} \sin\theta\, d\theta \int_0^{\infty} \left[-\left(\frac{2}{a_0 r}\right) + \left(\frac{1}{a_0^2}\right)\right] e^{-2r/a_0} r^2\, dr$

$$= -\left(\frac{2\hbar^2}{m_e a_0^3}\right) \int_0^{\infty} \left[-\left(\frac{2r}{a_0}\right) + \left(\frac{r^2}{a_0^2}\right)\right] e^{-2r/a_0}\, dr$$
$$= -\left(\frac{2\hbar^2}{m_e a_0^3}\right) \times \left(-\frac{a_0}{4}\right) \left[\int_0^{\infty} x^n e^{-ax}\, dx = \frac{n!}{a^{n+1}}\right] = \boxed{\frac{\hbar^2}{2m_e a_0^2}}$$

Inserting $a_0 = \dfrac{4\pi\varepsilon_0\hbar^2}{m_e e^2}$ [Chapter 13]

$$\langle T\rangle = \frac{e^2}{8\pi\varepsilon_0 a_0} = -\frac{1}{2}\langle V\rangle$$

P11.19 The quantity $\hat{\Omega}_1\hat{\Omega}_2 - \hat{\Omega}_2\hat{\Omega}_1$ [*Illustration*, p. 308] is referred to as the commutator of the operators $\hat{\Omega}_1$ and $\hat{\Omega}_2$. In obtaining the commutator it is necessary to realize that the operators operate on functions; thus, we form

$$\hat{\Omega}_1\hat{\Omega}_2 f(x) - \hat{\Omega}_2\hat{\Omega}_1 f(x)$$

(a) $$\frac{d}{dx}\hat{x} f(x) = \hat{x}\frac{df(x)}{dx} + f(x)$$
$$\hat{x}\frac{d}{dx} f(x) = x\frac{df(x)}{dx}$$
$$\left(\frac{d}{dx}\hat{x} - \hat{x}\frac{d}{dx}\right) f(x) = f(x)$$

Thus, $\left(\frac{d}{dx}\hat{x} - \hat{x}\frac{d}{dx}\right) = \boxed{1}$

(b) $$\frac{d}{dx}\hat{x}^2 f(x) = x^2 f'(x) + 2x f(x)$$
$$\hat{x}^2\frac{d}{dx} f(x) = x^2 f'(x)$$
$$\left(\frac{d}{dx}\hat{x}^2 - \hat{x}^2\frac{d}{dx}\right) f(x) = 2x f(x)$$

Thus, $\left(\frac{d}{dx}\hat{x}^2 - \hat{x}^2\frac{d}{dx}\right) = \boxed{2x}$

(c) $$p_x = \frac{\hbar}{i}\frac{d}{dx}$$

Therefore $a = \left(\hat{x} + \hbar\frac{d}{dx}\right)$ and $a^\dagger = \left(\hat{x} - \hbar\frac{d}{dx}\right)$

Then $aa^{\dagger}f(x) = \frac{1}{2}\left(\hat{x} + \hbar\frac{d}{dx}\right) \times \left(\hat{x} - \hbar\frac{d}{dx}\right) f(x)$

and $a^{\dagger}af(x) = \frac{1}{2}\left(\hat{x} - \hbar\frac{d}{dx}\right) \times \left(\hat{x} + \hbar\frac{d}{dx}\right) f(x)$

The terms in $\hat{x}^2$ and $\left(\frac{d}{dx}\right)^2$ obviously drop out when the difference is taken and are ignored in what follows; thus

$$aa^{\dagger}f(x) = \frac{1}{2}\left(-\hat{x}\hbar\frac{d}{dx} + \hbar\frac{d}{dx}x\right) f(x)$$

$$a^{\dagger}af(x) = \frac{1}{2}\left(x\hbar\frac{d}{dx} - \hbar\frac{d}{dx}x\right) f(x)$$

These expressions are the negative of each other, therefore

$$(aa^{\dagger} - a^{\dagger}a)f(x) = \hbar\frac{d}{dx}\hat{x}f(x) - \hbar\hat{x}\frac{d}{dx}f(x)$$

$$= \hbar\left(\frac{d}{dx}\hat{x} - \hat{x}\frac{d}{dx}\right) f(x) = \hbar f(x) \text{ [from (a)]}$$

Therefore, $(aa^{\dagger} - a^{\dagger}a) = \boxed{\hbar}$

Solutions to additional problems

P11.21 $\lambda_{max} = \frac{1.44\,\text{cm K}}{5T}$ [1]

$$= \frac{1.44\,\text{cm K}}{5(5800\,\text{K})} = 5.0 \times 10^{-5}\,\text{cm}\left(\frac{10^9\,\text{nm}}{10^2\,\text{cm}}\right)$$

$\lambda_{max} = \boxed{500\,\text{nm, blue-green}}$ [see Fig. 16.1 in the text]

P11.22 $I = aI + M = aI + \sigma T^4$ so $T = \left(\frac{I(1-a)}{\sigma}\right)^{1/4} = \left(\frac{(343\,\text{W m}^{-2}) \times (1-0.30)}{5.67 \times 10^{-8}\,\text{W m}^{-2}\,\text{K}^{-4}}\right)^{1/4}$

$$= \boxed{255\,\text{K}}$$

where I is the incoming energy flux, a the albedo (fraction of incoming radiation absorbed), M the excitance and σ the Stefan–Boltzmann constant. Wien's displacement law relates the temperature to the wavelength of the most intense radiation

$$T\lambda_{max} = c_2/5 \quad \text{so} \quad \lambda_{max} = \frac{c_2}{5T} = \frac{1.44\,\text{cm K}}{5(255\,\text{K})}$$

$$= 1.13 \times 10^{-3}\text{cm} = \boxed{11.3\ \ \text{m}} \text{ in the infrared.}$$

P11.23 **(a)** $CH_4(g) \rightarrow C(\text{graphite}) + 2H_2(g)$

$$\Delta_r G^{\ominus} = -\Delta_f G^{\ominus}(CH_4) = -(-50.72\,\text{kJ mol}^{-1}) = 50.72\,\text{kJ mol}^{-1} \quad \text{at } T$$

$$\Delta_r H^{\ominus} = -\Delta_f H^{\ominus}(CH_4) = -(-74.81\,\text{kJ mol}^{-1}) = 74.81\,\text{kJ mol}^{-1} \quad \text{at } T$$

We want to find the temperature at which $\Delta_r G^{\ominus}(T) = 0$. Below this temperature methane is stable with respect to decomposition into the elements. Above this temperature it is unstable. Assuming that the heat capacities are basically independent of temperature

$$\begin{aligned}\Delta_r C_p^{\ominus}(T) &\approx \Delta_r C_p^{\ominus}(T^{\ominus}) = [8.527 + 2(28.824) - 35.31]\,\mathrm{J\,K^{-1}\,mol^{-1}} \\ &\approx 30.865\,\mathrm{J\,K^{-1}\,mol^{-1}}\end{aligned}$$

$$\begin{aligned}\Delta_r H^{\ominus}(T) &= \Delta_r H^{\ominus}(T^{\ominus}) + \int_{T^{\ominus}}^{T} \Delta_r C_p^{\ominus}(T)\,\mathrm{d}T \text{ [2.45]} \\ &= \Delta_r H^{\ominus}(T^{\ominus}) + \Delta_r C_p^{\ominus} \times (T - T^{\ominus})\end{aligned}$$

$$\left(\frac{\partial}{\partial T}\left(\frac{\Delta_r G^{\ominus}}{T}\right)\right)_p = -\frac{\Delta_r H^{\ominus}}{T^2} \text{ [5.13]}$$

At constant pressure (the standard pressure)

$$\int_{T^{\ominus}}^{T} \mathrm{d}(\Delta_r G^{\ominus}/T) = -\int_{T^{\ominus}}^{T} \frac{\Delta_r H^{\ominus}}{T^2}\,\mathrm{d}T$$

$$\begin{aligned}\frac{\Delta_r G^{\ominus}(T)}{T} &= \frac{\Delta_r G^{\ominus}(T^{\ominus})}{T^{\ominus}} - \int_{T^{\ominus}}^{T} \frac{\Delta_r H^{\ominus}(T^{\ominus}) + \Delta_r C_p^{\ominus} \times (T - T^{\ominus})}{T^2}\,\mathrm{d}T \\ &= \frac{\Delta_r G^{\ominus}(T^{\ominus})}{T^{\ominus}} - [\Delta_r H^{\ominus}(T^{\ominus}) - \Delta_r C_p^{\ominus} \times T^{\ominus}] \int_{T^{\ominus}}^{T} \frac{1}{T^2}\,\mathrm{d}T - \Delta_r C_p^{\ominus} \int_{T^{\ominus}}^{T} \frac{1}{T}\,\mathrm{d}T \\ &= \frac{\Delta_r G^{\ominus}(T^{\ominus})}{T^{\ominus}} + [\Delta_r H^{\ominus}(T^{\ominus}) - \Delta_r C_p^{\ominus} \times T^{\ominus}] \times \left[\frac{1}{T} - \frac{1}{T^{\ominus}}\right] - \Delta_r C_p^{\ominus} \ln\left(\frac{T}{T^{\ominus}}\right)\end{aligned}$$

The value of T for which $\Delta_r G^{\ominus}(T) = 0$ can be determined by examination of a plot (Fig. 11.2) of $\dfrac{\Delta_r G^{\ominus}(T)}{T}$ against T.

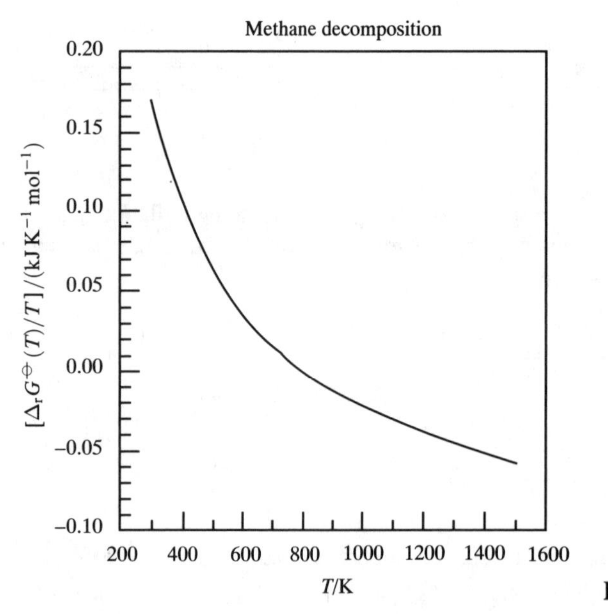

Figure 11.2

$$\frac{\Delta_r G^{\ominus}(\mathfrak{T})}{\mathfrak{T}} = 50.72\,\text{kJ mol}^{-1}/298.15\,\text{K} = 0.1701\,\text{kJ K}^{-1}\,\text{mol}^{-1}$$

$$\Delta_r H^{\ominus}(\mathfrak{T}) - \Delta_r C_p^{\ominus} \times \mathfrak{T} = 74.81\,\text{kJ mol}^{-1} - (30.865\,\text{J K}^{-1}\,\text{mol}^{-1}) \times (298\,\text{K}) \times \left(\frac{10^{-3}\,\text{kJ}}{\text{J}}\right)$$

$$= 65.61\,\text{kJ mol}^{-1}$$

$$\Delta_r C_p^{\ominus} = (30.865\,\text{J K}^{-1}\,\text{mol}^{-1}) \times \left(\frac{10^{-3}\,\text{kJ}}{\text{J}}\right) = 0.030\,865\,\text{kJ K}^{-1}\,\text{mol}^{-1}$$

With the estimate of constant $\Delta_r C_p^{\ominus}$, methane is unstable above 825 K.

(b) $$\lambda_{\text{max}} = \frac{\frac{1}{5}(1.44\,\text{cm K})}{T}\ [11.1]$$

$$\lambda_{\text{max}} = \frac{\frac{1}{5}(1.44\,\text{cm K})}{1000\,\text{K}} = 2.88 \times 10^{-4}\,\text{cm}\left(\frac{10^9\,\text{nm}}{10^2\,\text{cm}}\right)$$

$$\boxed{\lambda_{\text{max}}(1000\,\text{K}) = 2880\,\text{nm}}$$

(c) $$\text{Excitance ratio} = \frac{M(\text{brown dwarf})}{M(\text{Sun})} = \frac{\sigma T^4_{\text{brown dwarf}}}{\sigma T^4_{\text{Sun}}}\ [11.2b]$$

$$= \frac{(1000\,\text{K})^4}{(6000\,\text{K})^4} = \boxed{7.7 \times 10^{-4}}$$

$$\text{Energy density ratio} = \frac{\rho(\text{brown dwarf})}{\rho(\text{Sun})}$$

$$= \frac{\frac{8\pi hc}{\lambda^5}\left(\frac{1}{\mathrm{e}^{(hc/\lambda k T_{\text{brown dwarf}})} - 1}\right)}{\frac{8\pi hc}{\lambda^5}\left(\frac{1}{\mathrm{e}^{(hc/\lambda k T_{\text{Sun}})} - 1}\right)}\ [11.5]$$

$$= \frac{\mathrm{e}^{(hc/\lambda k T_{\text{Sun}})} - 1}{\mathrm{e}^{(hc/\lambda k T_{\text{brown dwarf}})} - 1}$$

The energy density ratio is a function of λ so we will calculate the ratio at λ_{max} of the brown dwarf.

$$\frac{hc}{\lambda_{\text{brown dwarf}} k} = \frac{(6.62 \times 10^{-34}\,\text{J s}) \times (3.00 \times 10^8\,\text{m s}^{-1})}{(2880 \times 10^{-9}\,\text{m}) \times (1.381 \times 10^{-23}\,\text{J K}^{-1})}$$

$$= 4998\,\text{K}$$

$$\text{Energy density ratio} = \frac{\mathrm{e}^{\frac{4998\,\text{K}}{T_{\text{Sun}}}} - 1}{\mathrm{e}^{\frac{4998}{T_{\text{brown dwarf}}}} - 1}$$

$$= \frac{\mathrm{e}^{\frac{4998}{6000}} - 1}{\mathrm{e}^{\frac{4998}{1000}} - 1} = \frac{1.300}{147}$$

$$= \boxed{8.8 \times 10^{-3}}$$

(d) The wavelength of visible radiation is between about 700 nm (red) and 420 nm (violet). (See text Fig. 16.1.)

$$\text{Fraction of visible energy density} = \frac{1}{aT^4}\left|\int_{700\,\text{nm}}^{420\,\text{nm}} \rho(\lambda)\,\mathrm{d}\lambda\right| \quad [11.2a, 11.5]$$

$$= \frac{c}{4\sigma T^4}\left|\int_{700\,\text{nm}}^{420\,\text{nm}} \rho(\lambda)\,\mathrm{d}\lambda\right|$$

As an estimate, let us suppose that $\rho(\lambda)$ doesn't vary too drastically in the visible at 1000 K. Then,

$$\left|\int_{700\,\text{nm}}^{420\,\text{nm}} \rho(\lambda)\,\mathrm{d}\lambda\right| \sim \rho(560\,\text{nm}) \times (700\,\text{nm} - 420\,\text{nm})$$

$$\sim \left(\frac{8\pi hc}{(560\times10^{-9}\,\text{m})^5}\right) \times \left(\frac{1}{\mathrm{e}^{\left(\left(\frac{4998\,\text{K}}{1000\,\text{K}}\right)\times\left(\frac{2880\,\text{nm}}{560\,\text{nm}}\right)\right)} - 1}\right) \times \left(280\times10^{-9}\,\text{m}\right)$$

$$= \frac{8\pi(6.626\times10^{-34}\,\text{J s})\times(3.00\times10^{8}\,\text{m s}^{-1})}{1.97\times10^{-25}\,\text{m}^4}\left(\frac{1}{\mathrm{e}^{25.70}-1}\right)$$

$$= 1.75\times10^{-10}\,\text{J m}^{-3}$$

$$\text{fraction of visible energy density} \sim \frac{(3.00\times10^{8}\,\text{m s}^{-1})\times(1.75\times10^{-10}\,\text{J m}^{-3})}{4(5.67\times10^{-8}\,\text{W m}^{-2}\,\text{K}^{-4})\times(1000\,\text{K})^4}$$

$$\sim \boxed{2.31\times10^{-7}}$$

Very little of the brown dwarf's radiation is in the visible. It doesn't shine brightly.

12 Quantum theory: techniques and applications

Solutions to exercises

E12.1 $E = \frac{n^2h^2}{8m_eL^2}$ [6]

$$\frac{h^2}{8m_eL^2} = \frac{(6.626 \times 10^{-34}\,\text{J s})^2}{(8) \times (9.109 \times 10^{-31}\,\text{kg}) \times (1.0 \times 10^{-9}\,\text{m})^2} = 6.02 \times 10^{-20}\,\text{J}$$

The conversion factors required are

$$E/(\text{kJ mol}^{-1}) = \frac{N_A}{10^3}E/\text{J}$$

$$1\,\text{eV} = 1.602 \times 10^{-19}\,\text{J}; \qquad 1\,\text{cm}^{-1} = 1.986 \times 10^{-23}\,\text{J}$$

(a) $$E_2 - E_1 = (4-1)\frac{h^2}{8m_eL^2} = (3) \times (6.02 \times 10^{-20}\,\text{J})$$
$$= 18.06 \times 10^{-20}\,\text{J}$$
$$= \boxed{1.81 \times 10^{-19}\,\text{J}}, \boxed{110\,\text{kJ mol}^{-1}}, \boxed{1.1\,\text{eV}}, \boxed{9100\,\text{cm}^{-1}}$$

(b) $$E_6 - E_5 = (36-25)\frac{h^2}{8m_eL^2} = \frac{11h^2}{8m_eL^2}$$
$$= (11) \times (6.02 \times 10^{-20}\,\text{J})$$
$$= \boxed{6.6 \times 10^{-19}\,\text{J}}, \boxed{400\,\text{kJ mol}^{-1}}, \boxed{4.1\,\text{eV}}, \boxed{33\,000\,\text{cm}^{-1}}$$

Comment. The energy level separations increase as n increases.

Question. For what value of n is $E_{n+1} - E_n$ for the system of this exercise equal to the ionization energy of the H-atom which is 13.6 eV?

E12.2 The wavefunctions are

$$\psi_n = \left(\frac{2}{L}\right)^{1/2} \sin\left(\frac{n\pi x}{L}\right) \text{ [7]}$$

The required probability is

$$P = \int_{0.49L}^{0.51L} \psi_n^2\,\text{d}x \approx \psi_n^2 \Delta x$$

(a) $$\psi_1^2 = \left(\frac{2}{L}\right)\sin^2\left(\frac{\pi x}{L}\right) = \left(\frac{2}{L}\right)\sin^2\left(\frac{\pi}{2}\right) [x \approx 0.50L] = \left(\frac{2}{L}\right) \quad \left[\sin\frac{\pi}{2} = 1\right]$$

$$P = \left(\frac{2}{L}\right) \times 0.02L = \boxed{0.04}$$

(b) $$\psi_2^2 = \left(\frac{2}{L}\right)\sin^2\left(\frac{2\pi x}{L}\right) = \left(\frac{2}{L}\right)\sin^2\pi = \boxed{0}$$

E12.3 The wavefunction for a particle in the state $n = 1$ in a square-well potential is

$$\psi_1 = \left(\frac{2}{L}\right)^{1/2} \sin\left(\frac{\pi x}{L}\right)$$

and

$$\hat{p} = \frac{\hbar}{\mathrm{i}}\frac{\mathrm{d}}{\mathrm{d}x}$$

$$\langle p\rangle = \int_0^L \psi_1^* \hat{p}\psi_1 \,\mathrm{d}x = \frac{2\hbar}{\mathrm{i}L}\int_0^L \sin\left(\frac{\pi x}{L}\right)\frac{\mathrm{d}}{\mathrm{d}x}\sin\left(\frac{\pi x}{L}\right)\mathrm{d}x$$

$$= \frac{2\pi\hbar}{\mathrm{i}L^2}\int_0^L \sin\left(\frac{\pi x}{L}\right)\cos\left(\frac{\pi x}{L}\right)\mathrm{d}x = \boxed{0}$$

$$\hat{p}^2 = -\hbar^2\frac{\mathrm{d}^2}{\mathrm{d}x^2}$$

$$\langle p^2\rangle = -\frac{2\hbar^2}{L}\int_0^L \sin\left(\frac{\pi x}{L}\right)\frac{\mathrm{d}^2}{\mathrm{d}x^2}\sin\left(\frac{\pi x}{L}\right)\mathrm{d}x = \left(\frac{2\hbar^2}{L}\right)\times\left(\frac{\pi}{L}\right)^2\int_0^L \sin^2 ax\,\mathrm{d}x \quad \left[a = \frac{\pi}{L}\right]$$

$$= \left(\frac{2\hbar^2}{L}\right)\times\left(\frac{\pi}{L}\right)^2\left(\frac{1}{2}x - \frac{1}{4a}\sin 2ax\right)\Bigg|_0^L = \left(\frac{2\hbar^2}{L}\right)\times\left(\frac{\pi}{L}\right)^2\times\left(\frac{L}{2}\right) = \boxed{\frac{h^2}{4L^2}}$$

Comment. The expectation value of $\hat{p}$ is zero because on average the particle moves to the left as often as the right.

E12.4

$$\psi_3 = \left(\frac{2}{L}\right)^{1/2}\sin\left(\frac{3\pi x}{L}\right)$$

$$P(x) \propto \psi_3^2 \propto \sin^2\left(\frac{3\pi x}{L}\right)$$

The maxima and minima in $P(x)$ correspond to $\frac{\mathrm{d}P(x)}{\mathrm{d}x} = 0$

$$\frac{\mathrm{d}P(x)}{\mathrm{d}x} \propto \psi^2 \propto \sin\left(\frac{3\pi x}{L}\right)\cos\left(\frac{3\pi x}{L}\right) \propto \sin\left(\frac{6\pi x}{L}\right) \quad [2\sin\alpha\cos\alpha = \sin 2\alpha]$$

$\sin\theta = 0$ when $\theta = \left(\frac{6\pi x}{L}\right) = n'\pi, n' = 0, 1, 2, \ldots$ which corresponds to $x = \frac{n'L}{6}, n' \le 6$. $n' = 0, 2, 4$, and 6 correspond to minima in ψ_3, leaving $n' = 1, 3$, and 5 for the maxima, that is

$$\boxed{x = \frac{L}{6}, \frac{L}{2} \text{ and } \frac{5L}{6}}$$

Comment. Maxima in ψ^2 correspond to maxima *and* minima in ψ itself, so one can also solve this exercise by finding all points where $\frac{\mathrm{d}\psi}{\mathrm{d}x} = 0$.

E12.5

$$E = (n_1^2 + n_2^2 + n_3^2)\times\left(\frac{h^2}{8mL^2}\right) \quad \text{[3-dimensional analogue of eqn 19]}$$

$$E_{111} = \frac{3h^2}{8mL^2}, \qquad 3E_{111} = \frac{9h^2}{8mL^2}$$

Hence, we require the values of n_1, n_2, and n_3 that make

$$n_1^2 + n_2^2 + n_3^2 = 9$$

Therefore, $(n_1, n_2, n_3) = (1, 2, 2)$, $(2, 1, 2)$, and $(2, 2, 1)$ and the degeneracy is $\boxed{3}$

Question. What is the smallest multiple of the lowest energy, E_{111} for which $E_{n_1n_2n_3}$ does not exist?

E12.6 $$E = (n_1^2 + n_2^2 + n_3^2) \times \left(\frac{h^2}{8mL^2}\right) = \frac{K}{L^2}, \quad K = (n_1^2 + n_2^2 + n_3^2) \times \left(\frac{h^2}{8m}\right)$$

$$\frac{\Delta E}{E} = \frac{\frac{K}{(0.9L)^2} - \frac{K}{L^2}}{\frac{K}{L^2}} = \frac{1}{0.81} - 1 = \boxed{0.23}, \quad \text{or} \quad \boxed{23 \text{ per cent}}$$

E12.7 $$E = \left(v + \frac{1}{2}\right)\hbar\omega, \quad \omega = \left(\frac{k}{m}\right)^{1/2} \quad [31]$$

The zero-point energy corresponds to $v = 0$; hence

$$E_0 = \frac{1}{2}\hbar\omega = \frac{1}{2}\hbar\left(\frac{k}{m}\right)^{1/2} = \left(\frac{1}{2}\right) \times (1.055 \times 10^{-34}\,\text{J s}) \times \left(\frac{155\,\text{N m}^{-1}}{2.33 \times 10^{-26}\,\text{kg}}\right)^{1/2}$$

$$= \boxed{4.30 \times 10^{-21}\,\text{J}}$$

E12.8 $$\Delta E = E_{v+1} - E_v = \left(v + 1 + \frac{1}{2}\right)\hbar\omega - \left(v + \frac{1}{2}\right)\hbar\omega = \hbar\omega = \hbar\left(\frac{k}{m}\right)^{1/2} \quad [31]$$

Hence $$k = m\left(\frac{\Delta E}{\hbar}\right)^2 = (1.33 \times 10^{-25}\,\text{kg}) \times \left(\frac{4.82 \times 10^{-21}\,\text{J}}{1.055 \times 10^{-34}\,\text{J s}}\right)^2 = \boxed{278\,\text{N m}^{-1}}$$

$[1\,\text{J} = 1\,\text{N m}]$

E12.9 The requirement for a transition to occur is that $\Delta E(\text{system}) = E(\text{photon})$.

$\Delta E(\text{system}) = \hbar\omega$ [Exercise 12.8(**a**)]

$$E(\text{photon}) = h\nu = \frac{hc}{\lambda}$$

Therefore, $$\frac{hc}{\lambda} = \frac{h\omega}{2\pi} = \left(\frac{h}{2\pi}\right) \times \left(\frac{k}{m}\right)^{1/2}$$

$$\lambda = 2\pi c\left(\frac{m}{k}\right)^{1/2} = (2\pi) \times (2.998 \times 10^8\,\text{m s}^{-1}) \times \left(\frac{1.673 \times 10^{-27}\,\text{kg}}{855\,\text{N m}^{-1}}\right)^{1/2}$$

$$= 2.63 \times 10^{-6}\,\text{m} = \boxed{2.63\,\mu\text{m}}$$

E12.10 Since $\lambda \propto m^{1/2}$, $\lambda_{\text{new}} = 2^{1/2}$, $\lambda_{\text{old}} = (2^{1/2}) \times (2.63\,\mu\text{m}) = \boxed{3.72\,\mu\text{m}}$

The change in wavelength is $\lambda_{\text{new}} - \lambda_{\text{old}} = 1.09\,\mu\text{m}$

E12.11 **(a)** $$\omega = \left(\frac{g}{l}\right)^{1/2} \quad \text{[elementary physics]}$$

$\Delta E = \hbar\omega$ [harmonic oscillator level separations, Exercise 12.8(**a**)]

$$= (1.055 \times 10^{-34}\,\text{J s}) \times \left(\frac{9.81\,\text{m s}^{-2}}{1\,\text{m}}\right)^{1/2} = \boxed{3.3 \times 10^{-34}\,\text{J}}$$

(b) $$\Delta E = h\nu = (6.626 \times 10^{-34}\,\text{J Hz}^{-1}) \times (5\,\text{Hz}) = \boxed{3.3 \times 10^{-33}\,\text{J}}$$

E12.12 The Schrödinger equation for the linear harmonic oscillator is

$$-\frac{\hbar^2}{2m}\frac{d^2\psi}{dx^2}+\frac{1}{2}kx^2\psi=E\psi \text{ [30]}$$

The ground-state wavefunction is

$$\psi_0=N_0 e^{-y^2/2}, \quad \text{where } y=\left(\frac{mk}{\hbar^2}\right)^{1/4}x[34]=\frac{x}{\alpha}$$

$$\text{with } \alpha=\left(\frac{\hbar^2}{mk}\right)^{1/4}=\left(\frac{\hbar^2}{m^2\omega^2}\right)^{1/4}; \quad k=\frac{\hbar^2}{m\alpha^4} \qquad \text{(a)}$$

Thus, $\psi_0=N_0 e^{-x^2/2\alpha^2}$

Performing the operations

$$\frac{d\psi_0}{dx}=\left(-\frac{1}{\alpha^2}x\right)\psi_0$$

$$\frac{d^2\psi_0}{dx^2}=\left(-\frac{1}{\alpha^2}x\right)\times\left(-\frac{1}{\alpha^2}x\right)\times\psi_0-\frac{1}{\alpha^2}\psi_0=\frac{x^2}{\alpha^4}\psi_0-\frac{1}{\alpha^2}\psi_0=\left(\frac{x^2}{\alpha^4}-\frac{1}{\alpha^2}\right)\psi_0$$

Thus,

$$-\frac{\hbar^2}{2m}\left(\frac{x^2}{\alpha^4}-\frac{1}{\alpha^2}\right)\psi_0+\frac{1}{2}kx^2\psi_0=E_0\psi_0$$

which implies

$$E_0=\frac{-\hbar^2}{2m}\left(\frac{x^2}{\alpha^4}-\frac{1}{\alpha^2}\right)+\frac{1}{2}kx^2 \qquad \text{(b)}$$

But E_0 is a constant, independent of x; therefore the terms which contain x must drop out, which is possible only if

$$-\frac{\hbar^2}{2m\alpha^4}+\frac{1}{2}k=0$$

which is consistent with $k=\dfrac{\hbar^2}{m\alpha^4}$ as in (a). What is left in (b) is

$$E_0=\frac{\hbar^2}{2m\alpha^2}=\hbar\omega \quad \left[\text{using } \omega=\left(\frac{k}{m}\right)^{1/2} \text{ and } k=\frac{\hbar^2}{m\alpha^4}\right]$$

Therefore, ψ_0 is a solution of the Schrödinger equation with energy $\frac{1}{2}\hbar\omega$.

E12.13 As described in Exercise 12.11(**b**), for the vibrations of a diatomic molecule $\mu=\dfrac{m}{2}$ must be substituted for m. Thus

$$E_0=\frac{1}{2}\hbar\omega=\frac{1}{2}\hbar\left(\frac{k}{\mu}\right)^{1/2}=\frac{1}{2}\hbar\left(\frac{2k}{m}\right)^{1/2}$$

$$m(^{35}\text{Cl})=34.96888\,\text{u}=(34.9688\,\text{u})\times(1.66054\times10^{-27}\,\text{kg/u})=5.807\times10^{-26}\,\text{kg}$$

$$E_0=\left(\frac{1.05457\times10^{-34}\,\text{J s}}{2}\right)\times\left(\frac{(2)\times(329\,\text{N m}^{-1})}{5.807\times10^{-26}\,\text{kg}}\right)^{1/2}=\boxed{5.61\times10^{-21}\,\text{J}}$$

E12.14 We require

$$\int \psi^* \psi \, d\tau = 1$$

that is

$$\int_0^{2\pi} N^2 e^{-im_l\phi} e^{im_l\phi} d\phi = \int_0^{2\pi} N^2 \, d\phi = 2\pi N^2 = 1$$

$$N^2 = \frac{1}{2\pi} \qquad N = \boxed{\left(\frac{1}{2\pi}\right)^{1/2}}$$

E12.15 Magnitude of angular momentum $= \{l(l+1)\}^{1/2}\hbar$ [67*a*]

Projection on arbitrary axis $= m_l\hbar$ [67*b*]

Thus,

$$\text{Magnitude} = (2^{1/2}) \times \hbar = \boxed{1.49 \times 10^{-34}\,\text{J s}}$$

$$\text{Possible projections} = \boxed{0}, \quad \pm\hbar = \boxed{0, \pm 1.05 \times 10^{-34}\,\text{J s}}$$

E12.16 The diagrams are drawn by forming a vector of length $\{j(j+1)\}^{1/2}$, with $j = s$ or l as appropriate, and with a projection m_j on the z-axis (see Fig. 12.1). Each vector represents the edge of a cone around the z-axis (that for $m_j = 0$ represents the side view of a disk perpendicular to z).

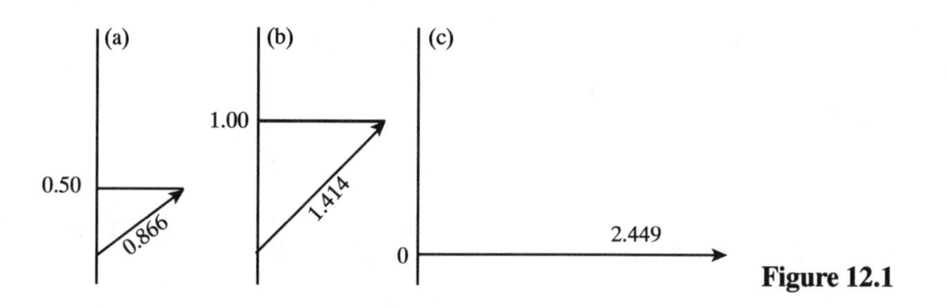

Figure 12.1

Solutions to problems

Solutions to numerical problems

P12.1 $E = \dfrac{n^2h^2}{8mL^2}, \quad E_2 - E_1 = \dfrac{3h^2}{8mL^2}$

We take $m(O_2) = (32.000) \times (1.6605 \times 10^{-27}\,\text{kg})$, and find

$$E_2 - E_1 = \frac{(3) \times (6.626 \times 10^{-34}\,\text{J s})^2}{(8) \times (32.00) \times (1.6605 \times 10^{-27}\,\text{kg}) \times (5.0 \times 10^{-2}\,\text{m})^2} = \boxed{1.24 \times 10^{-39}\,\text{J}}$$

We set $E = \dfrac{n^2h^2}{8mL^2} = \dfrac{1}{2}kT$ and solve for n.

From above $\dfrac{h^2}{8mL^2} = \dfrac{E_2 - E_1}{3} = 4.13 \times 10^{-40}\,\text{J}$; then

$$n^2 \times (4.13 \times 10^{-40}\,\text{J}) = \left(\tfrac{1}{2}\right) \times (1.381 \times 10^{-23}\,\text{J K}^{-1}) \times (300\,\text{K}) = 2.07 \times 10^{-21}\,\text{J}$$

We find $n = \left(\dfrac{2.07 \times 10^{-21}\,\text{J}}{4.13 \times 10^{-40}\,\text{J}}\right)^{1/2} = \boxed{2.2 \times 10^{9}}$

At this level,

$$E_n - E_{n-1} = \{n^2 - (n-1)^2\} \times \frac{h^2}{8mL^2} = (2n-1) \times \frac{h^2}{8mL^2} \approx (2n) \times \frac{h^2}{8mL^2}$$
$$= (4.4 \times 10^9) \times (4.13 \times 10^{-40}\,\text{J}) \approx \boxed{1.8 \times 10^{-30}\,\text{J}} \quad [\text{or } 1.1\ \ \text{J mol}^{-1}]$$

P12.3 $\omega = \left(\dfrac{k}{\mu}\right)^{1/2}$ [31, with μ in place of m]

Also, $\omega = 2\pi\nu = \dfrac{2\pi c}{\lambda} = 2\pi c\tilde{\nu}$

Therefore $k = \omega^2\mu = 4\pi^2c^2\tilde{\nu}^2\mu = \dfrac{4\pi^2c^2\tilde{\nu}^2 m_1 m_2}{m_1 + m_2}$

We draw up the following table using the information inside the back cover

	$^1H^{35}Cl$	$^1H^{81}Br$	$^1H^{127}I$	$^{12}C^{16}O$	$^{14}N^{16}O$
$\tilde{\nu}/\text{m}^{-1}$	299000	265000	231000	217000	190400
$10^{27}\,m_1/\text{kg}$	1.6735	1.6735	1.6735	19.926	23.253
$10^{27}\,m_2/\text{kg}$	58.066	134.36	210.72	26.560	26.560
$k/(\text{N m}^{-1})$	516	412	314	1902	1595

Therefore, the order of stiffness, is CO > NO > HCl > HBr > HI.

P12.4 $E = \dfrac{m_l^2\hbar^2}{2I}$ [49] $= \dfrac{m_l^2\hbar^2}{2mr^2}$ $[I = mr^2]$

$E_0 = 0 \quad [m_l = 0]$

$$E_1 = \frac{\hbar^2}{2mr^2} = \frac{(1.055 \times 10^{-34}\,\text{J s})^2}{(2) \times (1.008) \times (1.6605 \times 10^{-27}\,\text{kg}) \times (160 \times 10^{-12}\,\text{m})^2} = \boxed{1.30 \times 10^{-22}\,\text{J}}$$

$[1.96 \times 10^{11}\,\text{Hz}]$

The minimum angular momentum is $\boxed{\pm\hbar}$

Solutions to theoretical problems

P12.7 We assume that the barrier begins at $x = 0$ and that the barrier extends in the positive x direction.

(a) $P = \displaystyle\int_{\text{Barrier}} \psi^2\,\text{d}\tau = \int_0^{\infty} N^2\text{e}^{-2\kappa x}\,\text{d}x = \boxed{\frac{N^2}{2\kappa}}$

(b) $\langle x \rangle = \displaystyle\int_0^{\infty} x\psi^2\,\text{d}x = N^2 \int_0^{\infty} x\text{e}^{-2\kappa x}\,\text{d}x = \frac{N^2}{(2\kappa)^2} = \boxed{\frac{N^2}{4\kappa^2}}$

Question. Is N a normalization constant?

P12.9 $\langle E_K \rangle \equiv \langle T \rangle = \displaystyle\int_{-\infty}^{+\infty} \psi^*\hat{T}\psi\,\text{d}x$ with $\hat{T} \equiv \dfrac{\hat{p}^2}{2m}$ and $\hat{p} = \dfrac{\hbar}{\text{i}}\dfrac{\text{d}}{\text{d}x}$

$$\hat{T} = -\frac{\hbar^2}{2m}\frac{\text{d}^2}{\text{d}x^2} = -\frac{\hbar^2}{2m\alpha^2}\frac{\text{d}^2}{\text{d}y^2} = -\frac{1}{2}\hbar\omega\frac{\text{d}^2}{\text{d}y^2}, \quad \left[x = \alpha y,\ \alpha^2 = \frac{\hbar}{m\omega}\right]$$

which implies that

$$\hat{T}\psi = -\frac{1}{2}\hbar\,\omega\left(\frac{\mathrm{d}^2\psi}{\mathrm{d}y^2}\right)$$

We then use $\psi = NH\mathrm{e}^{-y^2/2}$, and obtain

$$\frac{\mathrm{d}^2\psi}{\mathrm{d}y^2} = N\frac{\mathrm{d}^2}{\mathrm{d}y^2}(H\mathrm{e}^{-y^2/2}) = N\{H'' - 2yH' - H + y^2H\}\mathrm{e}^{-y^2/2}$$

From Table 12.1

$$H_v'' - 2yH_v' = -2vH_v$$

$$y^2H_v = y\left(\tfrac{1}{2}H_{v+1} + vH_{v-1}\right) = \tfrac{1}{2}\left(\tfrac{1}{2}H_{v+2} + (v+1)H_v\right) + v\left(\tfrac{1}{2}H_v + (v-1)H_{v-2}\right)$$

$$= \tfrac{1}{4}H_{v+2} + v(v-1)H_{v-2} + \left(v+\tfrac{1}{2}\right)H_v$$

Hence, $\dfrac{\mathrm{d}^2\psi}{\mathrm{d}y^2} = N\left[\tfrac{1}{4}H_{v+2} + v(v-1)H_{v-2} - \left(v+\tfrac{1}{2}\right)H_v\right]\mathrm{e}^{-y^2/2}$

Therefore,

$$\langle T\rangle = N^2\left(-\frac{1}{2}\hbar\omega\right)\int_{-\infty}^{+\infty} H_v\left[\frac{1}{4}H_{v+2} + v(v-1)H_{v-2} - \left(v+\frac{1}{2}\right)H_v\right]\mathrm{e}^{-y^2}\,\mathrm{d}x$$

$[\mathrm{d}x = \alpha\,\mathrm{d}y]$

$$= \alpha N^2\left(-\tfrac{1}{2}\hbar\omega\right)\left[0 + 0 - \left(v+\tfrac{1}{2}\right)\pi^{1/2}2^v v!\right]$$

$$\left[\int_{-\infty}^{+\infty} H_v H_{v'}\mathrm{e}^{-y^2}\,\mathrm{d}y = 0 \quad \text{if } v' \neq v,\ \text{Table 12.1}\right]$$

$$= \boxed{\frac{1}{2}\left(v+\frac{1}{2}\right)\hbar\omega} \quad \left[N_v^2 = \frac{1}{\alpha\pi^{1/2}2^v v!},\ \text{Table 12.1}\right]$$

P12.11 **(a)**

$$\langle x\rangle = \int_0^L \left(\frac{2}{L}\right)^{1/2}\sin\left(\frac{n\pi x}{L}\right) x \left(\frac{2}{L}\right)^{1/2}\sin\left(\frac{n\pi x}{L}\right)\mathrm{d}x$$

$$= \left(\frac{2}{L}\right)\int_0^L x\sin^2 ax\,\mathrm{d}x \quad \left[a = \frac{n\pi}{L}\right]$$

$$= \left(\frac{2}{L}\right)\times\left(\frac{x^2}{4} - \frac{x\sin 2ax}{4a} - \frac{\cos 2ax}{8a^2}\right)\Big|_0^L = \left(\frac{2}{L}\right)\times\left(\frac{L^2}{4}\right)$$

$$= \frac{L}{2} \quad \text{[by symmetry also]}$$

$$\langle x^2\rangle = \frac{2}{L}\int_0^L x^2\sin^2 ax\,\mathrm{d}x = \left(\frac{2}{L}\right)\times\left[\frac{x^3}{6} - \left(\frac{x^2}{4a} - \frac{1}{8a^3}\right)\sin 2ax - \frac{x\cos 2ax}{4a^2}\right]\Big|_0^L$$

$$= \left(\frac{2}{L}\right)\times\left(\frac{L^3}{6} - \frac{L^3}{4n^2\pi}\right) = \frac{L^2}{3}\left(1 - \frac{1}{6n^2\pi^2}\right)$$

$$\Delta x = \left[\frac{L^2}{3}\left(1 - \frac{1}{6n^2\pi^2}\right) - \frac{L^2}{4}\right]^{1/2} = \boxed{L\left(\frac{1}{12} - \frac{1}{2\pi^2n^2}\right)^{1/2}}$$

$\langle p\rangle = 0$ [by symmetry, also see Exercise 12.3(a)]

$$\langle p^2\rangle = \frac{n^2h^2}{4L^2} \quad [\text{from } E = \frac{p^2}{2m}, \text{ also Exercise 12.3(a)}]$$

$$\Delta p = \left(\frac{n^2h^2}{4L^2}\right)^{1/2} = \boxed{\frac{nh}{2L}}$$

$$\Delta p\Delta x = \frac{nh}{2L} \times L\left(\frac{1}{12} - \frac{1}{2\pi^2n^2}\right)^{1/2} = \frac{nh}{2\sqrt{3}}\left(1 - \frac{1}{24\pi^2n^2}\right)^{1/2} > \frac{\hbar}{2}$$

(b) $\langle x\rangle = \alpha^2\int_{-\infty}^{+\infty}\psi^2 y\,\mathrm{d}y[x = \alpha y] = 0$ [by symmetry, y is an odd function]

$$\langle x^2\rangle = \frac{2}{k}\langle\frac{1}{2}kx^2\rangle = \frac{2}{k}\langle V\rangle$$

since $2\langle T\rangle = b\langle V\rangle[45, \langle T\rangle \equiv E_{\mathrm{K}}] = 2\langle V\rangle$ $\left[V = ax^b = \frac{1}{2}kx^2, b = 2\right]$

or $\langle V\rangle = \langle T\rangle = \frac{1}{2}\left(v + \frac{1}{2}\right)\hbar\omega$ [Problem 12.9]

$$\langle x^2\rangle = \left(v + \frac{1}{2}\right) \times \left(\frac{\hbar\omega}{k}\right) = \left(v + \frac{1}{2}\right) \times \left(\frac{\hbar}{\omega m}\right) = \left(v + \frac{1}{2}\right) \times \left(\frac{\hbar^2}{mk}\right)^{1/2} \quad [41]$$

$$\Delta x = \boxed{\left[\left(v + \frac{1}{2}\right)\frac{\hbar}{\omega m}\right]^{1/2}}$$

$\langle p\rangle = 0$ [by symmetry, or by noting that the integrand is always an odd function of x]

$\langle p^2\rangle = 2m\langle T\rangle = (2m) \times \left(\frac{1}{2}\right) \times \left(v + \frac{1}{2}\right) \times \hbar\omega$ [Problem 12.9]

$$\Delta p = \boxed{\left[\left(v + \frac{1}{2}\right)\hbar\omega m\right]^{1/2}}$$

$$\Delta p\Delta x = \left(v + \frac{1}{2}\right)\hbar \geq \frac{\hbar}{2}$$

Comment. Both results show a consistency with the uncertainty principle in the form $\Delta p\Delta q \geq \frac{\hbar}{2}$ as given is Section 11.6, eqn 46.

P12.12 $\mu \equiv \int\psi_{v'}x\psi_v\,\mathrm{d}x = \alpha^2\int\psi_{v'}y\psi_v\,\mathrm{d}y \quad [x = \alpha y]$

$$y\psi_v = N_v\left(\tfrac{1}{2}H_{v+1} + vH_{v-1}\right)\mathrm{e}^{-y^2/2} \text{ [Table 12.1]}$$

Hence

$$\mu = \alpha^2N_vN_{v'}\int\left(\tfrac{1}{2}H_{v'}H_{v+1} + vH_{v'}H_{v-1}\right)\mathrm{e}^{-y^2}\,\mathrm{d}y = 0 \quad \text{unless } v' = v \pm 1 \text{ [Table 12.1]}$$

For $v' = v + 1$

$$\mu = \frac{1}{2}\alpha^2N_vN_{v+1}\int H_{v+1}^2\mathrm{e}^{-y^2}\,\mathrm{d}y = \frac{1}{2}\alpha^2N_vN_{v+1}\pi^{1/2}2^{v+1}(v+1)! = \boxed{\alpha\left(\frac{v+1}{2}\right)^{1/2}}$$

For $v' = v - 1$

$$\mu = v\alpha^2N_vN_{v-1}\int H_{v-1}^2\mathrm{e}^{-y^2}\,\mathrm{d}y = v\alpha^2N_vN_{v-1}\pi^{1/2}2^{v-1}(v-1)! = \boxed{\alpha\left(\frac{v}{2}\right)^{1/2}}$$

No other values of v' result in a nonzero value for μ; hence, no other transitions are allowed.

P12.15 The Schrödinger equation is

$$-\frac{\hbar}{2m}\nabla^2\psi = E\psi \quad [61, \text{ with } V = 0]$$

$$\nabla^2\psi = \frac{1}{r}\frac{\partial^2(r\psi)}{\partial r^2} + \frac{1}{r^2}\Lambda^2\psi \text{ [Table 11.1]}$$

since r = constant, the first term is eliminated and the Schrödinger equation may be rewritten

$$-\frac{\hbar^2}{2mr^2}\Lambda^2\psi = E\psi \quad \text{or} \quad -\frac{\hbar^2}{2I}\Lambda^2\psi = E\psi \ [I = mr^2] \quad \text{or} \quad \Lambda^2\psi = -\frac{2IE\psi}{\hbar^2}$$

Now use $\psi = Y_{l,m_l}$ and see if they satisfy this equation.

(a) $\Lambda^2 Y_{0,0} = \boxed{0}\,[l = 0, m_l = 0]$, implying that $E = 0$
and angular momentum $= \boxed{0}$ [from $\{l(l+1)\}^{1/2}\hbar$]

(b) $\Lambda^2 Y_{2,-1} = -2(2+1)Y_{2,-1} \quad [l = 2]$, and hence

$$-2(2+1)Y_{2,-1} = -\frac{2IE}{\hbar^2}Y_{2,-1}, \quad \text{implying that} \quad \boxed{E = \frac{3\hbar^2}{I}}$$

and the angular momentum is $\{2(2+1)\}^{1/2}\hbar = \boxed{6^{1/2}\hbar}$

(c) $\Lambda^2 Y_{3,3} = -3(3+1)Y_{3,3} \quad [l = 3]$, and hence

$$-3(3+1)Y_{3,3} = -\frac{2IE}{\hbar^2}Y_{3,3}, \quad \text{implying that} \quad \boxed{E = \frac{6\hbar^2}{I}}$$

and the angular momentum is $\{3(3+1)\}^{1/2}\hbar = \boxed{2\sqrt{3}\hbar}$

P12.17 From the diagram in Fig. 12.2, $\cos\theta = \dfrac{m_l}{\{l(l+1)\}^{1/2}}$ and hence $\boxed{\theta = \arccos\dfrac{m_l}{\{l(l+1)\}^{1/2}}}$

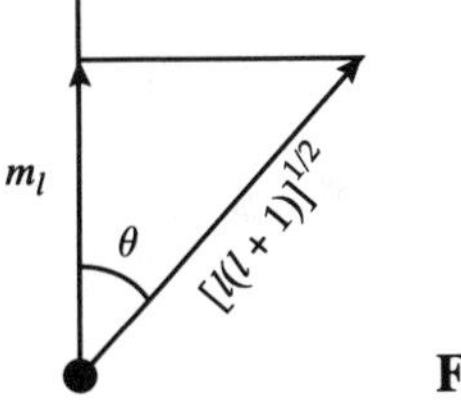

Figure 12.2

For an α electron, $m_s = +\frac{1}{2}$, $s = \frac{1}{2}$ and (with $m_l \to m_s$, $l \to s$)

$$\theta = \arccos\frac{\frac{1}{2}}{\left(\frac{3}{4}\right)^{1/2}} = \arccos\frac{1}{\sqrt{3}} = \boxed{54°44'}$$

The minimum angle occurs for $m_l = l$:

$$\lim_{l\to\infty}\theta_{\min} = \lim_{l\to\infty}\arccos\left(\frac{l}{\{l(l+1)\}^{1/2}}\right) = \lim_{l\to\infty}\arccos\frac{l}{l} = \arccos 1 = \boxed{0}$$

P12.19 $\boldsymbol{l} = \boldsymbol{r} \times \boldsymbol{p} = \begin{vmatrix} \boldsymbol{i} & \boldsymbol{j} & \boldsymbol{k} \\ \hat{x} & \hat{y} & \hat{z} \\ \hat{p}_x & \hat{p}_y & \hat{p}_z \end{vmatrix}$ [see any book treating the vector product of vectors]

$$= \boldsymbol{i}(\hat{y}\hat{p}_z - \hat{z}\hat{p}_y) + \boldsymbol{j}(\hat{z}\hat{p}_x - \hat{x}\hat{p}_z) + \boldsymbol{k}(\hat{x}\hat{p}_y - \hat{y}\hat{p}_x)$$

Therefore,

$$\hat{l}_x = (\hat{y}\hat{p}_z - \hat{z}\hat{p}_y) = \boxed{\frac{\hbar}{\mathrm{i}}\left(y\frac{\partial}{\partial z} - z\frac{\partial}{\partial y}\right)}$$

$$\hat{l}_y = (\hat{z}\hat{p}_x - \hat{x}\hat{p}_z) = \boxed{\frac{\hbar}{\mathrm{i}}\left(z\frac{\partial}{\partial x} - x\frac{\partial}{\partial z}\right)}$$

$$\hat{l}_z = (\hat{x}\hat{p}_y - \hat{y}\hat{p}_x) = \boxed{\frac{\hbar}{\mathrm{i}}\left(x\frac{\partial}{\partial y} - y\frac{\partial}{\partial x}\right)}$$

We have used $\hat{p}_x = \frac{\hbar}{\mathrm{i}}\frac{\partial}{\partial x}$, etc. The commutator of $\hat{l}_x$ and $\hat{l}_y$ is $(\hat{l}_x\hat{l}_y - \hat{l}_y\hat{l}_x)$. We note that the operations always imply operation on a function. We form

$$\begin{aligned}
\hat{l}_x\hat{l}_y f &= -\hbar^2\left(y\frac{\partial}{\partial z} - z\frac{\partial}{\partial y}\right)\left(z\frac{\partial}{\partial x} - x\frac{\partial}{\partial z}\right) f \\
&= -\hbar^2\left(yz\frac{\partial^2 f}{\partial z\partial x} + y\frac{\partial f}{\partial x} - yx\frac{\partial^2 f}{\partial z^2} - z^2\frac{\partial^2 f}{\partial y\partial x} + zx\frac{\partial^2 f}{\partial z\partial y}\right) \\
\hat{l}_y\hat{l}_x f &= -\hbar^2\left(z\frac{\partial}{\partial x} - x\frac{\partial}{\partial z}\right)\left(y\frac{\partial}{\partial z} - z\frac{\partial}{\partial y}\right) f \\
&= -\hbar^2\left(zy\frac{\partial^2 f}{\partial x\partial z} - z^2\frac{\partial^2 f}{\partial x\partial y} - xy\frac{\partial^2 f}{\partial z^2} + xz\frac{\partial^2 f}{\partial z\partial y} + x\frac{\partial f}{\partial y}\right)
\end{aligned}$$

Since multiplication and differentiation are each commutative, the results of the operation $\hat{l}_x\hat{l}_y$ and $\hat{l}_y\hat{l}_x$ differ only in one term. For $\hat{l}_y\hat{l}_x f$, $x\frac{\partial f}{\partial y}$ replaces $y\frac{\partial f}{\partial x}$. Hence, the commutator of the operations, $(\hat{l}_x\hat{l}_y - \hat{l}_y\hat{l}_x)$, is $-\hbar^2\left(y\frac{\partial}{\partial x} - x\frac{\partial}{\partial y}\right)$ or $\boxed{-\frac{\hbar}{\mathrm{i}}\hat{l}_z.}$

Comment. We also would find

$$(\hat{l}_y\hat{l}_z - \hat{l}_z\hat{l}_y) = -\frac{\hbar}{\mathrm{i}}\hat{l}_x \quad \text{and} \quad (\hat{l}_z\hat{l}_x - \hat{l}_x\hat{l}_z) = -\frac{\hbar}{\mathrm{i}}\hat{l}_y$$

P12.20 Upon making the operator substitutions

$$p_x = \frac{\hbar}{\mathrm{i}}\frac{\partial}{\partial x} \quad \text{and} \quad p_y = \frac{\hbar}{\mathrm{i}}\frac{\partial}{\partial y}$$

into $\hat{l}_z$ we find

$$\hat{l}_z = \frac{\hbar}{\mathrm{i}}\left(x\frac{\partial}{\partial y} - y\frac{\partial}{\partial x}\right)$$

But $\dfrac{\partial}{\partial\phi} = \dfrac{\partial x}{\partial\phi}\dfrac{\partial}{\partial x} + \dfrac{\partial y}{\partial\phi}\dfrac{\partial}{\partial y} + \dfrac{\partial z}{\partial\phi}\dfrac{\partial}{\partial z}$ which is the chain rule of partial differentiation.

$$\frac{\partial x}{\partial\phi} = \frac{\partial}{\partial\phi}(r\sin\theta\cos\phi) = -r\sin\theta\sin\phi = -y$$

$$\frac{\partial y}{\partial\phi} = \frac{\partial}{\partial\phi}(r\sin\theta\sin\phi) = r\sin\theta\cos\phi = x$$

$$\frac{\partial z}{\partial\phi} = 0$$

Thus,

$$\frac{\partial}{\partial\phi} = -y\frac{\partial}{\partial x} + x\frac{\partial}{\partial y}$$

Upon substitution,

$$\hat{l}_z = \frac{\hbar}{\mathrm{i}}\frac{\partial}{\partial\phi} = -\mathrm{i}\hbar\frac{\partial}{\partial\phi}$$

Solutions to additional problems

P12.21 The current is proportional to the probability of the electron getting to the probe tip; that probability is given by $|\psi|^2$, so

$$I \propto B^2 \mathrm{e}^{-2\kappa x}$$

The ratio of currents at two different distances is

$$\frac{I_1}{I_2} = \mathrm{e}^{-2\kappa(x_1-x_2)}$$

so

$$-2\kappa(x_1 - x_2) = \left(\frac{2(9.11\times10^{-31}\,\mathrm{kg})\times(2.0\,\mathrm{eV})\times(1.602\times10^{-19}\,\mathrm{J\,eV^{-1}})}{(1.0546\times10^{-34}\,\mathrm{J\,s})^2}\right)^{1/2} \times (0.10\times10^{-9}\,\mathrm{m}),$$

$$2\kappa(x_1 - x_2) = 0.72 \quad \text{and} \quad \frac{I_1}{I_2} = \boxed{0.49}$$

13 Atomic structure and atomic spectra

Solutions to exercises

E13.1 This is essentially the photoelectric effect [eqn 11.12 of Section 11.2] with the ionization energy of the ejected electron being the work function Φ.

$$h\nu = \tfrac{1}{2}m_e v^2 + I$$

$$I = h\nu - \frac{1}{2}m_e v^2 = (6.626 \times 10^{-34}\,\mathrm{J\,Hz^{-1}}) \times \left(\frac{2.998 \times 10^8\,\mathrm{m\,s^{-1}}}{58.4 \times 10^{-9}\,\mathrm{m}}\right)$$

$$- \left(\tfrac{1}{2}\right) \times (9.109 \times 10^{-31}\,\mathrm{kg}) \times (1.59 \times 10^6\,\mathrm{m\,s^{-1}})^2$$

$$= 2.25 \times 10^{-18}\,\mathrm{J}, \text{ corresponding to } \boxed{14.0\,\mathrm{eV}}$$

E13.2 $R_{2,0} \propto (2-\rho)\mathrm{e}^{-\rho/2}$ with $\rho = \dfrac{r}{a_0}$ [Table 13.1]

$$\frac{\mathrm{d}R}{\mathrm{d}r} = \frac{1}{a_0}\frac{\mathrm{d}R}{\mathrm{d}\rho} = \frac{1}{a_0}\left(-1 - 1 + \frac{1}{2}\rho\right)\mathrm{e}^{-\rho/2} = 0 \quad \text{when } \rho = 4$$

Hence, the wavefunction has an extremum at $r = \boxed{4a_0}$. Since $2-\rho < 0$, $\psi < 0$ and the extremum is a minimum $\left(\text{more formally: } \dfrac{\mathrm{d}^2\psi}{\mathrm{d}r^2} > 0 \text{ at } \rho = 4\right)$.

The second extremum is at $\boxed{r = 0}$. It is not a minimum and in fact is a physical maximum, though not one that can be obtained by differentiation. To see that it is maximum substitute $\rho = 0$ into $R_{2,0}$.

E13.3 The radial nodes correspond to $R_{3,0} = 0$. $R_{3,0} \propto 6 - 6\rho + \rho^2$ (Table 13.1); the radial nodes occur at

$$6 - 6\rho + \rho^2 = 0, \quad \text{or} \quad \rho = 3 \pm \sqrt{3} = 1.27 \text{ and } 4.73$$

Since $r = \dfrac{3\rho a_0}{2}$, the radial nodes occur at $\boxed{101\,\mathrm{pm} \text{ and } 376\,\mathrm{pm}}$

E13.4 $R_{1,0} = N\mathrm{e}^{-r/a_0}$

$$\int_0^\infty R^2 r^2\,\mathrm{d}r = 1 = \int_0^\infty N^2 r^2\,\mathrm{e}^{-2r/a_0}\,\mathrm{d}r = N^2 \times \frac{2!}{\left(\frac{2}{a_0}\right)^3} = 1 \quad \left[\int_0^\infty x^n \mathrm{e}^{-ax}\,\mathrm{d}x = \frac{n!}{a^{n+1}}\right]$$

$$N^2 = \frac{4}{a_0^3}, \qquad \boxed{N = \frac{2}{a_0^{3/2}}}$$

Thus,

$$R_{1,0} = 2\left(\frac{1}{a_0}\right)^{3/2}\mathrm{e}^{-r/a_0},$$

which agrees with Table 13.1.

E13.5 This exercise has already been solved in Problem 12.13 by use of the virial theorem. Here we will solve it by straightforward integration.

$$\psi_{1,0,0} = R_{1,0}Y_{0,0} = \left(\frac{1}{\pi a_0^3}\right)^{1/2}\mathrm{e}^{-r/a_0} \text{ [Tables 12.3 and 13.1]}$$

The potential energy operator is

$$V = -\frac{Ze^2}{4\pi\varepsilon_0} \times \left(\frac{1}{r}\right) = -k\left(\frac{1}{r}\right)$$

$$\langle V\rangle = -k\left\langle\frac{1}{r}\right\rangle \left[k = \frac{e^2}{4\pi\varepsilon_0}\right] = -k\int_0^\infty\int_0^\pi\int_0^{2\pi}\left(\frac{1}{\pi a_0^3}\right)\mathrm{e}^{-r/a_0}\left(\frac{1}{r}\right)\mathrm{e}^{-r/a_0}r^2\,\mathrm{d}r\sin\theta\,\mathrm{d}\theta\,\mathrm{d}\phi$$

$$= -k\times(4\pi)\times\left(\frac{1}{\pi a_0^3}\right)\int_0^\infty r\mathrm{e}^{-2r/a_0}\,\mathrm{d}r = -k\times\left(\frac{4}{a_0^3}\right)\times\left(\frac{a_0^2}{4}\right) = -k\left(\frac{1}{a_0}\right)$$

$$\left[\text{We have used } \int_0^\pi \sin\theta\,\mathrm{d}\theta = 2,\ \int_0^{2\pi}\mathrm{d}\phi = 2\pi,\ \text{and } \int_0^\infty x^n\,\mathrm{e}^{-ax}\,\mathrm{d}x = \frac{n!}{a^{n+1}}\right]$$

Hence,

$$\langle V\rangle = -\frac{e^2}{4\pi\varepsilon_0 a_0} = \boxed{2E_{1s}}$$

The kinetic energy operator is $-\frac{\hbar^2}{2\mu}\nabla^2$ [7]; hence

$$\langle E_{\mathrm{K}}\rangle \equiv \langle T\rangle = \int \psi_{1s}^*\left(-\frac{\hbar^2}{2\mu}\right)\nabla^2\psi_{1s}\,\mathrm{d}\tau$$

$$\nabla^2\psi_{1s} = \frac{1}{r}\frac{\partial^2(r\psi_{1s})}{\partial r^2} + \frac{1}{r^2}\Lambda^2\psi_{1s} \text{ [Problem 12.15]}$$

$$= \left(\frac{1}{\pi a_0^3}\right)^{1/2}\times\left(\frac{1}{r}\right)\times\left(\frac{\mathrm{d}^2}{\mathrm{d}r^2}\right)r\mathrm{e}^{-r/a_0}$$

$$[\Lambda^2\psi_{1s} = 0,\ \psi_{1s} \text{ contains no angular variables}]$$

$$= \left(\frac{1}{\pi a_0^3}\right)^{1/2}\left[-\left(\frac{2}{a_0 r}\right) + \left(\frac{1}{a_0^2}\right)\right]\mathrm{e}^{-r/a_0}$$

$$\langle T\rangle = -\left(\frac{\hbar^2}{2\mu}\right)\times\left(\frac{1}{\pi a_0^3}\right)\int_0^\infty\left[-\left(\frac{2}{a_0 r}\right)+\left(\frac{1}{a_0{}^2}\right)\right]\mathrm{e}^{-2r/a_0}r^2\,\mathrm{d}r\times\int_0^\pi\sin\theta\,\mathrm{d}\theta\int_0^{2\pi}\mathrm{d}\phi$$

$$= -\left(\frac{2\hbar^2}{\mu a_0{}^3}\right)\int_0^\infty\left[-\left(\frac{2r}{a_0}\right)\times\left(\frac{r^2}{a_0^2}\right)\right]\mathrm{e}^{-2r/a_0}\,\mathrm{d}r = -\left(\frac{2\hbar^2}{\mu a_0^3}\right)\times\left(-\frac{a_0}{4}\right) = \frac{\hbar^2}{2\mu a_0^2}$$

$$= \boxed{-E_{1s}}$$

Hence, $\langle T\rangle + \langle V\rangle = 2E_{1s} - E_{1s} = E_{1s}$

Comment. E_{1s} may also be written as

$$E_{1s} = -\frac{\mu e^4}{32\pi^2\varepsilon_0^2\hbar^2}$$

Question. Are the three different expressions for E_{1s} given in this exercise all equivalent?

E13.6 $P_{2s} = 4\pi r^2 \psi_{2s}^2$

$$\psi_{2s} = \frac{1}{2\sqrt{2}}\left(\frac{z}{a_0}\right)^{3/2} \times (2-\rho)e^{-\rho/2} \quad \left[\rho = \frac{Zr}{a_0}\right]$$

$$P_{2s} = 4\pi\left(\frac{a_0\rho}{Z}\right)^2 \times \left(\frac{1}{8}\right) \times \left(\frac{z}{a_0}\right)^3 (2-\rho)^2 e^{-\rho}$$

$$P_{2s} = k\rho^2(2-\rho)^2 e^{-\rho} \quad \left[k = \frac{\pi z}{2a_0} = \text{constant}\right]$$

The most probable value of r, or equivalently, ρ is where

$$\frac{d}{d\rho}\{\rho^2(2-\rho)^2 e^{-\rho}\} = 0$$

$$\propto \{2\rho(2-\rho)^2 - 2\rho^2(2-\rho) - \rho^2(2-\rho)^2\}e^{-\rho} = 0$$

$$\propto \rho(\rho-2)(\rho^2 - 6\rho + 4) = 0 \quad [e^{-\rho} \text{ is never zero, except as } \rho \to \infty]$$

Thus, $\rho^* = 0,\ \rho^* = 2,\ \rho^* = 3 \pm \sqrt{5}$

The principal (outermost) maximum is at $\rho^* = 3 + \sqrt{5}$

Hence, $r^* = (3+\sqrt{5})\dfrac{a_0}{Z} = \boxed{5.24\dfrac{a_0}{Z}}$

E13.7 Indentify l and use angular momentum $= \{l(l+1)\}^{1/2}\hbar$.

(a) $l = 0$, so angular momentum $= 0$ **(b)** $l = 0$, so angular momentum $= 0$

(c) $l = 2$, so angular momentum $= \sqrt{6}\hbar$

The total number of nodes is equal to $n - 1$ and the number of angular nodes is equal to l; hence the number of radial is equal to $n - l - 1$. We can draw up the following table

	1s	3s	3d
n, l	1, 0	3, 0	3, 2
Angular nodes	0	0	2
Radial nodes	0	2	0

E13.8 We use the Clebsch–Gordan series [46] in the form

$$j = l + s,\ l + s - l, \ldots, |l - s| \quad [\text{lower-case for a single electron}]$$

(a) $l = 2,\quad s = \frac{1}{2};\quad$ so $j = \boxed{\frac{5}{2}, \frac{3}{2}}$ **(b)** $l = 3,\quad s = \frac{1}{2};\quad$ so $j = \boxed{\frac{7}{2}, \frac{5}{2}}$

E13.9 The Clebsch–Gordan series for $\boxed{l = 1}$ and $s = \frac{1}{2}$ leads to $j = \frac{3}{2}$ and $\frac{1}{2}$

E13.10 The energies are $E = -\dfrac{hc\mathcal{R}_H}{n^2}$ [Table 13.2], and the orbital degeneracy g of an energy level of principal quantum number n is

$$g = \sum_{l=0}^{n-1}(2l-1) = 1 + 3 + 5 + \cdots + 2n - 1 = \frac{(1 + 2n - 1)n}{2} = n^2$$

(a) $E = -hc\mathcal{R}_H$ implies that $n = 1$, so $\boxed{g = 1}$ [the 1s orbital].

(b) $E = -\frac{hc\mathcal{R}_H}{9}$ implies that $n = 3$, so $\boxed{g = 9}$ (3s orbital, the three 3p orbitals, and the five 3d orbitals).

(c) $E = -\frac{hc\mathcal{R}_H}{25}$ implies that $n = 5$, so $\boxed{g = 25}$ (the 5s orbital, the three 5p orbitals, the five 5d orbitals, the seven 5f orbitals, the nine 5g orbitals).

E13.11 The letter D indicates that $L = 2$, the superscript 1 is the value of $2S + 1$, so $S = 0$ and the subscript 2 is the value of J. Hence, $\boxed{L = 2,\ S = 0,\ J = 2}$

E13.12 Here we use the probability density function ψ^2, rather than the radial distribution function P, since we are seeking the probability at a point, namely $\psi^2\,d\tau$

The probability density varies as

$$\psi^2 = \frac{1}{\pi a_0^3}e^{-2r/a_0} \text{ [From 20]}$$

Therefore, the maximum value is at $r = 0$ and ψ^2 is 50 per cent of the maximum when

$$e^{-2r/a_0} = 0.50$$

implying that $r = -\frac{1}{2}a_0 \ln 0.50$, which is at $r = \boxed{0.35a_0}$ [18 pm]

E13.13 The selection rules for a many-electron atom are given by the set [47]. For a single-electron transition these amount to Δn = any integer; $\Delta l = \pm 1$. Hence

(a) $2s \to 1s$; $\Delta l = 0$, $\boxed{\text{forbidden}}$ **(b)** $2p \to 1s$; $\Delta l = -1$, $\boxed{\text{allowed}}$

(c) $3d \to 2p$; $\Delta l = -1$, $\boxed{\text{allowed}}$

E13.14 For a given l there are $2l + 1$ values of m_l and hence $2l + 1$ orbitals. Each orbital may be occupied by two electrons. Hence the maximum occupancy is $2(2l + 1)$. Draw up the following table

	l	$2(2l+1)$		l	$2(2l+1)$
(a) 1s	0	$\boxed{2}$	**(c)** 3d	2	$\boxed{10}$
(b) 3p	1	$\boxed{6}$	**(d)** 6g	4	$\boxed{18}$

E13.15 **(a)** $1s^2 2s^2 2p^6 3s^2 3p^6 3d^8 = \boxed{[\text{Ar}]3d^8}$

(b) All subshells except 3d are filled and hence have no net spin. Applying Hund's rule to $3d^8$ shows that there are two unpaired spins. The paired spins do not contribute to the net spin, hence we consider only $s_1 = \frac{1}{2}$ and $s_2 = \frac{1}{2}$. The Clebsch–Gordan series [44, $l \to s$] produces

$$S = s_1 + s_2, \ldots, |s_1 - s_2|, \quad \text{hence} \quad \boxed{S = 1, 0}$$
$$M_S = -S, -S + 1, \ldots, S$$

For $S = 1$, $\boxed{M_S = -1, 0, +1}$

$S = 0$, $\boxed{M_S = 0}$

E13.16 Use the Clebsch–Gordan series in the form

$$S' = s_1 + s_2, s_1 + s_2 - 1, \ldots, |s_1 - s_2|$$

and

$$S = S' + s_1, S' + s_1 - 1, \ldots, |S' - s_1|$$

in succession. The multiplicity is $2S + 1$.

(a) $S = \frac{1}{2} + \frac{1}{2}, \frac{1}{2} - \frac{1}{2} = \boxed{1, 0}$ with multiplicities $\boxed{3, 1}$ respectively

(b) $S' = 1, 0$; then $S = \boxed{\frac{3}{2}, \frac{1}{2}}$ [from 1], $\boxed{\text{and } \frac{1}{2}}$ [from 0], with multiplicities $\boxed{4, 2, 2}$

E13.17 These electrons are not equivalent (different subshells), hence all the terms that arise from the vector model and the Clebsch–Gordan series are allowed (Example 13.6).

$$L = l_1 + l_2, \ldots, |l - l_2| \, [44] = 2 \text{ only}$$
$$S = s_1 + s_2, \ldots, |s_1 - s_2| = 1, 0$$

The allowed terms are then ${}^3\mathrm{D}$ and ${}^1\mathrm{D}$. The possible values of J are given by

$$J = L + S, \ldots, |L - S| \, [46] = 3,\ 2,\ 1 \text{ for } {}^3\mathrm{D} \text{ and } 2 \text{ for } {}^1\mathrm{D}$$

The allowed complete term symbols are then

$$\boxed{{}^3\mathrm{D}_3,\ {}^3\mathrm{D}_2,\ {}^3\mathrm{D}_1,\ {}^1\mathrm{D}_2}$$

The $\boxed{{}^3\text{D set of terms are the lower in energy}}$ [Hund's rule]

Comment. Hund's rule in the form given in the text does not allow the energies of the triplet terms to be distinguished. Experimental evidence indicates that ${}^3\mathrm{D}_1$ is lowest.

E13.18 Use the Clebsch–Gordan series in the form

$$J = L + S, L + S - 1, \ldots, |L - S|$$

The number of states (M_J values) is $2J + 1$ in each case.

(a) $L = 0, S = 0$; hence $\boxed{J = 0}$ and there is only $\boxed{1}$ state ($M_J = 0$)

(b) $L = 1, S = \frac{1}{2}$; hence $J = \boxed{\frac{3}{2}, \frac{1}{2}}$ (${}^2\mathrm{P}_{3/2},\ {}^2\mathrm{P}_{1/2}$) with 4, 2 states respectively.

(c) $L = 1, S = 1$; hence $J = \boxed{2, 1, 0}$ (${}^3\mathrm{P}_2,\ {}^3\mathrm{P}_1,\ {}^3\mathrm{P}_0$) with 5, 3, 1 states respectively.

E13.19 Closed shells and subshells do not contribute to either L or S and thus are ignored in what follows.

(a) Li[He]$2s^1$: $S = \frac{1}{2}, L = 0$; $J = \frac{1}{2}$, so the only term is $\boxed{{}^2\mathrm{S}_{1/2}}$

(b) Na[He]$3p^1$: $S = \frac{1}{2}, L = 1$; $J = \frac{3}{2}, \frac{1}{2}$, so the terms are $\boxed{{}^2\mathrm{P}_{3/2} \text{ and } {}^2\mathrm{P}_{1/2}}$

E13.20 $E = \mu_\mathrm{B}\mathcal{B}m_l$ [53]

implying that

$$\Delta E = E_{m_l+1} - E_{m_l} = \mu_\mathrm{B}\mathcal{B} = hc\tilde{\nu}$$

Therefore,

$$\mathcal{B} = \frac{hc\tilde{\nu}}{\mu_\mathrm{B}} = \frac{(6.626 \times 10^{-34}\,\mathrm{J\,s}) \times (2.998 \times 10^{10}\,\mathrm{cm\,s^{-1}}) \times (1.0\,\mathrm{cm^{-1}})}{9.274 \times 10^{-24}\,\mathrm{J\,T^{-1}}} = \boxed{2.1\,\mathrm{T}}$$

Solutions to problems

Solutions to numerical problems

P13.1 All lines in the hydrogen spectrum fit the Rydberg formula

$$\frac{1}{\lambda} = \mathcal{R}_H\left(\frac{1}{n_1^2} - \frac{1}{n_2^2}\right) \quad \left[1, \text{with } \tilde{\nu} = \frac{1}{\lambda}\right] \quad \mathcal{R}_H = 109\,677\,\text{cm}^{-1}$$

Find n_1 from the value of λ_{max}, which arises from the transition $n_1 + 1 \to n_1$

$$\frac{1}{\lambda_{max}\mathcal{R}_H} = \frac{1}{n_1^2} - \frac{1}{(n_1+1)^2} = \frac{2n_1+1}{n_1^2(n_1+1)^2}$$

$$\lambda_{max}\mathcal{R}_H = \frac{n_1^2(n_1+1)^2}{2n_1+1} = (12\,368 \times 10^{-9}\,\text{m}) \times (109\,677 \times 10^2\,\text{m}^{-1}) = 135.65$$

Since $n_1 = 1, 2, 3$, and 4 have already been accounted for, try $n_1 = 5, 6, \ldots$. With $n_1 = 6$ we get $\dfrac{n_1^2(n_1+1)^2}{2n_1+1} = 136$. Hence, the Humphreys series is $\boxed{n_2 \to 6}$ and the transitions are given by

$$\frac{1}{\lambda} = (109\,677\,\text{cm}^{-1}) \times \left(\frac{1}{36} - \frac{1}{n_2^2}\right), \quad n_2 = 7, 8, \ldots$$

and occur at 12 372 nm, 7503 nm, 5908 nm, 5129 nm, ..., 3908 nm (at $n_2 = 15$), converging to 3282 nm as $n_2 \to \infty$, in agreement with the quoted experimental result.

P13.3 A Lyman series corresponds to $n_1 = 1$; hence

$$\tilde{\nu} = \mathcal{R}_{Li^{2+}}\left(1 - \frac{1}{n^2}\right), \quad n = 2, 3, \ldots \quad \left[\tilde{\nu} = \frac{1}{\lambda}\right]$$

Therefore, if the formula is appropriate, we expect to find that $\tilde{\nu}\left(1 - \frac{1}{n^2}\right)^{-1}$ is a constant ($\mathcal{R}_{Li^{2+}}$). We therefore draw up the following table.

n	2	3	4
$\tilde{\nu}/\text{cm}^{-1}$	740 747	877 924	925 933
$\tilde{\nu}\left(1 - \frac{1}{n^2}\right)^{-1} / \text{cm}^{-1}$	987 663	987 665	987 662

Hence, the formula does describe the transitions, and $\boxed{\mathcal{R}_{Li^{2+}} = 987\,663\,\text{cm}^{-1}}$. The Balmer transitions lie at

$$\tilde{\nu} = \mathcal{R}_{Li^{2+}}\left(\frac{1}{4} - \frac{1}{n^2}\right) \quad n = 3, 4, \ldots$$

$$= (987\,663\,\text{cm}^{-1}) \times \left(\frac{1}{4} - \frac{1}{n^2}\right) = \boxed{137\,175\,\text{cm}^{-1}}, \boxed{185\,187\,\text{cm}^{-1}}, \ldots$$

The ionization energy of the ground-state ion is given by

$$\tilde{\nu} = \mathcal{R}_{Li^{2+}}\left(1 - \frac{1}{n^2}\right), \quad n \to \infty$$

and hence corresponds to

$$\tilde{\nu} = 987\,663\,\text{cm}^{-1}, \quad \text{or } \boxed{122.5\,\text{eV}}$$

P13.5 The ground term is $[\text{Ar}]4s^1\ {}^2\text{S}_{1/2}$ and the first excited is $[\text{Ar}]4p^1\ {}^2\text{P}$. The latter has two levels with $J = 1 + \frac{1}{2} = \frac{3}{2}$ and $J = 1 - \frac{1}{2} = \frac{1}{2}$ which are split by spin–orbit coupling (Section 13.8). Therefore, ascribe the transitions to $\boxed{{}^2\text{P}_{3/2} \rightarrow {}^2\text{S}_{1/2}}$ and $\boxed{{}^2\text{P}_{1/2} \rightarrow {}^2\text{S}_{1/2}}$ (since both are allowed). For these values of J, the splitting is equal to $\frac{3}{2}A$ (Example 13.5). Hence, since

$$(766.70 \times 10^{-7}\,\text{cm})^{-1} - (770.11 \times 10^{-7}\,\text{cm})^{-1} = 57.75\,\text{cm}^{-1}$$

we can conclude that $A = \boxed{38.50\,\text{cm}^{-1}}$

P13.7 The Rydberg constant for positronium ($\mathcal{R}_{\text{Ps}}$) is given by

$$\mathcal{R}_{\text{Ps}} = \frac{\mathcal{R}}{1 + \frac{m_e}{m_e}} = \frac{\mathcal{R}}{1+1} = \frac{1}{2}\mathcal{R} \quad [18; \text{ also Problem 13.6}; m(\text{positron}) = m_e]$$

$$= 54\,869\,\text{cm}^{-1} \quad [\mathcal{R} = 109\,737\,\text{cm}^{-1}]$$

Hence

$$\tilde{\nu} = \frac{1}{\lambda} = (54\,869\,\text{cm}^{-1}) \times \left(\frac{1}{4} - \frac{1}{n^2}\right), \quad n = 3, 4, \ldots$$

$$= \boxed{7621\,\text{cm}^{-1}}, \boxed{10\,288\,\text{cm}^{-1}}, \boxed{11\,522\,\text{cm}^{-1}}, \ldots$$

The binding energy of Ps is

$$E = -hc\mathcal{R}_{\text{Ps}}, \text{ corresponding to } (-)54\,869\,\text{cm}^{-1}$$

The ionization energy is therefore $54\,869\,\text{cm}^{-1}$, or $\boxed{6.80\,\text{eV}}$

Solutions to theoretical problems

P13.9 In each case we need to calculate $\langle r \rangle$. The radial wavefunctions (Table 13.1) rather than the radial distribution functions are appropriate for the purpose

$$\langle r \rangle_{2p} = \int_0^\infty R_{21} r R_{21} r^2\,\text{d}r \quad \left[\rho = \frac{2Zr}{na_0} = \frac{Zr}{a_0}\right]$$

$$= \left(\frac{Z}{a_0}\right)^3 \times \left(\frac{1}{2\sqrt{6}}\right)^2 \int_0^\infty r^3 \rho^2 \text{e}^{-\rho}\,\text{d}r \text{ [Table 13.1]}$$

$$= \left(\frac{Z}{a_0}\right)^3 \times \left(\frac{1}{24}\right) \times \left(\frac{a_0}{Z}\right)^4 \int_0^\infty \rho^5\,\text{e}^{-\rho}\,\text{d}\rho = \left(\frac{1}{24}\right) \times \left(\frac{a_0}{Z}\right) \times (5!) = \frac{5a_0}{Z}$$

$$\langle r \rangle_{2s} = \int_0^\infty R_{20} r R_{20} r^2\,\text{d}r = \left(\frac{Z}{a_0}\right)^3 \times \left(\frac{1}{8}\right) \times \left(\frac{a_0}{Z}\right)^4 \int_0^\infty \rho^3 (2-\rho)^2 \text{e}^{-\rho}\,\text{d}\rho$$

$$= \frac{a_0}{8Z} \int_0^\infty (4\rho^3 - 4\rho^4 + \rho^5)\,\text{e}^{-\rho}\,\text{d}\rho = \frac{a_0}{8Z}(4 \times 3! - 4 \times 4! + 5!) = \frac{6a_0}{Z}$$

Comment. We conclude that the $\boxed{2p \text{ orbital in hydrogen is on average closer}}$ to the nucleus. This is not necessarily true in heavier atoms where $E(2p) > E(2s)$.

P13.12 We use the p_x and p_y orbitals in the form (Section 13.2(f))

$$p_x = rf(r)\sin\theta\cos\phi \qquad p_y = rf(r)\sin\theta\sin\phi$$

and use $\cos\phi = \frac{1}{2}(\mathrm{e}^{\mathrm{i}\phi} + \mathrm{e}^{-\mathrm{i}\phi})$ and $\sin\phi = \frac{1}{2\mathrm{i}}(\mathrm{e}^{\mathrm{i}\phi} - \mathrm{e}^{-\mathrm{i}\phi})$ then

$$p_x = \frac{1}{2}rf(r)\sin\theta(\mathrm{e}^{\mathrm{i}\phi} + \mathrm{e}^{-\mathrm{i}\phi}) \qquad p_y = \frac{1}{2\mathrm{i}}rf(r)\sin\theta(\mathrm{e}^{\mathrm{i}\phi} - \mathrm{e}^{-\mathrm{i}\phi})$$

$$\hat{l}_z = \frac{\hbar}{\mathrm{i}}\frac{\partial}{\partial\phi} \text{ [Problem 12.20 and Section 12.6 and eqn 12.57]}$$

$$\hat{l}_z p_x = \frac{\hbar}{2}rf(r)\sin\theta\,\mathrm{e}^{\mathrm{i}\phi} - \frac{\hbar}{2}rf(r)\sin\theta\,\mathrm{e}^{-\mathrm{i}\phi} = \mathrm{i}\hbar p_y \neq \text{ constant} \times p_x$$

$$\hat{l}_z p_y = \frac{\hbar}{2\mathrm{i}}rf(r)\sin\theta\,\mathrm{e}^{\mathrm{i}\phi} + \frac{\hbar}{2\mathrm{i}}rf(r)\sin\theta\,\mathrm{e}^{-\mathrm{i}\phi} = -\mathrm{i}\hbar p_x \neq \text{ constant} \times p_y$$

Therefore, neither p_x nor p_y are eigenfunctions of $\hat{l}_z$. However, $\boxed{p_x + \mathrm{i}p_y \text{ and } p_x - \mathrm{i}p_y}$ are eigenfunctions

$$p_x + \mathrm{i}p_y = rf(r)\sin\theta\,\mathrm{e}^{\mathrm{i}\phi} \qquad p_x - \mathrm{i}p_y = rf(r)\sin\theta\,\mathrm{e}^{-\mathrm{i}\phi}$$

since both $\mathrm{e}^{\mathrm{i}\phi}$ and $\mathrm{e}^{-\mathrm{i}\phi}$ are eigenfunctions of $\hat{l}_z$ with eigenvalues $+h$ and $-h$.

P13.13 The general rule to use in deciding commutation properties is that operators having no variable in common will commute with each other. We first consider the commutation of $\hat{l}_z$ with the Hamiltonian. This is most easily solved in spherical polar coordinates.

$$\hat{l}_z = \frac{\hbar}{\mathrm{i}}\frac{\partial}{\partial\phi} \text{ [Problem 12.20 and Section 12.6 and eqn 12.57]}$$

$$H = -\frac{\hbar^2}{2\mu}\nabla^2 + V \text{ [7]} \quad V = -\frac{Ze^2}{4\pi\varepsilon_0 r}$$

Since V has no variable in common with $\hat{l}_z$, this part of the hamiltonian and $\hat{l}$ commute.

$$\nabla^2 = \text{terms in } r \text{ only} + \text{terms in } \theta \text{only} + \frac{1}{r^2\sin^2\theta}\frac{\partial^2}{\partial\phi^2} \quad \text{[Table 11.1]}$$

The terms in r only and θ only necessarily commute with $\hat{l}_z(\phi$ only). The final term in ∇^2 contains $\frac{\partial^2}{\partial\phi^2}$ which commutes with $\frac{\partial}{\partial\phi}$, since an operator necessarily commutes with itself. By symmetry we can deduce that if H commutes with $\hat{l}_z$ it must also commute with $\hat{l}_x$ and $\hat{l}_y$ since they are related to each other by a simple transformation of coordinates. This proves useful in establishing the commutation of l^2 and H. We form

$$\hat{l}^2 = \hat{\boldsymbol{l}}\cdot\hat{\boldsymbol{l}} = (\boldsymbol{i}\hat{l}_x + \boldsymbol{j}\hat{l}_y + \boldsymbol{k}\hat{l}_z)\cdot(\boldsymbol{i}\hat{l}_x + \boldsymbol{j}\hat{l}_y + \boldsymbol{k}\hat{l}_z) = \hat{l}_x^2 + \hat{l}_y^2 + \hat{l}_z^2$$

If H commutes with each of $\hat{l}_x, \hat{l}_y$, and $\hat{l}_z$ it must commute with $\hat{l}_x^2, \hat{l}_y^2$, and $\hat{l}_z^2$. Therefore it also commutes with $\hat{l}^2$. Thus H commutes with both $\hat{l}^2$ and $\hat{l}_z$.

Comment. As described at the end of Section 11.6, the physical properties associated with non-commuting operators cannot be simultaneously known with precision. However, since $H, \hat{l}^2$, and $\hat{l}_z$ commute we may simultaneously have exact knowledge of the energy, the total orbital angular momentum, and the projection of the orbital angular momentum along an arbitrary axis.

P13.14 $\psi_{1s} = \left(\frac{1}{\pi a_0^3}\right)^{1/2} e^{-r/a_0}$ [20]

The probability of the electron being within a sphere of radius r' is

$$\int_0^{r'}\int_0^{\pi}\int_0^{2\pi} \psi_{1s}^2 r^2\,dr\,\sin\theta\,d\theta\,d\phi$$

We set this equal to 0.90 and solve for r'. The integral over θ and ϕ gives a factor of 4π; thus

$$0.90 = \frac{4}{a_0^3}\int_0^{r'} r^2 e^{-2r/a_0}\,dr$$

$\int_0^{r'} r^2\, e^{-2r/a_0}\,dr$ is integrated by parts to yield

$$-\left.\frac{a_0 r^2 e^{-2r/a_0}}{2}\right|_0^{r'} + a_0\left[-\left.\frac{a_0 r e^{-2r/a_0}}{2}\right|_0^{r'} + \frac{a_0}{2}\left.\left(-\frac{a_0 e^{-2r/a_0}}{2}\right)\right|_0^{r'}\right]$$

$$= -\frac{a_0(r')^2 e^{-2r'/a_0}}{2} - \frac{a_0^2 r'}{2}e^{-2r'/a_0} - \frac{a_0^3}{4}e^{-2r'/a_0} + \frac{a_0^3}{4}$$

Multiplying by $\frac{4}{a_0^3}$ and factoring e^{-2r'/a_0}

$$0.90 = \left[-2\left(\frac{r'}{a_0}\right)^2 - 2\left(\frac{r'}{a_0}\right) - 1\right]e^{-2r'/a_0} + 1 \ \text{ or } \ 2\left(\frac{r'}{a_0}\right)^2 + 2\left(\frac{r'}{a_0}\right) + 1 = 0.10e^{2r'/a_0}$$

It is easiest to solve this numerically. It is seen that $\boxed{r' = 2.66\,a_0}$ satisfies the above equation.

P13.17 Refer to Problems 13.7 and 13.15 and their solutions.

$$\mu_{\rm H} = \frac{m_{\rm e}m_{\rm p}}{m_{\rm e}+m_{\rm p}} \approx m_{\rm e} \quad [m_{\rm p} = \text{mass of proton}]$$

$$\mu_{\rm Ps} = \frac{m_{\rm e}m_{\rm pos}}{m_{\rm e}+m_{\rm pos}} = \frac{m_{\rm e}}{2} \quad [m_{\rm pos} = \text{mass of positron} = m_{\rm e}]$$

$$a_0 = r(n=1) = \frac{4\pi\hbar^2\varepsilon_0}{e^2 m_{\rm e}} \ \text{[15 and Problem 13.15]}$$

To obtain $a_{\rm Ps}$, the radius of the first Bohr orbit of positronium, we replace $m_{\rm e}$ with $\mu_{\rm Ps} = \frac{m_{\rm e}}{2}$; hence,

$$\boxed{a_{\rm Ps} = 2a_0} = \frac{8\pi\hbar^2\varepsilon_0}{e^2 m_{\rm e}}$$

The energy of the first Bohr orbit of positronium is

$$E_{1,\rm Ps} = -hcR_{\rm Ps} = -\frac{hc}{2}R_\infty \ \text{[Problem 13.7]}$$

Thus, $\boxed{E_{1,\rm Ps} = \tfrac{1}{2}E_{1,\rm H}}$

Question. What modifications are required in these relations when the finite mass of the hydrogen nucleus is recognized?

Solutions to additional problems

P13.18 To determine the atomic diameters, we assume a body-centred cubic crystal array (Fig. 13.1) that has two atoms per unit cell. In this array there is an atom at each corner of the cube and one in the middle. The cubic diagonal is two atomic diameters (d) in length and equals $3^{1/2}$ times the length of the unit cell edge.

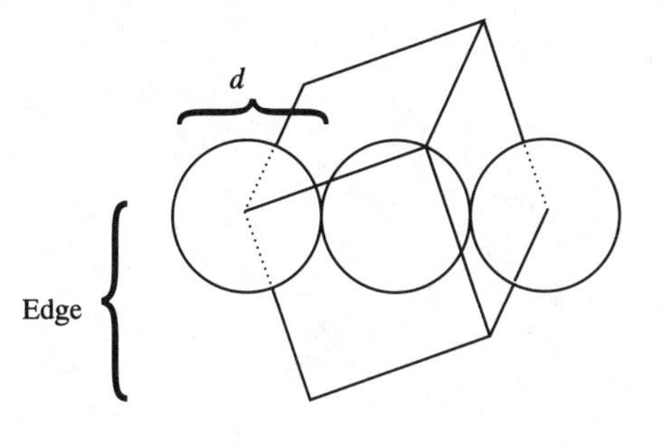

Figure 13.1

$$V_{\text{unit cell}} = \frac{(\text{edge})^3}{\text{unit cell}} = \left(\frac{V_{\text{m}}}{N_{\text{A}}}\right) \times \left(\frac{2\text{ atoms}}{\text{unit cell}}\right) = \left(\frac{M}{\rho N_{\text{A}}}\right) \times \left(\frac{2\text{ atoms}}{\text{unit cell}}\right)$$

$$\left(\frac{2d}{\sqrt{3}}\right)^3 \approx \frac{2M}{\rho N_{\text{A}}} \quad \text{or} \quad \boxed{d \simeq \frac{\sqrt{3}}{2}\left(\frac{2M}{\rho N_{\text{A}}}\right)^{1/3}} \quad \text{bcc array}$$

$$d_{\text{H}} = \frac{\sqrt{3}}{2}\left[\frac{2(1.008\,\text{g mol}^{-1}) \times \left(\frac{\text{m}^3}{10^6\text{cm}^3}\right)}{(0.071\,\text{g cm}^{-3}) \times (6.022 \times 10^{23}\,\text{mol}^{-1})}\right]^{1/3} = 3.14 \times 10^{-10}\,\text{m} = \boxed{0.314\,\text{nm}}$$

$$d_{\text{U}} = \frac{\sqrt{3}}{2}\left[\frac{2(238.0\,\text{g mol}^{-1}) \times \left(\frac{\text{m}^3}{10^6\text{cm}^3}\right)}{(18.95\,\text{g cm}^{-3}) \times (6.022 \times 10^{23}\,\text{mol}^{-1})}\right]^{1/3} = 3.01 \times 10^{-10}\,\text{m} = \boxed{0.301\,\text{nm}}$$

Therefore,

$$d_{\text{H}} \approx d_{\text{U}} \approx 0.3\,\text{nm}$$

It is reasonable to suppose that the radius of an atom can be approximated as equal to the average distance of a valence-shell electron from the nucleus. Substituting the effective nuclear charge (Z_{eff}) for the nuclear charge of a hydrogenic atom, the estimate becomes

$$\text{atomic radius} \approx \langle r \rangle_{n_{\text{valence shell}}} \approx \frac{(n_{\text{valence shell}})^2 a_0}{Z_{\text{eff}}}$$

This indicates that, when comparing atomic radii from left-to-right across a period of the periodic table, radii should decrease in proportion to (Z_{eff}). For a period of transition metals, the radii are approximately equal because the added electrons effectively screen the outermost electrons from the increased nuclear charge.

In a similar manner, the first ionization energy can be approximated with the proportionality

$$I \propto \frac{(Z_{\text{eff}})^2}{(n_{\text{valence}})^2}$$

This says that, in going across a period, I will vary more dramatically than does the atomic radius because of the $(Z_{\text{eff}})^2$ factor which provides greater variance than does $(Z_{\text{eff}})^{-1}$.

P13.19 E = kinetic energy + potential energy

$$E_{\text{classical}} = \frac{m_e v^2}{2} - \frac{Ze^2}{4\pi\varepsilon_0 a_0} \quad \text{where } v = \text{ electron speed}$$

Using the quantum energy for $n = 1$, $Z = 1$, and $\mu \approx m_e$

$$E_1 = -\frac{m_e e^4}{2(4\pi\varepsilon_0)^2\hbar^2} \; [13]$$
$$= -\frac{e^2}{8\pi\varepsilon_0 a_0} \; [15]$$

Therefore,

$$-\frac{e^2}{8\pi\varepsilon_0 a_0} = \frac{m_e v^2}{2} - \frac{e^2}{4\pi\varepsilon_0 a_0}$$

$$\text{or } v = \left(\frac{e^2}{4\pi\varepsilon_o m_e a_0}\right)^{1/2}$$

$$= \left(\frac{(1.602 \times 10^{-19}\,\text{C})^2}{(1.113 \times 10^{-10}\,\text{J}^{-1}\,\text{C}^2\,\text{m}^{-1}) \times (9.109 \times 10^{-31}\,\text{kg}) \times (5.29 \times 10^{-11}\,\text{m})}\right)^{1/2}$$

$$= 2.19 \times 10^6\,\text{m s}^{-1}\left(\frac{c}{3.00 \times 10^8\,\text{m s}^{-1}}\right)$$

$v = \boxed{0.00729c}$ The classical speed is 0.73 per cent of the speed of light.

$$\text{Electric field strength } \mathcal{E} = \frac{e}{4\pi\varepsilon_0 a_0^2} = \frac{1.602 \times 10^{-19}\,\text{C}}{(1.113 \times 10^{-10}\,\text{J}^{-1}\,\text{C}^2\,\text{m}^{-1}) \times (5.29 \times 10^{-11}\,\text{m})^2}$$
$$= \boxed{5.14 \times 10^{11}\,\text{V m}^{-1}}$$

We will also determine the magnetic field strength, $\mathcal{H}$, at the nucleus of the hydrogen atom. The magnetic field **B** at the nucleus is created by the current produced by the classical electron in the Bohr orbit for which $n = 1$ and $r = a_0$. Magnetic field strength at nucleus $\mathcal{H} = \frac{|\mathbf{B}|}{\mu_0} = \frac{\mathcal{B}}{\mu_0}$ where μ_0 is the vacuum permeability.

The electron in the classic Bohr orbit may be viewed as being electric current $I = e/\text{orbit period} = e/(2\pi a_0/v) = ev/(2\pi a_0)$ where v is the electron speed. The current creates the magnetic field and the Biot law relates the two (see Fig. 13.2)

$$\mathbf{dB} = \frac{\mu_0}{4\pi}\frac{I\,\mathbf{dl} \times \hat{\mathbf{r}}}{r^2}$$

Since we are interested in the magnetic field at the nucleus, note that Biot's law says that **B** will be perpendicular to the plane of the orbit (the vector cross-product says this). In terms of magnitudes alone the magnetic field is given by

$$\text{d}\mathcal{B} = \frac{\mu_0}{4\pi}\frac{I\,\text{d}l\sin\theta}{r^2} = \frac{\mu_0}{4\pi}\frac{I\,\text{d}l}{r^2} \quad (\theta = 90^\circ)$$

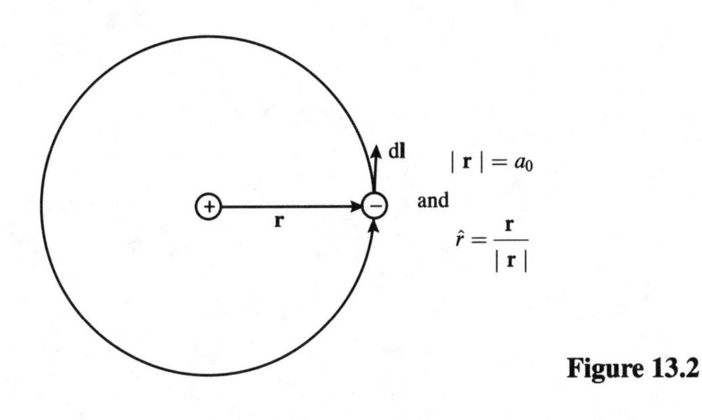

Figure 13.2

Since both I and r are constants for the orbit ($r = a_0$) this is easily integrated over one orbit.

$$\int_{\text{one orbit}} d\mathcal{B} = \frac{\mu_0 I}{4\pi a_0^2}\int_{\text{one orbit}} dl = \frac{\mu_0 I}{4\pi a_0^2}(2\pi a_0)$$

$$\mathcal{B} = \frac{\mu_0 I}{2a_0} = \frac{\mu_0 ev}{4\pi a_0^2} \quad \left[I = ev = \frac{ev}{2\pi a_0}\right]$$

$$= \frac{(4\pi \times 10^{-7}\,\text{J A}^{-2}\,\text{m}^{-1}) \times (1.602 \times 10^{-19}\,\text{C}) \times (2.19 \times 10^{6}\,\text{m s}^{-1})}{4\pi \times (5.29 \times 10^{-11}\,\text{m})^2} = \boxed{12.5\,\text{T}}$$

$$[\text{T} = \text{J A}^{-1}\,\text{m}^{-2}]$$

$$\text{magnetic field strength at nucleus} = \mathcal{H}_{\text{nucleus}} = \frac{\mathcal{B}}{\mu_0} = \frac{ev}{4\pi a_0^2}$$

$$= \frac{(1.6022 \times 10^{-19}\,\text{C}) \times (2.19 \times 10^{6}\,\text{m s}^{-1})}{4\pi(5.29 \times 10^{-11}\,\text{m})^2}$$

$$\mathcal{H}_{\text{nucleus}} = \boxed{9.98 \times 10^{6}\,\text{A m}^{-1}}$$

The electric field strength of the atom is about 5000 times larger than the dielectric strength of a mica capacitor. The magnetic field strength is the same order of magnitude as that in a superconducting magnet used in NMR experiments. (See Chapter 18.)

P13.21 **(a)** The emission line for the transitions $n+1 \to n$ is the first line of the Lyman series when $n = 1$, or the first line of the Balmer series when $n = 2$, or the first line of the Paschen series when $n = 3$, etc. Let $\tilde{\nu}$ be the wavenumber of this line.

Then,

$$\tilde{\nu} = \frac{|\Delta E|}{hc} = \frac{E_{n+1} - E_n}{hc}$$

For hydrogenic atoms $E_n = -\dfrac{C\mu Z^2}{n^2}$ where $C = \dfrac{e^4}{32\pi^2\varepsilon_0^2\hbar^2}$ [13]

$$\tilde{\nu} = \frac{C\mu Z^2}{hc}\left\{-\frac{1}{(n+1)^2} - \left(-\frac{1}{n^2}\right)\right\} = \frac{C\mu Z^2}{hc}\left\{\frac{2n+1}{n^2(n+1)^2}\right\}$$

$$\tilde{\nu}_{\text{He}} = \frac{4C\mu_{\text{He}}}{hc}\left\{\frac{2n+1}{n^2(n+1)^2}\right\}; \qquad \tilde{\nu}_{\text{H}} = \frac{C\mu_{\text{H}}}{hc}\left\{\frac{2n+1}{n^2(n+1)^2}\right\}$$

Therefore,

$$\gamma = \frac{\frac{1}{4}\tilde{\nu}_{\mathrm{He}} - \tilde{\nu}_{\mathrm{H}}}{\tilde{\nu}_{\mathrm{H}}} = \frac{\frac{1}{4}\left\{\frac{4C\mu_{\mathrm{He}}}{hc}\right\}\left\{\frac{2n+1}{n^2(n+1)^2}\right\} - \frac{C\mu_{\mathrm{H}}}{hc}\left\{\frac{2n+1}{n^2(n+1)^2}\right\}}{\frac{C\mu_{\mathrm{H}}}{hc}\left\{\frac{2n+1}{n^2(n+1)^2}\right\}}$$

$$\boxed{\gamma = \frac{\mu_{\mathrm{He}} - \mu_{\mathrm{H}}}{\mu_{\mathrm{H}}}}$$

(b) Using wavelength data

$$\tilde{\nu}_{\mathrm{H}} = \frac{1}{\lambda_{\mathrm{H}}} = \frac{1}{121.5664\,\mathrm{nm}}\left(\frac{10^9\,\mathrm{nm}}{10^2\,\mathrm{cm}}\right) = 82\,259.57\,\mathrm{cm}^{-1}$$

$$\tilde{\nu}_{\mathrm{He}^+} = \frac{1}{\lambda_{\mathrm{He}}} = \frac{1}{30.3779\,\mathrm{nm}}\left(\frac{10^9\,\mathrm{nm}}{10^2\,\mathrm{cm}}\right) = 329\,186.67\,\mathrm{cm}^{-1}$$

Therefore

$$\gamma_{2\to 1} = \frac{\frac{1}{4}(329\,186.67\,\mathrm{cm}^{-1}) - 82\,259.57\,\mathrm{cm}^{-1}}{82\,259.57\,\mathrm{cm}^{-1}} = 4.5098 \times 10^{-4}$$

$$\gamma = (\mu_{\mathrm{He}}/\mu_{\mathrm{H}}) - 1 \quad \text{or} \quad \frac{\mu_{\mathrm{H}}}{\mu_{\mathrm{He}}} = \frac{1}{1+\gamma} = 0.999\,549\,223$$

$$\frac{\mu_{\mathrm{H}}}{\mu_{\mathrm{He}}} = \left(\frac{m_{\mathrm{p}}m_{\mathrm{e}}}{m_{\mathrm{p}} + m_{\mathrm{e}}}\right)\left(\frac{4m_{\mathrm{p}} + m_{\mathrm{e}}}{4m_{\mathrm{p}}m_{\mathrm{e}}}\right) = \frac{4m_{\mathrm{p}} + m_{\mathrm{e}}}{4(m_{\mathrm{p}} + m_{\mathrm{e}})}$$

But, $m_{\mathrm{p}} \approx m_{\mathrm{H}} - m_{\mathrm{e}}$ so

$$\frac{\mu_{\mathrm{H}}}{\mu_{\mathrm{He}}} \approx \frac{4(m_{\mathrm{H}} - m_{\mathrm{e}}) + m_{\mathrm{e}}}{4m_{\mathrm{H}}} = \frac{4m_{\mathrm{H}} - 3m_{\mathrm{e}}}{4m_{\mathrm{H}}} = 1 - \frac{3}{4}\left(\frac{m_{\mathrm{e}}}{m_{\mathrm{H}}}\right)$$

Therefore,

$$\frac{m_{\mathrm{H}}}{m_{\mathrm{e}}} = \frac{3/4}{\left(1 - \frac{\mu_{\mathrm{H}}}{\mu_{\mathrm{He}}}\right)} = \frac{3}{4(1 - 0.999\,549\,223)} = \boxed{1663}$$

Using Rydberg constants $\mathcal{R}_{\mathrm{A}}$
Since $\mathcal{R} \propto \mu$,

$$\frac{\mu_{\mathrm{H}}}{\mu_{\mathrm{He}}} = \frac{\mathcal{R}_{\mathrm{H}}}{\mathcal{R}_{\mathrm{He}}} = \frac{109\,677.7\mathrm{cm}^{-1}}{109\,722.4\,\mathrm{cm}^{-1}} = 0.999\,592\,608$$

Therefore,

$$\frac{m_{\mathrm{H}}}{m_{\mathrm{e}}} = \frac{3}{4\left(1 - \frac{\mu_{\mathrm{H}}}{\mu_{\mathrm{He}}}\right)} = \frac{3}{4(1 - 0.999\,592\,608)} = \boxed{1841}$$

The value for the ratio calculated from the Rydberg constants is very close to the best modern accepted value of 1836. The ratio $m_{\mathrm{H}}/m_{\mathrm{e}}$ is very sensitive to errors in the (older) wavelength data and that may account for the discrepancy between the two values.

P13.22 $E_n = -\dfrac{hc\mathcal{R}_H}{n^2}$ where $\mathcal{R}_H = 109\,677\,\text{cm}^{-1}$ [13, 17]

For $n = 100$

$$\Delta E = E_{n+1} - E_n = -hc\mathcal{R}_H\left(\frac{1}{101^2} - \frac{1}{100^2}\right) = 1.97 \times 10^{-6}\, hc\mathcal{R}$$

$$\tilde{\nu} = \frac{\Delta E}{hc} = 1.97 \times 10^{-6}\mathcal{R} = \boxed{0.216\,\text{cm}^{-1}}$$

$$\langle r\rangle_{n,l} = n^2\left\{1 + \frac{1}{2}\left(1 - \frac{l(l+1)}{n^2}\right)\right\}\frac{a_0}{Z} \text{ [Example 13.2]}$$

$$\langle r\rangle_{100} \approx \frac{n^2 a_0}{Z} = 100^2 a_0 = 10^4 a_0 = \boxed{529\,\text{nm}}$$

$$I = E_\infty - E_n = -E_n = \frac{hc\mathcal{R}_H}{n^2}$$

$$I_{100} = 10^{-4} hc\mathcal{R}_H \quad \text{so} \quad \boxed{\frac{I_{100}}{hc} = 10.9677\,\text{cm}^{-1}}$$

At $\mathcal{T}$

$$\frac{k\mathcal{T}}{hc} = \frac{(1.38 \times 10^{-23}\,\text{J K}^{-1}) \times (298\,\text{K})\left(\frac{\text{m}}{10^2\,\text{cm}}\right)}{(6.63 \times 10^{-34}\,\text{J s}) \times (3.00 \times 10^8\,\text{m s}^{-1})} = 207\,\text{cm}^{-1}$$

so the thermal energy is readily available to ionize the state $n = 100$. Let v_{min} be the minimum speed required for collisional ionization. Then

$$\frac{1}{2}\frac{m_H v_{\text{min}}^2}{hc} = \frac{I_{100}}{hc}$$

$$v_{\text{min}} = \left[\frac{2hc}{m_H}\left(\frac{I_{100}}{hc}\right)\right]^{1/2}$$

$$= \sqrt{\frac{2(6.63 \times 10^{-34}\,\text{J s}) \times (3.00 \times 10^8\,\text{m s}^{-1}) \times (10.97\,\text{cm}^{-1})}{(1.008 \times 10^{-3}\,\text{kg mol}^{-1}) \times (6.022 \times 10^{23}\text{mol}^{-1})^{-1} \times \left(\frac{\text{m}}{10^2\,\text{cm}}\right)}}$$

$$\boxed{v_{\text{min}} = 511\,\text{m s}^{-1}} \quad \text{[very slow for an H atom]}$$

The radius of a Bohr orbit is $a_n \approx n^2 a_0$; hence the geometric cross-section $\pi a_n^2 \approx n^4 \pi a_0^2$. For $n = 1$ this is $8.8 \times 10^{-21}\,\text{m}^2$; for $n = 100$, it is $\boxed{8.8 \times 10^{-13}\,\text{m}^2}$. Thus a neutral H atom in its ground state is likely to pass right by the $n = 100$ Rydberg atom, leaving it undisturbed, since it is largely empty space.

The radial wavefunction for $n = 100$ will have 99 radial nodes and an extremely small amplitude above $\rho = \dfrac{r}{a_0} \approx 20$. For large values of n we expect the radial wavefunction [16] to be governed largely by the product of ρ^{n-1} and $e^{-\rho/2}$ and thus to approach a smoothly decreasing function of distance as the exponential will predominate over the power term.

14 Molecular structure

Solutions to exercises

E14.1 Refer to Fig. 14.31 of the text. Place two of the valence electrons in each orbital starting with the lowest energy orbital, until all valence electrons are used up. Apply Hund's rule to the filling of degenerate orbitals.

(a) Li_2 (2 electrons) $\boxed{1\sigma^2 \ b = 1}$ **(b)** Be_2 (4 electrons) $\boxed{1\sigma^2 2\sigma^{*2} \ b = 0}$

(c) C_2 (8 electrons) $\boxed{1\sigma^2 2\sigma^{*2} 1\pi^4 \ b = 2}$

E14.2 Note that CO and CN^- are isoelectronic with N_2 and that NO is isoelectronic with N_2^-, hence use Fig. 14.31 of the text.

(a) CO (10 electrons) $\boxed{1\sigma^2 2\sigma^{*2} 1\pi^4 3\sigma^2}$ $b = 3$

(b) NO (11 electrons) $\boxed{1\sigma^2 2\sigma^{*2} 1\pi^4 3\sigma^2 2\pi^{*1}}$ $b = 2.5$

(c) CN^- (10 electrons) $\boxed{1\sigma^2 2\sigma^{*2} 1\pi^4 3\sigma^2}$ $b = 3$

E14.3 B_2 (6 electrons): $1\sigma^2 2\sigma^{*2} 1\pi^2 \quad b = 1$

C_2 (8 electrons): $1\sigma^2 2\sigma^{*2} 1\pi^4 \quad b = 2$

The bond orders of B_2 and C_2 are respectively 1 and 2; so $\boxed{C_2}$ should have the greater bond dissociation enthalpy. The experimental values are approximately 4 eV and 6 eV respectively.

E14.4 We can use a version of Fig. 14.29 of the text, but with the energy levels of F lower than those of Xe as in Fig. 14.1.

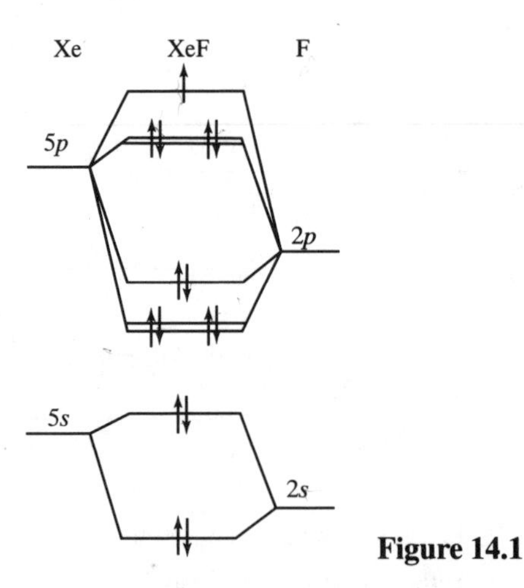

Figure 14.1

For XeF we insert 15 valence electrons. Since the bond order is increased when XeF^+ is formed from XeF (because an electron is removed from an antibonding orbital), XeF^+ will have a shorter bond length than XeF.

E14.5 Refer to Fig. 14.35 of the text.

(a) π^* is gerade, $\boxed{g}$

(b) g, u is inapplicable to a heteronuclear molecule, for it has no centre of inversion.

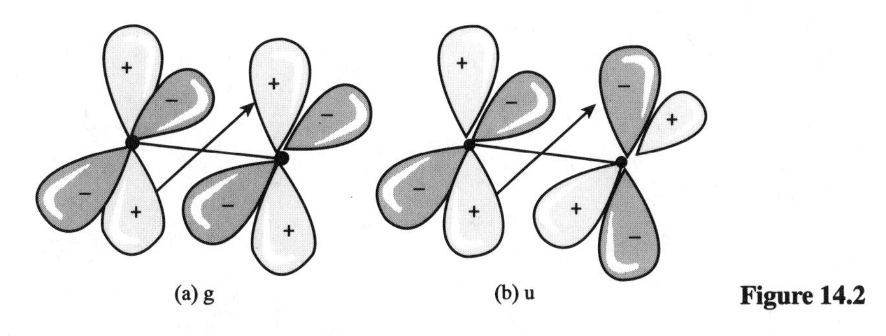

Figure 14.2

(c) A δ orbital (Fig. 14.2(a)) is gerade, $\boxed{\mathrm{g}}$ **(d)** A δ^* orbital (Fig. 14.2(b)) is ungerade, $\boxed{\mathrm{u}}$

E14.6 The left superscript is the value of $2S+1$, so $2S+1=2$ implies that $\boxed{S=\frac{1}{2}}$. The symbol Σ indicates that the total orbital angular momentum around the molecular axis is $\boxed{\text{zero}}$. The latter implies that the unpaired electron must be in a σ orbital. From Fig. 14.31 of the text, we predict the configuration of the ion to be $1\sigma^2 2\sigma^{*2} 1\pi^4 3\sigma^1$, which is in accord with the $^2\Sigma_g$ term symbol since 3σ is an even function (Fig. 14.35) and all lower energy orbitals are filled, leaving one unpaired electron; thus $S=\frac{1}{2}$.

E14.7 The electron configurations are used to determine the bond orders. Larger bond order corresponds qualitatively to shorter bond length.

The bond orders of NO and N_2 are 2.5 and 3, respectively (Exercises 14.1(**b**) and 14.2(**a**)); hence N_2 should have the shorter bond length. The experimental values are 115 pm and 110 pm, respectively.

E14.8 We need to demonstrate that $\int \psi^2\,\mathrm{d}\tau = 1$, where $\psi = \dfrac{s+\sqrt{2}p}{\sqrt{3}}$.

$$\int \psi^2\,\mathrm{d}\tau = \frac{1}{3}\int (s+\sqrt{2}p)^2\,\mathrm{d}\tau = \frac{1}{3}\int (s^2+2p^2+2\sqrt{2}sp)\,\mathrm{d}\tau = \frac{1}{3}(1+2+0) = 1$$

as $\int s^2\,\mathrm{d}\tau = 1$, $\int p^2\,\mathrm{d}\tau = 1$, and $\int sp\,\mathrm{d}\tau = 0$ [orthogonality]

E14.9 We evaluate $\int (\psi_A+\psi_B)(\psi_A-\psi_B)\,\mathrm{d}\tau$ and look at the result. If the integral is zero, then they are mutually orthogonal.

$$\int (\psi_A^2-\psi_B^2)\,\mathrm{d}\tau = 1-1 = \boxed{0}$$

Hence, they are orthogonal.

E14.10 **(a)** CO_2 is $\boxed{\text{linear}}$, either by VSEPR theory (two atoms attached to the central atom, no lone pairs on C), or by regarding the molecule as having a σ framework and π bonds between the C and O atoms.

(b) NO_2 is $\boxed{\text{nonlinear}}$, since it is isoelectronic with CO_2^-. The extra electron is 'half lone pair' and a bending agent. Alternatively, the extra electron is accommodated by the molecule bending so as to give the lone pair some s orbital character.

(c) NO_2^+ is $\boxed{\text{linear}}$, since it is isoelectronic with CO_2.

E14.11 The molecular orbitals of the fragments and the molecular orbitals that they form are shown in Fig. 14.3.

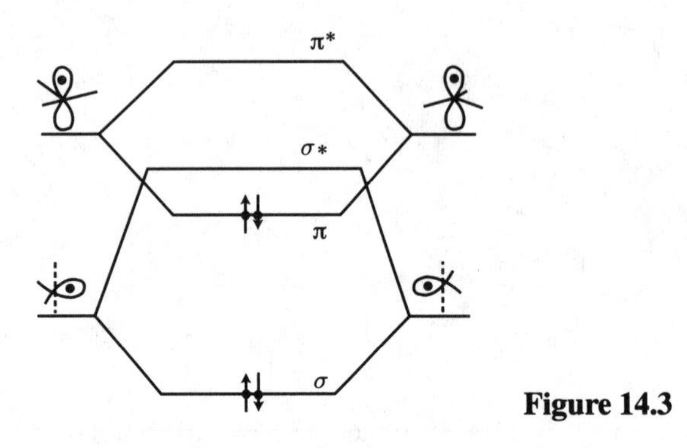

Figure 14.3

Comment. Note that the π-bonding orbital must be lower in energy than the σ-antibonding orbital for π-bonding to exist in ethene.

Question. Would the ethene molecule exist if the order of the energies of the π and σ^* orbitals were reversed?

E14.12 In setting up the secular determinant we use the approximation of Section 14.9

(a) $\begin{vmatrix} \alpha - E & \beta & 0 \\ \beta & \alpha - E & \beta \\ 0 & \beta & \alpha - E \end{vmatrix} = 0$ **(b)** $\begin{vmatrix} \alpha - E & \beta & \beta \\ \beta & \alpha - E & \beta \\ \beta & \beta & \alpha - E \end{vmatrix} = 0$

The atomic orbital basis is $1s_A$, $1s_B$, $1s_C$ in each case; in linear H_3 we ignore A, C overlap because A and C are not neighbouring atoms; in triangular H_3 we include it because they are.

Solutions to problems

Solutions to numerical problems

P14.2 Draw up the following table

R/a_0	0	1	2	3	4	5	6	7	8	9	10
S	1.000	0.858	0.586	0.349	0.189	0.097	0.047	0.022	0.010	0.005	0.002

The points are plotted in Fig. 14.4.

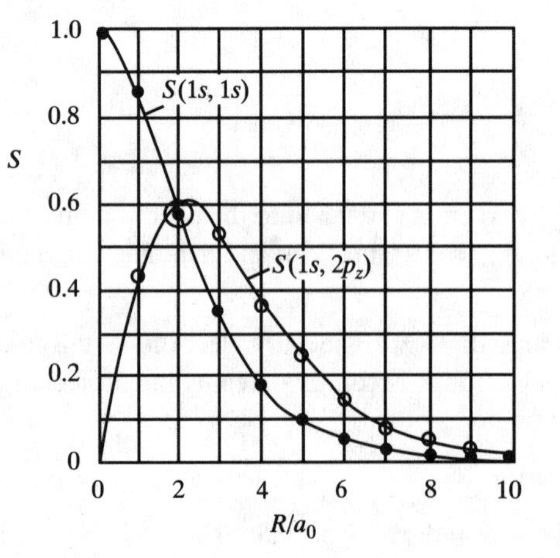

Figure 14.4

P14.3 The s orbital begins to spread into the region of negative amplitude of the p orbital. When their centres coincide, the region of positive overlap cancels the negative region. Draw up the following table

R/a_0	0	1	2	3	4	5	6	7	8	9	10
S	0	0.429	0.588	0.523	0.379	0.241	0.141	0.078	0.041	0.021	0.01

The points are plotted in Fig. 14.4. The maximum overlap occurs at $\boxed{R = 2.1a_0}$

P14.7 $E_H = E_1 = -hc\mathcal{R}_H$ [Section 13.2(b)]

Draw up the following table using the data in question and using

$$\frac{e^2}{4\pi\varepsilon_0 R} = \frac{e^2}{4\pi\varepsilon_0 a_0} \times \frac{a_0}{R} = \frac{e^2}{4\pi\varepsilon_0 \times (4\pi\varepsilon_0\hbar^2/m_e e^2)} \times \frac{a_0}{R}$$

$$= \frac{m_e e^4}{16\pi^2\varepsilon_0^2\hbar^2} \times \frac{a_0}{R} = E_h \times \frac{a_0}{R} \quad \left[E_h \equiv \frac{m_e e^4}{16\pi^2\varepsilon_0^2\hbar^2} = 2hc\mathcal{R}_H\right]$$

so that $\dfrac{\left(\frac{e^2}{4\pi\varepsilon_0 R}\right)}{E_h} = \dfrac{a_0}{R}$

R/a_0	0	1	2	3	4	∞
$\dfrac{\left(\frac{e^2}{4\pi\varepsilon_0 R}\right)}{E_h}$	∞	1	0.500	0.333	0.250	0
$(V_1 + V_2)/E_h$	2.000	1.465	0.843	0.529	0.342	0
$(E - E_H)/E_h$	∞	0.212	−0.031	−0.059	−0.038	0

The points are plotted in Fig. 14.5.

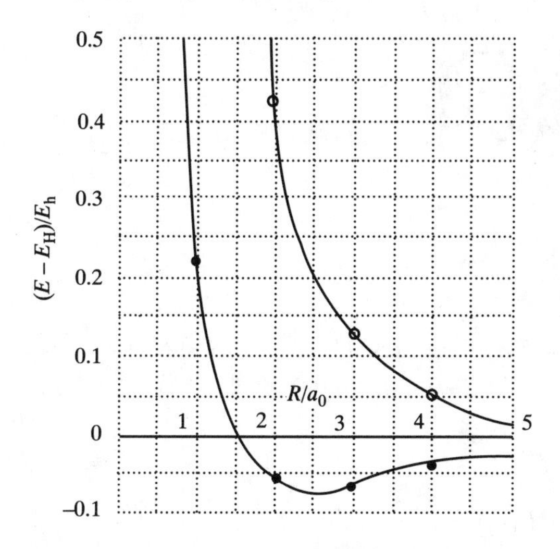

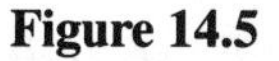
Figure 14.5

The minimum occurs at $R = 2.5a_0$, so $R = 130\,\text{pm}$. At the bond length

$$E - E_H = -0.07E_h = -1.91\,\text{eV}$$

Hence, the dissociation energy is predicted to be about $\boxed{1.9\,\text{eV}}$ and the equilibrium bond length about $\boxed{130\,\text{pm}}$

P14.8 We proceed as in Problem 14.7 and draw up the following table

R/a_0	0	1	2	3	4	∞
$\dfrac{\left(\frac{e^2}{4\pi\varepsilon_0 R}\right)}{E_h}$	∞	1	0.500	0.333	0.250	0
$(V_1 - V_2)/E_h$	0	−0.007	0.031	0.131	0.158	0
$(E - E_H)/E_h$	∞	1.049	0.425	0.132	0.055	0

The points are also plotted in Fig. 14.5. The contribution V_2 decreases rapidly because it depends on the overlap of the two orbitals.

P14.9 $E_n = \dfrac{n^2h^2}{8mL^2}$, $n = 1, 2, \ldots$ and $\psi_n = \left(\dfrac{2}{L}\right)^{1/2} \sin\left(\dfrac{n\pi x}{L}\right)$

Two electrons occupy each level (by the Pauli principle), and so butadiene (in which there are four π electrons) has two electrons in ψ_1 and two electrons in ψ_2

$$\psi_1 = \left(\frac{2}{L}\right)^{1/2} \sin\left(\frac{\pi x}{L}\right) \qquad \psi_2 = \left(\frac{2}{L}\right)^{1/2} \sin\left(\frac{2\pi x}{L}\right)$$

These orbitals are sketched in Fig. 14.6(a).

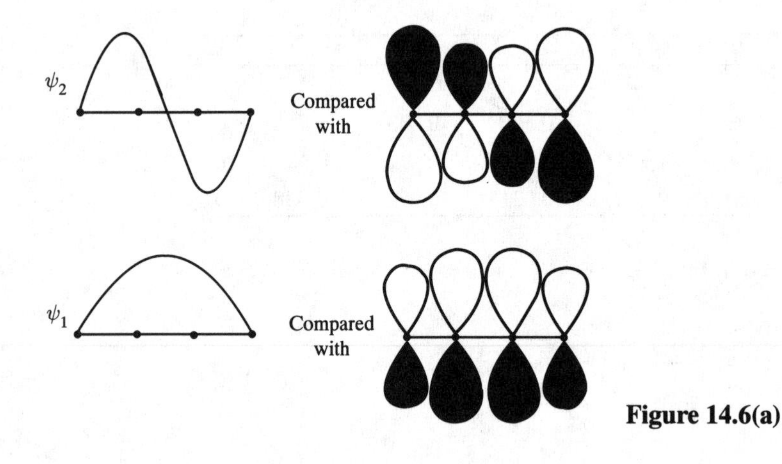

Figure 14.6(a)

The minimum excitation energy is

$$\Delta E = E_3 - E_2 = 5\left(\frac{h^2}{8m_eL^2}\right)$$

In $CH_2{=}CH{-}CH{=}CH{-}CH{=}CH{-}CH{=}CH_2$ there are eight π electrons to accommodate, so the HOMO will be ψ_4 and the LUMO ψ_5. From the particle-in-a-box solutions (Chapter 12)

$$\Delta E = E_5 - E_4 = (25 - 16)\frac{h^2}{8m_eL^2} = \frac{9h^2}{8m_eL^2}$$

$$= \frac{(9) \times (6.626 \times 10^{-34}\,\mathrm{J\,s})^2}{(8) \times (9.109 \times 10^{-31}\,\mathrm{kg}) \times (1.12 \times 10^{-9}\,\mathrm{m})^2} = 4.3 \times 10^{-19}\,\mathrm{J}$$

which corresponds to $\boxed{2.7\,\mathrm{eV}}$. The HOMO and LUMO are

$$\psi_n = \left(\frac{2}{L}\right)^{1/2} \sin\left(\frac{n\pi x}{L}\right)$$

with $n = 4, 5$ respectively; the two wavefunctions are sketched in Fig. 14.6(b).

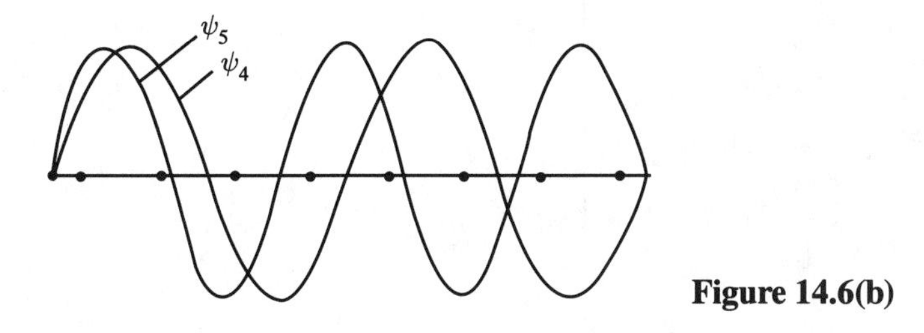

Figure 14.6(b)

Comment. It follows that

$$\lambda = \frac{hc}{\Delta E} = \frac{(6.626 \times 10^{-34}\,\text{J s}) \times (2.998 \times 10^{8}\,\text{m s}^{-1})}{4.3 \times 10^{-19}\,\text{J}} = 4.6 \times 10^{-7}\,\text{m}, \quad \text{or} \quad \boxed{460\,\text{nm}}$$

The wavelength 460 nm corresponds to blue light; so the molecule is likely to appear orange in white light (since blue is subtracted).

Solutions to theoretical problems

P14.11 We need to determine if $E_- + E_+ > 2E_H$.

$$\begin{aligned} E_- + E_+ &= -\frac{V_1 - V_2}{1 - S} + \frac{e^2}{4\pi\varepsilon_0 R} - \frac{V_1 + V_2}{1 + S} + \frac{e^2}{4\pi\varepsilon_0 R} + 2E_H \\ &= -\frac{\{(V_1 - V_2) \times (1 + S) + (1 - S) \times (V_1 + V_2)\}}{(1 - S) \times (1 + S)} + \frac{2e^2}{4\pi\varepsilon_0 R} + 2E_H \\ &= \frac{2(SV_2 - V_1)}{1 - S^2} + \frac{2e^2}{4\pi\varepsilon_0 R} + 2E_H \end{aligned}$$

The nuclear repulsion term is always positive, and always tends to raise the mean energy of the orbitals above E_H. The contribution of the first term is difficult to assess. Where $S \approx 0$, $SV_2 \approx 0$ and $V_1 \approx 0$, and its contribution is dominated by the nuclear repulsion term. Where $S \approx 1$, $SV_2 \approx V_1$ and once again the nuclear repulsion term is dominant. At intermediate values of S, the first term is negative, but of smaller magnitude than the nuclear repulsion term. Thus in all cases $E_- + E_+ > 2E_H$.

P14.13 The Walsh diagram is shown in Fig. 14.7.

The steep rise in energy of the $3a_1/1a_1''$ orbital arises from its loss of s character as the molecule becomes planar (120°).

(a) In NH_3 there are $5 + 3 = 8$ valence electrons to accommodate. This demands occupancy of the $3a_1/1a_2''$ orbital, and the lowest energy is obtained when the molecule is nonplanar with the configuration $2a_1^2 1e^4 3a_1^2$.

(b) CH_3^+ has only $4 + 3 - 1 = 6$ electrons. The $3a_1/1a_2''$ orbital is not occupied, and the lowest energy is attained with a planar molecule with configuration $2a_1'^2 1e'^4$.

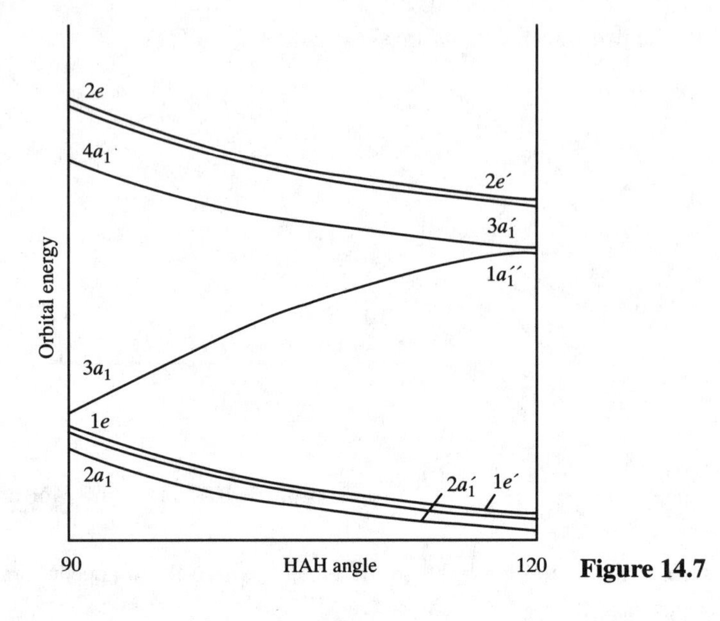

Figure 14.7

Solutions to additional problems

P14.16 The internuclear distance, $\langle r\rangle_n \approx n^2 a_0$, would be about twice the average distance ($\approx 1.06\times 10^6$ pm) of a hydrogenic electron from the nucleus when in the state $n = 100$. This distance is so large that each of the following estimates are applicable.

Resonance integral, $\beta \approx -\delta$ (where $\delta \approx 0$)
Overlap integral, $S \approx \varepsilon$ (where $\varepsilon \approx 0$)
Coulomb integral, $\alpha \approx E_{n=100}$ for atomic hydrogen

$$\text{Binding energy} = 2\{E_+ - E_{n=100}\} = 2\left\{\frac{\alpha+\beta}{1-S} - E_{n=100}\right\} = 2\{\alpha - E_{n=100}\} \approx 0$$

Vibrational force constant, $k \approx 0$ because of the weak binding energy. Rotational constant, $B = \dfrac{\hbar^2}{2hcI} = \dfrac{\hbar^2}{2hc\mu r_{AB}^2} \approx 0$ because r_{AB}^2 is so large.

The binding energy is so small that thermal energies would easily cause the Rydberg molecule to break apart. It is not likely to exist for much longer than a vibrational period.

P14.18 **(a)**
$$\begin{vmatrix} \alpha - E & \beta & \beta \\ \beta & \alpha - E & \beta \\ \beta & \beta & \alpha - E \end{vmatrix} = 0$$

$$(\alpha - E)\begin{vmatrix} \alpha - E & \beta \\ \beta & \alpha - E \end{vmatrix} - \beta\begin{vmatrix} \beta & \beta \\ \beta & \alpha - E \end{vmatrix} + \beta\begin{vmatrix} \beta & \alpha - E \\ \beta & \beta \end{vmatrix} = 0$$

$$(\alpha - E)\times\{(\alpha - E)^2 - \beta^2\} - \beta\{\beta(\alpha - E) - \beta^2\} + \beta\{\beta^2 - (\alpha - E)\beta\} = 0$$

$$(\alpha - E)\times\{(\alpha - E)^2 - \beta^2\} - 2\beta^2\{\alpha - E - \beta\} = 0$$

$$(\alpha - E)\times(\alpha - E - \beta)\times(\alpha - E + \beta) - 2\beta^2(\alpha - E - \beta) = 0$$

$$(\alpha - E - \beta) \times \{(\alpha - E) \times (\alpha - E + \beta) - 2\beta^2\} = 0$$
$$(\alpha - E - \beta) \times \{(\alpha - E) \times (\alpha - E + 2\beta) - \beta(\alpha - E) - 2\beta^2\} = 0$$
$$(\alpha - E - \beta) \times \{(\alpha - E) \times (\alpha - E + 2\beta) - \beta(\alpha - E + 2\beta)\} = 0$$
$$(\alpha - E - \beta) \times (\alpha - E + 2\beta) \times (\alpha - E - \beta) = 0$$

Therefore, the desired roots are $E = \boxed{\alpha - \beta,\ \alpha - \beta,\ \text{and}\ \alpha + 2\beta}$ The energy level diagram is shown in Fig. 14.8.

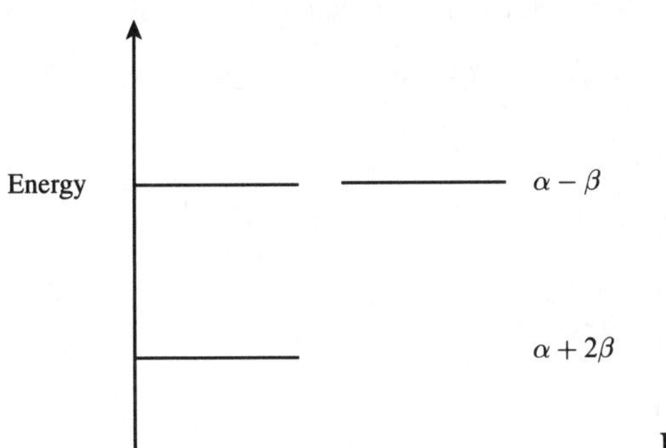

Figure 14.8

The binding energies are shown in the following table.

Species	Number of e^-	Binding energy
H_3^{2+}	1	$\alpha + 2\beta$
H_3^+	2	$2(\alpha + 2\beta) = 2\alpha + 4\beta$
H_3	3	$2(\alpha + 2\beta) + (\alpha - \beta) = 3\alpha + 3\beta$
H_3^-	4	$2(\alpha + 2\beta) + 2(\alpha - \beta) = 4\alpha + 2\beta$

(b)

$$H_3^+(g) \rightarrow 2H(g) + H^+(g) \qquad \Delta H_1 = 849\,\text{kJ mol}^{-1}$$
$$H^+(g) + H_2(g) \rightarrow H_3^+(g) \qquad \Delta H_2 = ?$$
$$H_2(g) \rightarrow 2H(g) \qquad \Delta H_3 = \{2(217.97) - 0\}\,\text{kJ mol}^{-1}$$

$$\Delta H_2 = \Delta H_3 - \Delta H_1 = \{2(217.97) - 849\}\,\text{kJ mol}^{-1}$$

$$\Delta H_2 = \boxed{-413\,\text{kJ mol}^{-1}}$$

This is only slightly less than the binding energy of H_2 ($435.94\,\text{kJ mol}^{-1}$)

(c) $2\alpha + 4\beta = -\Delta H_1 = -849\,\text{kJ mol}^{-1}$

$$\beta = \frac{-\Delta H_1 - 2\alpha}{4} \quad \text{where } \Delta H_1 = 849\,\text{kJ mol}^{-1}$$

Species	Binding energy
H_3^{2+}	$\alpha + 2\beta = -\dfrac{\Delta H_1}{2} = \boxed{-425\ \text{kJ mol}^{-1}}$
H_3^+	$2\alpha + 4\beta = -\Delta H_1 = \boxed{-849\ \text{kJ mol}^{-1}}$
H_3	$3\alpha + 3\beta = 3\left(\alpha - \dfrac{\Delta H_1 + 2\alpha}{4}\right) = 3\left(\dfrac{1}{2}\alpha - \dfrac{\Delta H_1}{4}\right) = \boxed{3(\alpha/2 - 212\ \text{kJ mol}^{-1})}$
H_3^-	$4\alpha + 2\beta = 4\alpha - \dfrac{\Delta H_1 + 2\alpha}{2} = 3\alpha - \dfrac{\Delta H_1}{2} = \boxed{3\alpha - 425\ \text{kJ mol}^{-1}}$

As α is a negative quantity, all four of these species are expected to be stable.

P14.19 The secular determinant for a cyclic species H_N^m has the form

$$\begin{array}{cccccccc} 1 & 2 & 3 & \cdots & \cdots & \cdots & N-1 & N \end{array}$$

$$\begin{vmatrix} x & 1 & 0 & \cdots & \cdots & \cdots & 0 & 1 \\ 1 & x & 1 & \cdots & \cdots & \cdots & 0 & 0 \\ 0 & 1 & x & 1 & \cdots & \cdots & 0 & 0 \\ 0 & 0 & 1 & x & 1 & \cdots & 0 & 0 \\ \vdots & \vdots & \vdots & \vdots & \vdots & \vdots & \vdots & \vdots \\ \vdots & \vdots & \vdots & \vdots & \vdots & \vdots & \vdots & 1 \\ 1 & 0 & 0 & 0 & 0 & \cdots & 1 & x \end{vmatrix} = 0$$

where $x = \dfrac{\alpha - E}{\beta}$ or $E = \alpha - \beta x$

Expanding the determinant, finding the roots of the polynomial, and solving for the total binding energy yields the following table. Note that $\alpha < 0$ and $\beta < 0$.

Species	Number of e^-	Permitted x	Total binding energy
H_4	4	$-2, 0, 0, 2$	$4\alpha + 4\beta$
H_5^+	4	$-2, \frac{1}{2}(1-\sqrt{5}), \frac{1}{2}(1-\sqrt{5}), \frac{1}{2}(1+\sqrt{5}), \frac{1}{2}(1+\sqrt{5})$	$4\alpha + (3+\sqrt{5})\beta$
H_5	5	$-2, \frac{1}{2}(1-\sqrt{5}), \frac{1}{2}(1-\sqrt{5}), \frac{1}{2}(1+\sqrt{5}), \frac{1}{2}(1+\sqrt{5})$	$5\alpha + \frac{1}{2}(5+3\sqrt{5})\beta$
H_5^-	6	$-2, \frac{1}{2}(1-\sqrt{5}), \frac{1}{2}(1-\sqrt{5}), \frac{1}{2}(1+\sqrt{5}), \frac{1}{2}(1+\sqrt{5})$	$6\alpha + (2+2\sqrt{5})\beta$
H_6	6	$-2, -1, -1, 1, 1, 2$	$6\alpha + 8\beta$
H_7^+	6	$-2, -1.248, -1.248, -1.248, 0.445, 0.445, 0.445$	$6\alpha + 8.992\beta$

$$H_4 \to 2H_2 \qquad \Delta_r U = 4(\alpha+\beta) - (4\alpha + 4\beta) = 0$$

$$H_5^+ \to H_2 + H_3^+ \qquad \Delta_r U = 2(\alpha+\beta) + (2\alpha + 4\beta) - (4\alpha + 5.236\beta) = 0.764\beta < 0$$

The above $\Delta_r U$ values indicate that H_4 and H_5^+ are unstable.

$$H_5^- \to H_2 + H_3^- \qquad \Delta_r U = 2(\alpha+\beta) + (4\alpha + 2\beta) - (6\alpha + 6.472\beta) = -2.472\beta > 0$$

$$H_6 \rightarrow 3H_2 \quad \Delta_r U = 6(\alpha + \beta) - (6\alpha + 8\beta)$$
$$= -2\beta > 0$$

$$H_7^+ \rightarrow 2H_2 + H_3^+ \quad \Delta_r U = 4(\alpha + \beta) + (2\alpha + 4\beta) - (6\alpha + 8.992\beta)$$
$$= -0.992\beta > 0$$

The $\Delta_r U$ values for H_5^-, H_6, and H_7^+ indicate that they are stable.

	Satisfies Hückel's $4n + 2$ low energy rule	
Species	Correct number of e^-	Stable
H_4, $4e^-$	No	No
H_5^+, $4e^-$	No	No
H_5^-, $6e^-$	Yes	Yes
H_6, $6e^-$	Yes	Yes
H_7^+, $6e^-$	Yes	Yes

Hückel's $4n + 2$ rule successfully predicts the stability of hydrogen rings.

15 Molecular symmetry

Solutions to exercises

E15.1 The elements, other than the identity E, are a C_3 axis and three vertical mirror planes σ_v. The symmetry axis passes through the C–Cl nuclei (Fig. 15.1). The mirror planes are defined by the three ClCH planes.

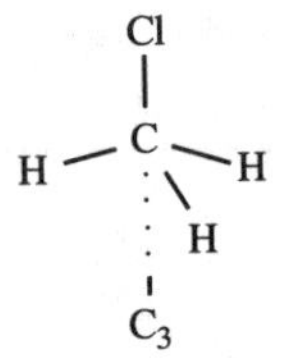

Figure 15.1

E15.2 Only molecules belonging to the groups C_n, C_{nv}, and C_s may be polar [Section 15.3(a)]; hence of the molecules listed only (a) pyridine and (b) nitroethane are polar.

E15.3 We refer to the character table for C_{4v} at the end of the *Data section*. We then use the procedure illustrated in Example 15.6, and draw up the following table of characters and their products

	E	$2C_4$	C_2	$2\sigma_v$	$2\sigma_d$	
$f_3 = p_z$	1	1	1	1	1	A_1
$f_2 = z$	1	1	1	1	1	A_1
$f_1 = p_x$	2	0	−2	0	0	E
$f_1 f_2 f_3$	2	0	−2	0	0	

The number of times that A_1 appears is 0 [since 2, 0, −2, 0, 0 are the characters of E itself], and so the integral is necessarily zero.

E15.4 We proceed as in Example 15.7, considering all three components of the electric dipole moment operator, $\boldsymbol{\mu}$.

Component of $\boldsymbol{\mu}$:		x			y			z	
A_1	1	1	1	1	1	1	1	1	1
$\Gamma(\mu)$	2	−1	0	2	−1	0	1	1	1
A_2	1	1	−1	1	1	−1	1	1	−1
$A_1\Gamma(\mu)A_2$	2	−1	0	2	−1	0	1	1	−1
		E			E			A_2	

Since A_1 is not present in any product, the transition dipole moment must be zero.

E15.5 We first determine how x and y individually transform under the operations of the group. Using these results we determine how the products xy transforms. The transform of xy is the product of the transforms of x and y.

Under each operation the functions transform as follows.

	E	C_2	C_4	σ_v	σ_d
x	x	$-x$	y	x	$-y$
y	y	$-y$	$-x$	$-y$	$-x$
xy	xy	xy	$-xy$	$-xy$	xy
χ	1	1	−1	−1	1

From the C_{4v} character table, we see that this set of characters belongs of B_2.

E15.6 In each molecule we must look for an improper rotation axis, perhaps in a disguised form ($S_1 = \sigma$, $S_2 = \mathrm{i}$) (Section 15.3). If present the molecule cannot be chiral. D_{2h} contains $\boxed{\mathrm{i}}$ and C_{3h} contains $\boxed{\sigma_h}$; therefore, molecules belonging to these point groups cannot be chiral. (Refer to Section 15.2.)

E15.7 In constructing the multiplication table it is convenient to consider the effects of the operation on an object or molecule belonging to that group. The molecule ion $[Pt(NH_2C_2H_4NH_2)_2]^{2+}$ belongs to D_2 (Fig. 15.2).

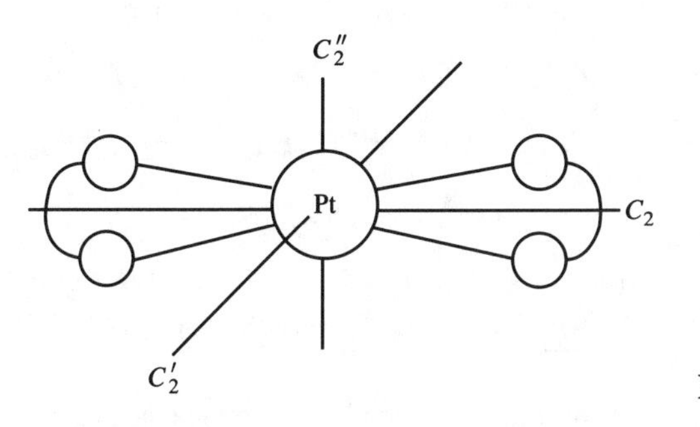

Figure 15.2

Alternatively, we may consider the effect of the operation on a point in space.

$$C_2(x, y, z) \rightarrow -x, \ -y, \ z$$
$$C_2'(x, y, z) \rightarrow x, \ -y, \ -z$$
$$C_2''(x, y, z) \rightarrow -x, \ y, \ -z$$

By inspection of the outsome of successive operations we can construct the following table

First operation :		E	C_2	C_2'	C_2''
	E	E	C_2	C_2'	C_2''
Second	C_2	C_2	E	C_2''	C_2'
operation	C_2'	C_2'	C_2''	E	C_2
	C_2''	C_2''	C_2'	C_2	E

E15.8 List the symmetry elements of the objects (the principal ones, not necessarily all the implied ones), then use the remarks in Section 15.2, and Fig. 15.3 below. Also refer to Figs 15.14 and 15.15 of the text.

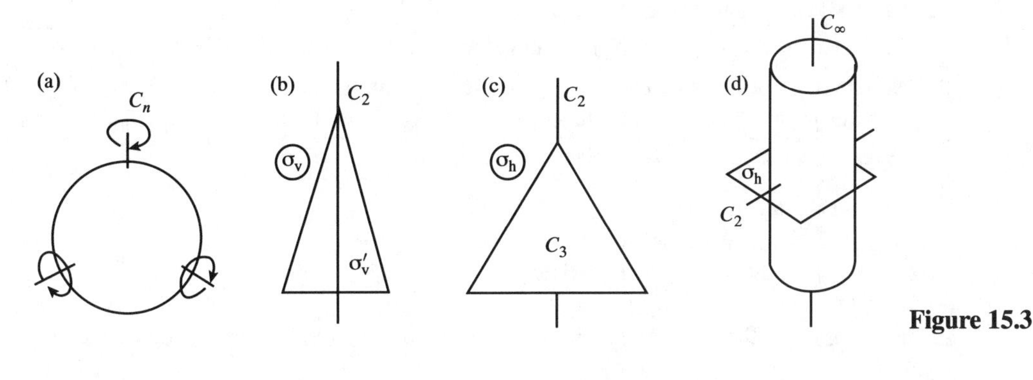

Figure 15.3

(a) Sphere: an infinite number of symmetry axes; therefore $\boxed{R_3}$

(b) Isosceles triangles: E, C_2, σ_v, and σ_v'; therefore $\boxed{C_{2v}}$

(c) Equilateral triangle: E, C_3, C_2, σ_h

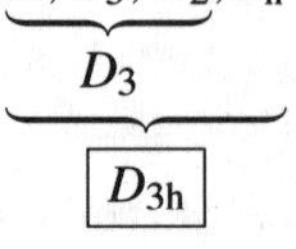

$\boxed{D_{3h}}$

(d) Cylinder: $E, C_\infty, C_2, \sigma_h$; therefore $\boxed{D_{\infty h}}$

E15.9 **(a)** NO_2: $E, C_2, \sigma_v, \sigma_v'$; $\boxed{C_{2v}}$ **(b)** N_2O: $E, C_\infty, C_2, \sigma_v$; $\boxed{C_{\infty v}}$

(c) $CHCl_3$: $E, C_3, 3\sigma_v$; $\boxed{C_{3v}}$ **(d)** $CH_2{=}CH_2$: $E, C_2, 2C_2', \sigma_h$: $\boxed{D_{2h}}$

(e) *cis*-CHBr=CHBr; $E, C_2, \sigma_v, \sigma_v'$; $\boxed{C_{2v}}$ **(f)** *trans*-CHCl=CHCl; E, C_2, σ_h, i; $\boxed{C_{2h}}$

E15.10 **(a)** *cis*-CHCl=CHCl; $E, C_2, \sigma_v, \sigma_v'$; $\boxed{C_{2v}}$ **(b)** *trans*-CHCl=CHCl; E, C_2, σ_h, i; $\boxed{C_{2h}}$

E15.11 **(a)** Only molecules belonging to the point groups C_n, C_{nv}, and C_s may be polar (Section 15.3(a)); hence of the molecules listed $\boxed{NO_2}$ (C_{2v}), $\boxed{N_2O}$ ($C_{\infty v}$), $\boxed{CHCl_3}$ (C_{3v}) and $\boxed{\textit{cis}\text{-CHBr=CHBr}}$ (C_{2v}) are polar.

(b) All the molecules listed possess an axis of improper rotation, S_n. (See the character tables at the end of the *Data section* which list the symmetry elements for the point groups involved. Note that a centre of inversion, i, is equivalent to S_2 and a mirror plane is equivalent to S_1.) Therefore, (Section 15.3(b)) $\boxed{\text{none}}$ of these molecules is chiral.

E15.12 Recall $p_x \propto x$, $p_y \propto y$, $p_z \propto z$, $d_{xy} \propto xy$, $d_{xz} \propto xz$, $d_{yz} \propto yz$, $d_{z^2} \propto z^2$, $d_{x^2-y^2} \propto x^2 - y^2$ (Section 13.2).

Now refer to the C_{2v} character table. The s orbital spans A_1 and the p orbitals of the central N atom span $A_1(p_z)$, $B_1(p_x)$, and $B_2(p_y)$. Therefore, $\boxed{\text{no orbitals}}$ span A_2, and hence $p_x(A) - p_x(B)$ is a nonbonding combination. If d orbitals are available, as they are in S of the SO_2 molecule, we could form a molecular orbital with $\boxed{d_{xy}}$, which is a basis for A_2.

E15.13 The electric dipole moment operator transforms as $x(B_1)$, $y(B_2)$, and $z(A_1)$ (C_{2v} character table). Transitions are allowed if $\int \psi_f^* \mu \psi_i \, d\tau$ is nonzero (Example 15.7), and hence are forbidden unless $\Gamma_f \times \Gamma(\mu) \times \Gamma_i$ contains A_1. Since $\Gamma_i = A_1$, this requires $\Gamma_f \times \Gamma(\mu) = A_1$. Since $B_1 \times B_1 = A_1$ and $B_2 \times B_2 = A_1$, and $A_1 \times A_1 = A_1$, x-polarized light may cause a transition to a B_1 term, y-polarized light to a B_2 term, and z-polarized light to an A_1 term.

E15.14 **(a)** The point group of benzene is D_{6h}. In D_{6h} μ spans $E_{1u}(x, y)$ and $A_{2u}(z)$, and the ground term is A_{1g}. Then, using $A_{2u} \times A_{1g} = A_{2u}$, $E_{1u} \times A_{1g} = E_{1u}$, $A_{2u} \times A_{2u} = A_{1g}$, and $E_{1u} \times E_{1u} = A_{1g} + A_{2g} + E_{2g}$, we conclude that the upper term is $\boxed{\text{either } E_{1u} \text{ or } A_{2u}}$

(b) Naphthalene belongs to D_{2h}. In D_{2h} itself, the components span $B_{3u}(x)$, $B_{2u}(y)$, and $B_{1u}(z)$ and the ground term is A_g. Hence, since $A_g \times \Gamma = \Gamma$ in this group, the upper terms are $\boxed{B_{3u}\ (x\text{-polarized})}$, $\boxed{B_{2u}\ (y\text{-polarized})}$, and $\boxed{B_{1u}\ (z\text{-polarized})}$

E15.15 We consider the integral

$$I = \int_{-a}^{a} f_1 f_2 \, d\theta = \int_{-a}^{a} \sin\theta \cos\theta \, d\theta$$

and hence draw up the following table for the effect of operations in the group C_s (see Fig. 15.4)

	E	σ_h
$f_1 = \sin\theta$	$\sin\theta$	$-\sin\theta$
$f_2 = \cos\theta$	$\cos\theta$	$\cos\theta$

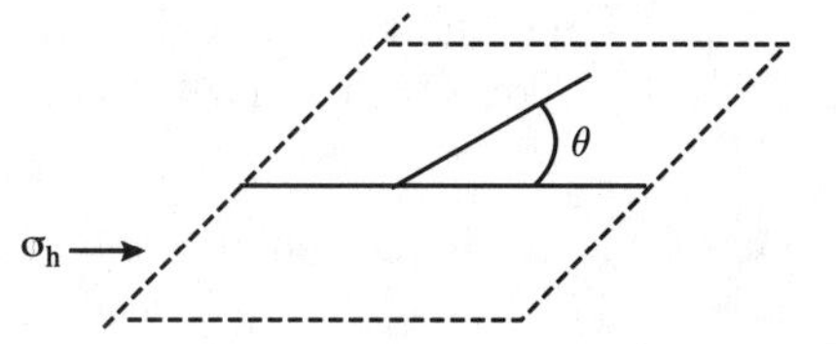

Figure 15.4

In terms of characters

	E	σ_h	
f_1	1	−1	A″
f_2	1	1	A′
f_1f_2	1	−1	A″

Solutions to problems

P15.1 **(a)** Staggered CH_3CH_3: $E, C_3, C_2, 3\sigma_d$; $\boxed{D_{3d}}$ [see Fig. 15.6b of the text]

(b) Chair C_6H_{12}: $E, C_3, C_2, 3\sigma_d$; $\boxed{D_{3d}}$

Boat C_6H_{12}: $E, C_2, \sigma_v, \sigma_v'$; $\boxed{C_{2v}}$

(c) B_2H_6: $E, C_2, 2C_2', \sigma_h$; $\boxed{D_{2h}}$

(d) $[Co(en)_3]^{3+}$: $E, 2C_3, 3C_2$; $\boxed{D_3}$

(e) Crown S_8: $E, C_4, C_2, 4C_2', 4\sigma_d, 2S_8$; $\boxed{D_{4d}}$

Only boat C_6H_{12} may be polar, since all the others are D point groups. Only $[Co(en)_3]^{3+}$ belongs to a group without an improper rotation axis ($S_1 = \sigma$), and hence is chiral.

P15.2 The operations are illustrated in Fig. 15.5. Note that $R^2 = E$ for all the operations of the groups, that $ER = RE = R$ always, and that $RR' = R'R$ for this group. Since $C_2\sigma_h = i$, $\sigma_h i = C_2$, and $iC_2 = \sigma_h$ we can draw up the following group multiplication table

	E	C_2	σ_h	i
E	E	C_2	σ_h	i
C_2	C_2	E	i	σ_h
σ_h	σ_h	i	E	C_2
i	i	σ_h	C_2	E

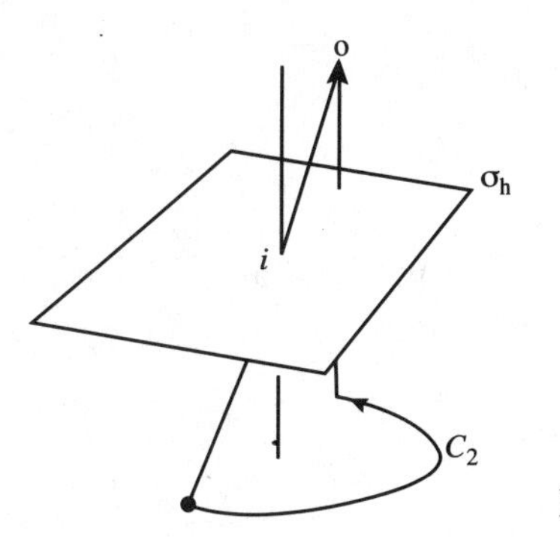

Figure 15.5

The $\boxed{trans\text{-}CHCl{=}CHCl}$ molecule belongs to the group C_{2h}.

Comment. Note that the multiplication table for C_{2h} can be put into a one-to-one correspondence with the multiplication table of D_2 obtained in Exercise 15.7. We say that they both belong to the same abstract group and are isomorphous.

Question. Can you find another abstract group of order 4 and obtain its multiplication table? There is only one other.

P15.4 Refer to Fig. 15.3 of the text. Place orbitals h_1 and h_2 on the H atoms and s, p_x, p_y, and p_z on the O atom. The z-axis is the C_2 axis; x lies perpendicular to σ_v', y lies perpendicular to σ_v. Then draw up the following table of the effect of the operations on the basis

	E	C_2	σ_v	σ_v'
h_1	h_1	h_2	h_2	h_1
h_2	h_2	h_1	h_1	h_2
s	s	s	s	s
p_x	p_x	$-p_x$	p_x	$-p_x$
p_y	p_y	$-p_y$	$-p_y$	p_y
p_z	p_z	p_z	p_z	p_z

Express the coloumns headed by each operations R in the form

$$(\text{new}) = D(R)(\text{original})$$

where $D(R)$ is the 6×6 representative of the operation R. We use the rules of matrix multiplication set out in *Justification* 15.1

(i) E: $(h_1, h_2, s, p_x, p_y, p_z) \leftarrow (h_1, h_2, s, p_x, p_y, p_z)$ is reproduced by the 6×6 unit matrix

(ii) C_2: $(h_2, h_1, s, -p_x, -p_y, p_z) \leftarrow (h_1, h_2, s, p_x, p_y, p_z)$ is reproduced by

$$D(C_2) = \begin{bmatrix} 0 & 1 & 0 & 0 & 0 & 0 \\ 1 & 0 & 0 & 0 & 0 & 0 \\ 0 & 0 & 1 & 0 & 0 & 0 \\ 0 & 0 & 0 & -1 & 0 & 0 \\ 0 & 0 & 0 & 0 & -1 & 0 \\ 0 & 0 & 0 & 0 & 0 & 1 \end{bmatrix}$$

(iii) σ_v: $(h_2, h_1, s, p_x, -p_y, p_z) \leftarrow (h_1, h_2, s, p_x, p_y, p_z)$ is reproduced by

$$D(\sigma_v) = \begin{bmatrix} 0 & 1 & 0 & 0 & 0 & 0 \\ 1 & 0 & 0 & 0 & 0 & 0 \\ 0 & 0 & 1 & 0 & 0 & 0 \\ 0 & 0 & 0 & 1 & 0 & 0 \\ 0 & 0 & 0 & 0 & -1 & 0 \\ 0 & 0 & 0 & 0 & 0 & 1 \end{bmatrix}$$

(iv) σ_v': $(h_1, h_2, s, -p_x, p_y, p_z) \leftarrow (h_1, h_2, s, p_x, p_y, p_z)$ is reproduced by

$$D(\sigma_v') = \begin{bmatrix} 1 & 0 & 0 & 0 & 0 & 0 \\ 0 & 1 & 0 & 0 & 0 & 0 \\ 0 & 0 & 1 & 0 & 0 & 0 \\ 0 & 0 & 0 & -1 & 0 & 0 \\ 0 & 0 & 0 & 0 & 1 & 0 \\ 0 & 0 & 0 & 0 & 0 & 1 \end{bmatrix}$$

(a) To confirm the correct representation of $C_2\sigma_v = \sigma_v'$ we write

$$D(C_2)D(\sigma_v) = \begin{bmatrix} 0 & 1 & 0 & 0 & 0 & 0 \\ 1 & 0 & 0 & 0 & 0 & 0 \\ 0 & 0 & 1 & 0 & 0 & 0 \\ 0 & 0 & 0 & -1 & 0 & 0 \\ 0 & 0 & 0 & 0 & -1 & 0 \\ 0 & 0 & 0 & 0 & 0 & 1 \end{bmatrix} \begin{bmatrix} 0 & 1 & 0 & 0 & 0 & 0 \\ 1 & 0 & 0 & 0 & 0 & 0 \\ 0 & 0 & 1 & 0 & 0 & 0 \\ 0 & 0 & 0 & 1 & 0 & 0 \\ 0 & 0 & 0 & 0 & -1 & 0 \\ 0 & 0 & 0 & 0 & 0 & 1 \end{bmatrix}$$

$$= \begin{bmatrix} 1 & 0 & 0 & 0 & 0 & 0 \\ 0 & 1 & 0 & 0 & 0 & 0 \\ 0 & 0 & 1 & 0 & 0 & 0 \\ 0 & 0 & 0 & -1 & 0 & 0 \\ 0 & 0 & 0 & 0 & 1 & 0 \\ 0 & 0 & 0 & 0 & 0 & 1 \end{bmatrix} = D(\sigma_v')$$

(b) Similarly, to confirm the correct representation of $\sigma_v\sigma_v' = C_2$, we write

$$\begin{bmatrix} 0 & 1 & 0 & 0 & 0 & 0 \\ 1 & 0 & 0 & 0 & 0 & 0 \\ 0 & 0 & 1 & 0 & 0 & 0 \\ 0 & 0 & 0 & 1 & 0 & 0 \\ 0 & 0 & 0 & 0 & -1 & 0 \\ 0 & 0 & 0 & 0 & 0 & 1 \end{bmatrix} \begin{bmatrix} 1 & 0 & 0 & 0 & 0 & 0 \\ 0 & 1 & 0 & 0 & 0 & 0 \\ 0 & 0 & 1 & 0 & 0 & 0 \\ 0 & 0 & 0 & -1 & 0 & 0 \\ 0 & 0 & 0 & 0 & 1 & 0 \\ 0 & 0 & 0 & 0 & 0 & 1 \end{bmatrix}$$

$$= \begin{bmatrix} 0 & 1 & 0 & 0 & 0 & 0 \\ 1 & 0 & 0 & 0 & 0 & 0 \\ 0 & 0 & 1 & 0 & 0 & 0 \\ 0 & 0 & 0 & -1 & 0 & 0 \\ 0 & 0 & 0 & 0 & -1 & 0 \\ 0 & 0 & 0 & 0 & 0 & 1 \end{bmatrix} = D(C_2)$$

(a) The characters of the representatives are the sums of their diagonal elements:

E	C_2	σ_v	σ_v'
6	0	2	4

(b) The characters are not those of any one irreducible representation, so the representation is reducible.

(c) The sum of the characters of the specified sum is

	E	C_2	σ_v	σ_v'
$3A_1$	3	3	3	3
B_1	1	−1	1	−1
$2B_2$	2	−2	−2	2
$3A_1 + B_1 + 2B_2$	6	0	2	4

which is the same as the original. Therefore the representation is $3A_1 + B_1 + 2B_2$.

P15.5 We examine how the operations of the C_{3v} group affect $l_z = xp_y - yp_x$ when applied to it. The transformation of $x,\ y$, and z, and by analogy p_x, p_y, and p_z, are as follows (see Fig. 15.6)

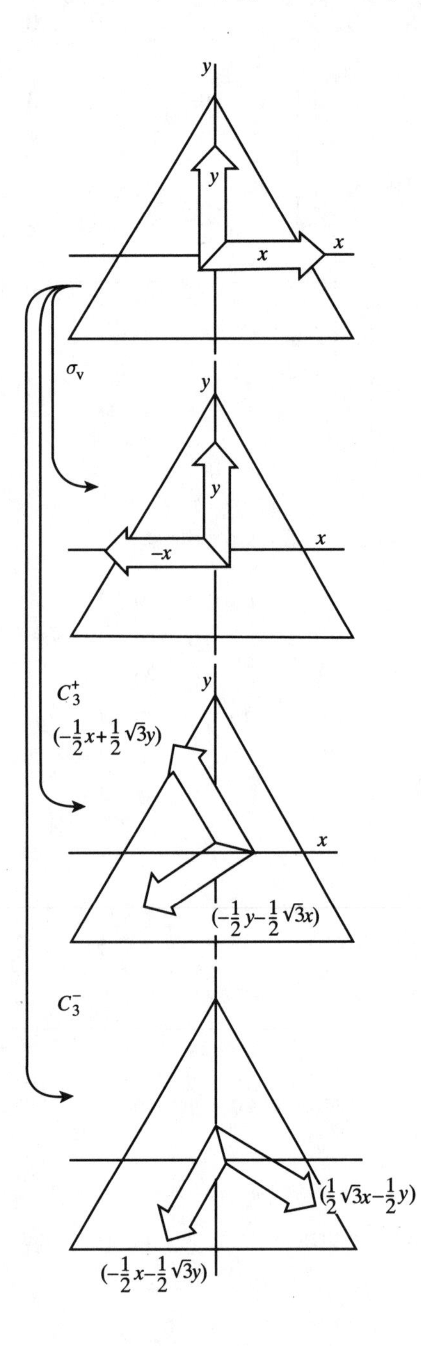

Figure 15.6

$$E(x, y, z) \to (x, y, z)$$
$$\sigma_v(x, y, z) \to (-x, y, z)$$
$$\sigma_v'(x, y, z) \to (x, -y, z)$$
$$\sigma_v''(x, y, z) \to (x, y, -z)$$

$$C_3^+(x, y, z) \to \left(-\tfrac{1}{2}x + \tfrac{1}{2}\sqrt{3}y, -\tfrac{1}{2}\sqrt{3}x - \tfrac{1}{2}y, z\right)$$
$$C_3^-(x, y, z) \to \left(-\tfrac{1}{2}x - \tfrac{1}{2}\sqrt{3}y, \tfrac{1}{2}\sqrt{3}x - \tfrac{1}{2}y, z\right)$$

The characters of all σ operations are the same, as are those of both C_3 operations (see the C_{3v} character table); hence we need consider only one operation in each class.

$$El_z = xp_y - yp_x = l_z$$
$$\sigma_v l_z = -xp_y + yp_x = -l_z \quad [(x, y, z) \to (-x, y, z)]$$
$$C_3^+ l_z = \left(-\tfrac{1}{2}x + \tfrac{1}{2}\sqrt{3}y\right) \times \left(-\tfrac{1}{2}\sqrt{3}p_x - \tfrac{1}{2}p_y\right) - \left(-\tfrac{1}{2}\sqrt{3}x - \tfrac{1}{2}y\right) \times \left(-\tfrac{1}{2}p_x + \tfrac{1}{2}\sqrt{3}p_y\right)$$
$$\left[(x, y, z) \to \left(-\tfrac{1}{2}x + \tfrac{1}{2}\sqrt{3}y, -\tfrac{1}{2}\sqrt{3}x - \tfrac{1}{2}y, z\right)\right]$$
$$= \tfrac{1}{4}(\sqrt{3}xp_x + xp_y - 3yp_x - \sqrt{3}yp_y - \sqrt{3}xp_x + 3xp_y - yp_x + \sqrt{3}yp_y)$$
$$= xp_y - yp_x = l_z$$

The representatives of E, σ_v, and C_3^+ are therefore all one-dimensional matrices with characters $1, -1, 1$ respectively. It follows that l_z is a basis for A_2 (see the C_{3v} character table).

P15.7 The multiplication table is

	1	σ_x	σ_y	σ_z
1	1	σ_x	σ_y	σ_z
σ_x	σ_x	1	$i\sigma_z$	$-i\sigma_y$
σ_y	σ_y	$-i\sigma_z$	1	$i\sigma_x$
σ_z	σ_z	$i\sigma_y$	$-i\sigma_x$	1

The matrices **do not form a group** since the products $i\sigma_z, i\sigma_y, i\sigma_x$ and their negatives are not among the four given matrices.

P15.10 **(a)** C_{2v}. The functions x^2, y^2, and z^2 are invariant under all operations of the group, and so $z(5z^2 - 3r^2)$ transforms as $z(A_1)$, $y(5y^2 - 3r^2)$ as $y(B_2)$, $x(5x^2 - 3r^2)$ as $x(B_1)$, and likewise for $z(x^2 - y^2)$, $y(x^2 - z^2)$, and $x(z^2 - y^2)$. The function xyz transforms as $B_1 \times B_2 \times A_1 = A_2$. Therefore, in group C_{2v}, $f \to$ $\boxed{2A_1 + A_2 + 2B_1 + 2B_2}$

(b) C_{3v}. In C_{3v}, z transforms as A_1, and hence so does z^3. From the C_{3v} character table, (x^2-y^2, xy) is a basis for E, and so $(xyz, z(x^2 - y^2))$ is a basis for $A_1 \times E = E$. The linear combinations $y(5y^2-3r^2)+5y(x^2-z^2) \propto y$ and $x(5x^2-3r^2)+5x(z^2-y^2) \propto x$ are a basis for E. Likewise, the two linear combinations orthogonal to these are another basis for E. Hence, in the group C_{3v}, $f \to$ $\boxed{A_1 + 3E}$

(c) T_d. Make the inspired guess that the f orbitals are a basis of dimension $3+3+1$, suggesting the decomposition $T + T + A$. Is the A representation A_1 or A_2? We see from the character table that the effect of S_4 discriminates between A_1 and A_2. Under S_4, $x \to y, y \to -x, z \to -z$, and so $xyz \to xyz$. The character is $\chi = 1$, and so xyz spans A_1. Likewise, $(x^3, y^3, z^3) \to (y^3, -x^3, -z^3)$ and $\chi = 0 + 0 - 1 = -1$. Hence, this trio spans T_2. Finally,

$$\{x(z^2 - y^2), y(z^2 - x^2), z(x^2 - y^2)\} \to \{y(z^2 - x^2), -x(z^2 - y^2), -z(y^2 - x^2)\}$$

resulting in $\chi = 1$, indicating T_1. Therefore, in T_d, $f \to$ $\boxed{A_1 + T_1 + T_2}$

(d) O_h. Anticipate an $A + T + T$ decomposition as in the other cubic group. Since x, y, and z all have odd parity, all the irreducible representatives will be u. Under S_4, $xyz \to xyz$ (as in (c)), and so the representation is A_{2u} (see the character table). Under S_4, $(x^3, y^3, z^3) \to (y^3, -x^3, -z^3)$, as before, and $\chi = -1$, indicating T_{1u}. In the same way, the remaining three functions span T_{2u}. Hence, in O_h, $f \rightarrow$ $\boxed{A_{2u} + T_{1u} + T_{2u}}$

(The shapes of the orbitals are shown in *Inorganic Chemistry*, D.F. Shriver, P.W. Atkins, and C.H. Langford, Oxford University Press and W.H. Freeman & Co (1994).)

The f orbitals will cluster into sets according to their irreducible representations. Thus **(a)** $f \rightarrow A_1 + T_1 + T_2$ in T_d symmetry, and there is one nondegenerate orbital and two sets of triply degenerate orbitals. **(b)** $f \rightarrow A_{2u} + T_{1u} + T_{2u}$, and the pattern of splitting (but not the order of energies) is the same.

P15.12 Refer to Fig. 15.7, and draw up the following table

	π_1	π_2	π_3	π_4	π_5	π_6	π_7	π_8	π_9	π_{10}	χ
E	π_1	π_2	π_3	π_4	π_5	π_6	π_7	π_8	π_9	π_{10}	10
C_2	π_5	π_6	π_7	π_8	π_1	π_2	π_3	π_4	π_{10}	π_9	0
σ_v	π_4	π_3	π_2	π_1	π_8	π_7	π_6	π_5	π_{10}	π_9	0
σ_v'	π_8	π_7	π_6	π_5	π_4	π_3	π_2	π_1	π_9	π_{10}	2

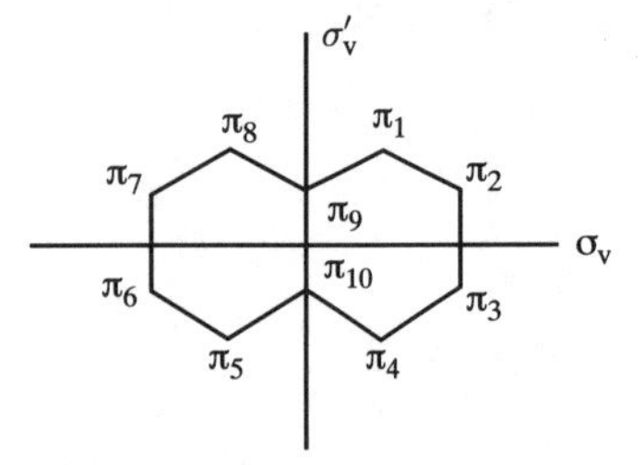

Figure 15.7

(χ is obtained from the number of unchanged orbitals.) The character set (10, 0, 0, 2) decomposes into $\boxed{3A_1 + 2A_2 + 2B_1 + 3B_2}$. Now form symmetry-adapted linear combinations as explained in Section 15.5

$\pi(A_1) = \pi_1 + \pi_4 + \pi_5 + \pi_8$ [from column 1]
$\pi(A_1) = \pi_2 + \pi_3 + \pi_6 + \pi_7$ [column 2]
$\pi(A_1) = \pi_9 + \pi_{10}$ [column 9]
$\pi(A_2) = \pi_1 + \pi_5 - \pi_4 - \pi_8$ [column 1]
$\pi(A_2) = \pi_2 + \pi_6 - \pi_3 - \pi_7$ [column 2]
(The other columns yield the same orbitals.)

$\pi(B_1) = \pi_1 - \pi_5 + \pi_4 - \pi_8$ [column 1]
$\pi(B_1) = \pi_2 - \pi_6 + \pi_3 - \pi_7$ [column 2]
$\pi(B_2) = \pi_1 - \pi_5 - \pi_4 + \pi_8$ [column 1]
$\pi(B_2) = \pi_2 - \pi_6 - \pi_3 + \pi_7$ [column 2]
$\pi(B_2) = \pi_9 - \pi_{10}$ [column 9]

Answers to additional problems

P15.16 **(a)** We work through the flow diagram in the text (Fig. 15.14). We note that this complex with freely rotating CF_3 groups is not linear, it has no C_n axes with $n > 2$, but it does have C_2 axes; in fact it has two C_2 axes perpendicular to whichever C_2 we call principal, and it has a σ_h. Therefore, the point group is $\boxed{D_{2h}}$

(b) The plane shown in Fig. 15.8 is a mirror plane so long as the CF_3 groups each have a CF bond in the plane. (i) If the CF_3 groups are staggered, then the Ag–CN axis is still a C_2 axis; however, there are no other C_2 axes. The Ag–CF_3 axis is an S_2 axis, though, which means that the Ag atom is at an inversion centre. Continuing with the flow diagram, there is a σ_h (the plane shown in the figure). So the point group is $\boxed{C_{2h}}$ (ii) If the CF_3 groups are eclipsed, then the axis through the Ag and perpendicular to the plane of the Ag bonds is still a C_2 axis; however, neither of the Ag bond axes is a C_2 axis. There is no σ_h, but there are two σ_v planes (the plane shown and the plane perpendicular to it and the Ag bond plane). So the point group is $\boxed{C_{2v}}$

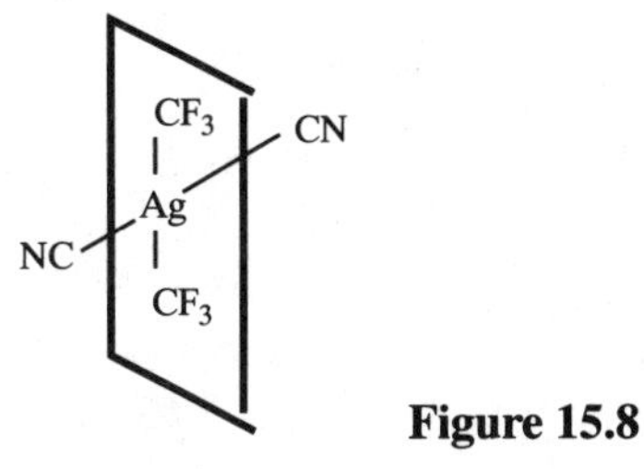

Figure 15.8

P15.18 For a photon to induce a spectroscopic transition, the transition moment $\langle \mu \rangle$ must be nonzero. The transition moment is the integral $\int \psi_f^* \mu \psi_i \, d\tau$, where the dipole moment operator has components proportional to the Cartesian coordinates. The integral vanishes unless the integrand, or at least some part of it, belongs to the totally symmetric representation of the molecule's point group. We can answer the first part of the question without reference to the character table, by considering the character of the integrand under inversion. Each component of μ has u character, but each state has g character; the integrand is $g \times g \times u = u$, so the integral vanishes and the transition is not allowed. However, if a vibration breaks the inversion symmetry, a look at the I character table shows that the components of μ have T_1 character. To find the character of the integrand, we multiply together the characters of its factors. For the transition to T_1

	E	$12C_5$	$12C_5^2$	$20C_3$	$15C_2$
A_1	1	1	1	1	1
$\mu(T_1)$	3	$\frac{1}{2}(1+\sqrt{5})$	$\frac{1}{2}(1-\sqrt{5})$	0	-1
T_1	3	$\frac{1}{2}(1+\sqrt{5})$	$\frac{1}{2}(1-\sqrt{5})$	0	-1
Integrand	9	$\frac{1}{2}(3+\sqrt{5})$	$\frac{1}{2}(3-\sqrt{5})$	0	1

The decomposition of the characters of the integrand into those of the irreducible representations is difficult to do by inspection, but when accomplished it is seen to contain A_1. Therefore the transition to T_1 would become allowed. It is easier to use the formula below which is obtained from what is referred to as the 'little orthogonality theorem' of group theory. (See the *Justification* in Section 15.5 of the 5th edition of this text.) The coefficient of A_1 in the integrand is given as

$$c_{A_1} = \frac{1}{h}\sum_C g(C)\chi(C) = \{9 + 12[\tfrac{1}{2}(3+\sqrt{5})] + 12[\tfrac{1}{2}(3-\sqrt{5})] + 20(0) + 15(1)\}/60 = 1$$

So the integrand contains A_1, and the $\boxed{\text{transition to } T_1 \text{ would become allowed}}$. For the transition to G

	E	$12C_5$	$12C_5^2$	$20C_3$	$15C_2$
A_1	1	1	1	1	1
$\mu(T_1)$	3	$\frac{1}{2}(1+\sqrt{5})$	$\frac{1}{2}(1-\sqrt{5})$	0	-1
G	4	-1	-1	1	0
Integrand	12	$-\frac{1}{2}(1+\sqrt{5})$	$-\frac{1}{2}(1-\sqrt{5})$	0	0

The little orthogonality theorem gives the coefficient of A_1 in the integrand as

$$c_{A_1} = \frac{1}{h}\sum_C g(C)\chi(C) = \{12 + 12[-\tfrac{1}{2}(1+\sqrt{5})]$$
$$+12[-\tfrac{1}{2}(1-\sqrt{5})] + 20(0) + 15(0)\}/60 = 0$$

So the integrand does not contain A_1, and the $\boxed{\text{transition to G would still be forbidden.}}$

P15.19 The shape of this molecule is shown in Fig. 15.9.

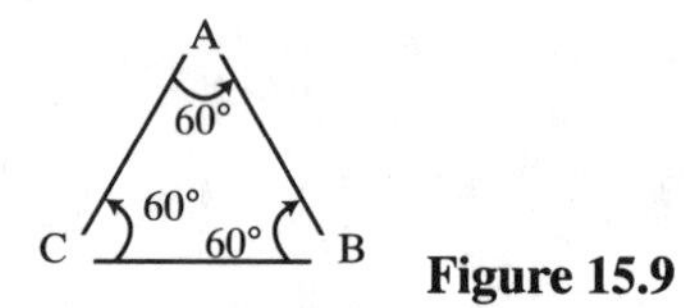

Figure 15.9

(a) Symmetry elements $\boxed{E, 2C_3, 3C_2, \sigma_h, 2S_3, 3\sigma_v}$

Point group $\boxed{D_{3h}}$

(b) $$D(E) = \begin{pmatrix} 1 & 0 & 0 \\ 0 & 1 & 0 \\ 0 & 0 & 1 \end{pmatrix} = D(\sigma_h)$$

$$D(C_3) = \begin{pmatrix} 0 & 0 & 1 \\ 1 & 0 & 0 \\ 0 & 1 & 0 \end{pmatrix}, \qquad D(C_3') = D^2(C_3) = \begin{pmatrix} 0 & 1 & 0 \\ 0 & 0 & 1 \\ 1 & 0 & 0 \end{pmatrix}$$

$D(S_3) = D(C_3), \qquad D(S_3') = D^2(S_3) = D(C_3')$

C_3' and S_3' are counter clockwise rotations.

σ_v is through A and perpendicular to B–C.

σ_v' is through B and perpendicular to A–C.

σ_v'' is through C and perpendicular to A–B.

$$D(\sigma_v) = \begin{pmatrix} 1 & 0 & 0 \\ 0 & 0 & 1 \\ 0 & 1 & 0 \end{pmatrix}, \qquad D(\sigma_v') = \begin{pmatrix} 0 & 0 & 1 \\ 0 & 1 & 0 \\ 1 & 0 & 0 \end{pmatrix},$$

$$D(\sigma_v'') = \begin{pmatrix} 0 & 1 & 0 \\ 1 & 0 & 0 \\ 0 & 0 & 1 \end{pmatrix}$$

$D(C_2) = D(\sigma_v), \qquad D(C_2') = D(\sigma_v'), \qquad D(C_2'') = D(\sigma_v''),$

(c) Example of elements of group multiplication table

$$D(C_3)D(C_2) = \begin{pmatrix} 0 & 0 & 1 \\ 1 & 0 & 0 \\ 0 & 1 & 0 \end{pmatrix}\begin{pmatrix} 1 & 0 & 0 \\ 0 & 0 & 1 \\ 0 & 1 & 0 \end{pmatrix}$$

$$= \begin{pmatrix} 0 & 1 & 0 \\ 1 & 0 & 0 \\ 0 & 0 & 1 \end{pmatrix} = D(\sigma_v'')$$

$$D(\sigma_v')D(\sigma_v) = \begin{pmatrix} 0 & 0 & 1 \\ 0 & 1 & 0 \\ 1 & 0 & 0 \end{pmatrix}\begin{pmatrix} 1 & 0 & 0 \\ 0 & 0 & 1 \\ 0 & 1 & 0 \end{pmatrix}$$

$$= \begin{pmatrix} 0 & 1 & 0 \\ 0 & 0 & 1 \\ 1 & 0 & 0 \end{pmatrix} = D(C_3')$$

D_{3h}	E	C_3	C_2	σ_v	σ_v'	σ_h	$\cdots$
E	E	C_3	C_2	σ_v	σ_v'	σ_h	$\cdots$
C_3	C_3	C_3'	σ_v''	σ_v''	σ_v	C_3	$\cdots$
C_2	C_2	σ_v'	E	E	C_3	C_2	$\cdots$
σ_v	σ_v	σ_v'	E	E	C_3	σ_v	$\cdots$
σ_v'	σ_v'	σ_v''	C_3'	C_3'	E	σ_v'	$\cdots$
σ_h	σ_h	C_3	C_2	σ_v	σ_v'	E	$\cdots$
$\vdots$	$\vdots$	$\vdots$	$\vdots$	$\vdots$	$\vdots$	$\vdots$	$\ddots$

(d) First, determine the number of *s* orbitals (the basis has three *s* orbitals) that have unchanged positions after application of each symmetry species of the D_{3h} point group.

D_{3h} :	E	$2C_3$	$3C_2$	σ_h	$2S_3$	$3\sigma_v$
Unchanged basis members	3	0	1	3	0	1

This is not one of the irreducible representations reported in the D_{3h} character table but inspection shows that it is identical to $A_1' + E'$. This allows us to conclude that the three *s* orbitals span $\boxed{A_1' + E'}$

Comment. The multiplication table in part (c) is not strictly speaking *the* group multiplication; it is instead the multiplication table for the matrix representations of the group in the basis under consideration.

16 Spectroscopy 1: rotational and vibrational spectra

Solutions to exercises

E16.1 The ratio of coefficients A/B is

$$\frac{A}{B}=\frac{8\pi h\nu^3}{c^3}\ [17]$$

The frequency is

$$\nu=\frac{c}{\lambda}\quad\text{so}\quad\frac{A}{B}=\frac{8\pi h}{\lambda^3}$$

(a) $$\frac{A}{B}=\frac{8\pi(6.626\times10^{-34}\,\mathrm{J\,s})}{(70.8\times10^{-12}\,\mathrm{m})^3}=\boxed{0.0469\,\mathrm{J\,m^{-3}\,s}}$$

(b) $$\frac{A}{B}=\frac{8\pi h}{\lambda^3}=\frac{8\pi(6.626\times10^{-34}\,\mathrm{J\,s})}{(500\times10^{-9}\,\mathrm{m})^3}=\boxed{1.33\times10^{-13}\,\mathrm{J\,m^{-3}\,s}}$$

(c) $$\frac{A}{B}=\frac{8\pi h}{\lambda^3}=8\pi h\tilde{\nu}^3=8\pi\left[6.62\times10^{-34}\,\mathrm{J\,s}\times3000\,\mathrm{cm^{-1}}\times(10^{-2}\,\mathrm{m^{-1}}/1\,\mathrm{cm^{-1}})\right]$$

$$=\boxed{4.50\times10^{-16}\,\mathrm{J\,m^{-3}\,s}}$$

Comment. Comparison of these ratios shows that the relative importance of spontaneous transitions decreases as the frequency decreases. The quotient $\dfrac{A}{B}$ has units. A unitless quotient is $\dfrac{A}{B\rho}$ with ρ given by eqn 13.

Question. What are the ratios $\dfrac{A}{B\rho}$ for the radiation of **(a)** through **(c)** and what additional conclusions can you draw from these results?

E16.2 NO is a linear rotor and we assume there is little centrifugal distortion; hence

$$F(J)=BJ(J+1)\ [37]$$

with $B=\dfrac{\hbar}{4\pi cI}$, $I=m_{\mathrm{eff}}R^2$ [Table 16.1], and

$$m_{\mathrm{eff}}=\frac{m_{\mathrm{N}}m_{\mathrm{O}}}{m_{\mathrm{N}}+m_{\mathrm{O}}}\ \text{[nuclide masses from inside back cover of the text]}$$

$$=\left(\frac{(14.003\,\mathrm{u})\times(15.995\,\mathrm{u})}{(14.003\,\mathrm{u})+(15.995\,\mathrm{u})}\right)\times(1.6605\times10^{-27}\,\mathrm{kg\,u^{-1}})=1.240\times10^{-26}\,\mathrm{kg}$$

Then, $I=(1.240\times10^{-26}\,\mathrm{kg})\times(1.15\times10^{-10}\,\mathrm{m})^2=1.64\bar{0}\times10^{-46}\,\mathrm{kg\,m^2}$

and $$B=\frac{1.0546\times10^{-34}\,\mathrm{J\,s}}{(4\pi)\times(2.998\times10^8\,\mathrm{m\,s^{-1}})\times(1.64\bar{0}\times10^{-46}\,\mathrm{kg\,m^2})}=170.\bar{7}\,\mathrm{m^{-1}}=1.70\bar{7}\,\mathrm{cm^{-1}}$$

The wavenumber of the $J=4\leftarrow3$ transition is

$$\tilde{\nu}=2B(J+1)[43]=8B[J=3]=(8)\times(1.70\bar{7}\,\mathrm{cm^{-1}})=13.6\,\mathrm{cm^{-1}}$$

The frequency is

$$\nu=\tilde{\nu}c=(13.6\bar{5}\,\mathrm{cm^{-1}})\times\left(\frac{10^2\,\mathrm{m^{-1}}}{1\,\mathrm{cm^{-1}}}\right)\times(2.998\times10^8\,\mathrm{m\,s^{-1}})=\boxed{4.09\times10^{11}\,\mathrm{Hz}}$$

Question. What is the percentage change in these calculated values if centrifugal distortion is included?

E16.3 **(a)** The wavenumber of the transition is related to the rotational constant by

$$hc\tilde{\nu} = \Delta E = hcB[J(J+1) - (J-1)J] = 2hcBJ \text{ [31, 33]}$$

where J refers to the upper state ($J = 3$). The rotational constant is related to molecular structure by

$$B = \frac{\hbar}{4\pi cI} \text{ [30]}$$

where I is moment of inertia, m_{eff} is the effective mass, and R is the bond length. Putting these expressions together yields

$$\tilde{\nu} = 2BJ = \frac{\hbar J}{2\pi cI} \quad \text{so} \quad I = \frac{\hbar J}{c\tilde{\nu}} = \frac{(1.0546 \times 10^{-34}\,\text{J s}) \times (3)}{2\pi(2.998 \times 10^{10}\,\text{cm s}^{-1}) \times (63.56\,\text{cm}^{-1})}$$

$$= \boxed{2.642 \times 10^{-47}\,\text{kg m}^2}$$

(b) The moment of inertia is related to the bond length by

$$I = m_{\text{eff}}R^2 \quad \text{so} \quad R = \sqrt{\frac{I}{m_{\text{eff}}}}$$

$$m_{\text{eff}}^{-1} = m_{\text{H}}^{-1} + m_{\text{Cl}}^{-1} = \frac{(1.0078\,\text{u})^{-1} + (34.9688\,\text{u})^{-1}}{1.66054 \times 10^{-27}\,\text{kg u}^{-1}} = 6.1477 \times 10^{26}\,\text{kg}^{-1}$$

$$\text{and } R = \sqrt{(6.1477 \times 10^{26}\,\text{kg}^{-1}) \times (2.642 \times 10^{-47}\,\text{kg m}^2)} = 1.274 \times 10^{-10}\,\text{m} = \boxed{127.4\,\text{pm}}$$

E16.4 If the spacing of lines is constant, the effects of centrifugal distortion are negligible. Hence we may use for the wavenumbers of the transitions

$$F(J) - F(J-1) = 2BJ \text{ [33]}$$

Since $J = 1, 2, 3, \ldots$, the spacing of the lines is $2B$

$$12.604\,\text{cm}^{-1} = 2B$$

$$B = 6.302\,\text{cm}^{-1} = 6.302 \times 10^2\,\text{m}^{-1}$$

$$I = \frac{\hbar}{4\pi cB} \text{ [Problem 16.3]} = m_{\text{eff}}R^2$$

$$\frac{\hbar}{4\pi c} = \frac{1.0546 \times 10^{-34}\,\text{J s}}{(4\pi) \times (2.9979 \times 10^8\,\text{m s}^{-1})} = 2.7993 \times 10^{-44}\,\text{kg m}$$

$$I = \frac{2.7993 \times 10^{-44}\,\text{kg m}}{6.302 \times 10^2\,\text{m}^{-1}} = \boxed{4.442 \times 10^{-47}\,\text{kg m}^2}$$

$$m_{\text{eff}} = \frac{m_{\text{Al}}m_{\text{H}}}{m_{\text{Al}} + m_{\text{H}}}$$

$$= \left(\frac{(26.98) \times (1.008)}{(26.98) + (1.008)}\right)\text{u} \times (1.6605 \times 10^{-27}\,\text{kg u}^{-1}) = 1.613\bar{6} \times 10^{-27}\,\text{kg}$$

$$R = \left(\frac{I}{m_{\text{eff}}}\right)^{1/2} = \left(\frac{4.442 \times 10^{-47}\,\text{kg m}^2}{1.6136 \times 10^{-27}\,\text{kg}}\right)^{1/2} = 1.659 \times 10^{-10}\,\text{m} = \boxed{165.9\,\text{pm}}$$

E16.5 $B = \dfrac{\hbar}{4\pi c I}$ [30], implying that $I = \dfrac{\hbar}{4\pi c B}$

Then, with $I = m_{\text{eff}} R^2$, $R = \left(\dfrac{\hbar}{4\pi m_{\text{eff}} c B}\right)^{1/2}$

We use $m_{\text{eff}} = \dfrac{m_1 m_2}{m_1 + m_2} = \dfrac{(126.904) \times (34.9688)}{(126.904) + (34.9688)}\,\text{u} = 27.4146\,\text{u}$

and hence obtain

$$R = \left(\frac{1.05457 \times 10^{-34}\,\text{J s}}{(4\pi) \times (27.4146) \times (1.66054 \times 10^{-27}\,\text{kg}) \times (2.99792 \times 10^{10}\,\text{cm s}^{-1}) \times (0.1142\,\text{cm}^{-1})}\right)^{1/2}$$

$$= \boxed{232.1\,\text{pm}}$$

E16.6 The determination of two unknowns requires data from two independent experiments and the equation which relates the unknowns to the experimental data. In this exercise two independently determined values of B for two isotopically different HCN molecules are used to obtain the moments of inertia of the molecules and from these, by use of the equation for the moment of inertia of linear triatomic rotors (Table 16.1), the interatomic distances R_{HC} and R_{CN} are calculated.

Rotational constants which are usually expressed in wavenumbers (cm^{-1}) are sometimes expressed in frequency units (Hz). The conversion between the two is

$$B/\text{Hz} = c \times B/\text{cm}^{-1} \quad [c \text{ in cm s}^{-1}]$$

Thus, $B(\text{in Hz}) = \dfrac{\hbar}{4\pi I}$ and $I = \dfrac{\hbar}{4\pi B}$

Let, $^1\text{H} = \text{H}$, $^2\text{H} = \text{D}$, $R_{\text{HC}} = R_{\text{DC}} = R$, $R_{\text{CN}} = R'$. Then

$$I(\text{HCN}) = \frac{1.05457 \times 10^{-34}\,\text{J s}}{(4\pi) \times (4.4316 \times 10^{10}\,\text{s}^{-1})} = 1.8937 \times 10^{-46}\,\text{kg m}^2$$

$$I(\text{DCN}) = \frac{1.05457 \times 10^{-34}\,\text{J s}}{(4\pi) \times (3.6208 \times 10^{10}\,\text{s}^{-1})} = 2.3178 \times 10^{-46}\,\text{kg m}^2$$

and from Table 16.1 with isotopic masses from the inside back cover

$$I(\text{HCN}) = m_{\text{H}} R^2 + m_{\text{N}} R'^2 - \frac{(m_{\text{H}} R - m_{\text{N}} R')^2}{m_{\text{H}} + m_{\text{C}} + m_{\text{N}}}$$

$$I(\text{HCN}) = \left[(1.0078 R^2) + (14.0031 R'^2) - \left(\frac{(1.0078 R - 14.0031 R')^2}{1.0078 + 12.0000 + 14.0031}\right)\right]\text{u}$$

Multiplying through by $m/\text{u} = (m_{\text{H}} + m_{\text{C}} + m_{\text{N}})/\text{u} = 27.0109$

$$27.0109 \times I(\text{HCN}) = \{27.0109 \times (1.0078 R^2 + 14.0031 R'^2) - (1.0078 R - 14.0031 R')^2\}\,\text{u}$$

or $\left(\dfrac{27.0109}{1.66054 \times 10^{-27}\,\text{kg}}\right) \times (1.8937 \times 10^{-46}\,\text{kg m}^2) = 3.0804 \times 10^{-18}\,\text{m}^2$

$$= \{27.0109 \times (1.0078 R^2 + 14.0031 R'^2) - (1.0078 R - 14.0031 R')^2\} \quad \text{(a)}$$

In a similar manner we find for DCN

$$\left(\frac{28.0172}{1.66054 \times 10^{-27}\,\text{kg}}\right) \times (2.3178 \times 10^{-46}\,\text{kg m}^2) = 3.9107 \times 10^{-18}\,\text{m}^2$$

$$= \{28.0172 \times (2.0141 R^2 + 14.0031 R'^2) - (2.0141 R - 14.0031 R')^2\} \quad \text{(b)}$$

Thus there are two simultaneous quadratic equations (a) and (b) to solve for R and R'. These equations are most easily solved by readily available computer programs or by successive approximations. The results are

$$R = 1.065 \times 10^{-10}\,\text{m} = \boxed{106.5\,\text{pm}} \quad \text{and} \quad R' = 1.156 \times 10^{-10}\,\text{m} = \boxed{115.6\,\text{pm}}$$

These values are easily verified by direct substitution into the equations and agree well with the accepted values $R_{\text{HC}} = 1.064 \times 10^{-10}$ m and $R_{\text{CN}} = 1.156 \times 10^{-10}$ m.

E16.7 The Stokes lines appear at

$$\tilde{\nu}(J+2 \leftarrow J) = \tilde{\nu}_i - 2B(2J+3)\ [48a] \quad \text{with } J = 0,\ \tilde{\nu} = \tilde{\nu}_i - 6B$$

Since $B = 1.9987\,\text{cm}^{-1}$ (Table 16.2), the Stokes line appears at

$$\tilde{\nu} = (20487) - (6) \times (1.9987\,\text{cm}^{-1}) = \boxed{20\,475\,\text{cm}^{-1}}$$

E16.8 The R branch obeys the relation

$$\tilde{\nu}_{\text{R}}(J) = \tilde{\nu} + 2B(J+1)\ [69c]$$

Hence, $\tilde{\nu}_{\text{R}}(2) = \tilde{\nu} + 6B = (2648.98) + (6) \times (8.465\,\text{cm}^{-1})$ [Table 16.2] $= \boxed{2699.77\,\text{cm}^{-1}}$

E16.9

$$\omega = 2\pi\nu = \left(\frac{k}{m}\right)^{1/2}$$

$$k = 4\pi^2\nu^2 m = 4\pi^2 \times (2.0\,\text{s}^{-1})^2 \times (1.0\,\text{kg}) = 1.6 \times 10^2\,\text{kg}\,\text{s}^{-2} = \boxed{1.6 \times 10^2\,\text{N}\,\text{m}^{-1}}$$

E16.10

$$\omega = \left(\frac{k}{m_{\text{eff}}}\right)^{1/2}\ [56]$$

The fractional difference is

$$\frac{\omega' - \omega}{\omega} = \frac{\left(\frac{k}{m'_{\text{eff}}}\right)^{1/2} - \left(\frac{k}{m_{\text{eff}}}\right)^{1/2}}{\left(\frac{k}{m_{\text{eff}}}\right)^{1/2}} = \frac{\left(\frac{1}{m'_{\text{eff}}}\right)^{1/2} - \left(\frac{1}{m_{\text{eff}}}\right)^{1/2}}{\left(\frac{1}{m_{\text{eff}}}\right)^{1/2}} = \left(\frac{m_{\text{eff}}}{m'_{\text{eff}}}\right)^{1/2} - 1$$

$$= \left(\frac{m(^{23}\text{Na})m(^{35}\text{Cl})\{m(^{23}\text{Na}) + m(^{37}\text{Cl})\}}{\{m(^{23}\text{Na}) + m(^{35}\text{Cl})\}m(^{23}\text{Na})m(^{37}\text{Cl})}\right)^{1/2} - 1$$

$$= \left(\frac{m(^{35}\text{Cl})}{m(^{37}\text{Cl})} \times \frac{m(^{23}\text{Na}) + m(^{37}\text{Cl})}{m(^{23}\text{Na}) + m(^{35}\text{Cl})}\right)^{1/2} - 1$$

$$= \left(\frac{34.9688}{36.9651} \times \frac{22.9898 + 36.9651}{22.9898 + 34.9688}\right)^{1/2} - 1 = -0.01089$$

Hence, the difference is $\boxed{1.089\ \text{per cent}}$

E16.11

$$\omega = \left(\frac{k}{m_{\text{eff}}}\right)^{1/2}\ [56]; \quad \omega = 2\pi\nu = 2\pi\left(\frac{c}{\lambda}\right) = 2\pi c\tilde{\nu}$$

Therefore, $k = m_{\text{eff}}\omega^2 = 4\pi^2 m_{\text{eff}} c^2\tilde{\nu}^2$, $m_{\text{eff}} = \frac{1}{2}m(^{35}\text{Cl})$

$$= (4\pi^2) \times \left(\frac{34.9688}{2}\right) \times (1.66054 \times 10^{-27}\,\text{kg}) \times [(2.997924 \times 10^{10}\,\text{cm}\,\text{s}^{-1}) \times (564.9\,\text{cm}^{-1})]^2$$

$$= \boxed{328.7\,\text{N}\,\text{m}^{-1}}$$

E16.12 Use the character table for the group C_{2v} (and see Example 16.6). The rotations span $A_2 + B_1 + B_2$. The translations span $A_1 + B_1 + B_2$. Hence the normal modes of vibration span the difference, $\boxed{4A_1 + A_2 + 2B_1 + 2B_2}$

Comment. A_1, B_1, and B_2 are infrared active; all modes are Raman active.

E16.13 Polar molecules show a pure rotational absorption spectrum. Therefore, select the polar molecules based on their well-known structures. Alternatively, determine the point groups of the molecules and use the rule that only molecules belonging to C_n, C_{nv}, and C_s may be polar, and in the case of C_n and C_{nv}, that dipole must lie along the rotation axis. Hence the polar molecules are

(b) HCl **(d)** CH_3Cl **(e)** CH_2Cl_2

Their point group symmetries are

(b) $C_{\infty v}$ **(d)** C_{3v} **(e)** $C_{2h}(\textit{trans})$, $C_{2v}(\textit{cis})$

Comment. Note that the *cis* form of CH_2Cl_2 is polar, but the *trans* form is not.

E16.14 See the *Illustration* on p. 478 of the text. Select those molecules in which a vibration gives rise to a change in dipole moment. It is helpful to write down the structural formulas of the compounds. The infrared active compounds are

(b) HCl **(c)** CO_2 **(d)** H_2O

Comment. A more powerful method for determining infrared activity based on symmetry considerations is described in Section 16.14. Also see Exercises 16.26–16.27.

E16.15 We select those molecules with an anisotropic polarizability. A practical rule to apply is that spherical rotors do not have anisotropic polarizabilities. Therefore **(c)** CH_4 is inactive. All others are active.

E16.16 A source approaching an observer appears to be emitting light of frequency

$$\nu' = \frac{\nu}{1 - \frac{v}{c}} \quad \text{[Section 16.3]}$$

Since $\nu \propto \frac{1}{\lambda}$, $\lambda_{\text{obs}} = \left(1 - \frac{v}{c}\right)\lambda$

$v = 80\ \text{km h}^{-1} = 22.\bar{2}\ \text{m s}^{-1}$. Hence,

$$\lambda_{\text{obs}} = \left(1 - \frac{22.\bar{2}\ \text{m s}^{-1}}{2.998 \times 10^8\ \text{m s}^{-1}}\right) \times (660\ \text{nm}) = \boxed{0.999\,999\,925 \times 660\ \text{nm}}$$

E16.17

$$\nu' = \frac{\nu}{1 + \frac{v}{c}} \quad \text{[Section 16.3]}$$

$$\text{or } \lambda_{\text{obs}} = \left(1 + \frac{v}{c}\right)\lambda \quad \left[\nu \propto \frac{1}{\lambda}\right]$$

$$v = \left(\frac{\lambda_{\text{obs}}}{\lambda} - 1\right)c = \left(\frac{706.5\ \text{nm}}{654.2\ \text{nm}} - 1\right) \times (2.998 \times 10^8\ \text{m s}^{-1}) = 2.4 \times 10^7\ \text{m s}^{-1}$$

$$= \boxed{2.4 \times 10^4\ \text{km s}^{-1}}$$

The broadening of the line is due to the local events (collisions) in the distant star. It is temperature-dependent and hence yields the surface temperature of the star.

$$\delta\lambda = \left(\frac{2\lambda}{c}\right) \times \left(\frac{2kT}{m}\ln 2\right)^{1/2} \quad [23], \quad \text{which implies that}$$

$$T = \left(\frac{m}{2k\ln 2}\right) \times \left(\frac{c\delta\lambda}{2\lambda}\right)^2$$

$$= \left(\frac{(48) \times (1.6605 \times 10^{-27}\,\text{kg})}{(2) \times (1.381 \times 10^{-23}\,\text{J K}^{-1}) \times (\ln 2)}\right)$$

$$\times \left(\frac{(2.998 \times 10^8\,\text{m s}^{-1}) \times (61.8 \times 10^{-12}\,\text{m})}{(2) \times (654.2 \times 10^{-9}\,\text{m})}\right)^2$$

$$= \boxed{8.4 \times 10^5\,\text{K}}$$

E16.18 $\delta\tilde{\nu} \approx \dfrac{5.31\,\text{cm}^{-1}}{\tau/\text{ps}}$ [25], implying that $\tau \approx \dfrac{5.31\,\text{ps}}{\delta\tilde{\nu}/\text{cm}^{-1}}$

(a) $\tau \approx \dfrac{5.31\,\text{ps}}{0.1} = \boxed{5\bar{3}\,\text{ps}}$ **(b)** $\tau \approx \dfrac{5.31\,\text{ps}}{1} = \boxed{5\,\text{ps}}$

E16.19 $\delta\tilde{\nu} \approx \dfrac{5.31\,\text{cm}^{-1}}{\tau/\text{ps}}$ [25]

(a) $\tau \approx 1.0 \times 10^{13}\,\text{s} = 0.10\,\text{ps}$, implying that $\delta\tilde{\nu} \approx \boxed{53\,\text{cm}^{-1}}$

(b) $\tau \approx (100) \times (1.0 \times 10^{-13}\,\text{s}) = 10\,\text{ps}$, implying that $\delta\tilde{\nu} \approx \boxed{0.53\,\text{cm}^{-1}}$

E16.20 We write, with $N' = N$ (upper state) and $N = N$ (lower state)

$$\frac{N'}{N} = e^{-h\nu/kT}\,[\text{from Boltzmann distribution}] = e^{-hc\tilde{\nu}/kT}$$

$$\frac{hc\tilde{\nu}}{k} = (1.4388\,\text{cm K}) \times (559.7\,\text{cm}^{-1})\,[\text{inside front cover}] = 805.3\,\text{K}$$

$$\frac{N(\text{upper})}{N(\text{lower})} = e^{-805.3\,\text{K}/T}$$

(a) $\dfrac{N(\text{upper})}{N(\text{lower})} = e^{-805.3/298} = \boxed{0.067}$ (1 : 15) **(b)** $\dfrac{N(\text{upper})}{N(\text{lower})} = e^{-805.3/500} = \boxed{0.20}$ (1 : 5)

E16.21 $\omega = \left(\dfrac{k}{m_{\text{eff}}}\right)^{1/2}$ [56], so $k = m_{\text{eff}}\omega^2 = 4\pi^2 m_{\text{eff}} c^2 \tilde{\nu}^2$

$$\frac{1}{m_{\text{eff}}} = \frac{1}{m_1} + \frac{1}{m_2}\ [55] \qquad m_{\text{eff}} = \frac{m_1 m_2}{m_1 + m_2}$$

$$m_{\text{eff}}(\text{H}^{19}\text{F}) = \frac{(1.0078) \times (18.9984)}{(1.0078) + (18.9984)}\,\text{u} = 0.9570\,\text{u}$$

$$m_{\text{eff}}(\text{H}^{35}\text{Cl}) = \frac{(1.0078) \times (34.9688)}{(1.0078) + (34.9688)}\,\text{u} = 0.9796\,\text{u}$$

$$m_{\text{eff}}(\text{H}^{81}\text{Br}) = \frac{(1.0078) \times (80.9163)}{(1.0078) + (80.9163)}\,\text{u} = 0.9954\,\text{u}$$

$$m_{\text{eff}}(\text{H}^{127}\text{I}) = \frac{(1.0078) \times (126.9045)}{(1.0078) + (126.9045)}\,\text{u} = 0.9999\,\text{u}$$

We draw up the following table

	HF	HCl	HBr	HI
$\tilde{\nu}/\mathrm{cm}^{-1}$	4141.3	2988.9	2649.7	2309.5
$m_{\mathrm{eff}}/\mathrm{u}$	0.9570	0.9796	0.9954	0.9999
$k/(\mathrm{N\,m}^{-1})$	$\boxed{967.0}$	$\boxed{515.6}$	$\boxed{411.8}$	$\boxed{314.2}$

Note the order of stiffness HF > HCl > HBr > HI.

Question. Which ratio, $\dfrac{k}{B(\mathrm{A-B})}$ or $\dfrac{\tilde{\nu}}{B(\mathrm{A-B})}$, where $B(\mathrm{A-B})$ are the bond enthalpies of Table 14.3, is the more nearly constant across the series of hydrogen halides? Why?

E16.22 Data on three transitions are provided. Only two are necessary to obtain the value of $\tilde{\nu}$ and x_e. The third datum can then be used to check the accuracy of the calculated values.

$$\Delta G(v=1\leftarrow 0)=\tilde{\nu}-2\tilde{\nu}x_e=1556.22\,\mathrm{cm}^{-1}\ [64]$$
$$\Delta G(v=2\leftarrow 0)=2\tilde{\nu}-6\tilde{\nu}x_e=3088.28\,\mathrm{cm}^{-1}\ [65]$$

Multiply the first equation by 3, then subtract the second.

$$\tilde{\nu}=(3)\times(1556.22\,\mathrm{cm}^{-1})-(3088.28\,\mathrm{cm}^{-1})=\boxed{1580.38\,\mathrm{cm}^{-1}}$$

Then from the first equation

$$x_e=\frac{\tilde{\nu}-1556.22\,\mathrm{cm}^{-1}}{2\tilde{\nu}}=\frac{(1580.38-1556.22)\,\mathrm{cm}^{-1}}{(2)\times(1580.38\,\mathrm{cm}^{-1})}=\boxed{7.644\times10^{-3}}$$

x_e data are usually reported as $x_e\tilde{\nu}$ which is

$$x_e\tilde{\nu}=12.08\,\mathrm{cm}^{-1}$$
$$\Delta G(v=3\leftarrow 0)=3\tilde{\nu}-12\tilde{\nu}x_e$$
$$=(3)\times(1580.38\,\mathrm{cm}^{-1})-(12)\times(12.08\,\mathrm{cm}^{-1})=4596.18\,\mathrm{cm}^{-1}$$

which is very close to the experimental value.

E16.23 $\Delta G_{v+1/2}=\tilde{\nu}-2(v+1)x_e\tilde{\nu}$ [64] where $\Delta G_{v+1/2}=G(v+1)-G(v)$

Therefore, since

$$\Delta G_{v+1/2}=(1-2x_e)\tilde{\nu}-2vx_e\tilde{\nu}$$

a plot of $\Delta G_{v+1/2}$ against v should give a straight line which gives $(1-2x_e)\tilde{\nu}$ from the intercept at $v=0$ and $-2x_e\tilde{\nu}$ from the slope. We draw up the following table

v	0	1	2	3	4
$G(v)/\mathrm{cm}^{-1}$	1481.86	4367.50	7149.04	9826.48	12399.8
$\Delta G_{v+1/2}/\mathrm{cm}^{-1}$	2885.64	2781.54	2677.44	2573.34	

The points are plotted in Fig. 16.1. The intercept lies at 2885.6 and the slope is $\dfrac{-312.3}{3}=-104.1$; hence $x_e\tilde{\nu}=52.1\,\mathrm{cm}^{-1}$.

Since $\tilde{\nu}-2x_e\tilde{\nu}=2885.6\,\mathrm{cm}^{-1}$, it follows that $\tilde{\nu}=2989.8\,\mathrm{cm}^{-1}$.

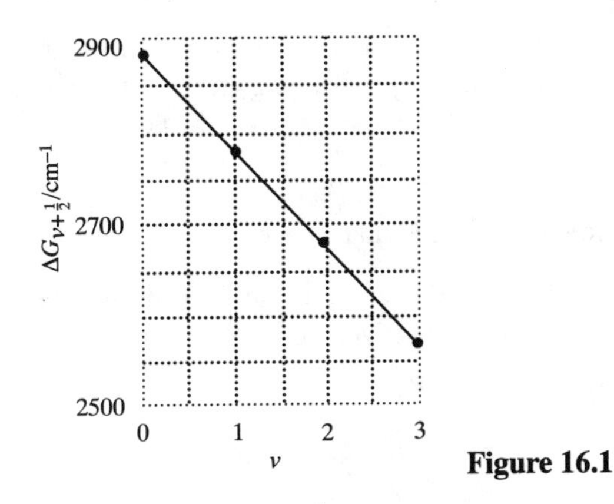

Figure 16.1

The dissociation energy may be obtained by assuming that the molecule is described by a Morse potential and that the constant D_e in the expression for the potential is an adequate first approximation for it. Then

$$D_e = \frac{\tilde{\nu}}{4x_e}[62] = \frac{\tilde{\nu}^2}{4x_e\tilde{\nu}} = \frac{(2989.8\,\mathrm{cm}^{-1})^2}{(4)\times(52.1\,\mathrm{cm}^{-1})} = 42.9\times10^3\,\mathrm{cm}^{-1}, \qquad 5.32\,\mathrm{eV}$$

However, the depth of the potential well D_e differs from D_0, the dissociation energy of the bond, by the zero-point energy; hence

$$\begin{aligned} D_0 &= D_e - \tfrac{1}{2}\tilde{\nu} \text{ [Fig. 16.36, p. 479 of text]} \\ &= (42.9\times10^3\mathrm{cm}^{-1}) - \left(\tfrac{1}{2}\right)\times(2889.8\,\mathrm{cm}^{-1}) \\ &= 41.5\times10^3\,\mathrm{cm}^{-1} = \boxed{5.15\,\mathrm{eV}} \end{aligned}$$

E16.24 The separation of lines is $4B$ [Section 16.7, eqns 48*a* and 48*b*], so $B = 0.2438\,\mathrm{cm}^{-1}$. Then we use

$$R = \left(\frac{\hbar}{4\pi m_{\mathrm{eff}}cB}\right)^{1/2}$$

with $m_{\mathrm{eff}} = \frac{1}{2}m(^{35}\mathrm{Cl}) = \left(\frac{1}{2}\right)\times(34.9688\,\mathrm{u}) = 17.4844\,\mathrm{u}$

Therefore

$$R = \left(\frac{1.05457\times10^{-34}\,\mathrm{J\,s}}{(4\pi)\times(17.4844)\times(1.6605\times10^{-27}\,\mathrm{kg})\times(2.9979\times10^{10}\,\mathrm{cm\,s}^{-1})\times(0.2438\,\mathrm{cm}^{-1})}\right)^{1/2}$$
$$= 1.989\times10^{-10}\,\mathrm{m} = \boxed{198.9\,\mathrm{pm}}$$

E16.25 The number of normal modes of vibration is given by (Section 16.14)

$$N_{\mathrm{vib}} = \begin{Bmatrix} 3N-5 \text{ for linear molecules} \\ 3N-6 \text{ for nonlinear molecules} \end{Bmatrix}$$

where N is the number of atoms in the molecule. Hence, since none of these molecules are linear,

(a) 3 **(b)** 6 **(c)** 12

Comment. Even for moderately sized molecules the number of normal modes of vibration is large and they are usually difficult to visualize.

E16.26 See Figs 16.47 (H_2O, bent) and 16.46 (CO_2, linear) of the text as well as Example 16.7 and the *Illustration* on p. 488. Decide which modes correspond to (i) a changing electric dipole moment, (ii) a changing polarizability, and take note of the exclusion rule (Sections 16.15 and 16.16).

(a) Nonlinear: all modes both infrared and Raman active.

(b) Linear: the symmetric stretch is infrared inactive but Raman active.

The antisymmetric stretch is infrared active and (by the exclusion rule) Raman inactive. The two bending modes are infrared active and therefore Raman inactive.

E16.27 The uniform expansion is depicted in Fig. 16.2.

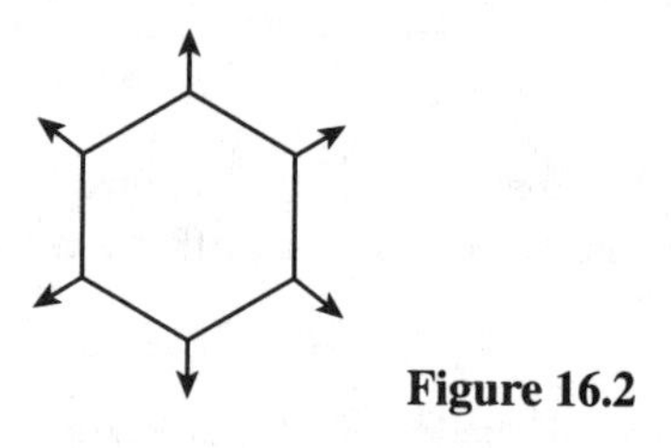

Figure 16.2

Benzene is centrosymmetric, and so the exclusion rule applies (Section 16.16). The mode is infrared inactive (symmetric breathing leaves the molecular dipole moment unchanged at zero), and therefore the mode may be $\boxed{\text{Raman active}}$ (and is). In group theoretical terms, the breathing mode has symmetry A_{1g} in D_{6h}, which is the point group for benzene, and the quadratic forms $x^2 + y^2$ and z^2 have this symmetry (see the character table for C_{6h}, a subgroup of D_{6h}). Hence, the mode is Raman active.

Solutions to problems

Solutions to numerical problems

P16.1

$$\frac{\delta\lambda}{\lambda} = \frac{2}{c}\left(\frac{2kT\ln 2}{m}\right)^{1/2} \quad [23]$$

$$= \left(\frac{2}{2.998\times 10^{8}\,\text{m s}^{-1}}\right) \times \left(\frac{(2)\times(1.381\times 10^{-23}\,\text{J K}^{-1})\times(298\,\text{K})\times(\ln 2)}{(m/\text{u})\times(1.6605\times 10^{-27}\,\text{kg})}\right)^{1/2}$$

$$= \frac{1.237\times 10^{-5}}{(m/\text{u})^{1/2}}$$

(a) For $^1H^{35}Cl$, $m \approx 36\,\text{u}$, so $\dfrac{\delta\lambda}{\lambda} \approx \boxed{2.1\times 10^{-6}}$

(b) For $^{127}I^{35}Cl$, $m \approx 162\,\text{u}$, so $\dfrac{\delta\lambda}{\lambda} \approx \boxed{9.7\times 10^{-7}}$

For the second part of the problem, we also need

$$\frac{\delta\tilde{\nu}}{\tilde{\nu}} = \frac{\delta\nu}{\nu} = \frac{2}{c}\left(\frac{2kT\ln 2}{m}\right)^{1/2} [23] = \frac{\delta\lambda}{\lambda} \quad \left[\frac{\delta\lambda}{\lambda} \ll 1\right]$$

(a) For HCl, $\nu(\text{rotation}) \approx 2Bc \approx (2)\times(10.6\,\text{cm}^{-1})\times(2.998\times 10^{10}\,\text{cm s}^{-1})$
$\approx 6.4\times 10^{11}\,\text{s}^{-1}$ or $6.4\times 10^{11}\,\text{Hz}$

Therefore, $\delta\nu(\text{rotation}) \approx (2.1\times 10^{-6})\times(6.4\times 10^{11}\,\text{Hz}) = \boxed{1.3\,\text{MHz}}$

$\tilde{\nu}(\text{vibration}) \approx 2991\,\text{cm}^{-1}$ [Table 16.2]; therefore

$$\delta\tilde{\nu}(\text{vibration}) \approx (2.1 \times 10^{-6}) \times (2991\,\text{cm}^{-1}) = \boxed{0.0063\,\text{cm}^{-1}}$$

(b) For ICl, $\nu(\text{rotation}) \approx (2) \times (0.1142\,\text{cm}^{-1}) \times (2.998 \times 10^{10}\,\text{cm s}^{-1}) \approx 6.8 \times 10^{9}\,\text{Hz}$

$\delta\nu(\text{rotation}) \approx (9.7 \times 10^{-7}) \times (6.8 \times 10^{9}\,\text{Hz}) = \boxed{6.6\,\text{kHz}}$

$\tilde{\nu}(\text{vibration}) \approx 384\,\text{cm}^{-1}$

$\delta\tilde{\nu}(\text{vibration}) \approx (9.7 \times 10^{-7}) \times (384\,\text{cm}^{-1}) \approx \boxed{0.0004\,\text{cm}^{-1}}$

Comment. ICl is a solid which melts at 27.2°C and has a significant vapour pressure at 25°C.

P16.3 Rotational line separations are $2B$ (in wavenumber units), $2Bc$ (in frequency units), and $(2B)^{-1}$ in wavelength units. Hence the transitions are separated by $\boxed{596\,\text{GHz}}$, $\boxed{19.9\,\text{cm}^{-1}}$, and $\boxed{0.503\,\text{mm}}$. Ammonia is a symmetric rotor (Section 16.4) and we know that

$$B = \frac{\hbar}{4\pi c I_{\perp}} \quad [36]$$

and from Table 16.1,

$$I_{\perp} = m_{\text{A}}R^2(1 - \cos\theta) + \left(\frac{m_{\text{A}}m_{\text{B}}}{m}\right) R^2(1 + 2\cos\theta)$$

$m_{\text{A}} = 1.6735 \times 10^{-27}\,\text{kg}$, $m_{\text{B}} = 2.3252 \times 10^{-26}\,\text{kg}$, and $m = 2.8273 \times 10^{-26}\,\text{kg}$ with $R = 101.4\,\text{pm}$ and $\theta = 106°47'$, which gives

$$\begin{aligned} I_{\perp} &= (1.6735 \times 10^{-27}\,\text{kg}) \times (101.4 \times 10^{-12}\,\text{m})^2 \times (1 - \cos 106°47') \\ &\quad + \left(\frac{(1.6735 \times 10^{-27}) \times (2.3252 \times 10^{-26}\,\text{kg}^2)}{2.8273 \times 10^{-26}\,\text{kg}}\right) \\ &\quad \times (101.4 \times 10^{-12}\,\text{m})^2 \times (1 + 2\cos 106°47') \\ &= 2.815\bar{8} \times 10^{-47}\,\text{kg m}^2 \end{aligned}$$

Therefore,

$$B = \frac{1.05457 \times 10^{-34}\,\text{J s}}{(4\pi) \times (2.9979 \times 10^{8}\,\text{m s}^{-1}) \times (2.815\bar{8} \times 10^{-47}\,\text{kg m}^2)} = 994.1\,\text{m}^{-1} = \boxed{9.941\,\text{cm}^{-1}}$$

which is in accord with the data.

P16.5 Rotation about any axis perpendicular to the C_6 axis may be represented in its essentials by rotation of the pseudolinear molecule in Fig. 16.3(a) about the x-axis in the figure.

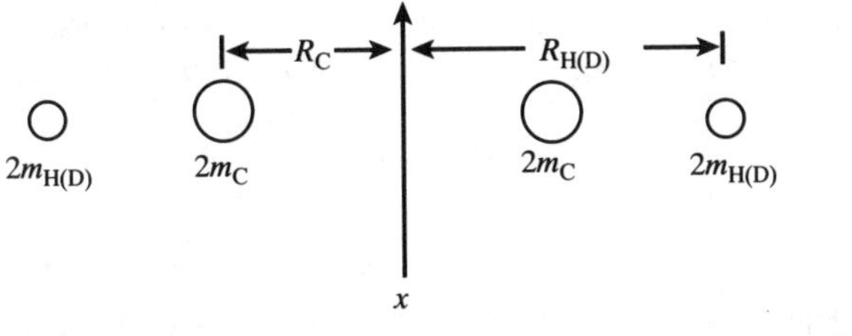

Figure 16.3(a)

The data allow for a determination of R_C and $R_{H(D)}$ which may be decomposed into R_{CC} and $R_{CH(D)}$.

$$I_H = 4m_H R_H^2 + 4m_C R_C^2 = 147.59 \times 10^{-47}\ \mathrm{kg\,m^2}$$
$$I_D = 4m_D R_D^2 + 4m_C R_C^2 = 178.45 \times 10^{-47}\ \mathrm{kg\,m^2}$$

Subtracting I_H from I_D (assume $R_H = R_D$) yields

$$4(m_D - m_H)R_H^2 = 30.86 \times 10^{-47}\ \mathrm{kg\,m^2}$$
$$4(2.0141\ \mathrm{u} - 1.0078\ \mathrm{u}) \times (1.66054 \times 10^{-27}\ \mathrm{kg\,u^{-1}}) \times (R_H^2) = 30.86 \times 10^{-47}\ \mathrm{kg\,m^2}$$
$$R_H^2 = 4.616\bar{9} \times 10^{-20}\ \mathrm{m^2} \qquad R_H = 2.149 \times 10^{-10}\ \mathrm{m}$$

$$R_C^2 = \frac{(147.59 \times 10^{-47}\ \mathrm{kg\,m^2}) - (4m_H R_H^2)}{4m_C}$$
$$= \frac{(147.59 \times 10^{-47}\ \mathrm{kg\,m^2}) - (4) \times (1.0078\ \mathrm{u}) \times (1.66054 \times 10^{-27}\ \mathrm{kg\,u^{-1}}) \times (4.616\bar{9} \times 10^{-20}\ \mathrm{m^2})}{(4) \times (12.011\ \mathrm{u}) \times (1.66054 \times 10^{-27}\ \mathrm{kg\,u^{-1}})}$$
$$= 1.4626 \times 10^{-20}\ \mathrm{m^2}$$

$$R_C = 1.209 \times 10^{-10}\ \mathrm{m}$$

Figure 16.3 (b) shows the relation between R_H, R_C, R_{CC}, and R_{CH}.

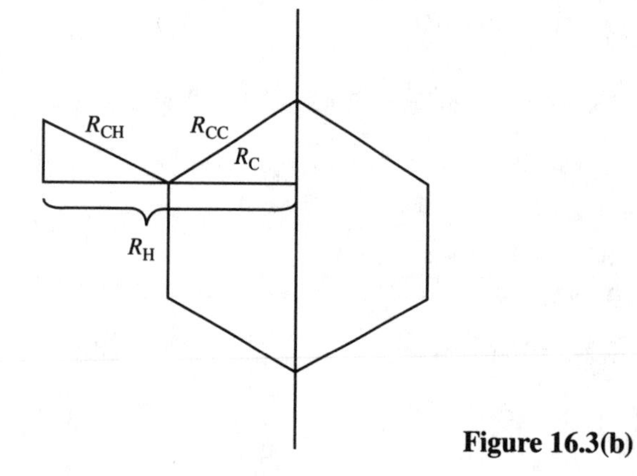

Figure 16.3(b)

$$R_{CC} = \frac{R_C}{\cos 30°} = \frac{1.209 \times 10^{-10}\ \mathrm{m}}{0.8660} = 1.396 \times 10^{-10}\ \mathrm{m} = \boxed{139.6\ \mathrm{pm}}$$
$$R_{CH} = \frac{R_H - R_C}{\cos 30°} = \frac{0.940 \times 10^{-10}}{0.8660} = 1.08\bar{5} \times 10^{-10} = \boxed{108.\bar{5}\ \mathrm{pm}}$$
$$R_{CD} = R_{CH}$$

Comment. These values are very close to the interatomic distances quoted by Herzberg in *Electronic spectra and electronic structure of polyatomic molecules*, p. 666 (*Further reading*, Chapter 17), which are 139.7 and 108.4 pm respectively.

P16.7 The Lewis structure is

$$[\ddot{\underset{..}{\mathrm{O}}}{=}\mathrm{N}{=}\ddot{\underset{..}{\mathrm{O}}}]^+$$

VSEPR indicates that the ion is $\boxed{\text{linear}}$ and has a centre of symmetry. The activity of the modes is consistent with the rule of mutual exclusion; none is both infrared and Raman active. These transitions may be compared to those for CO_2 (Fig. 16.46 of the text) and are consistent with them. The Raman active mode at $1400\,\text{cm}^{-1}$ is due to a symmetric stretch ($\tilde{\nu}_1$), that at $2360\,\text{cm}^{-1}$ to the antisymmetric stretch ($\tilde{\nu}_3$) and that at $540\,\text{cm}^{-1}$ to the two perpendicular bending modes ($\tilde{\nu}_2$). There is a combination band, $\tilde{\nu}_1 + \tilde{\nu}_3 = 3760\,\text{cm}^{-1} \approx 3735\,\text{cm}^{-1}$, which shows a weak intensity in the infrared.

P16.8 The separations between neighbouring lines are

$$20.81, 20.60, 20.64, 20.52, 20.34, 20.37, 20.26 \quad \text{mean: } 20.51\,\text{cm}^{-1}$$

Hence $B = \left(\frac{1}{2}\right) \times (20.51\,\text{cm}^{-1}) = 10.26\,\text{cm}^{-1}$ and

$$I = \frac{\hbar}{4\pi cB} = \frac{1.05457 \times 10^{-34}\,\text{J s}}{(4\pi) \times (2.99793 \times 10^{10}\,\text{cm s}^{-1}) \times (10.26\,\text{cm}^{-1})} = \boxed{2.728 \times 10^{-47}\,\text{kg m}^2}$$

$$R = \left(\frac{I}{m_{\text{eff}}}\right)^{1/2} \text{ [Table 16.1] with } m_{\text{eff}} = 1.6266 \times 10^{-27}\,\text{kg [Exercise 16.3(a)]}$$

$$= \left(\frac{2.728 \times 10^{-47}\,\text{kg m}^2}{1.6266 \times 10^{-27}\,\text{kg}}\right)^{1/2} = \boxed{129.5\,\text{pm}}$$

Comment. A more accurate value would be obtained by ascribing the variation of the separations to centrifugal distortion, and not taking a simple average. Alternatively, the effect of centrifugal distortion could be minimized by plotting the observed separations against J, fitting them to a smooth curve, and extrapolating that curve to $J = 0$. Since $B \propto \frac{1}{I}$ and $I \propto m_{\text{eff}}$, $B \propto \frac{1}{m_{\text{eff}}}$. Hence, the corresponding lines in $^2H^{35}Cl$ will lie at a factor

$$\frac{m_{\text{eff}}(^1H^{35}Cl)}{m_{\text{eff}}(^2H^{35}Cl)} = \frac{1.6266}{3.1624} = 0.5144$$

to low frequency of $^1H^{35}Cl$ lines. Hence, we expect lines at $\boxed{10.56, 21.11, 31.67, \ldots\,\text{cm}^{-1}}$

P16.10 From the equation for a linear rotor in Table 16.1 it is possible to show that $I_{\text{m}} = m_{\text{a}}m_{\text{c}}(R + R')^2 + m_{\text{a}}m_{\text{b}}R^2 + m_{\text{b}}m_{\text{c}}R'^2$.

Thus, $$I(^{16}O^{12}C^{32}S) = \left(\frac{m(^{16}O)m(^{32}S)}{m(^{16}O^{12}C^{32}S)}\right) \times (R + R')^2 + \left(\frac{m(^{12}C)\{m(^{16}O)R^2 + m(^{32}S)R'^2\}}{m(^{16}O^{12}C^{32}S)}\right)$$

$$I(^{16}O^{12}C^{34}S) = \left(\frac{m(^{16}O)m(^{34}S)}{m(^{16}O^{12}C^{34}S)}\right) \times (R + R')^2 + \left(\frac{m(^{12}C)\{m(^{16}O)R^2 + m(^{34}S)R'^2\}}{m(^{16}O^{12}C^{34}S)}\right)$$

$m(^{16}O) = 15.9949\,\text{u}$, $m(^{12}C) = 12.0000\,\text{u}$, $m(^{32}S) = 31.9721\,\text{u}$, and $m(^{34}S) = 33.9679\,\text{u}$. Hence,

$$I(^{16}O^{12}C^{32}S)/\text{u} = (8.5279) \times (R + R')^2 + (0.20011) \times (15.9949R^2 + 31.9721R'^2)$$
$$I(^{16}O^{12}C^{34}S)/\text{u} = (8.7684) \times (R + R')^2 + (0.19366) \times (15.9949R^2 + 33.9679R'^2)$$

The spectral data provides the experimental values of the moments of inertia based on the relation $\nu = 2cB(J + 1)$ [43] with $B = \frac{\hbar}{4\pi cI}$ [30]. These values are set equal to the above equations which are then solved for R and R'. The mean values of I obtained from the data are

$$I(^{16}O^{12}C^{32}S) = 1.37998 \times 10^{-45}\,\text{kg m}^2$$
$$I(^{16}O^{12}C^{34}S) = 1.41460 \times 10^{-45}\,\text{kg m}^2$$

Therefore, after conversion of the atomic mass units to kg, the equations we must solve are

$$1.37998\times10^{-45}\,\text{m}^2 = (1.4161\times10^{-26})\times(R+R')^2+(5.3150\times10^{-27}R^2) +(1.0624\times10^{-26}R'^2)$$

$$1.41460\times10^{-45}\,\text{m}^2 = (1.4560\times10^{-26})\times(R+R')^2+(5.1437\times10^{-27}R^2) +(1.0923\times10^{-26}R'^2)$$

These two equations may be solved for R and R'. They are tedious to solve, but straightforward. Exercise 16.6(b) illustrates the details of the solution. The outcome is $R = \boxed{116.28\,\text{pm}}$ and $R' = \boxed{155.97\,\text{pm}}$. These values may be checked by direct substitution into the equations.

Comment. The starting point of this problem is the actual experimental data on spectral line positions. Exercise 16.6(b) is similar to this problem; its starting points is, however, given values of the rotational constants B, which were themselves obtained from the spectral line positions. So the results for R and R' are expected to be essentially identical and they are.

Question. What are the rotational constants calculated from the data on the positions of the absorption lines?

P16.12 $V(R) = hcD_e\{1-e^{-a(R-R_e)}\}^2$ [61]

$$\tilde{\nu} = \frac{\omega}{2\pi c} = 936.8\,\text{cm}^{-1} \qquad x_e\tilde{\nu} = 14.15\,\text{cm}^{-1}$$

$$a = \left(\frac{m_{\text{eff}}}{2hcD_e}\right)^{1/2}\omega \quad x_e = \frac{\hbar a^2}{2m_{\text{eff}}\omega} \quad D_e = \frac{\tilde{\nu}}{4x_e}$$

$$m_{\text{eff}}(\text{RbH}) \approx \frac{(1.008)\times(85.47)}{(1.008)+(85.47)}\text{u} = 1.654\times10^{-27}\,\text{kg}$$

$$D_e = \frac{\tilde{\nu}^2}{4x_e\tilde{\nu}} = \frac{(936.8\,\text{cm}^{-1})^2}{(4)\times(14.15\,\text{cm}^{-1})} = 1550\bar{5}\,\text{cm}^{-1} \quad [1.92\,\text{eV}]$$

$$a = 2\pi\nu\left(\frac{m_{\text{eff}}}{2hcD_e}\right)^{1/2}[61] = 2\pi c\tilde{\nu}\left(\frac{m_{\text{eff}}}{2hcD_e}\right)^{1/2}$$

$$= (2\pi)\times(2.998\times10^{10}\,\text{cm s}^{-1})\times(936.8\,\text{cm}^{-1})$$

$$\times\left(\frac{1.654\times10^{-27}\,\text{kg}}{(2)\times(15505\,\text{cm}^{-1})\times(6.626\times10^{-34}\,\text{J s})\times(2.998\times10^{10}\,\text{cm s}^{-1})}\right)^{1/2}$$

$$= 9.144\times10^{9}\,\text{m}^{-1} = 9.144\,\text{nm}^{-1} = \frac{1}{0.1094\,\text{nm}}$$

Therefore, $\dfrac{V(R)}{hcD_e} = \{1-e^{-(R-R_e)/(0.1094\,\text{nm})}\}^2$

with $R_e = 236.7$ pm. We draw up the following table

R/pm	50	100	200	300	400	500	600	700	800
$V/(hcD_e)$	20.4	6.20	0.159	0.193	0.601	0.828	0.929	0.971	0.988

These points are plotted in Fig. 16.4 as the line labelled $J = 0$

For the second part, we note that $B \propto \dfrac{1}{R^2}$ and write

$$V_J^* = V + hcB_eJ(J+1)\times\left(\frac{R_e^2}{R^2}\right)$$

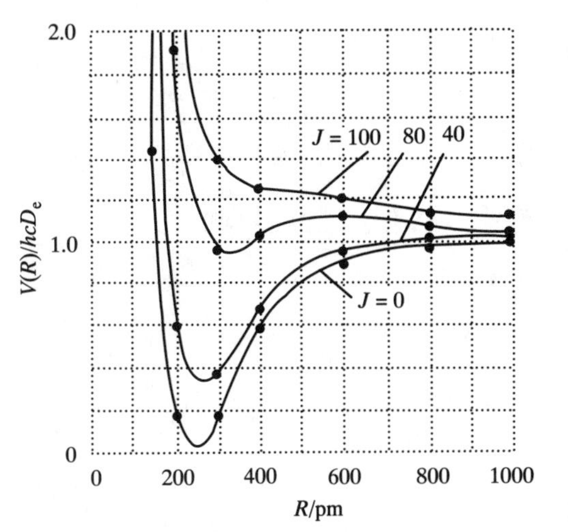

Figure 16.4

with B_e the equilibrium rotational constant, $B_e = 3.020\,\text{cm}^{-1}$.

We then draw up the following table using the values of V calculated above

R/pm	50	100	200	300	400	600	800	1000
$\dfrac{R_e}{R}$	4.73	2.37	1.18	0.79	0.59	0.39	0.30	0.24
$\dfrac{V}{hcD_e}$	20.4	6.20	0.159	0.193	0.601	0.929	0.988	1.000
$\dfrac{V_{40}^*}{hcD_e}$	27.5	7.99	0.606	0.392	0.713	0.979	1.016	1.016
$\dfrac{V_{80}^*}{hcD_e}$	48.7	13.3	1.93	0.979	1.043	1.13	1.099	1.069
$\dfrac{V_{100}^*}{hcD_e}$	64.5	17.2	2.91	1.42	1.29	1.24	1.16	1.11

These points are also plotted in Fig. 16.4.

Solutions to theoretical problems

P16.14 $N \propto g\mathrm{e}^{-E/kT}$ [Boltzmann distribution, Chapters 0 and 19]

$$N_J \propto g_J \mathrm{e}^{-Eg/kT} \propto (2J+1)\mathrm{e}^{-hcBJ(J+1)/kT} \quad [g_J = 2J+1 \text{ for a diatomic rotor}]$$

The maximum population occurs when

$$\frac{\mathrm{d}}{\mathrm{d}J}N_J \propto \left\{2 - (2J+1)^2 \times \left(\frac{hcB}{kT}\right)\right\} \mathrm{e}^{-hcBJ(J+1)/kT} = 0$$

and, since the exponential can never be zero at a finite temperature, when

$$(2J+1)^2 \times \left(\frac{hcB}{kT}\right) = 2$$

or when $J_{\max} = \boxed{\left(\dfrac{kT}{2hcB}\right)^{1/2} - \dfrac{1}{2}}$

For ICl, with $\dfrac{kT}{hc} = 207.22\,\text{cm}^{-1}$ (inside front cover)

$$J_{\text{max}} = \left(\frac{207.22\,\text{cm}^{-1}}{0.2284\,\text{cm}^{-1}}\right)^{1/2} - \frac{1}{2} = \boxed{30}$$

For a spherical rotor, $N_J \propto (2J+1)^2 e^{-hcBJ(J+1)/kT}$ $[g_J = (2J+1)^2]$
and the greatest population occurs when

$$\frac{dN_J}{dJ} \propto \left(8J + 4 - \frac{hcB(2J+1)^3}{kT}\right) e^{-hcBJ(J+1)/kT} = 0$$

which occurs when

$$4(2J+1) = \frac{hcB(2J+1)^3}{kT}$$

$$\text{or at } J_{\text{max}} = \boxed{\left(\frac{kT}{hcB}\right)^{1/2} - \frac{1}{2}}$$

$$\text{For } CH_4,\ J_{\text{max}} = \left(\frac{207.22\,\text{cm}^{-1}}{5.24\,\text{cm}^{-1}}\right)^{1/2} - \frac{1}{2} = \boxed{6}$$

Solutions to additional problems

P16.16 The rotational constant is related to the moment of inertia, which in turn is related to the internuclear separation

$$B = \frac{\hbar}{4\pi c I} = \frac{\hbar}{4\pi c m_{\text{eff}} R^2} \quad \text{so} \quad R = \left(\frac{\hbar}{4\pi c B m_{\text{eff}}}\right)^{1/2}$$

The effective mass is given by

$$m_{\text{eff}}^{-1} = m_1^{-1} + m_2^{-1} = (39.963\,\text{u})^{-1} + (19.992\,\text{u})^{-1} = 7.5043 \times 10^{-2}\,\text{u}^{-1}$$

$$\text{or} \quad m_{\text{eff}} = \frac{1.66054 \times 10^{-27}\,\text{kg u}^{-1}}{7.5043 \times 10^{-2}\,\text{u}^{-1}} = 2.2128 \times 10^{-26}\,\text{kg}$$

$$\text{then } R = \left(\frac{1.0546 \times 10^{-34}\,\text{J s}}{4\pi(2914.9 \times 10^6\,\text{s}^{-1}) \times (2.2128 \times 10^{-26}\,\text{kg})}\right)^{1/2} = \boxed{3.6071 \times 10^{-10}\,\text{m}}$$

The distortion constant is related to the fundamental vibrational wavenumber by

$$D_J = \frac{4B^3}{\tilde{\nu}^2} \quad \text{so} \quad \tilde{\nu} = \left(\frac{4B^3}{D_J}\right)^{1/2} = \left(\frac{4(cB)^3}{c^2(cD_J)}\right)^{1/2}$$

$$\tilde{\nu} = \left(\frac{4(2914.9 \times 10^6\,\text{s}^{-1})^3}{(2.998 \times 10^{10}\,\text{cm s}^{-1})^2 \times (231.01 \times 10^3\,\text{s}^{-1})}\right)^{1/2} = \boxed{21.84\,\text{cm}^{-1}}$$

The force constant is related to the vibrational frequency by

$$\omega = \left(\frac{k}{m_{\text{eff}}}\right)^{1/2} = 2\pi\nu = 2\pi c\tilde{\nu} \quad \text{so} \quad k = (2\pi c\tilde{\nu})^2 m_{\text{eff}}$$

$$k = [2\pi(2.998 \times 10^{10}\,\text{cm s}^{-1}) \times (21.84\,\text{cm}^{-1})]^2 \times (2.2128 \times 10^{-26}\,\text{kg}) = \boxed{0.3746\,\text{kg s}^{-2}}$$

P16.18 **(a)** In the harmonic approximation

$$D_e = D_0 + \tfrac{1}{2}\tilde{\nu} \quad \text{so} \quad \tilde{\nu} = 2(D_e - D_0)$$

$$\tilde{\nu} = \frac{2(1.51 \times 10^{-23}\,\text{J} - 2 \times 10^{-26}\,\text{J})}{(6.626 \times 10^{-34}\,\text{J s}) \times (2.998 \times 10^{8}\,\text{m s}^{-1})} = \boxed{152\,\text{m}^{-1}}$$

The force constant is related to the vibrational frequency by

$$\omega = \left(\frac{k}{m_{\text{eff}}}\right)^{1/2} = 2\pi\nu = 2\pi c\tilde{\nu} \quad \text{so} \quad k = (2\pi c\tilde{\nu})^2 m_{\text{eff}}$$

The effective mass is

$$m_{\text{eff}} = \tfrac{1}{2}m = \tfrac{1}{2}(4.003\,\text{u}) \times (1.66 \times 10^{-27}\,\text{kg u}^{-1}) = 3.32 \times 10^{-27}\,\text{kg}$$

$$k = [2\pi(2.998 \times 10^{8}\,\text{m s}^{-1}) \times (152\,\text{m}^{-1})]^2 \times (3.32 \times 10^{-27}\,\text{kg})$$

$$= \boxed{2.72 \times 10^{-4}\,\text{kg s}^{-2}}$$

The moment of inertia is

$$I = m_{\text{eff}} R_e^2 = (3.32 \times 10^{-27}\,\text{kg}) \times (297 \times 10^{-12}\,\text{m})^2 = \boxed{2.93 \times 10^{-46}\,\text{kg m}^2}$$

The rotational constant is

$$B = \frac{\hbar}{4\pi c I} = \frac{1.0546 \times 10^{-34}\,\text{J s}}{4\pi(2.998 \times 10^{8}\,\text{m s}^{-1}) \times (2.93 \times 10^{-46}\,\text{kg m}^2)} = \boxed{95.5\,\text{m}^{-1}}$$

(b) In the Morse potential

$$x_e = \frac{\tilde{\nu}}{4D_e} \quad \text{and} \quad D_e = D_0 + \tfrac{1}{2}(1 - \tfrac{1}{2}x_e)\tilde{\nu} = D_0 + \tfrac{1}{2}\left(1 - \tfrac{\tilde{\nu}}{8D_e}\right)\tilde{\nu}$$

This rearranges to a quadratic equation in $\tilde{\nu}$

$$\frac{\tilde{\nu}^2}{16D_e} - \tfrac{1}{2}\tilde{\nu} + D_e - D_0 = 0 \quad \text{so} \quad \tilde{\nu} = \frac{\frac{1}{2} - \sqrt{\left(\frac{1}{2}\right)^2 - \frac{4(D_e - D_0)}{16D_e}}}{2(16D_e)^{-1}}$$

$$\tilde{\nu} = 4D_e\left(1 - \sqrt{\frac{D_0}{D_e}}\right)$$

$$= \frac{4(1.51 \times 10^{-23}\,\text{J})}{(6.626 \times 10^{-34}\,\text{J s}) \times (2.998 \times 10^{8}\,\text{m s}^{-1})}\left(1 - \sqrt{\frac{2 \times 10^{-26}\,\text{J}}{1.51 \times 10^{-23}\,\text{J}}}\right)$$

$$= \boxed{293\,\text{m}^{-1}}$$

and $$x_e = \frac{(293\,\text{m}^{-1}) \times (6.626 \times 10^{-34}\,\text{J s}) \times (2.998 \times 10^{8}\,\text{m s}^{-1})}{4(1.51 \times 10^{-23}\,\text{J})} = \boxed{0.96}$$

P16.20 $E_J = J(J+1)hcB, \qquad g_J = 2J+1$

$$E_1 - E_0 = 2hcB = hc\left(\frac{1}{\lambda_{\text{shorter}}} - \frac{1}{\lambda_{\text{longer}}}\right)$$

$$B = \frac{1}{2}\left(\frac{1}{\lambda_{\text{shorter}}} - \frac{1}{\lambda_{\text{longer}}}\right) = \frac{1}{2}\left(\frac{1}{\lambda_{\text{shorter}}} - \frac{1}{\lambda_{\text{shorter}} + \Delta\lambda}\right)$$

$$= \frac{1}{2}\left(\frac{1}{\lambda_{\text{shorter}}}\right) \times \left(1 - \frac{1}{1 + \frac{\Delta\lambda}{\lambda_{\text{shorter}}}}\right)$$

$$= \frac{1}{2}\left(\frac{1}{387.5\text{nm}}\right) \times \left(1 - \frac{1}{1 + \frac{0.061}{387.5}}\right) \times \left(\frac{10^9\text{ nm}}{10^2\text{ cm}}\right)$$

$$B = \boxed{2.031\text{ cm}^{-1}}$$

$$\frac{E_1 - E_0}{k} = \frac{2hcB}{k} = \frac{2(6.626 \times 10^{-34}\text{ J s}) \times (3.00 \times 10^{10}\text{ cm s}^{-1}) \times (2.031\text{ cm}^{-1})}{1.381 \times 10^{-23}\text{ J K}^{-1}}$$

$$= 5.84\bar{7}\text{ K}$$

Intensity of $J' \leftarrow J$ absorption line $I_J \propto g_J e^{-E_J/kT}$

$$\frac{I_{\lambda_{\text{longer}}}}{I_{\lambda_{\text{shorter}}}} \simeq \frac{g_1 e^{-E_1/kT}}{g_0 e^{-E_0/kT}} = \frac{g_1}{g_0} e^{-(E_1 - E_0)/kT}$$

Solve for T

$$T = \left(\frac{E_1 - E_0}{k}\right) \times \left(\frac{1}{\ln\left(\frac{g_1 I_{\lambda_{\text{shorter}}}}{g_0 I_{\lambda_{\text{longer}}}}\right)}\right) = 5.84\bar{7}\text{ K}\left(\frac{1}{\ln(3 \times 4)}\right) = \boxed{2.35\text{ K}}$$

17 Spectroscopy 2: electronic transitions

Solutions to exercises

E17.1 The reduction in intensity obeys the Beer–Lambert law introduced in Chapter 16. It applies equally well to the spectroscopic methods of this chapter.

$$\log\frac{I}{I_0} = -\varepsilon[\mathrm{J}]l \text{ [16.9 and 16.10]}$$

$$= (-855\,\mathrm{L\,mol^{-1}\,cm^{-1}}) \times (3.25\times10^{-3}\,\mathrm{mol\,L^{-1}}) \times (0.25\,\mathrm{cm})$$

$$= -0.69\bar{5}$$

Hence, $\frac{I}{I_0} = 0.20$, and the reduction in intensity is $\boxed{80 \text{ per cent}}$

E17.2 $\log\frac{I}{I_0} = -\varepsilon[\mathrm{J}]l$ [16.9, 16.10]

Hence, $\varepsilon = -\frac{1}{[\mathrm{J}]l}\log\frac{I}{I_0} = -\frac{\log 0.201}{(1.11\times10^{-4}\,\mathrm{mol\,L^{-1}})\times(1.00\,\mathrm{cm})} = \boxed{6.28\times10^{3}\,\mathrm{L\,mol^{-1}\,cm^{-1}}}$

E17.3 $[\mathrm{J}] = -\frac{1}{\varepsilon l}\log\frac{I}{I_0}\text{[16.9, 16.10]} = \frac{-1}{(286\,\mathrm{L\,mol^{-1}\,cm^{-1}})\times(0.65\,\mathrm{cm})}\log(1-0.465)$

$= \boxed{1.5\,\mathrm{mmol\,L^{-1}}}$

E17.4 $A = \int \varepsilon\, d\tilde{\nu}$ [16.11]

The integral can be approximated by the area under the triangle [area $= \frac{1}{2}\times$ base $\times$ height]

$$A = \tfrac{1}{2}\times(43480 - 34480)\,\mathrm{cm^{-1}}\times(1.21\times10^{4}\,\mathrm{L\,mol^{-1}\,cm^{-1}})$$

$$= \boxed{5.44\times10^{7}\,\mathrm{L\,mol^{-1}\,cm^{-2}}}$$

E17.5 π-electrons in polyenes may be considered as particles in a one-dimensional box. Applying the Pauli exclusion principle, the N conjugated electrons will fill the levels, two electrons at a time, up to the level $n = \frac{N}{2}$. Since N is also the number of alkene carbon atoms. Nd is the length of the box, with d the carbon–carbon interatomic distance. Hence

$$E_n = \frac{n^2h^2}{8mN^2d^2}$$

where, for the lowest energy transition ($\Delta n = +1$)

$$\Delta E = h\nu = \frac{hc}{\lambda} = E_{(N/2)+1} - E_{(N/2)} = \frac{h^2(N+1)}{8md^2N^2}$$

Therefore, the larger N, the larger λ. Hence the absorption at 243 nm is due to the diene and that at 192 nm to the butene.

Question. How accurate is the formula derived above in predicting the wavelengths of the absorption maxima in these two compounds?

E17.6 $\varepsilon = -\frac{1}{[J]l}\log\frac{I}{I_0}$ [16.9, 16.10] with $l = 0.20\,\text{cm}$

We use this formula to draw up the following table

$[Br_2]/\text{mol L}^{-1}$	0.0010	0.0050	0.0100	0.0500	
I/I_0	0.814	0.356	0.127	3.0×10^{-5}	
$\varepsilon/(\text{L mol}^{-1}\,\text{cm}^{-1})$	447	449	448	452	mean: $44\overline{9}$

Hence, the molar absorption coefficient is $\varepsilon = \boxed{450\,\text{L mol}^{-1}\,\text{cm}^{-1}}$

E17.7 $\varepsilon = -\frac{1}{[J]l}\log\frac{I}{I_0}$ [16.9, 16.10] $= \frac{-1}{(0.010\,\text{mol L}^{-1})\times(0.20\,\text{cm})}\log 0.48 = \boxed{159\,\text{L mol}^{-1}\,\text{cm}^{-1}}$

$$T = \frac{I}{I_0} = 10^{-[J]\varepsilon l}$$

$$= 10^{(-0.010\,\text{mol L}^{-1})\times(159\,\text{L mol}^{-1}\,\text{cm}^{-1})\times(0.40)} = 10^{-0.63\overline{6}} = 0.23, \text{ or } \boxed{23 \text{ per cent}}$$

E17.8 $l = \frac{-1}{\varepsilon[J]}\log\frac{I}{I_0}$

For water, $[H_2O] \approx \frac{1.00\,\text{kg/L}}{18.02\,\text{g mol}^{-1}} = 55.5\,\text{mol L}^{-1}$

and $\varepsilon[J] = (55.5\,\text{M})\times(6.2\times10^{-5}\,\text{M}^{-1}\,\text{cm}^{-1}) = 3.4\times10^{-3}\,\text{cm}^{-1} = 0.34\,\text{m}^{-1}$, so $\frac{1}{\varepsilon[J]} = 2.9\,\text{nm}$

Hence, $l/\text{m} = -2.9\times\log\frac{I}{I_0}$

(a) $\frac{I}{I_0} = 0.5,\ l = -2.9\,\text{m}\times\log 0.5 = \boxed{0.9\,\text{m}}$ **(b)** $\frac{I}{I_0} = 0.1,\ l = -2.9\,\text{m}\times\log 0.1 = \boxed{3\,\text{m}}$

E17.9 We will make the same assumption as in Exercise 17.4**(a)** namely that the absorption curve can be approximated by a triangle. Refer to Fig. 17.1.

$$\mathcal{A} = \int \varepsilon\, d\tilde{\nu}\ [16.11]$$

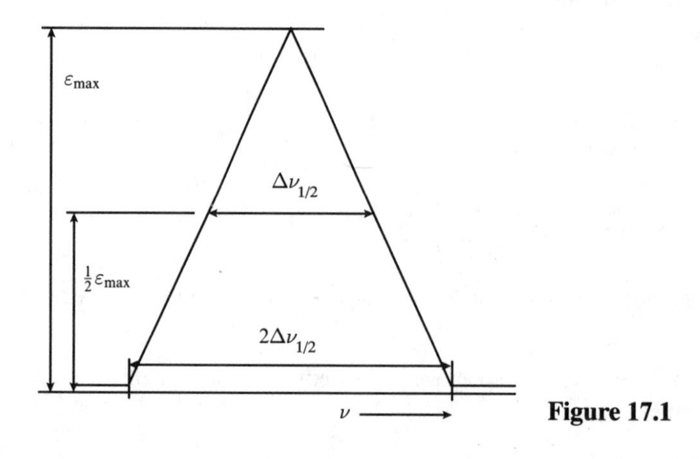

Figure 17.1

From the illustration,

$$\mathcal{A} = \tfrac{1}{2} \times \varepsilon_{\max} \times 2\Delta\tilde{\nu}_{1/2} \text{ [area} = \tfrac{1}{2} \times \text{height} \times \text{base]} = \varepsilon_{\max}\Delta\tilde{\nu}_{1/2}$$

$$\mathcal{A} = 5000\,\text{cm}^{-1} \times \varepsilon_{\max}$$

(a) $\mathcal{A} = 5000\,\text{cm}^{-1} \times 1 \times 10^4\,\text{L mol}^{-1}\,\text{cm}^{-1} = \boxed{5 \times 10^7\,\text{L mol}^{-1}\,\text{cm}^{-2}}$

(b) $\mathcal{A} = (5000\,\text{cm}^{-1}) \times (5 \times 10^2\,\text{L mol}^{-1}\,\text{cm}^{-1}) = \boxed{25 \times 10^5\,\text{L mol}^{-1}\,\text{cm}^{-2}}$

E17.10 The internuclear distance in H_2^+ is greater than that in H_2. The change in bond length and the corresponding shift in the molecular potential energy curves reduces the Franck–Condon factor for transitions between the two ground vibrational states. It creates a better overlap between $v = 2$ of H_2^+ and $v = 0$ of H_2, and so increases the Franck–Condon factor of that transition.

Solutions to problems

Solutions to numerical problems

P17.1 The potential energy curves for the $X^3\Sigma_g^-$ and $B^3\Sigma_u^-$ electronic states of O_2 are represented schematically in Fig. 17.2 along with the notation used to represent the energy separation of this problem.

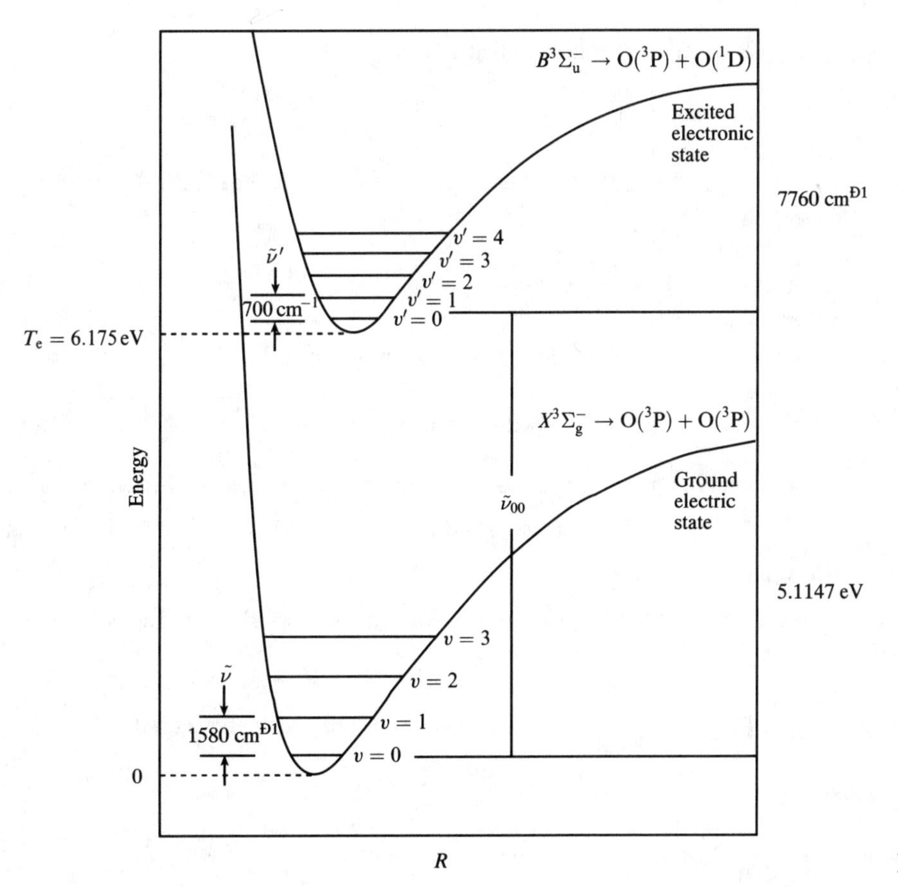

Figure 17.2

Curves for the other electronic state of O_2 are not shown. Ignoring rotational structure and anharmonicity we may write

$$\tilde{\nu}_{00} \approx T_e + \frac{1}{2}(\tilde{\nu}' - \tilde{\nu}) = 6.175\,\text{eV} \times \left(\frac{8065.5\,\text{cm}^{-1}}{1\,\text{eV}}\right) + \frac{1}{2}(700 - 1580)\,\text{cm}^{-1}$$

$$= \boxed{49\,36\bar{4}\,\text{cm}^{-1}}$$

Comment. Note that the selection rule $\Delta v = \pm 1$ does not apply to vibrational transitions between different electronic states.

Question. What is the percentage change in $\tilde{\nu}_{00}$ if the anharmonicity constants $x_e\tilde{\nu}$ (Section 16.11) $12.0730\,\text{cm}^{-1}$ and $8.002\,\text{cm}^{-1}$ for the ground and excited states, respectively, are included in the analysis?

P17.2 The energy of the dissociation products of the B state, $O(^3P)$ and $O(^1D)$, above the $v = 0$ state of the ground state is $7760\,\text{cm}^{-1} + 49\,363\,\text{cm}^{-1} = 57\,123\,\text{cm}^{-1}$. One of these products, $O(^1D)$, has energy $15\,870\,\text{cm}^{-1}$ above the energy of the ground-state atom, $O(^3P)$. Hence, the energy of two ground-state atoms, $2O(^3P)$, above the $v = 0$ state of the ground electronic state is $57\,123\,\text{cm}^{-1} - 15\,870\,\text{cm}^{-1} = 41\,253\,\text{cm}^{-1} = \boxed{5.1147\,\text{eV}}$. These energy relations are indicated (not to scale) in Fig. 17.2.

P17.4 We write $\varepsilon = \varepsilon_{max}e^{-x^2} = \varepsilon_{max}e^{-\tilde{\nu}^2/2\Gamma}$, the variable being $\tilde{\nu}$ and Γ being a constant. $\tilde{\nu}$ is measured from the band centre, at which $\tilde{\nu} = 0$. $\varepsilon = \frac{1}{2}\varepsilon_{max}$ when $\tilde{\nu}^2 = 2\Gamma \ln 2$.

Therefore, the width at half height is

$$\Delta\tilde{\nu}_{1/2} = 2 \times (2\Gamma \ln 2)^{1/2}, \quad \text{implying that} \quad \Gamma = \frac{\Delta\tilde{\nu}_{1/2}^2}{8 \ln 2}$$

Now we carry out the integration

$$\mathcal{A} = \int \varepsilon\,d\tilde{\nu} = \varepsilon_{max} \int_{-\infty}^{\infty} e^{-\tilde{\nu}^2/2\Gamma}\,d\tilde{\nu} = \varepsilon_{max}(2\Gamma\pi)^{1/2} \quad \left[\int_{-\infty}^{\infty} e^{-x^2}\,dx = \pi^{1/2}\right]$$

$$= \varepsilon_{max}\left(\frac{2\pi\,\Delta\tilde{\nu}_{1/2}^2}{8 \ln 2}\right)^{1/2} = \left(\frac{\pi}{4 \ln 2}\right)^{1/2} \varepsilon_{max}\Delta\tilde{\nu}_{1/2} = 1.0645\,\varepsilon_{max}\Delta\tilde{\nu}_{1/2}$$

From Fig. 17.44 of the text we estimate $\varepsilon_{max} \approx 9.5\,\text{L mol}^{-1}\,\text{cm}^{-1}$ and $\Delta\tilde{\nu}_{1/2} \approx 4760\,\text{cm}^{-1}$. Then

$$\mathcal{A} = 1.0645 \times (9.5\,\text{L mol}^{-1}\,\text{cm}^{-1}) \times (4760\,\text{cm}^{-1}) = \boxed{4.8 \times 10^4\,\text{L mol}^{-1}\,\text{cm}^{-2}}$$

The area under the curve on the printed page is about $1288\,\text{mm}^2$, each mm^2 corresponds to about $190.5\,\text{cm}^{-1} \times 0.189\,\text{L mol}^{-1}\,\text{cm}^{-1}$, and so $\int \varepsilon\,d\tilde{\nu} \approx 4.64 \times 10^4\,\text{L mol}^{-1}\,\text{cm}^{-2}$. The agreement with the calculated value above is good.

P17.6 The ratio of the transition probabilities of spontaneous emission to stimulated emission at a frequency ν is given by

$$A = \left(\frac{8\pi h\nu^3}{c^3}\right) B[16.17] = \frac{k}{\lambda^3}B, \text{ where } k \text{ is a constant and we have used } \nu = \frac{c}{\lambda}.$$

Thus at 400 nm

$$A(400) = \frac{k}{(400)^3}B(400), \quad \text{and at 500 nm} \quad A(500) = \frac{k}{(500)^3}B(500)$$

Then, $\dfrac{A(500)}{A(400)} = \left(\dfrac{(400)^3}{(500)^3}\right) \times \left(\dfrac{B(500)}{B(400)}\right) = \left(\dfrac{64}{125}\right) \times 10^{-5} = 5 \times 10^{-6}$

Lifetimes and half-lives are inversely proportional to transition probabilities (rate constants) and hence

$$t_{1/2}(\mathrm{T} \to \mathrm{S}) = \frac{1}{5 \times 10^{-6}} t_{1/2}(\mathrm{S}^* \to \mathrm{S}) = (2 \times 10^5) \times (1.0 \times 10^{-9}\ \mathrm{s}) = \boxed{2 \times 10^{-4}\ \mathrm{s}}$$

P17.8 The valence electron configuration of NO is $\boxed{(1\sigma)^2(2\sigma^*)^2(1\pi)^4(3\sigma)^2(2\pi^*)^1}$. The data refer to the kinetic energies of the ejected electrons, and so the ionization energies are 16.52 eV, 15.65 eV, and 9.21 eV. The 16.52 eV line refers to ionization of 3σ electron, and the 15.65 eV line (with its long vibrational progression) to ionization of a 1π electron. The 9.21 eV line refers to the ionization of the least strongly attached electron, that is, $2\pi^*$.

Solutions to theoretical problems

P17.10 **(a)** Ethene (ethylene) belongs to D_{2h}. In this group the x, y, and z components of the dipole moment transform as B_{3u}, B_{2u}, and B_{1u} respectively. [See a more extensive set of character tables than in the text.] The π orbital is B_{1u} (like z, the axis perpendicular to the plane) and π^* is B_{3g}. Since $B_{3g} \times B_{1u} = B_{2u}$ and $B_{2u} \times B_{2u} = A_{1g}$, the transition is $\boxed{\text{allowed}}$ (and is y-polarized).

(b) Regard the CO group with its attached groups as locally C_{2v}. The dipole moment has components that transform as $A_1(z)$, $B_1(x)$, and $B_2(y)$, with the z-axis along the C=O direction and x perpendicular to the R_2CO plane. The n orbital is p_y (in the R_2CO plane), and hence transforms as B_2. The π^* orbital is p_x (perpendicular to the R_2CO plane), and hence transforms as B_1. Since $\Gamma_f \times \Gamma_i = B_1 \times B_2 = A_2$, but no component of the dipole moment transforms as A_2, the transition is $\boxed{\text{forbidden}}$

P17.12 The fluorescence spectrum gives the vibrational splitting of the lower state. The wavelengths stated correspond to the wavenumbers 22 730, 24 390, 25 640, 27 030 $\mathrm{cm^{-1}}$, indicating spacings of 1660, 1250, and 1390 $\mathrm{cm^{-1}}$. The absorption spectrum spacing gives the separation of the vibrational levels of the upper state. The wavenumbers of the absorption peaks are 27 800, 29 000, 30 300, and 32 800 $\mathrm{cm^{-1}}$. The vibrational spacings are therefore 1200, 1300, and 2500 $\mathrm{cm^{-1}}$.

P17.15 $\mu = -eSR$ [given]

$$S = \left[1 + \frac{R}{a_0} + \frac{1}{3}\left(\frac{R}{a_0}\right)^2\right] \mathrm{e}^{-R/a_0} \text{ [Problem 14.3]}$$

$$f = \frac{8\pi^2 m_e \nu}{3he^2}\mu^2 = \frac{8\pi^2 m_e \nu}{3h} R^2 S^2 = \frac{8\pi^2 m_e \nu a_0^2}{3h}\left(\frac{RS}{a_0}\right)^2 = \boxed{\left(\frac{RS}{a_0}\right)^2 f_0}$$

We then draw up the following table

R/a_0	0	1	2	3	4	5	6	7	8
f/f_0	0	0.737	1.376	1.093	0.573	0.233	0.080	0.024	0.007

These points are plotted in Fig. 17.3.

The maximum in f occurs at the maximum of RS

$$\frac{\mathrm{d}}{\mathrm{d}R}(RS) = S + R\frac{\mathrm{d}S}{\mathrm{d}R} = \left[1 + \frac{R}{a_0} - \frac{1}{3}\left(\frac{R}{a_0}\right)^3\right]\mathrm{e}^{-R/a_0} = 0 \quad \text{at } R = R^*$$

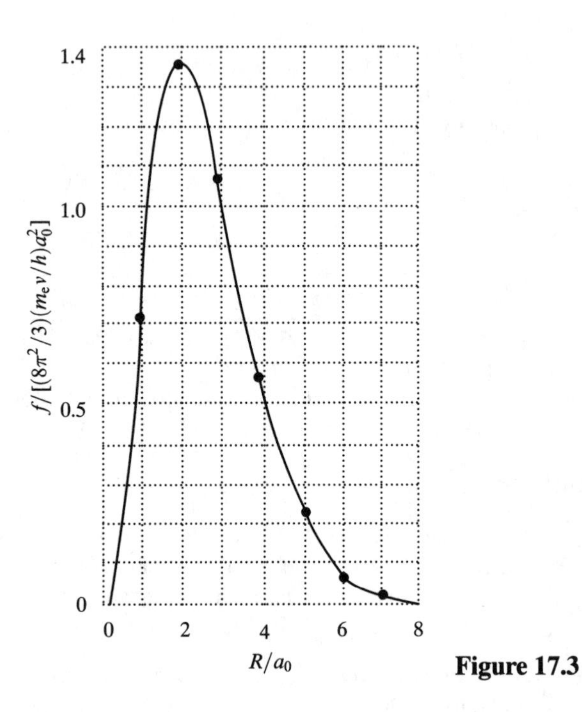

Figure 17.3

That is, $1 + \dfrac{R^*}{a_0} - \dfrac{1}{3}\left(\dfrac{R^*}{a_0}\right)^3 = 0$

This equation may be solved either numerically or analytically (see Abramowitz and Stegun, *Handbook of mathematical functions*, Section 3.8.2), and $R^* = 2.10380a_0$

As $R \to 0$, the transition becomes $s \to s$, which is forbidden. As $R \to \infty$, the electron is confined to a single atom because its wavefunction does not extend to the other.

Solutions to additional problems

P17.16 $V = \text{volume of crystal} = \pi r^2 \cdot l = \pi \left(\dfrac{0.50\,\text{cm}}{2}\right)^2 \times (5.0\,\text{cm}) \quad = 0.98\,\text{cm}^3$

$$\text{Maximum mass of excited ions} \simeq \left(\frac{0.050\,\text{g Cr}^{3+}}{100\,\text{g crystal}}\right) \times \left(\frac{3.97\,\text{g crystal}}{\text{cm}^3}\right) \times (0.98\,\text{cm}^3)$$
$$\simeq 1.95 \times 10^{-3}\,\text{g Cr}^{3+}$$

$$\text{Maximum number of photon emissions} \simeq (1.95 \times 10^{-3}\,\text{g Cr}^{3+}) \times \left(\frac{1\,\text{mol}}{52.0\,\text{g Cr}}\right) \times \left(\frac{6.022 \times 10^{23}}{\text{mol}}\right)$$
$$\simeq 2.26 \times 10^{19}$$

$$\text{Maximum total energy of emissions} \simeq 2.26 \times 10^{19}\frac{hc}{\lambda}$$
$$\simeq \frac{2.26 \times 10^{19}\,(6.626 \times 10^{-34}\,\text{J s}) \times (3.00 \times 10^{8}\,\text{m s}^{-1})}{694.3 \times 10^{-9}\,\text{m}}$$
$$\simeq 6.46\,\text{J}$$

$$\text{Maximum power} = \frac{\text{Maximum emitted energy}}{\Delta t} = \frac{6.46\,\text{J}}{100 \times 10^{-9}\,\text{s}}$$

$$\text{Maximum power} \approx \boxed{65\,\text{MW}}$$

P17.17 For a photon to induce a spectroscopic transition, the transition moment $\langle\mu\rangle$ must be nonzero. The Laporte selection rule forbids transitions that involve no change in parity. So transitions to the Π_u states are forbidden. (Note. These states may not even be reached by a vibronic transition, for these molecules have only one vibrational mode and it is centrosymmetric.)

We will judge transitions to the other states with the assistance of the $D_{\infty h}$ character table. The transition moment is the integral $\int \psi_f^* \mu \psi_i \, d\tau$, where the dipole moment operator has components proportional to the Cartesian coordinates. The integral vanishes unless the integrand, or at least some part of it, belongs to the totally symmetric representation of the molecule's point group. To find the character of the integrand, we multiply together the characters of its factors. Note that the μ_z has the same symmetry species as the ground state, namely A_{1u}, and the product of the ground state and μ_z has the A_{1g} symmetry species; since the symmetry species are mutually orthogonal, only a state with A_{1g} symmetry can be reached from the ground state with z-polarized light. The ${}^2\Sigma_g^+$ state is such a state, so $\boxed{{}^2\Sigma_g^+ \leftarrow {}^2\Sigma_u^+ \text{ is allowed}}$ That leaves x- or y- polarized transitions to the ${}^2\Pi_g$ states to consider.

	E	$\infty C_2'$	$2C_\phi$	i	$\infty\sigma_v$	$2S_\phi$
$\Sigma_u^+(A_{1u})$	1	-1	1	-1	1	-1
$\mu_{x \text{ or } y}(E_{1u})$	2	0	$2\cos\phi$	-2	0	$2\cos\phi$
$\Pi_g(E_{1g})$	2	0	$2\cos\phi$	2	0	$-2\cos\phi$
Integrand	4	0	$4\cos^2\phi$	4	0	$4\cos^2\phi$

The little orthogonality theorem (see the solution to Problem 15.18) gives the coefficient of A_{1g} in the integrand as

$$c_{A_{1g}} = (1/h)\Sigma_C g(C)\chi(C) = [4 + 0 + 2(4\cos^2\phi) + 4 + 0 + 2(4\cos^2\phi)]/\infty = 0.$$

So the integrand does not contain A_{1g}, and the $\boxed{\text{transition to } {}^2\Pi_g \text{ would be forbidden}}$

P17.19 **(a)** The integrated absorption coefficient is (specializing to a triangular lineshape)

$$\mathcal{A} = \int \varepsilon(\tilde{\nu})\,d\tilde{\nu} = (1/2)\varepsilon_{\max}\Delta\tilde{\nu}$$

$$= (1/2) \times (150\,\text{L mol}^{-1}\,\text{cm}^{-1}) \times (34\,483 - 31\,250)\,\text{cm}^{-1},$$

$$\mathcal{A} = \boxed{2.42 \times 10^5\,\text{L mol}^{-1}\,\text{cm}^{-2}}$$

(b) The concentration of gas under these conditions is

$$c = \frac{n}{V} = \frac{p}{RT} = \frac{2.4\,\text{Torr}}{(62.364\,\text{Torr L mol}^{-1}\,\text{K}^{-1}) \times (373\,\text{K})} = 1.03 \times 10^{-4}\,\text{mol L}^{-1}$$

Over 99 per cent of these gas molecules are monomers, so we take this concentration to be that of CH_3I. (If 1 of every 100 of the original monomers turned to dimers, each produces 0.5 dimers; remaining monomers represent 99 of 99.5 molecules.) Beer's law states

$$A = \varepsilon cl = (150\,\mathrm{L\,mol^{-1}\,cm^{-1}}) \times (1.03 \times 10^{-4}\,\mathrm{mol\,L^{-1}}) \times (12.0\,\mathrm{cm}) = \boxed{0.185}$$

(c) The concentration of gas under these conditions is

$$c = \frac{n}{V} = \frac{p}{RT} = \frac{100\,\mathrm{Torr}}{(62.364\,\mathrm{Torr\,L\,mol^{-1}\,K^{-1}}) \times (373\,\mathrm{K})} = 4.30 \times 10^{-3}\,\mathrm{mol\,L^{-1}}$$

Since 18 per cent of these CH_3I units are in dimers (forming 9 per cent as many molecules as were originally present as monomers), the monomer concentration is only 82/91 of this value or $3.87 \times 10^{-3}\,\mathrm{mol\,L^{-1}}$. Beer's law is

$$A = \varepsilon cl = (150\,\mathrm{L\,mol^{-1}\,cm^{-1}}) \times (3.87 \times 10^{-3}\,\mathrm{mol\,L^{-1}}) \times (12.0\,\mathrm{cm}) = \boxed{6.97}$$

If this absorbance were measured, the molar absorption coefficient inferred from it without consideration of the dimerization would be

$$\begin{aligned}\varepsilon &= A/cl = 6.97/((4.30 \times 10^{-1}\,\mathrm{mol\,L^{-1}}) \times (12.0\,\mathrm{cm})) \\ &= \boxed{135\,\mathrm{L\,mol^{-1}\,cm^{-1}}}\end{aligned}$$

an apparent drop of 10 per cent compared to the low-pressure value.

P17.22 In Fig. 17.4

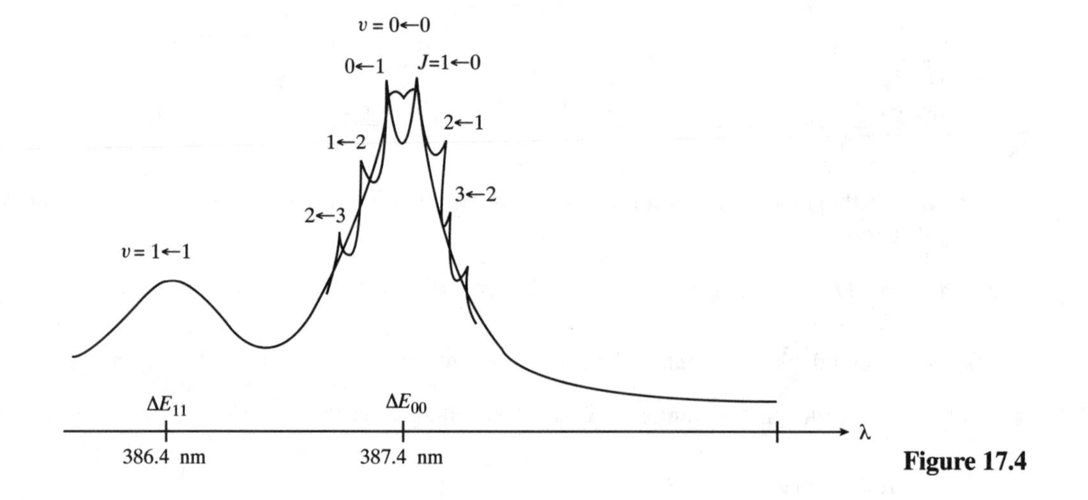

Figure 17.4

$$\Delta E_{11} = \frac{hc}{\lambda_{11}} = \frac{hc}{386.4\,\mathrm{nm}} = 5.1409 \times 10^{-19}\,\mathrm{J} = 3.2087\,\mathrm{eV}$$

and

$$\Delta E_{00} = \frac{hc}{\lambda_{00}} = \frac{hc}{387.6\,\mathrm{nm}} = 5.1250 \times 10^{-19}\,\mathrm{J} = 3.1987\,\mathrm{eV}$$

Energy of excited singlet, S_1: $E_1(v, J) = V_1 + (v + 1/2)\tilde{\nu}_1 hc + J(J+1)\tilde{B}_1 hc$

Energy of ground singlet, S_0: $E_0(v, J) = V_0 + (v + 1/2)\tilde{\nu}_0 hc + J(J+1)\tilde{B}_0 hc$

The midpoint of the 0–0 band corresponds to the forbidden Q branch ($\Delta J = 0$) with $J = 0$ and $v = 0 \leftarrow 0$.

$$\Delta E_{00} = E_1(0,0) - E_0(0,0) = (V_1 - V_0) + \tfrac{1}{2}(\tilde{\nu}_1 - \tilde{\nu}_0)hc \qquad (1)$$

The midpoint of the 1–1 band corresponds to the forbidden Q branch ($\Delta J = 0$) with $J = 0$ and $v = 1 \leftarrow 1$.

$$\Delta E_{11} = E_1(1,0) - E_0(1,0) = (V_1 - V_0) + \tfrac{3}{2}(\tilde{\nu}_1 - \tilde{\nu}_0)hc \qquad (2)$$

Multiplying eqn 1 by three and subtracting eqn 2 gives

$$3\Delta E_{00} - \Delta E_{11} = 2(V_1 - V_0)$$

$$\begin{aligned} V_1 - V_0 &= \tfrac{1}{2}(3\Delta E_{00} - \Delta E_{11}) \\ &= \tfrac{1}{2}\{3(5.1250) - (5.1409)\}10^{-19}\ \text{J} \\ &= 5.1171 \times 10^{-19}\ \text{J} = \boxed{3.1938\ \text{eV}} \end{aligned} \qquad (3)$$

This is the potential energy difference between S_0 and S_1.

Equations (1) and (3) may be solved for $\tilde{\nu}_1 - \tilde{\nu}_0$.

$$\begin{aligned} \tilde{\nu}_1 - \tilde{\nu}_0 &= 2\{\Delta E_{00} - (V_1 - V_0)\} \\ &= 2\{5.1250 - 5.1171\}10^{-19}\ \text{J}/hc \\ &= 1.5800 \times 10^{-21}\ \text{J} = 0.009\,861\,5\ \text{eV} = \boxed{79.538\ \text{cm}^{-1}} \end{aligned}$$

The $\tilde{\nu}_1$ value can be determined by analyzing the band head data for which $J + 1 \leftarrow J$.

$$\begin{aligned} \Delta E_{10}(J) &= E_1(0, J) - E_0(1, J+1) \\ &= V_1 - V_0 + \tfrac{1}{2}(\tilde{\nu}_1 - 3\tilde{\nu}_0)hc + J(J+1)\tilde{B}_1 hc - (J+1) \times (J+2)\tilde{B}_0 hc \end{aligned}$$

$$\Delta E_{00}(J) = V_1 - V_0 + \tfrac{1}{2}(\tilde{\nu}_1 - \tilde{\nu}_0)hc + J(J+1)\tilde{B}_1 hc - (J+1) \times (J+2)\tilde{B}_0 hc$$

Therefore,

$$\Delta E_{00}(J) - \Delta E_{10}(J) = \tilde{\nu}_0 hc$$

$$\Delta E_{00}(J_{\text{head}}) = \frac{hc}{388.3\ \text{nm}} = 5.1158 \times 10^{-19}\ \text{J}$$

$$\Delta E_{10}(J_{\text{head}}) = \frac{hc}{421.6\ \text{nm}} = 4.7117 \times 10^{-19}\ \text{J}$$

$$\begin{aligned} \tilde{\nu}_0 &= \frac{\Delta E_{00}(J) - \Delta E_{10}(J)}{hc} \\ &= \frac{(5.1158 - 4.7117) \times 10^{-19}\ \text{J}}{hc} \\ &= \frac{4.0410 \times 10^{-20}\ \text{J}}{hc} = 0.25222\ \text{eV} = \boxed{2034.3\ \text{cm}^{-1}} \end{aligned}$$

$$\begin{aligned} \tilde{\nu}_1 &= \tilde{\nu}_0 + 79.538\ \text{cm}^{-1} \\ &= (2034.3 + 79.538)\ \text{cm}^{-1} = \boxed{2113.8\ \text{cm}^{-1} = \frac{4.1990 \times 10^{-20}\ \text{J}}{hc}} \end{aligned}$$

$$\frac{I_{1\text{–}1}}{I_{0\text{–}0}} \approx \frac{\mathrm{e}^{-E_1(1,0)/kT_{\text{eff}}}}{\mathrm{e}^{-E_1(0,0)/kT_{\text{eff}}}} = \mathrm{e}^{-(E_1(1,0)-E_1(0,0))/kT_{\text{eff}}}$$

$$\approx \mathrm{e}^{-hc\tilde{\nu}_1/kT_{\text{eff}}}$$

$$\ln\left(\frac{I_{1\text{–}1}}{I_{0\text{–}0}}\right) = -\frac{hc\tilde{\nu}_1}{kT_{\text{eff}}}$$

$$T_{\text{eff}} = \frac{hc\tilde{\nu}_1}{k\ln\left(\frac{I_{0\text{–}0}}{I_{1\text{–}1}}\right)} = \frac{4.1990 \times 10^{-20}\ \text{J}}{(1.38066 \times 10^{-23}\ \text{J K}^{-1})\ln(10)} = \boxed{1321\ \text{K}}$$

The relative population of the $v = 0$ and $v = 1$ vibrational states is the inverse of the relative intensities of the transitions from those states; hence $\frac{1}{0.1} = \boxed{10}$

It would seem that with such a high effective temperature more than eight of the rotational levels of the S_1 state should have a significant population. But the spectra of molecules in comets are never as clearly resolved as those obtained in the laboratory and that is most probably the reason why additional rotational structure does not appear in these spectra.

18 Spectroscopy 3: magnetic resonance

Solutions to exercises

E18.1 The resonance frequency is equal to the Larmor frequency of the proton and is given by

$$\nu = \nu_{\rm L} = \frac{\gamma\mathcal{B}}{2\pi}\ [4] \quad \text{with } \gamma = \frac{g_I\mu_{\rm N}}{\hbar}\ [2]$$

hence $\nu = \dfrac{g_I\mu_{\rm N}\mathcal{B}}{h} = \dfrac{(5.5857)\times(5.0508\times10^{-27}\,{\rm J\,T^{-1}})\times(14.1\,{\rm T})}{6.626\times10^{-34}\,{\rm J\,s}} = \boxed{600\,{\rm MHz}}$

E18.2 $E_{m_I} = -\gamma\hbar\mathcal{B}m_I[3] = -g_I\mu_{\rm N}\mathcal{B}m_I\ [2, \gamma\hbar = g_I\mu_{\rm N}]$

$$m_I = \frac{3}{2}, \frac{1}{2}, -\frac{1}{2}, -\frac{3}{2}$$

$$E_{m_I} = (-0.4289)\times(5.051\times10^{-27}\,{\rm J\,T^{-1}})\times(7.500\,{\rm T}\times m_I) = \boxed{-1.625\times10^{-26}\,{\rm J}\times m_I}$$

E18.3 The energy level separation is

$$\Delta E = h\nu \quad \text{where } \nu = \frac{\gamma\mathcal{B}}{2\pi}\ [5]$$

So

$$\nu = \frac{(6.73\times10^{7}\,{\rm T^{-1}\,s^{-1}})\times14.4\,{\rm T}}{2\pi} = 1.54\times10^{8}\,{\rm s^{-1}}$$

$$= 1.54\times10^{8}\,{\rm Hz} = \boxed{154\,{\rm MHz}}$$

E18.4 **(a)** As shown in Exercise 18.1 a 600-MHz NMR spectrometer operates in a magnetic field of 14.1 T. Thus

$$\Delta E = \gamma\hbar\mathcal{B} = h\nu_{\rm L} = h\nu \quad \text{at resonance}$$

$$= (6.626\times10^{-34}\,{\rm J\,s})\times(6.00\times10^{8}\,{\rm s^{-1}}) = \boxed{3.98\times10^{-25}\,{\rm J}}$$

(b) A 600-MHz NMR spectrometer means 600-MHz is the resonance frequency for protons for which the magnetic field is 14.1 T. In high-field NMRs it is the field not the frequency which is fixed, so for the deuteron

$$\nu = \frac{g_I\mu_{\rm N}\mathcal{B}}{h}\ [\text{Exercise 18.1}]$$

$$= \frac{(0.8575)\times(5.051\times10^{-27}\,{\rm J\,T^{-1}})\times(14.1\,{\rm T})}{6.626\times10^{-34}\,{\rm J\,s}} = 9.22\times10^{7}\,{\rm Hz} = 92.2\,{\rm MHz}$$

$$\Delta E = h\nu = (6.626\times10^{-34}\,{\rm J\,s})\times(9.21\bar{6}\times10^{7}\,{\rm s^{-1}}) = \boxed{6.11\times10^{-26}\,{\rm J}}$$

Thus the separation in energy is larger for the proton **(a)**.

E18.5 This is similar to Exercise 18.4**(a)(b)**. There the energy difference was calculated for $|\Delta m_I| = 1$, here it is for $|\Delta m_I| = 2$. That is

$$\Delta E = 2g_I\mu_{\rm N}\mathcal{B} = (2)\times(0.4036)\times(5.051\times10^{-27}\,{\rm J\,T^{-1}})\times(15.00\,{\rm T})$$

$$= \boxed{6.116\times10^{-26}\,{\rm J}}$$

E18.6 In all cases the selection rule $\Delta m_I = \pm1$ is applied; hence

$$\mathcal{B} = \frac{h\nu}{g_I\mu_{\rm N}} = \frac{6.626\times10^{-34}\,{\rm J\,Hz^{-1}}}{5.0508\times10^{-27}\,{\rm J\,T^{-1}}}\times\frac{\nu}{g_I}$$

$$= (1.3119\times10^{-7})\times\frac{(\nu/{\rm Hz})}{g_I}{\rm T} = (0.13119)\times\frac{(\nu/{\rm MHz})}{g_I}{\rm T}$$

We can draw up the following table

$\mathcal{B}$/T	(a) ^{1}H	(b) ^{2}H	(c) ^{13}C
g_I	5.5857	0.85745	1.4046
(i) 250 MHz	5.87	38.3	23.4
(ii) 500 MHz	11.7	76.6	46.8

Comment. Magnetic fields above 20 T have not yet been obtained for use in NMR spectrometers. As discussed in the solution to Exercise 18.4**(a)**(**b**), it is the field, not the frequency, that is fixed in high-field NMR spectrometers. Thus an NMR spectrometer that is called a 500-MHz spectrometer refers to the resonance frequency for protons and has a magnetic field fixed at 11.7 T.

Question. What are the resonance frequencies of these nuclei in 250-MHz and 500-MHz spectrometers? See Exercise 18.4**(a)**(**b**).

E18.7 The ground state has

$$m_I = +\frac{1}{2} = \alpha \text{ spin}, \qquad m_I = -\frac{1}{2} = \beta \text{ spin [3]}$$

Hence, with

$$\delta N = N_\alpha - N_\beta$$

$$\frac{\delta N}{N} = \frac{N_\alpha - N_\beta}{N_\alpha + N_\beta} = \frac{N_\alpha - N_\alpha e^{-\Delta E/kT}}{N_\alpha + N_\alpha e^{-\Delta E/kT}} \quad [\textit{Justification } 18.1]$$

$$= \frac{1 - e^{-\Delta E/kT}}{1 + e^{-\Delta E/kT}} \approx \frac{1 - (1 - \Delta E/kT)}{1+1} \approx \frac{\Delta E}{2kT} = \frac{g_I \mu_N \mathcal{B}}{2kT} \quad [\text{for } \Delta E \ll kT]$$

That is, $\frac{\delta N}{N} \approx \frac{g_I \mu_N \mathcal{B}}{2kT} = \frac{(5.5857) \times (5.0508 \times 10^{-27}\,\text{J T}^{-1}) \times (\mathcal{B})}{(2) \times (1.38066 \times 10^{-23}\,\text{J K}^{-1}) \times (298\,\text{K})} \approx 3.43 \times 10^{-6}\,\mathcal{B}/\text{T}$

(a) $\mathcal{B} = 0.3\,\text{T}, \qquad \delta N/N = \boxed{1 \times 10^{-6}}$

(b) $\mathcal{B} = 1.5\,\text{T}, \qquad \delta N/N = \boxed{5.1 \times 10^{-6}}$

(c) $\mathcal{B} = 10\,\text{T}, \qquad \delta N/N = \boxed{3.4 \times 10^{-5}}$

E18.8 $\delta N \approx \frac{N g_I \mu_N \mathcal{B}}{2kT} [\text{Exercise } 18.7] = \frac{Nh\nu}{2kT}$

Thus, $\delta N \propto \nu$

$$\frac{\delta N(600\,\text{MHz})}{\delta N(60\,\text{MHz})} = \frac{600\,\text{MHz}}{60\,\text{MHz}} = \boxed{10}$$

This ratio is not dependent on the nuclide as long as the approximation $\Delta E \ll kT$ holds (Exercise 18.7**(a)**).

E18.9 $\mathcal{B}_{\text{loc}} = (1 - \sigma)\mathcal{B}$ [8]

$$|\Delta \mathcal{B}_{\text{loc}}| = |(\Delta\sigma)|\mathcal{B} \approx |[\delta(\text{CH}_3) - \delta(\text{CHO})]|\mathcal{B} \quad \left[|\Delta\sigma| \approx \left|\frac{\nu - \nu_0}{\nu_0}\right|\right]$$

$$= |(2.20 - 9.80)| \times 10^{-6}\mathcal{B} = 7.60 \times 10^{-6}\mathcal{B}$$

(a) $\mathcal{B} = 1.5\,\text{T}, \qquad |\Delta\mathcal{B}_{\text{loc}}| = 7.60 \times 10^{-6} \times 1.5\,\text{T} = \boxed{11\,\mu\text{T}}$

(b) $\mathcal{B} = 15\,\text{T}, \qquad |\Delta\mathcal{B}_{\text{loc}}| = \boxed{110\,\mu\text{T}}$

E18.10 $\nu - \nu_0 = \nu_0\delta \times 10^{-6}$ [10]

$$|\Delta\nu| \equiv (\nu - \nu_0)(\text{CHO}) - (\nu - \nu_0)(\text{CH}_3)$$
$$= \nu(\text{CHO}) - \nu(\text{CH}_3)$$
$$= \nu_0[\delta(\text{CHO}) - \delta(\text{CH}_3)] \times 10^{-6}$$
$$= (9.80 - 2.20) \times 10^{-6}\nu_0 = 7.60 \times 10^{-6}\nu_0$$

(a) $\nu_0 = 250\,\text{MHz}, \qquad |\Delta\nu| = 7.60 \times 10^{-6} \times 250\,\text{MHz} = 1.90\,\text{kHz}$

(b) $\nu_0 = 500\,\text{MHz}, \qquad |\Delta\nu| = 3.80\,\text{kHz}$

(a) The spectrum is shown in Fig. 18.1 with the value of $|\Delta\nu|$ as calculated above.

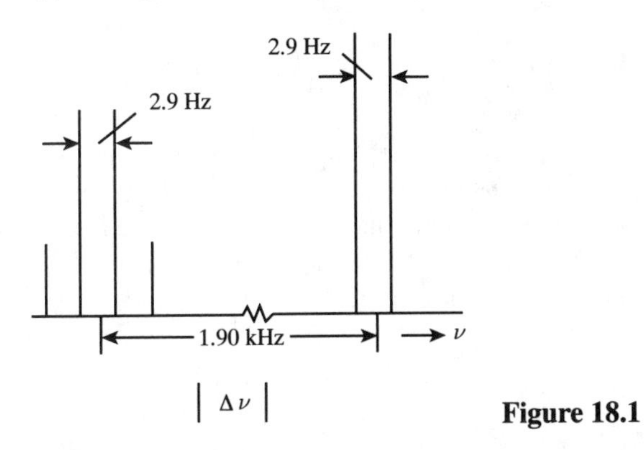

Figure 18.1

(b) When the frequency is changed to 500 MHz, the $|\Delta\nu|$ changes to 3.80 kHz. The fine structure (the splitting within groups) remains the same as spin–spin splitting is unaffected by the strength of the applied field. However, the intensity of the lines increases by a factor of 2 because $\delta N/N \propto \nu$ (Exercise 18.8(**a**)).

The observed splitting pattern is that of an AX_3 (or A_3X) species, the spectrum of which is described in Section 18.4.

E18.11 $\tau \approx \dfrac{\sqrt{2}}{\pi\Delta\nu}$ [32, with $\delta\nu$ written as $\Delta\nu$]

$\Delta\nu = \nu_0(\delta' - \delta) \times 10^{-6}$ [Exercise 18.10(**a**)]

Then $\tau \approx \dfrac{\sqrt{2}}{\pi\nu_0(\delta' - \delta) \times 10^{-6}}$

$$\approx \frac{\sqrt{2}}{(\pi) \times (250 \times 10^6\,\text{Hz}) \times (5.2 - 4.0) \times 10^{-6}} \approx 1.5 \times 10^{-3}\,\text{s}$$

Therefore, the signals merge when the lifetime of each isomer is less than about 1.5 ms, corresponding to a conversion rate of about $\boxed{6.7 \times 10^2\,\text{s}^{-1}}$

E18.12 The four equivalent ^{19}F nuclei ($I = \frac{1}{2}$) give a single line. However, the ^{10}B nucleus ($I = 3$, 19.6 per cent abundant) splits this line into $2 \times 3 + 1 = 7$ lines and the ^{11}B nucleus ($I = \frac{3}{2}$, 80.4 per cent abundant) into $2 \times \frac{3}{2} + 1 = 4$ lines. The splitting arising from the ^{11}B nucleus will be larger than that arising from the ^{10}B nucleus (since its magnetic moment is larger, by a factor of 1.5, Table 18.1). Moreover, the total intensity of the four lines due to the ^{11}B nuclei will be greater (by a factor of $80.4/19.6 \approx 4$) than the total intensity of the seven lines due to the ^{10}B nuclei. The individual line intensities will be in the ratio $\frac{7}{4} \times 4 = 7$ ($\frac{4}{7}$ the number of lines and about four times as abundant). The spectrum is sketched in Fig. 18.2.

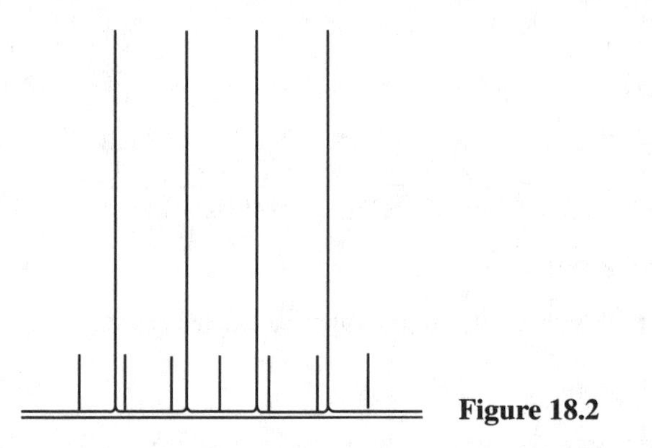

Figure 18.2

E18.13 The A, M, and X resonances lie in distinctively different groups. The A resonance is split into a 1 : 2 : 1 triplet by the M nuclei, and each line of that triplet is split into a 1 : 4 : 6 : 4 : 1 quintet by the X nuclei, (with $J_{AM} > J_{AX}$). The M resonance is split into a 1 : 3 : 3 : 1 quartet by the A nuclei and each line is split into a quintet by the X nuclei (with $J_{AM} > J_{MX}$). The X resonance is split into a quartet by the A nuclei and then each line is split into a triplet by the M nuclei (with $J_{AX} > J_{MX}$). The spectrum is sketched in Fig. 18.3.

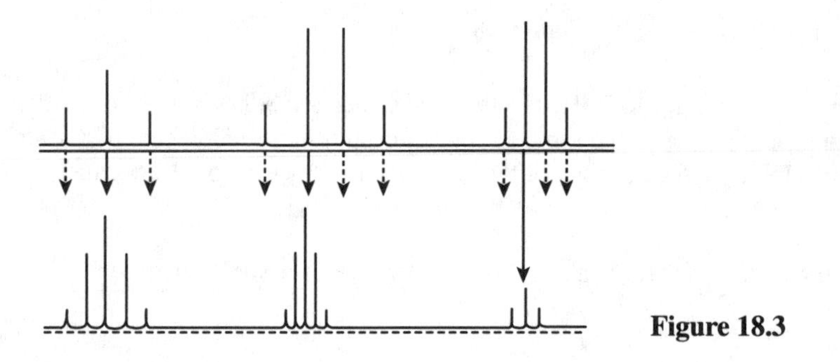

Figure 18.3

E18.14 **(a)** If there is rapid rotation about the axis, the H nuclei are both chemically and magnetically equivalent.

(b) Since $J_{cis} \neq J_{trans}$, the H nuclei are chemically but not magnetically equivalent.

E18.15 Analogous to precession of the magnetization vector in the laboratory frame due to the presence of $\mathcal{B}_0$ that is

$$\nu_L = \frac{\gamma \mathcal{B}_0}{2\pi} \quad [4],$$

there is a precession in the rotating frame, due to the presence of $\mathcal{B}_1$, namely

$$\nu_L = \frac{\gamma \mathcal{B}_1}{2\pi} \quad \text{or} \quad \omega_1 = \gamma \mathcal{B}_1 \quad [\omega = 2\pi\nu]$$

Since ω is an angular frequency, the angle through which the magnetization vector rotates is

$$\theta = \gamma \mathcal{B}_1 t = \frac{g_I \mu_N}{\hbar} \mathcal{B}_1 t$$

and $\mathcal{B}_1 = \dfrac{\theta \hbar}{g_I \mu_N t} = \dfrac{(\frac{\pi}{2}) \times (1.055 \times 10^{-34}\,\mathrm{J\,s})}{(5.586) \times (5.051 \times 10^{-27}\,\mathrm{J\,T^{-1}}) \times (1.0 \times 10^{-5}\,\mathrm{s})} = \boxed{5.9 \times 10^{-4}\,\mathrm{T}}$

A 180° pulse requires $2 \times 10\,\mu s = \boxed{20\,\mu s}$

E18.16 **(a)** $\mathcal{B} = \dfrac{h\nu}{g_I \mu_N} = \dfrac{(6.626 \times 10^{-34}\,\mathrm{J\,Hz^{-1}}) \times (9 \times 10^9\,\mathrm{Hz})}{(5.5857) \times (5.051 \times 10^{-27}\,\mathrm{J\,T^{-1}})} = \boxed{2 \times 10^2\,\mathrm{T}}$

(b) $\mathcal{B} = \dfrac{h\nu}{g_e \mu_B} = \dfrac{(6.626 \times 10^{-34}\,\mathrm{J\,Hz^{-1}}) \times (300 \times 10^6\,\mathrm{Hz})}{(2.0023) \times (9.274 \times 10^{-24}\,\mathrm{J\,T^{-1}})} = \boxed{10\,\mathrm{mT}}$

Comment. Because of the sizes of these magnetic fields neither experiment seems feasible.

Question. What frequencies are required to observe electron resonance in the magnetic field of a 300 MHz NMR magnet and nuclear resonance in the field of a 9 GHz ($g = 2.00$) ESR magnet? Are these experiments feasible?

E18.17 $g = \dfrac{h\nu}{\mu_B \mathcal{B}}$ [38]

We shall often need the value

$$\frac{h}{\mu_B} = \frac{6.62608 \times 10^{-34}\,\mathrm{J\,Hz^{-1}}}{9.27402 \times 10^{-24}\,\mathrm{J\,T^{-1}}} = 7.14478 \times 10^{-11}\,\mathrm{T\,Hz^{-1}}$$

Then, in this case

$$g = \frac{(7.14478 \times 10^{-11}\,\mathrm{T\,Hz^{-1}}) \times (9.2231 \times 10^9\,\mathrm{Hz})}{329.12 \times 10^{-3}\,\mathrm{T}} = \boxed{2.0022}$$

E18.18 $a = \mathcal{B}(\text{line } 3) - \mathcal{B}(\text{line } 2) = \mathcal{B}(\text{line } 2) - \mathcal{B}(\text{line } 1)$

$$\left.\begin{aligned} \mathcal{B}_3 - \mathcal{B}_2 &= (334.8 - 332.5)\,\mathrm{mT} = 2.3\,\mathrm{mT} \\ \mathcal{B}_2 - \mathcal{B}_1 &= (332.5 - 330.2)\,\mathrm{mT} = 2.3\,\mathrm{mT} \end{aligned}\right\} \quad a = \boxed{2.3\,\mathrm{mT}}$$

Use the centre line to calculate g

$$g = \frac{h\nu}{\mu_B \mathcal{B}} = (7.14478 \times 10^{-11}\,\mathrm{T\,Hz^{-1}}) \times \frac{9.319 \times 10^9\,\mathrm{Hz}}{332.5 \times 10^{-3}\,\mathrm{T}} = \boxed{2.002\bar{5}}$$

E18.19 The centre of the spectrum will occur at 332.5 mT. Proton 1 splits the line into two components with separation 2.0 mT and hence at 332.5 ± 1.0 mT. Proton 2 splits these two hyperfine lines into two, each with separation 2.6 mT, and hence the lines occur at $332.5 \pm 1.0 \pm 1.3$ mT. The spectrum therefore consists of four lines of $\boxed{\text{equal intensity}}$ at the fields $\boxed{330.2\,\mathrm{mT},\ 332.2\,\mathrm{mT},\ 332.8\,\mathrm{mT},\ 334.8\,\mathrm{mT}}$

E18.20 We construct Fig. 18.4(a) for CH_3 and Fig. 18.4(b) for CD_3. The predicted intensity distribution is determined by counting the number of overlapping lines of equal intensity from which the hyperfine line is constructed.

E18.21 $\mathcal{B} = \dfrac{h\nu}{g\mu_B} = \dfrac{7.14478 \times 10^{-11}}{2.0025}\,\mathrm{T\,Hz^{-1}} \times \nu$ [Exercise 18.17(**a**)] $= 35.68\,\mathrm{mT} \times (\nu/\mathrm{GHz})$

(a) $\nu = 9.302\,\mathrm{GHz}$, $\quad \mathcal{B} = \boxed{331.9\,\mathrm{mT}}$

(b) $\nu = 33.67\,\mathrm{GHz}$, $\quad \mathcal{B} = 1201\,\mathrm{mT} = \boxed{1.201\,\mathrm{T}}$

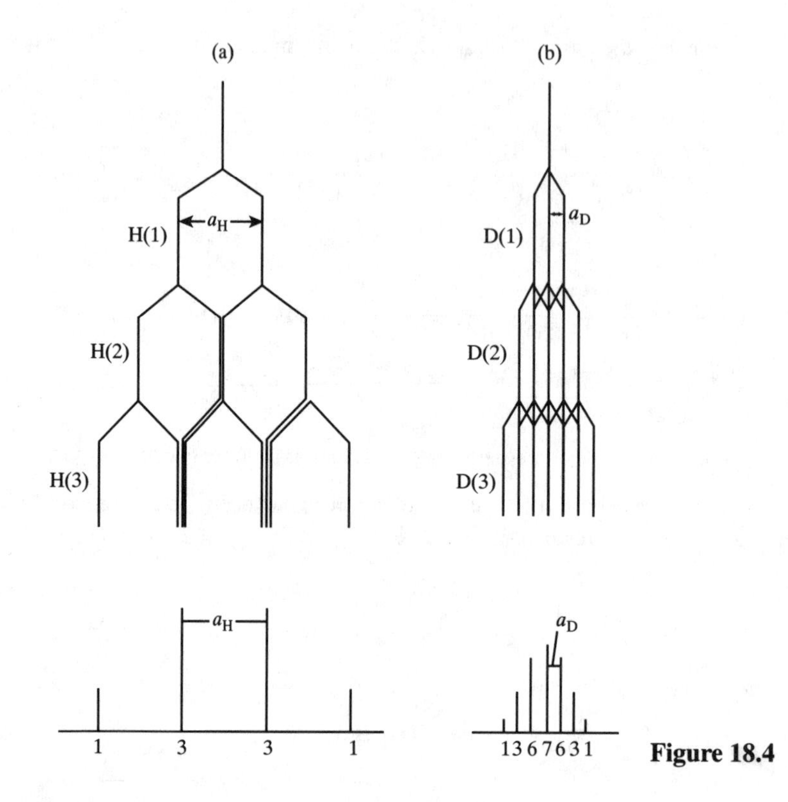

Figure 18.4

E18.22 Since the number of hyperfine lines arising from a nucleus of spin I is $2I + 1$, we solve $2I + 1 = 4$ and find that $\boxed{I = \frac{3}{2}}$

Comment. Four lines of equal intensity could also arise from two inequivalent nuclei with $I = \frac{1}{2}$.

E18.23 The X nucleus produces six lines of equal intensity. The pair of H nuclei in XH_2 split each of these lines into a 1 : 2 : 1 triplet (Fig. 18.5(a)). The pair of D nuclei ($I = 1$) in XD_2 split each line into a 1 : 2 : 3 : 2 : 1 quintet (Fig. 18.5(b)). The total number of hyperfine lines observed is then $6 \times 3 = 18$ in XH_2 and $6 \times 5 = 30$ in XD_2.

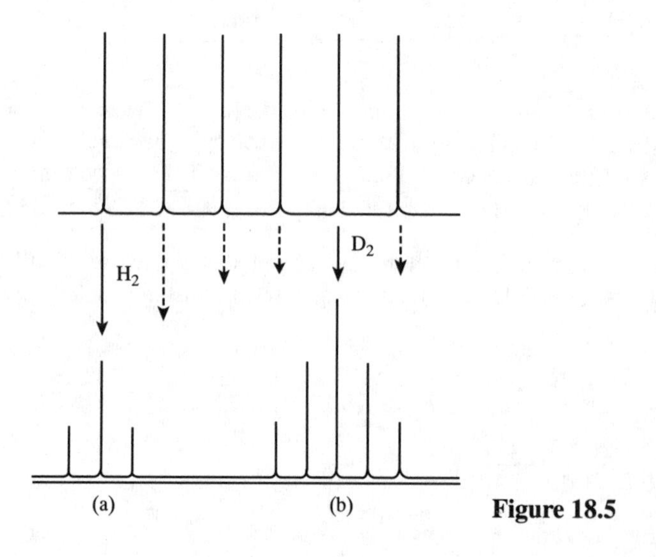

Figure 18.5

Solutions to problems

Solutions to numerical problems

P18.1 $g_I = -3.8260$ (Table 18.1)

$$\mathcal{B} = \frac{h\nu}{g_I\mu_N} = \frac{(6.626\times10^{-34}\,\mathrm{J\,Hz^{-1}})\times\nu}{(-)(3.8260)\times(5.0508\times10^{-27}\,\mathrm{J\,T^{-1}})} = 3.429\times10^{-8}(\nu/\mathrm{Hz})\,\mathrm{T}$$

Therefore, with $\nu = 300\,\mathrm{MHz}$,

$$\mathcal{B} = (3.429\times10^{-8})\times(300\times10^{6}\,\mathrm{T}) = \boxed{10.3\,\mathrm{T}}$$

$$\frac{\delta N}{N} \approx \frac{g_I\mu_N\mathcal{B}}{2kT} \quad \text{[Exercise 18.7(a)]}$$

$$= \frac{(-3.8260)\times(5.0508\times10^{-27}\,\mathrm{J\,T^{-1}})\times(10.3\,\mathrm{T})}{(2)\times(1.381\times10^{-23}\,\mathrm{J\,K^{-1}})\times(298\,\mathrm{K})} = \boxed{2.42\times10^{-5}}$$

Since $g_I < 0$ (as for an electron, the magnetic moment is antiparallel to its spin), the $\boxed{\beta}$ state $\left(m_I = -\tfrac{1}{2}\right)$ lies lower.

P18.3

$$g = \frac{h\nu}{\mu_B\mathcal{B}}[38] = \frac{(7.14478\times10^{-11}\,\mathrm{T})\times(\nu/\mathrm{Hz})}{\mathcal{B}}$$

$$= \frac{(7.14478\times10^{-11}\,\mathrm{T})\times(9.302\times10^{9})}{\mathcal{B}} = \frac{0.66461}{\mathcal{B}/T}$$

$$g_{\parallel} = \frac{0.66461}{0.33364} = \boxed{1.992} \qquad g_{\perp} = \frac{0.66461}{0.33194} = \boxed{2.002}$$

P18.5

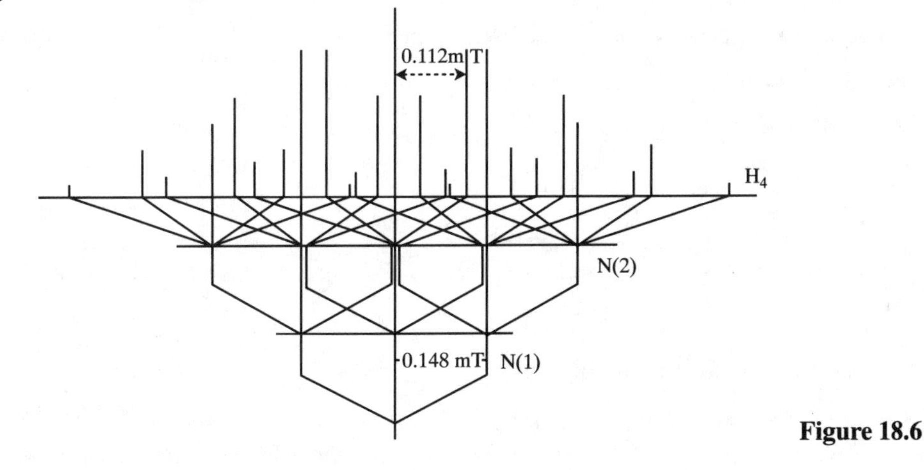

Figure 18.6

Construct the spectrum by taking into account first the two equivalent ^{14}N splitting (producing a $\boxed{1:2:3:2:1 \text{ quintet}}$) and then the splitting of each of these lines into a $\boxed{1:4:6:4:1 \text{ quintet}}$ by the four equivalent protons. The resulting 25-line spectrum is shown in Fig. 18.6. Note that Pascal's triangle does not apply to the intensities of the quintet due to ^{14}N, but does apply to the quintet due to the protons.

Solutions to theoretical problems

P18.7 $\mathcal{B}_{loc} = -\frac{\gamma\hbar\mu_0 m_I}{4\pi R^3}(1-3\cos^2\theta)\,[34] = \frac{g_I\mu_N\mu_0}{4\pi R^3} \quad [m_I = +\frac{1}{2}, \theta = 0,\ \gamma\hbar = g_I\mu_N]$

which rearranges to

$$R = \left(\frac{g_I\mu_N\mu_0}{4\pi\mathcal{B}}\right)^{1/3} = \left(\frac{(5.5857)\times(5.0508\times10^{-27}\,\mathrm{J\,T^{-1}})\times(4\pi\times10^{-7}\,\mathrm{T^2\,J^{-1}\,m^3})}{(4\pi)\times(0.715\times10^{-3}\,\mathrm{T})}\right)^{1/3}$$

$$= (3.946\times10^{-30}\,\mathrm{m^3})^{1/3} = \boxed{158\,\mathrm{pm}}$$

P18.8 $\langle\mathcal{B}_{nucl}\rangle = \frac{-g_I\mu_N\mu_0 m_I}{4\pi R^3}\frac{\int_0^{\theta_{max}}(1-3\cos^2\theta)\sin\theta\,\mathrm{d}\theta}{\int_0^{\theta_{max}}\sin\theta\,\mathrm{d}\theta}$

The denominator is the normalization constant, and ensures that the total probability of being between 0 and θ_{max} is 1.

$$= \frac{-g_I\mu_N\mu_0 m_I}{4\pi R^3}\frac{\int_1^{x_{max}}(1-3x^2)\,\mathrm{d}x}{\int_1^{x_{max}}\mathrm{d}x} \quad [x_{max} = \cos\theta_{max}]$$

$$= \frac{-g_I\mu_N\mu_0 m_I}{4\pi R^3}\times\frac{x_{max}(1-x_{max}^2)}{x_{max}-1} = \boxed{\frac{+g_I\mu_N\mu_0 m_I}{4\pi R^3}(\cos^2\theta_{max}+\cos\theta_{max})}$$

If $\theta_{max} = \pi$ (complete rotation), $\cos\theta_{max} = -1$ and $\langle\mathcal{B}_{nucl}\rangle = 0$. If $\theta_{max} = 30°$, $\cos^2\theta_{max} + \cos\theta_{max} = 1.616$, and

$$\langle\mathcal{B}_{nucl}\rangle = \frac{(5.5857)\times(5.0508\times10^{-27}\,\mathrm{J\,T^{-1}})\times(4\pi\times10^{-7}\,\mathrm{T^2\,J^{-1}\,m^3})\times(1.616)}{(4\pi)\times(1.58\times10^{-10}\,\mathrm{m})^3\times(2)}$$

$$= \boxed{0.58\,\mathrm{mT}}$$

P18.10 We have seen (Problem 18.9) that, if $G \propto \cos\omega_0 t$, then $I(\omega) \propto \frac{1}{[1+(\omega_0-\omega)^2\tau^2]}$ which peaks at $\omega \approx \omega_0$. Therefore, if

$$G(t) \propto a\cos\omega_1 t + b\cos\omega_2 t$$

we can anticipate that

$$I(\omega) \propto \frac{a}{1+(\omega_1-\omega)^2\tau^2} + \frac{b}{1+(\omega_2-\omega)^2\tau^2}$$

and explicit calculation shows this to be so. Therefore, $I(\omega)$ consists of two absorption lines, one peaking at $\omega \approx \omega_1$ and the other at $\omega \approx \omega_2$.

Solutions to additional problems

P18.11 The envelopes of maxima and minima of the curve are determined by T_2 through eqn 25, but the time interval between the maxima of this decaying curve corresponds to the reciprocal of the frequency difference $\Delta\nu$ between the pulse frequency ν_0 and the Larmor frequency ν_L, that is $\Delta\nu = |\nu_0 - \nu_L|$

$$\Delta\nu = \frac{1}{0.10\,\mathrm{s}} = 10\,\mathrm{s^{-1}} = 10\,\mathrm{Hz}$$

Therefore the Larmor frequency is $\boxed{300\times10^6\,\mathrm{Hz}\pm10\,\mathrm{Hz}}$

According to eqn 25 the intensity of the maxima in the FID curve decays exponentially as

$$\mathrm{e}^{-t/T_2}$$

Therefore T_2 corresponds to the time at which the intensity has been reduced to $1/e$ of the original value. In the text figure, this corresponds to a time slightly before the fourth maximum has occurred, or about $\boxed{0.29\,\text{s}}$

P18.12 The three rotational conformations of $F_2BrC{-}CBrCl_2$ are shown in Fig. 18.7. In conformation I, the two F atoms are equivalent. However, in conformations II and III they are non-equivalent. At low temperature, the molecular residence time in conformation I is longer (because this conformation has the lowest repulsive energy of the large bromine atoms) than that of conformations II and III, which have equal residence times. With its longer residence time, we expect that the NMR signal intensity of conformation I should be stronger and we can conclude that it is the low-temperature singlet. It is a singlet because equivalent atoms do not have detectable spin–spin couplings.

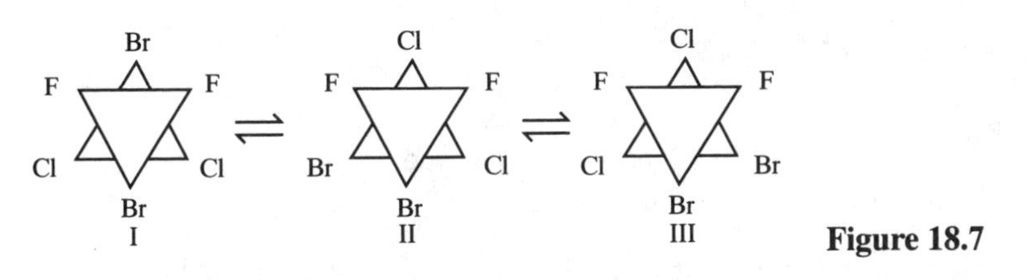

Figure 18.7

The fluorines of conformations II and III are non-equivalent, so their coupling is observed at low-temperature. Fluorine has a nuclear spin of $1/2$, so we expect a doublet for each fluorine. These are observed with strong geminal coupling of 160 Hz. As temperature increases, the rate of rotation between II and III increases and the two fluorines become equivalent in these conformations, though remaining distinct from I. The doublets collapse to singlets. With a further temperature increase to $-30°$C, and above, the rate of rotation about the C$-$C bond becomes so rapid that the residence times of the three conformations become equal. The very short residence times produce an average NMR signal that is a singlet and the fluorines appear totally equivalent.

The spectra shown in text Fig. 18.50 for conformations II and III show both spin–spin coupling and differences in chemical shift. The spin–spin splitting is 160 Hz. The difference in chemical shift can be estimated from the separation between the doublet centres, Δ

$$\Delta = (J^2 + \delta\nu^2)^{1/2}$$

Δ is estimated from the figure to be 210 Hz. This yields for $\delta\nu$, the chemical shift,

$$\delta\nu = (\Delta^2 - J^2)^{1/2}$$
$$= (210^2 - 160^2)^{1/2}\,\text{Hz} \approx 140\,\text{Hz}$$

Collapse to a single line will occur when the rate of interconversion satisfies

$$k \approx \frac{1}{\tau} \approx \frac{\pi\Delta}{\sqrt{2}}\ [32]$$

$$k = \frac{\pi \times 200\,\text{s}^{-1}}{\sqrt{2}} \approx \boxed{4 \times 10^2\,\text{s}^{-1}}$$

The relative intensities, I, of the lines at $-80°$C can be used to estimate the energy difference $(E_{\text{II}} - E_{\text{I}})$ between conformation I and conformations II and III. We assume that the relative intensities of the lines are proportional to the populations of conformers and that these populations follow the Boltzmann distribution (Chapters 0 and 19). Then

$$\frac{I_{\text{I}}}{I_{\text{II}}} = \frac{e^{-E_{\text{I}}/RT}}{e^{-E_{\text{II}}/RT}} = e^{(E_{\text{II}}-E_{\text{I}})/RT}$$

$$E_{\text{II}} - E_{\text{I}} = RT \ln\left(\frac{I_{\text{I}}}{I_{\text{II}}}\right) = 8.314\ \text{J K}^{-1}\ \text{mol}^{-1} \times (273 - 80)\ \text{K} \ln(10)$$

$$= 3.7 \times 10^3\ \text{J mol}^{-1} = \boxed{3.7\ \text{kJ mol}^{-1}}$$

This energy difference is not, however, the rotational energy barrier between the rotational isomers. The latter can be estimated from the rate of interconversion between the isomers as a function of temperature. That rate of interconversion is roughly $4 \times 10^2\ \text{s}^{-1}$ at $-30°\text{C}$. At $-60°\text{C}$, as estimated from the line width at that temperature [16.25], it is roughly 1/3 of that value, or $\sim 1.3 \times 10^2\ \text{s}^{-1}$. Assuming that the rate of interconversion satisfies an Arrhenius type of behaviour, $k \propto e^{-E_{\text{a}}/RT}$, where E_{a} is the rotational energy barrier,

$$\frac{k(-30°\text{C})}{k(-60°\text{C})} = 3 = e^{\left\{-\frac{E_{\text{a}}}{R}\left(\frac{1}{243\ \text{K}} - \frac{1}{213\ \text{K}}\right)\right\}}$$

$$E_{\text{a}} = \frac{R \ln 3}{\left(\frac{1}{213\ \text{K}} - \frac{1}{243\ \text{K}}\right)} = 1.6 \times 10^4\ \text{J mol}^{-1} = \boxed{16\ \text{kJ mol}^{-1}}$$

This value is typical of the rotational barriers observed in compounds of this kind.

P18.14 (a) The Karplus equation [21] for ${}^3J_{\text{HH}}$ is a linear equation in $\cos\phi$ and $\cos 2\phi$. The experimentally determined equation for ${}^3J_{\text{SnSn}}$ is a linear equation in ${}^3J_{\text{HH}}$. In general, if $F(f)$ is linear in f, and if $f(x)$ is linear in x, then $F(x)$ is linear. So we expect ${}^3J_{\text{SnSn}}$ to be linear in $\cos\phi$ and $\cos 2\phi$. This is demonstrated in **(b)**.

(b) $${}^3J_{\text{SnSn}}/\text{Hz} = 78.86({}^3J_{\text{HH}}/\text{Hz}) + 27.84$$

Inserting the Karplus equation for ${}^3J_{\text{HH}}$ we obtain

${}^3J_{\text{SnSn}}/\text{Hz} = 78.86\{A + B\cos\phi + C\cos 2\phi\} + 27.84$. Using $A = 7$, $B = -1$, and $C = 5$, we obtain

$${}^3J_{\text{SnSn}}/\text{Hz} = \boxed{580 - 79\cos\phi + 395\cos 2\phi}$$

The plot of ${}^3J_{\text{SnSn}}$ is shown in Fig. 18.8.

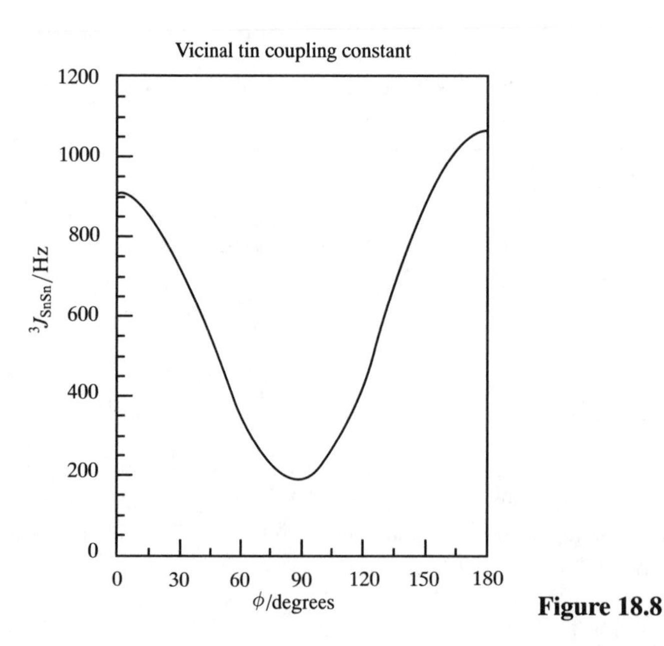

Figure 18.8

(c) A staggered configuration (Fig. 18.9) with the $SnMe_3$ groups *trans* to each other is the preferred configuration. The $SnMe_3$ repulsions are then at a minimum.

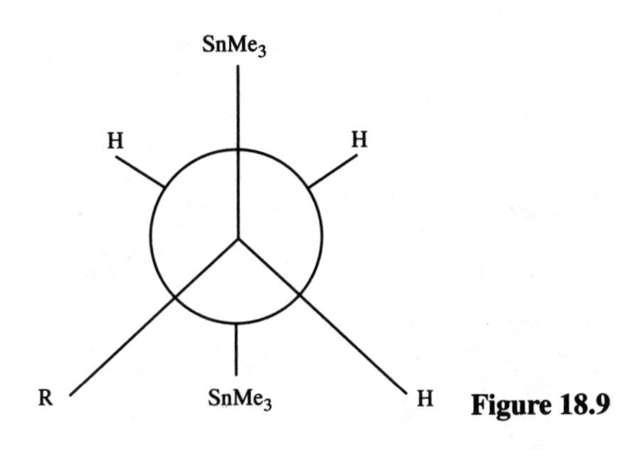

Figure 18.9

19 Statistical thermodynamics: the concepts

Solution to exercises

E19.1 $n_i = \dfrac{N\mathrm{e}^{-\beta\varepsilon_i}}{q} \quad \left[6, \text{ with } q = \sum_j \mathrm{e}^{-\beta\varepsilon_j}\right]$

Hence, $\dfrac{n_2}{n_1} = \dfrac{\mathrm{e}^{-\beta\varepsilon_2}}{\mathrm{e}^{-\beta\varepsilon_1}} = \mathrm{e}^{-\beta(\varepsilon_2-\varepsilon_1)} = \mathrm{e}^{-\beta\Delta\varepsilon} = \mathrm{e}^{-\Delta\varepsilon/kT} \quad \left[\beta = \dfrac{1}{kT}\right]$

as $T \to \infty$, $\dfrac{n_2}{n_1} = \mathrm{e}^{-0} = \boxed{1}$

E19.2 $q = \dfrac{V}{\Lambda^3} = \left(\dfrac{2\pi m}{h^2\beta}\right)^{3/2} V\,[22] = \left(\dfrac{2\pi mkT}{h^2}\right)^{3/2} V$

$$= \left(\frac{(2\pi) \times (120 \times 10^{-3}\,\mathrm{kg\,mol^{-1}}) \times (1.381 \times 10^{-23}\,\mathrm{J\,K^{-1}}) \times T}{(6.022 \times 10^{23}\,\mathrm{mol^{-1}}) \times (6.626 \times 10^{-34}\,\mathrm{J\,s})^2}\right)^{3/2} \times (2.00 \times 10^{-6}\,\mathrm{m^3})$$

(a) $T = 300\,\mathrm{K}, \quad q = (4.94 \times 10^{23}) \times (300)^{3/2} = \boxed{2.57 \times 10^{27}}$

(b) $T = 600\,\mathrm{K}, \quad q = (4.94 \times 10^{23}) \times (600)^{3/2} = \boxed{7.26 \times 10^{27}}$

E19.3 $q = \dfrac{V}{\Lambda^3}[22]$, implying that $\dfrac{q}{q'} = \left(\dfrac{\Lambda'}{\Lambda}\right)^3$

However, as $\Lambda \propto \dfrac{1}{m^{1/2}}$, $\dfrac{q}{q'} = \left(\dfrac{m}{m'}\right)^{3/2}$

Therefore, $\dfrac{q(\mathrm{D_2})}{q(\mathrm{H_2})} = 2^{3/2} = \boxed{2.83} \quad (m(\mathrm{D_2}) = 2m(\mathrm{H_2}))$

E19.4 $q = \sum_{\text{levels}} g_j \mathrm{e}^{-\beta\varepsilon_j}[12] = 3 + (\mathrm{e}^{-\beta\varepsilon_1}) + (3\mathrm{e}^{-\beta\varepsilon_2})$

$\beta\varepsilon = \dfrac{hc\tilde{\nu}}{kT} = \dfrac{1.4388(\tilde{\nu}/\mathrm{cm^{-1}})}{T/\mathrm{K}}$ [inside front cover]

Therefore, $q = 3+(\mathrm{e}^{-(1.4388)\times(3500)/1900})+(3\mathrm{e}^{-(1.4388)\times(4700)/1900}) = 3+0.0706+0.085 = \boxed{3.156}$

E19.5 $E = -\dfrac{N}{q}\dfrac{\mathrm{d}q}{\mathrm{d}\beta}[25] = -\dfrac{N}{q}\dfrac{\mathrm{d}}{\mathrm{d}\beta}(3 + \mathrm{e}^{-\beta\varepsilon_1} + 3\mathrm{e}^{-\beta\varepsilon_2}) = -\dfrac{N}{q}(-\varepsilon_1\mathrm{e}^{-\beta\varepsilon_1} - 3\varepsilon_2\mathrm{e}^{-\beta\varepsilon_2})$

$$= \frac{Nhc}{q}(\tilde{\nu}_1\mathrm{e}^{-\beta hc\tilde{\nu}_1} + 3\tilde{\nu}_2\mathrm{e}^{-\beta hc\tilde{\nu}_2})$$

$$= \left(\frac{N_\mathrm{A}hc}{3.156}\right) \times \left(\tilde{\nu}_1\mathrm{e}^{(-hc\tilde{\nu}_1)}/kT + 3\tilde{\nu}_2\mathrm{e}^{-(hc\tilde{\nu}_2)/kT}\right)$$

$$= \left(\frac{N_\mathrm{A}hc}{3.156}\right) \times \left(0 + 3500\,\mathrm{cm^{-1}} \times \mathrm{e}^{-(1.4388\times3500)/1900} + 3 \times 4700\,\mathrm{cm^{-1}} \times \mathrm{e}^{-(1.4388\times4700)/1900}\right)$$

$$= N_\mathrm{A}hc \times (204.9\,\mathrm{cm^{-1}}) = \boxed{2.45\,\mathrm{kJ\,mol^{-1}}}$$

E19.6 $\frac{n_i}{N} = \frac{e^{-\beta\varepsilon_i}}{q}$ [6]

Therefore, $\frac{n_{ex}}{n_g} = \frac{e^{-\beta\varepsilon_{ex}}}{e^{-\beta\varepsilon_g}} = e^{-\beta\varepsilon} \quad [\varepsilon = \varepsilon_{ex} - \varepsilon_g]$

Solving for β, $\beta = \frac{1}{\varepsilon}\ln\frac{n_g}{n_{ex}}$ or $T = \frac{\varepsilon/k}{\ln\left(\frac{n_g}{n_{ex}}\right)}$

and $T = \frac{\left(\frac{hc\tilde{\nu}}{k}\right)}{\ln\left(\frac{n_g}{n_{ex}}\right)} = \frac{(1.4388\,\text{cm K}) \times (540\,\text{cm}^{-1})}{\ln\left(\frac{0.90}{0.10}\right)} = \boxed{354\,\text{K}}$

E19.7 $q = \sum_i e^{-\beta\varepsilon_i} = \boxed{1 + e^{-2\mu_B\beta\mathcal{B}}}$ [energies measured from lower state]

$$\langle\varepsilon\rangle = \frac{E}{N} = -\frac{1}{q}\frac{dq}{d\beta}[25] = \boxed{\frac{2\mu_B \mathcal{B}\, e^{-2\mu_B\beta\mathcal{B}}}{1 + e^{-2\mu_B\beta\mathcal{B}}}}$$

We write $x = 2\mu_B\beta\mathcal{B}$, then $\frac{\langle\varepsilon\rangle}{2\mu_B\mathcal{B}} = \frac{e^{-x}}{1+e^{-x}} = \frac{1}{e^x+1}$

This function is plotted in Fig. 19.1. For the partition function we plot

$$q = 1 + e^{-x}$$

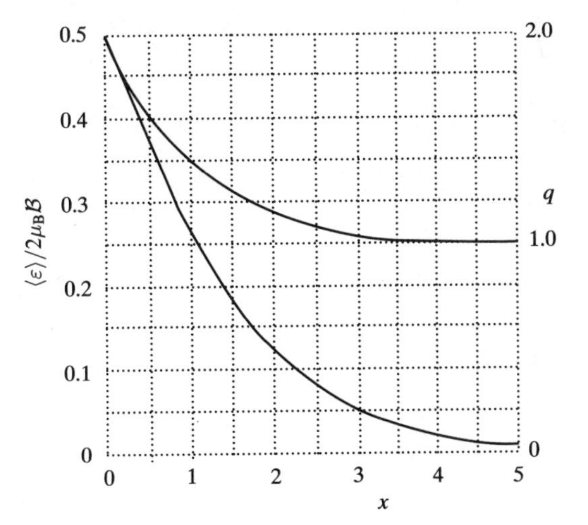

Figure 19.1

The relative populations are

$$\frac{n_+}{n_-} = e^{-\beta\Delta\varepsilon}[\text{Exercise 19.1}] = e^{-x}$$

$$x = 2\mu_B\beta\mathcal{B} = \frac{(2) \times (9.274 \times 10^{-24}\,\text{J T}^{-1}) \times (1.0)\text{T}}{(1.381 \times 10^{-23}\,\text{J K}^{-1})T} = 1.343/(T/\text{K})$$

(a) $T = 4\,\text{K}$, $\frac{n_+}{n_-} = e^{-1.343/4} = \boxed{0.71}$ **(b)** $T = 298\,\text{K}$, $\frac{n_+}{n_-} = e^{-1.343/298} = \boxed{0.996}$

E19.8 The energy separation is $\varepsilon = k \times (10\,\text{K})$

(a) $\dfrac{n_1}{n_0} = e^{-\beta(\varepsilon_1-\varepsilon_0)}$[Exercise 19.1] $= e^{-\beta\varepsilon} = e^{-10/(T/\text{K})}$

(1) $T = 1.0\,\text{K};\quad \dfrac{n_1}{n_0} = e^{-10} = \boxed{\bar{5} \times 10^{-5}}$

(2) $T = 10\,\text{K};\quad \dfrac{n_1}{n_0} = e^{-1.0} = \boxed{0.4}$

(3) $T = 100\,\text{K};\quad \dfrac{n_1}{n_0} = e^{-0.100} = \boxed{0.905}$

(b) $q = \sum_j g_j e^{-\varepsilon_j/kT} = e^0 + e^{-1.0} = \boxed{1.4}$

(c) $E = -\dfrac{N_\text{A}}{q}\dfrac{dq}{d\beta}$ [25]

$q = 1 + e^{-10\,\text{K}\times k\beta}$

$$E = -\frac{N_\text{A}}{q}\left\{-(10\,\text{K}) \times k e^{-(10\,\text{K}\times k\beta)}\right\} = \frac{(10\,\text{K}) \times R}{1 + e^{(10\,\text{K}\times k\beta)}} = \frac{(10\,\text{K}) \times R}{1 + e^{10/(T/\text{K})}}$$

At $T = 10\,\text{K}$, $\quad E = \dfrac{(10\,\text{K}) \times R}{1 + e} = \dfrac{(10\,\text{K}) \times (8.314\,\text{J K}^{-1}\,\text{mol}^{-1})}{3.718} = \boxed{22\,\text{J mol}^{-1}}$

(d) $C_V = \left(\dfrac{\partial U}{\partial T}\right)_V = \dfrac{dE}{dT} \qquad \dfrac{d}{dT} = -\dfrac{1}{kT^2}\dfrac{d}{d\beta}$

$$= -\frac{1}{kT^2}\frac{d}{d\beta}\left(\frac{(10\,\text{K}) \times R}{1 + e^{(10\,\text{K}\times k\beta)}}\right) = R\left(\frac{e}{(1+e)^2}\right) = 0.19\bar{7}R = \boxed{1.6\,\text{J K}^{-1}\,\text{mol}^{-1}}$$

(e) $S = \dfrac{U - U(0)}{T} + N_\text{A}k\ln q = \dfrac{E}{T} + R\ln q$

$$= \left(\frac{22.3\bar{6}\,\text{J mol}^{-1}}{10\,\text{K}}\right) + (R\ln 1.3\overline{68}) = \boxed{4.8\,\text{J K}^{-1}\,\text{mol}^{-1}}$$

E19.9 $\dfrac{n_1}{n_0} = e^{-\beta\varepsilon} = e^{-hc\tilde{\nu}/kT}$

$\tilde{\nu} = 2991\,\text{cm}^{-1}$ [Table 16.2, assume $^1\text{H}^{35}\text{Cl}$]

$\dfrac{1}{e} = e^{-hc\tilde{\nu}/kT}$

$-1 = \dfrac{-hc\tilde{\nu}}{kT}$

$T = \dfrac{hc\tilde{\nu}}{k} = (1.4388\,\text{cm K}) \times (2991\,\text{cm}^{-1}) = \boxed{4303\,\text{K}}$

Comment. Vibrational energy levels are large compared to kT at room temperature which is $207\,\text{cm}^{-1}$ at 298 K. Thus high temperatures are required to achieve substantial population in excited vibrational states.

Question. If thermal decomposition of HCl occurs when 1 per cent of HCl molecules find themselves in a vibrational state of energy corresponding to the bond dissociation energy ($431\,\text{kJ mol}^{-1}$), what temperature is required? Assume ε is constant at $2991\,\text{cm}^{-1}$ and do not take the result too seriously.

E19.10 $S_m^\ominus = R\ln\left(\dfrac{e^{5/2}kT}{p^\ominus\Lambda^3}\right)$ [47b with $p = p^\ominus$]

$$\Lambda = \frac{h}{(2\pi mkT)^{1/2}} = \frac{6.626\times10^{-34}\,\mathrm{J\,s}}{[(2\pi)\times(20.18)\times(1.6605\times10^{-27}\,\mathrm{kg})\times(1.381\times10^{-23}\,\mathrm{J\,K^{-1}}T)]^{1/2}}$$

$$= \frac{3.886\times10^{-10}\,\mathrm{m}}{(T/\mathrm{K})^{1/2}}$$

$$S_m^\ominus = R\ln\left(\frac{(e^{5/2})\times(1.381\times10^{-23}\,\mathrm{J\,K^{-1}}T)}{(1\times10^5\,\mathrm{Pa})\times(3.886\times10^{-10}\,\mathrm{m})^3}\right)\times\left(\frac{T}{\mathrm{K}}\right)^{3/2} = R\ln(28.67)\times(T/\mathrm{K})^{5/2}$$

(a) $T = 200\,\mathrm{K},\quad S_m^\ominus = (8.314\,\mathrm{J\,K^{-1}\,mol^{-1}})\times\ln(28.67)\times(200)^{5/2} = \boxed{138\,\mathrm{J\,K^{-1}\,mol^{-1}}}$

(b) $T = 298.15\,\mathrm{K},\quad S_m^\ominus = (8.314\,\mathrm{J\,K^{-1}\,mol^{-1}})\times\ln(28.67)\times(298.15)^{5/2}$

$$= \boxed{146\,\mathrm{J\,K^{-1}\,mol^{-1}}}$$

E19.11 $q = \dfrac{1}{1-e^{-\beta\varepsilon}}$ [Example 19.4] $= \dfrac{1}{1-e^{-hc\beta\tilde{\nu}}}$

$$hc\beta\tilde{\nu} = \frac{(1.4388\,\mathrm{cm\,K})\times(560\,\mathrm{cm^{-1}})}{500\,\mathrm{K}} = 1.611$$

Therefore, $q = \dfrac{1}{1-e^{-1.611}} = 1.249$

The internal energy due to vibrational excitation is

$$U - U(0) = \frac{N\varepsilon e^{-\beta\varepsilon}}{1-e^{-\beta\varepsilon}}\text{[Example 19.4]} = \frac{Nhc\tilde{\nu}e^{-hc\tilde{\nu}\beta}}{1-e^{-hc\tilde{\nu}\beta}} = \frac{Nhc\tilde{\nu}}{e^{hc\tilde{\nu}\beta}-1}$$

$$= (0.249)\times(Nhc)\times(560\,\mathrm{cm^{-1}})$$

and hence

$$\frac{S_m}{N_A k} = \frac{U-U(0)}{N_A kT} + \ln q[35] = (0.249)\times\left(\frac{hc}{kT}\right)\times(560\,\mathrm{cm^{-1}}) + \ln(1.249)$$

$$= \left(\frac{(0.249)\times(1.4388\,\mathrm{K\,cm})\times(560\,\mathrm{cm^{-1}})}{500\,\mathrm{K}}\right) + \ln(1.249) = 0.401 + 0.222 = 0.623$$

Hence, $S_m = 0.623R = \boxed{5.18\,\mathrm{J\,K^{-1}\,mol^{-1}}}$

E19.12 **[(a)]** Yes; He atoms indistinguishable and non-localized.

[(b)] Yes; CO molecules indistinguishable and non-localized.

(c) No; CO molecules can be identified by their locations.

[(d)] Yes; H_2O molecules indistinguishable and non-localized.

Solutions to problems

Solutions to numerical problems

P19.2 $q = \dfrac{V}{\Lambda^3},\qquad \Lambda = \dfrac{h}{(2\pi mkT)^{1/2}}\quad\left[22,\ \beta = \dfrac{1}{kT}\right]$

and hence

$$\begin{aligned}T &= \left(\frac{h^2}{2\pi mk}\right) \times \left(\frac{q}{V}\right)^{2/3} \\ &= \left(\frac{(6.626 \times 10^{-34}\,\mathrm{J\,s})^2}{(2\pi) \times (39.95) \times (1.6605 \times 10^{27}\,\mathrm{kg}) \times (1.381 \times 10^{-23}\,\mathrm{J\,K^{-1}})}\right) \\ &\quad \times \left(\frac{10}{1.0 \times 10^{-6}\,\mathrm{m^3}}\right)^{2/3} \\ &= \boxed{3.5 \times 10^{-15}\,\mathrm{K}}\ \text{[a very low temperature]}\end{aligned}$$

The exact partition function in one dimension is

$$q = \sum_{n=1}^{\infty} \mathrm{e}^{-(n^2-1)h^2\beta/8mL^2}$$

For an Ar atom in a cubic box of side 1.0 cm,

$$\begin{aligned}\frac{h^2\beta}{8mL^2} &= \frac{(6.626 \times 10^{-34}\,\mathrm{J\,s})^2}{(8) \times (39.95) \times (1.6605 \times 10^{-27}\,\mathrm{kg}) \times (1.381 \times 10^{-23}\,\mathrm{J\,K^{-1}}) \times (3.5 \times 10^{-15}\,\mathrm{K}) \times (1.0 \times 10^{-2}\,\mathrm{m})^2} \\ &= 0.17\bar{1}\end{aligned}$$

Then $q = \sum_{n=1}^{\infty} \mathrm{e}^{-0.17\bar{1}(n^2-1)} = 1.00 + 0.60 + 0.25 + 0.08 + 0.02 + \cdots = 1.95$

The partition function for motion in three dimensions is therefore $q = (1.95)^3 = \boxed{7.41}$

Comment. Temperatures as low as 3.5×10^{-15} K have never been achieved. However, a temperature of 2×10^{-8} K has been attained by adiabatic nuclear demagnetization (Section 4.5).

Question. Does the integral approximation apply at 2×10^{-8} K?

P19.3 (a) $q = \sum_j g_j \mathrm{e}^{-\beta\varepsilon_j}\,[12] = \sum_j g_j \mathrm{e}^{-hc\beta\tilde{\nu}_j}$

We use $hc\beta = \dfrac{1}{207\,\mathrm{cm^{-1}}}$ at 298 K and $\dfrac{1}{3475\,\mathrm{cm^{-1}}}$ at 5000 K. Therefore,

(i) $q = 5 + \mathrm{e}^{-4707/207} + 3\mathrm{e}^{-4751/207} + 5\mathrm{e}^{-10559/207}$

$= (5) + (1.3 \times 10^{-10}) + (3.2 \times 10^{-10}) + (3.5 \times 10^{-22}) = \boxed{5.00}$

(ii) $q = 5+\mathrm{e}^{-4707/3475}+3\mathrm{e}^{-4751/3475}+5\mathrm{e}^{-10559/3475} = (5)+(0.26)+(0.76)+(0.24) = \boxed{6.26}$

(b) $p_j = \dfrac{g_j\mathrm{e}^{-\beta\varepsilon_j}}{q} = \dfrac{g_j\mathrm{e}^{-hc\beta\tilde{\nu}_j}}{q}$ [10, with degeneracy g_j included]

Therefore, $p_0 = \dfrac{5}{q} = \boxed{1.00}$ at 298 K and $\boxed{0.80}$ at 5000 K

$p_2 = \dfrac{3\mathrm{e}^{-4751/207}}{5.00} = \boxed{6.5 \times 10^{-11}}$ at 298 K

$p_2 = \dfrac{3\mathrm{e}^{-4751/3475}}{6.26} = \boxed{0.12}$ at 5000 K

(c) $S_m = \dfrac{U_m - U_m(0)}{T} + Nk \ln q$ [35]

We need $U_m - U_m(0)$, and evaluate it by explicit summation

$$U_m - U_m(0) = E = \frac{N_A}{q}\sum_j g_j\varepsilon_j e^{-\beta\varepsilon_j} \quad [24 \text{ with degeneracy } g_j \text{ included}]$$

In terms of wavenumber units

(i) $\dfrac{U_m - U_m(0)}{N_A hc} = \dfrac{1}{5.00}\{0 + 4707\,\text{cm}^{-1} \times e^{-4707/207} + \cdots\} = 4.32 \times 10^{-7}\,\text{cm}^{-1}$

(ii) $\dfrac{U_m - U_m(0)}{N_A hc} = \dfrac{1}{6.26}\{0 + 4707\,\text{cm}^{-1} \times e^{-4707/3475} + \cdots\} = 1178\,\text{cm}^{-1}$

Hence, at 298 K

$$U_m - U_m(0) = 5.17 \times 10^{-6}\,\text{J mol}^{-1}$$

and at 5000 K

$$U_m - U_m(0) = 14.10\,\text{kJ mol}^{-1}$$

It follows that

(i) $S_m = \left(\dfrac{5.17 \times 10^{-6}\,\text{J mol}^{-1}}{298\,\text{K}}\right) + (8.314\,\text{J K}^{-1}\,\text{mol}^{-1}) \times (\ln 5.00)$

$= \boxed{13.38\,\text{J K}^{-1}\,\text{mol}^{-1}}$ [essentially $R \ln 5$]

(ii) $S_m = \left(\dfrac{14.09 \times 10^{3}\,\text{J mol}^{-1}}{5000\,\text{K}}\right) + (8.314\,\text{J K}^{-1}\,\text{mol}^{-1}) \times (\ln 6.26) = \boxed{18.07\,\text{J K}^{-1}\,\text{mol}^{-1}}$

P19.5 $q = \sum_j g_j e^{-\beta\varepsilon_j}\,[12] = \sum_j g_j e^{-hc\beta\tilde{\nu}_j}$

$$p_i = \frac{g_i e^{-\beta\varepsilon_i}}{q}[10] = \frac{g_i e^{-hc\beta\tilde{\nu}_i}}{q}$$

We measure energies from the lower states, and write

$$q = 2 + 2e^{-hc\beta\tilde{\nu}} = 2 + 2e^{-(1.4388\times121.1)/(T/\text{K})} = 2 + 2e^{-174.2/(T/\text{K})}$$

This function is plotted in Fig. 19.2.

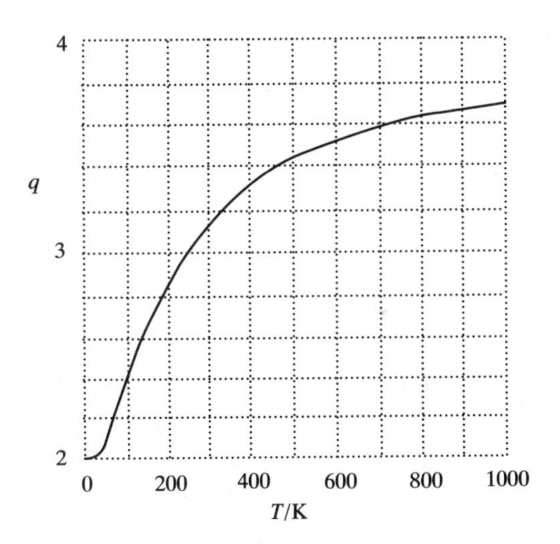

Figure 19.2

(a) At 300 K

$$p_0 = \frac{2}{q} = \frac{1}{1+e^{-174.2/300}} = \boxed{0.64}$$

$$p_1 = 1 - p_0 = \boxed{0.36}$$

(b) The electronic contribution to U_m in wavenumber units is

$$\frac{U_m - U_m(0)}{N_A hc} = -\frac{1}{hcq}\frac{dq}{d\beta}[25] = \frac{2\tilde{\nu}e^{-hc\beta\tilde{\nu}}}{q}$$

$$= \frac{(121.1\,\text{cm}^{-1}) \times (e^{-174.2/300})}{1+e^{-174.2/300}} = 43.45\,\text{cm}^{-1}$$

which corresponds to $\boxed{0.52\,\text{kJ mol}^{-1}}$

For the electronic contribution to the molar entropy, we need q and $U_m - U_m(0)$ at 500 K as well as at 300 K. These are

	300K	500K
$U_m - U_m(0)$	$0.518\,\text{kJ mol}^{-1}$	$0.599\,\text{kJ mol}^{-1}$
q	3.120	3.412

Then we form

$$S_m = \frac{U_m - U_m(0)}{T} + R\ln q\ [35]$$

At 300 K $\quad S_m = \left(\frac{518\,\text{J mol}^{-1}}{300\,\text{K}}\right) + (8.314\,\text{J K}^{-1}\,\text{mol}^{-1}) \times (\ln 3.120) = 11.2\,\text{J K}^{-1}\,\text{mol}^{-1}$

At 500 K $\quad S_m = \left(\frac{599\,\text{J mol}^{-1}}{500\,\text{K}}\right) + (8.314\,\text{J K}^{-1}\,\text{mol}^{-1}) \times (\ln 3.412) = 11.4\,\text{J K}^{-1}\,\text{mol}^{-1}$

Solutions to theoretical problems

P19.7 **(a)** $\quad W = \frac{N!}{n_1!n_2!\cdots}[1] = \frac{5!}{0!5!0!0!0!} = \boxed{1}$

(b) We draw up the following table

0	ε	2ε	3ε	4ε	5ε	$W = \frac{N!}{n_1!n_2!\cdots}$
4	0	0	0	0	1	5
3	1	0	0	1	0	20
3	0	1	1	0	0	20
2	2	0	1	0	0	30
2	1	2	0	0	0	30
1	3	1	0	0	0	20
0	5	0	0	0	0	1

The most probable configurations are $\boxed{\{2, 2, 0, 1, 0, 0\}}$ and $\boxed{\{2, 1, 2, 0, 0, 0\}}$ jointly.

P19.10 **(a)** $\quad q = \sum_j g_j e^{-\beta\varepsilon_j} = 1 + 3e^{-\beta\varepsilon} = \boxed{1 + 3e^{-\varepsilon/kT}}$

at $T = \frac{\varepsilon}{k}$, $q = 1 + 3e^{-1} = 2.104$

(b) $$U_m - U_m(0) = E = -\frac{N_A}{q}\frac{dq}{d\beta} = \frac{N_A}{q}(3\varepsilon e^{-\beta\varepsilon})$$
$$= \frac{N_A}{q}(3RTe^{-1}) = \frac{3RT}{2.104\,e} = \boxed{0.5245RT}$$

A numerical value cannot be obtained for the energy without specific knowledge of the temperature, but that is not required for the heat capacity on the entropy.

$$C_V = \left(\frac{\partial U_m}{\partial T}\right)_V = \left(\frac{\partial E}{\partial T}\right)_V$$

Since $\frac{d}{dT} = \frac{d\beta}{dT} \times \frac{d}{d\beta} = -\frac{1}{kT^2}\frac{d}{d\beta} = -k\beta^2\frac{d}{d\beta}$

$$C_V = -k\beta^2\left(\frac{\partial E}{\partial \beta}\right)_V = -k\beta^2(3\varepsilon N_A)\frac{\partial}{\partial\beta}\left(\frac{e^{-\beta\varepsilon}}{q}\right)$$
$$= -k\beta^2(3\varepsilon N_A)\frac{\partial}{\partial\beta}\left(\frac{e^{-\beta\varepsilon}}{1+3e^{-\beta\varepsilon}}\right)$$
$$= -k\beta^2(3\varepsilon N_A)\left[\frac{(1+3e^{-\beta\varepsilon})\times(-\varepsilon)e^{-\beta\varepsilon} - e^{-\beta\varepsilon}(-3\varepsilon e^{-\beta\varepsilon})}{(1+3e^{-\beta\varepsilon})^2}\right]$$
$$= -k\beta^2(3\varepsilon N_A)\left[\frac{-\varepsilon e^{-\beta\varepsilon} - 3\varepsilon e^{-2\beta\varepsilon} + 3\varepsilon e^{-2\beta\varepsilon}}{(1+3e^{-\beta\varepsilon})^2}\right]$$
$$= -k\beta^2(3\varepsilon N_A)\left[\frac{-\varepsilon e^{-\beta\varepsilon}}{(1+3e^{-\beta\varepsilon})^2}\right]$$
$$= \frac{3R\varepsilon^2 e^{-\beta\varepsilon}}{(kT)^2(1+3e^{-\beta\varepsilon})^2}$$

For $\varepsilon = kT$, $C_V = \frac{3Re^{-1}}{(1+3e^{-1})^2} = \frac{3R}{e(1+\frac{3}{e})^2} = \boxed{2.074\,\text{J K}^{-1}\,\text{mol}^{-1}}$

Note that taking the deriative of $0.5245RT$ with regard to T does not give the correct answer. That is because the temperature dependence of q is not taken into account by that process.

$$\frac{\partial}{\partial T}(0.5245RT) = 0.5245R = 4.361\,\text{J K}^{-1}\,\text{mol}^{-1}$$

and this is not the correct value.

The calculation of S does not require taking another derivative, so we can use $E = 0.5245RT$

$$S_m = \frac{E}{T} + R\ln q = 0.5245\,R + R\ln(2.104) = \boxed{10.55\,\text{J K}^{-1}\,\text{mol}^{-1}}$$

P19.11 (a) $U - U(0) = -N\frac{d\ln q}{d\beta}$ [27*b*], with $q = \frac{1}{1-e^{-\beta\varepsilon}}$ [15]

$$\frac{d\ln q}{d\beta} = \frac{1}{q}\frac{dq}{d\beta} = \frac{-\varepsilon e^{-\beta\varepsilon}}{1-e^{-\beta\varepsilon}}$$

$$a\varepsilon = \frac{U-U(0)}{N} = \frac{\varepsilon e^{(-\beta\varepsilon)}}{1-e^{-\beta\varepsilon}} = \frac{\varepsilon}{e^{\beta\varepsilon}-1}$$

Hence, $e^{\beta\varepsilon} = \dfrac{1+a}{a}$, implying that, $\beta = \dfrac{1}{\varepsilon}\ln\left(1+\dfrac{1}{a}\right)$

For a mean energy of ε, $a = 1$, $\beta = \dfrac{1}{\varepsilon}\ln 2$, implying that

$$T = \frac{\varepsilon}{k\ln 2}\ln 2 = (50\,\text{cm}^{-1}) \times \left(\frac{hc}{k\ln 2}\right) = \boxed{104\,\text{K}}$$

(b) $$q = \frac{1}{1-e^{-\beta\varepsilon}} = \frac{1}{1-\left(\frac{a}{1+a}\right)} = \boxed{1+a}$$

(c) $$\frac{S}{Nk} = \frac{U-U(0)}{NkT} + \ln q\,[35] = a\beta\varepsilon + \ln q$$

$$= a\ln\left(1+\frac{1}{a}\right) + \ln(1+a) = a\ln(1+a) - a\ln a + \ln(1+a)$$

$$= \boxed{(1+a)\ln(1+a) - a\ln a}$$

When the mean energy is ε, $a = 1$ and then $\boxed{\dfrac{S}{Nk} = 2\ln 2}$

Solutions to additional problems

P19.14 Number of configurations of combined system, $W = W_1W_2$

$$W = (10^{20}) \times (2 \times 10^{20}) = \boxed{2 \times 10^{40}}$$

$$S = k\ln W\,[30]; \qquad S_1 = k\ln W_1; \qquad S_2 = k\ln W_2$$

$$S = k\ln(2 \times 10^{40}) = k\{\ln 2 + 40\ln 10\} = 92.8\,k$$

$$= 92.8 \times (1.381 \times 10^{-23}\,\text{J K}^{-1}) = \boxed{1.282 \times 10^{-21}\,\text{J K}^{-1}}$$

$$S_1 = k\ln(10^{20}) = k\{20\ln 10\} = 46.1\,k$$

$$= 46.1 \times (1.381 \times 10^{-23}\,\text{J K}^{-1}) = \boxed{0.637 \times 10^{-21}\,\text{J K}^{-1}}$$

$$S_2 = k\ln(2 \times 10^{20}) = k\{\ln 2 + 20\ln 10\} = 46.7\,k$$

$$= 46.7 \times (1.381 \times 10^{-23}\,\text{J K}^{-1}) = \boxed{0.645 \times 10^{-21}\,\text{J K}^{-1}}$$

These results are significant in that they show that the statistical mechanical entropy is an additive property consistent with the thermodynamic result. That is, $S = S_1 + S_2 = (0.637 \times 10^{-21} + 0.645 \times 10^{-21})\,\text{J K}^{-1} = 1.282 \times 10^{-21}\,\text{J K}^{-1}$

P19.15 Although He is a liquid at these temperatures ($T_b = 4.22\,\text{K}$), we will test it as if it were a perfect gas with no interaction potential.

$$p_i = \frac{N_i}{N} = g_i e^{-\beta\varepsilon_i}/q\ [10\text{–}12]$$

$$\varepsilon_i = \frac{h^2}{8mX^2}\{n_x^2 + n_y^2 + n_z^2\}\ [19]; \qquad q = \frac{V}{\Lambda^3}; \quad \Lambda = h\left(\frac{\beta}{2\pi m}\right)^{1/2}\ [22]$$

Ground state $n_x = n_y = n_z = 1$; $g = 1$

First excited state

$$\left.\begin{array}{ll} n_x = n_y = 1; & n_z = 2 \\ n_x = n_z = 1; & n_y = 2 \\ n_y = n_z = 1; & n_x = 2 \end{array}\right\} g = 3$$

$$q = \frac{V}{\Lambda^3} = \frac{V}{h^3}(2\pi mkT)^{3/2}$$

$$= \frac{(1\,\text{cm}^3) \times \left(\frac{1\,\text{m}^3}{10^6\,\text{cm}^3}\right) \times \left[2\pi(1.381 \times 10^{-23}\,\text{J K}^{-1})\right]^{+3/2} \times (mT)^{+3/2}}{(6.626 \times 10^{-34}\,\text{J s})^3}$$

$$= 2.28 \times 10^{60}\,\text{kg}^{-3/2}\,\text{K}^{-3/2}(mT)^{3/2}$$

$$\beta\varepsilon_{\text{1st excited}} = \left(\frac{1}{kT}\right) \times \left(\frac{h^2}{8mX^2}\right)\ (6)$$

$$= \frac{6(6.626 \times 10^{-34}\,\text{J s})^2}{8(1.381 \times 10^{-23}\,\text{J K}^{-1}) \times (0.01\,\text{m})^2}\frac{1}{mT}$$

$$= \frac{2.38 \times 10^{-40}\,\text{kg K}}{mT}$$

$$p_{\text{1st excited}} = \frac{3e^{-\left(\frac{2.38\times10^{-40}\,\text{kg K}}{mT}\right)}}{(2.78 \times 10^{60}\,\text{kg}^{-3/2}\,\text{K}^{-3/2}) \times (mT)^{3/2}}$$

Isotope	m/kg	T/K	$p_{\text{1st excited}}$	Occupancy $= pN = 10^{22}p$
^{4}He	6.64×10^{-27}	0.0010	6.30×10^{-17}	6.30×10^5
		2.0	7.04×10^{-22}	7
		4.0	2.49×10^{-22}	2
^{3}He	5.01×10^{-27}	0.0010	9.63×10^{-17}	9.63×10^5
		2.0	1.08×10^{-21}	11
		4.0	3.81×10^{-22}	4

These results may at first seem to contradict the expected common sense result that the populations of excited states increase as the temperature increases, but the energy separations of these states is so small that even a slight increase in temperature promotes the particles to much higher quantum states.

P19.17 $S = k \ln W$ [30]

Therefore,

$$\left(\frac{\partial S}{\partial U}\right)_V = \frac{k}{W}\left(\frac{\partial W}{\partial U}\right)_V$$

or

$$\left(\frac{\partial W}{\partial U}\right)_V = \frac{W}{k}\left(\frac{\partial S}{\partial U}\right)_V$$

But from eqn 5.4

$$\left(\frac{\partial U}{\partial S}\right)_V = T$$

So,

$$\left(\frac{\partial S}{\partial U}\right)_V = \frac{1}{T}$$

then

$$\left(\frac{\partial W}{\partial U}\right)_V = \frac{W}{k}\left(\frac{1}{T}\right)$$

Therefore,

$$\begin{aligned}\frac{\Delta W}{W} &\approx \frac{\Delta U}{kT} \\ &= \frac{100 \times 10^3\,\mathrm{J}}{(1.381 \times 10^{-23}\,\mathrm{J\,K^{-1}}) \times 298\,\mathrm{K}} \\ &= \boxed{2.4 \times 10^{25}}\end{aligned}$$

P19.19 **(a)** Total entropy, $S = S_1 + S_2 = (5.69 + 11.63)\,\mathrm{J\,K^{-1}} = 17.32\,\mathrm{J\,K^{-1}}$

$$\begin{aligned}W &= \mathrm{e}^{S/k} = \mathrm{e}^{17.32\,\mathrm{J\,K^{-1}}/1.381\times10^{-23}\,\mathrm{J\,K^{-1}}}\ [30] \\ &= \mathrm{e}^{1.254\times10^{24}} = 10^{5.44\times10^{23}}\end{aligned}$$

(b) Total entropy, $S = 2\,\mathrm{mol}(9.03\,\mathrm{J\,K^{-1}\,mol^{-1}}) = 18.06\,\mathrm{J\,K^{-1}}$

$$\begin{aligned}W &= \mathrm{e}^{S/k} = \mathrm{e}^{18.06\,\mathrm{J\,K^{-1}}/1.381\times10^{-23}\,\mathrm{J\,K^{-1}}} \\ &= \mathrm{e}^{1.31\times10^{24}} = 10^{5.69\times10^{23}}\end{aligned}$$

The final temperature is not the average because the molar heat capacity of graphite increases with temperature. At 298 K it is $8.54\,\mathrm{J\,K^{-1}\,mol^{-1}}$, whereas at 498 K it is $14.64\,\mathrm{J\,K^{-1}\,mol^{-1}}$.

(c) At constant internal energy and volume the condition for spontaneity is $\boxed{\Delta S_{U,V} > 0}$. Since $W_{(\mathbf{b})} > W_{(\mathbf{a})}$, the process part **(b)** is $\boxed{\text{spontaneous}}$

P19.21 At equilibrium $\dfrac{N(r)/V}{N(r_0)/V} = \mathrm{e}^{-\{V(r)-V(r_0)\}/kT}\,[6]$

Since $V(r) = -GMm/r,\ V(\infty) = 0$ and [Note: $V(r)$ is potential energy, V is volume]

$$\frac{N(\infty)/V}{N(r_0)/V} = \mathrm{e}^{V(r_0)/kT}$$

which says that $N(\infty)/V \propto \mathrm{e}^{V(r_0)/kT}$ = constant. This is obviously not the current distribution for planetary atmospheres where $\lim_{r\to\infty} N(r)/V = 0$. Consequently, we may conclude that the earth's atmosphere, or any other planetary atmosphere, cannot be at equilibrium.

20 Statistical thermodynamics: the machinery

Solutions to exercises

E20.1 $C_{V,\mathrm{m}} = \frac{1}{2}(3 + \nu_{\mathrm{R}}^* + 2\nu_{\mathrm{V}}^*)R$ [42]
with a mode active if $T > \theta_{\mathrm{M}}$.

(a) $\nu_{\mathrm{R}}^* = 2,\ \nu_{\mathrm{V}}^* \approx 0;$ hence $C_{V,\mathrm{m}} = \frac{1}{2}(3+2)R = \boxed{\frac{5}{2}R}$ [experimental: $3.4R$]

(b) $\nu_{\mathrm{R}}^* = 3,\ \nu_{\mathrm{V}}^* \approx 0;$ hence $C_{V,\mathrm{m}} = \frac{1}{2}(3+3)R = \boxed{3R}$ [experimental: $3.2R$]

(c) $\nu_{\mathrm{R}}^* = 3,\ \nu_{\mathrm{V}}^* \approx 0;$ hence $C_{V,\mathrm{m}} = \frac{1}{2}(3+3)R = \boxed{3R}$ [experimental: $8.8R$]

Comment. Data from the books by Herzberg (see *Further reading*, Chapters 16 and 17) give for the vibrational wavenumbers

I_2 $\quad \tilde{\nu} = 214\ \mathrm{cm^{-1}} \approx 207\ \mathrm{cm^{-1}}$ [kT at $T = 298$ K]
CH_4 $\quad$ all greater than $1300\ \mathrm{cm^{-1}}$
C_6H_6 $\quad$ 4 less than $207\ \mathrm{cm^{-1}}$ [kT at $T = 298$ K]

Thus, we expect the vibrational mode of I_2 to contribute significantly to $C_{V,\mathrm{m}}$, and hence $C_{V,\mathrm{m}} > \dfrac{5}{2}R$.

We expect little vibrational contribution to $C_{V,\mathrm{m}}$ for methane; hence $C_{V,\mathrm{m}} \approx 3R$. For benzene, there is a lot of vibrational contribution; hence $C_{V,\mathrm{m}} \gg 3R$.

E20.2 Assuming that all rotational modes are active we can draw up the following table for $C_{V,\mathrm{m}}$, $C_{p,\mathrm{m}}$, and γ with and without active vibrational modes.

	$C_{V,\mathrm{m}}$	$C_{p,\mathrm{m}}$	γ	Exptl	
$NH_3(\nu_{\mathrm{V}}^* = 0)$	$3R$	$4R$	1.33	1.31	closer
$NH_3(\nu_{\mathrm{V}}^* = 6)$	$9R$	$10R$	1.11		
$CH_4(\nu_{\mathrm{V}}^* = 0)$	$3R$	$4R$	1.33	1.31	closer
$CH_4(\nu_{\mathrm{V}}^* = 9)$	$12R$	$13R$	1.08		

The experimental values are obtained from Table 2.6 assuming $C_{p,\mathrm{m}} = C_{V,\mathrm{m}} + R$. It is clear from the comparison in the above table that the vibrational modes are not active. This is confirmed by the experimental vibrational wavenumbers (see Herzberg references in *Further reading*, Chapters 16 and 17) all of which are much greater than kT at 298 K.

E20.3
$$q^{\mathrm{R}} = \frac{0.6950}{\sigma} \times \frac{T/\mathrm{K}}{(B/\mathrm{cm^{-1}})} \text{ [Table 20.4]}$$
$$= \frac{(0.6950) \times (T/\mathrm{K})}{10.59}\ [\sigma = 1] = 0.06563(T/\mathrm{K})$$

(a) $q^{\mathrm{R}} = (0.06563) \times (298) = \boxed{19.6}$ **(b)** $q^{\mathrm{R}} = (0.06563) \times (523) = \boxed{34.3}$

E20.4 Look for the rotational subgroup of the molecule (the group of the molecule composed only of the identity and the rotational elements, and assess its order).

(a) CO. Full group $C_{\infty\mathrm{v}}$; subgroup C_1; hence $\sigma = \boxed{1}$

(b) O_2. Full group $D_{\infty\mathrm{h}}$; subgroup C_2; hence $\sigma = \boxed{2}$

(c) H_2S. Full group $C_{2\mathrm{v}}$; subgroup C_2; hence $\sigma = \boxed{2}$

(d) SiH_4. Full group T_d; subgroup T; hence $\sigma = \boxed{12}$

(e) $CHCl_3$. Full group C_{3v}; subgroup C_3; hence $\sigma = \boxed{3}$

See the references in the *Further reading* for Chapter 15 for a more complete set of character tables including those of the rotational subgroups.

E20.5

$$q^{R} = \frac{1.0270}{\sigma} \frac{(T/K)^{3/2}}{(ABC/cm^{-3})^{1/2}} \text{ [Table 20.4]}$$
$$= \frac{1.0270 \times 298^{3/2}}{(2) \times (27.878 \times 14.509 \times 9.287)^{1/2}} [\sigma = 2] = \boxed{43.1}$$

The high-temperature approximation is valid if $T > \theta_R$, where

$$\theta_R = \frac{hc(ABC)^{1/3}}{k},$$
$$= \frac{(6.626 \times 10^{-34}\,\text{J s}) \times (2.998 \times 10^{10}\,\text{cm s}^{-1}) \times [(27.878) \times (14.509) \times (9.287)\,\text{cm}^{-3}]^{1/3}}{1.38 \times 10^{-23}\,\text{J K}^{-1}}$$
$$= \boxed{22.36\,\text{K}}$$

E20.6 $q^{R} = 43.1$ [Exercise 20.5]

All the rotational modes of water are fully active at 25°C (Example 20.6 and Exercise 20.5); therefore

$$U_m^R - U_m^R(0) = E^R = \tfrac{3}{2}RT$$
$$S_m^R = \frac{E^R}{T} + R \ln q^R \text{ [Table 20.1]}$$
$$= \tfrac{3}{2}R + R \ln 43.1 = \boxed{43.76\,\text{J K}^{-1}\,\text{mol}^{-1}}$$

Comment. Division of q^R by $N_A!$ is not required for the internal contributions; internal motions may be thought of as localized (distinguishable). It is the overall canonical partition function, which is a product of internal and external contributions, that is divided by $N!$ (Table 20.1).

E20.7 **(a)** For a spherical rotor (Section 16.5)

$$E = hcBJ(J+1) \text{ [16.31]} \quad [B = 5.2412\,\text{cm}^{-1} \text{ for } CH_4]$$

and the degeneracy is $g(J) = (2J+1)^2$. Hence

$$q \approx \frac{1}{\sigma}\sum_J (2J+1)^2 e^{-\beta hcBJ(J+1)}$$

which is analogous for spherical rotors to eqn 14 for q^R for linear rotors.

$$hcB\beta = \frac{(1.4388\,\text{K}) \times (5.2412)}{T} = \frac{7.5410}{T/\text{K}}, \qquad \sigma = 12$$
$$q = \frac{1}{12}\sum_J (2J+1)^2 e^{-7.5410\,J(J+1)/(T/\text{K})}$$
$$= \frac{1}{12}(1.0000 + 8.5561 + 21.480 + 36.173 + \cdots)$$
$$= \frac{1}{12} \times 443.427 = \boxed{36.95} \text{ at 298 K}$$

The sum converged after 20 terms.

Similarly, at 500 K

$$q = \frac{1}{12}(1.0000 + 8.7326 + 22.8370 + 40.8880 + \cdots) = \frac{1}{12} \times 960.96 = \boxed{80.08}$$

The sum converged after 24 terms.

(Note that the results are still approximate because the symmetry number is a valid corrector only at high temperatures. To get exact values of q we should do a detailed analysis of the rotational states allowed by the Pauli principle.)

(b) $$q \approx \frac{1.0270}{\sigma} \times \frac{(T/\mathrm{K})^{3/2}}{(B/\mathrm{cm}^{-1})^{3/2}} \text{ [Table 20.4, } A = B = C]$$

$$= \frac{1.0270}{12} \times \frac{(T/\mathrm{K})^{3/2}}{(5.2412)^{3/2}} = 7.133 \times 10^{-3} \times (T/\mathrm{K})^{3/2}$$

At 298 K, $q = 7.133 \times 10^{-3} \times 298^{3/2} = \boxed{36.7}$

At 500 K, $q = 7.133 \times 10^{-3} \times 500^{3/2} = \boxed{79.7}$

The difference in this case is small.

E20.8 $$q^{\mathrm{R}} = \frac{kT}{\sigma hcB} \text{ [17]}, \quad B = \frac{\hbar}{4\pi cI}, \quad I = \mu R^2$$

Hence $q = \dfrac{8\pi^2 kTI}{\sigma h^2} = \dfrac{8\pi^2 kT\mu R^2}{\sigma h^2}$

For O_2, $\mu = \frac{1}{2}m(\mathrm{O}) = \frac{1}{2} \times 16.00\,\mathrm{u} = 8.00\,\mathrm{u}$ and $\sigma = 2$; therefore

$$q = \frac{(8\pi^2) \times (1.381 \times 10^{-23}\,\mathrm{J\,K^{-1}}) \times (300\,\mathrm{K}) \times (8.00) \times (1.6605 \times 10^{-27}\,\mathrm{kg}) \times (1.21 \times 10^{-10}\,\mathrm{m})^2}{(2) \times (6.626 \times 10^{-34}\,\mathrm{J\,s})^2}$$

$$= \boxed{72.5}$$

E20.9 $$C_{V,\mathrm{m}}/R = f^2, \quad f = \left(\frac{\theta_{\mathrm{V}}}{T}\right) \times \left(\frac{\mathrm{e}^{-\theta_{\mathrm{V}}/2T}}{1 - \mathrm{e}^{-\theta_{\mathrm{V}}/T}}\right) \text{ [41]}; \quad \theta = \frac{hc\tilde{\nu}}{k}$$

We write $x = \dfrac{\theta_{\mathrm{V}}}{T}$; then $C_{V,\mathrm{m}}/R = \dfrac{x^2\mathrm{e}^{-x}}{(1 - \mathrm{e}^{-x})^2}$

This function is plotted in Fig. 20.1. For the acetylene (ethyne) calculation, use the expression above for each mode. We draw up the following table using $kT/hc = 207\,\mathrm{cm}^{-1}$ at 298 K and $348\,\mathrm{cm}^{-1}$ at 500 K, and $\theta_{\mathrm{V}}/T = hc\tilde{\nu}/kT$.

	x		$C_{V,\mathrm{m}}/R$	
$\tilde{\nu}/\mathrm{cm}^{-1}$	298 K	500 K	298 K	500 K
612	2.96	1.76	0.505	0.777
612	2.96	1.76	0.505	0.777
729	3.52	2.09	0.389	0.704
729	3.52	2.09	0.389	0.704
1974	9.54	5.67	0.007	0.112
3287	15.88	9.45	3.2×10^{-5}	0.007
3374	16.30	9.70	2.2×10^{-5}	0.006

The heat capacity of the molecule is the sum of these contributions, namely

(a) $1.796R = \boxed{14.93\,\mathrm{J\,K^{-1}\,mol^{-1}}}$ at 298 K and **(b)** $3.086R = \boxed{25.65\,\mathrm{J\,K^{-1}\,mol^{-1}}}$ at 500 K.

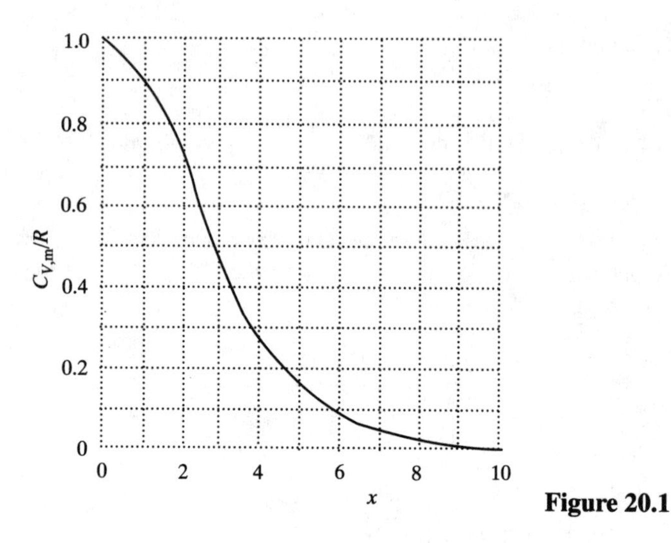

Figure 20.1

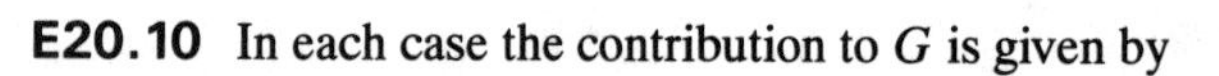

E20.10 In each case the contribution to G is given by

$$G - G(0) = -nRT \ln q \text{ [Table 20.1, also see Comment to Exercise 20.6]}$$

Therefore, we first evaluate q^{R} and q^{V}.

$$\begin{aligned} q^{\mathrm{R}} &= \frac{0.6950}{\sigma} \frac{T/\mathrm{K}}{B/\mathrm{cm}^{-1}} \text{ [Table 20.4, } \sigma = 2] \\ &= \frac{(0.6950) \times (298)}{(2) \times (0.3902)} = 265 \\ q^{\mathrm{V}} &= \left(\frac{1}{1 - \mathrm{e}^{-a}}\right) \times \left(\frac{1}{1 - \mathrm{e}^{-b}}\right)^2 \times \left(\frac{1}{1 - \mathrm{e}^{-c}}\right) \text{ [Table 20.4]} \end{aligned}$$

with

$$\begin{aligned} a &= \frac{(1.4388) \times (1388.2)}{298} = 6.70\bar{2} \\ b &= \frac{(1.4388) \times (667.4)}{298} = 3.22\bar{2} \\ c &= \frac{(1.4388) \times (2349.2)}{298} = 11.3\bar{4} \end{aligned}$$

Hence

$$q^{\mathrm{V}} = \frac{1}{1 - \mathrm{e}^{-6.702}} \times \left(\frac{1}{1 - \mathrm{e}^{-3.222}}\right)^2 \times \frac{1}{1 - \mathrm{e}^{-11.34}} = 1.08\bar{6}$$

Therefore, the rotational contribution to the molar Gibbs energy is

$$\begin{aligned} -RT \ln q^{\mathrm{R}} &= -8.314\,\mathrm{J\,K^{-1}\,mol^{-1}} \times 298\,\mathrm{K} \times \ln 265 \\ &= \boxed{-13.8\,\mathrm{kJ\,mol^{-1}}} \end{aligned}$$

and the vibrational contribution is

$$-RT \ln q^{\mathrm{V}} = -8.314\,\mathrm{J\,K^{-1}\,mol^{-1}} \times 298\,\mathrm{K} \times \ln 1.08\bar{6} = \boxed{-0.20\,\mathrm{kJ\,mol^{-1}}}$$

E20.11 $q = \sum_j g_j e^{-\beta\varepsilon_j}, \quad g_j = 2J+1$

$$= 4 + 2e^{-\beta\varepsilon} \quad [g(^2P_{3/2}) = 4, \; g(^2P_{1/2}) = 2]$$

$$U - U(0) = -\frac{N}{q}\frac{dq}{d\beta} = \frac{N\varepsilon e^{-\beta\varepsilon}}{2 + e^{-\beta\varepsilon}}$$

$$C_V = \left(\frac{\partial U}{\partial T}\right)_V = -k\beta^2\left(\frac{\partial U}{\partial \beta}\right)_V = \frac{2R(\varepsilon\beta)^2 e^{-\beta\varepsilon}}{(2 + e^{-\beta\varepsilon})^2} \quad [N = N_A]$$

(a) Therefore, since at 500 K $\beta\varepsilon = 2.53\bar{5}$

$$C_{V,m}/R = \frac{(2) \times (2.53\bar{5})^2 \times (e^{-2.53\bar{5}})}{(2 + e^{-2.53\bar{5}})^2} = \boxed{0.236}$$

(b) At 900 K, when $\beta\varepsilon = 1.408$,

$$C_{V,m}/R = \frac{(2) \times (1.408)^2 \times (e^{-1.408})}{(2 + e^{-1.408})^2} = \boxed{0.193}$$

Comment. $C_{V,m}$ is smaller at 900 K than at 500 K, for then the temperature is higher than the peak in the 'two-level' heat capacity curve.

E20.12 We assume that the upper eight of the $\left(2 \times \frac{9}{2} + 1\right) = 9$ spin–orbit states of the ion lies at an energy much greater than kT at 1 K; hence, since the spin degeneracy of Co^{2+} is 4 (the ion is a spin quartet), $q = 4$. The contribution to the entropy is

$$R \ln q = (8.314\,\mathrm{J\,K^{-1}\,mol^{-1}}) \times (\ln 4) = \boxed{11.5\,\mathrm{J\,K^{-1}\,mol^{-1}}}$$

E20.13 In each case $S_m = R \ln s$ [54]. Therefore,

(a) $S_m = R \ln 3 = 1.1R = \boxed{9\,\mathrm{J\,K^{-1}\,mol^{-1}}}$ **(b)** $S_m = R \ln 5 = 1.6R = \boxed{13\,\mathrm{J\,K^{-1}\,mol^{-1}}}$

(c) $S_m = R \ln 6 = 1.8R = \boxed{15\,\mathrm{J\,K^{-1}\,mol^{-1}}}$

E20.14 $S = k \ln W$ [19.30]

$$= k \ln 4^N = Nk \ln 4$$

$$= (5 \times 10^8) \times (1.38 \times 10^{-23}\,\mathrm{J\,K^{-1}}) \times \ln 4 = \boxed{9.57 \times 10^{-15}\,\mathrm{J\,K^{-1}}}$$

Comment. Even for a molecule as large as DNA the residual molecular entropy is small compared to normal entropies for macroscopic systems.

E20.15 We use eqn 59 with $X = \mathrm{I}, \; X_2 = \mathrm{I_2}, \; \Delta E_0 = D_0$.

$$D_0 = D_e - \frac{1}{2}\tilde{\nu} = 1.5422\,\mathrm{eV} \times \frac{8065.5\,\mathrm{cm^{-1}}}{1\,\mathrm{eV}} - 107.18\,\mathrm{cm^{-1}}$$

$$= 1.2331 \times 10^4\,\mathrm{cm^{-1}} = 1.475 \times 10^5\,\mathrm{J\,mol^{-1}}$$

$$K = \left(\frac{q^{\ominus 2}_{\mathrm{I,m}}}{q^{\ominus}_{\mathrm{I_2,m}} N_A}\right) e^{-\Delta E_0/RT} \quad [59]$$

$$q^{\ominus}_{\mathrm{I,m}} = q^{\mathrm{T}}_{\mathrm{m}}(\mathrm{I})q^{\mathrm{E}}(\mathrm{I}), \quad q^{\mathrm{E}}(\mathrm{I}) = 4$$

$$q^{\ominus}_{\mathrm{I_2,m}} = q^{\mathrm{T}}_{\mathrm{m}}(\mathrm{I_2})q^{\mathrm{R}}(\mathrm{I_2})q^{\mathrm{V}}(\mathrm{I_2})q^{\mathrm{E}}(\mathrm{I_2}), \quad q^{\mathrm{E}}(\mathrm{I_2}) = 1$$

$$\frac{q_{\rm m}^{\rm T}}{N_{\rm A}} = 2.561 \times 10^{-2}(T/{\rm K})^{5/2} \times (M/{\rm g\,mol^{-1}})^{3/2} \text{ [Table 20.4]}$$

$$\frac{q_{\rm m}^{\rm T}({\rm I_2})}{N_{\rm A}} = 2.561 \times 10^{-2} \times 1000^{5/2} \times 253.8^{3/2} = 3.27 \times 10^{9}$$

$$\frac{q_{\rm m}^{\rm T}({\rm I})}{N_{\rm A}} = 2.561 \times 10^{-2} \times 1000^{5/2} \times 126.9^{3/2} = 1.16 \times 10^{9}$$

$$q^{\rm R}({\rm I_2}) = \frac{0.6950}{\sigma} \times \frac{T/{\rm K}}{B/{\rm cm^{-1}}} = \frac{1}{2} \times 0.6950 \times \frac{1000}{0.0373} = 931\bar{6}$$

$$q^{\rm V}({\rm I_2}) = \frac{1}{1-{\rm e}^{-a}}, \quad a = 1.4388\frac{\tilde{\nu}/{\rm cm^{-1}}}{T/{\rm K}} \text{ [Table 20.4]}$$

$$= \frac{1}{1-{\rm e}^{-1.4388\times 214.36/1000}} = 3.77$$

$$K = \frac{(1.16 \times 10^{9} \times 4)^2{\rm e}^{-17.741}}{(3.27 \times 10^{9}) \times (9316) \times (3.77)} = \boxed{3.70 \times 10^{-3}}$$

Solutions to problems

Solutions to numerical problems

P20.1 $q^{\rm E} = \sum_j g_j {\rm e}^{-\beta\varepsilon_j} = 2 + 2{\rm e}^{-\beta\varepsilon}, \quad \varepsilon = \Delta\varepsilon = 121.1\,{\rm cm^{-1}}$

$$U_{\rm m} - U_{\rm m}(0) = -\frac{N_{\rm A}}{q^{\rm E}}\left(\frac{\partial q^{\rm E}}{\partial \beta}\right)_V [28] = \frac{2N_{\rm A}\varepsilon{\rm e}^{-\beta\varepsilon}}{q^{\rm E}}$$

$$C_{V,{\rm m}} = -k\beta^2\left(\frac{\partial U_{\rm m}}{\partial \beta}\right)_V [37]$$

Let $x = \beta\varepsilon$, then $d\beta = \frac{1}{\varepsilon}dx$

$$C_{V,{\rm m}} = -N_{\rm A}k\left(\frac{x}{\varepsilon}\right)^2 \times (\varepsilon)^2\frac{\partial}{\partial x}\left(\frac{{\rm e}^{-x}}{1+{\rm e}^{-x}}\right) = R\left(\frac{x^2{\rm e}^{-x}}{(1+{\rm e}^{-x})^2}\right)$$

Therefore

$$C_{V,{\rm m}}/R = \frac{x^2{\rm e}^{-x}}{(1+{\rm e}^{-x})^2}, \quad x = \beta\varepsilon$$

We then draw up the following table

$T/{\rm K}$	50	298	500
$(kT/hc)/{\rm cm^{-1}}$	34.8	207	348
x	3.48	0.585	0.348
$C_{V,{\rm m}}/R$	$\boxed{0.351}$	$\boxed{0.079}$	$\boxed{0.029}$
$C_{V,{\rm m}}/({\rm J\,K^{-1}\,mol^{-1}})$	2.91	0.654	0.244

Comment. Note that the double degeneracies do not affect the results because the two factors of 2 in q cancel when U is formed. In the range of temperature specified, the electronic contribution to the heat capacity decreases with increasing temperature.

P20.3 The energy expression for a particle on a ring is

$$E = \frac{\hbar^2 m_l^2}{2I} \text{ [12.49]}$$

Therefore

$$q = \sum_{m=-\infty}^{\infty} e^{-m^2\hbar^2/2IkT} = \sum_{m=-\infty}^{\infty} e^{-\beta\hbar^2 m^2/2I}$$

The summation may be approximated by an integration

$$q \approx \frac{1}{\sigma}\int_{-\infty}^{\infty} e^{-m^2\hbar^2/2IkT}\,dm = \frac{1}{\sigma}\left(\frac{2IkT}{\hbar^2}\right)^{1/2}\int_{-\infty}^{\infty} e^{-x^2}\,dx$$
$$\approx \frac{1}{\sigma}\left(\frac{2\pi IkT}{\hbar^2}\right)^{1/2}$$

$$U - U(0) = -\frac{N}{q}\frac{\partial q}{\partial \beta} = \frac{N}{2\beta} = \frac{1}{2}NkT = \frac{1}{2}RT \quad (N = N_A)$$

$$C_{V,m} = \left(\frac{\partial U_m}{\partial T}\right)_V = \frac{1}{2}R = \boxed{4.2\,\mathrm{J\,K^{-1}\,mol^{-1}}}$$

$$S_m = \frac{U_m - U_m(0)}{T} + R\ln q \text{ [Table 20.1]}$$
$$= \frac{1}{2}R + R\ln\frac{1}{\sigma}\left(\frac{2\pi IkT}{\hbar^2}\right)^{1/2}$$
$$= \frac{1}{2}R + R\ln\frac{1}{3}\left(\frac{(2\pi)\times(5.341\times10^{-47}\,\mathrm{kg\,m^2})\times(1.381\times10^{-23}\,\mathrm{J\,K^{-1}})\times(298)}{(1.055\times10^{-34}\,\mathrm{J\,s})^2}\right)^{1/2}$$
$$= \frac{1}{2}R + 1.31R = 1.81R, \text{ or } \boxed{15\,\mathrm{J\,K^{-1}\,mol^{-1}}}$$

P20.5 The absorption lines are the values of $\{E(J+1) - E(J)\}/hc$ for $J = 0, 1, \ldots$ Therefore, we can reconstruct the energy levels from the data; they are

$$\frac{E_J}{hc} = \sum_{J'=0}^{J-1}\{E(J'+1) - E(J')\}/hc$$

Using $hc\beta = \dfrac{hc}{kT} = 207.223\,\mathrm{cm^{-1}}$ [inside front cover]

$$q = \sum_{J=0}^{\infty}(2J+1)e^{-\beta E(J)}$$
$$= 1 + 3e^{-21.19/207.223} + 5e^{-(21.19+42.37)/207.223} + 7e^{-(21.19+42.37+63.56)/207.223} + \cdots$$
$$= 1 + 2.708 + 3.679 + 3.790 + 3.237 + \cdots = \boxed{19.89}$$

P20.7 $H_2O + DCl \rightleftharpoons HDO + HCl$

$$K = \frac{q^{\ominus}(HDO)q^{\ominus}(HCl)}{q^{\ominus}(H_2O)q^{\ominus}(DCl)} e^{-\beta\Delta_r E_0} \text{ [56, with } \Delta_r E_0 \text{ in joules]}$$

The ratio of translational partition functions (Table 20.4) is

$$\frac{q_m^T(HDO)q_m^T(HCl)}{q_m^T(H_2O)q_m^T(DCl)} = \left(\frac{M(HDO)M(HCl)}{M(H_2O)M(DCl)}\right)^{3/2} = \left(\frac{19.02 \times 36.46}{18.02 \times 37.46}\right)^{3/2} = 1.041$$

The ratio of rotational partition functions is

$$\frac{q^R(HDO)q^R(HCl)}{q^R(H_2O)q^R(DCl)} = 2 \times \frac{(27.88 \times 14.51 \times 9.29)^{1/2} \times 5.449}{(23.38 \times 9.102 \times 6.417)^{1/2} \times 10.59} = 1.707$$

($\sigma = 2$ for H_2O; $\sigma = 1$ for the other molecules).

The ratio of vibrational partition functions is

$$\frac{q^V(HDO)q^V(HCl)}{q^V(H_2O)q^V(DCl)} = \frac{q(2726.7)q(1402.2)q(3707.5)q(2991)}{q(3656.7)q(1594.8)q(3755.8)q(2145)} = Q$$

where

$$q(x) = \frac{1}{1 - e^{-1.4388 \times x/(T/K)}}$$

$$\frac{\Delta_r E_0}{hc} = \frac{1}{2}\{(2726.7 + 1402.2 + 3707.5 + 2991) - (3656.7 + 1594.8 + 3755.8 + 2145)\}\,\text{cm}^{-1} = -162\,\text{cm}^{-1}$$

Therefore, $K = 1.041 \times 1.707 \times Q \times e^{1.4388 \times 162/(T/K)} = 1.777\,Qe^{233/(T/K)}$

We then draw up the following table (using a computer)

T/K	100	200	300	400	500	600	700	800	900	1000
K	18.3	5.70	3.87	3.19	2.85	2.65	2.51	2.41	2.34	2.29

and specifically $K = \boxed{3.89}$ at **(a)** 298 K and $\boxed{2.41}$ at **(b)** 800 K.

Solutions to theoretical problems

P20.8 A Sackur–Tetrode type of equation describes the translational entropy of the gas. Here

$$q^T = q_x^T q_y^T \quad \text{with } q_x^T = \left(\frac{2\pi m X^2}{\beta h^2}\right)^{1/2} \quad [19.18]$$

Therefore,

$$q^T = \left(\frac{2\pi m}{\beta h^2}\right) XY = \frac{2\pi m\sigma}{\beta h^2}, \quad \sigma = XY$$

$$U_m - U_m(0) = -\frac{N_A}{q}\left(\frac{\partial q}{\partial \beta}\right) = RT \text{ [or by equipartition]}$$

$$S_{\mathrm{m}} = \frac{U_{\mathrm{m}} - U_{\mathrm{m}}(0)}{T} + R(\ln q_{\mathrm{m}} - \ln N_{\mathrm{A}} + 1) \quad \left[\text{Table 20.1}, q_{\mathrm{m}} = \frac{q}{n}\right]$$

$$= R + R\ln\left(\frac{\mathrm{e}q_{\mathrm{m}}}{N_{\mathrm{A}}}\right) = R\ln\left(\frac{\mathrm{e}^2 q_{\mathrm{m}}}{N_{\mathrm{A}}}\right)$$

$$= \boxed{R\ln\left(\frac{2\pi \mathrm{e}^2 m\sigma_{\mathrm{m}}}{h^2 N_{\mathrm{A}}\beta}\right)} \quad \left[\sigma_{\mathrm{m}} = \frac{\sigma}{n}\right]$$

Since in three dimensions

$$S_{\mathrm{m}} = R\ln\left\{\mathrm{e}^{5/2}\left(\frac{2\pi m}{h^2\beta}\right)^{3/2}\frac{V_{\mathrm{m}}}{N_{\mathrm{A}}}\right\} \quad \text{[Sackur–Tetrode equation]}$$

The entropy of condensation is the difference

$$\Delta S_{\mathrm{m}} = R\ln\frac{\mathrm{e}^2(2\pi m/h^2\beta)\times(\sigma_{\mathrm{m}}/N_{\mathrm{A}})}{\mathrm{e}^{5/2}(2\pi m/h^2\beta)^{3/2}\times(V_{\mathrm{m}}/N_{\mathrm{A}})}$$

$$= \boxed{R\ln\left\{\left(\frac{\sigma_{\mathrm{m}}}{V_{\mathrm{m}}}\right)\times\left(\frac{h^2\beta}{2\pi m\mathrm{e}}\right)^{1/2}\right\}}$$

P20.10 **(a)**

$$U - U(0) = -\frac{N}{q}\frac{\partial q}{\partial\beta} = -\frac{N}{q}\sum_j \varepsilon_j \mathrm{e}^{-\beta\varepsilon_j} = \frac{NkT}{q}\dot{q}$$

$$= \boxed{nRT\left(\frac{\dot{q}}{q}\right)}$$

$$C_V = \left(\frac{\partial U}{\partial T}\right)_V = \frac{\mathrm{d}\beta}{\mathrm{d}T}\left(\frac{\partial U}{\partial\beta}\right)_V = \frac{1}{kT^2}\frac{\partial}{\partial\beta}\left(\frac{N}{q}\sum_j \varepsilon_j \mathrm{e}^{-\beta\varepsilon_j}\right)$$

$$= \left(\frac{N}{kT^2}\right)\times\left[\frac{1}{q}\sum_j \varepsilon_j^2 \mathrm{e}^{-\beta\varepsilon_j} + \frac{1}{q^2}\left(\frac{\partial q}{\partial\beta}\right)\sum_j \varepsilon_j \mathrm{e}^{-\beta\varepsilon_j}\right]$$

$$= \left(\frac{N}{kT^2}\right)\times\left[\frac{1}{q}\sum_j \varepsilon_j^2 \mathrm{e}^{-\beta\varepsilon_j} - \frac{1}{q^2}\left(\sum_j \varepsilon_j \mathrm{e}^{-\beta\varepsilon_j}\right)^2\right]$$

$$= \left(\frac{N}{kT^2}\right)\times\left[\frac{k^2T^2\ddot{q}}{q} - \frac{k^2T^2}{q^2}\dot{q}^2\right]$$

$$= \boxed{nR\left\{\frac{\ddot{q}}{q} - \left(\frac{\dot{q}}{q}\right)^2\right\}}$$

$$S = \frac{U - U(0)}{T} + nR\ln\left(\frac{q}{N} + 1\right) = \boxed{nR\left(\frac{\dot{q}}{q} + \ln\frac{\mathrm{e}q}{N}\right)}$$

(b) At 5000 K, $\frac{kT}{hc} = 3475\ \mathrm{cm}^{-1}$. We form the sums

$$q = \sum_j \mathrm{e}^{-\beta\varepsilon_j} = 1 + \mathrm{e}^{-21870/3475} + 3\mathrm{e}^{-21870/3475} + \cdots = 1.0167$$

$$\dot{q} = \sum_j \frac{\varepsilon_j}{kT}\mathrm{e}^{-\beta\varepsilon_j} = \frac{hc}{kT}\sum_j \tilde{\nu}_j \mathrm{e}^{-\beta\varepsilon_j}$$

$$= \left(\frac{1}{3475}\right) \times \{0 + 21850\,\mathrm{e}^{-21850/3475} + 3 \times 21870\,\mathrm{e}^{-21870/3475} + \cdots\} = 0.1057$$

$$\ddot{q} = \sum_j \left(\frac{\varepsilon_j}{kT}\right)^2 \mathrm{e}^{-\beta\varepsilon_j} = \left(\frac{hc}{kT}\right)^2 \sum_j \tilde{\nu}_j^2 \mathrm{e}^{-\beta\varepsilon_j}$$

$$= \left(\frac{1}{3475}\right)^2 \times \{0 + 21850^2\,\mathrm{e}^{-21850/3475} + 3 \times 21870^2\,\mathrm{e}^{-21870/3475} + \cdots\} = 0.6719$$

The electronic contribution to the molar constant-volume heat capacity is

$$C_{V,\mathrm{m}} = R\left\{\frac{\ddot{q}}{q} - \left(\frac{\dot{q}}{q}\right)^2\right\}$$

$$= 8.314\,\mathrm{J\,K^{-1}\,mol^{-1}} \times \left\{\frac{0.6719}{1.0167} - \left(\frac{0.1057}{1.0167}\right)^2\right\} = \boxed{5.41\,\mathrm{J\,K^{-1}\,mol^{-1}}}$$

P20.12 $c_\mathrm{s} = \left(\dfrac{\gamma RT}{M}\right)^{1/2}, \quad \gamma = \dfrac{C_{p,\mathrm{m}}}{C_{V,\mathrm{m}}}, \qquad C_{p,\mathrm{m}} = C_{V,\mathrm{m}} + R$

(a) $C_{V,\mathrm{m}} = \frac{1}{2}R(3 + \nu_\mathrm{R}^* + 2\nu_\mathrm{V}^*) = \frac{1}{2}R(3+2) = \frac{5}{2}R$

$C_{p,\mathrm{m}} = \frac{5}{2}R + R = \frac{7}{2}R$

$\gamma = \dfrac{7}{5} = 1.40;$ hence $\boxed{c_\mathrm{s} = \left(\dfrac{1.40RT}{M}\right)^{1/2}}$

(b) $C_{V,\mathrm{m}} = \dfrac{1}{2}R(3+2) = \dfrac{5}{2}R, \qquad \gamma = 1.40, \qquad \boxed{c_\mathrm{s} = \left(\dfrac{1.40RT}{M}\right)^{1/2}}$

(c) $C_{V,\mathrm{m}} = \frac{1}{2}R(3+3) = 3R$

$C_{p,\mathrm{m}} = 3R + R = 4R, \qquad \gamma = \dfrac{4}{3}, \qquad \boxed{c_\mathrm{s} = \left(\dfrac{4RT}{3M}\right)^{1/2}}$

For air, $M \approx 29\,\mathrm{g\,mol^{-1}}$, $T \approx 298\,\mathrm{K}$, $\gamma = 1.40$

$$c_\mathrm{s} = \left(\frac{(1.40) \times (2.48\,\mathrm{kJ\,mol^{-1}})}{29 \times 10^{-3}\,\mathrm{mol^{-1}}}\right)^{1/2} = \boxed{350\,\mathrm{m\,s^{-1}}}$$

Solutions to additional problems

P20.14 The molar entropy is given by

$$S_\mathrm{m} = \frac{U_\mathrm{m} - U_\mathrm{m}(0)}{T} + R\left(\ln\frac{q_\mathrm{m}}{N_\mathrm{A}} - 1\right) \quad \text{where} \quad \frac{U_\mathrm{m} - U_\mathrm{m}(0)}{T} = -N_\mathrm{A}\left(\frac{\partial \ln q}{\partial \beta}\right)_V \quad \text{[Table 20.1]}$$

and $\dfrac{q_\mathrm{m}}{N_\mathrm{A}} = \dfrac{q_\mathrm{m,tr}}{N_\mathrm{A}} q_\mathrm{rot}\, q_\mathrm{vib}\, q_\mathrm{elec}$

The energy term $U_m - U_m(0)$ works out to be

$$U_m - U_m(0) = N_A[\langle\varepsilon_{tr}\rangle + \langle\varepsilon_{rot}\rangle + \langle\varepsilon_{vib}\rangle + \langle\varepsilon_{elec}\rangle] \text{ [Table 20.4]}$$

Translation:

$$\frac{q^\circ_{m,tr}}{N_A} = \frac{kT}{p^\circ\Lambda^3} = 2.561\times10^{-2}(T/\text{K})^{5/2}\times(M/\text{g mol}^{-1})^{3/2} \text{ [Table 20.4]}$$

$$= 2.561\times10^{-2}\times(298)^{5/2}\times(38.00)^{3/2}$$

$$= 9.20\times10^6 \quad\text{and}\quad \langle\varepsilon_{tr}\rangle = 3/2kT$$

Rotation of a linear molecule

$$q_{rot} = \frac{kT}{\sigma hcB} = \frac{0.6950}{\sigma}\times\frac{T/\text{K}}{B/\text{cm}^{-1}} \text{ [Table 20.4]}$$

The rotational constant is

$$B = \frac{\hbar}{4\pi cI} = \frac{\hbar}{4\pi c\mu R^2}$$

$$= \frac{(1.0546\times10^{-34}\,\text{J s})\times(6.022\times10^{23}\,\text{mol}^{-1})}{4\pi(2.998\times10^{10}\,\text{cm s}^{-1})\times(1/2\times19.00\times10^{-3}\,\text{kg mol}^{-1})\times(190.0\times10^{-12}\,\text{m})^2}$$

$$= 0.4915\,\text{cm}^{-1} \quad\text{so}\quad q_{rot} = \frac{0.6950}{2}\times\frac{298}{0.4915} = 210.7$$

Also $\langle\varepsilon_{rot}\rangle = kT$

Vibration

$$q_{vib} = \frac{1}{1-e^{-hc\tilde{\nu}/kT}} = \frac{1}{1-\exp\left(\frac{-1.4388(\tilde{\nu}/\text{cm}^{-1})}{T/\text{K}}\right)} = \frac{1}{1-\exp\left(\frac{-1.4388(450.0)}{298}\right)}$$

$$= 1.129$$

$$\langle\varepsilon_{vib}\rangle = \frac{hc\tilde{\nu}}{e^{hc\tilde{\nu}/kT}-1} = \frac{(6.626\times10^{-34}\,\text{J s})\times(2.998\times10^{10}\,\text{cm s}^{-1})\times(450.0\,\text{cm}^{-1})}{\exp\left(\frac{1.4388(450.0)}{298}\right)-1}$$

$$= 1.149\times10^{-21}\,\text{J}$$

The Boltzmann factor for the lowest-lying electronic excited state is

$$\exp\left(\frac{-(1.609\,\text{eV})\times(1.602\times10^{-19}\,\text{J eV}^{-1})}{(1.381\times10^{-23}\,\text{J K}^{-1})\times(298\,\text{K})}\right) = 6\times10^{-28}$$

so we may take q_{elec} to equal the degeneracy of the ground state, namely 2 and $\langle\varepsilon_{elec}\rangle$ to be zero. Putting it all together yields

$$\frac{U_m - U_m(0)}{T} = \frac{N_A}{T}\left(\tfrac{3}{2}kT + kT + 1.149\times10^{-21}\,\text{J}\right) = \tfrac{5}{2}R + \tfrac{N_A(1.149\times10^{-21}\,\text{J})}{T}$$

$$= (2.5)\times(8.3145\,\text{J mol}^{-1}\,\text{K}^{-1}) + \frac{(6.022\times10^{23}\,\text{mol}^{-1})\times(1.149\times10^{-21}\,\text{J})}{298\,\text{K}}$$

$$= 23.11\,\text{J mol}^{-1}\,\text{K}^{-1}$$

$$R\left(\ln\frac{q_m}{N_A} - 1\right) = (8.3145\,\text{J mol}^{-1}\,\text{K}^{-1})\times\{\ln[(9.20\times10^6)\times(210.7)\times(1.129)\times(2)] - 1\}$$

$$= 176.3\,\text{J mol}^{-1}\,\text{K}^{-1} \quad\text{and}\quad S^\circ_m = \boxed{199.4\,\text{J mol}^{-1}\,\text{K}^{-1}}$$

P20.16 The vibrational temperature is defined by

$$k\theta_V = hc\tilde{\nu},$$

so a vibration with θ_V less than 1000 K has a wavenumber less than

$$\tilde{\nu} = \frac{k\theta_V}{hc} = \frac{(1.381\times10^{-23}\,\mathrm{J\,K^{-1}})\times(1000\,\mathrm{K})}{(6.626\times10^{-34}\,\mathrm{J\,s})\times(2.998\times10^{10}\,\mathrm{cm\,s^{-1}})} = 695.2\,\mathrm{cm^{-1}}$$

There are seven such wavenumbers listed among those for C_{60} : two T_{1u}, a T_{2u}, a G_u, and three H_u. The number of modes involved, ν_V^*, must take into account the degeneracy of these vibrational energies

$$\nu_V^* = 2(3) + 1(3) + 1(4) + 3(5) = \boxed{28}$$

The molar heat capacity of a molecule is roughly

$$C_{V,\mathrm{m}} = \tfrac{1}{2}(3 + \nu_R^* + 2\nu_V^*)R\ [42] = \tfrac{1}{2}(3+3+2\times28)R = 31R = 31(8.3145\,\mathrm{J\,mol^{-1}\,K^{-1}})$$
$$= \boxed{258\,\mathrm{J\,mol^{-1}\,K^{-1}}}$$

P20.18 The standard molar Gibbs energy is given by

$$G_\mathrm{m}^\circ - G_\mathrm{m}^\circ(0) = RT\ln\frac{q_\mathrm{m}^\circ}{N_A}\ [10]\quad \text{where } \frac{q_\mathrm{m}^\circ}{N_A} = \frac{q_\mathrm{m,tr}^\circ}{N_A}q_\mathrm{rot}q_\mathrm{vib}q_\mathrm{elec}$$

Translation

$$\frac{q_\mathrm{m,tr}^\circ}{N_A} = 2.561\times10^{-2}\times(T/\mathrm{K})^{5/2}\times(M/\mathrm{g\,mol^{-1}})^{3/2}$$
$$= 2.561\times10^{-2}\times(200.0)^{5/2}\times(102.9)^{3/2} = 1.512\times10^{7}$$

Rotation of a nonlinear molecule

$$q_\mathrm{rot} = \frac{1}{\sigma}\left(\frac{kT}{hc}\right)^{3/2}\left(\frac{\pi}{ABC}\right)^{1/2} = \frac{1.0270}{\sigma}\times\frac{(T/\mathrm{K})^{3/2}}{(ABC/\mathrm{cm^{-3}})^{1/2}}$$
$$= \frac{1.0270}{2}\times\frac{[(200.0)\times(2.998\times10^{10}\,\mathrm{cm\,s^{-1}})]^{3/2}}{[(13109.4)\times(2409.8)\times(2139.7)\times(10^6\,\mathrm{s^{-1}})^3/\mathrm{cm^{-3}}]^{1/2}} = 2.900\times10^{4}$$

Vibration

$$q_\mathrm{vib}^{(1)} = \frac{1}{1-\exp\left(\frac{-1.4388(\tilde{\nu}/\mathrm{cm^{-1}})}{T/\mathrm{K}}\right)} = \frac{1}{1-\exp\left(\frac{-1.4388(753)}{200.0}\right)} = 1.004$$

$$q_\mathrm{vib}^{(2)} = \frac{1}{1-\exp\left(\frac{-1.4388(542)}{200.0}\right)} = 1.021$$

$$q_\mathrm{vib}^{(3)} = \frac{1}{1-\exp\left(\frac{-1.4388(310)}{200.0}\right)} = 1.120$$

$$q_\mathrm{vib}^{(4)} = \frac{1}{1-\exp\left(\frac{-1.4388(127)}{200.0}\right)} = 1.670$$

$$q_{\text{vib}}^{(5)} = \frac{1}{1 - \exp\left(\frac{-1.4388(646)}{200.0}\right)} = 1.010$$

$$q_{\text{vib}}^{(6)} = \frac{1}{1 - \exp\left(\frac{-1.4388(419)}{200.0}\right)} = 1.052$$

$$q_{\text{vib}} = \prod_{i=1}^{6} q_{\text{vib}}^{(i)} = 2.037$$

Putting it all together yields

$$\begin{aligned} G_{\text{m}}^{\circ} - G_{\text{m}}^{\circ}(0) &= (8.3145\,\text{J}\,\text{mol}^{-1}\,\text{K}^{-1}) \times (200.0\,\text{K}) \times \ln(1.512 \times 10^{7}) \\ &\quad \times (2.900 \times 10^{4}) \times (2.037) \times (1) \\ G_{\text{m}}^{\circ} - G_{\text{m}}^{\circ}(0) &= 4.576 \times 10^{4}\,\text{J}\,\text{mol}^{-1} = \boxed{45.76\,\text{kJ}\,\text{mol}^{-1}} \end{aligned}$$

21 Diffraction techniques

Solutions to exercises

E21.1 There are four equivalent lattice points in the fcc unit cell. One way of choosing them is shown by the positions of the Cl^- ions in Fig. 21.1 (which is similar to Fig. 21.28 in the text). The three lattice points equivalent to $(\frac{1}{2}, 0, 0)$ are $\boxed{(1, \frac{1}{2}, 0)}$, $\boxed{(1, 0, \frac{1}{2})}$, and $\boxed{(\frac{1}{2}, \frac{1}{2}, \frac{1}{2})}$. Figure 21.1 shows location of the atoms in the fcc unit cell of NaCl. The tinted circles are Na^+; the open circles are Cl^-.

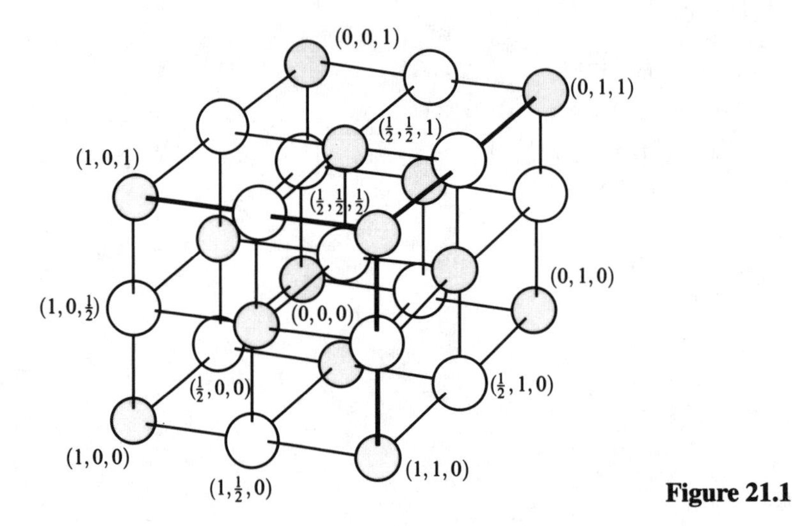

Figure 21.1

Comment. The positions of the other Cl^- ions in Fig. 21.1 do not correspond to lattice points of the unit cell shown, as they are generated by full unit cell translations, and hence belong to neighbouring unit cells.

Question. What Na^+ positions define the unit cell of NaCl in Fig. 21.1? What lattice points are equivalent to (0, 0, 0)?

E21.2 The planes are sketched in Fig. 21.2. Expressed in multiples of the unit cell distances the planes are labelled (2, 3, 2) and (2, 2, ∞). Their Miller indices are the reciprocals of these multiples with all fractions cleared, thus

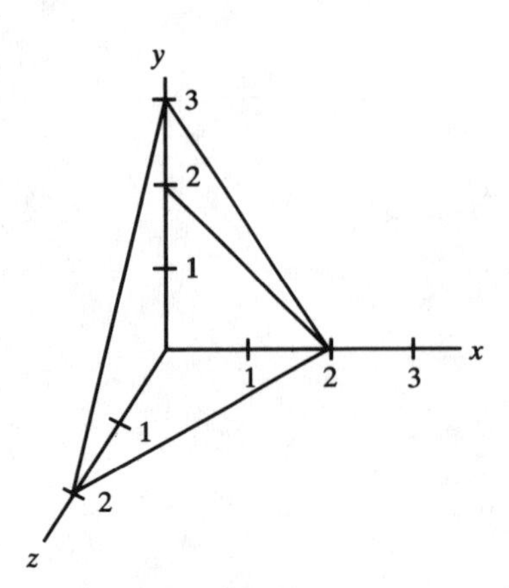

Figure 21.2

$$(2,3,2) \rightarrow \left(\tfrac{1}{2},\tfrac{1}{3},\tfrac{1}{2}\right) \rightarrow (3,2,3) \quad \text{[multiply by 6]}$$
$$(2,2,\infty) \rightarrow \left(\tfrac{1}{2},\tfrac{1}{2},0\right) \rightarrow (1,1,0) \quad \text{[multiply by 2]}$$

Dropping the commas, the planes are written $\boxed{(3\,2\,3)}$ and $\boxed{(1\,1\,0)}$

E21.3 $d_{khl} = \dfrac{a}{(h^2+k^2+l^2)^{1/2}}$ [2]

Therefore, $d_{111} = \dfrac{a}{3^{1/2}} = \dfrac{432\,\text{pm}}{3^{1/2}} = \boxed{249\,\text{pm}}$ $\quad d_{211} = \dfrac{a}{6^{1/2}} = \dfrac{432\,\text{pm}}{6^{1/2}} = \boxed{176\,\text{pm}}$

$d_{100} = a = \boxed{432\,\text{pm}}$

E21.4 $\lambda = 2d\sin\theta[5] = (2)\times(99.3\,\text{pm})\times(\sin 20.85°) = \boxed{70.7\,\text{pm}}$

Comment. Knowledge of the type of crystal is not needed to complete this exercise.

E21.5 Refer to Fig. 21.26 of the text. Systematic absences correspond to $h+k+l=$ odd. Hence the first three lines are from planes (1 1 0), (2 0 0), and (2 1 1).

$$\sin\theta_{hkl} = (h^2+k^2+l^2)^{1/2} \times \left(\frac{\lambda}{2a}\right) \text{ [6]}$$

In a bcc unit cell, the body diagonal of the cube is $4R$ where R is the atomic radius. The relationship of the side of the unit cell to R is therefore (using the Pythagorean theorem twice)

$$(4R)^2 = a^2 + 2a^2 = 3a^2 \quad \text{or} \quad a = \frac{4R}{3^{1/2}}$$

This can be seen from Fig. 21.3.

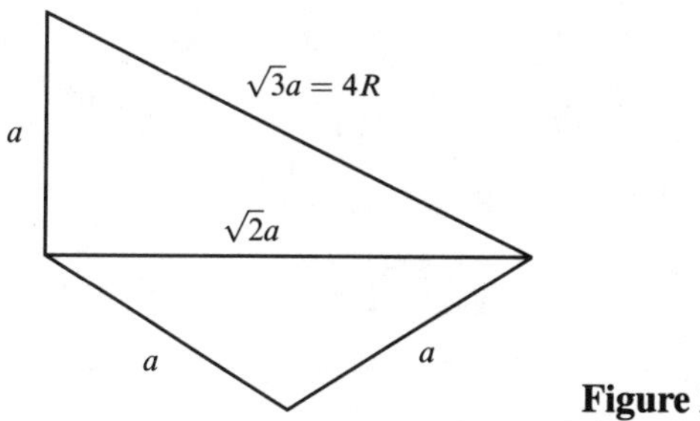

Figure 21.3

$$a = \frac{4\times 126\,\text{pm}}{3^{1/2}} = 291\,\text{pm}$$

$$\frac{\lambda}{2a} = \frac{58\,\text{pm}}{(2)\times(291\,\text{pm})} = 0.099\bar{7}$$

$\sin\theta_{110} = \sqrt{2}\times(0.099\bar{7}) = 0.14\bar{1}$ $\quad 2\theta_{110} = \boxed{16°}$

$\sin\theta_{200} = (2)\times(0.099\bar{7}) = 0.19\bar{9}$ $\quad 2\theta_{200} = \boxed{23°}$

$\sin\theta_{211} = \sqrt{6}\times(0.099\bar{7}) = 0.24\bar{4}$ $\quad 2\theta_{211} = \boxed{28°}$

E21.6 $\theta = \arcsin \dfrac{\lambda}{2d}$ [5, $\arcsin \equiv \sin^{-1}$]

$$\Delta\theta = \arcsin \frac{\lambda_1}{2d} - \arcsin \frac{\lambda_2}{2d} = \arcsin\left(\frac{154.051\,\text{pm}}{(2)\times(77.8\,\text{pm})}\right) - \arcsin\left(\frac{154.433\,\text{pm}}{(2)\times(77.8\,\text{pm})}\right)$$

$$= -1.07^\circ = -0.0187\,\text{rad}$$

The angle θ in radians is related to the distances D of the reflection line from the centre of the pattern by $\theta = \dfrac{D}{2R}$; hence

$$D = 2R\theta = (2)\times(5.74\,\text{cm})\times(0.0187) = \boxed{0.215\,\text{cm}}$$

E21.7 A tetragonal unit cell, as shown in Fig. 21.9 of the text, has $a = b \neq c$. Therefore

$$V = (651\,\text{pm})\times(651\,\text{pm})\times(934\,\text{pm}) = \boxed{3.96\times10^{-28}\,\text{m}^3}$$

E21.8 $\rho = \dfrac{\text{mass of unit cell}}{\text{volume of unit cell}} = \dfrac{m}{V}$

$m = nM = \dfrac{N}{N_A}M$ [N is the number of formula units per unit cell]

Then, $\rho = \dfrac{NM}{VN_A}$

and $N = \dfrac{\rho V N_A}{M}$

$$= \frac{(3.9\times10^6\,\text{g}\,\text{m}^{-3})\times(634)\times(784)\times(516\times10^{-36}\,\text{m}^3)\times(6.022\times10^{23}\,\text{mol}^{-1})}{154.77\,\text{g}\,\text{mol}^{-1}} = 3.9$$

Therefore, $\boxed{N=4}$ and the true calculated density (in the absence of defects) is

$$\rho = \frac{(4)\times(154.77\,\text{g}\,\text{mol}^{-1})}{(634)\times(784)\times(516\times10^{-30}\,\text{cm}^3)\times(6.022\times10^{23})\,\text{mol}^{-1}} = \boxed{4.01\,\text{g}\,\text{cm}^{-3}}$$

E21.9 $d_{hkl} = \left[\left(\dfrac{h}{a}\right)^2 + \left(\dfrac{k}{b}\right)^2 + \left(\dfrac{l}{c}\right)^2\right]^{-1/2}$ [3]

$$d_{411} = \left[\left(\frac{4}{812}\right)^2 + \left(\frac{1}{947}\right)^2 + \left(\frac{1}{637}\right)^2\right]^{-1/2}\,\text{pm} = \boxed{190\,\text{pm}}$$

E21.10 Since the reflection at 32.6° is (220), we know that

$$d_{220} = \frac{\lambda}{2\sin\theta}[5] = \frac{154\,\text{pm}}{2\sin 32.6} = 143\,\text{pm}$$

and hence, since $d_{220} = \dfrac{a}{(2^2+2^2)^{1/2}}[1] = \dfrac{a}{8^{1/2}}$

it follows that $a = (8^{1/2})\times(143\,\text{pm}) = 404\,\text{pm}$

The indices of the other reflections are obtained from

$$(h^2+k^2+l^2) = \left(\frac{a}{d_{hkl}}\right)^2[1] = \left(\frac{(a)\times 2\sin\theta}{\lambda}\right)^2 \quad \text{[using eqn 5]}$$

We draw up the following table

θ	$a^2\left(\frac{2\sin\theta}{\lambda}\right)^2$	$h^2+k^2+l^2$	(hkl)	a/pm
19.4	3.04	3	$\boxed{(111)}$	402
22.5	4.03	4	$\boxed{(200)}$	402
32.6	7.99	8	(220)	404
39.4	11.09	11	$\boxed{(311)}$	402

The values of a in the final column are obtained from

$$a = \left(\frac{\lambda}{2\sin\theta}\right) \times (h^2+k^2+l^2)^{1/2}$$

and average to 402 pm.

E21.11 $$\theta_{hkl} = \arcsin\frac{\lambda}{2d_{hkl}} \text{(from eqn 5)} = \arcsin\left\{\frac{\lambda}{2}\left[\left(\frac{h}{2}\right)^2+\left(\frac{k}{b}\right)^2+\left(\frac{l}{c}\right)^2\right]^{1/2}\right\} \text{[from eqn 3]}$$

$$= \arcsin\left\{77\left[\left(\frac{h}{542}\right)^2+\left(\frac{k}{917}\right)^2+\left(\frac{l}{645}\right)^2\right]^{1/2}\right\}$$

Therefore,

$$\theta_{100} = \arcsin\left(\frac{77}{542}\right) = \boxed{8.17^\circ} \qquad \theta_{010} = \arcsin\left(\frac{77}{917}\right) = \boxed{4.82^\circ}$$

$$\theta_{111} = \arcsin\left\{77\times\left[\left(\frac{1}{542}\right)^2+\left(\frac{1}{917}\right)^2+\left(\frac{1}{645}\right)^2\right]^{1/2}\right\} = \arcsin\frac{77}{378} = \boxed{11.75^\circ}$$

E21.12 From the discussion of systematic absences (Section 21.4 and Fig. 21.26 of the text) we can conclude that the unit cell is $\boxed{\text{face-centred cubic}}$

E21.13 $$F_{hkl} = \sum_j f_j e^{2\pi i(hx_j+ky_j+lz_j)} \text{ [12]}$$

with $f_j = \frac{1}{8}f$ (each atom is shared by eight cells). Therefore,

$$F_{hkl} = \frac{1}{8}f\{1+e^{2\pi ih}+e^{2\pi ik}+e^{2\pi il}+e^{2\pi i(h+k)}+e^{2\pi i(h+l)}+e^{2\pi i(k+l)}+e^{2\pi i(h+k+l)}\}$$

However all the exponential terms are unity since, h, k, and l are all integers and

$$e^{i\theta} = \cos\theta + i\sin\theta[\theta = 2\pi h, 2\pi k, \ldots] = \cos\theta = 1$$

Therefore, $F_{hkl} = \boxed{f}$

E21.14 The hatched area is $h \times 2R = 3^{1/2}R \times 2R = 2\sqrt{3}R^2$ where $h = 2R\cos 30^\circ$. The net number of cylinders in a hatched area is 1, and the area of the cylinder's base is πR^2. The volume of the prism (of which the hatched area is the base) is $2\sqrt{3}R^2L$, and the volume occupied by the cylinders is $\pi R^2 L$. Hence, the packing fraction is

$$f = \frac{\pi R^2 L}{2\sqrt{3}R^2 L} = \frac{\pi}{2\sqrt{3}} = \boxed{0.9069}$$

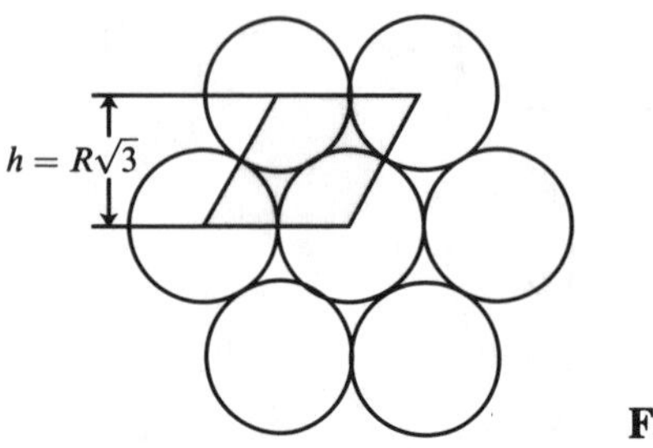

Figure 21.4

E21.15 For sixfold coordination see Fig. 21.5.

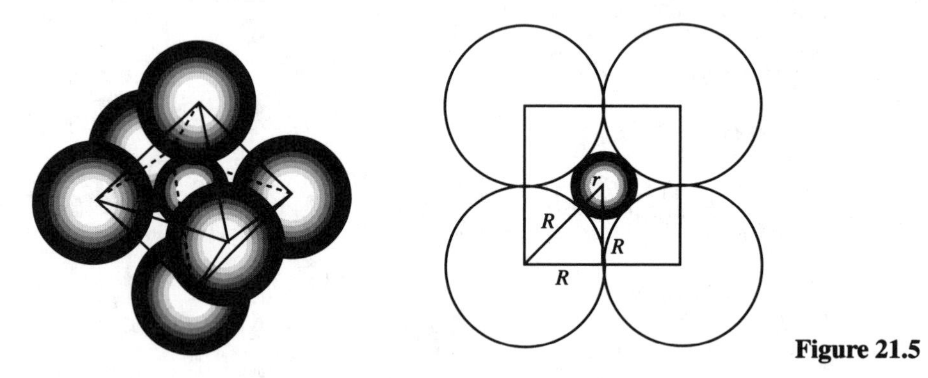

Figure 21.5

We assume that the larger spheres of radius R touch each other and that they also touch the smaller interior sphere. Hence, by the Pythagorean theorem

$$(R+r)^2 = 2(R)^2 \quad \text{or} \quad \left(1+\frac{r}{R}\right)^2 = 2$$

Thus, $\dfrac{r}{R} = \boxed{0.414}$

E21.16 The radius ratios determined in Exercises 21.15(a) and 21.15(b) correspond to the smallest value of the radius of the interior cation, since any smaller value would tend to bring the anions closer and increase their interionic repulsion and at the same time decrease the attractions of cation and anion.

(a) $\dfrac{r_+}{r_-} = 0.414$ [result of Exercise 21.15(a)]

$r_+(\text{smallest}) = (0.414) \times (140\,\text{pm})[\text{Table 21.3}] = \boxed{58.0\,\text{pm}}$

(b) $\dfrac{r_+}{r_-} = 0.732$ [result of Exercise 21.15(b)]

$r_+(\text{smallest}) = (0.732) \times (140\,\text{pm}) = \boxed{102\,\text{pm}}$

Comment. As is evident from the data in Table 21.3 larger values than these do not preclude the occurrence of coordination number 6.

E21.17 Figure 21.32 in the text shows the diamond structure. Figure 21.6(a) below is an easier to visualize form of the structure which shows the unit cell of diamond.

The number of carbon atom in the unit cell is $\left(8 \times \frac{1}{8}\right) + \left(6 \times \frac{1}{2}\right) + (4 \times 1) = 8$ ($\frac{1}{8}$ for a corner atom, $\frac{1}{2}$ for a face-centred atom, and 1 for an atom entirely in the cell). The positions of the atoms are $(0, 0, 0)$, $\left(\frac{1}{2}, \frac{1}{2}, 0\right)$, $\left(\frac{1}{2}, 0, \frac{1}{2},\right)$, $\left(0, \frac{1}{2}, \frac{1}{2}\right)$, $\left(\frac{1}{4}, \frac{1}{4}, \frac{1}{4}\right)$, $\left(\frac{1}{4}, \frac{3}{4}, \frac{3}{4}\right)$, $\left(\frac{3}{4}, \frac{1}{4}, \frac{3}{4}\right)$, and $\left(\frac{3}{4}, \frac{3}{4}, \frac{1}{4}\right)$ as indicated in Fig. 21.6(b).

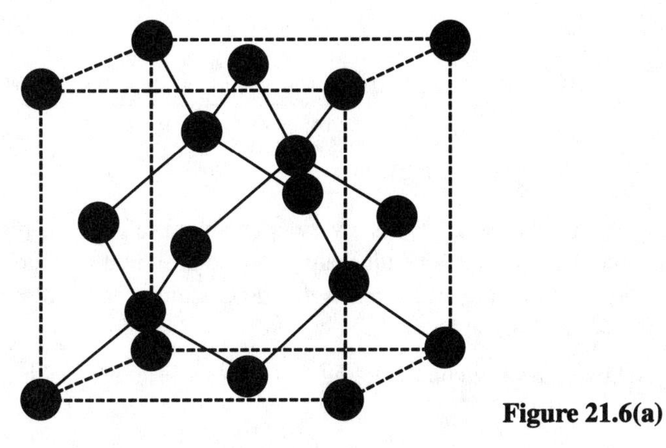

Figure 21.6(a)

The fractions in Fig. 21.6(b) denote height above the base in units of the cube edge, a. Two atoms that touch lie along the body diagonal at (0, 0, 0) and $\left(\frac{1}{4}, \frac{1}{4}, \frac{1}{4}\right)$. Hence the distance $2r$ is one-fourth of the body diagonal which is $\sqrt{3}a$ in a cube. That is

$$2r = \frac{\sqrt{3}a}{4}$$

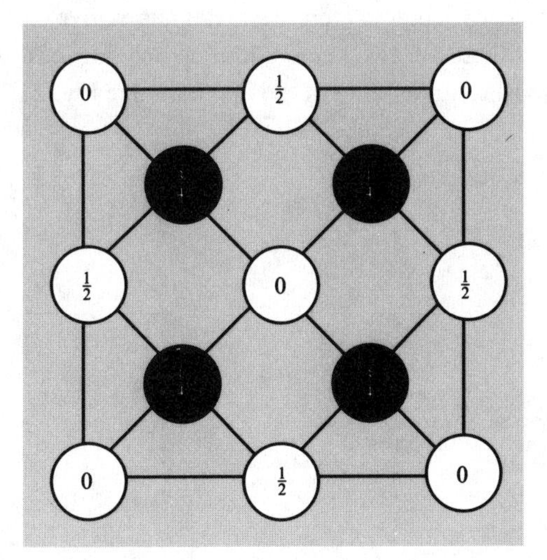

Figure 21.6(b)

The packing fraction is $\dfrac{\text{volume of atoms}}{\text{volume of unit cell}} = \dfrac{8V_a}{a^3} = \dfrac{(8) \times \frac{4}{3}\pi r^3}{\left(\frac{8r}{\sqrt{3}}\right)^3} = \boxed{0.340}$

E21.18 As demonstrated in *Justification* 21.3 of the text, close-packed spheres fill 0.7404 of the total volume of the crystal. Therefore 1 cm^3 of close-packed carbon atoms would contain

$$\frac{0.74040\,\text{cm}^3}{\left(\frac{4}{3}\pi r^3\right)} = 3.838 \times 10^{23}\ \text{atoms}$$

$$\left(r = \left(\frac{154.45}{2}\right)\text{pm} = 77.225\,\text{pm} = 77.225 \times 10^{-10}\,\text{cm}\right)$$

Hence the close-packed density would be

$$\rho = \frac{\text{mass in } 1\,\text{cm}^3}{1\,\text{cm}^3} = \frac{(3.838 \times 10^{23}\,\text{atom}) \times (12.01\,\text{u/atom}) \times (1.6605 \times 10^{-24}\,\text{g}\,\text{u}^{-1})}{1\,\text{cm}^3}$$
$$= \boxed{7.654\,\text{g}\,\text{cm}^{-3}}$$

The diamond structure (solution to Exercise 21.17) is a very open structure which is dictated by the tetrahedral bonding of the carbon atoms. As a result many atoms that would be touching each other in a normal fcc structure do not in diamond; for example, the C atom in the centre of a face does not touch the C atoms at the corners of the face.

E21.19 The volume change is a result of two partially counteracting factors: (1) different packing fraction (f), and (2) different radii.

$$\frac{V(\text{bcc})}{V(\text{hcp})} = \frac{f(\text{hcp})}{f(\text{bcc})} \times \frac{v(\text{bcc})}{v(\text{hcp})}$$
$f(\text{hcp}) = 0.7404,\quad f(\text{bcc}) = 0.6802$ [*Justification* 21.3 and Problem 21.13]
$$\frac{V(\text{bcc})}{V(\text{hcp})} = \frac{0.7405}{0.6802} \times \frac{(142.5)^3}{(145.8)^3} = 1.016$$

Hence there is an $\boxed{\text{expansion}}$ of 1.6 per cent.

E21.20 Draw points corresponding to the vectors joining each pair of atoms. Heavier atoms give more intense contribution than light atoms. Remember that there are two vectors joining any pair of atoms ($\overrightarrow{AB}$ and $\overleftarrow{AB}$); don't forget the AA zero vectors for the centre point of the diagram. See Fig. 21.7.

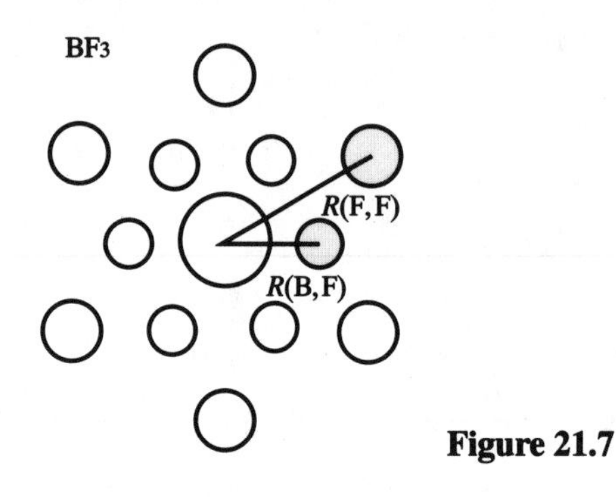

Figure 21.7

E21.21 $$\lambda = \frac{h}{p} = \frac{h}{mv}$$

Hence, $$v = \frac{h}{m\lambda} = \frac{6.626 \times 10^{-34}\,\text{J}\,\text{s}}{(1.675 \times 10^{-27}\,\text{kg}) \times (50 \times 10^{-12}\,\text{m})} = \boxed{7.9\,\text{km}\,\text{s}^{-1}}$$

E21.22 Combine $E = \frac{1}{2}kT$ and $E = \frac{1}{2}mv^2$ with $m^2v^2 = \frac{h^2}{\lambda^2}$ [de Broglie relation]; then

$$E = \frac{h^2}{2m\lambda^2} \quad \text{and} \quad E = e\Delta\phi$$

Therefore, $$\Delta\phi = \frac{h^2}{2me\lambda^2} = \frac{(6.626 \times 10^{-34}\,\text{J}\,\text{s})^2}{(2) \times (9.109 \times 10^{-31}\,\text{kg}) \times (1.602 \times 10^{-19}\,\text{C}) \times (18 \times 10^{-12}\,\text{m})^2}$$
$$= \boxed{4.6\,\text{kV}} \quad (1\,\text{J} = 1\,\text{C}\,\text{V})$$

E21.23 The maxima and minima are determined by $\sin sR$ [19]. For the maxima, $\sin sR = 1$, and sR satisfies

$$sR = (4n+1)\frac{\pi}{2} \quad n = 0, 1, 2, 3, \ldots$$

Combining this relation with

$$s = \frac{4\pi}{\lambda}\sin\frac{1}{2}\theta$$

yields $\sin\frac{1}{2}\theta = \dfrac{(4n+1)\lambda}{8R}$

For $n = 0$ (the first maximum), for neutrons

$$\sin\frac{1}{2}\theta = \frac{\lambda}{8R} = \frac{80\,\text{pm}}{(8)\times(198.75\,\text{pm})} = 0.050\bar{3} \quad \theta = \boxed{5.8°}$$

For electrons

$$\sin\frac{1}{2}\theta = \frac{\lambda}{8R} = \frac{4\,\text{pm}}{(8)\times(198.75\,\text{pm})} = 2.\overline{52}\times 10^{-3} \quad \theta = \boxed{0.3°}$$

For the minima, $\sin sR = -1$, and sR satisfies

$$sR = (4n+3)\frac{\pi}{2} \quad n = 0, 1, 2, 3, \ldots$$

This yields

$$\sin\frac{1}{2}\theta = \frac{(4n+3)\lambda}{8R}$$

For $n = 0$ (the first minimum), for neutrons

$$\sin\frac{1}{2}\theta = \frac{3\lambda}{8R} = \frac{(3)\times(80\,\text{pm})}{(8)\times(198.75\,\text{pm})} = 0.15\bar{1} \quad \boxed{\theta = 17°}$$

For electrons

$$\sin\frac{1}{2}\theta = \frac{3\lambda}{8R} = \frac{(3)\times(4\,\text{pm})}{(8)\times(198.75\,\text{pm})} = 7.\bar{5}\times 10^{-3} \quad \boxed{\theta = 0.9°}$$

Comment. The maxima and minima are widely separated in neutron diffraction but not in electron diffraction. Camera design is therefore different for neutron and electron diffraction.

Solutions to problems

Solutions to numerical problems

P21.2 A large separation between the sixth and seventh lines relative to the separation between the fifth and sixth lines is characteristic of a $\boxed{\text{simple (primitive) cubic lattice}}$. This is readily seen without indexing the lines. The conclusion that the unit cell is simple cubic is then confirmed by the presence of reflections from (100) planes.

$$d_{100} = a[1] = \frac{\lambda}{2\sin\theta} \text{ [5]}$$

$$a = \frac{154\,\text{pm}}{(2)\times(0.225)} = \boxed{342\,\text{pm}}$$

P21.3 See Fig. 21.28 of the text or Fig. 21.1 of this manual. The length of an edge in the fcc lattice of these compounds is

$$a = 2(r_+ + r_-)$$

Then

(1) $a(\text{NaCl}) = 2(r_{\text{Na}^+} + r_{\text{Cl}^-}) = 562.8\,\text{pm}$ (2) $a(\text{KCl}) = 2(r_{\text{K}^+} + r_{\text{Cl}^-}) = 627.7\,\text{pm}$
(3) $a(\text{NaBr}) = 2(r_{\text{Na}^+} + r_{\text{Br}^-}) = 596.2\,\text{pm}$ (4) $a(\text{KBr}) = 2(r_{\text{K}^+} + r_{\text{Br}^-}) = 658.6\,\text{pm}$

If the ionic radii of all the ions are constant then

$$(1) + (4) = (2) + (3)$$
$$(1) + (4) = (562.8 + 658.6)\,\text{pm} = 1221.4\,\text{pm}$$
$$(2) + (3) = (627.7 + 596.2)\,\text{pm} = 1223.9\,\text{pm}$$

The difference is slight; $\boxed{\text{hence the data support}}$ the constancy of the radii of the ions.

P21.5 For the three given reflections

$$\sin 19.076° = 0.32682 \qquad \sin 22.171° = 0.37737 \qquad \sin 32.256° = 0.53370$$

For cubic lattices $\sin\theta_{hkl} = \dfrac{\lambda(h^2+k^2+l^2)^{1/2}}{2a}$ [6]

First consider the possibility of simple cubic; the first three reflections are (100), (110), and (111). (See Fig. 21.26 of the text.)

$$\frac{\sin\theta(100)}{\sin\theta(110)} = \frac{1}{\sqrt{2}} \neq \frac{0.32682}{0.37737} \quad \text{[not simple cubic]}$$

Consider next the possibility of body-centred cubic; the first three reflections are (110), (200), and (211).

$$\frac{\sin\theta(110)}{\sin\theta(200)} = \frac{\sqrt{2}}{\sqrt{4}} = \frac{1}{\sqrt{2}} \neq \frac{0.32682}{0.37737} \quad \text{[not bcc]}$$

Consider finally face-centred cubic; the first three reflections are (111), (200), and (220)

$$\frac{\sin\theta(111)}{\sin\theta(200)} = \frac{\sqrt{3}}{\sqrt{4}} = 0.86603$$

which compares very favourably to $\dfrac{0.32682}{0.37737} = 0.86605$. Therefore, the lattice is $\boxed{\text{face-centred cubic}}$

This conclusion may easily be confirmed in the same manner using the second and third reflection.

$$a = \frac{\lambda}{2\sin\theta}(h^2+k^2+l^2)^{1/2}[6] = \left(\frac{154.18\,\text{pm}}{(2)\times(0.32682)}\right) \times \sqrt{3} = \boxed{408.55\,\text{pm}}$$

$$\rho = \frac{NM}{N_A V}[\text{Exercise 21.8(a)}] = \frac{(4)\times(107.87\,\text{g mol}^{-1})}{(6.0221\times10^{23}\,\text{mol}^{-1})\times(4.0855\times10^{-8}\,\text{cm})^3}$$

$$= \boxed{10.507\,\text{g cm}^{-3}}$$

This compares favourably to the value listed in the *Data section.*

P21.7 $\lambda = 2a\sin\theta_{100}$ as $d_{100} = a$

Therefore, $a = \dfrac{\lambda}{2\sin\theta_{100}}$ and

$$\frac{a(\text{KCl})}{a(\text{NaCl})} = \frac{\sin\theta_{100}(\text{NaCl})}{\sin\theta_{100}(\text{KCl})} = \frac{\sin 6°0'}{\sin 5°23'} = 1.114$$

Therefore, $a(\text{KCl}) = (1.114)\times(564\,\text{pm}) = \boxed{628\,\text{pm}}$

The relative densities calculated from these unit cell dimensions are

$$\frac{\rho(\text{KCl})}{\rho(\text{NaCl})} = \left(\frac{M(\text{KCl})}{M(\text{NaCl})}\right)\times\left(\frac{a(\text{NaCl})}{a(\text{KCl})}\right)^3 = \left(\frac{74.55}{58.44}\right)\times\left(\frac{564\,\text{pm}}{628\,\text{pm}}\right)^3 = 0.924$$

Experimentally

$$\frac{\rho(\text{KCl})}{\rho(\text{NaCl})} = \frac{1.99\,\text{g cm}^{-3}}{2.17\,\text{g cm}^{-3}} = 0.917$$

and the measurements $\boxed{\text{are broadly consistent}}$

P21.9 $\theta(100\,\text{K}) = 22°2'25''$, $\theta(300\,\text{K}) = 21°57'59''$

$\sin\theta(100\,\text{K}) = 0.37526$, $\sin\theta(300\,\text{K}) = 0.37406$

$$\frac{\sin\theta(300\,\text{K})}{\sin\theta(100\,\text{K})} = 0.99681 = \frac{a(100\,\text{K})}{a(300\,\text{K})} \quad \text{[see Problem 21.7]}$$

$$a(300\,\text{K}) = \frac{\lambda\sqrt{3}}{2\sin\theta} = \frac{(154.062\,\text{pm})\times\sqrt{3}}{(2)\times(0.37406)} = 356.67\,\text{pm}$$

$$a(100\,\text{K}) = (0.99681)\times(356.67\,\text{pm}) = 355.53\,\text{pm}$$

$$\frac{\delta a}{a} = \frac{356.67 - 355.53}{355.53} = 3.206\times10^{-3}$$

$$\frac{\delta V}{V} = \frac{356.67^3 - 355.53^3}{355.53^3} = 9.650\times10^{-3}$$

$$\alpha_{\text{volume}} = \frac{1}{V}\frac{\delta V}{\delta T} = \frac{9.560\times10^{-3}}{200\,\text{K}} = \boxed{4.8\times10^{-5}\,\text{K}^{-1}}$$

$$\alpha_{\text{linear}} = \frac{1}{a}\frac{\delta a}{\delta T} = \frac{3.206\times10^{-3}}{200\,\text{K}} = \boxed{1.6\times10^{-5}\,\text{K}^{-1}}$$

Solutions to theoretical problems

P21.11 Consider, for simplicity, the two-dimensional lattice and planes shown in Fig. 21.8.

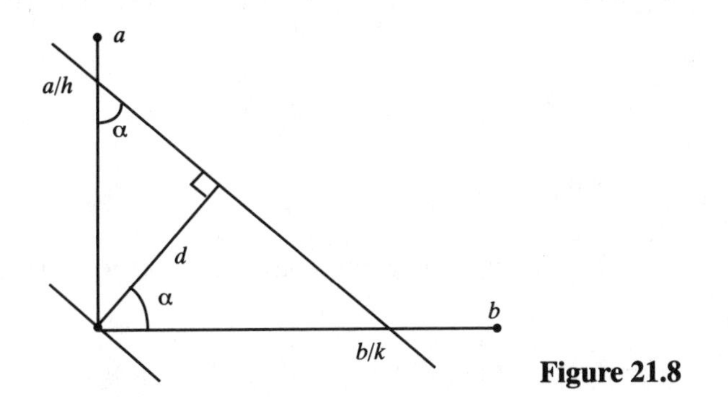

Figure 21.8

The (hk) planes cut the a and b axes at $\frac{a}{h}$ and $\frac{b}{k}$, and we have

$$\sin\alpha = \frac{d}{(a/h)} = \frac{hd}{a}, \qquad \cos\alpha = \frac{d}{(b/k)} = \frac{kd}{b}$$

Then, since $\sin^2\alpha + \cos^2\alpha = 1$, we can write

$$\left(\frac{hd}{a}\right)^2 + \left(\frac{kd}{b}\right)^2 = 1$$

and therefore

$$\frac{1}{d^2} = \left(\frac{h}{a}\right)^2 + \left(\frac{k}{b}\right)^2$$

The argument extends by analogy (or further trigonometry) to three dimensions, to give

$$\boxed{\frac{1}{d^2} = \left(\frac{h}{a}\right)^2 + \left(\frac{k}{b}\right)^2 + \left(\frac{l}{c}\right)^2}$$

P21.13 $f = \frac{NV_a}{V_c}$

where N is the number of atoms in each unit cell, V_a their individual volumes, and V_c the volume of the unit cell itself. Refer to Fig. 21.9.

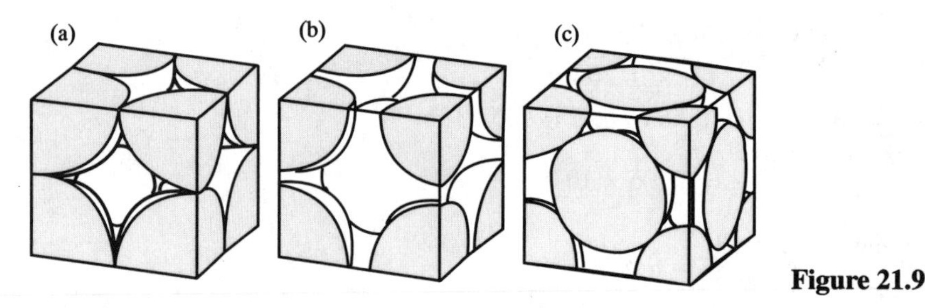

Figure 21.9

(a) $N = 1, \quad V_a = \frac{4}{3}\pi R^3, \quad V_c = (2R)^3$

$$f = \frac{\left(\frac{4}{3}\pi R^3\right)}{(2R)^3} = \frac{\pi}{6} = \boxed{0.5236}$$

(b) $N = 2, \quad V_a = \frac{4}{3}\pi R^3, \quad V_c = \left(\frac{4R}{\sqrt{3}}\right)^3$ [body diagonal of a unit cube is $\sqrt{3}$]

$$f = \frac{(2)\times\frac{4}{3}\pi R^3}{\left(\frac{4R}{\sqrt{3}}\right)^3} = \frac{\pi\sqrt{3}}{8} = \boxed{0.6802}$$

(c) $N = 4, \quad V_a = \frac{4}{3}\pi R^3, \quad V_c = (2\sqrt{2}R)^3$

$$f = \frac{(4)\times\frac{4}{3}\pi R^3}{(2\sqrt{2}R)^3} = \frac{\pi}{3\sqrt{2}} = \boxed{0.7405}$$

P21.14 The four values of $hx + ky + lz$ that occur in the exponential functions in F have the values 0, $\frac{5}{2}$, 3, and $\frac{7}{2}$, and so

$$F_{hkl} \propto 1 + e^{5i\pi} + e^{6i\pi} + e^{7i\pi} = 1 - 1 + 1 - 1 = \boxed{0}$$

Solutions to additional problems

P21.17 The volume per unit cell is

$$V = abc = (3.6881\,\text{nm}) \times (0.9402\,\text{nm}) \times (1.7652\,\text{nm}) = 6.121\,\text{nm}^3 = 6.121 \times 10^{-21}\,\text{cm}^3$$

The mass per unit cell is 8 times the mass of the formula unit, $RuN_2C_{28}H_{44}S_4$, for which the molar mass is

$$M = \{101.07 + 2(14.007) + 28(12.011) + 44(1.008) + 4(32.066)\}\,\text{g mol}^{-1} = 638.01\,\text{g mol}^{-1}$$

The density is

$$\rho = \frac{m}{V} = \frac{8M}{N_A V} = \frac{8(638.01\,\text{g mol}^{-1})}{(6.022 \times 10^{23}\,\text{mol}^{-1}) \times (6.121 \times 10^{-21}\,\text{cm}^3)} = \boxed{1.385\,\text{g cm}^{-3}}$$

The osmium analogue has a molar mass of $727.1\,\text{g mol}^{-1}$. If the volume of the crystal changes negligibly with the substitution, then the densities of the complexes are in proportion to their molar masses

$$\rho_{\text{Os}} = \frac{727.1}{638.01}(1.385\,\text{g cm}^{-3}) = \boxed{1.578\,\text{g cm}^{-3}}$$

P21.19 $$f = 4\pi \int_0^\infty \rho(r)\frac{\sin kr}{kr} r^2\,\text{d}r$$

where

$$k = \frac{4\pi}{\lambda}\sin\theta$$

$$\rho(r) = |\psi_{1s}|^2 = \frac{Z^3}{\pi a_0^3}\text{e}^{-2Zr/a_0}$$

$$f = \frac{4Z^3}{ka_0^3}\int_0^\infty r\text{e}^{-2Zr/a_0}\sin kr\,\text{d}r$$

$$= \left(\frac{4Z^3}{ka_0^3}\right)\left[\left\{\frac{r\text{e}^{-2Zr/a_0}\left(\left(-\frac{2Z}{a_0}\right)\sin kr - k\cos kr\right)}{\left(-\frac{2Z}{a_0}\right)^2 + k^2}\right\}\right.$$

$$\left.-\left\{\frac{\text{e}^{-2Zr/a_0}\left[\left(\left(-\frac{2Z}{a_0}\right)^2 - k^2\right)\sin kr - 2\left(-\frac{2Z}{a_0}\right)k\cos kr\right]}{\left[\left(-\frac{2Z}{a_0}\right)^2 + k^2\right]^2}\right\}\right]_{r=0}^{r=\infty}$$

$$= \left(\frac{4Z^3}{ka_0^3}\right)\left[\frac{\left(\frac{4Zk}{a_0}\right)}{\left[\left(\frac{2Z}{a_0}\right)^2 + k^2\right]^2}\right] = \frac{16Z^4}{a_0^4}\left[\frac{1}{\left(\frac{2Z}{a_0}\right)^4\left[1 + \left(\frac{ka_0}{2Z}\right)^2\right]^2}\right] = \frac{1}{\left[1 + \left(\frac{ka_0}{2Z}\right)^2\right]^2}$$

$$\boxed{f = \frac{1}{\left[1 + \left(\frac{2\pi a_0}{\lambda Z}\sin\theta\right)^2\right]^2}}$$

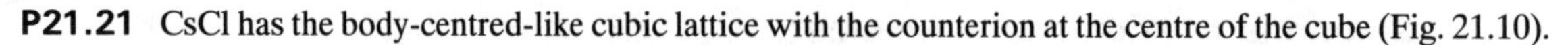

P21.21 CsCl has the body-centred-like cubic lattice with the counterion at the centre of the cube (Fig. 21.10).

$$F_{hkl} = \sum_i f_i e^{i\phi_i}, \quad \phi_i = 2\pi(hx_i + ky_i + lz_i)$$

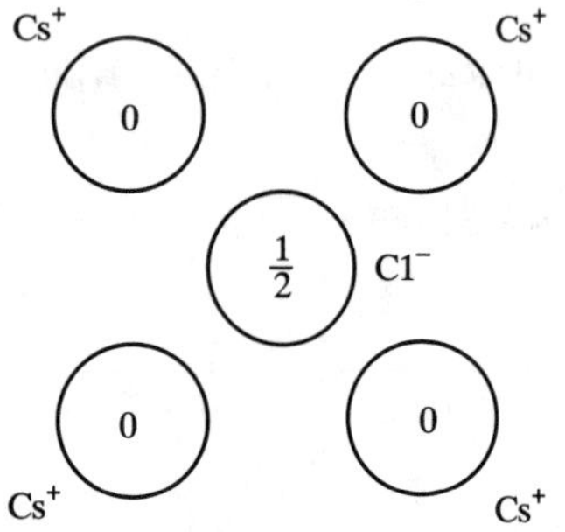

Figure 21.10

$$\sum_{\substack{i \text{ corner} \\ \text{atoms}}}^{8} f_i e^{i\phi_i} = \frac{1}{8}(f^+) \sum_{\substack{i \text{ corner} \\ \text{atoms}}}^{8} e^{i\phi_i} = \frac{1}{8}(8) f^+ = f^+$$

$$\sum_{\substack{i \text{ center} \\ \text{atom}}}^{1} f_i e^{i\phi_i} = (f^-) e^{\left[2\pi i\left(\frac{h}{2}+\frac{k}{2}+\frac{l}{2}\right)\right]} = (f^-) e^{\pi i(h+k+l)}$$

$$= (f^-)\cos[\pi(h+k+l)] = (f^-)(-1)^{(h+k+l)}$$

$$(F^-) = (-1)^{(h+k+l)}$$

Therefore

$$F_{hkl} = f^+ + (-1)^{(h+k+l)} f^-$$

If $h + k + l$ is odd, $F_{hkl} = f^+ - f^-$. If $h + k + l$ is even, $F_{hkl} = f^+ + f^-$

$$F_{100} = f^+ - f^-; \qquad F_{110} = f^+ + f^-; \qquad F_{200} = f^+ + f^-$$

Therefore, for CsCl, assuming that the atomic scattering factors are given by the number of electrons

$$F_{100} = f(Cs^+) - f(Cl^-) = 54 - 18 = \boxed{36}$$
$$F_{110} = f(Cs^+) + f(Cl^-) = 54 + 18 = \boxed{72}$$
$$F_{200} = f(Cs^+) + f(Cl^-) = 54 + 18 = \boxed{72}$$

In this case, reflections from planes with hkl odd are diminished in intensity, but not absent.

22 The electric and magnetic properties of molecules

Solutions to exercises

E22.1 A molecule with a centre of symmetry may not be polar. Therefore $ClF_3(D_{3h})$ may not be polar as the group D_{3h} contains C_2 and σ_h (equivalent to i). Molecules belonging to the groups C_n and C_{nv} may be polar (Section 15.3); therefore $\boxed{O_3}$ (C_{2v}) is polar as well as $\boxed{H_2O_2}$ (C_2) except in one configuration when the two O–H bonds are at 180° to each other. But nearly free rotation about the O–O bond makes the average dipole zero.

E22.2 Polarizability, dipole moment, and molar polarization are related by

$$P_m = \left(\frac{N_A}{3\varepsilon_0}\right) \times \left(\alpha + \frac{\mu^2}{3kT}\right) \; [16]$$

In order to solve for α, it is first necessary to obtain μ from the temperature variation of P_m.

$$\alpha + \frac{\mu^2}{3kT} = \frac{3\varepsilon_0 P_m}{N_A}$$

Therefore, $\left(\frac{\mu^2}{3k}\right) \times \left(\frac{1}{T} - \frac{1}{T'}\right) = \left(\frac{3\varepsilon_0}{N_A}\right) \times (P - P') \quad (P \text{ at } T, P' \text{ at } T')$

and hence

$$\mu^2 = \frac{\left(\frac{9\varepsilon_0 k}{N_A}\right) \times (P - P')}{\frac{1}{T} - \frac{1}{T'}}$$

$$= \frac{(9) \times (8.854 \times 10^{-12}\,\mathrm{J^{-1}\,C^2\,m^{-1}}) \times (1.381 \times 10^{-23}\,\mathrm{J\,K^{-1}}) \times (70.62 - 62.47) \times 10^{-6}\,\mathrm{m^3\,mol^{-1}}}{(6.022 \times 10^{23}\,\mathrm{mol^{-1}}) \times \left(\frac{1}{351.0\,\mathrm{K}} - \frac{1}{423.2\,\mathrm{K}}\right)}$$

$$= 3.06\bar{4} \times 10^{-59}\,\mathrm{C^2 m^2}$$

and hence $\mu = \boxed{5.5 \times 10^{-30}\,\mathrm{C\,m}}$, or 1.7 D

$$\text{Then } \alpha = \frac{3\varepsilon_0 P_m}{N_A} - \frac{\mu^2}{3kT} = \frac{(3) \times (8.854 \times 10^{-12}\,\mathrm{J^{-1}\,C^2\,m^{-1}}) \times (70.62 \times 10^{-6}\,\mathrm{m^3\,mol^{-1}})}{6.022 \times 10^{23}\,\mathrm{mol^{-1}}} - \frac{3.06\bar{4} \times 10^{-59}\,\mathrm{C^2\,m^2}}{(3) \times (1.381 \times 10^{-23}\,\mathrm{J\,K^{-1}}) \times (351.0\,\mathrm{K})}$$

$$= \boxed{1.01 \times 10^{-39}\,\mathrm{J^{-1}\,C^2 m^2}}$$

Corresponding to $\alpha' = \frac{\alpha}{4\pi\varepsilon_0}[10] = \boxed{9.1 \times 10^{-24}\,\mathrm{cm^3}}$

E22.3

$$\frac{\varepsilon_r - 1}{\varepsilon_r + 2} = \frac{\rho P_m}{M}[15] = \frac{(1.89\,\mathrm{g\,cm^{-3}}) \times (27.18\,\mathrm{cm^3\,mol^{-1}})}{92.45\,\mathrm{g\,mol^{-1}}} = 0.556$$

Hence, $\varepsilon_r = \frac{(1) + (2) \times (0.556)}{1 - 0.556} = \boxed{4.8}$

E22.4

$$n_r = (\varepsilon_r)^{1/2} \; [19]$$

$$\frac{\varepsilon_r - 1}{\varepsilon_r + 2} = \frac{N\alpha}{3\varepsilon_0} \; [17]; \quad N = \frac{\rho N_A}{M}$$

Therefore,

$$\alpha = \left(\frac{3\varepsilon_0 M}{\rho N_A}\right) \times \left(\frac{n_r^2 - 1}{n_r^2 + 2}\right) = \left(\frac{(3) \times (8.854 \times 10^{-12}\,\mathrm{J^{-1}\,C^2\,m^{-1}}) \times (267.8\,\mathrm{g\,mol^{-1}})}{(3.32 \times 10^6\,\mathrm{g\,m^{-3}}) \times (6.022 \times 10^{23}\,\mathrm{mol^{-1}})}\right)$$

$$\times \left(\frac{1.732^2 - 1}{1.732^2 + 2}\right) = \boxed{1.42 \times 10^{-39}\,\mathrm{J^{-1}\,C^2\,m^2}}$$

and $\alpha' = \boxed{1.28 \times 10^{-23}\,\mathrm{cm^3}}$

E22.5 $\mu = qR \quad (q = be, b = \text{bond order})$

For example, $\mu_{\text{ionic}}(\text{C–O}) = (1.602 \times 10^{-19}\,\mathrm{C}) \times (1.43 \times 10^{-10}\,\mathrm{m}) = 22.9 \times 10^{-30}\,\mathrm{C\,m} = 6.86\,\mathrm{D}$

Then, percentage ionic character $= \dfrac{\mu_{\text{obs}}}{\mu_{\text{ionic}}} \times 100$ per cent

$\Delta\chi$ values are based on Pauling electronegativities.

We draw up the following table

Bond	$\mu_{\text{obs}}/\mathrm{D}$	$\mu_{\text{ionic}}/\mathrm{D}$	per cent	$\Delta\chi$
C–O	1.2	6.86	$\boxed{17}$	0.8
C=O	2.7	11.72	$\boxed{23}$	0.8

There is $\boxed{\text{no correlation}}$ based on this set of bonds between the same two atoms, but in general there is a qualitative correlation for bonds between different atoms.

Comment. There are other contributions to the observed dipole moment besides the term qR. These are a result of the delocalization of the charge distribution in the bond orbitals.

E22.6 Refer to Fig. 22.2 of the text, and add moments vectorially.

Use $\mu = 2\mu_1 \cos\frac{1}{2}\theta$ [4*b*].

(a) *p*-xylene: the resultant is zero, so $\mu = \boxed{0}$

(b) *o*-xylene: $\mu = (2) \times (0.4\,\mathrm{D}) \times \cos 30° = \boxed{0.7\,\mathrm{D}}$

(c) *m*-xylene: $\mu = (2) \times (0.4\,\mathrm{D}) \times \cos 60° = \boxed{0.4\,\mathrm{D}}$

The *p*-xylene molecule belongs to the group D_{2h}, and so it is necessarily nonpolar.

E22.7

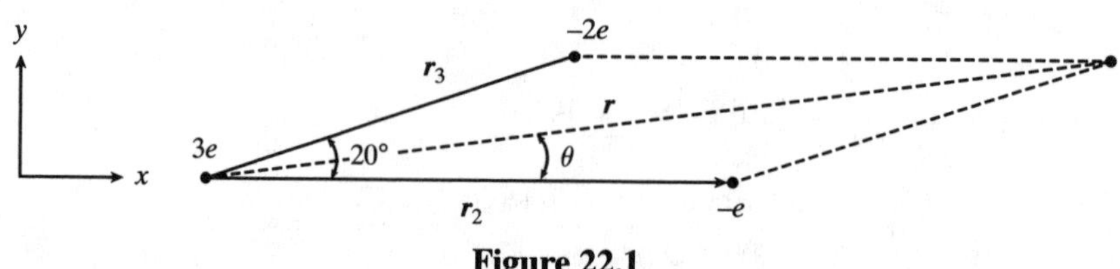

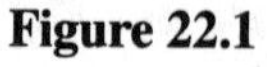
Figure 22.1

The dipole moment is the vector sum (see Fig. 22.1)

$$\mu = \sum_i q_i \boldsymbol{r}_i = 3e(0) - e\boldsymbol{r}_2 - 2e\boldsymbol{r}_3$$

$$\boldsymbol{r}_2 = \mathbf{i}x_2, \qquad \boldsymbol{r}_3 = \mathbf{i}x_3 + \mathbf{j}y_3$$

$$x_2 = +0.32\,\text{nm}$$
$$x_3 = r_3 \cos 20^\circ = (+0.23\,\text{nm}) \times (0.940) = 0.21\overline{6}\,\text{nm}$$
$$y_3 = r_3 \sin 20^\circ = (+0.23\,\text{nm}) \times (0.342) = 0.078\overline{7}\,\text{nm}$$

The components of the vector sum are the sums of the components. That is (with all distances in nm)

$$\mu_x = -ex_2 - 2ex_3 = -(e) \times \{(0.32) + (2) \times (0.21\overline{6})\} = -(e) \times (0.752\,\text{nm})$$
$$\mu_y = -2ey_3 = -(e) \times (2) \times (0.078\overline{7}) = -(e) \times (0.1574\,\text{nm})$$
$$\mu = (\mu_x^2 + \mu_y^2)^{1/2} = (e) \times (0.76\overline{8}\,\text{nm}) = (1.602 \times 10^{-19}\,\text{C}) \times (0.76\overline{8} \times 10^{-9}\,\text{m})$$
$$= 1.2\overline{3} \times 10^{-28}\,\text{C m} = \boxed{37\,\text{D}}$$

The angle that μ makes with x-axis is given by

$$\cos\theta = \frac{|\mu_x|}{\mu} = \frac{0.752}{0.768}; \qquad \boxed{\theta = 11.7^\circ}$$

E22.8 $\mu^* = \alpha\mathcal{E}[8] = 4\pi\varepsilon_0\alpha'\mathcal{E}\ [10]$

$$= (4\pi) \times (8.854 \times 10^{-12}\,\text{J}^{-1}\,\text{C}^2\,\text{m}^{-1}) \times (1.48 \times 10^{-30}\,\text{m}^3) \times (1.0 \times 10^5\,\text{V m}^{-1})$$
$$= 1.6 \times 10^{-35}\,\text{C m} \quad [1\,\text{J} = 1\,\text{C V}]$$

which corresponds to $\boxed{4.9\,\mu\text{D}}$

E22.9 The solution to Exercise 22.4 showed that

$$\alpha = \left(\frac{3\varepsilon_0 M}{\rho N_A}\right) \times \left(\frac{n_r^2 - 1}{n_r^2 + 2}\right) \quad \text{or} \quad \alpha' = \left(\frac{3M}{4\pi\rho N_A}\right) \times \left(\frac{n_r^2 - 1}{n_r^2 + 2}\right)$$

which may be solved for n_r to yield

$$n_r = \left(\frac{\beta' + 2\alpha'}{\beta' - \alpha'}\right)^{1/2} \quad \text{with } \beta' = \frac{3M}{4\pi\rho N_A}$$

$$\beta' = \frac{(3) \times (18.02\,\text{g mol}^{-1})}{(4\pi) \times (0.99707 \times 10^6\,\text{g m}^{-3}) \times (6.022 \times 10^{23}\,\text{mol}^{-1})} = 7.165 \times 10^{-30}\,\text{m}^3$$

$$n_r = \left(\frac{(7.165) + (2) \times (1.5)}{(7.165) - (1.5)}\right)^{1/2} = \boxed{1.34}$$

There is little or no discrepancy to be explained!

E22.10 $\dfrac{\varepsilon_r - 1}{\varepsilon_r + 2} = \left(\dfrac{\rho N_A}{3\varepsilon_0 M}\right) \times \left(\alpha + \dfrac{\mu^2}{3kT}\right)$ [16, with 15]

Hence, $\varepsilon_r = \dfrac{1 + 2x}{1 - x}$ with $x = \left(\dfrac{\rho N_A}{3\varepsilon_0 M}\right) \times \left(\alpha + \dfrac{\mu^2}{3kT}\right)$

$$x = \left(\frac{(1.173 \times 10^6\,\text{g m}^{-3}) \times (6.022 \times 10^{23}\,\text{mol}^{-1})}{(3) \times (8.854 \times 10^{-12}\,\text{J}^{-1}\,\text{C}^2\,\text{m}^{-1}) \times (112.6\,\text{g mol}^{-1})}\right)$$
$$\times \left[(4\pi) \times (8.854 \times 10^{-12}\,\text{J}^{-1}\,\text{C}^2\,\text{m}^{-1}) \times (1.23 \times 10^{-29}\,\text{m}^3) + \left(\frac{[(1.57) \times (3.336 \times 10^{-30}\,\text{C m})]^2}{(3) \times (1.381 \times 10^{-23}\,\text{J K}^{-1}) \times (298.15\,\text{K})}\right)\right]$$
$$= 0.848$$

Therefore, $\varepsilon_r = \dfrac{(1) + (2) \times (0.848)}{1 - 0.848} = \boxed{18}$

E22.11 We start with the equation derived in *Justification* 22.4

$$\Delta\theta = (n_R - n_L) \times \left(\frac{2\pi l}{\lambda}\right)$$

From the construction in Fig. 22.2 we see that the angle of rotation of the plane of polarization is $\delta = \dfrac{\Delta\theta}{2}$.

$$\delta = (n_R - n_L) \times \left(\frac{\pi l}{\lambda}\right)$$

$$(n_R - n_L) = \frac{\delta\lambda}{\pi l} = \frac{\delta\lambda}{180° \times l} = \frac{(250°) \times (5.00 \times 10^{-7}\,\text{m})}{(180°) \times (0.10\,\text{m})} = \boxed{6.9 \times 10^{-6}}$$

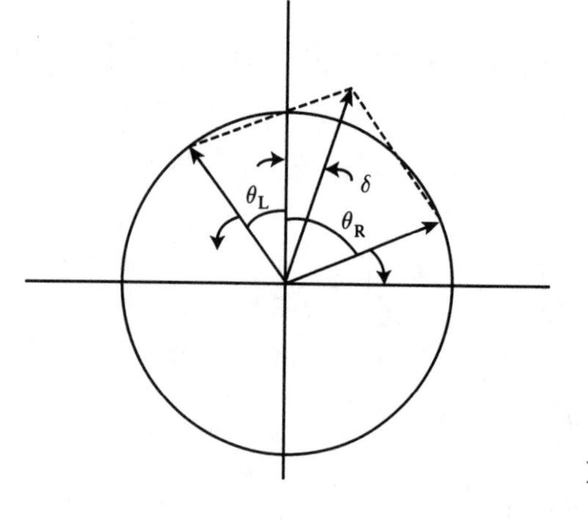

Figure 22.2

E22.12 $m = g_e\{S(S+1)\}^{1/2}\mu_B$ [43, with S in place of s]

Therefore, since $m = 3.81\mu_B$

$$S(S+1) = \left(\tfrac{1}{4}\right) \times (3.81)^2 = 3.63, \quad \text{implying that} \quad S = 1.47$$

Since $S \approx \frac{3}{2}$, there must be $\boxed{\text{three}}$ unpaired spins.

E22.13 $\chi_m = \chi V_m[36] = \dfrac{\chi M}{\rho} = \dfrac{(-7.2 \times 10^{-7}) \times (78.11\,\text{g mol}^{-1})}{0.879\,\text{g cm}^{-3}} = \boxed{-6.4 \times 10^{-5}\,\text{cm}^3\,\text{mol}^{-1}}$

E22.14 We need to compare the experimentally determined expression for χ_m to the theoretical expression

$$\chi_m = \frac{N_A g_e^2 \mu_0 \mu_B^2 S(S+1)}{3kT} \quad [44]$$

where in making the comparison we are assuming spin-only magnetism. Inserting the constants we obtain (*Illustration*, Section 22.7)

$$\chi_m = (6.3001 \times 10^{-6}\,\text{m}^3\,\text{K mol}^{-1}) \times \left(\frac{S(S+1)}{T}\right) = \frac{1.22 \times 10^{-5}\,\text{m}^3\,\text{K mol}^{-1}}{T}$$

Therefore, $S(S+1) = \dfrac{1.22 \times 10^{-5}}{6.3001 \times 10^{-6}} = 1.94 \approx 2$ or $S = 1$

and the number of unpaired electrons is $\boxed{2}$.

The problem of the Lewis structure is resolved in molecular orbital theory which shows that it is possible to have simultaneously a double bond and two unpaired electrons. See Section 14.5(f).

Comment. The discrepancy between 1.94 and 2 in $S(S+1)$ can probably be accounted for by allowing for some orbital contribution to the magnetic moment of O_2. The assumption of spin-only magnetism is not exact.

E22.15 $\chi_m(\text{theor}) = \dfrac{N_A g_e^2 \mu_0 \mu_B^2 S(S+1)}{3kT}$ [44]

The molar susceptibility is given by

$$\chi_m = \frac{N_A g_e^2 \mu_0 \mu_B^2 S(S+1)}{3kT} \quad \text{so} \quad S(S+1) = \frac{3kT\chi_m}{N_A g_e^2 \mu_0 \mu_B^2}$$

$$S(S+1) = \frac{3(1.381 \times 10^{-23}\,\text{J K}^{-1}) \times (294.53\,\text{K}) \times (0.1463 \times 10^{-6}\,\text{m}^3\,\text{mol}^{-1})}{(6.022 \times 10^{23}\,\text{mol}^{-1}) \times (2.0023)^2 \times (4\pi \times 10^{-7}\,\text{T}^2\,\text{J}^{-1}\,\text{m}^3) \times (9.27 \times 10^{-24}\,\text{J T}^{-1})^2}$$
$$= 6.84$$

so

$$S^2 + S - 6.841 = 0 \quad \text{and} \quad S = \frac{-1 + \sqrt{1 + 4(6.841)}}{2} = 2.163$$

corresponding to $\boxed{4.326}$ effective unpaired spins. The theoretical number is $\boxed{5}$ corresponding to the $3d^5$ electronic configuration of Mn^{2+}.

Comment. The discrepancy between the two values is accounted for by an antiferromagnetic interaction between the spins which alters χ_m from the form of eqn 44.

E22.16 $\chi_m = (6.3001) \times \left(\dfrac{S(S+1)}{T/\text{K}} \text{cm}^3\,\text{mol}^{-1}\right)$ [*Illustration*, Section 22.7]

Since Cu(II) is a d^9 species, it has one unpaired spin, and so $S = s = \frac{1}{2}$. Therefore,

$$\chi_m = \frac{(6.3001) \times \left(\frac{1}{2}\right) \times \left(\frac{3}{2}\right)}{298}\,\text{cm}^3\,\text{mol}^{-1} = \boxed{+0.016\,\text{cm}^3\,\text{mol}^{-1}}$$

E22.17 The magnitude of the orientational energy is given by

$$g_e \mu_B M_S \mathcal{B} \quad \text{with } M_S = S = 1$$

Setting this equal to kT and solving for $\mathcal{B}$

$$\mathcal{B} = \frac{kT}{g_e \mu_B} = \frac{(1.38 \times 10^{-23}\,\text{J K}^{-1}) \times (298\,\text{K})}{(2.00) \times (9.27 \times 10^{-24}\,\text{J T}^{-1})} = \boxed{222\,\text{T}}$$

Comment. This is an enormous magnetic field and it is a measure of the strength of the internal magnetic fields required for spin alignment in ferromagnetic and antiferromagnetic materials in which such alignments occur.

Solutions to problems

Solutions to numerical problems

P22.1 The positive (H) end of the dipole will lie closer to the (negative) anion. The electric field generated by a dipole is

$$\mathcal{E} = \left(\frac{\mu}{4\pi\varepsilon_0}\right) \times \left(\frac{2}{r^3}\right) \text{ [23]}$$

$$= \frac{(2) \times (1.85) \times (3.34 \times 10^{-30}\,\mathrm{C\,m})}{(4\pi) \times (8.854 \times 10^{-12}\,\mathrm{J^{-1}\,C^2\,m^{-1}}) \times r^3} = \frac{1.11 \times 10^{-19}\,\mathrm{V\,m^{-1}}}{(r/\mathrm{m})^3} = \frac{1.11 \times 10^{8}\,\mathrm{V\,m^{-1}}}{(r/\mathrm{nm})^3}$$

(a) $\mathcal{E} = \boxed{1.1 \times 10^8\,\mathrm{V\,m^{-1}}}$ when $r = 1.0\,\mathrm{nm}$

(b) $\mathcal{E} = \dfrac{1.11 \times 10^8\,\mathrm{V\,m^{-1}}}{0.3^3} = \boxed{4 \times 10^9\,\mathrm{V\,m^{-1}}}$ for $r = 0.3\,\mathrm{nm}$

(c) $\mathcal{E} = \dfrac{1.11 \times 10^8\,\mathrm{V\,m^{-1}}}{30^3} = \boxed{4\,\mathrm{kV\,m^{-1}}}$ for $r = 30\,\mathrm{nm}$

P22.3 The equations relating dipole moment and polarizability volume to the experimental quantities ε_r and ρ are

$$P_m = \left(\frac{M}{\rho}\right) \times \left(\frac{\varepsilon_r - 1}{\varepsilon_r + 2}\right) \text{ [15] and } P_m = \frac{4\pi}{3}N_A\alpha' + \frac{N_A\mu^2}{9\varepsilon_0 kT} \quad [16, \text{ with } \alpha = 4\pi\varepsilon_0\alpha']$$

Therefore, we draw up the following table (with $M = 119.4\,\mathrm{g\,mol^{-1}}$)

$\theta/°\mathrm{C}$	−80	−70	−60	−40	−20	0	20
T/K	193	203	213	233	253	273	293
$\dfrac{1000}{T/\mathrm{K}}$	5.18	4.93	4.69	4.29	3.95	3.66	3.41
ε_r	3.1	3.1	7.0	6.5	6.0	5.5	5.0
$\dfrac{\varepsilon_r - 1}{\varepsilon_r + 2}$	0.41	0.41	0.67	0.65	0.63	0.60	0.57
$\rho/\mathrm{g\,cm^{-3}}$	1.65	1.64	1.64	1.61	1.57	1.53	1.50
$P_m/(\mathrm{cm^3\,mol^{-1}})$	29.8	29.9	48.5	48.0	47.5	46.8	45.4

P_m is plotted against $\dfrac{1}{T}$ in Fig. 22.3.

The (dangerously unreliable) intercept is ≈ 30 and the slope is $\approx 4.5 \times 10^3$. It follows that

$$\alpha' = \frac{(3) \times (30\,\mathrm{cm^3\,mol^{-1}})}{(4\pi) \times (6.022 \times 10^{23}\,\mathrm{mol^{-1}})} = \boxed{1.2 \times 10^{-23}\,\mathrm{cm^3}}$$

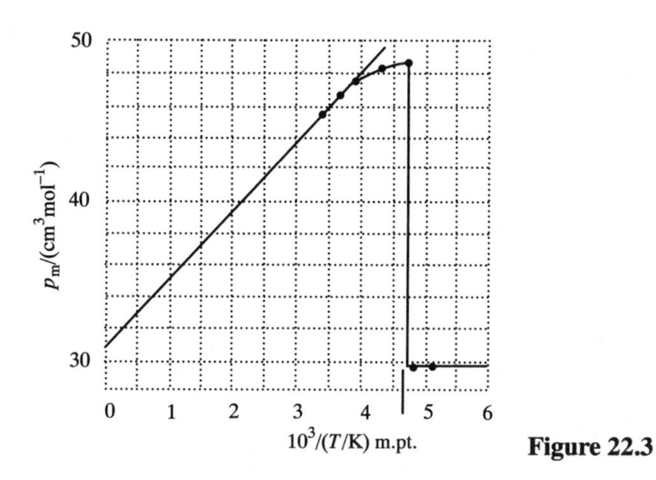

Figure 22.3

To determine μ we need

$$\mu = \left(\frac{9\varepsilon_0 k}{N_A}\right)^{1/2} \times (\text{slope} \times \text{cm}^3\,\text{mol}^{-1}\,\text{K})^{1/2}$$

$$= \left(\frac{(9) \times (8.854 \times 10^{-12}\,\text{J}^{-1}\,\text{C}^2\,\text{m}^{-1}) \times (1.381 \times 10^{-23}\,\text{J}\,\text{K}^{-1})}{6.022 \times 10^{-23}\,\text{mol}^{-1}}\right)^{1/2}$$

$$\times (\text{slope} \times \text{cm}^3\,\text{mol}^{-1}\text{K})^{1/2}$$

$$= (4.275 \times 10^{-29}\,\text{C}) \times \left(\frac{\text{mol}}{\text{K}\,\text{m}}\right)^{1/2} \times (\text{slope} \times \text{cm}^3\,\text{mol}^{-1}\,\text{K})^{1/2}$$

$$= (4.275 \times 10^{-29}\,\text{C}) \times (\text{slope} \times \text{cm}^3\,\text{m}^{-1})^{1/2} = (4.275 \times 10^{-29}\,\text{C}) \times (\text{slope} \times 10^{-6}\,\text{m}^2)^{1/2}$$

$$= (4.275 \times 10^{-32}\,\text{C}\,\text{m}) \times (\text{slope})^{1/2} = (1.282 \times 10^{-2}\,\text{D}) \times (\text{slope})^{1/2}$$

$$= (1.282 \times 10^{-2}\,\text{D}) \times (4.5 \times 10^3)^{1/2} = \boxed{0.86\,\text{D}}$$

The sharp decrease in P_m occurs at the freezing point of chloroform (−63°C), indicating that the dipole reorientation term no longer contributes. Note that P_m for the solid corresponds to the extrapolated, dipole-free, value of P_m, so the extrapolation is less hazardous than it looks.

P22.5 $P_m = \frac{4\pi}{3} N_A \alpha' + \frac{N_A \mu^2}{9\varepsilon_0 kT}$ [16, with $\alpha = 4\pi\varepsilon_0\alpha'$]

Therefore, draw up the following table

T/K	292.2	309.0	333.0	387.0	413.0	446.0
$\frac{1000}{T/\text{K}}$	3.42	3.24	3.00	2.58	2.42	2.24
$P_m/(\text{cm}^3\,\text{mol}^{-1})$	57.57	55.01	51.22	44.99	42.51	39.59

The points are plotted in Fig. 22.4.

The extrapolated (least squares) intercept lies at $5.65\,\text{cm}^3\,\text{mol}^{-1}$ (not shown in the figure), and the least squares slope is $1.52 \times 10^4\,\text{cm}^3\,\text{K}^{-1}\,\text{mol}^{-1}$. It follows that

$$\alpha' = \frac{3P_m(\text{at intercept})}{4\pi N_A} = \frac{3 \times 5.65\,\text{cm}^3\,\text{mol}^{-1}}{4\pi \times 6.022 \times 10^{23}\,\text{mol}^{-1}}$$

$$= \boxed{2.24 \times 10^{-24}\,\text{cm}^3}$$

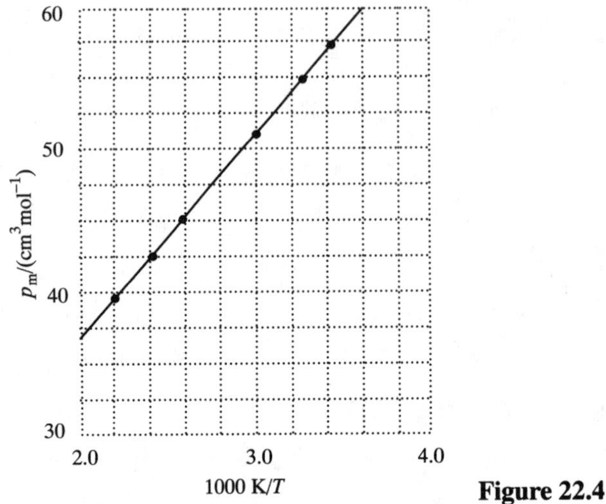

Figure 22.4

$\mu = 1.282 \times 10^{-2}\,\mathrm{D} \times (1.52 \times 10^4)^{1/2}$ [from Problem 22.3] $= \boxed{1.58\,\mathrm{D}}$

The high-frequency contribution to the molar polarization, P'_m, at 273 K may be calculated from the refractive index

$$P'_m = \left(\frac{M}{\rho}\right) \times \left(\frac{\varepsilon_r - 1}{\varepsilon_r + 2}\right) [15] = \left(\frac{M}{\rho}\right) \times \left(\frac{n_r^2 - 1}{n_r^2 + 2}\right)$$

Assuming that ammonia under these conditions (1.00 atm pressure assumed) can be considered a perfect gas, we have

$$\rho = \frac{pM}{RT}$$

$$\text{and } \frac{M}{\rho} = \frac{RT}{p} = \frac{82.06\,\mathrm{cm^3\,atm\,K^{-1}\,mol^{-1}} \times 273\,\mathrm{K}}{1.00\,\mathrm{atm}} = 2.24 \times 10^4\,\mathrm{cm^3\,mol^{-1}}$$

$$\text{Then } P'_m = 2.24 \times 10^4\,\mathrm{cm^3\,mol^{-1}} \times \left\{\frac{(1.000379)^2 - 1}{(1.000379)^2 + 2}\right\} = \boxed{5.66\,\mathrm{cm^3\,mol^{-1}}}$$

If we assume that the high-frequency contribution to P_m remains the same at 292.2 K then we have

$$\frac{N_A\mu^2}{q\varepsilon_0 kT} = P_m - P'_m = (57.57 - 5.66)\,\mathrm{cm^3\,mol^{-1}}$$

$$= 51.91\,\mathrm{cm^3\,mol^{-1}} = 5.191 \times 10^{-5}\,\mathrm{m^3\,mol^{-1}}$$

Solving for μ we have

$$\mu = \left(\frac{9\varepsilon_0 k}{N_A}\right)^{1/2} T^{1/2}(P_m - P'_m)^{1/2}$$

The factor $\left(\frac{9\varepsilon_0 k}{N_A}\right)^{1/2}$ has been calculated in Problem 22.3 and is $4.275 \times 10^{-29}\,\mathrm{C} \times \left(\frac{\mathrm{mol}}{\mathrm{K\,m}}\right)^{1/2}$

$$\text{Therefore } \mu = 4.275 \times 10^{-29}\,\mathrm{C} \times \left(\frac{\mathrm{mol}}{\mathrm{K\,m}}\right)^{1/2} \times (292.2\,\mathrm{K})^{1/2} \times (5.191 \times 10^{-5})^{1/2}\left(\frac{\mathrm{m^3}}{\mathrm{mol}}\right)^{1/2}$$

$$= 5.26 \times 10^{-30}\,\mathrm{C\,m} = \boxed{1.58\,\mathrm{D}}$$

The agreement is exact!

Solutions to theoretical problems

P22.7 **(a)** Consider the arrangement shown in Fig. 22.5(a). There are a total of $3 \times 3 = 9$ Coulombic interactions at the distances shown. The total potential energy of interaction of the two quadrupoles is

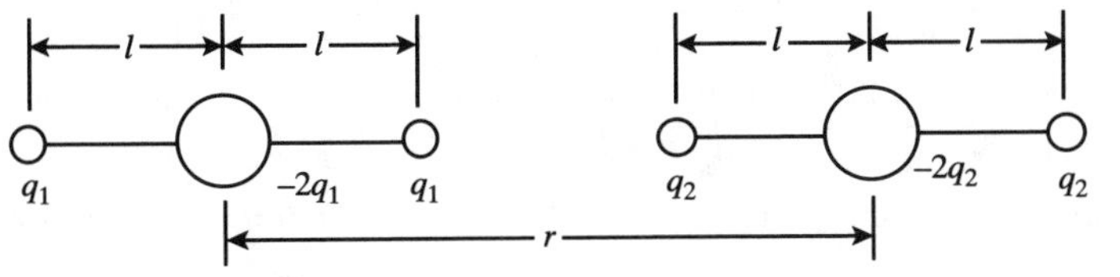

Figure 22.5(a)

$$V = \frac{q_1 q_2}{4\pi\varepsilon_0} \times \left[\left(\frac{1}{r} - \frac{2}{r-l} + \frac{1}{r-2l}\right) - 2\left(\frac{1}{r+l} - \frac{2}{r} + \frac{1}{r-l}\right) + \left(\frac{1}{r+2l} - \frac{2}{r+l} + \frac{1}{r}\right)\right]$$

$$= \frac{q_1 q_2}{4\pi\varepsilon_0 r} \times \left[\left(1 - \frac{2}{1-\lambda} + \frac{1}{1-2\lambda}\right) - 2\left(\frac{1}{1+\lambda} - 2 + \frac{1}{1-\lambda}\right) + \left(\frac{1}{1+2\lambda} - \frac{2}{1+\lambda} + 1\right)\right] \quad \left(\lambda = \frac{l}{r} \ll 1\right)$$

Expand each term using

$$\frac{1}{1+x} = 1 - x + x^2 - x^3 + x^4 - \cdots$$

and keep up to λ^4 (the preceding terms cancel). The result is

$$V = \frac{q_1 q_2}{4\pi\varepsilon_0 r} \times 24\lambda^4 = \frac{6 q_1 q_2 l^4}{\pi\varepsilon_0 r^5}$$

Define the quadrupole moments of the two distributions as

$$Q_1 = q_1 l^2, \qquad Q_2 = q_2 l^2$$

and hence obtain $\boxed{V = \dfrac{6 Q_1 Q_2}{\pi\varepsilon_0} \times \dfrac{1}{r^5}}$

(b) Consider Fig. 22.5(b). There are three different distances, r, r', and r''. Three interactions are at r, four at r', and two at r''.

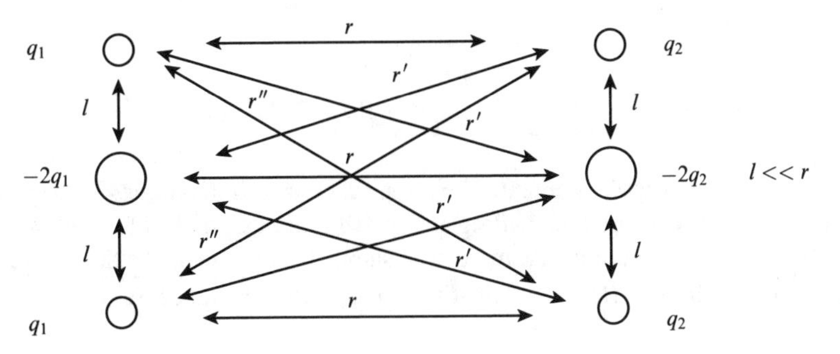

Figure 22.5(b)

$$r' = (r^2 + l^2)^{1/2} = r(1+\lambda^2)^{1/2} \approx r\left(1 + \frac{\lambda^2}{2} - \frac{\lambda^4}{8} + \cdots\right)$$
$$r'' = (r^2 + 4l^2)^{1/2} = r(1+4\lambda^2)^{1/2} \approx r(1 + 2\lambda^2 - 2\lambda^4 + \cdots)$$
$$V = \frac{q_1 q_2}{4\pi\varepsilon_0} \times \left[\left(\frac{1}{r} - \frac{2}{r'} + \frac{1}{r''}\right) - 2\left(\frac{2}{r'} - \frac{4}{r} + \frac{2}{r'}\right) + \left(\frac{1}{r''} - \frac{2}{r'} + \frac{1}{r}\right)\right]$$
$$= \left(\frac{2q_1 q_2}{4\pi\varepsilon_0}\right) \times \left(\frac{3}{r} - \frac{4}{r'} + \frac{1}{r''}\right) = \left(\frac{2q_1 q_2}{4\pi\varepsilon_0 r}\right) \times \left(3 - 4\frac{r}{r'} + \frac{r}{r''}\right)$$

Substituting for r' and r'' in terms of r and λ from above we obtain (dropping terms beyond λ^4)

$$V = V_0\left(3 - \frac{4}{\left(1 + \frac{\lambda^2}{2} - \frac{\lambda^4}{8}\right)} + \frac{1}{(1 + 2\lambda^2 - 2\lambda^4)}\right) \quad \left[V_0 = \frac{2q_1 q_2}{4\pi\varepsilon_0 r}\right]$$
$$= V_0\left[3 - 4\left(1 - \frac{\lambda^2}{2} + \frac{\lambda^4}{8} + \frac{\lambda^4}{4}\right) + (1 - 2\lambda^2 + 2\lambda^4 + 4\lambda^4)\right]$$

The terms in λ^0 and λ^2 cancel leaving

$$V = V_0\left(6 - \frac{3}{2}\right)\lambda^4 = \frac{9}{2}V_0\lambda^4 = \frac{9q_1 q_2 \lambda^4}{4\pi\varepsilon_0 r} = \frac{9q_1 q_2 l^4}{4\pi\varepsilon_0 r^5} = \boxed{\frac{9Q_1 Q_2}{4\pi\varepsilon_0 r^5}}$$

P22.8 Exercise 22.4 showed

$$\alpha = \left(\frac{3\varepsilon_0 M}{\rho N_A}\right) \times \left(\frac{n_r^2 - 1}{n_r^2 + 2}\right) \quad \text{or} \quad \alpha' = \left(\frac{3M}{4\pi\rho N_A}\right) \times \left(\frac{n_r^2 - 1}{n_r^2 + 2}\right)$$

Therefore, $\dfrac{n_r^2 - 1}{n_r^2 + 2} = \dfrac{4\pi\alpha' N_A \rho}{3M}$

$$\text{Solving for } n_r,\ n_r = \left(\frac{1 + \frac{8\pi\alpha'\rho N_A}{3M}}{1 - \frac{4\pi\alpha'\rho N_A}{3M}}\right)^{1/2} = \left(\frac{1 + \frac{8\pi\alpha' p}{3kT}}{1 - \frac{4\pi\alpha' p}{3kT}}\right)^{1/2} \quad \left[\text{for a gas, } \rho = \frac{M}{V_m} = \frac{Mp}{RT}\right]$$
$$\approx \left[\left(1 + \frac{8\pi\alpha' p}{3kT}\right) \times \left(1 + \frac{4\pi\alpha' p}{3kT}\right)\right]^{1/2} \quad \left[\frac{1}{1-x} \approx 1 + x\right]$$
$$\approx \left(1 + \frac{12\pi\alpha' p}{3kT} + \cdots\right)^{1/2} \approx 1 + \frac{2\pi\alpha' p}{kT} \quad \left[(1+x)^{1/2} \approx 1 + \frac{1}{2}x\right]$$

Hence, $\boxed{n_r = 1 + \text{const.} \times p}$, with constant $= \boxed{\dfrac{2\pi\alpha'}{kT}}$ From the first line above,

$$\alpha' = \left(\frac{3M}{4\pi N_A \rho}\right) \times \left(\frac{n_r^2 - 1}{n_r^2 + 2}\right) = \boxed{\left(\frac{3kT}{4\pi p}\right) \times \left(\frac{n_r^2 - 1}{n_r^2 + 2}\right)}$$

P22.10 The dimers should have a zero dipole moment. The strong molecular interactions in the pure liquid probably break up the dimers and produce hydrogen-bonded groups of molecules with a chain-like structure. In very dilute benzene solutions, the molecules should behave much like those in the gas and should tend to form planar dimers. Hence the relative permittivity $\boxed{\text{should decrease}}$ as the dilution increases.

P22.11 Consider a single molecule surrounded by $N - 1(\approx N)$ others in a container of volume V. The number of molecules in a spherical shell of thickness dr at a distance r is $4\pi r^2 \times \frac{N}{V}\,dr$. Therefore, the interaction energy is

$$u = \int_a^R 4\pi r^2 \times \left(\frac{N}{V}\right) \times \left(\frac{-C_6}{r^6}\right) dr = \frac{-4\pi NC_6}{V}\int_a^R \frac{dr}{r^4}$$

where R is the radius of the container and d the molecular diameter (the distance of closest approach). Therefore,

$$u = \left(\frac{4\pi}{3}\right) \times \left(\frac{N}{V}\right)(C_6) \times \left(\frac{1}{R^3} - \frac{1}{d^3}\right) \approx \frac{-4\pi NC_6}{3Vd^3}$$

because $d \ll R$. The mutual pairwise interaction energy of all N molecules is $U = \frac{1}{2}Nu$ (the $\frac{1}{2}$ appears because each pair must be counted only once, i.e. A with B but not A with B and B with A). Therefore,

$$U = \boxed{\frac{-2\pi N^2 C_6}{3Vd^3}}$$

For a van der Waals gas, $\frac{n^2 a}{V^2} = \left(\frac{\partial U}{\partial V}\right)_T = \frac{2\pi N^2 C_6}{3V^2 d^3}$

and therefore $a = \boxed{\frac{2\pi N_A^2 C_6}{3d^3}}$ $\quad [N = nN_A]$

P22.14 Refer to Fig. 22.6(a).

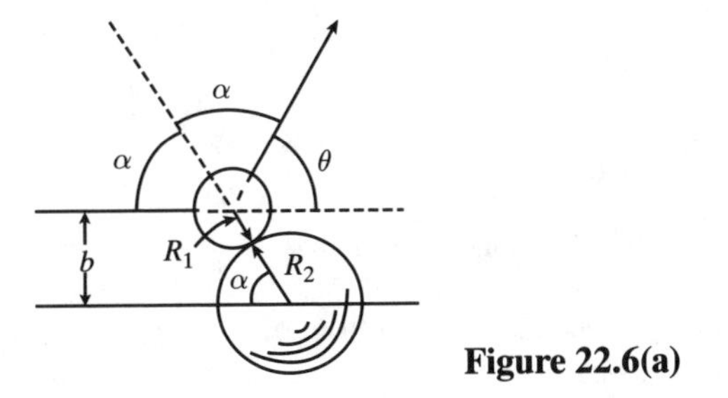

Figure 22.6(a)

The scattering angle is $\theta = \pi - 2\alpha$ if specular reflection occurs in the collision (angle of impact equal to angle of departure from the surface). For $b \le R_1 + R_2$, $\sin\alpha = \frac{b}{R_1 + R_2}$.

$$\theta = \begin{cases} \pi - 2\arcsin\left(\dfrac{b}{R_1 + R_2}\right) & b \le R_1 + R_2 \\ 0 & b > R_1 + R_2 \end{cases}$$

The function is plotted in Fig. 22.6(b).

P22.16 $\xi = \frac{-e^2}{6m_e}\langle r^2\rangle$

$$\langle r^2\rangle = \int_0^\infty r^2\psi^2\,d\tau \quad \text{with } \psi = \left(\frac{1}{\pi a_0^3}\right)^{1/2} e^{-r/a_0}$$

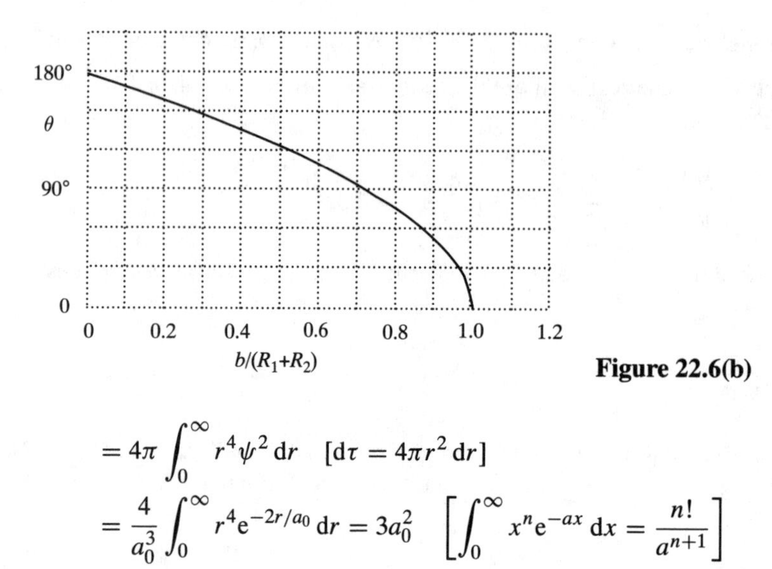

Figure 22.6(b)

$$= 4\pi \int_0^\infty r^4 \psi^2 \, dr \quad [d\tau = 4\pi r^2 \, dr]$$

$$= \frac{4}{a_0^3} \int_0^\infty r^4 e^{-2r/a_0} \, dr = 3a_0^2 \quad \left[\int_0^\infty x^n e^{-ax} \, dx = \frac{n!}{a^{n+1}}\right]$$

Therefore, $\boxed{\xi = \dfrac{-e^2 a_0^2}{2m_e}}$

Then, since $\chi_m = N_A \mu_0 \xi$ [41, $m = 0$]

$$\chi_m = \boxed{\frac{-N_A \mu_0 e^2 a_0^2}{2m_e}}$$

Solutions to additional problems

P22.20 **(a)** The depth of the well in energy units is

$$\varepsilon = hcD_e = \boxed{1.51 \times 10^{-23}\ \text{J}}$$

The distance at which the potential is zero is given by

$$R_e = 2^{1/6} r_0 \quad \text{so} \quad r_0 = R_e 2^{-1/6} = 2^{-1/6}(297\ \text{pm}) = \boxed{265\ \text{pm}}$$

(b) In Fig. 22.7 both potentials were plotted with respect to the bottom of the well, so the Lennard-Jones potential is the usual L-J potential plus ε.

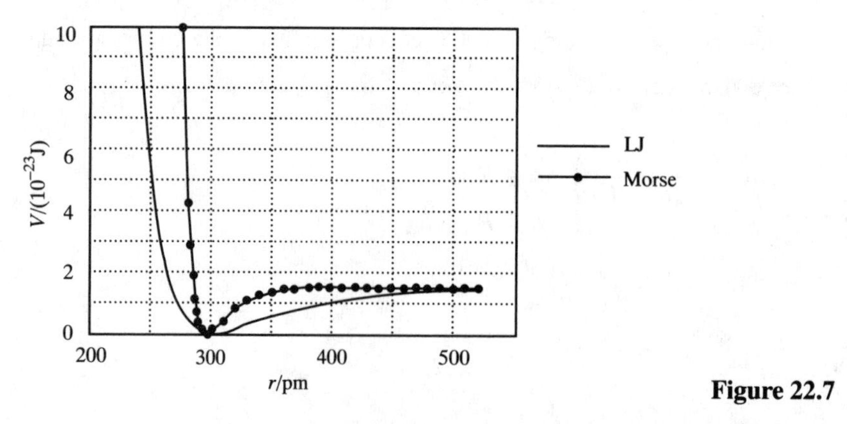

Figure 22.7

Note that the Lennard-Jones potential has a much softer repulsive branch than the Morse.

P22.21 The molar magnetic susceptibility is given by

$$\chi_m = \frac{N_A g_e^2 \mu_0 \mu_B^2 S(S+1)}{3kT} = 6.3001 \times \frac{S(S+1)}{T/K} \text{ cm}^3 \text{ mol}^{-1} \text{ [}\textit{Illustration}\text{, Section 22.7]}$$

$$\text{For } S = 2, \ \chi_m = \frac{(6.3001) \times (2) \times (2+1)}{298} \text{ cm}^3 \text{ mol}^{-1} = \boxed{0.127 \text{ cm}^3 \text{ mol}^{-1}}$$

$$\text{For } S = 3, \ \chi_m = \frac{(6.3001) \times (3) \times (3+1)}{298} \text{ cm}^3 \text{ mol}^{-1} = \boxed{0.254 \text{ cm}^3 \text{ mol}^{-1}}$$

$$\text{For } S = 4, \ \chi_m = \frac{(6.3001) \times (4) \times (4+1)}{298} \text{ cm}^3 \text{ mol}^{-1} = \boxed{0.423 \text{ cm}^3 \text{ mol}^{-1}}$$

Instead of a single value of S, we use an average weighted by the Boltzmann factor

$$\exp\left(\frac{-50 \times 10^3 \text{ J mol}^{-1}}{(8.3145 \text{ J mol}^{-1} \text{ K}^{-1}) \times (298 \text{ K})}\right) = 1.7 \times 10^{-9}$$

Thus the $S = 2$ and $S = 4$ forms are present in negligible quantities compared to the $S = 3$ form. The compound's susceptibility, then, is that of the $S = 3$ form, namely $\boxed{0.254 \text{ cm}^3 \text{ mol}^{-1}}$

23 Macromolecules and colloids

Solutions to exercises

E23.1 Equal amounts imply equal numbers of molecules; hence

$$\overline{M}_n = \frac{N_1M_1 + N_2M_2}{N}[1] = \frac{n_1M_1 + n_2M_2}{n} = \tfrac{1}{2}(M_1 + M_2) \quad \left[n_1 = n_2 = \tfrac{1}{2}n\right]$$

$$= \frac{62+78}{2}\,\text{kg mol}^{-1} = \boxed{70\,\text{kg mol}^{-1}}$$

$$\overline{M}_w = \frac{m_1M_1 + m_2M_2}{m}[2] = \frac{n_1M_1^2 + n_2M_2^2}{n_1M_1 + n_2M_2} = \frac{M_1^2 + M_2^2}{M_1 + M_2} \quad [n_1 = n_2]$$

$$= \frac{62^2+78^2}{62+78}\,\text{kg mol}^{-1} = \boxed{71\,\text{kg mol}^{-1}}$$

E23.2 $$R_{\text{rms}} = N^{1/2}l[36] = (700)^{1/2} \times (0.90\,\text{nm}) = \boxed{24\,\text{nm}}$$

E23.3 $$R_g = \frac{N^{1/2}l}{\sqrt{6}}[37] \quad N = 6\left(\frac{R_g}{l}\right)^2 = (6) \times \left(\frac{7.3\,\text{nm}}{0.154\,\text{nm}}\right)^2 = \boxed{1.4 \times 10^4}$$

E23.4 The repeating unit (monomer) of polyethylene is $(-CH_2-CH_2-)$ which has a molar mass of 28 g mol^{-1}. The number of repeating units, N, is therefore

$$N = \frac{280\,000\,\text{g mol}^{-1}}{28\,\text{g mol}^{-1}}$$

$$= 1.00 \times 10^4; \qquad l = 2R(\text{C–C}) \left[\text{Add } \tfrac{1}{2} \text{ bond on each side of monomer}\right]$$

$$R_c = Nl[35] = 2 \times (1.00 \times 10^4) \times (154\,\text{pm}) = 3.08 \times 10^6\,\text{pm} = \boxed{3.08 \times 10^{-6}\,\text{m}}$$

$$R_{\text{rms}} = N^{1/2} \times l[36] = 2 \times (1.00 \times 10^4)^{1/2} \times (154\,\text{pm})$$

$$= 3.08 \times 10^4\,\text{pm} = \boxed{3.08 \times 10^{-8}\,\text{m}}$$

E23.5 The effective mass of the particles is

$$m_{\text{eff}} = bm = (1 - \rho v_s)m[9] = m - \rho v_s m = v\rho_p - v\rho = v(\rho_p - \rho)$$

where v is the particle volume, ρ_p is the particle density. Equating the forces

$$m_{\text{eff}} r\omega^2 = fs = 6\pi\eta as \quad [13,\ a = \text{particle radius}]$$

or $v(\rho_p - \rho)r\omega^2 = 6\pi\eta as$ or $\frac{4}{3}\pi a^3(\rho_p - \rho)r\omega^2 = 6\pi\eta as$

Solving for s, $s = \dfrac{2a^2(\rho_p - \rho)r\omega^2}{9\eta}$

Thus, the relative rates of sedimentation are $\dfrac{s_2}{s_1} = \dfrac{a_2^2}{a_1^2} = 10^2 = \boxed{100}$

E23.6 See the solution to Exercise 23.5(a). In place of force $= m_{\text{eff}} r\omega^2$ we have force $= m_{\text{eff}} g$

The rest of the analysis is similar, leading to

$$s = \frac{2a^2(\rho_p - \rho)g}{9\eta} = \frac{(2) \times (2.0 \times 10^{-5}\,\text{m})^2 \times (1750 - 1000)\,\text{kg m}^{-3} \times (9.81\,\text{m s}^{-2})}{(9) \times (8.9 \times 10^{-4}\,\text{kg m}^{-1}\,\text{s}^{-1})}$$

$$= \boxed{7.3 \times 10^{-4}\,\text{m s}^{-1}}$$

E23.7 The data yield the number-average molar mass using

$$\overline{M}_n = \frac{SRT}{bD}[16] = \frac{SRT}{(1 - \rho v_s)D}\ [9,\ \text{for } b]$$

$$= \frac{(4.48 \times 10^{-13}\,\text{s}) \times (8.314\,\text{J K}^{-1}\,\text{mol}^{-1}) \times (293\,\text{K})}{[(1) - (0.9982 \times 10^3\,\text{kg m}^3) \times (0.749 \times 10^{-3}\,\text{m}^3\,\text{kg}^{-1})] \times (6.9 \times 10^{-11}\,\text{m}^2\,\text{s}^{-1})}$$

$$= \boxed{63\,\text{kg mol}^{-1}}$$

E23.8 $$\overline{M}_n = \frac{SRT}{bD}[16] = \frac{(3.2 \times 10^{-13}\,\text{s}) \times (8.314\,\text{J K}^{-1}\,\text{mol}^{-1}) \times (293\,\text{K})}{[(1) - (0.656) \times (1.06)] \times (8.3 \times 10^{-11}\,\text{m}^2\,\text{s}^{-1})} = \boxed{31\,\text{kg mol}^{-1}}$$

E23.9 **(a)** Osmometry gives the number-average molar mass, so

$$\overline{M}_n = \frac{N_1M_1 + N_2M_2}{N_1 + N_2} = \frac{\left(\frac{m_1}{M_1}\right)M_1 + \left(\frac{m_2}{M_2}\right)M_2}{\left(\frac{m_1}{M_1}\right) + \left(\frac{m_2}{M_2}\right)} = \frac{m_1 + m_2}{\left(\frac{m_1}{M_1}\right) + \left(\frac{m_2}{M_2}\right)}$$

$$= \frac{100\,\text{g}}{\left(\frac{30\,\text{g}}{30\,\text{kg mol}^{-1}}\right) + \left(\frac{70\,\text{g}}{15\,\text{kg mol}^{-1}}\right)}\ [\text{assume 100 g of solution}] = \boxed{18\,\text{kg mol}^{-1}}$$

(b) Light-scattering gives the mass-average molar mass, so

$$\overline{M}_w = \frac{m_1M_1 + m_2M_2}{m_1 + m_2} = (0.30) \times (30) + (0.70) \times (15)\,\text{kg mol}^{-1} = \boxed{20\,\text{kg mol}^{-1}}$$

E23.10 The difference in sodium concentration is

$$[\text{Na}^+]_\text{L} - [\text{Na}^+]_\text{R} = \frac{\nu[\text{P}][\text{Na}^+]_\text{L}}{2[\text{Cl}^-] + \nu[\text{P}]}$$

Charge balance also demands that

$$[\text{Na}^+]_\text{L} + [\text{Na}^+]_\text{R} = 2[\text{Cl}^-] + \nu[\text{P}]$$

Solving these equations for $2[\text{Cl}^-] + \nu[\text{P}]$ and setting the expressions equal yields

$$[\text{Na}^+]_\text{L} + [\text{Na}^+]_\text{R} = \frac{\nu[\text{P}][\text{Na}^+]_\text{L}}{[\text{Na}^+]_\text{L} - [\text{Na}^+]_\text{R}}$$

so $[\text{Na}^+]_\text{L}^2 - [\text{Na}^+]_\text{R}^2 = \nu[\text{P}][\text{Na}^+]_\text{L}$ and $[\text{Na}^+]_\text{L}^2 - \nu[\text{P}][\text{Na}^+]_\text{L} - [\text{Na}^+]_\text{R}^2 = 0$

Application of the quadratic formula yields

$$[\text{Na}^+]_\text{L} = \frac{\nu[\text{P}] + \sqrt{\nu^2[\text{P}]^2 + 4[\text{Na}^+]_\text{R}^2}}{2} = \frac{1}{2}\nu[\text{P}]\left(1 + \sqrt{1 + \frac{4[\text{Na}^+]_\text{R}^2}{\nu^2[\text{P}]^2}}\right)$$

The concentration of the polyanion is

$$[\mathrm{P}] = \frac{1.00 \times 10^{-3}\,\mathrm{kg}/0.100\,\mathrm{L}}{100\,\mathrm{kg\,mol^{-1}}} = 1.00 \times 10^{-4}\,\mathrm{mol\,L^{-1}}$$

$$\text{so } [\mathrm{Na^+}]_\mathrm{L} = (1/2) \times (20) \times (1.00 \times 10^{-4}\,\mathrm{mol\,L^{-1}}) \times \left(1 + \sqrt{1 + \frac{4(0.0010)^2}{[20(1.00 \times 10^{-4})]^2}}\right)$$

$$[\mathrm{Na^+}]_\mathrm{L} = \boxed{2.4 \times 10^{-4}\,\mathrm{mol\,L^{-1}}}$$

E23.11 $$[\mathrm{Cl^-}]_\mathrm{L} - [\mathrm{Cl^-}]_\mathrm{R} = \frac{-\nu[\mathrm{P}][\mathrm{Cl^-}]_\mathrm{L}}{[\mathrm{Cl^-}]_\mathrm{L} + [\mathrm{Cl^-}]_\mathrm{R}}\ [8]$$

For simplicity, write $[\mathrm{Cl^-}]_\mathrm{L} = L$, $[\mathrm{Cl^-}]_\mathrm{R} = R$, and $[\mathrm{P}] = P$. Then, since $\nu = 1$,

$$L - R = \frac{-PL}{L+R} \quad \text{implying that} \quad L^2 - R^2 = -PL$$

Suppose an amount n mol of $\mathrm{Cl^-}$ ions migrates from the right-hand compartment to the left, L becomes n M and R changes from 0.030 M to $\left\{\frac{(2) \times (0.030) - n}{2}\right\}$M (since its volume is $2L$). Therefore, at equilibrium

$$n^2 + Pn - \left(\frac{0.060 - n}{2}\right)^2 = 0 \quad \text{with } P = 0.100$$

This quadratic equation solves to $n = 6.7 \times 10^{-3}$; therefore, at equilibrium,

$$[\mathrm{Cl^-}]_\mathrm{L} = \boxed{6.7\,\mathrm{mmol\,L^{-1}}}$$

E23.12 The number of solute molecules with potential energy E is proportional to $\mathrm{e}^{-E/kT}$; hence

$$c \propto N \propto \mathrm{e}^{-E/kT} \quad E = \tfrac{1}{2}m_\mathrm{eff}r^2\omega^2$$

Therefore, $c \propto \mathrm{e}^{Mb\omega^2r^2/2RT}$ $[m_\mathrm{eff} = bm,\ M = mN_\mathrm{A}]$

$$\ln c = \text{const.} + \frac{Mb\omega^2r^2}{2RT} \quad [b = 1 - \rho v_\mathrm{s}]$$

and slope of $\ln c$ against r^2 is equal to $\frac{Mb\omega^2}{2RT}$. Therefore

$$M = \frac{2RT \times \text{slope}}{b\omega^2} = \frac{(2) \times (8.314\,\mathrm{J\,K^{-1}\,mol^{-1}}) \times (300\,\mathrm{K}) \times (729 \times 10^4\,\mathrm{m^{-2}})}{(1 - 0.997 \times 0.61) \times \left(\frac{(2\pi)\times(50000)}{60\,\mathrm{s}}\right)^2}$$

$$= \boxed{3.4 \times 10^3\,\mathrm{kg\,mol^{-1}}}$$

E23.13 The centrifugal force acting is $F = mr\omega^2$, and by Newton's second law of motion, $F = ma$; hence

$$a = r\omega^2 = 4\pi^2 r\nu^2 = 4\pi^2 \times (6.0 \times 10^{-2}\,\mathrm{m}) \times \left(\frac{80 \times 10^3}{60\,\mathrm{s}}\right)^2 = 4.2\bar{1} \times 10^6\,\mathrm{m\,s^{-2}}$$

Then, since $g = 9.81\,\mathrm{m\,s^{-2}}$, $a = \boxed{4.3 \times 10^5\,g}$

E23.14 Washing a cotton shirt disturbs the secondary structure of the cellulose. The water tends to break the hydrogen bonds between the cellulose chains by forming its own hydrogen bonds to the chains. Upon drying, the hydrogen bonds between chains reform, but in a random manner, causing wrinkles. The wrinkles are removed by moistening the shirt, which again breaks the hydrogen bonds, making the fibre more flexible and plastically deformable. A hot iron shapes the cloth and causes the water to evaporate so new hydrogen bonds are formed between the chains while they are held in place by the pressure of the iron.

Solutions to problems

Solutions to numerical problems

P23.1 $$\frac{\Pi}{c} = \frac{RT}{\overline{M}_n}\left(1 + B\frac{c}{\overline{M}_n} + \cdots\right) \quad \text{[Example 23.2, with } \Pi = \rho g h\text{]}$$

Therefore, to determine $\overline{M}_n$ and B we need to plot Π/c against c. We draw up the following table

$c/(\mathrm{g\,L^{-1}})$	1.21	2.72	5.08	6.60
$(\Pi/c)/(\mathrm{Pa/g\,L^{-1}})$	111	118	129	136

The points are plotted in Fig. 23.1.

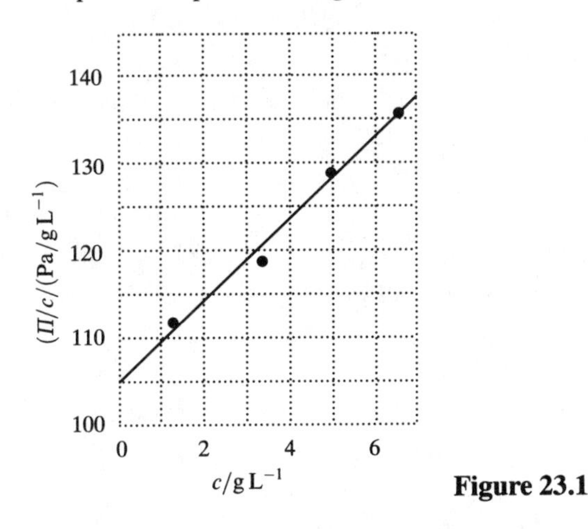

Figure 23.1

A least-squares analysis gives an intercept of $105.\overline{4}$ and a slope of 4.64. It follows that

$$\frac{RT}{\overline{M}_n} = 105.\overline{4}\,\mathrm{Pa\,g^{-1}\,L} = 105.\overline{4}\,\mathrm{Pa\,kg^{-1}\,m^3}$$

and hence that $$\overline{M}_n = \frac{(8.314\,\mathrm{J\,K^{-1}\,mol^{-1}}) \times (293\,\mathrm{K})}{105.\overline{4}\,\mathrm{Pa\,kg^{-1}\,m^3}} = \boxed{23.1\,\mathrm{kg\,mol^{-1}}}$$

The slope of the graph is equal to $\dfrac{RTB}{\overline{M}_n^2}$, so

$$\frac{RTB}{\overline{M}_n^2} = 4.64\,\mathrm{Pa\,g^{-2}\,L^2} = 4.64\,\mathrm{Pa\,kg^{-2}\,m^6}$$

Therefore, $$B = \frac{(23.1\,\mathrm{kg\,mol^{-1}})^2 \times (4.64\,\mathrm{Pa\,kg^{-2}\,m^6})}{(8.314\,\mathrm{J\,K^{-1}\,mol^{-1}}) \times (293\,\mathrm{K})} = \boxed{1.02\,\mathrm{m^3\,mol^{-1}}}$$

P23.3 $[\eta] = \lim_{c \to 0} \left(\dfrac{\eta/\eta^* - 1}{c} \right)$ [19]

We see that the intercept of a plot of the right-hand side against c, extrapolated to $c = 0$, gives $[\eta]$. We begin by constructing the following table using $\eta^* = 0.985\,\mathrm{g\,m^{-1}\,s^{-1}}$

$c/(\mathrm{g\,L^{-1}})$	1.32	2.89	5.73	9.17
$\left(\dfrac{\eta/\eta^* - 1}{c} \right) \Big/ (\mathrm{L\,g^{-1}})$	0.0731	0.0755	0.0771	0.0825

The points are plotted in Fig. 23.2. The least-squares intercept is at 0.0716, so $[\eta] = \boxed{0.0716\,\mathrm{L\,g^{-1}}}$

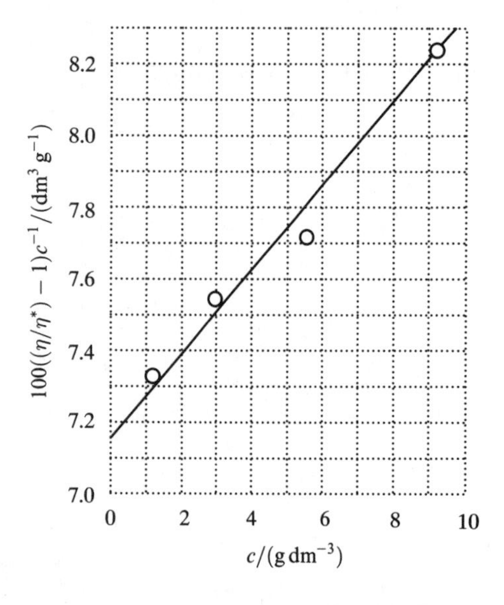

Figure 23.2

P23.4 $S = \dfrac{s}{r\omega^2}$ [11]

Since $s = \dfrac{\mathrm{d}r}{\mathrm{d}t}$, $\dfrac{s}{r} = \dfrac{1}{r}\dfrac{\mathrm{d}r}{\mathrm{d}t} = \dfrac{\mathrm{d}\ln r}{\mathrm{d}t}$

and if we plot $\ln r$ against t, the slope gives S through

$$S = \frac{1}{\omega^2}\frac{\mathrm{d}\ln r}{\mathrm{d}t}$$

The data are as follows

t/min	15.5	29.1	36.4	58.2
r/cm	5.05	5.09	5.12	5.19
$\ln(r/\mathrm{cm})$	1.619	1.627	1.633	1.647

The points are plotted in Fig. 23.3.

The least-squares slope is 6.62×10^{-4}, so

$$S = \frac{6.62 \times 10^{-4}\,\mathrm{min^{-1}}}{\omega^2} = \frac{(6.62 \times 10^{-4}) \times \left(\frac{1}{60}\right)\mathrm{s^{-1}}}{\left(2\pi \times \frac{4.5 \times 10^4}{60\,\mathrm{s}}\right)^2} = 4.9\bar{7} \times 10^{-13}\,\mathrm{s} \quad \text{or} \quad \boxed{5.0\,\mathrm{Sv}}$$

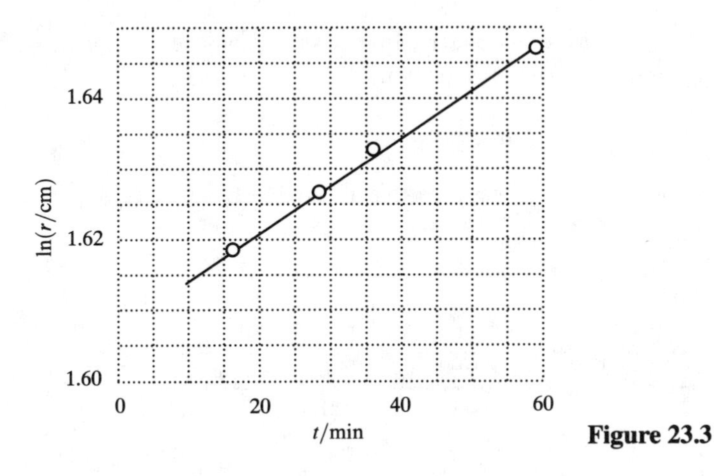

Figure 23.3

P23.7 $$[\mathrm{Na}^+]_\mathrm{L} - [\mathrm{Na}^+]_\mathrm{R} = \frac{\nu[\mathrm{P}][\mathrm{Na}^+]_\mathrm{L}}{[\mathrm{Na}^+]_\mathrm{L} + [\mathrm{Na}^+]_\mathrm{R}} \ [6]$$

Therefore, writing $[\mathrm{Na}^+]_\mathrm{L} = L$ and $[\mathrm{Na}^+]_\mathrm{R} = R$, and setting $\nu = 2$

$$(L + R) \times (L - R) = 2[\mathrm{P}]L$$

Suppose an amount $2n$ mol Na^+ migrate from the left- to the right-hand compartments to reach equilibrium, then L changes from $(0.030+0.010)$ M to $(0.040-n)$ M and R changes from 0.0050 M to $(0.0050 + n)$ M. We must therefore solve

$$(0.045) \times (0.035 - 2n) = (0.030) \times (0.040 - n)$$

which gives $n = 6.25 \times 10^{-3}$. Therefore, the concentration of Na^+ ions at equilibrium are $L =$ 0.034 M, $R = 0.011$ M. The potential difference across the membrane is therefore

$$E = \frac{RT}{F} \ln \frac{R}{L} = \frac{(8.314\,\mathrm{J\,K^{-1}\,mol^{-1}}) \times (300\,\mathrm{K})}{96.485\,\mathrm{kC\,mol^{-1}}} \ln \frac{0.011}{0.034} = \boxed{-29\,\mathrm{mV}}$$

P23.8 $$\Pi = RT[\mathrm{P}](1 + B[\mathrm{P}])$$

$$B = \frac{\nu^2[\mathrm{Cl}^-]_\mathrm{R}}{4[\mathrm{Cl}^-]^2 + 2\nu[\mathrm{Cl}^-]_\mathrm{L}[\mathrm{P}]} \ [7]$$

When $4[\mathrm{Cl}^-] \gg \nu[\mathrm{P}]$ as is assumed here, the expression for B reduces to

$$B = \frac{\nu^2[\mathrm{Cl}^-]_\mathrm{R}}{4[\mathrm{Cl}^-]^2} \approx \frac{\nu^2}{4[\mathrm{Cl}^-]}$$

$$B = \frac{400}{(4) \times (0.020\,\mathrm{M})} = 5 \times 10^3\,\mathrm{L\,mol^{-1}} = \boxed{5\,\mathrm{m^3\,mol^{-1}}}$$

This value of B is comparable to the values calculated for non-electrolyte solutions (Example 23.2), and so the two effects are comparable in this case.

P23.10 We need to determine the intrinsic viscosity from a plot of $\dfrac{\left(\frac{\eta}{\eta^*}\right) - 1}{c/(\mathrm{g\,L^{-1}})}$ against c, extrapolated to $c = 0$ as in Example 23.5. Then from the relation

$$[\eta] = K\overline{M}_\mathrm{V}^a \ [21]$$

with K and a from Table 23.3, the viscosity average molar mass $\overline{M}_V$ may be calculated. η/η^* values are determined from the times of flow using the relation

$$\frac{\eta}{\eta^*} = \frac{t}{t^*} \times \frac{\rho}{\rho^*} \approx \frac{t}{t^*}$$

noting that in the limit as $c \to 0$ it becomes exact. As explained in Example 23.5, $[\eta]$ can also be determined from the limit of $\frac{1}{c} \ln\left(\frac{\eta}{\eta^*}\right)$ as $c \to 0$.

We draw up the following table

$c/(\text{g L}^{-1})$	0.000	2.22	5.00	8.00	10.00
t/s	208.2	248.1	303.4	371.8	421.3
$\frac{\eta}{\eta^*}$	—	1.192	1.457	1.786	2.024
$\frac{100\left[\left(\frac{\eta}{\eta^*}\right)-1\right]}{c/(\text{g L}^{-1})}$	—	8.63	9.15	9.82	10.24
$\ln\left(\frac{\eta}{\eta^*}\right)$	—	0.1753	0.3766	0.5799	0.7048
$\frac{100\ln\left(\frac{\eta}{\eta^*}\right)}{c/(\text{g L}^{-1})}$	—	7.89	7.52	7.24	7.05

The points are plotted in Fig. 23.4.

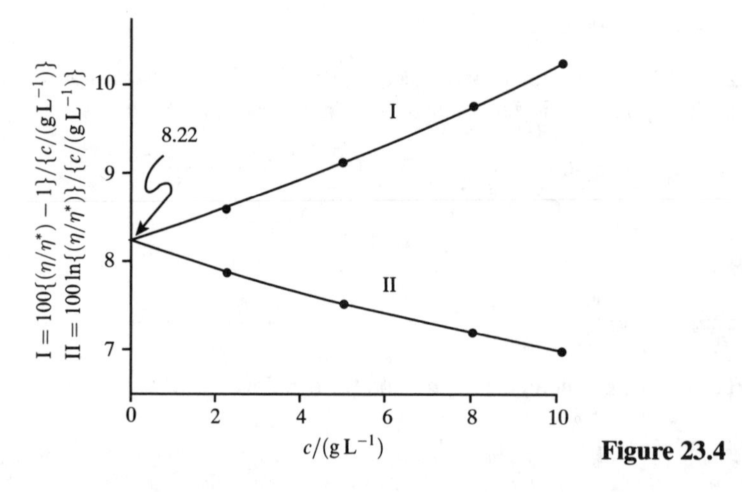

Figure 23.4

The intercept as determined from the simultaneous extrapolation of both plots is $0.0822\,\text{L g}^{-1}$.

$$\overline{M}_V = \left(\frac{[\eta]}{K}\right)^{1/a} = \left(\frac{0.0822\,\text{L g}^{-1}}{9.5 \times 10^{-6}\,\text{L g}^{-1}}\right)^{1/0.74} = \boxed{2.1 \times 10^5\,\text{g mol}^{-1}}$$

Comment. This value differs markedly in molar mass from the sample of polystyrene in toluene described in Example 23.5.

P23.12 We follow the procedure of Example 23.5. Also compare to Problems 23.3 and 23.10.

$$[\eta] = \lim_{c\to 0}\left(\frac{\eta/\eta^* - 1}{c}\right) \quad \text{and} \quad [\eta] = K\overline{M}_V^a \quad [\text{with } K \text{ and } a \text{ from Table 23.3}]$$

We draw up the following table using $\eta^* = 0.647 \times 10^{-3}\,\mathrm{kg\,m^{-1}\,s^{-1}}$

$c/(\mathrm{g}/100\,\mathrm{cm}^3)$	0	0.2	0.4	0.6	0.8	1.0
$\eta/(10^{-3}\,\mathrm{kg\,m^{-1}\,s^{-1}})$	0.647	0.690	0.733	0.777	0.821	0.865
$\left(\dfrac{\eta/\eta^* - 1}{c}\right)\Big/(100\,\mathrm{cm^3\,g^{-1}})$		0.333	0.332	0.335	0.336	0.337

The values are plotted in Fig. 23.5, and extrapolated to 0.330.

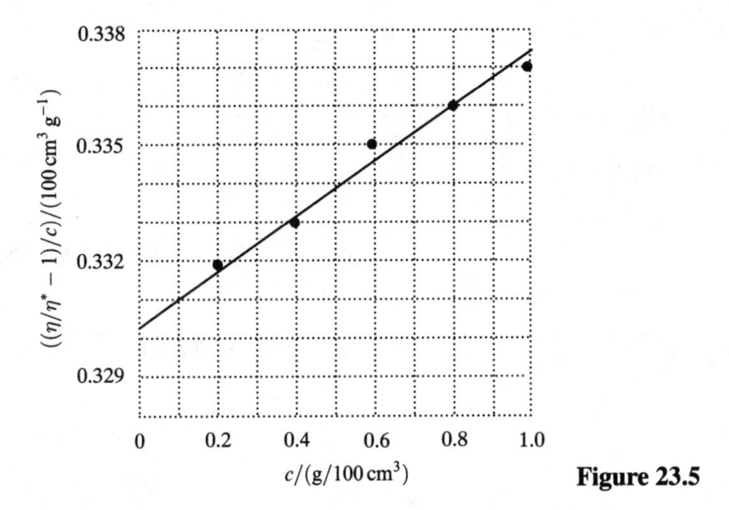

Figure 23.5

Hence $[\eta] = (0.330) \times (100\,\mathrm{cm^3\,g^{-1}}) = 33.0\,\mathrm{cm^3\,g^{-1}}$

$$\text{and } M_V = \left(\frac{33.0\,\mathrm{cm^3\,g^{-1}}}{8.3 \times 10^{-2}\,\mathrm{cm^3\,g^{-1}}}\right)^{1/0.50} = 158 \times 10^3$$

That is, $M = \boxed{158\,\mathrm{kg\,mol^{-1}}}$

P23.14 Assume the solute particles are solid spheres and see how well R_g calculated on the basis of that assumption agrees with experimental values.

$$R_g = (0.05690) \times \{(v_s/\mathrm{cm^3\,g^{-1}}) \times (M/\mathrm{g\,mol^{-1}})\}^{1/3}\,\mathrm{nm} \quad \text{[Problem 23.13]}$$

and draw up the following table

	$M/(\mathrm{g\,mol^{-1}})$	$v_s/(\mathrm{cm^3\,g^{-1}})$	$(R_g/\mathrm{nm})_{\mathrm{calc}}$	$(R_g/\mathrm{nm})_{\mathrm{expt}}$
Serum albumin	66×10^3	0.752	2.09	2.98
Bushy stunt virus	10.6×10^6	0.741	11.3	12.0
DNA	4×10^6	0.556	7.43	117.0

Therefore, serum albumin and bushy stunt virus resemble solid spheres, but DNA does not.

P23.16 $$\rho = \frac{m(\text{unit cell})}{V(\text{unit cell})} = \frac{(2)\times\frac{M(\mathrm{CH_2CH_2})}{N_\mathrm{A}}}{abc}$$
$$= \frac{(2)\times(28.05\,\mathrm{g\,mol^{-1}})}{(6.022\times10^{23}\,\mathrm{mol^{-1}})\times[(740\times493\times253)\times10^{-36}]\,\mathrm{m^3}}$$
$$= 1.01\times10^{6}\,\mathrm{g\,m^{-3}} = \boxed{1.01\,\mathrm{g\,cm^{-3}}}$$

Solutions to theoretical problems

P23.18 $G = U - TS - tl$ [given]

Hence $\mathrm{d}G = \mathrm{d}U - T\,\mathrm{d}S - S\,\mathrm{d}T - l\,\mathrm{d}t - t\,\mathrm{d}l = T\,\mathrm{d}S + t\,\mathrm{d}l - T\,\mathrm{d}S - S\,\mathrm{d}T - l\,\mathrm{d}t - t\,\mathrm{d}l = \boxed{-S\,\mathrm{d}T - l\,\mathrm{d}t}$

$A = U - TS = G + tl$

Hence $\mathrm{d}A = \mathrm{d}G + t\,\mathrm{d}l + l\,\mathrm{d}t = -S\,\mathrm{d}T - l\,\mathrm{d}t + t\,\mathrm{d}l + l\,\mathrm{d}t = \boxed{-S\,\mathrm{d}T + t\,\mathrm{d}l}$

Since $\mathrm{d}G$ and $\mathrm{d}A$ are both exact differentials

$$\left(\frac{\partial S}{\partial l}\right)_T = -\left(\frac{\partial t}{\partial T}\right)_l \quad \text{and} \quad \left(\frac{\partial S}{\partial t}\right)_T = \left(\frac{\partial l}{\partial T}\right)_t$$

Since $\mathrm{d}U = T\,\mathrm{d}S + t\,\mathrm{d}l$ [given],

$$\left(\frac{\partial U}{\partial l}\right)_T = T\left(\frac{\partial S}{\partial l}\right)_T + t = \boxed{-T\left(\frac{\partial t}{\partial T}\right)_l + t} \quad \text{[Maxwell relation, above]}$$

P23.21 **(a)** $$R_\mathrm{rms}^2 = \int_0^\infty R^2 f\,\mathrm{d}R \;[36]$$

$$f = 4\pi\left(\frac{a}{\pi^{1/2}}\right)^3 R^2 \mathrm{e}^{-a^2R^2}\;[34], \quad a = \left(\frac{3}{2Nl^2}\right)^{1/2}$$

Therefore,

$$R_\mathrm{rms}^2 = 4\pi\left(\frac{a}{\pi^{1/2}}\right)^3\int_0^\infty R^4\mathrm{e}^{-a^2R^2}\,\mathrm{d}R = 4\pi\left(\frac{a}{\pi^{1/2}}\right)^3\times\left(\frac{3}{8}\right)\times\left(\frac{\pi}{a^{10}}\right)^{1/2} = \frac{3}{2a^2} = Nl^2$$

Hence, $R_\mathrm{rms} = \boxed{lN^{1/2}}$

(b) $$R_\mathrm{mean} = \int_0^\infty Rf\,\mathrm{d}r = 4\pi\left(\frac{a}{\pi^{1/2}}\right)^3\int_0^\infty R^3\mathrm{e}^{-a^2R^2}\,\mathrm{d}R$$
$$= 4\pi\left(\frac{a}{\pi^{1/2}}\right)^3\times\left(\frac{1}{2a^4}\right) = \frac{2}{a\pi^{1/2}} = \boxed{\left(\frac{8N}{3\pi}\right)^{1/2} l}$$

(c) Set $\dfrac{\mathrm{d}f}{\mathrm{d}R} = 0$ and solve for R

$$\frac{\mathrm{d}f}{\mathrm{d}R} = 4\pi\left(\frac{a}{\pi^{1/2}}\right)^3\{2R - 2a^2R^3\}\mathrm{e}^{-a^2R^2} = 0 \quad \text{when } a^2R^2 = 1$$

Therefore, the most probable separation is

$$R^* = \frac{1}{a} = \boxed{l\left(\frac{2}{3}N\right)^{1/2}}$$

When $N = 4000$ and $l = 154\,\mathrm{pm}$,

(a) $R_\mathrm{rms} = \boxed{9.74\,\mathrm{nm}}$ **(b)** $R_\mathrm{mean} = \boxed{8.97\,\mathrm{nm}}$ **(c)** $R^* = \boxed{7.95\,\mathrm{nm}}$

Solutions to additional problems

P23.22 The relationship [21] between $[\eta]$ and $\overline{M}_V$ can be transformed into a linear one

$$\ln[\eta] = \ln K + a \ln M_V$$

so a plot of $\ln[\eta]$ versus $\ln M_V$ will have a slope of a and a y-intercept of $\ln K$. The transformed data and plot are shown below (Fig. 23.6)

$\overline{M}_V/(\mathrm{kg\,mol^{-1}})$	10.0	19.8	106	249	359	860	1800	5470	9720	56 800
$[\eta]/(\mathrm{cm^3\,g^{-1}})$	8.90	11.9	28.1	44.0	51.2	77.6	113.9	195	275	667
$\ln \overline{M}_V/(\mathrm{kg\,mol^{-1}})$	2.30	2.99	4.66	5.52	5.88	6.76	7.50	8.61	9.18	10.9
$\ln[\eta]/(\mathrm{cm^3\,g^{-1}})$	2.19	2.48	3.34	3.78	3.94	4.35	4.74	5.27	5.62	6.50

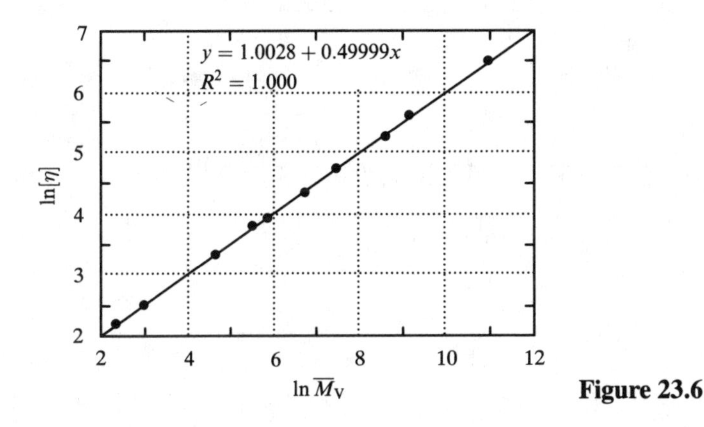

Figure 23.6

Thus $a = \boxed{0.500}$ and $K = \mathrm{e}^{1.0028}\,\mathrm{cm^3\,g^{-1}\,kg^{-1/2}\,mol^{1/2}} = \boxed{2.73\,\mathrm{cm^3\,g^{-1}\,kg^{-1/2}\,mol^{1/2}}}$

Solving for $\overline{M}_V$ yields

$$\overline{M}_V = \left(\frac{[\eta]}{K}\right)^{1/a} = \left(\frac{100\,\mathrm{cm^3\,g^{-1}}}{2.73\,\mathrm{cm^3\,g^{-1}\,kg^{-1/2}\,mol^{1/2}}}\right)^2 = \boxed{1.34\times10^3\,\mathrm{kg\,mol^{-1}}}$$

P23.25 We use eqn 7.38 in the form given in Example 23.2 with $\Pi = \rho g h$, then

$$\frac{\Pi}{c} = \frac{RT}{\overline{M}_n}\left(1 + \frac{B}{\overline{M}_n}c\right) = \frac{RT}{\overline{M}_n} + \frac{RTB}{\overline{M}_n^2}c$$

where c is the mass concentration of the polymer. Therefore plot Π/c against c. The intercept gives $RT/\overline{M}_n$ and the slope gives $RTB/\overline{M}_n^2$.

The transformed data to plot are given in the table

$c/(\mathrm{mg\,cm^{-3}})$	1.33	2.10	4.52	7.18	9.87
$(\Pi/c)/(\mathrm{N\,m^{-2}\,mg^{-1}\,cm^3})$	$22.5\overline{6}$	$24.2\overline{9}$	$29.2\overline{0}$	$34.2\overline{6}$	$39.5\overline{1}$

The plot is shown in Fig. 23.7. The intercept is $20.0\overline{9}\,\mathrm{N\,m^{-2}/(mg\,cm^{-3})}$. The slope is $1.974\,\mathrm{N\,m^{-2}/(mg\,cm^{-3})^2}$. Therefore

$$\overline{M}_n = \frac{RT}{20.0\overline{9}\,\mathrm{N\,m^{-2}/(mg\,cm^{-3})}}$$

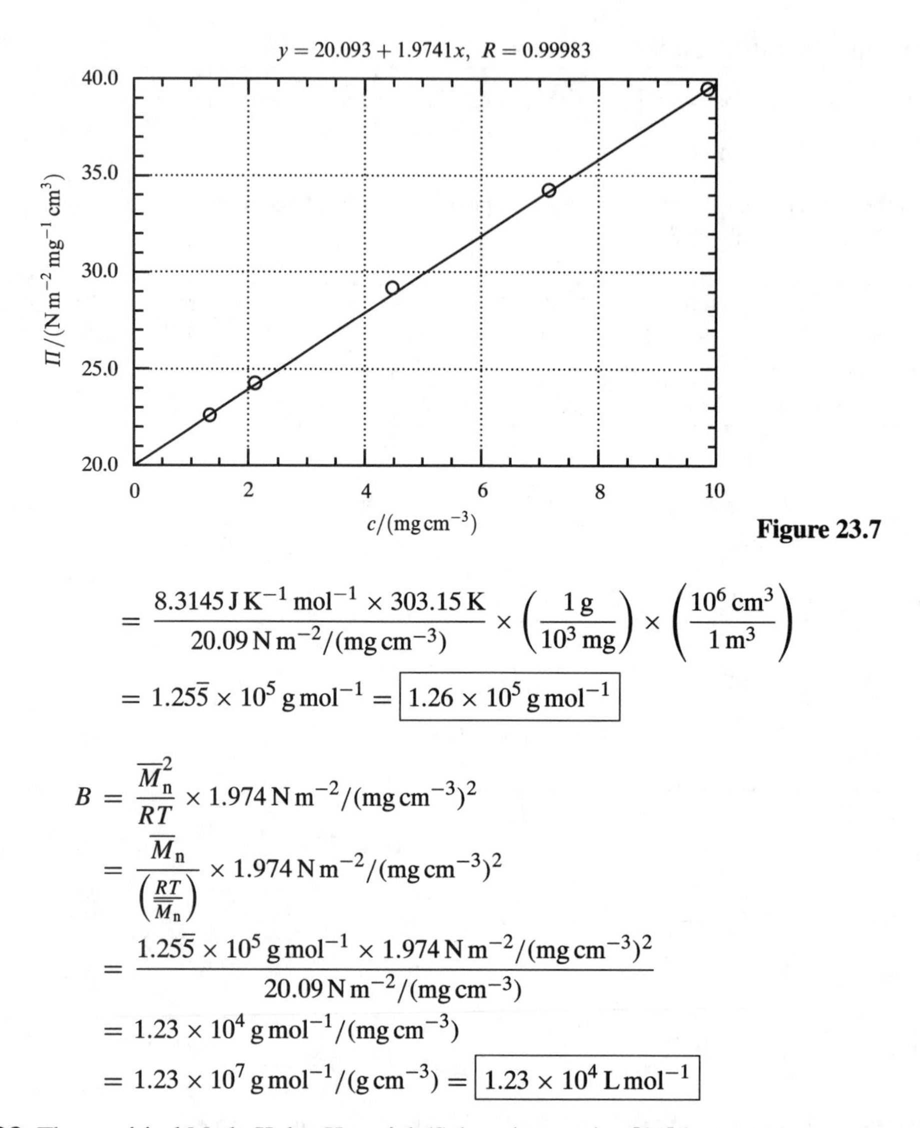

Figure 23.7

$$= \frac{8.3145\,\mathrm{J\,K^{-1}\,mol^{-1}} \times 303.15\,\mathrm{K}}{20.09\,\mathrm{N\,m^{-2}/(mg\,cm^{-3})}} \times \left(\frac{1\,\mathrm{g}}{10^3\,\mathrm{mg}}\right) \times \left(\frac{10^6\,\mathrm{cm^3}}{1\,\mathrm{m^3}}\right)$$

$$= 1.25\bar{5} \times 10^5\,\mathrm{g\,mol^{-1}} = \boxed{1.26 \times 10^5\,\mathrm{g\,mol^{-1}}}$$

$$\begin{aligned} B &= \frac{\overline{M}_\mathrm{n}^2}{RT} \times 1.974\,\mathrm{N\,m^{-2}/(mg\,cm^{-3})^2} \\ &= \frac{\overline{M}_\mathrm{n}}{\left(\frac{RT}{\overline{M}_\mathrm{n}}\right)} \times 1.974\,\mathrm{N\,m^{-2}/(mg\,cm^{-3})^2} \\ &= \frac{1.25\bar{5} \times 10^5\,\mathrm{g\,mol^{-1}} \times 1.974\,\mathrm{N\,m^{-2}/(mg\,cm^{-3})^2}}{20.09\,\mathrm{N\,m^{-2}/(mg\,cm^{-3})}} \\ &= 1.23 \times 10^4\,\mathrm{g\,mol^{-1}/(mg\,cm^{-3})} \\ &= 1.23 \times 10^7\,\mathrm{g\,mol^{-1}/(g\,cm^{-3})} = \boxed{1.23 \times 10^4\,\mathrm{L\,mol^{-1}}} \end{aligned}$$

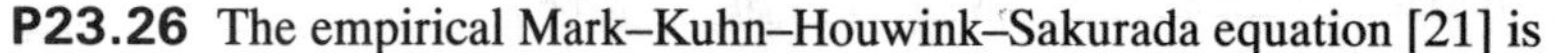

P23.26 The empirical Mark–Kuhn–Houwink–Sakurada equation [21] is

$$[\eta] = K\overline{M}_\mathrm{V}^a$$

As the constant a may be non-integral the molar mass here is to be interpreted as unitless, that is, as $\overline{M}_\mathrm{V}/(\mathrm{g\,mol^{-1}})$. The units of K are then the same as $[\eta]$.

We fit the data to the above equation and obtain K and a from the fitting procedure. The plot is shown in Fig. 23.8.

$$\boxed{K = 0.0117\,\mathrm{cm^3\,g^{-1}}}$$

$$\boxed{a = 0.717}$$

This value for a is not much different from that for polystryene in benzene listed in Table 23.3. This is somewhat surprising as one would expect both the K and a values to be solvent-dependent. THF is not chemically similar to benzene. On the other hand, benzene and toluene are very much alike, yet the values of K and a as determined in Example 23.5 are markedly different from those in Table 23.3 for polystyrene in benzene.

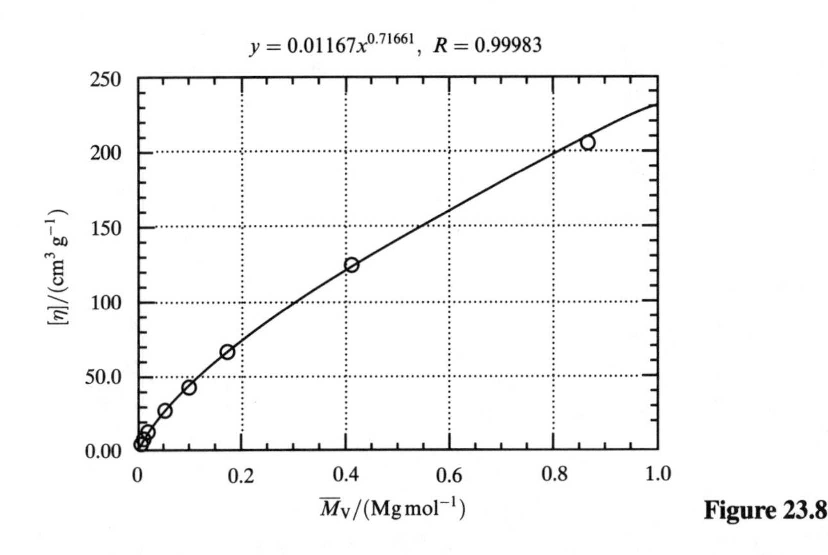

Figure 23.8

P23.27 The procedure is that described in Problem 23.26. The data are fitted to the Mark–Kuhn–Houwink–Sakurada equation (Fig. 23.9)

$$[\eta] = K\overline{M}_{\mathrm{V}}^{a}$$

The values obtained for the parameters are

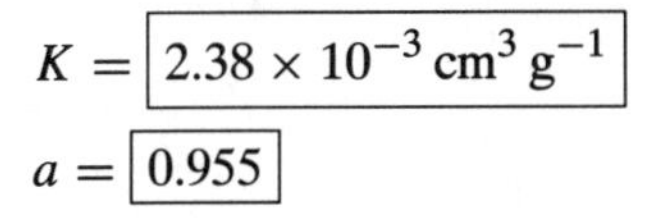

$$K = \boxed{2.38 \times 10^{-3}\,\mathrm{cm^3\,g^{-1}}}$$

$$a = \boxed{0.955}$$

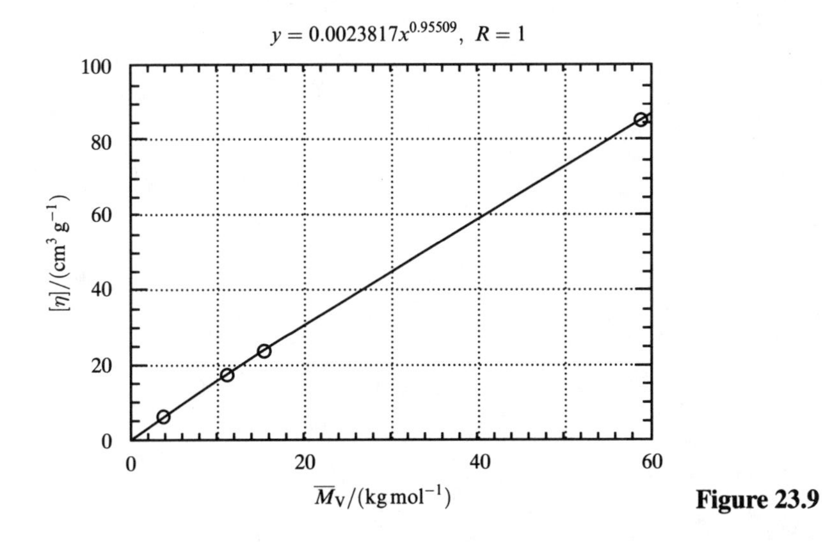

Figure 23.9

This K value is smaller than any in Table 23.3 or that in Problem 23.26. The value for a is quite close to 1. When $a = 1$ exactly, the molar mass, $\overline{M}_{\mathrm{V}}$ corresponds to the weight average molar mass, $\overline{M}_{w}$.

The magnitude of the constant a reflects the stiffness of the polymer chain as a result of π-orbital interactions between heterocyclic rings.

Part 3: Change

24 Molecules in motion

Solutions to exercises

E24.1 We first calculate, Z_W, the number of collisions per unit area per unit time; the number of collisions is then Z_W multiplied by the area of the surface and the time.

$$Z_W = \frac{p}{(2\pi mkT)^{1/2}} \,[3]$$

$$= \frac{90\,\text{Pa}}{[(2\pi) \times (39.95) \times (1.6605 \times 10^{-27}\,\text{kg}) \times (1.381 \times 10^{-23}\,\text{J K}^{-1}) \times (500\,\text{K})]^{1/2}}$$

$$= 1.7 \times 10^{24}\,\text{m}^{-2}\,\text{s}^{-1}$$

Therefore, the number of collisions is

$$N = (1.7 \times 10^{24}\,\text{m}^{-2}\,\text{s}^{-1}) \times (2.5 \times 3.0 \times 10^{-6}\,\text{m}^2) \times (15\,\text{s}) = \boxed{1.9 \times 10^{20}}$$

Comment. Equation 3 in the form $p = Z_W(2\pi mkT)^{1/2}$ is considered the molecular explanation of pressure and is used in this form in Example 24.2.

Question. How many collisions are there per second on the walls of a room with dimensions 3 m × 5 m × 5 m with 'air' molecules at 25°C and 1.00 atm?

E24.2

$$\Delta m = Z_W A_0 m \Delta t \text{ [Example 24.2]} = \frac{pA_0 m\Delta t}{(2\pi mkT)^{1/2}} = pA_0\Delta t\left(\frac{m}{2\pi kT}\right)^{1/2}$$

$$= pA_0\Delta t\left(\frac{M}{2\pi RT}\right)^{1/2}$$

From the data, with $A_0 = \pi r^2$,

$$\Delta m = (0.835\,\text{Pa}) \times (\pi) \times (1.25 \times 10^{-3}\,\text{m})^2 \times (7.20 \times 10^3\,\text{s}) \times \left(\frac{260 \times 10^{-3}\,\text{kg mol}^{-1}}{(2\pi) \times (8.314\,\text{J K}^{-1}\,\text{mol}^{-1}) \times (400\,\text{K})}\right)^{1/2}$$

$$= 1.04 \times 10^{-4}\,\text{kg}, \quad \text{or} \quad \boxed{104\,\text{mg}}$$

Question. For the same solid shaped in the form of a sphere of radius 0.050 m and suspended in a vacuum, what will be the mass loss in 2.00 h? *Hint*. Make any reasonable approximations.

E24.3

$$J_z = -\kappa\frac{dT}{dz}[9] = \left(\frac{-0.163\,\text{mJ cm}^{-2}\,\text{s}^{-1}}{\text{K cm}^{-1}}\right) \times (-2.5\,\text{K m}^{-1}) \text{ [Table 24.1]}$$

$$= (0.41\,\text{mJ cm}^{-2}\,\text{s}^{-1}) \times (\text{cm/m}) = 0.41 \times 10^{-2}\,\text{mJ cm}^{-2}\,\text{s}^{-1} = \boxed{4.1 \times 10^{-2}\,\text{J m}^{-2}\,\text{s}^{-1}}$$

E24.4 The thermal conductivity, κ, is a function of the mean free path, λ, which in turn is a function of the collision cross-section, σ. Hence, reversing the order, σ can be obtained from κ.

$$\kappa = \frac{1}{3}\lambda\bar{c}C_{V,\text{m}}[\text{A}] \,[16]$$

$$\bar{c} = \left(\frac{8RT}{\pi M}\right)^{1/2} [2] \quad \text{and} \quad \lambda = \frac{kT}{2^{1/2}\sigma p}[1] = \frac{V}{2^{1/2}\sigma nN_A} = \frac{1}{2^{1/2}\sigma N_A[\text{A}]}$$

Hence,

$$[\mathrm{A}]\lambda\bar{c} = \left(\frac{8RT}{\pi M}\right)^{1/2} \times \left(\frac{1}{2^{1/2}\sigma N_{\mathrm{A}}}\right) = \left(\frac{4RT}{\pi M}\right)^{1/2} \times \left(\frac{1}{\sigma N_{\mathrm{A}}}\right)$$

and so

$$\kappa = \left(\frac{1}{3\sigma N_{\mathrm{A}}}\right) \times \left(\frac{4RT}{\pi M}\right)^{1/2} C_{V,\mathrm{m}} = \left(\frac{1}{3\sigma N_{\mathrm{A}}}\right) \times \left(\frac{4RT}{\pi M}\right)^{1/2}$$

$$\times \frac{3}{2}R\left[C_{V,\mathrm{m}} = \tfrac{3}{2}R\right] = \left(\frac{k}{2\sigma}\right) \times \left(\frac{4RT}{\pi M}\right)^{1/2}$$

$$\sigma = \left(\frac{k}{2\kappa}\right) \times \left(\frac{4RT}{\pi M}\right)^{1/2} = \left(\frac{1.318 \times 10^{-23}\,\mathrm{J\,K^{-1}}}{(2) \times (0.0465\,\mathrm{J\,s^{-1}\,K^{-1}\,m^{-1}})}\right)$$

$$\times \left(\frac{(4) \times (8.314\,\mathrm{J\,K^{-1}\,mol^{-1}}) \times (273\,\mathrm{K})}{(\pi) \times (20.2 \times 10^{-3}\,\mathrm{kg\,mol^{-1}})}\right)^{1/2}$$

$$= \boxed{5.6 \times 10^{-20}\,\mathrm{m^2}} \quad \text{or} \quad 0.056\,\mathrm{nm^2}$$

The experimental value is $0.24\,\mathrm{nm^2}$.

Question. What approximations inherent in the equations used in the solution to this exercise are the likely cause of the factor of 4 difference between the experimental and calculated values of the collision cross-section for neon?

E24.5 $J_z(\text{energy}) = -\kappa\dfrac{\mathrm{d}T}{\mathrm{d}z}$ [9]

This is the rate of energy transfer per unit area. For an area A

Rate of energy transfer: $\dfrac{\mathrm{d}E}{\mathrm{d}t} = AJ_z = \kappa A\dfrac{\mathrm{d}T}{\mathrm{d}z}$

Therefore, with $\kappa \approx 0.241\,\mathrm{mJ\,cm^{-2}\,s^{-1}/(K\,cm^{-1})}$ [Table 24.1]

$$\frac{\mathrm{d}E}{\mathrm{d}t} \approx \left(\frac{0.241\,\mathrm{mJ\,cm^{-2}\,s^{-1}}}{\mathrm{K\,cm^{-1}}}\right) \times (1.0 \times 10^4\,\mathrm{cm^2}) \times \left(\frac{35\,\mathrm{K}}{5.0\,\mathrm{cm}}\right)$$

$$\approx 17 \times 10^3\,\mathrm{mJ\,s^{-1}} = 17\,\mathrm{J\,s^{-1}}, \quad \text{or} \quad \boxed{17\,\mathrm{W}}$$

Therefore, a $\boxed{17\,\mathrm{W}}$ heater is required.

E24.6 The pressure change follows the equation

$$p = p_0\mathrm{e}^{-t/\tau}, \quad \tau = \left(\frac{2\pi m}{kT}\right)^{1/2} \times \left(\frac{V}{A_0}\right) \text{ [Example 24.1]}$$

Therefore, the time required for the pressure to fall from p_0 to p is

$$t = \tau \ln\frac{p_0}{p}$$

Consequently for two different gases at the same initial and final pressures

$$\frac{t'}{t} = \frac{\tau'}{\tau} = \left(\frac{M'}{M}\right)^{1/2}$$

and hence $M' = \left(\frac{t'}{t}\right)^2 \times M = \left(\frac{52}{42}\right)^2 \times (28.02\,\mathrm{g\,mol^{-1}}) = \boxed{43\,\mathrm{g\,mol^{-1}}}$

Comment. The actual value of CO_2 is $44.01\,\mathrm{g\,mol^{-1}}$

E24.7 $t = \tau \ln \frac{p_0}{p}, \quad \tau = \left(\frac{2\pi m}{kT}\right)^{1/2} \times \left(\frac{V}{A_0}\right)$ [Example 24.1 and Exercise 24.6(a)]

$$\text{Since } \tau = \left(\frac{2\pi M}{RT}\right)^{1/2} \times \left(\frac{V}{A_0}\right) = \left(\frac{(2\pi) \times (32.0 \times 10^{-3}\,\mathrm{kg\,mol^{-1}})}{(8.314\,\mathrm{J\,K^{-1}\,mol^{-1}}) \times (298\,\mathrm{K})}\right)^{1/2}$$

$$\times \left(\frac{3.0\,\mathrm{m^3}}{\pi \times (1.0 \times 10^{-4}\,\mathrm{m})^2}\right) = 8.6 \times 10^5\,\mathrm{s}$$

we find that $t = (8.6 \times 10^5) \times \ln\left(\frac{0.80}{0.70}\right) = \boxed{1.1 \times 10^5\,\mathrm{s}}$ (30 h)

E24.8 $\eta = \frac{1}{3} m\lambda\bar{c} N_A[\mathrm{A}] \quad [21,\ M = mN_A]$

$$\lambda\bar{c}[\mathrm{A}] = \left(\frac{4RT}{\pi M}\right)^{1/2} \times \left(\frac{1}{\sigma N_A}\right) \text{ [Exercise 24.4(a)]}$$

$$\text{Therefore, } \eta = \left(\frac{m}{3\sigma}\right) \times \left(\frac{4RT}{\pi M}\right)^{1/2}$$

$$\text{and } \sigma = \left(\frac{m}{3\eta}\right) \times \left(\frac{4RT}{\pi M}\right)^{1/2}$$

$$= \left(\frac{(20.2) \times (1.6605 \times 10^{-27}\,\mathrm{kg})}{(3) \times (2.98 \times 10^{-5}\,\mathrm{kg\,m^{-1}\,s^{-1}})}\right) \times \left(\frac{(4) \times (8.314\,\mathrm{J\,K^{-1}\,mol^{-1}}) \times (273\,\mathrm{K})}{\pi \times (20.2 \times 10^{-3}\,\mathrm{kg\,mol^{-1}})}\right)^{1/2}$$

$$= \boxed{1.42 \times 10^{-19}\,\mathrm{m^2}} \quad \text{or} \quad 0.142\,\mathrm{nm^2}$$

E24.9 $\frac{\mathrm{d}V}{\mathrm{d}t} = \frac{(p_1^2 - p_2^2)\pi r^4}{16 l\eta p_0}$ [23]

which rearranges to

$$p_1^2 = p_2^2 + \left(\frac{16 l\eta p_0}{\pi r^4}\right) \times \left(\frac{\mathrm{d}V}{\mathrm{d}t}\right)$$

$$= p_2^2 + \left(\frac{(16) \times (8.50\,\mathrm{m}) \times (1.76 \times 10^{-5}\,\mathrm{kg\,m^{-1}\,s^{-1}}) \times (1.00 \times 10^5\,\mathrm{Pa})}{\pi \times (5.0 \times 10^{-3}\,\mathrm{m})^4}\right) \times \left(\frac{9.5 \times 10^2\,\mathrm{m^3}}{3600\,\mathrm{s}}\right)$$

$$= p_2^2 + (3.22 \times 10^{10}\,\mathrm{Pa^2}) = (1.00 \times 10^5)^2\,\mathrm{Pa^2} + (3.22 \times 10^{10}\,\mathrm{Pa^2}) = 4.22 \times 10^{10}\,\mathrm{Pa^2}$$

Hence, $p_1 = \boxed{205\,\mathrm{kPa}}$ (2.05 bar)

E24.10 $\eta = \frac{1}{3} m\lambda\bar{c} N_A[\mathrm{A}]$ [21, with $M = mN_A$] $= \left(\frac{m}{3\sigma}\right) \times \left(\frac{4RT}{\pi M}\right)^{1/2}$

$$= \left(\frac{(29) \times (1.6605 \times 10^{-27}\,\mathrm{kg})}{(3) \times (0.40 \times 10^{-18}\,\mathrm{m^2})}\right) \times \left(\frac{(4) \times (8.314\,\mathrm{J\,K^{-1}\,mol^{-1}}) \times T}{\pi \times (29 \times 10^{-3}\,\mathrm{kg\,mol^{-1}})}\right)^{1/2}$$

$$= (7.7 \times 10^{-7}\,\mathrm{kg\,m^{-1}\,s^{-1}}) \times (T/\mathrm{K})^{1/2}$$

(a) At $T = 273\,\text{K}$, $\eta = 1.3 \times 10^{-5}\,\text{kg m}^{-1}\,\text{s}^{-1}$, or $\boxed{130\,\mu\text{P}}$

(b) At $T = 298\,\text{K}$, $\eta = \boxed{130\,\mu\text{P}}$ **(c)** At $T = 1000\,\text{K}$, $\eta = \boxed{240\,\mu\text{P}}$

E24.11 $$\kappa = \frac{1}{3}\lambda\bar{c}C_{V,\text{m}}[\text{A}]\ [16] = \left(\frac{k}{2\sigma}\right) \times \left(\frac{4RT}{\pi M}\right)^{1/2} \quad [\text{Exercise 24.4(a)}]$$

$$= \left(\frac{1.381 \times 10^{-23}\,\text{J K}^{-1}}{(2) \times (\sigma/\text{nm}^2) \times 10^{-18}\,\text{m}^{-2}}\right) \times \left(\frac{(4) \times (8.314\,\text{J K}^{-1}\,\text{mol}^{-1}) \times (300\,\text{K})}{\pi \times (M/\text{g mol}^{-1}) \times 10^{-3}\,\text{kg mol}^{-1}}\right)^{1/2}$$

$$= \frac{1.23 \times 10^{-2}\,\text{J K}^{-1}\,\text{m}^{-1}\,\text{s}^{-1}}{(\sigma/\text{nm}^2) \times (M/\text{g mol}^{-1})^{1/2}}$$

(a) For Ar, $$\kappa = \frac{1.23 \times 10^{-2}\,\text{J K}^{-1}\,\text{m}^{-1}\,\text{s}^{-1}}{(0.36) \times (39.95)^{1/2}} = \boxed{5.4\,\text{mJ K}^{-1}\,\text{m}^{-1}\,\text{s}^{-1}}$$

(b) For He, $$\kappa = \frac{1.23 \times 10^{-2}\,\text{J K}^{-1}\,\text{m}^{-1}\,\text{s}^{-1}}{(0.21) \times (4.00)^{1/2}} = \boxed{29\,\text{mJ K}^{-1}\,\text{m}^{-1}\,\text{s}^{-1}}$$

The rate of flow of energy as heat is [Exercise 24.5(a)]

$$kA\frac{\text{d}T}{\text{d}z} = \kappa \times (100 \times 10^{-4}\,\text{m}^2) \times (150\,\text{K m}^{-1}) = (1.50\,\text{K m}) \times \kappa$$

$$= 8.1\,\text{mJ s}^{-1} = \boxed{8.1\,\text{mW}} \text{ for Ar, } 44\,\text{mJ s}^{-1} = \boxed{44\,\text{mW}} \text{ for He}$$

E24.12 $$\frac{\text{d}V}{\text{d}t} \propto \frac{1}{\eta}\ [23]$$

which implies

$$\frac{\eta(\text{CO}_2)}{\eta(\text{Ar})} = \frac{\tau(\text{CO}_2)}{\tau(\text{Ar})} = \frac{55\,\text{s}}{83\,\text{s}} = 0.66\bar{3}$$

Therefore, $\eta(\text{CO}_2) = 0.66\bar{3} \times \eta(\text{Ar}) = \boxed{13\bar{8}\,\mu\text{P}}$

For the molecular diameter of CO_2 we use

$$\sigma = \left(\frac{m}{3\eta}\right) \times \left(\frac{3RT}{\pi M}\right)^{1/2} \quad [\text{Exercise 24.8(a)}]$$

$$= \left(\frac{(44.01) \times (1.6605 \times 10^{-27}\,\text{kg})}{(3) \times (1.38 \times 10^{-5}\,\text{kg m}^{-1}\,\text{s}^{-1})}\right) \times \left(\frac{(4) \times (8.314\,\text{J K}^{-1}\,\text{mol}^{-1}) \times (298\,\text{K})}{\pi \times (44.01 \times 10^{-3}\,\text{kg mol}^{-1})}\right)^{1/2}$$

$$= 4.7 \times 10^{-19}\,\text{m}^2 \approx \pi d^2$$

therefore $$d \approx \left(\frac{1}{\pi} \times (4.7 \times 10^{-19}\,\text{m}^2)\right)^{1/2} = \boxed{390\,\text{pm}}$$

E24.13 $$\kappa = \frac{1}{3}\lambda\bar{c}C_{V,\text{m}}[\text{A}]\ [16], \quad \bar{c} = \left(\frac{8RT}{\pi M}\right)^{1/2}$$

$$\lambda = \frac{kT}{2^{1/2}\sigma p}[1] = \frac{1}{2^{1/2}\sigma N_\text{A}[\text{A}]} \quad \left[\frac{p}{kT} = N_\text{A}[\text{A}]\right]$$

Therefore, $$\kappa = \frac{\bar{c}C_{V,\text{m}}}{(3) \times (2^{1/2})\sigma N_\text{A}}$$

For argon $M \approx 39.95\,\mathrm{g\,mol^{-1}}$

$$\bar{c} = \left(\frac{(8) \times (8.314\,\mathrm{J\,K^{-1}\,mol^{-1}}) \times (298\,\mathrm{K})}{\pi \times (39.95 \times 10^{-3}\,\mathrm{kg\,mol^{-1}})}\right)^{1/2} = 397\,\mathrm{m\,s^{-1}}$$

$$\kappa = \frac{(397\,\mathrm{m\,s^{-1}}) \times (12.5\,\mathrm{J\,K^{-1}\,mol^{-1}})}{(3) \times (2^{1/2}) \times (0.36 \times 10^{-18}\,\mathrm{m^2}) \times (6.022 \times 10^{23}\,\mathrm{mol^{-1}})} = \boxed{5.4 \times 10^{-3}\,\mathrm{J\,K^{-1}\,m^{-1}\,s^{-1}}}$$

Comment. This calculated value does not agree well with the value of κ listed in Table 24.1.

Question. Can the differences between the calculated and experimental values of κ be accounted for by the difference in temperature (298 K here, 273 K in Table 24.1)? If not, what might be responsible for the difference?

E24.14 $$D = \frac{1}{3}\lambda\bar{c}\,[11] = \left(\frac{2}{3p\sigma}\right) \times \left(\frac{k^3T^3}{\pi m}\right)^{1/2}$$

$$= \left(\frac{2}{(3p) \times (0.36 \times 10^{-18}\,\mathrm{m^2})}\right) \times \left(\frac{(1.381 \times 10^{-23}\,\mathrm{J\,K^{-1}})^3 \times (298\,\mathrm{K})^3}{\pi \times (39.95) \times (1.6605 \times 10^{-27}\,\mathrm{kg})}\right)^{1/2}$$

$$= \frac{1.07\,\mathrm{m^2\,s^{-1}}}{(p/\mathrm{Pa})}$$

Therefore, **(a)** at 1 Pa, $D = \boxed{1.1\,\mathrm{m^2\,s^{-1}}}$, **(b)** at 100 kPa, $D = \boxed{1.1 \times 10^{-5}\,\mathrm{m^2\,s^{-1}}}$, and **(c)** at 10 M Pa, $D = \boxed{1.1 \times 10^{-7}\,\mathrm{m^2\,s^{-1}}}$

The flux due to diffusion is

$$J = -D\left(\frac{\mathrm{dN}}{\mathrm{d}z}\right)\ [8]$$

Dividing both sides by the Avogadro constant converts the flux to number of moles per unit area per second. Thus

$$J = -D\frac{\mathrm{d}c}{\mathrm{d}x}[66] = -D\frac{\mathrm{d}}{\mathrm{d}x}\left(\frac{n}{V}\right)$$

$$= -\left(\frac{D}{RT}\right)\frac{\mathrm{d}p}{\mathrm{d}x}\quad \text{[perfect gas law]}$$

The negative sign indicates flow from high pressure to low. For a pressure gradient of $0.10\,\mathrm{atm\,cm^{-1}}$

$$J = \left[\frac{D/(\mathrm{m^2\,s^{-1}})}{(8.3145\,\mathrm{J\,K^{-1}\,mol^{-1}} \times 298\,\mathrm{K})}\right] \times (0.10\,\mathrm{atm\,cm^{-1}} \times 100\,\mathrm{cm\,m^{-1}} \times 1.01 \times 10^5\,\mathrm{Pa\,atm^{-1}})$$

$$= (4.1 \times 10^2\,\mathrm{mol\,m^{-2}\,s^{-1}}) \times (D/(\mathrm{m^2\,s^{-1}}))$$

(a) $J = (4.1 \times 10^2\,\mathrm{mol\,m^{-2}\,s^{-1}}) \times 1.07 = \boxed{4.4 \times 10^2\,\mathrm{mol\,m^{-2}\,s^{-1}}}$

(b) $J = (4.1 \times 10^2\,\mathrm{mol\,m^{-2}\,s^{-1}}) \times 1.07 \times 10^{-5} = \boxed{4.4 \times 10^{-3}\,\mathrm{mol\,m^{-2}\,s^{-1}}}$

(c) $J = (4.1 \times 10^2\,\mathrm{mol\,m^{-2}\,s^{-1}}) \times 1.07 \times 10^{-7} = \boxed{4.4 \times 10^{-5}\,\mathrm{mol\,m^{-2}\,s^{-1}}}$

E24.15 Molar ionic conductivity is related to mobility by

$$\lambda = zuF\ [46]$$

$$= 1 \times 7.91 \times 10^{-8}\,\mathrm{m^2\,s^{-1}\,V^{-1}} \times 96\,485\,\mathrm{C\,mol^{-1}}$$

$$= \boxed{7.63 \times 10^{-3}\,\mathrm{S\,m^2\,mol^{-1}}}$$

E24.16 $s = u\mathcal{E}$ [44] $\mathcal{E} = \dfrac{\Delta\phi}{l}$

Therefore,

$$s = u\left(\frac{\Delta\phi}{l}\right) = (7.92 \times 10^{-8}\,\text{m}^2\,\text{s}^{-1}\,\text{V}^{-1}) \times \left(\frac{35.0\,\text{V}}{8.00 \times 10^{-3}\,\text{m}}\right)$$

$$= 3.47 \times 10^{-4}\,\text{m}\,\text{s}^{-1}, \quad \text{or} \quad \boxed{347\,\mu\text{m}\,\text{s}^{-1}}$$

E24.17 $t_+^\circ = \dfrac{u_+}{u_+ + u_-}$ [53] $= \dfrac{4.01 \times 10^{-4}\,\text{cm}^2\,\text{s}^{-1}\,\text{V}^{-1}}{(4.01 + 8.09) \times 10^{-4}\,\text{cm}^2\,\text{s}^{-1}\,\text{V}^{-1}}$ (Table 24.5) $= \boxed{0.331}$

E24.18 The basis for the solution is Kohlrausch's law of independent of ions [32]. Switching counterions does not affect the mobility of the remaining other ion at infinite dilution.

$$\Lambda_\text{m}^\circ = \nu_+\lambda_+ + \nu_-\lambda_- \; [32]$$
$$\Lambda_\text{m}^\circ(\text{KCl}) = \lambda(\text{K}^+) + \lambda(\text{Cl}^-) = 14.99\,\text{mS}\,\text{m}^2\,\text{mol}^{-1}$$
$$\Lambda_\text{m}^\circ(\text{KNO}_3) = \lambda(\text{K}^+) + \lambda(\text{NO}_3^-) = 14.50\,\text{mS}\,\text{m}^2\,\text{mol}^{-1}$$
$$\Lambda_\text{m}^\circ(\text{AgNO}_3) = \lambda(\text{Ag}^+) + \lambda(\text{NO}_3^-) = 13.34\,\text{mS}\,\text{m}^2\,\text{mol}^{-1}$$

Hence, $\Lambda_\text{m}^\circ(\text{AgCl}) = \Lambda_\text{m}^\circ(\text{AgNO}_3) + \Lambda_\text{m}^\circ(\text{KCl}) - \Lambda_\text{m}^\circ(\text{KNO}_3)$

$$= (13.34 + 14.99 - 14.50)\,\text{mS}\,\text{m}^2\,\text{mol}^{-1} = \boxed{13.83\,\text{mS}\,\text{m}^2\,\text{mol}^{-1}}$$

Question. How well does this result agree with the value calculated directly from the data of Table 24.4?

E24.19 $u = \dfrac{\lambda}{zF}$ [46]

$$u(\text{Li}^+) = \frac{3.87\,\text{mS}\,\text{m}^2\,\text{mol}^{-1}}{9.6485 \times 10^4\,\text{C}\,\text{mol}^{-1}} = 4.01 \times 10^{-5}\,\text{mS}\,\text{C}^{-1}\,\text{m}^2$$

$$= \boxed{4.01 \times 10^{-8}\,\text{m}^2\,\text{V}^{-1}\,\text{s}^{-1}} \; (1\,\text{C}\,\Omega = 1\,\text{A}\,\text{s}\,\Omega = 1\,\text{V}\,\text{s})$$

$$u(\text{Na}^+) = \frac{5.01\,\text{mS}\,\text{m}^2\,\text{mol}^{-1}}{9.6485 \times 10^4\,\text{C}\,\text{mol}^{-1}} = \boxed{5.19 \times 10^{-8}\,\text{m}^2\,\text{V}^{-1}\,\text{s}^{-1}}$$

$$u(\text{K}^+) = \frac{7.35\,\text{mS}\,\text{m}^2\,\text{mol}^{-1}}{9.6485 \times 10^4\,\text{C}\,\text{mol}^{-1}} = \boxed{7.62 \times 10^{-8}\,\text{m}^2\,\text{V}^{-1}\,\text{s}^{-1}}$$

E24.20 $D = \dfrac{uRT}{zF}$ [70] $= \dfrac{(7.40 \times 10^{-8}\,\text{m}^2\,\text{s}^{-1}\,\text{V}^{-1}) \times (8.314\,\text{J}\,\text{K}^{-1}) \times (298\,\text{K})}{9.6485 \times 10^4\,\text{C}\,\text{mol}^{-1}} = \boxed{1.90 \times 10^{-9}\,\text{m}^2\,\text{s}^{-1}}$

E24.21 Equation [82] gives the mean square distance travelled in any one dimension. We need the distance travelled from a point in any direction. The distinction here is the distinction between one-dimensional and three-dimensional diffusion. The mean square three-dimensional distance can be obtained from the one-dimensional mean square distance since motions in the three directions are independent. Since

$$r^2 = x^2 + y^2 + z^2 \; \text{[Pythagorean theorem]}$$
$$\langle r^2\rangle = \langle x^2\rangle + \langle y^2\rangle + \langle z^2\rangle = 3\langle x^2\rangle \; \text{[independent motion]}$$
$$= 3 \times 2Dt \; [82 \text{ for } \langle x^2\rangle] = 6Dt$$

Therefore, $t = \dfrac{\langle r^2\rangle}{6D} = \dfrac{(5.0 \times 10^{-3}\,\text{m})^2}{(6) \times (3.17 \times 10^{-9}\,\text{m}^2\,\text{s}^{-1})} = \boxed{1.3 \times 10^3\,\text{s}}$

E24.22 $a = \dfrac{kT}{6\pi\eta D}$ [74 and Example 24.6]

$$= \frac{(1.381 \times 10^{-23}\,\mathrm{J\,K^{-1}}) \times (298\,\mathrm{K})}{(6\pi) \times (1.00 \times 10^{-3}\,\mathrm{kg\,m^{-1}\,s^{-1}}) \times (5.2 \times 10^{-10}\,\mathrm{m^2\,s^{-1}})} = 4.2\times10^{-10}\,\mathrm{m}, \text{ or } \boxed{420\,\mathrm{pm}}$$

E24.23 The Einstein–Smoluchowski equation [84] relates the diffusion constant to the unit jump distance and time

$$D = \frac{\lambda^2}{2\tau}\ [84] \quad \text{so} \quad \tau = \frac{\lambda^2}{2D}$$

If the jump distance is about one molecular diameter, or two effective molecular radii, then the jump distance can be obtained by use of the Stokes–Einstein equation [74]

$$D = \frac{kT}{6\pi\eta a} = \frac{kT}{3\pi\eta\lambda} \quad \text{so} \quad \lambda = \frac{kT}{3\pi\eta D}$$

$$\text{and } \tau = \frac{(kT)^2}{18(\pi\eta)^2 D^3} = \frac{[(1.381 \times 10^{-23}\,\mathrm{J\,K^{-1}}) \times (298\,\mathrm{K})]^2}{18[\pi(0.601 \times 10^{-3}\,\mathrm{kg\,m^{-1}\,s^{-1}})]^2 \times (2.13 \times 10^{-9}\,\mathrm{m^2\,s^{-1}})^3}$$

$$= 2.73 \times 10^{-11}\,\mathrm{s} = \boxed{27\,\mathrm{ps}}$$

Comment. In the strictest sense we are again (cf. Exercise 24.21(a)) dealing with three-dimensional diffusion here. However, since we are assuming that only one jump occurs, it is probably an adequate approximation to use an equation derived for one-dimensional diffusion. For three-dimensional diffusion the equation analogous to eqn 84 is

$$\tau = \frac{\lambda^2}{6D}$$

Question. Can you derive this equation? *Hint*. Use an analysis similar to that described in the solution to Exercise 24.21(a).

E24.24 For three-dimensional diffusion we use the equation analogous to eqn 82 derived in Exercise 24.21(a), that is

$$\langle r^2\rangle = 6Dt$$

$$\langle r^2\rangle^{1/2} = [(6) \times (2.13 \times 10^{-9}\,\mathrm{m^2\,s^{-1}}) \times (1.0\,\mathrm{s})]^{1/2} = \boxed{113\,\mu\mathrm{m}}$$

[Data from Table 24.7]

E24.25 $t = \dfrac{\langle r^2\rangle}{6D}$ [Exercise 24.21(a)]

(a) Iodine: $t = \dfrac{(1.0 \times 10^{-3}\,\mathrm{m})^2}{(6) \times (2.13 \times 10^{-9}\,\mathrm{m^2\,s^{-1}})} = \boxed{78\,\mathrm{s}}$

(b) Since $t \propto \langle r^2\rangle$, for a 10-fold increase in distance

$$\text{Iodine: } t = \boxed{7.8 \times 10^3\,\mathrm{s}}$$

Comment. In the solution above we have assumed that the distances given are root mean square distances. For mean distances the results would be slightly different. For mean distances the three-dimensional analogue of eqn 81 should be used.

Question. Can you derive the three-dimensional analogue of eqn 81? *Hint*. Use an analysis similar to that of the solution to Exercise 24.21(a). What times does it take for the iodine molecules to drift mean distances of (a) 1 mm and (b) 1 cm from their starting point?

Solutions to problems

Solutions to numerical problems

P24.1 For order of magnitude calculations we restrict our assumed values to powers of 10 of the base units. Thus

$$\rho = 1\,\mathrm{g\,cm^{-3}} = 1 \times 10^{3}\,\mathrm{kg\,m^{-3}}$$
$$\eta(\mathrm{air}) = 1 \times 10^{-5}\,\mathrm{kg\,m^{-1}\,s^{-1}}$$ [See comment and question below.]

We need the diffusion constant

$$D = \frac{kT}{6\pi\eta a}$$

a is calculated from the volume of the virus which is assumed to be spherical

$$V = \frac{m}{\rho} \approx \frac{(1 \times 10^{5}\,\mathrm{u}) \times (1 \times 10^{-27}\,\mathrm{kg\,u^{-1}})}{1 \times 10^{3}\,\mathrm{kg\,m^{3}}} \approx 1 \times 10^{-25}\,\mathrm{m^{3}}$$

$$V = \frac{4}{3}\pi a^{3}$$

$$a \approx \left(\frac{V}{4}\right)^{1/3} \approx \left(\frac{1 \times 10^{-25}\,\mathrm{m^{3}}}{4}\right)^{1/3} \approx 1 \times 10^{-8}\,\mathrm{m}$$

$$D \approx \left(\frac{(1 \times 10^{-23}\,\mathrm{J\,K^{-1}}) \times (300\,\mathrm{K})}{(6\pi) \times (1 \times 10^{-5}\,\mathrm{kg\,m^{-1}\,s^{-1}}) \times (1 \times 10^{-8}\,\mathrm{m})}\right) \approx 1 \times 10^{-9}\,\mathrm{m^{2}\,s^{-1}}$$

For three-dimensional diffusion

$$t = \frac{\langle r^{2}\rangle}{6D} \approx \frac{1\,\mathrm{m^{2}}}{1 \times 10^{-8}\,\mathrm{m^{2}\,s^{-1}}} \approx \boxed{10^{8}\,\mathrm{s}}$$

Therefore it does not seem likely that a cold could be caught by the process of diffusion.

Comment. In a Fermi calculation only those values of physical quantities that can be determined by scientific common sense should be used. Perhaps the value for $\eta(\mathrm{air})$ used above does not fit that description.

Question. Can you obtain the value of $\eta(\mathrm{air})$ by a Fermi calculation based on the relation in Table 24.1?

P24.3 The number of molecules that escape in unit time is the number per unit time that would have collided with a wall section of area A equal to the area of the small hole. That is,

$$\frac{\mathrm{d}N}{\mathrm{d}t} = -Z_{\mathrm{W}}A = \frac{-Ap}{(2\pi mkT)^{1/2}} \quad [3]$$

where p is the (constant) vapour pressure of the solid. The change in the number of molecules inside the cell in an interval Δt is therefore $\Delta N = -Z_{\mathrm{W}}A\Delta t$, and so the mass loss is

$$\Delta w = \Delta Nm = -Ap\left(\frac{m}{2\pi kT}\right)^{1/2}\Delta t = -Ap\left(\frac{M}{2\pi RT}\right)^{1/2}\Delta t$$

Therefore, the vapour pressure of the substance in the cell is

$$\boxed{p = \left(\frac{-\Delta w}{A\Delta t}\right) \times \left(\frac{2\pi RT}{M}\right)^{1/2}}$$

For the vapour pressure of germanium

$$p = \left(\frac{4.3 \times 10^{-8}\,\text{kg}}{\pi \times (5.0 \times 10^{-4}\,\text{m})^2 \times (7200\,\text{s})}\right) \times \left(\frac{(2\pi) \times (8.314\,\text{J K}^{-1}\,\text{mol}^{-1}) \times (1273\,\text{K})}{72.6 \times 10^{-3}\,\text{kg mol}^{-1}}\right)^{1/2}$$

$$= 7.3 \times 10^{-3}\,\text{Pa}, \quad \text{or} \quad \boxed{7.3\,\text{mPa}}$$

P24.5 Radioactive decay follows first-order kinetics (Chapter 25); hence the two contributions to the rate of change of the number of helium atoms are

$$\frac{dN}{dt} = k_r[\text{Bk}] \text{ (Radioactive decay)} \qquad \frac{dN}{dt} = -Z_W[\text{A}] \quad \text{[Problem 24.3]}$$

Therefore, the total rate of change is

$$\frac{dN}{dt} = k_r[\text{Bk}] - Z_W A \quad \text{with } Z_W = \frac{p}{(2\pi mkT)^{1/2}}$$

$$[\text{Bk}] = [\text{Bk}]_0 e^{-k_r t} \quad \text{and} \quad p = \frac{nRT}{V} = \frac{nN_A kT}{V} = \frac{NkT}{V}$$

Therefore, the pressure of helium inside the container obeys

$$\frac{dp}{dt} = \frac{kT}{V}\frac{dN}{dt} = \frac{kk_rT}{V}[\text{Bk}]_0 e^{-k_r t} - \frac{\left(\frac{pAkT}{V}\right)}{(2\pi mkT)^{1/2}}$$

If we write $a = \dfrac{kk_rT[\text{Bk}]_0}{V}$, $b = \left(\dfrac{A}{V}\right) \times \left(\dfrac{kT}{2\pi m}\right)^{1/2}$, the rate equation becomes

$$\frac{dp}{dt} = ae^{-k_r t} - bp, \quad p = 0 \text{ at } t = 0$$

which is a first-order linear differential equation with the solution

$$p = \left(\frac{a}{k_r - b}\right) \times (e^{-bt} - e^{-k_r t})$$

Since $[\text{Bk}] = \frac{1}{2}[\text{Bk}]_0$ when $t = 4.4\,\text{h}$, it follows from the radioactive decay law ($[\text{Bk}] = [\text{Bk}]_0 e^{-k_r t}$) that (Chapter 25)

$$k_r = \frac{\ln 2}{(4.4) \times (3600\,\text{s})} = 4.4 \times 10^{-5}\,\text{s}^{-1}$$

We also know that $[\text{Bk}]_0 = \left(\dfrac{1.0 \times 10^{-3}\,\text{g}}{244\,\text{g mol}^{-1}}\right) \times (6.022 \times 10^{23}\,\text{mol}^{-1}) = 2.5 \times 10^{18}$

Then, $a = \dfrac{kk_rT[\text{Bk}]_0}{V} = \dfrac{(1.381 \times 10^{-23}\,\text{J K}^{-1}) \times (4.4 \times 10^{-5}\,\text{s}^{-1}) \times (298\,\text{K}) \times (2.5 \times 10^{18})}{1.0 \times 10^{-6}\,\text{m}^3}$

$$= 0.45\,\text{Pa s}^{-1}$$

and $b = \left(\dfrac{\pi \times (2.0 \times 10^{-6}\,\text{m})^2}{1.0 \times 10^{-6}\,\text{m}^3}\right) \times \left(\dfrac{(1.381 \times 10^{-23}\,\text{J K}^{-1}) \times (298\,\text{K})}{(2\pi) \times (4.0) \times (1.6605 \times 10^{-27}\,\text{kg})}\right)^{1/2} = 3.9 \times 10^{-3}\,\text{s}^{-1}$

Hence, $p = \left(\dfrac{0.45\,\text{Pa s}^{-1}}{[(4.4 \times 10^{-5}) - (3.9 \times 10^{-3})]\,\text{s}^{-1}}\right) \times (e^{-3.9\times10^{-3}(t/\text{s})} - e^{-4.4\times10^{-5}(t/\text{s})})$

$$= (120\,\text{Pa}) \times (e^{-4.4\times10^{-5}(t/\text{s})} - e^{-3.9\times10^{-3}(t/\text{s})})$$

(a) $t = 1\,\text{h}$, $p = (120\,\text{Pa}) \times (e^{-0.16} - e^{-14}) = \boxed{100\,\text{Pa}}$

(b) $t = 10\,\text{h}$, $p = (120\,\text{Pa}) \times (e^{-1.6} - e^{140}) = \boxed{24\,\text{Pa}}$

P24.7 $\kappa \propto \dfrac{1}{R}$ [29, and the discussion above 29]

Because both solutions are aqueous their conductivities include a contribution of $76\,\text{mS m}^{-1}$ from the water. Therefore,

$$\frac{\kappa(\text{acid soln})}{\kappa(\text{KCl soln})} = \frac{\kappa(\text{acid}) + \kappa(\text{water})}{\kappa(\text{KCl}) + \kappa(\text{water})} = \frac{R(\text{KCl soln})}{R(\text{acid soln})} = \frac{33.21\,\Omega}{300.0\,\Omega}$$

Hence, $\kappa(\text{acid}) = \{\kappa(\text{KCl}) + \kappa(\text{water})\} \times \left(\dfrac{33.21}{300.0}\right) - \kappa(\text{water}) = 53\,\text{mS m}^{-1}$

$$\Lambda_\text{m} = \frac{\kappa}{c} = \frac{53\,\text{mS m}^{-1}}{1.00 \times 10^5\,\text{mol m}^{-3}} = \boxed{5.3 \times 10^{-4}\,\text{mS m}^2\,\text{mol}^{-1}}$$

P24.8 $\Lambda_\text{m} = \Lambda_\text{m}^\circ - \mathcal{K}c^{1/2}$ [31], $\Lambda_\text{m} = \dfrac{C}{cR}$ [30]

where $C = 20.63\,\text{m}^{-1}$ ($C = \kappa^* R^*$, where κ^* and R^* are the conductivity and resistance, of a standard solution respectively,)

Therefore, we draw up the following table

c/M	0.0005	0.001	0.005	0.010	0.020	0.050
$(c/\text{M})^{1/2}$	0.224	0.032	0.071	0.100	0.141	0.224
R/Ω	3314	1669	342.1	174.1	89.08	37.14
$\Lambda_\text{m}/(\text{mS m}^2\,\text{mol}^{-1})$	12.45	12.36	12.06	11.85	11.58	11.11

The values of Λ_m are plotted against $c^{1/2}$ in Fig. 24.1.

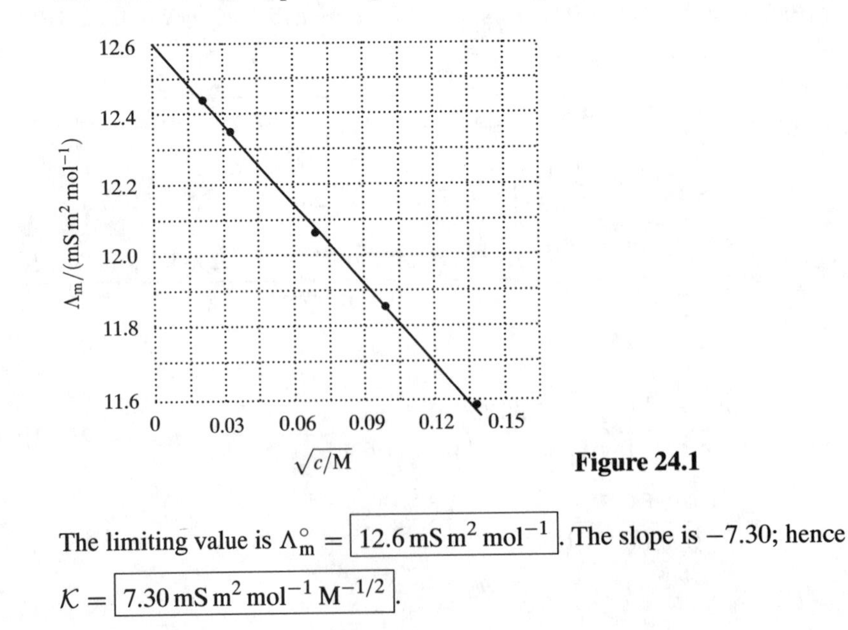

Figure 24.1

The limiting value is $\Lambda_\text{m}^\circ = \boxed{12.6\,\text{mS m}^2\,\text{mol}^{-1}}$. The slope is -7.30; hence $\mathcal{K} = \boxed{7.30\,\text{mS m}^2\,\text{mol}^{-1}\,\text{M}^{-1/2}}$.

(a) $\Lambda_m = (5.01 + 7.68)\,\text{mS m}^2\,\text{mol}^{-1} - (+7.30\,\text{mS m}^2\,\text{mol}^{-1}) \times (0.010)^{1/2}$

$= \boxed{11.96\,\text{mS m}^2\,\text{mol}^{-1}}$

(b) $\kappa = c\Lambda_m = (10\,\text{mol m}^{-3}) \times (11.96\,\text{mS m}^2\,\text{mol}^{-1}) = 119.6\,\text{mS m}^2\,\text{m}^{-3} = \boxed{119.6\,\text{mS m}^{-1}}$

(c) $R = \dfrac{C}{\kappa} = \dfrac{20.63\,\text{m}^{-1}}{119.6\,\text{mS m}^{-1}} = \boxed{172.5\,\Omega}$

P24.10 $s = u\mathcal{E}$ [44] with $\mathcal{E} = \dfrac{10\,\text{V}}{1.00\,\text{cm}} = 10\,\text{V cm}^{-1}$

$s(\text{Li}^+) = (4.01 \times 10^{-4}\,\text{cm}^2\,\text{s}^{-1}\,\text{V}^{-1}) \times (10\,\text{V cm}^{-1}) = \boxed{4.0 \times 10^{-3}\,\text{cm s}^{-1}}$

$s(\text{Na}^+) = (5.19 \times 10^{-4}\,\text{cm}^2\,\text{s}^{-1}\,\text{V}^{-1}) \times (10\,\text{V cm}^{-1}) = \boxed{5.2 \times 10^{-3}\,\text{cm s}^{-1}}$

$s(\text{K}^+) = (7.62 \times 10^{-4}\,\text{cm}^2\,\text{s}^{-1}\,\text{V}^{-1}) \times (10\,\text{V cm}^{-1}) = \boxed{7.6 \times 10^{-3}\,\text{cm s}^{-1}}$

$t = \dfrac{d}{s}$ with $d = 1.0\,\text{cm}$:

$t(\text{Li}^+) = \dfrac{1.0\,\text{cm}}{4.0 \times 10^{-3}\,\text{cm s}^{-1}} = \boxed{250\,\text{s}}, \quad t(\text{Na}^+) = \boxed{190\,\text{s}}, \quad t(\text{K}^+) = \boxed{130\,\text{s}}$

(a) For the distance moved during a half-cycle, write

$$d = \int_0^{1/2\nu} s\,\text{d}t = \int_0^{1/2\nu} u\mathcal{E}\,\text{d}t = u\mathcal{E}_0 \int_0^{1/2\nu} \sin(2\pi\nu t)\,\text{d}t \quad [\mathcal{E} = \mathcal{E}_0 \sin(2\pi\nu t)]$$

$$= \frac{u\mathcal{E}_0}{\pi\nu} = \frac{u \times (10\,\text{V cm}^{-1})}{\pi \times (1.0 \times 10^3\,\text{s}^{-1})} \text{ [assume } \mathcal{E}_0 = 10\,\text{V]} = 3.18 \times 10^{-3}\,u\,\text{V s cm}^{-1}$$

That is, $d/\text{cm} = (3.18 \times 10^{-3}) \times (u/\text{cm}^2\,\text{V}^{-1}\,\text{s}^{-1})$. Hence,

$d(\text{Li}^+) = (3.18 \times 10^{-3}) \times (4.0 \times 10^{-4}\,\text{cm}) = \boxed{1.3 \times 10^{-6}\,\text{cm}}$

$d(\text{Na}^+) = \boxed{1.7 \times 10^{-6}\,\text{cm}}, \quad d(\text{K}^+) = \boxed{2.4 \times 10^{-6}\,\text{cm}}$

(b) These correspond to about $\boxed{43}$, $\boxed{55}$, and $\boxed{81}$ solvent molecule diameters respectively.

P24.12 $t = \dfrac{zcVF}{I\Delta t} = \dfrac{zcAFl}{I\Delta t}$ [56]

$$= \left(\frac{(21\,\text{mol m}^{-3}) \times (\pi) \times (2.073 \times 10^{-3}\,\text{m})^2 \times (9.6485 \times 10^4\,\text{C mol}^{-1})}{18.2 \times 10^{-3}\,\text{A}}\right) \times \left(\frac{l}{\Delta t}\right)$$

$$= (1.50 \times 10^3\,\text{m}^{-1}\,\text{s}) \times \left(\frac{l}{\Delta t}\right) = (1.50) \times \left(\frac{l/\text{mm}}{\Delta t/\text{s}}\right)$$

Then we draw up the following table

$\Delta t/\text{s}$	200	400	600	800	1000
l/mm	64	128	192	254	318
t_+	0.48	0.48	0.48	0.48	0.48
$t_- = 1 - t_+$	0.52	0.52	0.52	0.52	0.52

Hence, we conclude that $t_+ = \boxed{0.48}$ and $t_- = \boxed{0.52}$. For the mobility of K^+ we use

$$t_+ = \frac{\lambda_+}{\Lambda_m^\circ}[54] = \frac{u_+ F}{\Lambda_m^\circ}\ [46]$$

to obtain

$$u_+ = \frac{t_+ \Lambda_m^\circ}{F} = \frac{(0.48) \times (149.9\,\mathrm{S\,cm^2\,mol^{-1}})}{9.6485 \times 10^4\,\mathrm{C\,mol^{-1}}} = \boxed{7.5 \times 10^{-4}\,\mathrm{cm^2\,s^{-1}\,V^{-1}}}$$

$$\lambda_+ = t_+ \Lambda_m^\circ[54] = (0.48) \times (149.9\,\mathrm{S\,cm^2\,mol^{-1}}) = \boxed{72\,\mathrm{S\,cm^2\,mol^{-1}}}$$

P24.14 $$\mathcal{F} = -\frac{RT}{c} \times \frac{dc}{dx}\ [65]$$

$$\frac{dc}{dx} = \frac{(0.05 - 0.10)\,\mathrm{M}}{0.10\,\mathrm{m}} = -0.50\,\mathrm{M\,m^{-1}}\ \text{[linear gradation]}$$

$$RT = 2.48 \times 10^3\,\mathrm{J\,mol^{-1}} = 2.48 \times 10^3\,\mathrm{N\,m\,mol^{-1}}$$

(a) $$\mathcal{F} = \left(\frac{-2.48\,\mathrm{kN\,m\,mol^{-1}}}{0.10\,\mathrm{M}}\right) \times (-0.50\,\mathrm{M\,m^{-1}}) = \boxed{12\,\mathrm{kN\,mol^{-1}}},\ \boxed{2.1 \times 10^{-20}\,\mathrm{N\,molecule^{-1}}}$$

(b) $$\mathcal{F} = \left(\frac{-2.48\,\mathrm{kN\,m\,mol^{-1}}}{0.075\,\mathrm{M}}\right) \times (-0.50\,\mathrm{M\,m^{-1}}) = \boxed{17\,\mathrm{kN\,mol^{-1}}},\ \boxed{2.8 \times 10^{-20}\,\mathrm{N\,molecule^{-1}}}$$

(c) $$\mathcal{F} = \left(\frac{-2.48\,\mathrm{kN\,m\,mol^{-1}}}{0.05\,\mathrm{M}}\right) \times (-0.50\,\mathrm{M\,m^{-1}}) = \boxed{25\,\mathrm{kN\,mol^{-1}}},\ \boxed{4.1 \times 10^{-20}\,\mathrm{N\,molecule^{-1}}}$$

P24.15 $$D = \frac{uRT}{zF}\ [70] \quad \text{and} \quad a = \frac{ze}{6\pi\eta u}\ [45]$$

$$D = \frac{(8.314\,\mathrm{J\,K^{-1}\,mol^{-1}}) \times (298.15\,\mathrm{K}) \times u}{9.6485 \times 10^4\,\mathrm{C\,mol^{-1}}} = 2.569 \times 10^{-2}\,\mathrm{V} \times u$$

so $D/(\mathrm{cm^2\,s^{-1}}) = (2.569 \times 10^{-2}) \times u/(\mathrm{cm^2\,s^{-1}\,V^{-1}})$

$$a = \frac{1.602 \times 10^{-19}\,\mathrm{C}}{(6\pi) \times (0.891 \times 10^{-3}\,\mathrm{kg\,m^{-1}\,s^{-1}}) \times u}$$

$$= \frac{9.54 \times 10^{-18}\,\mathrm{C\,kg^{-1}\,m\,s}}{u} = \frac{9.54 \times 10^{-18}\,\mathrm{V^{-1}\,m^3\,s^{-1}}}{u} \quad (1\,\mathrm{J} = 1\,\mathrm{C\,V},\ 1\,\mathrm{J} = 1\,\mathrm{kg\,m^2\,s^{-2}})$$

and so $a/\mathrm{m} = \dfrac{9.54 \times 10^{-14}}{u/\mathrm{cm^2\,s^{-1}\,V^{-1}}}$

and therefore $a/\mathrm{pm} = \dfrac{9.54 \times 10^{-2}}{u/\mathrm{cm^2\,s^{-1}\,V^{-1}}}$

We can now draw up the following table using data from Table 24.5

	Li^+	Na^+	K^+	Rb^+
$10^4 u/(\mathrm{cm^2\,s^{-1}\,V^{-1}})$	4.01	5.19	7.62	7.92
$10^5 D/\mathrm{cm^2}$	1.03	1.33	1.96	2.04
a/pm	238	184	125	120

The ionic radii themselves (i.e. their crystallographic radii) are

	Li^+	Na^+	K^+	Rb^+
r_+/pm	59	102	138	149

and it would seem that K^+ and Rb^+ have effective hydrodynamic radii that are smaller than their ionic radii. The effective hydrodynamic and ionic volumes of Li^+ and Na^+ are $\frac{4\pi}{3}\pi a^3$ and $\frac{4\pi}{3}\pi r_+^3$ respectively, and so the volumes occupied by hydrating water molecules are

$$\textbf{(a)}\quad Li^+: \Delta V = \left(\frac{4\pi}{3}\right) \times (212^3 - 59^3) \times 10^{-36}\,\text{m}^3 = 5.5\bar{6} \times 10^{-29}\,\text{m}^3$$

$$\textbf{(b)}\quad Na^+: \Delta V = \left(\frac{4\pi}{3}\right) \times (164^3 - 102^3) \times 10^{-36}\,\text{m}^3 = 2.1\bar{6} \times 10^{-29}\,\text{m}^3$$

The volume occupied by a single H_2O molecule is approximately $\left(\frac{4\pi}{3}\right) \times (150\,\text{pm})^3 = 1.4 \times 10^{-29}\,\text{m}^3$.

Therefore, Li^+ has about $\boxed{\text{four}}$ firmly attached H_2O molecules whereas Na^+ has only $\boxed{\text{one to two}}$ (according to this analysis).

P24.17 This is essentially one-dimensional diffusion and therefore eqn 79 applies.

$$c = \frac{n_0 e^{-x^2/4Dt}}{A(\pi Dt)^{1/2}} \quad [79]$$

and we know that $n_0 = \left(\frac{10\,\text{g}}{342\,\text{g mol}^{-1}}\right) = 0.0292\,\text{mol}$

$$A = \pi R^2 = 19.6\,\text{cm}^2, \qquad D = 5.21 \times 10^{-6}\,\text{cm}^2\,\text{s}^{-1} \text{ [Table 24.7]}$$

$$A(\pi Dt)^{1/2} = (19.6\,\text{cm}^2) \times [(\pi) \times (5.21 \times 10^{-6}\,\text{cm}^2\,\text{s}^{-1}) \times (t)]^{1/2}$$

$$= 7.93 \times 10^{-2}\,\text{cm}^3 \times (t/\text{s})^{1/2}$$

$$\frac{x^2}{4Dt} = \frac{25\,\text{cm}^2}{(4) \times (5.21 \times 10^{-6}\,\text{cm}^2\,\text{s}^{-1}) \times t} = \frac{1.20 \times 10^6}{(t/\text{s})}$$

Therefore, $c = \left(\frac{0.0292\,\text{mol} \times 10^{22}}{(7.93 \times 10^{-2}\,\text{cm}^3) \times (t/\text{s})^{1/2}}\right) \times e^{-1.20\times 10^6/(t/\text{s})} = (369\,\text{M}) \times \left(\frac{e^{-1.20\times 10^6/(t/\text{s})}}{(t/\text{s})^{1/2}}\right)$

$$\textbf{(a)}\quad t = 10\,\text{s}, \qquad c = (369\,\text{M}) \times \left(\frac{e^{-1.2\times 10^5}}{10^{1/2}}\right) \approx \boxed{0}$$

$$\textbf{(b)}\quad t = 1\,\text{yr} = 3.16 \times 10^7\,\text{s}, \qquad c = (369\,\text{M}) \times \left(\frac{e^{-0.038}}{(3.16 \times 10^7)^{1/2}}\right) = \boxed{0.063\,\text{M}}$$

Comment. This problem illustrates the extreme slowness of diffusion through typical macroscopic distances; however, it is rapid enough through distances comparable to the dimensions of a cell. Compare to Exercise 24.25(b).

Solutions to theoretical problems

P24.20 $p(x) = \dfrac{N!}{\left\{\frac{1}{2}(N+s)\right\}!\left\{\frac{1}{2}(N-s)\right\}!2^N}$ [*Justification* 24.9], $s = \dfrac{x}{\lambda}$

$$p(6d) = \frac{N!}{\left\{\frac{1}{2}(N+6)\right\}!\left\{\frac{1}{2}(N-6)\right\}!2^N}$$

(a) $N = 4,\ p(6\lambda) = \boxed{0}$ $(m! = \infty \text{ for } m < 0)$

(b) $N = 6,\quad p(6\lambda) = \dfrac{6!}{6!0!2^6} = \dfrac{1}{2^6} = \dfrac{1}{64} = \boxed{0.016}$

(c) $N = 12,\quad p(6\lambda) = \dfrac{12!}{9!3!2^{12}} = \dfrac{12 \times 11 \times 10}{3 \times 2 \times 2^{12}} = \boxed{0.054}$

(NB $0! = 1$)

Solutions to additional problems

P24.22 Kohlrausch's law states that the molar conductance of a strong electrolyte varies with the square root of concentration

$$\Lambda_m = \Lambda_m^\circ - Kc^{1/2}$$

Therefore, a plot of Λ_m versus $c^{1/2}$ should be a straight line with y-intercept Λ_m°. The data and plot (Fig. 24.2) are shown below

NaI			KI		
$c/(\text{mmol L}^{-1})$	$c^{1/2}$	$\Lambda_m/(\text{S cm}^2\ \text{mol}^{-1})$	$c/(\text{mmol L}^{-1})$	$c^{1/2}$	$\Lambda_m/(\text{S cm}^2\ \text{mol}^{-1})$
32.02	5.659	50.26	17.68	4.205	42.45
20.28	4.503	51.99	10.88	3.298	45.91
12.06	3.473	54.01	7.19	2.68	47.53
8.64	2.94	55.75	2.67	1.63	51.81
2.85	1.69	57.99	1.28	1.13	54.09
1.24	1.11	58.44	0.83	0.91	55.78
0.83	0.91	58.67	0.19	0.44	57.42

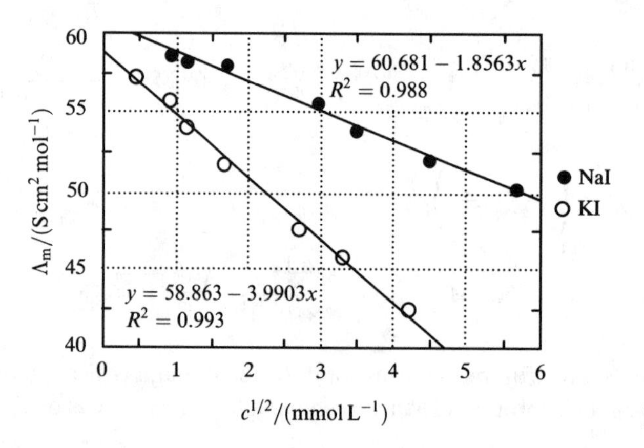

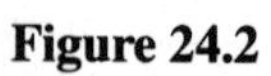
Figure 24.2

Thus $\Lambda_m^\circ(\text{NaI}) = \boxed{60.7\,\text{S cm}^2\,\text{mol}^{-1}}$ and $\Lambda_m^\circ(\text{KI}) = \boxed{58.9\,\text{S cm}^2\,\text{mol}^{-1}}$

Since these two electrolytes have a common anion, the difference in conductances is due to the cations

$$\lambda^\circ(\text{Na}^+) - \lambda^\circ(\text{K}^+) = \Lambda_m^\circ(\text{NaI}) - \Lambda_m^\circ(\text{KI}) = \boxed{1.8\,\text{S cm}^2\,\text{mol}^{-1}}$$

The analogous quantities in water are

$$\Lambda_m^\circ(\text{NaI}) = \lambda(\text{Na}^+) + \lambda(\text{I}^-) = (50.10 + 76.8)\,\text{S cm}^2\,\text{mol}^{-1} = \boxed{126.9\,\text{S cm}^2\,\text{mol}^{-1}}$$

$$\Lambda_m^\circ(\text{KI}) = \lambda(\text{K}^+) + \lambda(\text{I}^-) = (73.50 + 76.8)\,\text{S cm}^2\,\text{mol}^{-1} = \boxed{150.3\,\text{S cm}^2\,\text{mol}^{-1}}$$

$$\lambda^\circ(\text{Na}^+) - \lambda^\circ(\text{K}^+) = (50.10 - 73.50)\,\text{S cm}^2\,\text{mol}^{-1} = \boxed{-23.4\,\text{S cm}^2\,\text{mol}^{-1}}$$

The ions are considerably more mobile in water than in this solvent. Also, the differences between Na^+ and K^+ are minimized and even inverted compared to water.

P24.24 The diffusion coefficient for a perfect gas is

$$D = \tfrac{1}{3}\lambda\bar{c} \quad \text{where } \lambda = (2^{1/2}\sigma\mathcal{N})^{-1} \text{ where } \mathcal{N} \text{ is number density.}$$

The mean speed is

$$\bar{c} = \left(\frac{8kT}{\pi m}\right)^{1/2} = \left(\frac{8(1.381\times10^{-23}\,\text{J K}^{-1})\times(10^4\,\text{K})}{\pi(1\,\text{u})\times(1.66\times10^{-27}\,\text{kg u}^{-1})}\right)^{1/2} = 1.46\times10^4\,\text{m s}^{-1}$$

$$\text{So } D = \frac{\bar{c}}{3\sigma\mathcal{N}2^{1/2}} = \frac{1.46\times10^4\,\text{m s}^{-1}}{3(0.21\times10^{-18}\,\text{m}^2)\times(1\times(10^{-2}\,\text{m})^{-3})2^{1/2}} = \boxed{1.\bar{6}\times10^{16}\,\text{m}^2\,\text{s}^{-1}}$$

The thermal conductivity is

$$\kappa = \frac{\bar{c}C_{V,\text{m}}}{3\sigma N_A 2^{1/2}} = \frac{(1.46\times10^4\,\text{m s}^{-1})\times(20.784 - 8.3145)\,\text{J K}^{-1}\,\text{mol}^{-1}}{3(0.21\times10^{-18}\,\text{m}^2)\times(6.022\times10^{23}\,\text{mol}^{-1})2^{1/2}}$$

$$\kappa = \boxed{0.3\underline{4}\,\text{J K}^{-1}\,\text{m}^{-1}\,\text{s}^{-1}}$$

Comment. The validity of these calculations is in doubt because the kinetic theory of gases assumes the Maxwell–Boltzmann distribution, essentially an equilibrium distribution. In such a dilute medium, the timescales on which particles exchange energy by collision make an assumption of equilibrium unwarranted. It is especially dubious considering that atoms are more likely to interact with photons from stellar radiation than with other atoms.

25 The rates of chemical reactions

Solutions to exercises

E25.1 $v = \frac{1}{\nu_J}\frac{d[J]}{dt}$ [1] so $\frac{d[J]}{dt} = \nu_J v$

The reaction has the form

$0 = 3C + D - A - 2B$

Rate of formation of C $= 3v = \boxed{3.0\,\text{mol L}^{-1}\,\text{s}^{-1}}$

Rate of formation of D $= v = \boxed{1.0\,\text{mol L}^{-1}\text{s}^{-1}}$

Rate of consumption of A $= v = \boxed{1.0\,\text{mol L}^{-1}\text{s}^{-1}}$

Rate of consumption of B $= 2v = \boxed{2.0\,\text{mol L}^{-1}\text{s}^{-1}}$

E25.2 $v = \frac{1}{\nu_J}\frac{d[J]}{dt}$ [1]

For the reaction $2A + B \rightarrow 2C + 3D$, $\nu_C = +2$; hence

$v = \frac{1}{2} \times (1.0\,\text{M s}^{-1}) = \boxed{0.50\,\text{mol L}^{-1}\text{s}^{-1}}$

Rate of formation of D $= 3v = \boxed{1.5\,\text{mol L}^{-1}\,\text{s}^{-1}}$

Rate of consumption of A $= 2v = \boxed{1.0\,\text{mol L}^{-1}\,\text{s}^{-1}}$

Rate of consumption of B $= v = \boxed{0.50\,\text{mol L}^{-1}\,\text{s}^{-1}}$

E25.3 The rate is expressed in $\text{mol}^{-1}\text{L}^{-1}\text{s}^{-1}$; therefore

$\text{mol L}^{-1}\text{s}^{-1} = [k] \times (\text{mol L}^{-1}) \times (\text{mol L}^{-1})$ $\quad [[k] = \text{units of } k]$

requires the units to be $\boxed{\text{L mol}^{-1}\,\text{s}^{-1}}$

(a) Rate of formation of A $= v = \boxed{k[A][B]}$ **(b)** Rate of consumption of C $= 3v = \boxed{3k[A][B]}$

E25.4 $\frac{d[C]}{dt} = k[A][B][C]$

$v = \frac{1}{\nu_J}\frac{d[J]}{dt}$ with $\nu_J = \nu_C = 2$

Therefore $v = \frac{1}{2}\frac{d[C]}{dt} = \boxed{\frac{1}{2}k[A][B][C]}$

The units of k, $[k]$, must satisfy

$\text{mol L}^{-1}\text{s}^{-1} = [k] \times (\text{mol L}^{-1}) \times (\text{mol L}^{-1}) \times (\text{mol L}^{-1})$

Therefore, $[k] = \boxed{\text{L}^2\,\text{mol}^{-2}\,\text{s}^{-1}}$

E25.5 For A $\rightarrow$ Products

$v_A = -\frac{d[A]}{dt} = k[A]^a$

Since concentration and partial pressure are proportional to each other we may write

$v_A = -\frac{dp_A}{dt} = kp_A^a$

and $\dfrac{v_{A,1}}{v_{A,2}} = \dfrac{p_{A,1}^a}{p_{A,2}^a} = \left(\dfrac{p_{A,1}}{p_{A,2}}\right)^a$

Taking logarithms

$$\log\left(\frac{v_{A,1}}{v_{A,2}}\right) = a\log\left(\frac{p_{A,1}}{p_{A,2}}\right)$$

$$a = \frac{\log\left(\frac{v_{A,1}}{v_{A,2}}\right)}{\log\left(\frac{p_{A,1}}{p_{A,2}}\right)} = \frac{\log\left(\frac{1.07}{0.76}\right)}{\log\left(\frac{0.95}{0.80}\right)} = 1.9\overline{9}$$

Hence, the reaction is $\boxed{\text{second-order}}$

Comment. Knowledge of the initial pressure is not required for the solution to this exercise.

E25.6 The general expression for the half-life of a reaction of the type A → P is

$$t_{1/2} = \frac{2^{n-1}-1}{(n-1)k[A]_0^{n-1}}\ \text{[Table 25.3]} = f(n,k)[A]_0^{1-n}$$

where $f(n,k) = \dfrac{2^{n-1}-1}{(n-1)k}$. Then

$$\log t_{1/2} = \log f + (1-n)\log p_0 \quad [p_0 \propto [A]_0]$$

Hence, $\log\left(\dfrac{t_{1/2}(p_{0,1})}{t_{1/2}(p_{0,2})}\right) = (1-n)\log\left(\dfrac{p_{0,1}}{p_{0,2}}\right) = (n-1)\log\left(\dfrac{p_{0,2}}{p_{0,1}}\right)$

or $(n-1) = \dfrac{\log\left(\frac{410}{880}\right)}{\log\left(\frac{169}{363}\right)} = 0.999 \approx 1$

Therefore, $\boxed{n=2}$ in agreement with the result of Exercise 25.5(a).

E25.7 $2N_2O_5 \rightarrow 4NO_2 + O_2 \quad v = k[N_2O_5]$

Therefore, rate of consumption of $N_2O_5 = 2v = 2k[N_2O_5]$ [1]

$$\frac{d[N_2O_5]}{dt} = -2k[N_2O_5]$$

$$[N_2O_5] = [N_2O_5]_0 e^{-2kt}$$

which implies that $t = \dfrac{1}{2k}\ln\dfrac{[N_2O_5]_0}{[N_2O_5]}$

and therefore that $t_{1/2} = \dfrac{1}{2k}\ln 2 = \dfrac{\ln 2}{(2)\times(3.38\times10^{-5}\,\text{s}^{-1})} = \boxed{1.03\times10^4\,\text{s}}$

Since the partial pressure of N_2O_5 is proportional to its concentration,

$$p(N_2O_5) = p_0(N_2O_5)e^{-2kt}$$

(a) $p(N_2O_5) = (500\,\text{Torr}) \times \left(e^{-(6.76\times10^{-5})\times10}\right) = \boxed{499.\overline{7}\,\text{Torr}}$

(b) $p(N_2O_5) = (500\,\text{Torr}) \times \left(e^{-(6.76\times10^{-5})\times600}\right) = \boxed{480\,\text{Torr}}$

E25.8 We use $kt = \dfrac{1}{[\mathrm{B}]_0 - [\mathrm{A}]_0} \ln\left\{\left(\dfrac{[\mathrm{B}]}{[\mathrm{B}]_0}\right) \Big/ \left(\dfrac{[\mathrm{A}]}{[\mathrm{A}]_0}\right)\right\}$ [15]

(a) The stoichiometry of the reaction requires that when $\Delta[\mathrm{A}] = (0.050 - 0.020)\,\mathrm{mol\,L^{-1}} = 0.030\,\mathrm{mol\,L^{-1}}$, $\Delta[\mathrm{B}] = 0.030\,\mathrm{mol\,L^{-1}}$ as well. Thus $[\mathrm{B}] = 0.080\,\mathrm{mol\,L^{-1}} - 0.030\,\mathrm{mol\,L^{-1}} = 0.050\,\mathrm{mol\,L^{-1}}$ when $[\mathrm{A}] = 0.20\,\mathrm{mol\,L^{-1}}$. Thus,

$$kt = \left(\frac{1}{(0.080 - 0.050)\,\mathrm{mol\,L^{-1}}}\right) \ln\left\{\left(\frac{0.050}{0.080}\right) \Big/ \left(\frac{0.020}{0.050}\right)\right\}$$

$$k \times 1.0\,\mathrm{h} = 14.88\,\mathrm{L\,mol^{-1}}$$

$$k = 14.\bar{9}\,\mathrm{L\,mol^{-1}\,h^{-1}} = \boxed{4.1 \times 10^{-3}\,\mathrm{L\,mol^{-1}\,s^{-1}}}$$

(b) The half-life with respect to A is the time required for [A] to fall to $0.025\,\mathrm{mol\,L^{-1}}$. We solve eqn 15 for t

$$t_{1/2}(\mathrm{A}) = \left(\frac{1}{(14.\bar{9}\,\mathrm{L\,mol^{-1}\,h^{-1}}) \times (0.030\,\mathrm{mol\,L^{-1}})}\right) \times \ln\left\{\left(\frac{0.055}{0.080}\right) \Big/ 0.50\right\}$$

$$= 0.71\bar{2}\,\mathrm{h} = \boxed{2.6 \times 10^3\,\mathrm{s}}$$

Similarly, $t_{1/2}(\mathrm{B}) = \left(\dfrac{1}{0.44\bar{7}\,\mathrm{h^{-1}}}\right) \ln\left\{0.50 \Big/ \left(\dfrac{0.010}{0.050}\right)\right\} = 2.0\bar{5}\,\mathrm{h} = \boxed{7.4 \times 10^3\,\mathrm{s}}$

Comment. This exercise illustrates that there is no unique half-life for reactions other than those of the type $\mathrm{A} \rightarrow \mathrm{P}$.

E25.9 **(a)** For a second-order reaction, denoting the units of k by $[k]$

$$\mathrm{mol\,L^{-1}\,s^{-1}} = [k] \times (\mathrm{mol\,L^{-1}})^2, \quad \text{therefore} \quad [k] = \boxed{\mathrm{L\,mol^{-1}\,s^{-1}}}$$

For a third-order reaction

$$\mathrm{mol\,L^{-1}\,s^{-1}} = [k] \times (\mathrm{mol\,L^{-1}})^3, \quad \text{therefore} \quad [k] = \boxed{\mathrm{L^2\,mol^{-2}\,s^{-1}}}$$

(b) For a second-order reaction

$$\mathrm{kPa\,s^{-1}} = [k] \times \mathrm{kPa}^2, \quad \text{therefore} \quad [k] = \boxed{\mathrm{kPa^{-1}\,s^{-1}}}$$

For a third-order reaction

$$\mathrm{kPa\,s^{-1}} = [k] \times \mathrm{kPa}^3, \quad \text{therefore} \quad [k] = \boxed{\mathrm{kPa^{-2}\,s^{-1}}}$$

E25.10 $[^{14}\mathrm{C}] = [^{14}\mathrm{C}]_0 \mathrm{e}^{-kt}$ [10*b*], $\quad k = \dfrac{\ln 2}{t_{1/2}}$

Solving for t, $t = \dfrac{1}{k} \ln \dfrac{[^{14}\mathrm{C}]_0}{[^{14}\mathrm{C}]} = \dfrac{t_{1/2}}{\ln 2} \ln \dfrac{[^{14}\mathrm{C}]_0}{[^{14}\mathrm{C}]} = \left(\dfrac{5730\,\mathrm{y}}{\ln 2}\right) \times \ln\left(\dfrac{1.00}{0.72}\right) = \boxed{27\bar{2}0\,\mathrm{y}}$

E25.11 For a reaction of the type $\mathrm{A} + \mathrm{B} \rightarrow$ products we use

$$kt = \left(\frac{1}{[\mathrm{B}]_0 - [\mathrm{A}]_0}\right) \ln\left\{\left(\frac{[\mathrm{B}]}{[\mathrm{B}]_0}\right) \Big/ \left(\frac{[\mathrm{A}]}{[\mathrm{A}]_0}\right)\right\} \quad [15]$$

Introducing $[B] = [B]_0 - x$ and $[A] = [A]_0 - x$ and rearranging we obtain

$$kt = \left(\frac{1}{[B]_0 - [A]_0}\right) \ln\left(\frac{[A]_0([B]_0 - x)}{([A]_0 - x)[B]_0}\right)$$

Solving for x

$$x = \frac{[A]_0[B]_0\left\{e^{k([B]_0-[A]_0)t} - 1\right\}}{[B]_0 e^{([B]_0-[A]_0)kt} - [A]_0} = \frac{(0.050) \times (0.100\,\text{mol L}^{-1}) \times \left\{e^{(0.100-0.050)\times 0.11 \times t/\text{s}} - 1\right\}}{(0.100) \times \left\{e^{(0.100-0.050)\times 0.11\times t/\text{s}}\right\} - 0.050}$$

$$= \frac{(0.100\,\text{mol L}^{-1}) \times (e^{5.5\times 10^{-3}\,t/\text{s}} - 1)}{2e^{5.5\times 10^{-3}\,t/\text{s}} - 1}$$

(a) $$x = \frac{(0.100\,\text{mol L}^{-1}) \times (e^{0.055} - 1)}{2e^{0.055} - 1} = 5.1 \times 10^{-3}\,\text{mol L}^{-1}$$

which implies that $[\text{NaOH}] = (0.050 - 0.0051)\,\text{mol L}^{-1} = \boxed{0.045\,\text{mol L}^{-1}}$ and

$[CH_3COOC_2H_5] = (0.100 - 0.0051)\,\text{mol L}^{-1} = \boxed{0.095\,\text{mol L}^{-1}}$

(b) $$x = \frac{(0.100\,\text{mol L}^{-1}) \times (e^{3.3} - 1)}{2e^{3.3} - 1} = 0.049\,\text{mol L}^{-1}$$

Hence, $[\text{NaOH}] = (0.050 - 0.049)\,\text{mol L}^{-1} = \boxed{0.001\,\text{mol L}^{-1}}$

$[CH_3COOC_2H_5] = (0.100 - 0.049)\,\text{mol L}^{-1} = \boxed{0.051\,\text{mol L}^{-1}}$

E25.12 The rate of consumption of A is

$$\frac{d[A]}{dt} = -2k[A]^2 \quad [\nu_A = -2]$$

which integrates to $\dfrac{1}{[A]} - \dfrac{1}{[A]_0} = 2kt$ [12*b* with *k* repalced by 2*k*]

Therefore, $$t = \frac{1}{2k}\left(\frac{1}{[A]} - \frac{1}{[A]_0}\right) = \left(\frac{1}{(2) \times (3.50 \times 10^{-4}\,\text{L mol}^{-1}\,\text{s}^{-1})}\right)$$

$$\times \left(\frac{1}{0.011\,\text{mol L}^{-1}} - \frac{1}{0.260\,\text{mol L}^{-1}}\right) = \boxed{1.24 \times 10^5\,\text{s}}$$

E25.13 $\ln k = \ln A - \dfrac{E_a}{RT}$ [24] $\qquad \ln k' = \ln A - \dfrac{E_a}{RT'}$

Hence, $$E_a = \frac{R\ln\left(\frac{k'}{k}\right)}{\left(\frac{1}{T} - \frac{1}{T'}\right)} = \frac{(8.314\,\text{J K}^{-1}\,\text{mol}^{-1}) \times \ln\left(\frac{1.38\times 10^{-2}}{2.80\times 10^{-3}}\right)}{\frac{1}{303\,\text{K}} - \frac{1}{323\,\text{K}}} = \boxed{64.9\,\text{kJ mol}^{-1}}$$

For *A*, we use

$$A = k \times e^{E_a/RT}\,[26] = (2.80 \times 10^{-3}\,\text{mol L}^{-1}\,\text{s}^{-1}) \times e^{64.9\times 10^3/(8.314\times 303)}$$

$$= \boxed{4.32 \times 10^8\,\text{mol L}^{-1}\,\text{s}^{-1}}$$

E25.14 We assume a pre-equilibrium (as the first step is fast), and write

$$K = \frac{[A]^2}{[A_2]}, \quad \text{implying that} \quad [A] = K^{1/2}[A_2]^{1/2}$$

The rate-determining step then gives

$$v = \frac{d[P]}{dt} = k_2[A][B] = \boxed{k_2 K^{1/2}[A_2]^{1/2}[B]} = k_{\text{eff}}[A_2]^{1/2}[B]$$

where $k_{\text{eff}} = k_2 K^{1/2}$.

E25.15 The rate of change of [A] is

$$\frac{d[A]}{dt} = -k[A]^n$$

Hence, $\displaystyle\int_{[A]_0}^{[A]} \frac{d[A]}{[A]^n} = -k\int_0^t dt = -kt$

Therefore, $kt = \left(\dfrac{1}{n-1}\right) \times \left(\dfrac{1}{[A]^{n-1}} - \dfrac{1}{[A]_0^{n-1}}\right)$

and $kt_{1/2} = \left(\dfrac{1}{n-1}\right) \times \left(\dfrac{2^{n-1}}{[A]_0^{n-1}} - \dfrac{1}{[A]_0^{n-1}}\right) = \left(\dfrac{2^{n-1}-1}{n-1}\right) \times \left(\dfrac{1}{[A]_0^{n-1}}\right)$ [as in Table 25.3]

Hence, $\boxed{t_{1/2} \propto \dfrac{1}{[A]_0^{n-1}}}$

E25.16 Maximum velocity $= k_b[E]_0$ [paragraph following eqn 52]; also

$$\frac{d[P]}{dt} = k[E]_0, \quad k = \frac{k_b[S]}{K_M + [S]} \ [50]$$

Therefore, since $v = \dfrac{k_b[S][E]_0}{K_M + [S]}$

we know that

$$k_b[E]_0 = \left(\frac{K_M + [S]}{[S]}\right) v = \left(\frac{0.035 + 0.110}{0.110}\right) \times (1.15 \times 10^{-3}\,\text{mol L}^{-1}\,\text{s}^{-1})$$

$$= \boxed{1.52 \times 10^{-3}\,\text{mol L}^{-1}\,\text{s}^{-1}}$$

E25.17 $\dfrac{1}{k} = \dfrac{1}{k_a p_A} + \dfrac{k_a'}{k_a k_b}$ [analogous to eqn 65]

Therefore, for two different pressures we have

$$\frac{1}{k} - \frac{1}{k'} = \frac{1}{k_a}\left(\frac{1}{p} - \frac{1}{p'}\right)$$

and hence $k_a = \dfrac{\left(\frac{1}{p} - \frac{1}{p'}\right)}{\left(\frac{1}{k} - \frac{1}{k'}\right)} = \dfrac{\left(\frac{1}{12\,\text{Pa}} - \frac{1}{1.30\times 10^3\,\text{Pa}}\right)}{\left(\frac{1}{2.10\times 10^{-5}\,\text{s}^{-1}} - \frac{1}{2.50\times 10^{-4}\,\text{s}^{-1}}\right)} = 1.9 \times 10^{-6}\,\text{Pa}^{-1}\,\text{s}^{-1}$, or

$\boxed{1.9\,\text{MPa}^{-1}\,\text{s}^{-1}}$

E25.18 $NH_4^+(aq) + H_2O(l) \rightleftharpoons NH_3(aq) + H_3O^+(aq) \quad pK_a = 9.25$

$$NH_3(aq) + H_2O(l) \underset{k'}{\overset{k}{\rightleftharpoons}} NH_4^+(aq) + OH^-(aq)$$

$$pK_b = pK_w - pK_a = 14.00 - 9.25 = 4.75$$

Therefore, $K_b = \dfrac{k}{k'} = 10^{-4.75} = 1.78 \times 10^{-5}\,\text{mol L}^{-1}$

and $k = k'K_b = (1.78 \times 10^{-5}\,\text{mol L}^{-1}) \times (4.0 \times 10^{10}\,\text{L mol}^{-1}\text{s}^{-1}) = \boxed{7.1 \times 10^5\,\text{s}^{-1}}$

$$\frac{1}{\tau} = k + k'([\mathrm{NH_4^+}] + [\mathrm{OH^-}])\ [\text{Example 25.4}]$$

$$= k + 2k'K_b^{1/2}[\mathrm{NH_3}]^{1/2} \quad [[\mathrm{NH_4^+}] = [\mathrm{OH^-}] = (K_b[\mathrm{NH_3}])^{1/2}]$$

$$= (7.1 \times 10^5\,\mathrm{s^{-1}}) + (2) \times (4.0 \times 10^{10}\,\mathrm{L\,mol^{-1}\,s^{-1}}) \times (1.78 \times 10^{-5}\,\mathrm{mol\,L^{-1}})^{1/2}$$

$$\times (0.15\,\mathrm{mol\,L^{-1}})^{1/2}$$

$$= 1.31 \times 10^8\,\mathrm{s^{-1}}; \quad \text{hence} \quad \boxed{\tau = 7.61\,\mathrm{ns}}$$

Comment. The rate constant k corresponds to the pseudo-first-order protonation of NH_3 in excess water and hence has the units s^{-1}. Therefore, K_b in this problem must be assigned the units mol L^{-1} to obtain proper cancellation of units in the equation for $\frac{1}{\tau}$.

Solutions to problems

Solutions to numerical problems

P25.1 A simple but practical approach is to make an initial guess at the order by observing whether the half-life of the reaction appears to depend on concentration. If it does not, the reaction is first-order; if it does, it may be second-order. Examination of the data shows that the first half-life is roughly 45 minutes, but that the second is about double the first. (Compare the $0 \rightarrow 50.0$ minute data to the $50.0 \rightarrow 150$ minute data.) Therefore, assume second-order and confirm by a plot of $\frac{1}{[\mathrm{A}]}$ against time.

We draw up the following table (A = NH_4CNO)

t/min	0	20.0	50.0	65.0	150
m(urea)/g	0	7.0	12.1	13.8	17.7
m(A)/g	22.9	15.9	10.8	9.1	5.2
$[\mathrm{A}]/(\mathrm{mol\,L^{-1}})$	0.382	0.265	0.180	0.152	0.0866
$\frac{1}{[\mathrm{A}]}/(\mathrm{L\,mol^{-1}})$	2.62	3.77	5.56	6.59	11.5

The data are plotted in Fig. 25.1 and fit closely to a straight line. Hence, the reaction is $\boxed{\text{second-order}}$ The rate constant is the slope. A least-squares fit gives $\boxed{k = 0.059\bar{4}\,\mathrm{L\,mol^{-1}\,min^{-1}}}$ At 300 min $[\mathrm{A}] = 0.049\,\mathrm{mol\,L^{-1}}$. These calculations were performed on an inexpensive hand-held calculator which is pre-programmed to do linear regression (and other kinds too). The value of [A] at 300 min is provided automatically by the calculator. It could be obtained by

$$\frac{1}{[\mathrm{A}]} = kt + \frac{1}{[\mathrm{A}]_0} \quad [12b]$$

$$\text{or } [\mathrm{A}] = \frac{[\mathrm{A}]_0}{kt[\mathrm{A}]_0 + 1} = \frac{0.382\,\mathrm{mol\,L^{-1}}}{(0.059\bar{4}) \times (300) \times (0.382) + 1} = 0.049\,\mathrm{mol\,L^{-1}}$$

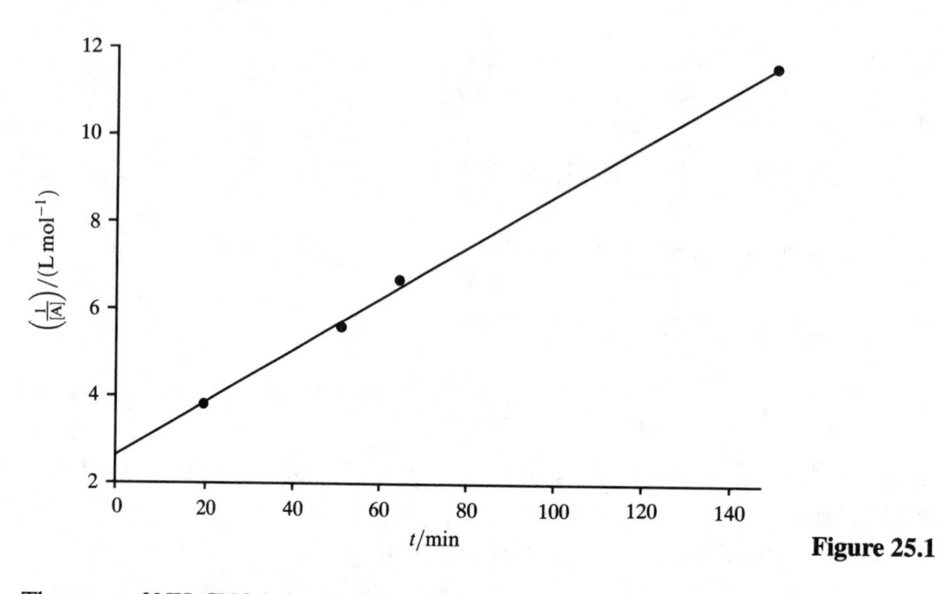

Figure 25.1

The mass of NH_4CNO left after 300 minutes is

$$\text{mass} = (0.048\overline{9}\,\mathrm{mol\,L^{-1}}) \times (1.00\,\mathrm{L}) \times (60.06\,\mathrm{g\,mol^{-1}}) = \boxed{2.94\,\mathrm{g}}$$

P25.3 The procedure adopted in the solutions to Problems 25.1 and 25.2 is employed here. Examination of the data indicates that the half-life is independent of concentration and that the reaction is therefore first- order. That is confirmed by a plot of $\ln\left(\frac{[A]}{[A]_0}\right)$ against time.

We draw up the following table (A = nitrile)

$t/(10^3\,\mathrm{s})$	0	2.00	4.00	6.00	8.00	10.00	12.00
$[A]/(\mathrm{mol\,L^{-1}})$	1.10	0.86	0.67	0.52	0.41	0.32	0.25
$\frac{[A]}{[A]_0}$	1	0.78	0.61	0.47	0.37	0.29	0.23
$\ln\left(\frac{[A]}{[A]_0}\right)$	0	−0.246	−0.496	−0.749	−0.987	− 1.235	−1.482

A least-squares fit to a linear equation gives $k = \boxed{1.2\overline{3} \times 10^{-4}\,\mathrm{s^{-1}}}$ with a correlation coefficient of 1.000.

P25.5 As described in Example 25.5, if the rate constant obeys the Arrhenius equation [24], a plot of $\ln k$ against $\frac{1}{T}$ should yield a straight line with slope $\frac{-E_a}{R}$. However, since data are available only at three temperatures, we use the two-point method, that is

$$\ln\frac{k_2}{k_1} = -\frac{E_a}{R}\left(\frac{1}{T_2} - \frac{1}{T_1}\right)$$

which yields $E_a = \dfrac{-R\ln\left(\frac{k_2}{k_1}\right)}{\left(\frac{1}{T_2} - \frac{1}{T_1}\right)}$

For the pair $\theta = 0°C$ and $40°C$,

$$E_a = \frac{-R\ln\left(\frac{576}{2.46}\right)}{\left(\frac{1}{313\,\mathrm{K}} - \frac{1}{273\,\mathrm{K}}\right)} = 9.69 \times 10^4\,\mathrm{J\,mol^{-1}}$$

For the pair $\theta = 20°C$ and $40°C$,

$$E_a = \frac{-R\ln\left(\frac{576}{45.1}\right)}{\left(\frac{1}{313\,\mathrm{K}} - \frac{1}{293\,\mathrm{K}}\right)} = 9.71 \times 10^4\,\mathrm{J\,mol^{-1}}$$

The agreement of these values of E_a indicates that the rate constant data fits the Arrhenius equation and that the activation energy is $\boxed{9.70 \times 10^4\,\mathrm{J\,mol^{-1}}}$

P25.6 We have, since both reactions are first-order

$$-\frac{\mathrm{d}[\mathrm{A}]}{\mathrm{d}t} = k_1[\mathrm{A}] + k_2[\mathrm{A}]$$

or $\frac{\mathrm{d}x}{\mathrm{d}t} = k_1([\mathrm{A}]_0 - x) + k_2([\mathrm{A}]_0 - x)\ [x = [\mathrm{A}]_0 - [\mathrm{A}]] = (k_1 + k_2) \times ([\mathrm{A}]_0 - x)$

which integrates to $(k_1 + k_2)t = \ln\frac{[\mathrm{A}]_0}{[\mathrm{A}]_0 - x}$

Solving for x, $x = [\mathrm{A}]_0\left(1 - \mathrm{e}^{-(k_1+k_2)t}\right)$

For reaction (2) above $x_2 = [\mathrm{A}]_0(1 - \mathrm{e}^{-k_2 t})$

At the start of the reaction, $t = 0$ and x and x_2 are both zero as well. When $t \to \infty$ we may expand both exponentials

$$x = [\mathrm{A}]_0(k_1 + k_2)t \qquad x_2 = [\mathrm{A}]_0 k_2 t$$

The fraction of the ketene formed is

$$\frac{x_2}{x} = \frac{k_2}{k_1 + k_2} = \frac{4.65\,\mathrm{s^{-1}}}{(3.74\,\mathrm{s^{-1}}) + (4.65\,\mathrm{s^{-1}})} = 0.554$$

The maximum percentage yield is then $\boxed{55.4\text{ per cent}}$

Comment. If a substance reacts by parallel processes of the same order, then the ratio of the amounts of products will be constant and independent of the extent of the reaction, no matter what the order.

Question. Can you demonstrate the truth of the statement made in the above comment?

P25.8 The data for this experiment do not extend much beyond one half-life. Therefore the half-life method of predicting the order of the reaction as described in the solutions to Problems 25.1 and 25.2 cannot be used here. However, a similar method based on three-quarters lives will work. Analogous to the derivation leading to eqn 11, we may write

$$kt_{3/4} = -\ln\frac{\frac{3}{4}[\mathrm{A}]_0}{[\mathrm{A}]_0} = -\ln\frac{3}{4} = \ln\frac{4}{3} = 0.288$$

or $t_{3/4} = \frac{0.288}{k}$

and we see that the three-quarters life is also independent of concentration for a first-order reaction. Examination of the data shows that the first three-quarters life is about 80 min ($0.237\,\mathrm{mol\,L^{-1}}$) and by interpolation the second is also about 80 min (0.178 mol L^{-1}). Therefore the reaction is first-order and the rate constant is approximately

$$k = \frac{0.288}{t_{3/4}} \approx \frac{0.288}{80\,\text{min}} = 3.6 \times 10^{-3}\,\text{min}^{-1}$$

A least-squares fit of the data to the first-order integrated rate law [10b] gives the slightly more accurate result, $\boxed{k = 3.65 \times 10^{-3}\,\text{min}^{-1}}$ The average lifetime is calculated from

$$\frac{[\text{A}]}{[\text{A}]_0} = \text{e}^{-kt}\ [10b]$$

which has the form of a distribution function. The ratio $\frac{[\text{A}]}{[\text{A}]_0}$ is the fraction of sucrose molecules which have lived to time t. The average lifetime is then

$$\langle t \rangle = \frac{\int_0^\infty t\text{e}^{-kt}\,\text{d}t}{\int_0^\infty \text{e}^{-kt}\,\text{d}t} = \frac{1}{k} = \boxed{274\,\text{min}}$$

The denominator ensures normalization of the distribution function.

Comment. The average lifetime is also called the relaxation time. Compare to eqn 23. Note that the average lifetime is not the half-life. The latter is 190 minutes. Also note that $2 \times t_{3/4} \neq t_{1/2}$

P25.10 If the reaction is first-order the concentrations obey

$$\ln\left(\frac{[\text{A}]}{[\text{A}]_0}\right) = -kt\ [10b]$$

and, since pressures and concentrations of gases are proportional, the pressures should obey

$$\ln\frac{p_0}{p} = kt$$

and $\frac{1}{t}\ln\frac{p_0}{p}$ should be a constant. We test this by drawing up the following table

p_0/Torr	200	200	400	400	600	600
t/s	100	200	100	200	100	200
p/Torr	186	173	373	347	559	520
$10^4\left(\frac{1}{t/\text{s}}\right)\ln\frac{p_0}{p}$	7.3	7.3	7.0	7.1	7.1	7.2

The values in the last row of the table are virtually constant, and so (in the pressure range spanned by the data) the reaction has $\boxed{\text{first-order kinetics}}$ with $k = \boxed{7.2 \times 10^{-4}\,\text{s}^{-1}}$

P25.11 $\text{A} + \text{B} \rightarrow \text{P}, \quad \frac{\text{d[P]}}{\text{d}t} = k[\text{A}]^m[\text{B}]^n$

and for a short interval δt,

$$\delta[\text{P}] \approx k[\text{A}]^m[\text{B}]^n\delta t$$

Therefore, since $\delta[\text{P}] = [\text{P}]_t - [\text{P}]_0 = [\text{P}]_t$,

$$\frac{[\text{P}]}{[\text{A}]} = k[\text{A}]^{m-1}[\text{B}]^n\delta t$$

$\dfrac{[\text{Chloropropane}]}{[\text{Propene}]}$ independent of [Propene] implies that $m = 1$.

$$\frac{[\text{Chloropropane}]}{[\text{HCl}]} = \begin{cases} p(\text{HCl}) & 10 & 7.5 & 5.0 \\ & 0.05 & 0.03 & 0.01 \end{cases}$$

These results suggest that the ratio is roughly proportional to p^2, and therefore that $m = 3$ when A is identified with HCl. The rate law is therefore

$$\frac{\mathrm{d}[\text{Chloropropane}]}{\mathrm{d}t} = k[\text{Propane}][\text{HCl}]^3$$

and the reaction is $\boxed{\text{first-order}}$ in propene and $\boxed{\text{third-order}}$ in HCl.

P25.12 $2\text{HCl} \rightleftharpoons (\text{HCl})_2, \quad K_1 \qquad [(\text{HCl})_2] = K_1[\text{HCl}]^2$

$\text{HCl} + \text{CH}_3\text{CH}{=}\text{CH}_2 \rightleftharpoons \text{Complex} \quad K_2$

$[\text{Complex}] = K_2[\text{HCl}][\text{CH}_3\text{CH}{=}\text{CH}_2]$

$(\text{HCl})_2 + \text{Complex} \rightarrow \text{CH}_3\text{CHClCH}_3 + 2\text{HCl} \quad k$

$\text{rate} = k[(\text{HCl})_2][\text{Complex}] = kK_2[(\text{HCl})_2][\text{HCl}][\text{CH}_3\text{CH}{=}\text{CH}_2]$

$= \boxed{kK_2K_1[\text{HCl}]^3[\text{CH}_3\text{CH}{=}\text{CH}_2]}$

Use infrared spectroscopy to search for $(\text{HCl})_2$.

P25.14 $$E_\text{a} = \frac{R\ln\left(\frac{k'}{k}\right)}{\left(\frac{1}{T} - \frac{1}{T'}\right)} \quad \text{[Exercise 25.13(a) from eqn 24]}$$

We then draw up the following table

T/K	300.3	300.3	341.2
T'/K	341.2	392.2	392.2
$10^{-7}k/(\text{L mol}^{-1}\,\text{s}^{-1})$	1.44	1.44	3.03
$10^{-7}k'/(\text{L mol}^{-1}\,\text{s}^{-1})$	3.03	6.9	6.9
$E_\text{a}/(\text{kJ mol}^{-1})$	15.5	16.7	18.0

The mean is $\boxed{16.7\,\text{kJ mol}^{-1}}$ For A, use

$$A = k\mathrm{e}^{E_\text{a}/RT}$$

and draw up the following table

T/K	300.3	341.2	392.2
$10^{-7}k/(\text{L mol}^{-1}\,\text{s}^{-1})$	1.44	3.03	6.9
E_a/RT	6.69	5.89	5.12
$10^{-10}A/(\text{L mol}^{-1}\,\text{s}^{-1})$	1.16	1.10	1.16

The mean is $\boxed{1.14 \times 10^{10}\,\text{L mol}^{-1}\,\text{s}^{-1}}$

P25.16 $$\frac{\mathrm{d}[\text{P}]}{\mathrm{d}t} = \frac{k_\text{b}[\text{E}]_0[\text{S}]}{K_\text{M} + [\text{S}]} \quad [50]$$

Write $v = \dfrac{\mathrm{d}[\text{P}]}{\mathrm{d}t}$, then $\dfrac{1}{v} = \left(\dfrac{1}{k_\text{b}[\text{E}]_0}\right) + \left(\dfrac{K_\text{M}}{k_\text{b}[\text{E}]_0}\right) \times \left(\dfrac{1}{[\text{S}]}\right)$

We therefore draw up the following table

$10^3[\mathrm{S}]/\mathrm{mol\,L^{-1}}$	50	17	10	5	2
$\dfrac{1}{[\mathrm{S}]/\mathrm{mol\,L^{-1}}}$	20.0	58.8	100	200	500
$v/(\mathrm{mm^3\,min^{-1}})$	16.6	12.4	10.1	6.6	3.3
$\dfrac{1}{v/\mathrm{mm^3\,min^{-1}}}$	0.0602	0.0806	0.0990	0.152	0.303

The points are plotted in Fig. 25.2.

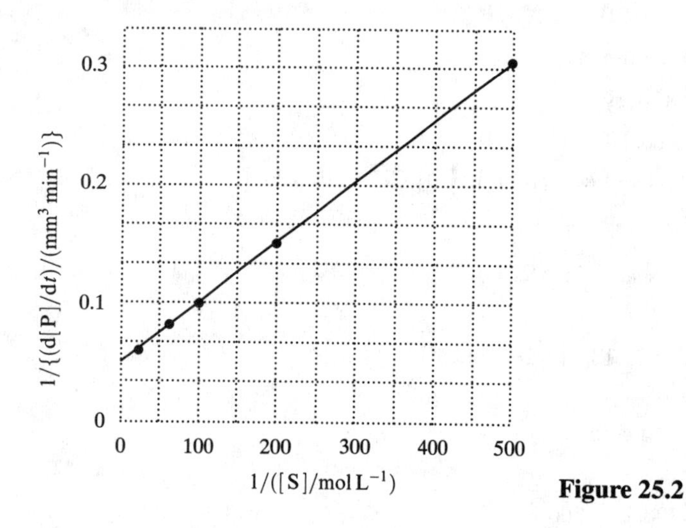

Figure 25.2

The intercept lies at 0.050, which implies that $\dfrac{1}{k_\mathrm{b}[\mathrm{E}]_0} = 0.050\,\mathrm{mm^{-3}\,min}$. The slope is 5.06×10^{-4}, which implies that

$$\frac{K_\mathrm{M}}{k_\mathrm{b}[\mathrm{E}]_0} = 5.06 \times 10^{-4}\,\mathrm{mm^{-3}\,min\,mol\,L^{-1}}$$

and therefore that $K_\mathrm{M} = \dfrac{5.06 \times 10^{-4}\,\mathrm{mm^{-3}\,min\,mol\,L^{-1}}}{0.050\,\mathrm{mm^{-3}\,min}} = \boxed{0.010\,\mathrm{mol\,L^{-1}}}$

Solutions to theoretical problems

P25.17 $\mathrm{A} \rightleftharpoons \mathrm{B}$

$$\frac{\mathrm{d[A]}}{\mathrm{d}t} = -k[\mathrm{A}] + k'[\mathrm{B}] \quad \frac{\mathrm{d[B]}}{\mathrm{d}t} = -k'[\mathrm{B}] + k[\mathrm{A}]$$

$[\mathrm{A}] + [\mathrm{B}] = [\mathrm{A}]_0 + [\mathrm{B}]_0$ at all times.

Therefore, $[\mathrm{B}] = [\mathrm{A}]_0 + [\mathrm{B}]_0 - [\mathrm{A}]$

$$\frac{\mathrm{d[A]}}{\mathrm{d}t} = -k[\mathrm{A}] + k'\{[\mathrm{A}]_0 + [\mathrm{B}]_0 - [\mathrm{A}]\} = -(k+k')[\mathrm{A}] + k'([\mathrm{A}]_0 + [\mathrm{B}]_0)$$

The solution is $[\mathrm{A}] = \dfrac{k'([\mathrm{A}]_0 + [\mathrm{B}]_0) + (k[\mathrm{A}]_0 - k'[\mathrm{B}]_0)\mathrm{e}^{-(k+k')t}}{k+k'}$

The final composition is found by setting $t = \infty$

$$[A]_\infty = \left(\frac{k'}{k+k'}\right) \times ([A]_0 + [B]_0)$$

$$[B]_\infty = [A]_0 + [B]_0 - [A]_\infty = \left(\frac{k}{k+k'}\right) \times ([A]_0 + [B]_0)$$

Note that $\boxed{\dfrac{[B]_\infty}{[A]_\infty} = \dfrac{k}{k'}}$

P25.19 $\dfrac{d[A]}{dt} = -2k[A]^2[B], \quad 2A + B \rightarrow P$

(a) Let $[P] = x$ at t, then $[A] = A_0 - 2x$ and $[B] = B_0 - x$. Therefore,

$$\frac{d[A]}{dt} = -2\frac{dx}{dt} = -2k(A_0 - 2x)^2 \times (B_0 - x)$$

$$\frac{dx}{dt} = k(A_0 - 2x)^2 \times \left(\frac{1}{2}A_0 - x\right) = \frac{1}{2}k(A_0 - 2x)^3$$

$$\frac{1}{2}kt = \int_0^x \frac{dx}{(A_0 - 2x)^3} = \frac{1}{4} \times \left[\left(\frac{1}{A_0 - 2x}\right)^2 - \left(\frac{1}{A_0}\right)^2\right]$$

Therefore, $\boxed{kt = \dfrac{2x(A_0 - x)}{A_0^2(A_0 - 2x)^2}}$

(b) $\dfrac{dx}{dt} = k(A_0 - 2x)^2 \times (B_0 - x) = k(A_0 - 2x)^2 \times (A_0 - x) \quad \left[B_0 = 2 \times \dfrac{1}{2}A_0 = A_0\right]$

$$kt = \int_0^x \frac{dx}{(A_0 - 2x)^2 \times (A_0 - x)}$$

We proceed by the method of partial fractions (which is employed in the general case too), and look for the concentrations α, β, and γ in

$$\frac{1}{(A_0 - 2x)^2 \times (A_0 - x)} = \frac{\alpha}{(A_0 - 2x)^2} + \frac{\beta}{A_0 - 2x} + \frac{\gamma}{A_0 - x}$$

which requires that

$$\alpha(A_0 - x) + \beta(A_0 - 2x) \times (A_0 - x) + \gamma(A_0 - 2x)^2 = 1$$

$$(A_0\alpha + A_0^2\beta + A_0^2\gamma) - (\alpha + 3\beta A_0 + 4\gamma A_0)x + (2\beta + 4\gamma)x^2 = 1$$

This must be true for all x; therefore

$$A_0\alpha + A_0^2\beta + A_0^2\gamma = 1$$
$$\alpha + 3A_0\beta + 4A_0\gamma = 0$$
$$2\beta + 4\gamma = 0$$

These solve to give $\alpha = \dfrac{2}{A_0}$, $\beta = \dfrac{-2}{A_0^2}$, and $\gamma = \dfrac{1}{A_0^2}$

Therefore,

$$kt = \int_0^x \left(\frac{(2/A_0)}{(A_0 - 2x)^2} - \frac{(2/A_0^2)}{A_0 - 2x} + \frac{(1/A_0^2)}{A_0 - x} \right) dx$$

$$= \left(\frac{(1/A_0)}{A_0 - 2x} + \frac{1}{A_0^2}\ln(A_0 - 2x) - \frac{1}{A_0^2}\ln(A_0 - x) \right)\Bigg|_0^x$$

$$= \boxed{\left(\frac{2x}{A_0^2(A_0 - 2x)} \right) + \left(\frac{1}{A_0^2} \right) \ln\left(\frac{A_0 - 2x}{A_0 - x} \right)}$$

P25.22 The mechanism considered is

$$\mathrm{E} + \mathrm{S} \underset{k_a'}{\overset{k_a}{\rightleftharpoons}} (\mathrm{ES}) \underset{k_b'}{\overset{k_b}{\rightleftharpoons}} \mathrm{P} + \mathrm{E}$$

We apply the steady-state approximation to [(ES)].

$$\frac{d[\mathrm{ES}]}{dt} = k_a[\mathrm{E}][\mathrm{S}] - k_a'[(\mathrm{ES})] - k_b[(\mathrm{ES})] + k_b'[\mathrm{E}][\mathrm{P}] = 0$$

Substituting $[\mathrm{E}] = [\mathrm{E}]_0 - [(\mathrm{ES})]$ we obtain

$$k_a([\mathrm{E}]_0 - [(\mathrm{ES})])[\mathrm{S}] - k_a'[(\mathrm{ES})] - k_b[(\mathrm{ES})] + k_b'([\mathrm{E}]_0 - [(\mathrm{ES})])[\mathrm{P}] = 0$$
$$(-k_a[\mathrm{S}] - k_a' - k_b - k_b'[\mathrm{P}])[(\mathrm{ES})] + k_a[\mathrm{E}]_0[\mathrm{S}] - k_b'[\mathrm{E}]_0[\mathrm{P}] = 0$$

$$[(\mathrm{ES})] = \frac{k_a[\mathrm{E}]_0[\mathrm{S}] + k_b'[\mathrm{E}]_0[\mathrm{P}]}{k_a[\mathrm{S}] + k_a' + k_b + k_b'[\mathrm{P}]} = \frac{[\mathrm{E}]_0[\mathrm{S}] + \left(\frac{k_b'}{k_a}\right)[\mathrm{E}]_0[\mathrm{P}]}{K_M + [\mathrm{S}] + \left(\frac{k_b'}{k_a}\right)[\mathrm{P}]} \quad \left[K_M = \frac{k_a' + k_b}{k_a} \right]$$

Then,

$$\frac{d[\mathrm{P}]}{dt} = k_b[(\mathrm{ES})] - k_b'[\mathrm{P}][\mathrm{E}] = k_b \frac{[\mathrm{E}]_0[\mathrm{S}] + \left(\frac{k_b'}{k_a}\right)[\mathrm{E}]_0[\mathrm{P}]}{K_M + [\mathrm{S}] + \left(\frac{k_b'}{k_a}\right)[\mathrm{P}]} - k_b'[\mathrm{P}] \times \left([\mathrm{E}]_0 - \frac{[\mathrm{E}]_0[\mathrm{S}] + \left(\frac{k_b'}{k_a}\right)[\mathrm{E}]_0[\mathrm{P}]}{K_M + [\mathrm{S}] + \left(\frac{k_b'}{k_a}\right)[\mathrm{P}]} \right)$$

$$= \frac{k_b\left[[\mathrm{E}]_0[\mathrm{S}] + \left(\frac{k_b'}{k_a}\right)[\mathrm{E}]_0[\mathrm{P}]\right] - k_b'[\mathrm{E}]_0[\mathrm{P}]K_M}{K_M + [\mathrm{S}] + \left(\frac{k_b'}{k_a}\right)[\mathrm{P}]}$$

Substituting for K_M in the numerator and rearranging

$$\boxed{\frac{d[\mathrm{P}]}{dt} = \frac{k_b[\mathrm{E}]_0[\mathrm{S}] + \left(\frac{k_a'k_b'}{k_a}\right)[\mathrm{E}]_0[\mathrm{P}]}{K_M + [\mathrm{S}] + \left(\frac{k_b'}{k_a}\right)[\mathrm{P}]}}$$

For large concentrations of substrate, such that $[\mathrm{S}] \gg K_M$ and $[\mathrm{S}] \gg [\mathrm{P}]$,

$$\boxed{\frac{d[\mathrm{P}]}{dt} = k_b[\mathrm{E}]_0}$$

which is the same as the unmodified mechanism. For [S] $\gg K_M$, but [S] $\approx$ [P]

$$\boxed{\frac{d[P]}{dt} = k_b[E]_0\left\{\frac{[S]-(k/k_b)[P]}{[S]+(k/k_a')[P]}\right\}} \qquad k = \frac{k_a'k_b'}{k_a}$$

$$\text{For } [S] \to 0, \quad \frac{d[P]}{dt} = \frac{-k_a'k_b'[E]_0[P]}{k_a'+k_b+k_b'[P]} = \boxed{\frac{-k_a'[E]_0[P]}{k_P+[P]}}$$

where $k_P = \dfrac{k_a'+k_b}{k_b'}$

Comment. The negative sign in the expression for $\dfrac{d[P]}{dt}$ for the case $[S] \to 0$ is to be interpreted to mean that the mechanism in this case is the reverse of the mechanism for the case $[P] \to 0$. The roles of P and S are interchanged.

Question. Can you demonstrate the last statement in the comment above?

Solutions to additional problems

P25.24

$$E_a = -R\frac{d\ln(k)}{d(1/T)} \quad [24, 25]$$

$$= -R\ln(10)\frac{d\log(k)}{d(1/T)} = -R\ln(10)\frac{d}{d(1/T)}\left\{11.75 - \frac{5488}{T/K}\right\}$$

$$= -R\ln(10)\{-5488\,K\} = \{8.3145\,J\,K^{-1}\,mol^{-1}\}\ln(10)\{5488\,K\}$$

$$\boxed{E_a = 105\,kJ\,mol^{-1}}$$

$$\Delta_r G^{\ominus} = -RT\ln K \quad [9.18]$$

$$= -RT\ln(10)\log K$$

At 298.15 K

$$\Delta_r G^{\ominus} = -(8.3145\,J\,K^{-1}\,mol^{-1}) \times (298.15\,K)\ln(10)\left\{-1.36 + \frac{1794}{298.15}\right\}$$

$$\boxed{\Delta_r G^{\ominus} = -26.6\,kJ\,mol^{-1}}$$

$$\Delta_r H^{\ominus} = -R\frac{d\ln(K)}{d(1/T)} \quad [9.24]$$

$$= -R\ln(10)\frac{d\log(K)}{d(1/T)}$$

$$= -R\ln(10)\frac{d}{d(1/T)}\left\{-1.36 + \frac{1794}{T/K}\right\}$$

$$= -R\ln(10)\{1794\,K\} = -\{8.3145\,J\,K^{-1}\,mol^{-1}\}\ln(10)\{1794\,K\}$$

$$\boxed{\Delta_r H^{\ominus} = -34.3\,kJ\,mol^{-1}}$$

The reaction is

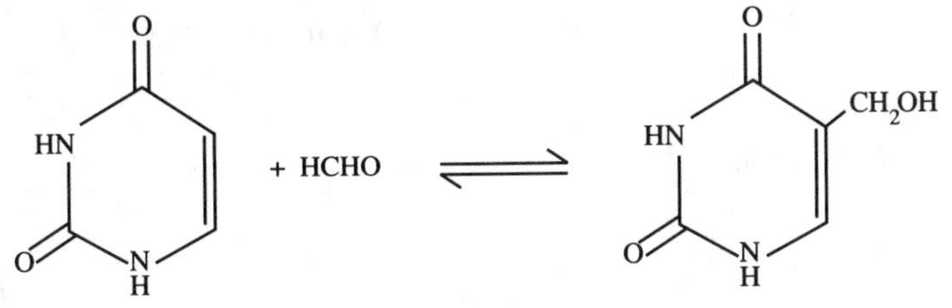

O
HN
O N H
+ HCHO ⇌
O
CH_2OH
HN
O N H

The equations for the rate constant k and the equilibrium constant K were obtained under conditions corresponding to the biological standard state (pH $= 7$, $p = 1\,\text{bar}$; Section 9.6). Thus the values of $\Delta_r G$ calculated from the equation for K are $\Delta_r G^{\oplus}$ values which differ significantly from $\Delta_r G^{\ominus}(\text{pH} = 1, p = 1\,\text{bar})$. Prebiotic conditions are more likely to be near pH $= 7$ than pH $= 1$, so we expect that the reaction will still be favourable ($K \gg 1$) thermodynamically.

Because $\Delta_r G = \Delta_r G^{\oplus} + RT \ln Q$ [9.10] and since we might expect $Q < 1$ in a prebiotic environment, $\Delta_r G < \Delta_r G^{\oplus}$. But, as shown in the calculation above, $\Delta_r G^{\oplus}$ is rather large and negative ($-26.6\,\text{kJ mol}^{-1}$), so we expect it will still be large and negative under the prebiotic conditions; hence the reaction will be spontaneous for these conditions. We expect that $\Delta_r H \approx \Delta_r H^{\oplus}$ because enthalpy changes largely reflect bond breakage and bond formation energies.

A plot of the equation for the rate constant k is shown in Fig. 25.3(a) and that for the equilibrium constant in Fig. 25.3(b). From a kinetic point of view the reaction becomes more favourable at higher temperatures; from a thermodynamic point of view it becomes less favourable, but $K \gg 1$ at all temperatures.

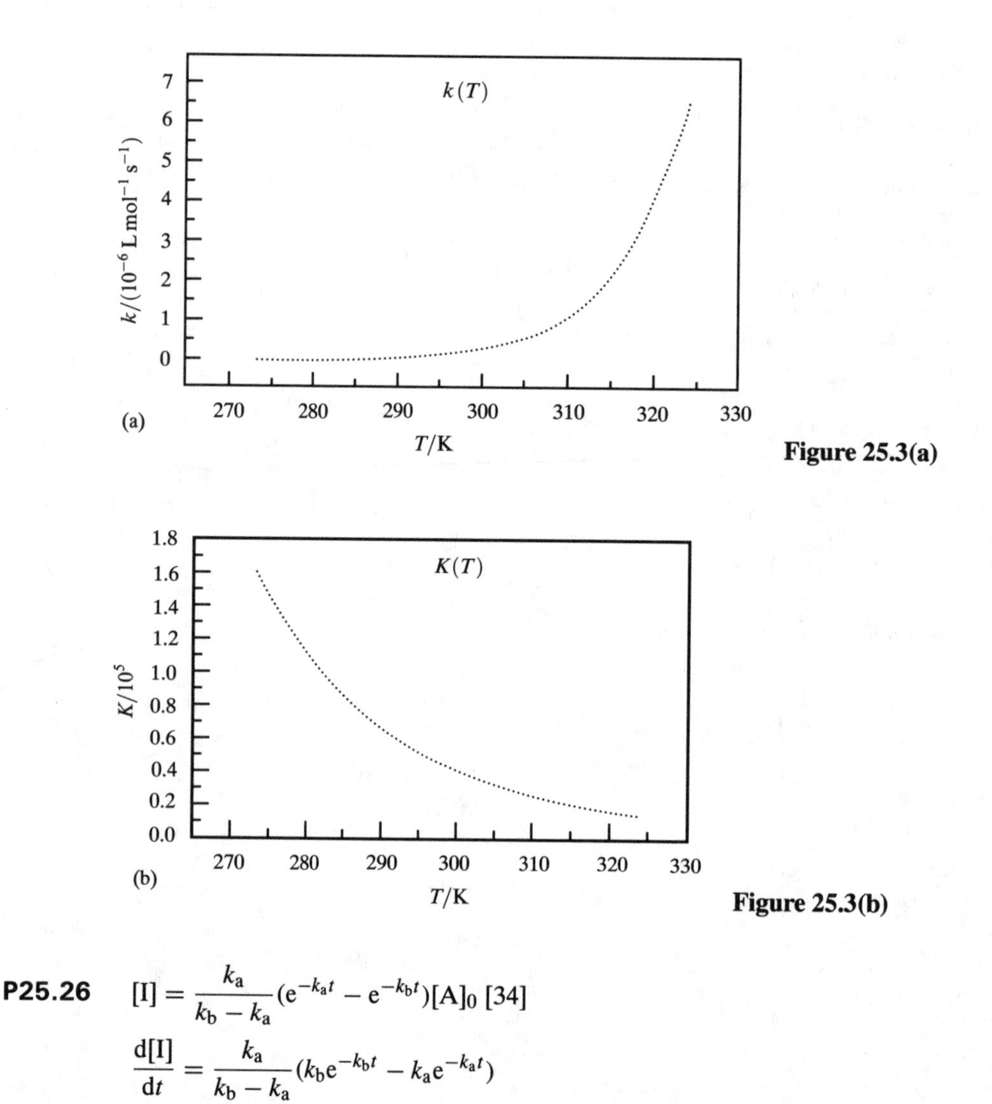

Figure 25.3(a)

Figure 25.3(b)

P25.26 $[\text{I}] = \dfrac{k_a}{k_b - k_a}(\mathrm{e}^{-k_a t} - \mathrm{e}^{-k_b t})[\text{A}]_0$ [34]

$$\frac{\mathrm{d}[\text{I}]}{\mathrm{d}t} = \frac{k_a}{k_b - k_a}(k_b \mathrm{e}^{-k_b t} - k_a \mathrm{e}^{-k_a t})$$

[I] reaches a maximum when $d[I]/dt = 0$. This occurs when t satisfies the equation

$$k_b e^{-k_b t_{max}} - k_a e^{-k_a t_{max}} = 0$$

$$k_b e^{-k_b t_{max}} \left(1 - \frac{k_a}{k_b} e^{-(k_a - k_b) t_{max}}\right) = 0$$

$$1 - \frac{k_a}{k_b} e^{-(k_a - k_b) t_{max}} = 0$$

$$e^{-(k_a - k_b) t_{max}} = k_b / k_a$$

$$-(k_a - k_b) t_{max} = \ln(k_b / k_a)$$

$$t_{max} = \frac{\ln(k_b/k_a)}{(k_b - k_a)} = \frac{(k_a/k_b)\ln(k_a/k_b)}{k_a\left(\frac{k_a}{k_b} - 1\right)}$$

For $k_a = 1.0\,\text{min}^{-1}$, the times at which [I] is a maximum are

k_a/k_b	5	1	0.5
t_{max}/min	2.01	1	0.693

The evaluation of t_{max} when $k_a/k_b = 1$ requires special care. Imagine $k_a/k_b > 0$ and take $\lim_{k_a/k_b \to 1} (t_{max})$. In this limit the value of $\left(\frac{k_a}{k_b} - 1\right)$ in the denominator becomes very small (call this value x) and can be viewed as being part of the Taylor series expansion of $\ln(1 + x)$

$$\ln(1 + x) = x - \frac{x^2}{2} + \frac{x^3}{3} + \cdots \approx x$$

$$\lim_{k_a/k_b \to 1} (t_{max}) = \frac{1}{k_a} \lim_{k_a/k_b \to 1} \left\{\frac{(k_a/k_b)\ln(k_a/k_b)}{\ln(k_a/k_b)}\right\}$$

$$= \frac{1}{k_a}$$

Plots of $\frac{[I]}{[A]_0}$ for $k_a/k_b = 5$, 1, and 0.5 are shown in Fig. 25.4.

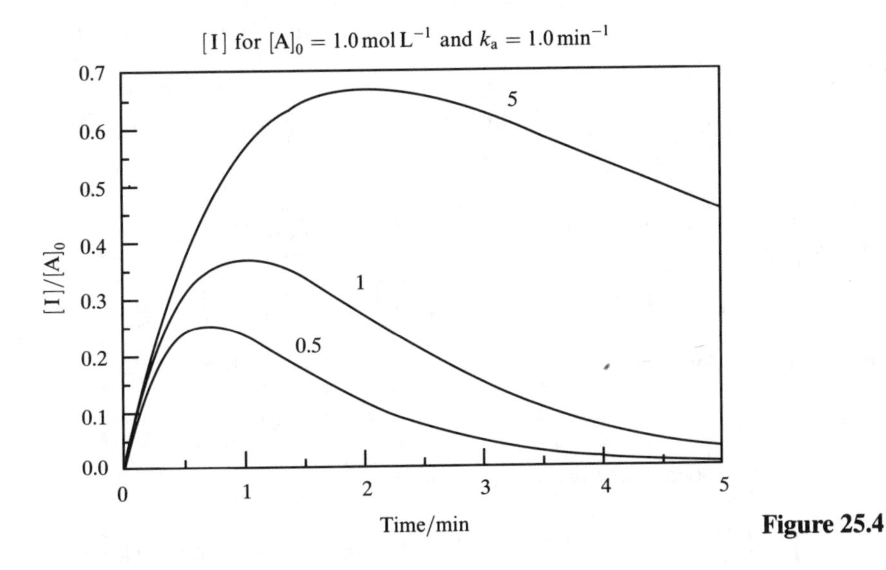

Figure 25.4

P25.28 The Arrhenius expression for the rate constant is

$$k = Ae^{-E_a/RT} \quad [26] \quad \text{so } \ln k = \ln A - E_a/RT \ [24]$$

A plot of $\ln k$ versus $1/T$ will have slope $-E_a/R$ and y- intercept $\ln A$. The transformed data and plot (Fig. 25.5) follow

T/K	295	295	223	218	213	206	200	195
$10^{-6}k/(\mathrm{L\,mol^{-1}\,s^{-1}})$	3.70	3.55	0.494	0.452	0.379	0.295	0.241	0.217
$\ln k/(\mathrm{L\,mol^{-1}\,s^{-1}})$	15.12	15.08	13.11	13.02	12.85	12.59	12.39	12.29
K/T	0.00339	0.00339	0.00448	0.00459	0.00469	0.00485	0.00500	0.00513

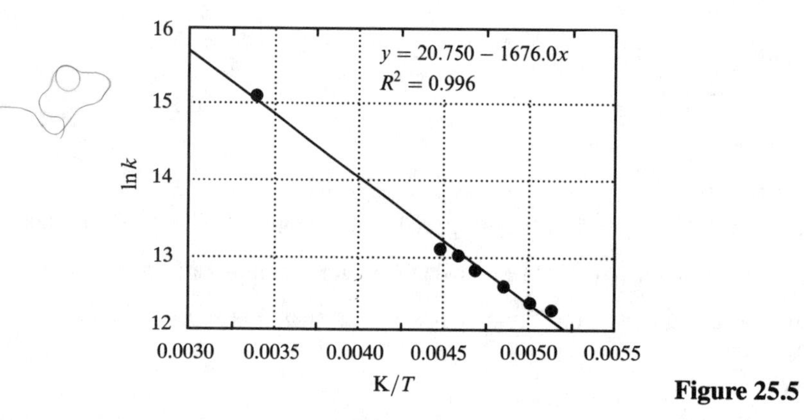

Figure 25.5

So $E_a = -(8.3145\,\mathrm{J\,K^{-1}mol^{-1}}) \times (-1676\,\mathrm{K}) = 1.39 \times 10^4\,\mathrm{J\,mol^{-1}} = \boxed{13.9\,\mathrm{kJ\,mol^{-1}}}$

and $A = e^{20.750}\,\mathrm{L\,mol^{-1}\,s^{-1}} = \boxed{1.03 \times 10^9\,\mathrm{L\,mol^{-1}\,s^{-1}}}$

P25.30 The rate of reaction [1] for this reaction is

$$v = k[\mathrm{Cl}][\mathrm{O_3}]$$

(a) $v = (1.7\times10^{10}\,\mathrm{L\,mol^{-1}s^{-1}})\exp(-260\,\mathrm{K}/220\,\mathrm{K})\times(5\times10^{-17}\,\mathrm{mol\,L^{-1}})\times(8\times10^{-9}\,\mathrm{mol\,L^{-1}})$

$= \boxed{2.\bar{1} \times 10^{-15}\,\mathrm{mol\,L^{-1}\,s^{-1}}}$

(b) $v = (1.7\times10^{10}\,\mathrm{L\,mol^{-1}s^{-1}})\exp(-260\,\mathrm{K}/270\,\mathrm{K})\times(3\times10^{-15}\,\mathrm{mol\,L^{-1}})\times(8\times10^{-11}\,\mathrm{mol\,L^{-1}})$

$= \boxed{1.\bar{6} \times 10^{-15}\,\mathrm{mol\,L^{-1}\,s^{-1}}}$

P25.32 The relation between equilibrium and rate constant [Section 27.6(a)] is

$$K = \frac{k}{k'} = \exp\left(\frac{-\Delta_r H^{\ominus}}{RT}\right)\exp\left(\frac{-\Delta_r S^{\ominus}}{R}\right) = \left(\frac{A}{A'}\right)\exp\left(\frac{E'_a - E_a}{RT}\right)$$

Setting the temperature-dependent parts equal yields

$$\Delta_r H^{\ominus} = E_a - E'_a = [-4.2 - (53.3)]\,\mathrm{kJ\,mol^{-1}} = -57.5\,\mathrm{kJ\,mol^{-1}}$$

Setting the temperature-independent parts equal yields

$$\exp\left(\frac{\Delta_r S^{\ominus}}{R}\right) = \left(\frac{A}{A'}\right) \quad \text{so} \quad \Delta_r S^{\ominus} = R\ln\left(\frac{A}{A'}\right) = (8.3145\,\mathrm{J\,K^{-1}\,mol^{-1}})\ln\left(\frac{1.0\times10^9}{1.4\times10^{11}}\right)$$

$$\Delta_r S^{\ominus} = -41.1\,\mathrm{J\,K^{-1}\,mol^{-1}}$$

The thermodynamic quantities of the reaction are related to standard molar quantities

$$\Delta_r H^{\ominus} = \Delta_f H^{\ominus}(\mathrm{C_2H_6}) + \Delta_f H^{\ominus}(\mathrm{Br}) - \Delta_f H^{\ominus}(\mathrm{C_2H_5}) - \Delta_f H^{\ominus}(\mathrm{HBr})$$

so $\Delta_f H^{\ominus}(\mathrm{C_2H_5}) = \Delta_f H^{\ominus}(\mathrm{C_2H_6}) + \Delta_f H^{\ominus}(\mathrm{Br}) - \Delta_f H^{\ominus}(\mathrm{HBr}) - \Delta_r H^{\ominus}$

$$\Delta_f H^{\ominus}(\mathrm{C_2H_5}) = [(-84.68) + 111.88 - (-36.40) - (-57.5)]\,\mathrm{kJ\,mol^{-1}} = \boxed{121.2\,\mathrm{kJ\,mol^{-1}}}$$

$$S_m^{\ominus}(\mathrm{C_2H_5}) = [229.60 + 175.02 - 198.70 - (-41.1)]\,\mathrm{J\,mol^{-1}\,K^{-1}} = \boxed{247.0\,\mathrm{J\,K^{-1}\,mol^{-1}}}$$

$$\begin{aligned}\Delta_f G^{\ominus}(\mathrm{C_2H_5}) &= [-32.82 + 82.396 - (-53.45)]\,\mathrm{kJ\,mol^{-1}} - \Delta_r G^{\ominus}\\ &= 103.03\,\mathrm{kJ\,mol^{-1}} - \Delta_r G^{\ominus}\end{aligned}$$

$$\begin{aligned}\Delta_r G^{\ominus} &= \Delta_r H^{\ominus} - T\Delta_r S^{\ominus} = -57.5\,\mathrm{kJ\,mol^{-1}} - (298\,\mathrm{K}) \times (-41.1 \times 10^{-3}\,\mathrm{kJ\,K^{-1}\,mol^{-1}})\\ &= -45.3\,\mathrm{kJ\,mol^{-1}}\end{aligned}$$

$$\Delta_f G^{\ominus}(\mathrm{C_2H_5}) = [103.03 - (-45.3)]\,\mathrm{kJ\,mol^{-1}} = \boxed{148.3\,\mathrm{kJ\,mol^{-1}}}$$

26 The kinetics of complex reactions

Solutions to exercises

In the following exercises and problems, it is recommended that rate constants are labelled with the number of the step in the proposed reaction mechanism and that any reverse steps are labelled similarly but with a prime.

E26.1 We assume that the steady-state approximation applies to [O] (but see the question below). Then

$$\frac{d[O]}{dt} = 0 = k_1[O_3] - k_1'[O][O_2] - k_2[O][O_3]$$

Solving for [O],

$$[O] = \frac{k_1[O_3]}{k_1'[O_2] + k_2[O_3]}$$

$$\text{Rate} = -\frac{1}{2}\frac{d[O_3]}{dt}$$

$$\frac{d[O_3]}{dt} = -k_1[O_3] + k_1'[O][O_2] - k_2[O][O_3]$$

Substituting for [O] from above

$$\frac{d[O_3]}{dt} = -k_1[O_3] + \frac{k_1[O_3](k_1'[O_2] - k_2[O_3])}{k_1'[O_2] + k_2[O_3]}$$

$$= \frac{-k_1[O_3](k_1'[O_2] + k_2[O_3]) + k_1[O_3](k_1'[O_2] - k_2[O_3])}{k_1'[O_2] + k_2[O_3]} = \frac{-2k_1k_2[O_3]^2}{k_1'[O_2] + k_2[O_3]}$$

$$\text{Rate} = \boxed{\frac{k_1k_2[O_3]^2}{k_1'[O_2] + k_2[O_3]}}$$

Question. Can you determine the rate law expression if the first step of the proposed mechanism is a rapid pre-equilibrium? Under what conditions does the rate expression above reduce to the latter?

E26.2 The steady-state expressions are now

$$k_2[NO_2][NO_3] - k_3[NO][NO_3] = 0$$

$$k_1[N_2O_5] - k_1'[NO_2][NO_3] - k_2[NO_2][NO_3] - k_3[NO][NO_3] = 0$$

$$\frac{d[N_2O_5]}{dt} = -k_1[N_2O_5] + k_1'[NO_2][NO_3]$$

From the steady-state equations

$$k_3[NO][NO_3] = k_2[NO_2][NO_3]$$

$$[NO_2][NO_3] = \frac{k_1}{k_1' + 2k_2}[N_2O_5]$$

Substituting, $$\frac{d[N_2O_5]}{dt} = -k_1[N_2O_5] + \frac{k_1k_1'}{k_1' + 2k_2}[N_2O_5] = \frac{-2k_1k_2}{k_1' + 2k_2}[N_2O_5]$$

$$\text{Rate} = \frac{k_1k_2}{k_1' + 2k_2}[N_2O_5] = k[N_2O_5]$$

E26.3 At 800 K, the branching-chain explosion occurs between $\boxed{0.16\,\text{kPa and } 4.0\,\text{kPa}}$

E26.4 Number of photons absorbed $= \Phi^{-1} \times$ number of molecules that react [Section 26.3]. Therefore,

$$\text{Number absorbed} = \frac{(1.14 \times 10^{-3}\,\text{mol}) \times (6.022 \times 10^{23}\,\text{einstein}^{-1})}{2.1 \times 10^{2}\,\text{mol einstein}^{-1}} = \boxed{3.3 \times 10^{18}}$$

E26.5 For a source of power P and wavelength λ, the amount of photons (n_λ) generated in a time t is

$$n_\lambda = \frac{Pt}{h\nu N_A} = \frac{P\lambda t}{hcN_A} = \frac{(100\,\text{W}) \times (45) \times (60\,\text{s}) \times (490 \times 10^{-9}\,\text{m})}{(6.626 \times 10^{-34}\,\text{J s}) \times (2.998 \times 10^{8}\,\text{m s}^{-1}) \times (6.022 \times 10^{23}\,\text{mol}^{-1})}$$
$$= 1.11\,\text{mol}$$

The amount of photons absorbed is 60 per cent of this incident flux, or 0.664 mol. Therefore,

$$\Phi = \frac{0.344\,\text{mol}}{0.664\,\text{mol}} = \boxed{0.518}$$

Alternatively, expressing the amount of photons in einsteins [1 mol photons = 1 einstein], $\Phi = 0.518\,\text{mol einstein}^{-1}$.

E26.6
$$\frac{d[A^-]}{dt} = k_1[AH][B] - k_2[A^-][BH^+] - k_3[A^-][AH] = 0$$

Therefore, $[A^-] = \boxed{\dfrac{k_1[AH][B]}{k_2[BH^+] + k_3[AH]}}$

and the rate of formation of product is

$$\frac{d[P]}{dt} = k_3[AH][A^-] = \boxed{\frac{k_1k_3[AH]^2[B]}{k_2[BH^+] + k_3[AH]}}$$

E26.7 $\boxed{\text{Step1: initiation}}$ [radicals formed]; $\boxed{\text{Steps 2 and 3: propagation}}$ [new radicals formed]; $\boxed{\text{Step 4: termination}}$ [non-radical product formed].

$$\frac{d[AH]}{dt} = -k_a[AH] - k_c[AH][B]$$

$$\text{(i)}\ \frac{d[A]}{dt} = k_a[AH] - k_b[A] + k_c[AH][B] - k_d[A][B] \approx 0$$

$$\text{(ii)}\ \frac{d[B]}{dt} = k_b[A] - k_c[AH][B] - k_d[A][B] \approx 0$$

$$\left.\begin{aligned}&\text{(i + ii) } [A][B] = \left(\frac{k_a}{2k_d}\right)[AH]\\ &\text{(i − ii) } [A] = \left(\frac{k_a + 2k_c[B]}{2k_b}\right)[AH]\end{aligned}\right\}$$

Then, solving for [A]

$$[A] = k[AH],\ k = \left(\frac{k_a}{4k_b}\right) \times \left[1 + \left(1 + \frac{8k_bk_c}{k_ak_d}\right)^{1/2}\right]$$

from which it follows that

$$[B] = \frac{k_a[AH]}{2k_d[A]} = \frac{k_a}{2kk_d}$$

and hence that $\dfrac{d[AH]}{dt} = -k_a[AH] - \left(\dfrac{k_a k_c}{2kk_d}\right)[AH] = \boxed{-k_{eff}[AH]}$

with $\boxed{k_{eff} = k_a + \dfrac{k_a k_c}{2kk_d}}$

Solutions to problems

Solutions to numerical problems

P26.1

$$UO_2^{2+} + h\nu \rightarrow (UO_2^{2+})^*$$
$$(UO_2^{2+})^* + (COOH)_2 \rightarrow UO_2^{2+} + H_2O + CO_2 + CO$$
$$2MnO_4^- + 5(COOH)_2 + 6H^+ \rightarrow 10CO_2 + 8H_2O + 2Mn^{2+}$$

$17.0\,cm^3$ of $0.212\,M\ KMnO_4$ is equivalent to

$$\tfrac{5}{2} \times (17.0\,cm^3) \times (0.212\,mol\,L^{-1}) = 9.01 \times 10^{-3}\,mol\ (COOH)_2$$

The initial sample contained $5.232\,g\ (COOH)_2$, corresponding to

$$\frac{5.232\,g}{90.04\,g\,mol^{-1}} = 5.81 \times 10^{-2}\,mol\ (COOH)_2$$

Therefore, $(5.81\times10^{-2}\,mol)-(9.01\times10^{-3}\,mol) = 4.91\times10^{-2}\,mol$ of the acid has been consumed. A quantum efficiency 0.53 implies that the amount of photons absorbed must have been

$$\frac{4.91 \times 10^{-2}\,mol}{0.53} = 9.3 \times 10^{-2}\,mol$$

Since the exposure was for 300 s, the rate of incidence of photons was

$$\frac{9.3 \times 10^{-2}\,mol}{300\,s} = 3.1 \times 10^{-4}\,mol\,s^{-1}$$

Since 1 mol photons = 1 einstein, the incident rate is $\boxed{3.1 \times 10^{-4}\,einstein\,s^{-1}}$ or $\boxed{1.9 \times 10^{20}\,s^{-1}}$

P26.3

$$H + NO_2 \rightarrow OH + NO \quad k_1 = 2.9 \times 10^{10}\,L\,mol^{-1}\,s^{-1}$$
$$OH + OH \rightarrow H_2O + O \quad k_2 = 1.55 \times 10^{9}\,L\,mol^{-1}\,s^{-1}$$
$$O + OH \rightarrow O_2 + H \quad k_3 = 1.1 \times 10^{10}\,L\,mol^{-1}\,s^{-1}$$
$$[H]_0 = 4.5 \times 10^{-10}\,mol\,cm^{-3} \qquad [NO_2]_0 = 5.6 \times 10^{-10}\,mol\,cm^{-3}$$
$$\frac{d[O]}{dt} = k_2[OH]^2 + k_3[O][OH] \qquad \frac{d[O_2]}{dt} = k_3[O][OH]$$
$$\frac{d[OH]}{dt} = k_1[H][NO_2] - 2k_2[OH]^2 - k_3[O][OH] \qquad \frac{d[NO_2]}{dt} = -k_1[H][NO_2]$$
$$\frac{d[H]}{dt} = k_3[O][OH] - k_1[H][NO_2]$$

These equations serve to show how even a simple sequence of reactions leads to a complicated set of nonlinear differential equations. Since we are interested in the time behaviour of the composition we may not invoke the steady-state assumption. The only thing left is to use a computer and to integrate the equations numerically. The outcome of this is the set of curves shown in Fig. 26.1 (they have been sketched from the original reference). The similarity to an $A \rightarrow B \rightarrow C$ scheme should be noticed (and expected), and the general features can be analysed quite simply in terms of the underlying reactions.

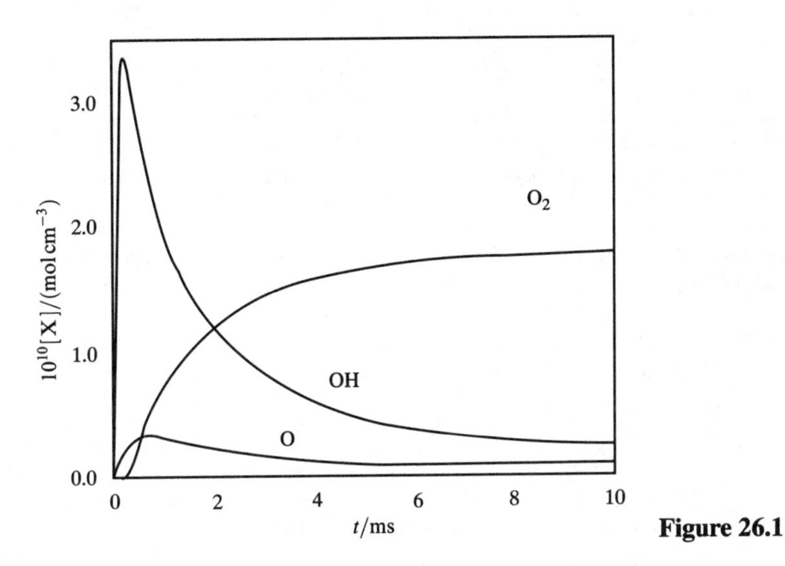

Figure 26.1

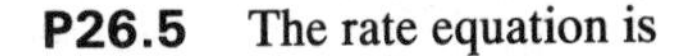

P26.5 The rate equation is

$$\frac{dN}{dt} = bN - dN$$

which has the solution

$$\boxed{N(t) = N_0 e^{(b-d)t} = N_0 e^{kt}}$$

A least-squares fit to the above data gives

$$N_0 = 0.484 \times 10^9 \approx 0.5 \times 10^9$$
$$k = 9.19 \times 10^{-3}\,\text{y}^{-1}$$
$$R^2 = \text{(coefficient of determination)} = 0.983$$
$$\text{Standard error of estimate} = 0.130 \times 10^9$$

Thus, this model of population growth for the planet as a whole fits the data fairly well.

Comment. Despite the fact that the Malthusian model seems to fit the (admittedly crude) population data it has been much criticized. An alternative rate equation that takes into account the carrying capacity K of the planet is due to Verhulst (1836). This rate equation is

$$\frac{dN}{dt} = kN\left(1 - \frac{N}{K}\right)$$

Question. Does the Verhulst model fit our limited data any better?

Solutions to theoretical problems

P26.6 $$\frac{d[P]}{dt} = k[A]^2[P]$$

$$[A] = A_0 - x, \quad [P] = P_0 + x, \quad \frac{d[P]}{dt} = \frac{dx}{dt} = k(A_0 - x)^2(P_0 + x)$$

$$\int_0^x \frac{dx}{(A_0 - x)^2(P_0 + x)} = kt$$

Solve the integral by partial fractions

$$\frac{1}{(A_0-x)^2(P_0+x)} = \frac{\alpha}{(A_0-x)^2} + \frac{\beta}{A_0-x} + \frac{\gamma}{P_0+x}$$
$$= \frac{\alpha(P_0+x)+\beta(A_0-x)(P_0+x)+\gamma(A_0-x)^2}{(A_0-x)^2(P_0+x)}$$

$$\left.\begin{aligned} P_0\alpha + A_0P_0\beta + A_0^2\gamma &= 1 \\ \alpha + (A_0-P_0)\beta - 2A_0\gamma &= 0 \\ -\beta+\gamma &= 0 \end{aligned}\right\}$$

This set of simultaneous equations solves to

$$\alpha = \frac{1}{A_0+P_0}, \qquad \beta = \gamma = \frac{\alpha}{A_0+P_0}$$

Therefore,

$$\begin{aligned} kt &= \left(\frac{1}{A_0+P_0}\right)\int_0^x \left[\left(\frac{1}{A_0-x}\right)^2 + \left(\frac{1}{A_0+P_0}\right)\left(\frac{1}{A_0-x}+\frac{1}{P_0+x}\right)\right]\mathrm{d}x \\ &= \left(\frac{1}{A_0+P_0}\right)\left\{\left(\frac{1}{A_0-x}\right) - \left(\frac{1}{A_0}\right) + \left(\frac{1}{A_0+P_0}\right)\left[\ln\left(\frac{A_0}{A_0-x}\right) + \ln\left(\frac{P_0+x}{P_0}\right)\right]\right\} \\ &= \left(\frac{1}{A_0+P_0}\right)\left[\left(\frac{x}{A_0(A_0-x)}\right) + \left(\frac{1}{A_0+P_0}\right)\ln\left(\frac{A_0(P_0+x)}{(A_0-x)P_0}\right)\right] \end{aligned}$$

Therefore, with $y = \dfrac{x}{A_0}$ and $p = \dfrac{P_0}{A_0}$,

$$A_0(A_0+P_0)kt = \boxed{\left(\frac{y}{1-y}\right) + \left(\frac{1}{1+p}\right)\ln\left(\frac{p+y}{p(1-y)}\right)}$$

The maximum rate occurs at

$$\frac{\mathrm{d}v_\mathrm{P}}{\mathrm{d}t} = 0, \qquad v_\mathrm{P} = k[\mathrm{A}]^2[\mathrm{P}]$$

and hence at the solution of

$$\begin{aligned} &2k\left(\frac{\mathrm{d}[\mathrm{A}]}{\mathrm{d}t}\right)[\mathrm{A}][\mathrm{P}] + k[\mathrm{A}]^2\frac{\mathrm{d}[\mathrm{P}]}{\mathrm{d}t} = 0 \\ &-2k[\mathrm{A}][\mathrm{P}]v_\mathrm{P} + k[\mathrm{A}]^2 v_\mathrm{P} = 0 \quad [\text{as } v_\mathrm{A} = -v_\mathrm{P}] \\ &k[\mathrm{A}]([\mathrm{A}] - 2[\mathrm{P}])v_\mathrm{P} = 0 \end{aligned}$$

That is, the rate is a maximum when $[\mathrm{A}] = 2[\mathrm{P}]$, which occurs at

$$A_0 - x = 2P_0 + 2x, \quad \text{or} \quad x = \tfrac{1}{3}(A_0 - 2P_0); \quad y = \tfrac{1}{3}(1-2p)$$

Substituting this condition into the integrated rate law gives

$$A_0(A_0+P_0)kt_{\mathrm{max}} = \left(\frac{1}{1+p}\right)\left(\frac{1}{2}(1-2p) + \ln\frac{1}{2p}\right)$$

or $\boxed{(A_0+P_0)^2 kt_{\mathrm{max}} = \tfrac{1}{2} - p - \ln 2p}$

P26.9 $$\frac{d[Cr(CO)_5]}{dt} = I - k_2[Cr(CO)_5][CO] - k_3[Cr(CO)_5][M] + k_4[Cr(CO)_5M] = 0 \text{ [steady state]}$$

Hence, $[Cr(CO)_5] = \dfrac{I + k_4[Cr(CO)_5M]}{k_2[CO] + k_3[M]}$

$$\frac{d[Cr(CO)_5M]}{dt} = k_3[Cr(CO)_5][M] - k_4[Cr(CO)_5M]$$

Substituting for $[Cr(CO)_5]$ from above,

$$\frac{d[Cr(CO)_5M]}{dt} = \frac{k_3I[M] - k_2k_4[Cr(CO)_5M][CO]}{k_2[CO] + k_3[M]} = -f[Cr(CO)_5M]$$

if $f = \boxed{\dfrac{k_2k_4[CO]}{k_2[CO] + k_3[M]}}$

and we have taken $k_3I[M] \ll k_2k_4[Cr(CO)_5M][CO]$. Therefore,

$$\frac{1}{f} = \frac{1}{k_4} + \frac{k_3[M]}{k_2k_4[CO]}$$

and a graph of $\dfrac{1}{f}$ against [M] should be a straight line.

P26.11 $\langle \overline{M} \rangle_N = \dfrac{M}{1-p}$ [Example 26.5 with $\langle \overline{M} \rangle_N = \langle n \rangle M$]

The probability P_n that a polymer consists of n monomers is equal to the probability that it has $n-1$ reacted end groups and one unreacted end group. The former probability is p^{n-1}; the latter $1-p$. Therefore, the total probability of finding an n-mer is

$$P_n = p^{n-1}(1-p)$$

$$\langle M^2 \rangle_N = M^2\langle n^2 \rangle = M^2\sum_n n^2 P_n = M^2(1-p)\sum_n n^2 p^{n-1} = M^2(1-p)\frac{d}{dp}p\frac{d}{dp}\sum_n p^n$$

$$= M^2(1-p)\frac{d}{dp}p\frac{d}{dp}(1-p)^{-1} = \frac{M^2(1+p)}{(1-p)^2}$$

We see that $\langle n^2 \rangle = \dfrac{1+p}{(1-p)^2}$

and that $\langle M^2 \rangle_N - \langle \overline{M} \rangle_{N^2} = M^2\left(\dfrac{1+p}{(1-p)^2} - \dfrac{1}{(1-p)^2}\right) = \dfrac{pM^2}{(1-p)^2}$

Hence, $\boxed{\delta M = \dfrac{p^{1/2}M}{1-p}}$

The time dependence is obtained from

$$p = \frac{kt[A]_0}{1 + kt[A]_0} \quad [26]$$

and $\dfrac{1}{1-p} = 1 + kt[A]_0$ [Example 26.5]

Hence $\dfrac{p^{1/2}}{1-p} = p^{1/2}(1 + kt[A]_0) = \{kt[A]_0(1 + kt[A]_0)\}^{1/2}$

and $\delta M = \boxed{M\{kt[A]_0(1 + kt[A]_0)\}^{1/2}}$

P26.13 $A \rightarrow 2R \quad \mathcal{I}$

$A + R \rightarrow R + B \quad k_2$

$R + R \rightarrow R_2 \quad k_3$

$$\frac{d[A]}{dt} = \boxed{-\mathcal{I} - k_2[A][R]}, \quad \frac{d[R]}{dt} = 2\mathcal{I} - 2k_3[R]^2 = 0$$

The latter implies that $[R] = \left(\frac{\mathcal{I}}{k_3}\right)^{1/2}$, and so

$$\frac{d[A]}{dt} = \boxed{-\mathcal{I} - k_2\left(\frac{\mathcal{I}}{k_3}\right)^{1/2}[A]}$$

$$\frac{d[B]}{dt} = k_2[A][R] = k_2\left(\frac{\mathcal{I}}{k_3}\right)^{1/2}[A]$$

Therefore, only the combination $\frac{k_2}{k_3^{1/2}}$ may be determined if the reaction attains a steady state.

Comment. If the reaction can be monitored at short enough times so that termination is negligible compared to initiation, then $[R] \approx 2\mathcal{I}t$ and $\frac{d[B]}{dt} \approx 2k_2\mathcal{I}t\,[A]$. So monitoring B sheds light on just k_2.

P26.16 Write the differential equations for [X] and [Y]

$$\text{(i)}\ \frac{d[X]}{dt} = k_a[A][X] - k_b[X][Y]$$

$$\text{(ii)}\ \frac{d[Y]}{dt} = k_b[X][Y] - k_c[Y]$$

and express them as finite-difference equations

$$\text{(i)}\ X(t_{i+1}) = X(t_i) + k_a[A]X(t_i)\Delta t - k_b X(t_i)Y(t_i)\Delta t$$

$$\text{(ii)}\ Y(t_{i+1}) = Y(t_i) - k_c Y(t_i)\Delta t + k_b X(t_i)Y(t_i)\Delta t$$

and iterate for different values of [A], X(0), and Y(0). For the steady state,

$$\text{(i)}\ \frac{d[X]}{dt} = k_a[A][X] - k_b[X][Y] = 0$$

$$\text{(ii)}\ \frac{d[Y]}{dt} = k_b[X][Y] - k_c[Y] = 0$$

which solve to

$$\text{(i)}\ k_b[X] = k_c \qquad \text{(ii)}\ k_a[A] = k_b[Y]$$

Hence, $\boxed{[X] = \frac{k_c}{k_b}}$ $\boxed{[Y] = \frac{k_a[A]}{k_b}}$

P26.18 The description of the progress of infectious diseases can be represented by the mechanism

$S \rightarrow I \rightarrow R$

Only the $\boxed{\text{first step is autocatalytic}}$ as indicated in the first rate expression. If the three rate equations are added

$$\frac{dS}{dt} + \frac{dI}{dt} + \frac{dR}{dt} = 0$$

and, hence, there is no change with time of the total population, that is

$$\mathrm{S}(t) + \mathrm{I}(t) + \mathrm{R}(t) = N$$

Whether the infection spreads or dies out is determined by

$$\frac{\mathrm{dI}}{\mathrm{d}t} = r\mathrm{SI} - a\mathrm{I}$$

At $t = 0$, $\mathrm{I} = \mathrm{I}(0) = \mathrm{I}_0$. Since the process is autocatalytic $\mathrm{I}(0) \neq 0$.

$$\left(\frac{\mathrm{dI}}{\mathrm{d}t}\right)_{t=0} = \mathrm{I}_0(r\mathrm{S}_0 - a)$$

If $a > r\mathrm{S}_0$, $\left(\frac{\mathrm{dI}}{\mathrm{d}t}\right)_{t=0} < 0$, and the infection dies out. If $a < r\mathrm{S}$, $\left(\frac{\mathrm{dI}}{\mathrm{d}t}\right)_{t=0} > 0$ and the infection spreads (an epidemic). Thus

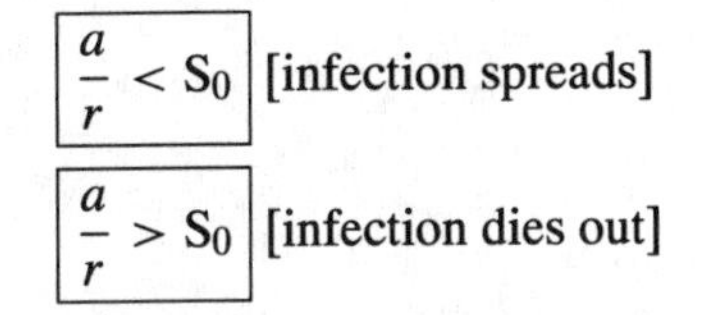

$$\boxed{\frac{a}{r} < \mathrm{S}_0} \ \text{[infection spreads]}$$

$$\boxed{\frac{a}{r} > \mathrm{S}_0} \ \text{[infection dies out]}$$

Solutions to additional problems

P26.19 The roles are

	Reaction	Role
(1)	$N_2O \rightarrow N_2 + O$	initiation
(2)	$O + SiH_4 \rightarrow SiH_3 + OH$	propagation [or transfer]
(3)	$OH + SiH_4 \rightarrow SiH_3 + H_2O$	propagation [or transfer]
(4)	$SiH_3 + N_2O \rightarrow SiH_3O + N_2$	propagation
(5)	$SiH_3O + SiH_4 \rightarrow SiH_3OH + SiH_3$	propagation
(6)	$SiH_3 + SiH_3O \rightarrow (H_3Si)_2O$	termination

The rate of silane consumption is

$$\frac{\mathrm{d}[\mathrm{SiH_4}]}{\mathrm{d}t} = -k_2[\mathrm{SiH_4}][\mathrm{O}] - k_3[\mathrm{SiH_4}][\mathrm{OH}] - k_5[\mathrm{SiH_3O}][\mathrm{SiH_4}]$$

Steady-state approximation (SSA) for O

$$\frac{\mathrm{d}[\mathrm{O}]}{\mathrm{d}t} = k_1[\mathrm{N_2O}] - k_2[\mathrm{SiH_4}][\mathrm{O}] \approx 0 \quad \text{so} \quad [\mathrm{O}] = \frac{k_1[\mathrm{N_2O}]}{k_2[\mathrm{SiH_4}]}$$

SSA for OH

$$\frac{\mathrm{d}[\mathrm{OH}]}{\mathrm{d}t} = k_2[\mathrm{SiH_4}][\mathrm{O}] - k_3[\mathrm{OH}][\mathrm{SiH_4}] \approx 0 = k_1[\mathrm{N_2O}] - k_3[\mathrm{OH}][\mathrm{SiH_4}]$$

$$\text{so} \quad [\mathrm{OH}] = \frac{k_1[\mathrm{N_2O}]}{k_3[\mathrm{SiH_4}]}$$

SSA for SiH_3O and SiH_3

$$\frac{\mathrm{d}[\mathrm{SiH_3O}]}{\mathrm{d}t} = k_4[\mathrm{SiH_3}][\mathrm{N_2O}] - k_5[\mathrm{SiH_3O}][\mathrm{SiH_4}] - k_6[\mathrm{SiH_3O}][\mathrm{SiH_3}] = 0$$

$$\begin{aligned}\frac{\mathrm{d}[\mathrm{SiH_3}]}{\mathrm{d}t} &= k_2[\mathrm{SiH_4}][\mathrm{O}] + k_3[\mathrm{SiH_4}][\mathrm{OH}] - k_4[\mathrm{SiH_3}][\mathrm{N_2O}] \\ &\quad + k_5[\mathrm{SiH_3O}][\mathrm{SiH_4}] - k_6[\mathrm{SiH_3O}][\mathrm{SiH_3}] \\ &= 2k_1[\mathrm{N_2O}] - k_4[\mathrm{SiH_3}][\mathrm{N_2O}] + k_5[\mathrm{SiH_3O}][\mathrm{SiH_4}] - k_6[\mathrm{SiH_3O}][\mathrm{SiH_3}] \approx 0\end{aligned}$$

Adding these expressions together yields

$$0 = 2k_1[\mathrm{N_2O}] - 2k_6[\mathrm{SiH_3O}][\mathrm{SiH_3}] \quad \text{so} \quad [\mathrm{SiH_3}] = \frac{k_1[\mathrm{N_2O}]}{k_6[\mathrm{SiH_3O}]}$$

and subtracting them gives

$$0 = 2k_1[\mathrm{N_2O}] - k_4[\mathrm{SiH_3}][\mathrm{N_2O}] + k_5[\mathrm{SiH_3O}][\mathrm{SiH_4}]$$

Solve for $[\mathrm{SiH_3O}]$

$$0 = 2k_1[\mathrm{N_2O}] - \frac{k_1k_4[\mathrm{N_2O}]^2}{k_6[\mathrm{SiH_3O}]} + k_5[\mathrm{SiH_3O}][\mathrm{SiH_4}]$$

$$= 2k_1k_6[\mathrm{SiH_3O}][\mathrm{N_2O}] - k_1k_4[\mathrm{N_2O}]^2 + k_5k_6[\mathrm{SiH_3O}]^2[\mathrm{SiH_4}]$$

$$[\mathrm{SiH_3O}] = \frac{-2k_1k_6[\mathrm{N_2O}] \pm (4k_1^2k_6^2[\mathrm{N_2O}]^2 + 4k_1k_4k_5k_6[\mathrm{N_2O}]^2[\mathrm{SiH_4}])^{1/2}}{2k_5k_6[\mathrm{SiH_4}]}$$

$$= \frac{k_1[\mathrm{N_2O}]}{k_5[\mathrm{SiH_4}]}\left[-1 + \left(1 + \frac{k_4k_5[\mathrm{SiH_4}]}{k_1k_6}\right)^{1/2}\right]$$

If k_1 is small, then

$$[\mathrm{SiH_3O}] \approx \frac{k_1[\mathrm{N_2O}]}{k_5[\mathrm{SiH_4}]}\left(\frac{k_4k_5[\mathrm{SiH_4}]}{k_1k_6}\right)^{1/2} = [\mathrm{N_2O}]\left(\frac{k_1k_4}{k_5k_6[\mathrm{SiH_4}]}\right)^{1/2}$$

Putting it all together yields

$$\frac{\mathrm{d}[\mathrm{SiH_4}]}{\mathrm{d}t} = -2k_1[\mathrm{N_2O}] - k_5[\mathrm{SiH_4}][\mathrm{N_2O}]\left(\frac{k_1k_4}{k_5k_6[\mathrm{SiH_4}]}\right)^{1/2} \approx \boxed{-\left(\frac{k_1k_4k_5}{k_6}\right)^{1/2}[\mathrm{N_2O}][\mathrm{SiH_4}]^{1/2}}$$

P26.21 (a) Note that $[\mathrm{A}]_0 - x = \dfrac{[\mathrm{A}]_0}{1 + \left(\frac{x}{[\mathrm{A}]_0 - x}\right)}$

as can be seen by dividing both sides by $[\mathrm{A}]_0 - x$. The ratio $x/([\mathrm{A}]_0 - x)$ is given in the problem.

Test of first-order rate law. The plot of $\ln([\mathrm{A}]_0 - x)$ against t is linear with slope $-k$ and intercept $\ln[\mathrm{A}]_0$. The regression fit [Fig. 26.2(a)] of $\ln([\mathrm{A}]_0 - x)$ against t gives the following results for X = H in the $\mathrm{XC_6H_4HgCl}$ formula.

$R = 0.986\,324$

Slope $= -3.44 \times 10^{-3}\ \mathrm{min^{-1}}$ standard deviation $= 2.34 \times 10^{-4}\ \mathrm{min^{-1}}$

Intercept $= \ln([\mathrm{A}]_0/(\mathrm{mmol\,L^{-1}})) = 4.532$ standard deviation $= 0.031$

Test of second-order rate law. The plot of $1/([\mathrm{A}]_0 - x)$ against t is linear with slope k and intercept $1/[\mathrm{A}]_0$. The regression fit of $1/([\mathrm{A}]_0 - x)$ against t gives the following results for X = H in the $\mathrm{XC_6H_4HgCl}$ formula.

$R = 0.999\,691$

Slope $= 5.46 \times 10^{-5}\ (\mathrm{mmol\,L^{-1}})^{-1}\ \mathrm{min^{-1}}$

standard deviation $= 7.84 \times 10^{-7}\ (\mathrm{mmol\,L^{-1}})^{-1}\ \mathrm{min^{-1}}$

Intercept $= 1.01 \times 10^{-2} (\mathrm{mmol\,L^{-1}})^{-1}$

standard deviation $= 1.02 \times 10^{-4}\ (\mathrm{mmol\,L^{-1}})^{-1}$

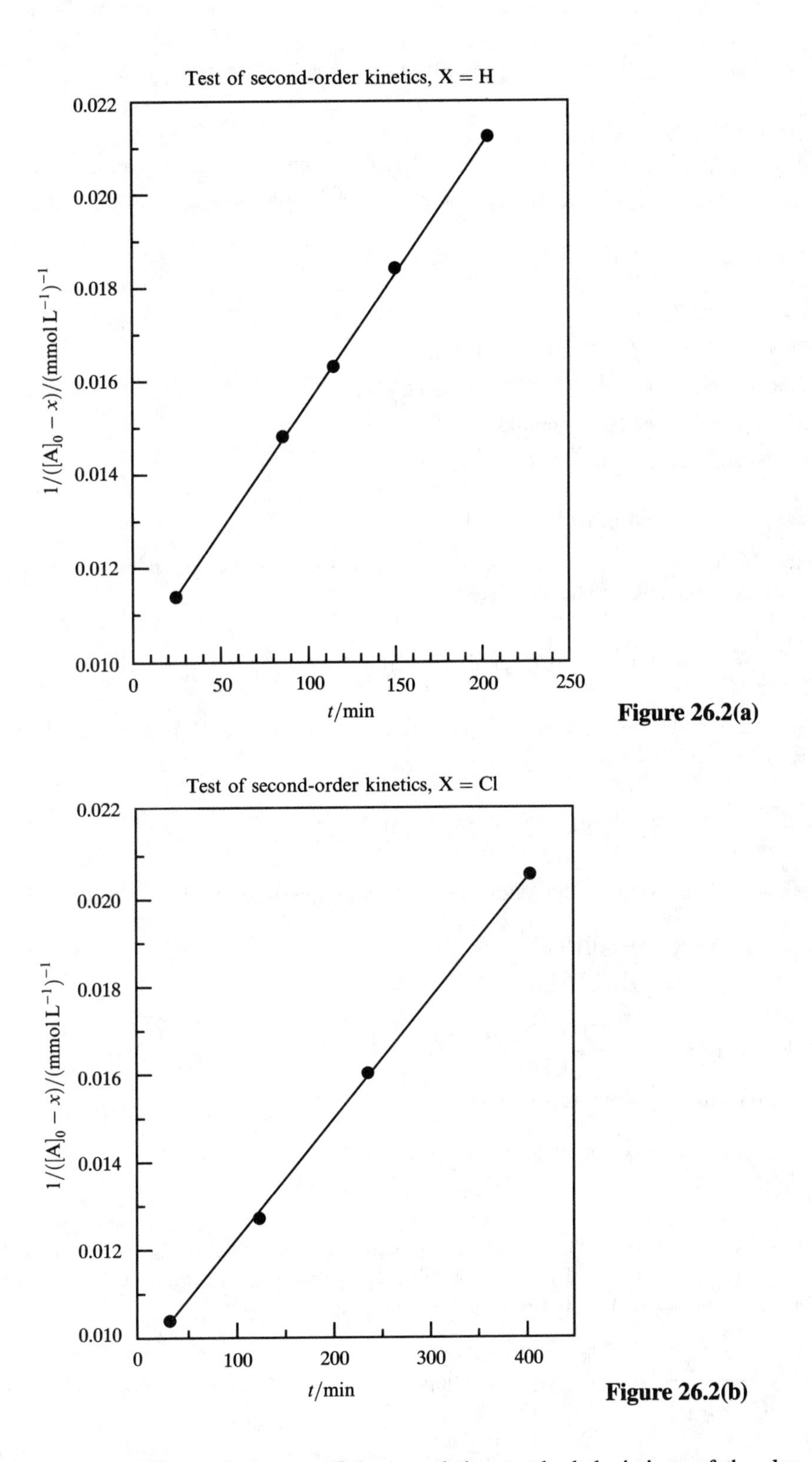

Figure 26.2(a)

Figure 26.2(b)

Both the correlation coefficients and the standard deviations of the slopes indicate that the regression plot of $1/([A]_0 - x)$ against t is the better fit. We conclude that the rate law is second-order when X = H.

For X = Cl the regression fit [Fig. 26.2(b)] of $\ln([A]_0 - x)$ against t gives

$R = 0.995\,421$
Slope $= -1.807 \times 10^{-3}\ \mathrm{min}^{-1}$ standard deviation $= 1.23 \times 10^{-4}\ \mathrm{min}^{-1}$
Intercept $= \ln([A]_0/(\mathrm{mmol\,L^{-1}})) = 4.597$ standard deviation $= 0.030$

while the regression fit of $1/([A]_0 - x)$ against t gives

$R = 0.999\,697$
Slope $= 2.716 \times 10^{-5}\ (\mathrm{mmol\,L^{-1}})^{-1}\,\mathrm{min}^{-1}$
standard deviation $= 4.72 \times 10^{-7}\ (\mathrm{mmol\,L^{-1}})\,\mathrm{min}^{-1}$
Intercept $= 9.51 \times 10^{-3}\ (\mathrm{mmol\,L^{-1}})^{-1}$
standard deviation $= 1.15 \times 10^{-4}\ (\mathrm{mmol\,L^{-1}})^{-1}$

Once again the **second-order** fit is the better.

(b) Since the third step is very fast while the second is slow, we may consider the second step of the proposed mechanism to be rate-determining.

$$\frac{d[\mathrm{ArHgCl}]}{dt} = -k_2[(\mathrm{ArRh(CO)_2})_2][\mathrm{ArHgCl}] + k_{-2}[\mathrm{Ar_2Rh(CO)_2HgCl}]$$

The second term to the right is very small because the fast step 3 prevents the build-up of $[\mathrm{Ar_2Rh(CO)_2HgCl}]$. Thus,

$$\frac{d[\mathrm{ArHgCl}]}{dt} = -k_2[(\mathrm{ArRh(CO)_2})_2][\mathrm{ArHgCl}]$$

If we assume that step 1 is fast enough so as to be at quasi-equilibrium, we may write that

$$K_1 \approx \frac{[(\mathrm{ArRh(CO)_2})_2][\mathrm{HgCl_2}]}{[(\mathrm{ClRh(CO)_2})_2][\mathrm{ArHgCl}]}$$

or $$(\mathrm{ArRh(CO)_2})_2 \approx \frac{K_1[(\mathrm{ClRh(CO)_2})_2][\mathrm{ArHgCl}]}{[\mathrm{HgCl_2}]}$$

Substitution into the rate expression gives

$$\frac{d[\mathrm{ArHgCl}]}{dt} = -\frac{k_2K_1[(\mathrm{ClRh(CO)_2})_2][\mathrm{ArHgCl}]^2}{[\mathrm{HgCl_2}]}$$

If the concentrations of both the catalyst and $HgCl_2$ remain constant, the rate will be second-order in [ArHgCl] as observed. Step 3 assumes that the catalyst concentration remains constant. The limited solubility of $HgCl_2$ keeps this concentration constant.

P26.22 (a)

$2\mathrm{NO} \rightarrow \mathrm{N_2O} + \mathrm{O}$	k_a	initiation
$\mathrm{O} + \mathrm{NO} \rightarrow \mathrm{O_2} + \mathrm{N}$	k_b	propagation
$\mathrm{N} + \mathrm{NO} \rightarrow \mathrm{N_2} + \mathrm{O}$	k_c	propagation
$2\mathrm{O} + \mathrm{M} \rightarrow \mathrm{O_2} + \mathrm{M}$	k_d	termination
$\mathrm{O_2} + \mathrm{M} \rightarrow 2\mathrm{O} + \mathrm{M}$	k_{-d}	initiation

(b) $$\frac{d[\mathrm{NO}]}{dt} = -2k_a[\mathrm{NO}]^2 - k_b[\mathrm{O}][\mathrm{NO}] - k_c[\mathrm{N}][\mathrm{NO}]$$

To determine the steady-state concentration of N, $[N]_{SS}$, write the rate expression for $d[N]/dt$ and set it equal to zero.

$$\frac{d[N]}{dt} = k_b[O][NO] - k_c[N]_{SS}[NO] = 0$$

$$[N]_{SS} = \frac{k_b}{k_c}[O]$$

Substitution of $[N]_{SS}$ into the expression of $d[NO]/dt$ indicates that, under steady-state conditions for [N], $v_b = v_c$ and

$$\boxed{\frac{d[NO]}{dt} = -2k_a[NO]^2 - 2k_b[O][NO]}$$

[N] steady-state conditions.

If the propagation step is much more rapid than initiation, the last term predominates

$$\boxed{\frac{d[NO]}{dt} = -2k_b[O][NO]}$$

[N] is in steady-state and initiation is very slow.

If oxygen atoms and molecules are in equilibrium

$$2O + M \underset{k_{-d}}{\overset{k_d}{\rightleftharpoons}} O_2 + M$$

$$K_{O/O_2} = \frac{k_d}{k_{-d}} = \frac{[O_2]}{[O]^2}$$

$$[O] = \left(\frac{k_{-d}[O_2]}{k_d}\right)^{1/2}$$

Substitution into the previous rate expression yields

$$\boxed{\frac{d[NO]}{dt} = -2k_b\left(\frac{k_{-d}}{k_d}\right)^{1/2}[O_2]^{1/2}[NO]}$$

[N] is in steady-state; initiation is very slow; atomic and molecular oxygen are in equilibrium.

(c) Since $k \propto e^{-E_a/RT}$ where E_a is the activation energy, we may write the individual rate constants in the form $k_i \propto e^{-E_i/RT}$ where the subscript 'a' has been dropped and 'i' represents the ith elementary step with activation energy E_i. Substitution of such expressions into the last equation of part **(b)** yields

$$\frac{d[NO]}{dt} \propto e^{-E_b/RT}\left(\frac{e^{-E_{-d}/RT}}{e^{-E_d/RT}}\right)^{1/2}[O_2]^{1/2}[NO]$$

$$\propto e^{-\left(\frac{-\left(E_b+\frac{1}{2}E_{-d}-\frac{1}{2}E_d\right)}{RT}\right)}[O_2]^{1/2}[NO]$$

We conclude that the effective activation energy, $E_{a,\text{eff}}$, is given by

$$\boxed{E_{a,\text{eff}} = E_b + \tfrac{1}{2}E_{-d} - \tfrac{1}{2}E_d}$$

(d) Using the estimate that activation energies are approximately equal to the bond energies that must be broken

$$\begin{aligned} E_{a,\text{eff}} &\approx B(\text{NO}) + \tfrac{1}{2}B(\text{O}_2) - \tfrac{1}{2}B(\text{O}) \\ &\approx 630.57\,\text{kJ mol}^{-1} + \tfrac{1}{2}(498.36\,\text{kJ mol}^{-1}) - \tfrac{1}{2}(0) \\ &\approx 879.75\,\text{kJ mol}^{-1} \end{aligned}$$

where this is the unimolecular bond-breakage estimate of activation energies.

The previous estimate of $E_{a,\text{eff}}$ may be much too high because the activation energy of step (b) is probably being greatly overestimated. A more realistic estimate of E_b would be that difference between the energies of the NO bond that must be broken and the O_2 bond that is formed. Then,

$$\begin{aligned} E_{a,\text{eff}} &\approx \{B(\text{NO}) - B(\text{O}_2)\} + \tfrac{1}{2}B(\text{O}_2) - \tfrac{1}{2}B(\text{O}) \\ &\approx B(\text{NO}) - \tfrac{1}{2}B(\text{O}_2) \\ &\approx 630.57\,\text{kJ mol}^{-1} - \tfrac{1}{2}(498.36\,\text{kJ mol}^{-1}) \\ &\approx \boxed{381.39\,\text{kJ mol}^{-1}} \end{aligned}$$

The energy of the activated complex of step (b) is the difference between bond-breakage and bond-formation energies.

It is interesting to compare these estimates with the value based upon E_i values that have been determined by experiment

$$\begin{aligned} E_{a,\text{eff}} &= (161\,\text{kJ mol}^{-1}) + \tfrac{1}{2}(493\,\text{kJ mol}^{-1}) - \tfrac{1}{2}(14\,\text{kJ mol}^{-1}) \\ &= 401\,\text{kJ mol}^{-1} \end{aligned}$$

This value is based upon experimental activation energies for the elementary steps.

(e) We now eliminate the assumption of O/O_2 equilibrium and assume that both [N] and [O] are at steady-state values. From part (b), $[\text{N}_{\text{SS}}] = k_b[\text{O}]_{\text{SS}}/k_c$

$$\frac{d[\text{O}]}{dt} = k_a[\text{NO}]^2 - k_b[\text{O}]_{\text{SS}}[\text{NO}] + k_c[\text{N}]_{\text{SS}}[\text{NO}] - 2k_d[\text{O}]_{\text{SS}}^2[\text{M}] + 2k_{-d}[\text{O}_2][\text{M}] = 0$$

$$k_a[\text{NO}]^2 - k_b[\text{O}]_{\text{SS}}[\text{NO}] + k_b[\text{O}]_{\text{SS}}[\text{NO}] - 2k_d[\text{O}]_{\text{SS}}^2[\text{M}] + 2k_{-d}[\text{O}_2][\text{M}] = 0$$

$$k_a[\text{NO}]^2 - 2k_d[\text{O}]_{\text{SS}}^2[\text{M}] + 2k_{-d}[\text{O}_2][\text{M}] = 0$$

At very low values of $[O_2]$ the last term is negligible so that

$$[\text{O}]_{\text{SS}} \approx \left(\frac{k_a}{2k_d[\text{M}]}\right)^{1/2}[\text{NO}]$$

Substitution of the expressions for $[\text{N}]_{\text{SS}}$ and $[\text{O}]_{\text{SS}}$ into the expression for $d[\text{NO}]/dt$ (top of part(**b**)) gives

$$\begin{aligned} \frac{d[\text{NO}]}{dt} &= -2k_a[\text{NO}]^2 - k_b[\text{O}]_{\text{SS}}[\text{NO}] - k_c\left(\frac{k_b[\text{O}]_{\text{SS}}}{k_c}\right)[\text{NO}] \\ &= -2k_a[\text{NO}]^2 - 2k_b[\text{O}]_{\text{SS}}[\text{NO}] \\ &= -2k_a[\text{NO}]^2 - 2k_b\left(\frac{k_a}{2k_d[\text{M}]}\right)^{1/2}[\text{NO}]^2 \end{aligned}$$

If propagation is much more rapid than initiation so that $k_b\left(\frac{k_a}{2k_d[\mathrm{M}]}\right)^{1/2} \gg k_a$, this expression becomes

$$\boxed{\frac{d[\mathrm{NO}]}{dt} = -2k_b\left(\frac{k_a}{2k_d[\mathrm{M}]}\right)^{1/2}[\mathrm{NO}]^2}$$

(f) $\mathrm{NO} + \mathrm{O_2} \rightarrow \mathrm{O} + \mathrm{NO_2}$ $\quad k_e$ initiation

$$\frac{d[\mathrm{NO}]}{dt} = -2k_a[\mathrm{NO}]^2 - k_b[\mathrm{O}][\mathrm{NO}] - k_c[\mathrm{N}][\mathrm{NO}] - k_e[\mathrm{NO}][\mathrm{O_2}]$$

if the conversion has proceeded to the extent that $[\mathrm{O_2}]$ has become significant and $k_e[\mathrm{NO}][\mathrm{O_2}] \gg 2k_a[\mathrm{NO}]^2$

$$\frac{d[\mathrm{NO}]}{dt} = -k_b[\mathrm{O}][\mathrm{NO}] - k_c[\mathrm{N}][\mathrm{NO}] - k_e[\mathrm{NO}][\mathrm{O_2}]$$

Applying the steady-state approximation to both [N] and [O] gives $[\mathrm{N}]_{\mathrm{SS}} = k_b[\mathrm{O}]_{\mathrm{SS}}/k_c$ and

$$\frac{d[\mathrm{O}]}{dt} = k_a[\mathrm{NO}]^2 - k_b[\mathrm{O}]_{\mathrm{SS}}[\mathrm{NO}] + k_c[\mathrm{N}]_{\mathrm{SS}}[\mathrm{NO}] - 2k_d[\mathrm{O}]_{\mathrm{SS}}^2[\mathrm{M}] + 2k_{-d}[\mathrm{O}^2][\mathrm{M}]$$

$$+ k_e[\mathrm{O_2}][\mathrm{NO}] = 0 \text{ and } -k_b[\mathrm{O}]_{\mathrm{SS}}[\mathrm{NO}] + k_c[\mathrm{N}]_{\mathrm{SS}}[\mathrm{NO}] = 0$$

Thus $2k_d[\mathrm{O}]_{\mathrm{SS}}^2[\mathrm{M}] - 2k_{-d}[\mathrm{O_2}][\mathrm{M}] - k_e[\mathrm{O_2}][\mathrm{NO}] = 0$

At high concentrations of $\mathrm{O_2}$, species 'M' is likely to be $\mathrm{O_2}$ and $k_d[\mathrm{O}]_{\mathrm{SS}}^2[\mathrm{M}] \gg k_b[\mathrm{O}]_{\mathrm{SS}}[\mathrm{NO}]$. The value of k_{-d} is so small that it can be neglected.

$$2k_d[\mathrm{O}]_{\mathrm{SS}}^2[\mathrm{M}] \approx k_e[\mathrm{O_2}][\mathrm{NO}]$$

$$[\mathrm{O}]_{\mathrm{SS}} = \left(\frac{k_e}{2k_d[\mathrm{M}]}\right)^{1/2}[\mathrm{O_2}]^{1/2}[\mathrm{NO}]^{1/2}$$

Substitution of $[\mathrm{N}]_{\mathrm{SS}}$ and $[\mathrm{O}]_{\mathrm{SS}}$ into the expression for $d[\mathrm{NO}]/dt$ gives

$$\begin{aligned}\frac{d[\mathrm{NO}]}{dt} &= -k_b[\mathrm{O}]_{\mathrm{SS}}[\mathrm{NO}] - k_c[\mathrm{N}]_{\mathrm{SS}}[\mathrm{NO}] - k_e[\mathrm{NO}][\mathrm{O_2}] \\ &= -2k_b[\mathrm{O}]_{\mathrm{SS}}[\mathrm{NO}] - k_e[\mathrm{NO}][\mathrm{O_2}] \\ &= -2k_b\left(\frac{k_e}{2k_d[\mathrm{M}]}\right)^{1/2}[\mathrm{O_2}]^{1/2}[\mathrm{NO}]^{3/2} - k_e[\mathrm{NO}][\mathrm{O_2}]\end{aligned}$$

If propagation is much more rapid than initiation, the expression becomes

$$\boxed{\frac{d[\mathrm{NO}]}{dt} = -2k_b\left(\frac{k_e}{2k_d[\mathrm{M}]}\right)^{1/2}[\mathrm{O_2}]^{1/2}[\mathrm{NO}]^{3/2}}$$

$$E_{\mathrm{a,eff}} = E_b + \tfrac{1}{2}E_e - \tfrac{1}{2}E_d$$

Using the experimental values of E_i, $E_{\mathrm{a,eff}}$ is estimated to be given by

$$\begin{aligned}E_{\mathrm{a,eff}} &= 161\,\mathrm{kJ\,mol^{-1}} + \tfrac{1}{2}(198\,\mathrm{kJ\,mol^{-1}}) - \tfrac{1}{2}(14\,\mathrm{kJ\,mol^{-1}}) \\ &= \boxed{253\,\mathrm{kJ\,mol^{-1}}}\end{aligned}$$

This value is consistent with the low range of the experimental values of $E_{\mathrm{a,eff}}$, whereas the value found in part **(b)** is consistent with the high experimental values.

P26.24
$$\frac{d[HI]}{dt} = 2k_b[I\cdot]^2[H_2] \tag{1}$$

$$\frac{d[I\cdot]}{dt} = 2k_a[I_2] - 2k_a'[I\cdot]^2 - 2k_b[I\cdot]^2[H_2]$$

In the steady-state approximation for $[I\cdot]$

$$\frac{d[I\cdot]}{dt} = 0 = 2k_a[I_2] - 2k_a'[I\cdot]_{SS}^2 - 2k_b[I\cdot]_{SS}^2[H_2]$$

$$[I\cdot]_{SS}^2 = \frac{k_a}{k_a' + k_b[H_2]}[I_2] \tag{2}$$

Substitution of (2) into (1) gives

$$\frac{d[HI]}{dt} = \frac{2k_bk_a[I_2][H_2]}{k_a' + k_b[H_2]}$$

This simple rate law is observed when step (b) is rate-determining so that step (a) is a rapid equilibrium and $[I\cdot]$ is in an approximate steady state. This is equivalent to $k_b[H_2] \ll k_a'$ and hence,

$$\frac{d[HI]}{dt} = 2k_bK[I_2][H_2]$$

27 Molecular reaction dynamics

Solutions to exercises

E27.1 $z = \frac{2^{1/2}\sigma\bar{c}p}{kT}$ [1.31 with 1.28]

and $\bar{c} = \left(\frac{8RT}{\pi M}\right)^{1/2}$ [Example 1.6] $= \left(\frac{8kT}{\pi m}\right)^{1/2}$

Therefore, $z = \frac{4\sigma p}{(\pi mkT)^{1/2}}$ with $\sigma \approx \pi d^2 \approx 4\pi R^2$

The collision frequency z gives the number of collisions made by a single molecule. We can obtain the *total* collision frequency, the rate of collisions between all the molecules in the gas, by multiplying z by $\frac{1}{2}N$ (the factor $\frac{1}{2}$ ensures that the A . . . A′ and A′ . . . A collisions are counted as one). Therefore the collision density, Z, the total number of collisions per unit time per unit volume, is

$$Z_{AA} = \frac{\frac{1}{2}zN}{V} = \frac{\sigma\bar{c}}{2^{1/2}}\left(\frac{N}{V}\right)^2$$

Introducing the expression for $\bar{c}$ above

$$Z_{AA} = \sigma\left(\frac{4kT}{\pi m}\right)^{1/2} \times \left(\frac{N}{V}\right)^2 = \sigma\left(\frac{4kT}{\pi m}\right)^{1/2} \times \left(\frac{p}{kT}\right)^2 \quad [N/V = p/kT]$$

We express these equations in the form

$$z = \frac{(16\pi R^2) \times (1.00 \times 10^5\,\text{Pa})}{\{(\pi) \times (M/\text{g mol}^{-1}) \times (1.6605 \times 10^{-27}\,\text{kg}) \times (1.381 \times 10^{-23}\,\text{J K}^{-1}) \times (298.15\,\text{K})\}^{1/2}}$$

$$= \frac{(1.08 \times 10^{30}\,\text{m}^{-2}\,\text{s}^{-1}) \times R^2}{(M/\text{g mol}^{-1})^{1/2}} = \frac{1.08 \times 10^6 \times (R/\text{pm})^2\,\text{s}^{-1}}{(M/\text{g mol}^{-1})^{1/2}}$$

$$Z_{AA} = 4\pi R^2\left(\frac{(4) \times (1.381 \times 10^{-23}\,\text{J K}^{-1}) \times (298.15\,\text{K})}{(\pi) \times (M/\text{g mol}^{-1}) \times (1.6605 \times 10^{-27}\,\text{kg})}\right)^{1/2}$$

$$\times \left(\frac{1.00 \times 10^5\,\text{Pa}}{(1.381 \times 10^{-23}\,\text{J K}^{-1}) \times (298.15\,\text{K})}\right)^2$$

$$= \frac{(1.32 \times 10^{55}\,\text{m}^{-5}\,\text{s}^{-1}) \times R^2}{(M/\text{g mol}^{-1})^{1/2}} = \frac{1.35 \times 10^{31}(R/\text{pm})^2}{(M/\text{g mol}^{-1})^{1/2}}\,\text{m}^{-3}\,\text{s}^{-1}$$

For NH_3, $R = 190\,\text{pm}$, $M = 17\,\text{g mol}^{-1}$

$$z = \frac{(1.08 \times 10^6) \times (190^2\,\text{s}^{-1})}{17^{1/2}} = \boxed{9.5 \times 10^9\,\text{s}^{-1}}$$

$$Z_{AA} = \frac{(1.32 \times 10^{31}) \times (190^2\,\text{m}^{-3}\,\text{s}^{-1})}{17^{1/2}} = \boxed{1.2 \times 10^{35}\,\text{m}^{-3}\,\text{s}^{-1}}$$

For the percentage increase at constant volume, use

$$\frac{1}{z}\frac{\text{d}z}{\text{d}T} = \frac{1}{\bar{c}}\frac{\text{d}\bar{c}}{\text{d}T} = \frac{1}{2T}, \qquad \left(\frac{1}{Z}\right)\left(\frac{\text{d}Z}{\text{d}T}\right) = \frac{1}{2T}$$

Therefore, $\dfrac{\delta z}{z} \approx \dfrac{\delta T}{2T}$ and $\dfrac{\delta T}{Z} \approx \dfrac{\delta T}{2T}$

and since $\dfrac{\delta T}{T} = \dfrac{10\,\text{K}}{298\,\text{K}} = 0.034$, both z and Z increase by about $\boxed{1.7\text{ per cent}}$

E27.2 In each case use $f = \text{e}^{-E_\text{a}/RT}$

(a)

$$\frac{E_\text{a}}{RT} = \frac{10 \times 10^3\,\text{J mol}^{-1}}{(8.314\,\text{J K}^{-1}\,\text{mol}^{-1}) \times (300\,\text{K})} = 4.01, \qquad f = \text{e}^{-4.01} = \boxed{0.018}$$

$$\frac{E_\text{a}}{RT} = \frac{10 \times 10^3\,\text{J mol}^{-1}}{(8.314\,\text{J K}^{-1}\,\text{mol}^{-1}) \times (1000\,\text{K})} = 1.20, \qquad f = \text{e}^{-1.20} = \boxed{0.30}$$

(b)

$$\frac{E_\text{a}}{RT} = \frac{100 \times 10^3\,\text{J mol}^{-1}}{(8.314\,\text{J K}^{-1}\,\text{mol}^{-1}) \times (300\,\text{K})} = 40.1, \qquad f = \text{e}^{-40.1} = \boxed{3.9 \times 10^{-18}}$$

$$\frac{E_\text{a}}{RT} = \frac{100 \times 10^3\,\text{J mol}^{-1}}{(8.314\,\text{J K}^{-1}\,\text{mol}^{-1}) \times (1000\,\text{K})} = 12.0, \qquad f = \text{e}^{-12.0} = \boxed{6.0 \times 10^{-6}}$$

E27.3 The percentage increase is

$$(100) \times \left(\frac{\delta f}{f}\right) \approx (100) \times \left(\frac{\text{d}f}{\text{d}T}\right) \times \left(\frac{\delta T}{f}\right) \approx \frac{100 E_\text{a}}{RT^2}\delta T$$

(a) $E_\text{a} = 10\,\text{kJ mol}^{-1}, \qquad \delta T = 10\,\text{K}$

$$(100)\left(\frac{\delta f}{f}\right) = \frac{(100) \times (10 \times 10^3\,\text{J mol}^{-1}) \times (10\,\text{K})}{(8.314\,\text{J K}^{-1}\,\text{mol}^{-1}) \times (T^2)}$$

$$= \frac{1.20 \times 10^6}{(T/\text{K})^2} = \begin{cases} \boxed{13\text{ per cent}} \text{ at } 300\,\text{K} \\ \boxed{1.2\text{ per cent}} \text{ at } 1000\,\text{K} \end{cases}$$

(b) $E_\text{a} = 100\,\text{kJ mol}^{-1}, \qquad \delta T = 10\,\text{K}$

$$(100)\left(\frac{\delta f}{f}\right) = \frac{1.20 \times 10^7}{(T/\text{K})^2} = \begin{cases} \boxed{130\text{ per cent}} \text{ at } 300\,\text{K} \\ \boxed{12\text{ per cent}}\ 1000\,\text{K} \end{cases}$$

E27.4 $k_2 = \sigma \left(\dfrac{8kT}{\pi\mu}\right)^{1/2} N_\text{A}\text{e}^{-E_\text{a}/RT}$ [16 with 9]

The activation energy E_a to be used in this formula is related to the experimental activation by

$$\begin{aligned} E_\text{a} &= E_\text{a}^\text{exp} - \tfrac{1}{2}RT \text{ [Footnote 1, p. 823]} \\ &= (1.71 \times 10^5\,\text{J mol}^{-1}) - \left(\tfrac{1}{2}\right) \times (8.314\,\text{J K}^{-1}\,\text{mol}^{-1}) \times (650\,\text{K}) \\ &= 1.68\overline{3} \times 10^5\,\text{J mol}^{-1} \end{aligned}$$

$$\text{e}^{-E_\text{a}/RT} = \text{e}^{-1.68\overline{3}\times10^5\,\text{J mol}^{-1}/(8.314\,\text{J K}^{-1}\,\text{mol}^{-1}\times 650\,\text{K})} = 2.9\overline{9} \times 10^{-14}$$

$$\left(\frac{8kT}{\pi\mu}\right)^{1/2} = \left(\frac{(8) \times (1.381 \times 10^{-23}\,\text{J K}^{-1}) \times (650\,\text{K})}{(\pi) \times (3.32 \times 10^{-27}\,\text{kg})}\right)^{1/2} = 2.62\overline{3} \times 10^3\,\text{m s}^{-1}$$

$$\begin{aligned} k_2 &= (0.36 \times 10^{-18}\,\text{m}^2) \times (2.62\overline{3} \times 10^3\,\text{m s}^{-1}) \times (6.022 \times 10^{23}\,\text{mol}^{-1}) \times (2.9\overline{9} \times 10^{-14}) \\ &= 1.7 \times 10^{-5}\,\text{m}^3\,\text{mol}^{-1}\,\text{s}^{-1} = \boxed{1.7 \times 10^{-2}\,\text{L mol}^{-1}\,\text{s}^{-1}} \end{aligned}$$

Comment. Estimates of collision cross-sections are notoriously variable. For the $H_2 + I_2$ reaction they have ranged from $0.28\,\text{nm}^2$ to $0.50\,\text{nm}^2$. However, that factor alone will not account for the differences between theoretical and experimental values of k_2. See Example 27.1.

E27.5 $k_d = 4\pi R^* DN_A$ [27]

$D = D_A + D_B = 2 \times 5 \times 10^{-9}\,\text{m}^2\,\text{s}^{-1} = 1 \times 10^{-8}\,\text{m}^2\,\text{s}^{-1}$

$k_d = (4\pi) \times (0.4 \times 10^{-9}\,\text{m}) \times (1 \times 10^{-8}\,\text{m}^2\,\text{s}^{-1}) \times (6.02 \times 10^{23}\,\text{mol}^{-1}) = 3 \times 10^7\,\text{m}^3\,\text{mol}^{-1}\,\text{s}^{-1}$

$= \boxed{3 \times 10^{10}\,\text{L}\,\text{mol}^{-1}\,\text{s}^{-1}}$

E27.6

$$k_d = \frac{8RT}{3\eta}\ [35] = \frac{(8) \times (8.314 \times \text{J}\,\text{K}^{-1}\,\text{mol}^{-1}) \times (298\,\text{K})}{3\eta}$$

$$= \frac{6.61 \times 10^3\,\text{J}\,\text{mol}^{-1}}{\eta} = \frac{6.61 \times 10^3\,\text{kg}\,\text{m}^2\,\text{s}^{-2}\,\text{mol}^{-1}}{(\eta/\text{kg}\,\text{m}^{-1}\text{s}^{-1}) \times \text{kg}\,\text{m}^{-1}\,\text{s}^{-1}} = \frac{6.61 \times 10^3\,\text{m}^3\,\text{mol}^{-1}\,\text{s}^{-1}}{(\eta/\text{kg}\,\text{m}^{-1}\,\text{s}^{-1})}$$

$$= \frac{6.61 \times 10^6\,\text{M}^{-1}\,\text{s}^{-1}}{(\eta/\text{kg}\,\text{m}^{-1}\,\text{s}^{-1})} = \frac{6.61 \times 10^9\,\text{M}^{-1}\,\text{s}^{-1}}{(\eta/\text{cP})}$$

(a) Water, $\eta = 1.00\,\text{cP}$,

$$k_d = \frac{6.61 \times 10^9}{1.00}\,\text{L}\,\text{mol}^{-1}\,\text{s}^{-1} = 6.61 \times 10^9\,\text{L}\,\text{mol}^{-1}\,\text{s}^{-1}$$

$$= \boxed{6.61 \times 10^6\,\text{m}^3\,\text{mol}^{-1}\,\text{s}^{-1}}$$

(b) Pentane, $\eta = 0.22\,\text{cP}$,

$$k_d = \frac{6.61 \times 10^9}{0.22}\,\text{L}\,\text{mol}^{-1}\,\text{s}^{-1} = 3.0 \times 10^{10}\,\text{L}\,\text{mol}^{-1}\,\text{s}^{-1}$$

$$= \boxed{3.0 \times 10^7\,\text{m}^3\,\text{mol}^{-1}\,\text{s}^{-1}}$$

E27.7

$$k_d = \frac{8RT}{3\eta} = \frac{(8) \times (8.314\,\text{J}\,\text{K}^{-1}\,\text{mol}^{-1}) \times (298\,\text{K})}{(3) \times (0.89 \times 10^{-3}\,\text{kg}\,\text{m}^{-1}\,\text{s}^{-1})}$$

$$= 7.4 \times 10^6\,\text{m}^3\,\text{mol}^{-1}\,\text{s}^{-1} = \boxed{7.4 \times 10^9\,\text{L}\,\text{mol}^{-1}\,\text{s}^{-1}}$$

Since this reaction is elementary bimolecular it is second-order; hence

$$t_{1/2} = \frac{1}{k_d[\text{A}]_0}\ [\text{Table 25.3}]$$

$$= \frac{1}{(7.4 \times 10^9\,\text{L}\,\text{mol}^{-1}\,\text{s}^{-1}) \times (1.0 \times 10^{-3}\,\text{mol}\,\text{L}^{-1})} = \boxed{1.4 \times 10^{-7}\,\text{s}}$$

E27.8 $P = \dfrac{\sigma^*}{\sigma}$ [Section 27.1(c)]

For the mean collision cross-section, write $\sigma_A = \pi d_A{}^2$, $\sigma_B = \pi d_B{}^2$, and $\sigma = \pi d^2$, with $d = \frac{1}{2}(d_A + d_B)$

$$\sigma = \tfrac{1}{4}\pi(d_A + d_B)^2 = \tfrac{1}{4}\pi(d_A{}^2 + d_B{}^2 + 2d_A d_B)$$

$$= \tfrac{1}{4}(\sigma_A + \sigma_B + 2\sigma_A{}^{1/2}\sigma_B{}^{1/2})$$

$$= \tfrac{1}{4}\{0.95 + 0.65 + 2 \times (0.95 \times 0.65)^{1/2}\}\,\text{nm}^2 = 0.793\,\text{nm}^2$$

Therefore, $P \approx \dfrac{9.2 \times 10^{-22}\,\text{m}^2}{0.793 \times 10^{-18}\,\text{m}^2} = \boxed{1.2 \times 10^{-3}}$

E27.9 Since the reaction is assumed to be elementary bimolecular, it is necessarily second-order; hence

$$\frac{d[P]}{dt} = k_2[A][B]$$

$$k_2 = 4\pi R^* D N_A [27] = 4\pi R^*(D_A + D_B)N_A$$

$$= \frac{2kTN}{3\eta}(R_A + R_B) \times \left(\frac{1}{R_A} + \frac{1}{R_B}\right)$$

$$= \frac{2RT}{3\eta}(R_A + R_B) \times \left(\frac{1}{R_A} + \frac{1}{R_B}\right)$$

$$= \frac{(2) \times (8.314\,\mathrm{J\,K^{-1}\,mol^{-1}}) \times (313\,\mathrm{K})}{(3) \times (2.37 \times 10^{-3}\,\mathrm{kg\,m^{-1}\,s^{-1}})} \times (294 + 825) \times \left(\frac{1}{294} + \frac{1}{825}\right)$$

$$= 3.8 \times 10^6\,\mathrm{mol^{-1}\,m^3\,s^{-1}} = 3.8 \times 10^9\,\mathrm{L\,mol^{-1}\,s^{-1}}$$

Therefore, the initial rate is

$$\frac{d[P]}{dt} = (3.8 \times 10^9\,\mathrm{L\,mol^{-1}\,s^{-1}}) \times (0.150\,\mathrm{mol\,L^{-1}}) \times (0.330\,\mathrm{mol\,L^{-1}})$$

$$= \boxed{1.9 \times 10^8\,\mathrm{mol\,L^{-1}\,s^{-1}}}$$

Comment. If eqn 35 is used in place of eqn 27, $k_2 = 2.9 \times 10^9\,\mathrm{L\,mol^{-1}\,s^{-1}}$ which yields $\frac{d[P]}{dt} = 1.4 \times 10^8\,\mathrm{mol\,L^{-1}\,s^{-1}}$. In this case the approximation that led to eqn 35 results in a difference of ~30 per cent.

E27.10 For reactions in solution the relation between energy and enthalpy of activation is

$$\Delta^{\ddagger}H = E_a - RT$$

[See, for example, the 5th edition of this textbook.]

$$k_2 = B\,e^{\Delta^{\ddagger}S/R}e^{-\Delta^{\ddagger}H/RT}, \quad B = \left(\frac{kT}{h}\right) \times \left(\frac{RT}{p^{\ominus}}\right)\ [64]$$

$$= B\,e^{\Delta^{\ddagger}S/R}e^{-E_a/RT}\,e = A\,e^{-E_a/RT}$$

Therefore, $A = e\,B\,e^{\Delta^{\ddagger}S/R}$, implying that $\Delta^{\ddagger}S = R\left(\ln\frac{A}{B} - 1\right)$

Therefore, since $E_a = 8681\,\mathrm{K} \times R$

$$\Delta^{\ddagger}H = E_a - RT = (8681\,\mathrm{K} - 303\,\mathrm{K})R$$

$$= (8378\,\mathrm{K}) \times (8.314\,\mathrm{J\,K^{-1}\,mol^{-1}}) = \boxed{69.7\,\mathrm{kJ\,mol^{-1}}}$$

$$B = \frac{(1.381 \times 10^{-23}\,\mathrm{J\,K^{-1}}) \times (303\,\mathrm{K})}{6.626 \times 10^{-34}\,\mathrm{J\,s}} \times \frac{(8.314\,\mathrm{J\,K^{-1}\,mol^{-1}}) \times (303\,\mathrm{K})}{10^5\,\mathrm{Pa}}$$

$$= 1.59 \times 10^{11}\,\mathrm{m^3\,mol^{-1}\,s^{-1}} = 1.59 \times 10^{14}\,\mathrm{L\,mol^{-1}\,s^{-1}}$$

and hence $\Delta^{\ddagger}S = R\left[\ln\left(\frac{2.05 \times 10^{13}\,\mathrm{L\,mol^{-1}\,s^{-1}}}{1.59 \times 10^{14}\,\mathrm{L\,mol^{-1}\,s^{-1}}}\right) - 1\right]$

$$= 8.314\,\mathrm{J\,K^{-1}\,mol^{-1}} \times (-3.05) = \boxed{-25\,\mathrm{J\,K^{-1}\,mol^{-1}}}$$

E27.11 $\Delta^{\ddagger}H = E_a - RT$ [Exercise 27.10(a)]

$$\Delta^{\ddagger}H = (9134\,\text{K} - 303\,\text{K}) \times (8.314\,\text{J K}^{-1}\,\text{mol}^{-1}) = +73.4\,\text{kJ mol}^{-1}$$

$$\Delta^{\ddagger}S = R\left(\ln\frac{A}{B} - 1\right) \text{ [Exercise 27.10(a)]}$$

with $B = \left(\frac{kT}{h}\right) \times \left(\frac{RT}{p^{\ominus}}\right)$ [64] $= 1.59 \times 10^{14}\,\text{L mol}^{-1}\,\text{s}^{-1}$ at 30°C

Therefore, $\Delta^{\ddagger}S = 8.314\,\text{J K}^{-1}\,\text{mol}^{-1} \times \left[\ln\left(\frac{7.78 \times 10^{14}}{1.59 \times 10^{14}}\right) - 1\right] = +4.9\,\text{J K}^{-1}\,\text{mol}^{-1}$

Hence, $\Delta^{\ddagger}G = \Delta^{\ddagger}H - T\Delta^{\ddagger}S = \{(73.4) - (303) \times (4.9 \times 10^{-3})\}\,\text{kJ mol}^{-1} = \boxed{+71.9\,\text{kJ mol}^{-1}}$

E27.12 $\Delta^{\ddagger}H = E_a - 2RT$ [expression above eqn 65]

$$= \{(56.8) - (2) \times (8.314 \times 10^{-3}) \times (338)\}\,\text{kJ mol}^{-1} = 51.2\,\text{kJ mol}^{-1}$$

$k_2 = A\,\text{e}^{-E_a/RT}$ implies that

$$A = k_2\,\text{e}^{E_a/RT} = 7.84 \times 10^{-3}\,\text{kPa}^{-1}\,\text{s}^{-1} \times \text{e}^{58.6\times 10^3/(8.314\times 338)}$$
$$= 4.70\bar{5} \times 10^{6}\,\text{kPa}^{-1}\,\text{s}^{-1} = 4.70\bar{5} \times 10^{3}\,\text{Pa}^{-1}\,\text{s}^{-1}$$

In terms of molar concentrations

$$v = k_2 p_A p_B = k_2 (RT)^2 [\text{A}][\text{B}]$$

and instead of $\frac{\text{d}p_A}{\text{d}t} = -k_2 p_A p_B$

we have $\frac{\text{d}[\text{A}]}{\text{d}t} = -k_2 RT[\text{A}][\text{B}]$

and hence use

$$A = (4.70\bar{5} \times 10^{3}\,\text{Pa}^{-1}\,\text{s}^{-1}) \times (8.314\,\text{J K}^{-1}\,\text{mol}^{-1}) \times (338\,\text{K}) = 1.32\bar{2} \times 10^{7}\,\text{m}^3\,\text{mol}^{-1}\,\text{s}^{-1}$$

Then $B = \frac{kT}{h} \times \frac{RT}{p^{\ominus}}[64] = \frac{(1.381 \times 10^{-23}) \times (338\,\text{K})}{6.626 \times 10^{-34}\,\text{J s}} \times \frac{(8.314\,\text{J K}^{-1}\,\text{mol}^{-1}) \times (338\,\text{K})}{10^5\,\text{Pa}}$

$$= 1.98 \times 10^{11}\,\text{m}^3\,\text{s}^{-1}\,\text{mol}^{-1}$$

and $\Delta^{\ddagger}S = R\left[\ln\left(\frac{A}{B}\right) - 2\right]$ [66] $= (8.314\,\text{J K}^{-1}\,\text{mol}^{-1}) \times \left\{\ln\left(\frac{1.32\bar{2} \times 10^{7}}{1.98 \times 10^{11}}\right) - 2\right\}$

$$= \boxed{-96.6\,\text{J K}^{-1}\,\text{mol}^{-1}}$$

and hence $\Delta^{\ddagger}G = \Delta^{\ddagger}H - T\,\Delta^{\ddagger}S = [(51.2) - (338) \times (-96.6 \times 10^{-3})]\,\text{kJ mol}^{-1}$

$$= +83.9\,\text{kJ mol}^{-1}$$

E27.13 $k_2 = N_A\sigma^*\left(\frac{8kT}{\pi\mu}\right)^{1/2} \text{e}^{-\Delta E_0/RT}$ [58]

The pre-exponential factor is

$$A = N_A\sigma^*\left(\frac{8kT}{\pi\mu}\right)^{1/2}$$

Therefore, $\dfrac{A}{B} = \left(\dfrac{N_A\sigma^* hp^{\ominus}}{kT \times RT}\right) \times \left(\dfrac{8kT}{\pi\mu}\right)^{1/2} = \dfrac{8^{1/2}\sigma^* hp^{\ominus}}{(\pi\mu k^3T^3)^{1/2}}$

For identical particles, $\mu = \frac{1}{2}m$, so

$$\frac{A}{B} = \frac{4\sigma^* hp^{\ominus}}{(\pi m k^3T^3)^{1/2}}$$

$$= \frac{(4) \times (0.4 \times 10^{-18}\,\text{m}^2) \times (6.626 \times 10^{-34}\,\text{J s}) \times (10^5\,\text{Pa})}{\{(\pi) \times (50) \times (1.6605 \times 10^{-27}\,\text{kg}) \times (1.381 \times 10^{-23}\,\text{J K}^{-1} \times 300\,\text{K})^3\}^{1/2}}$$

$$= 7.78 \times 10^{-4}$$

and hence $\Delta^{\ddagger}S = R\left[\ln\left(\dfrac{A}{B}\right) - 2\right]$ [from eqn 66] $= 8.314\,\text{J K}^{-1}\,\text{mol}^{-1}\{\ln 7.78 \times 10^{-4} - 2\}$

$$= \boxed{-76\,\text{J K}^{-1}\,\text{mol}^{-1}}$$

E27.14 $B = \left(\dfrac{kT}{h}\right) \times \left(\dfrac{RT}{p^{\ominus}}\right)$ [64]

$$= \left(\frac{(1.381 \times 10^{-23}\,\text{J K}^{-1}) \times (298.15\,\text{K})}{6.626 \times 10^{-34}\,\text{J s}}\right) \times \left(\frac{(8.314\,\text{J K}^{-1}\,\text{mol}^{-1}) \times (298.15\,\text{K})}{10^5\,\text{Pa}}\right)$$

$$= 1.540 \times 10^{11}\,\text{m}^3\,\text{mol}^{-1}\,\text{s}^{-1} = 1.540 \times 10^{14}\,\text{L mol}^{-1}\,\text{s}^{-1}$$

Therefore

(a) $\Delta^{\ddagger}S = R\left[\ln\left(\dfrac{4.6 \times 10^{12}}{1.540 \times 10^{14}}\right) - 2\right] = \boxed{-45.8\,\text{J K}^{-1}\,\text{mol}^{-1}}$

(b) $\Delta^{\ddagger}H = E_a - 2RT = \{(10.0) - (2) \times (2.48)\}\,\text{kJ mol}^{-1} = \boxed{+5.0\,\text{kJ mol}^{-1}}$

(c) $\Delta^{\ddagger}G = \Delta^{\ddagger}H - T\Delta^{\ddagger}S = \{(5.0) - (298.15) \times (-45.8 \times 10^{-3})\}\,\text{kJ mol}^{-1} = \boxed{+18.7\,\text{kJ mol}^{-1}}$

E27.15 If cleavage of a C–D or C–H bond is involved in the rate-determining step then use

$$\frac{k(\text{C–D})}{k(\text{C–H})} = \text{e}^{-\lambda}, \quad \lambda = \left(\frac{\hbar k_f^{1/2}}{2kT}\right) \times \left(\frac{1}{\mu_{\text{CH}}^{1/2}} - \frac{1}{\mu_{\text{CD}}^{1/2}}\right) \text{ [60]}$$

and see if this accounts for the difference.

$$\mu_{\text{CD}} \approx \frac{2 \times 12}{2 + 12}\text{u} = 1.71\,\text{u}$$

$$\mu_{\text{CH}} \approx \frac{1 \times 12}{1 + 12}\text{u} = 0.92\,\text{u}$$

$$\lambda \approx \left(\frac{(1.054 \times 10^{-34}\,\text{J s}) \times (450\,\text{N m}^{-1})^{-1/2}}{(2) \times (1.381 \times 10^{-23}\,\text{J K}^{-1}) \times (298\,\text{K})}\right) \times \left(\frac{1}{0.92^{1/2}} - \frac{1}{1.71^{1/2}}\right) \times \left(\frac{1}{(1.6605 \times 10^{-27}\,\text{kg})^{1/2}}\right)$$

$$\approx 1.85$$

Hence, $\dfrac{k_2(\text{D})}{k_2(\text{H})} = \text{e}^{-1.85} = \boxed{0.156}$

That is, $k_2(\text{H}) \approx 6.4 \times k_2(\text{D})$, in reasonable accord with the data.

E27.16 $\log k_2 = \log k_2^\circ + 2Az_Az_BI^{1/2}$ [73]
Hence

$$\log k_2^\circ = \log k_2 - 2Az_Az_BI^{1/2}$$
$$= (\log 12.2) - (2) \times (0.509) \times (1) \times (-1) \times (0.0525^{1/2}) = 1.32$$
$$k_2^\circ = \boxed{20.9\,\mathrm{L^2\,mol^{-2}\,min^{-1}}}$$

Solutions to problems

Solutions to numerical problems

P27.2 Draw up the following table as the basis of an Arrhenius plot

T/K	600	700	800	1000
$10^3\,\mathrm{K}/T$	1.67	1.43	1.25	1.00
$k/(\mathrm{cm^3\,mol^{-1}\,s^{-1}})$	4.6×10^2	9.7×10^3	1.3×10^5	3.1×10^6
$\ln(k/\mathrm{cm^3\,mol^{-1}\,s^{-1}})$	6.13	9.18	11.8	14.9

The points are plotted in Fig. 27.1.

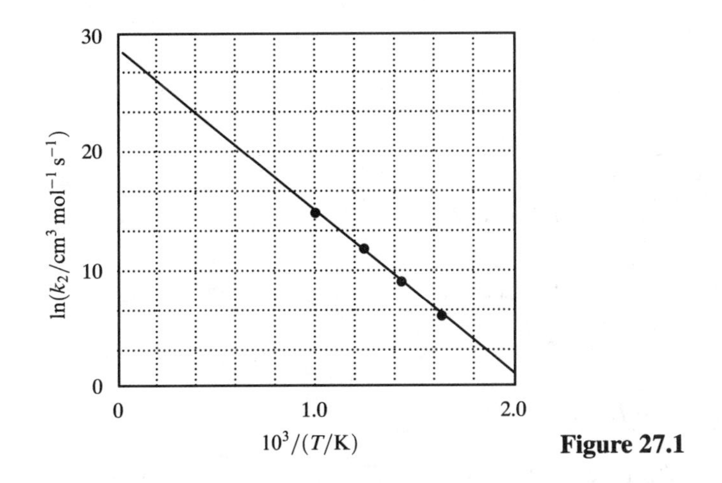

Figure 27.1

The least-squares intercept is at 28.3, which implies that

$$A/(\mathrm{cm^3\,mol^{-1}\,s^{-1}}) = \mathrm{e}^{28.3} = 2.0 \times 10^{12}$$

From $A = N_A\sigma^*\left(\dfrac{8kT}{\pi\mu}\right)^{1/2}$ [Exercise 27.13(a)]

$$\sigma^* = \frac{A_{\mathrm{exptl}}}{N_A(8kT/\pi\mu)^{1/2}} \quad \text{with } \mu = \tfrac{1}{2}m(\mathrm{NO_2})$$
$$= \left(\frac{A_{\mathrm{exptl}}}{4N_A}\right)\left(\frac{\pi m}{kT}\right)^{1/2} = \left(\frac{2.0 \times 10^6\,\mathrm{m^3\,mol^{-1}\,s^{-1}}}{(4) \times (6.022 \times 10^{23}\,\mathrm{mol^{-1}})}\right)$$
$$\times \left(\frac{(\pi) \times (46\,\mathrm{u}) \times (1.6605 \times 10^{-27}\,\mathrm{kg\,u^{-1}})}{(1.381 \times 10^{-23}\,\mathrm{J\,K^{-1}}) \times (750\,\mathrm{K})}\right)^{1/2}$$

$$= 4.0 \times 10^{-21}\,\text{m}^2 \quad \text{or} \quad \boxed{4.0 \times 10^{-3}\,\text{nm}^2}$$

$$P = \frac{\sigma^*}{\sigma} = \frac{4.0 \times 10^{-3}\,\text{nm}^2}{0.60\,\text{nm}^2} = \boxed{0.007}$$

P27.4 Draw up the following table for an Arrhenius plot

$\theta/°\text{C}$	−24.82	−20.73	−17.02	−13.00	−8.95
T/K	248.33	252.42	256.13	260.15	264.20
$10^3/(T/\text{K})$	4.027	3.962	3.904	3.844	3.785
$\ln(k/\text{s}^{-1})$	−9.01	−8.37	−7.73	−7.07	−6.55

The points are plotted in Fig.27.2.

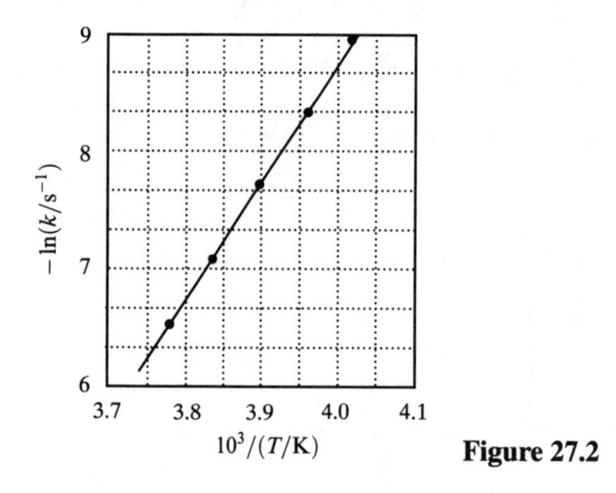

Figure 27.2

A least-squares fit of the data yields the intercept +32.6 at $\dfrac{1}{T} = 0$ and slope $-10.33 \times 10^3\,\text{K}$. The former implies that $\ln\left(\dfrac{A}{\text{s}^{-1}}\right) = 32.6$, and hence that $A = 1.4 \times 10^{14}\,\text{s}^{-1}$. The slope yields $\dfrac{E_\text{a}}{R} = 10.33 \times 10^3\,\text{K}$, and hence $E_\text{a} = \boxed{85.9\,\text{kJ mol}^{-1}}$

In solution $\Delta^{\ddagger}H = E_\text{a} - RT$ [See, for example, the fifth edition of this textbook], so at −20°C

$$\Delta^{\ddagger}H = (85.9\,\text{kJ mol}^{-1}) - (8.314\,\text{J K}^{-1}\,\text{mol}^{-1}) \times (253\,\text{K})$$
$$= \boxed{83.8\,\text{kJ mol}^{-1}}$$

We assume that the reaction is first-order for which, by analogy to Section 27.5

$$K^{\ddagger} = K = \frac{kT}{h\nu}\overline{K}$$

and $k_1 = k^{\ddagger}\,K^{\ddagger} = \nu \times \dfrac{kT}{h\nu} \times \overline{K}$

with $\Delta^{\ddagger}G = -RT\,\ln\overline{K}$

Therefore, $k_1 = A\,\text{e}^{-E_\text{a}/RT} = \dfrac{kT}{h}\,\text{e}^{-\Delta^{\ddagger}G/RT} = \dfrac{kT}{h}\,\text{e}^{\Delta^{\ddagger}S/R}\,\text{e}^{-\Delta^{\ddagger}H/RT}$

and hence we can identify $\Delta^{\ddagger}S$ by writing

$$k_1 = \frac{kT}{h}\,e^{\Delta^\ddagger S/R}\,e^{-E_a/RT}\,e = A\,e^{-E_a/RT}$$

and hence obtain

$$\Delta^\ddagger S = R\left[\ln\left(\frac{hA}{kT}\right) - 1\right]$$

$$= 8.314\,\mathrm{J\,K^{-1}\,mol^{-1}} \times \left[\ln\left(\frac{(6.626\times10^{-34}\,\mathrm{J\,s})\times(1.4\times10^{14}\,\mathrm{s^{-1}})}{(1.381\times10^{-23}\,\mathrm{J\,K^{-1}})\times(253\,\mathrm{K})}\right) - 1\right]$$

$$= \boxed{+19.1\,\mathrm{J\,K^{-1}mol^{-1}}}$$

Therefore, $\Delta^\ddagger G = \Delta^\ddagger H - T\Delta^\ddagger S = 83.8\,\mathrm{kJ\,mol^{-1}} - 253\,\mathrm{K}\times 19.1\,\mathrm{J\,K^{-1}\,mol^{-1}}$

$$= \boxed{+79.0\,\mathrm{kJ\,mol^{-1}}}$$

P27.5 $\log k_2 = \log k_2^\circ + 2Az_Az_BI^{1/2}$ with $A = 0.509\,(\mathrm{mol\,L^{-1}})^{-1/2}$ [73]

This expression suggests that we should plot $\log k$ against $I^{1/2}$ and determine z_B from the slope, since we know that $|z_A| = 1$. We draw up the following table

$I/(\mathrm{mol\,L^{-1}})$	0.0025	0.0037	0.0045	0.0065	0.0085
$(I/(\mathrm{mol\,L^{-1}}))^{1/2}$	0.050	0.061	0.067	0.081	0.092
$\log(k_2/(\mathrm{L\,mol^{-1}\,s^{-1}}))$	0.021	0.049	0.064	0.072	0.100

These points are plotted in Fig. 27.3.

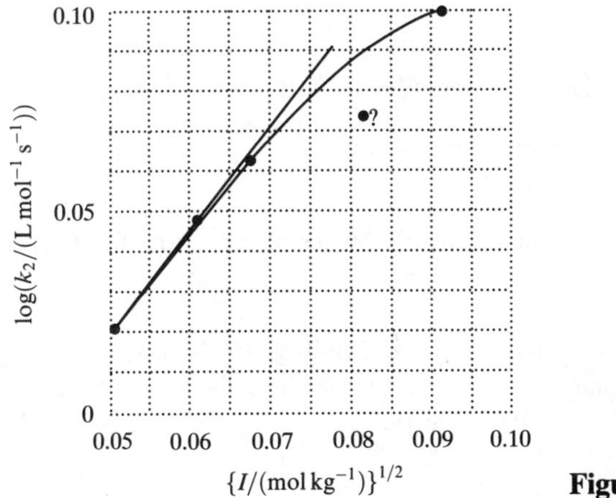

Figure 27.3

The slope of the limiting line in Fig. 27.3 is ≈2.5. Since this slope is equal to $2\,Az_Az_B\times(\mathrm{mol\,L^{-1}})^{1/2} = 1.018\,z_Az_B$, we have $z_Az_B \approx 2.5$. But $|z_A| = 1$, and so $|z_B| = 2$. Furthermore, z_A and z_B have the same sign because $z_Az_B > 0$. (The data refer to I^- and $S_2O_8^{2-}$.)

Solutions to theoretical problems

P27.8 $[J]^* = k\int_0^t [J]\,e^{-kt}\,dt + [J]\,e^{-kt}$ [40]

$$\frac{\partial[J]^*}{\partial t} = k[J]\,e^{-kt} + \frac{\partial[J]}{\partial t}\,e^{-kt} - k[J]\,e^{-kt} = \left(\frac{\partial[J]}{\partial t}\right)e^{-kt}$$

$$\frac{\partial^2[\mathrm{J}]^*}{\partial x^2} = k\int_0^t \left(\frac{\partial^2[\mathrm{J}]}{\partial x^2}\right) \mathrm{e}^{-kt}\,\mathrm{d}t + \left(\frac{\partial^2[\mathrm{J}]}{\partial x^2}\right) \mathrm{e}^{-kt}$$

Then, since

$$D\frac{\partial^2[\mathrm{J}]}{\partial x^2} = \frac{\partial[\mathrm{J}]}{\partial t}\ [39,\ k=0]$$

we find that

$$D\frac{\partial^2[\mathrm{J}]^*}{\partial x^2} = k\int_0^t \left(\frac{\partial[\mathrm{J}]}{\partial t}\right) \mathrm{e}^{-kt}\,\mathrm{d}t + \left(\frac{\partial[\mathrm{J}]}{\partial t}\right) \mathrm{e}^{-kt}$$

$$= k\int_0^t \left(\frac{\partial[\mathrm{J}]^*}{\partial t}\right) \mathrm{d}t + \frac{\partial[\mathrm{J}]^*}{\partial t} = k[\mathrm{J}]^* + \frac{\partial[\mathrm{J}]^*}{\partial t}$$

which rearranges to eqn 39. When $t = 0$, $[\mathrm{J}]^* = [\mathrm{J}]$, and so the same initial conditions are satisfied. (The same boundary conditions are also satisfied.)

P27.11 $K_\mathrm{a} = \dfrac{[\mathrm{H}^+][\mathrm{A}^-]}{[\mathrm{HA}]\gamma_\mathrm{HA}}\gamma_\pm^{\,2} \approx \dfrac{[\mathrm{H}^+][\mathrm{A}^-]\gamma_\pm^{\,2}}{[\mathrm{HA}]}$

Therefore, $[\mathrm{H}^+] = \dfrac{[\mathrm{HA}]\,K_\mathrm{a}}{[\mathrm{A}^-]\,\gamma_\pm^{\,2}}$

and $\log[\mathrm{H}^+] = \log K_\mathrm{a} + \log\dfrac{[\mathrm{HA}]}{[\mathrm{A}^-]} - 2\log\gamma_\pm = \log K_\mathrm{a} + \log\dfrac{[\mathrm{HA}]}{[\mathrm{A}^-]} + 2AI^{1/2}$

Write $v = k_2[\mathrm{H}^+][\mathrm{B}]$

then $\log v = \log(k_2[\mathrm{B}]) + \log[\mathrm{H}^+]$

$$= \log(k_2[\mathrm{B}]) + \log\frac{[\mathrm{HA}]}{[\mathrm{A}^-]} + 2AI^{1/2} + \log K_\mathrm{a}$$

$$= \log v^\circ + 2AI^{1/2}, \quad v^\circ = k_2\frac{[\mathrm{B}][\mathrm{HA}]K_\mathrm{a}}{[\mathrm{A}^-]}$$

That is, the logarithm of the rate should depend linearly on the square root of the ionic strength,

$$\boxed{\log v \propto I^{1/2}}$$

P27.12 The structure of the activated complex is shown in Fig. 27.4(a). Its geometry is that of an asymmetric rotor, like H_2O, and thus has three principal moments of inertia about the axes labelled A, B, and C in Fig. 27.4(a). They are [see eqn 16.26, also Example 16.1]

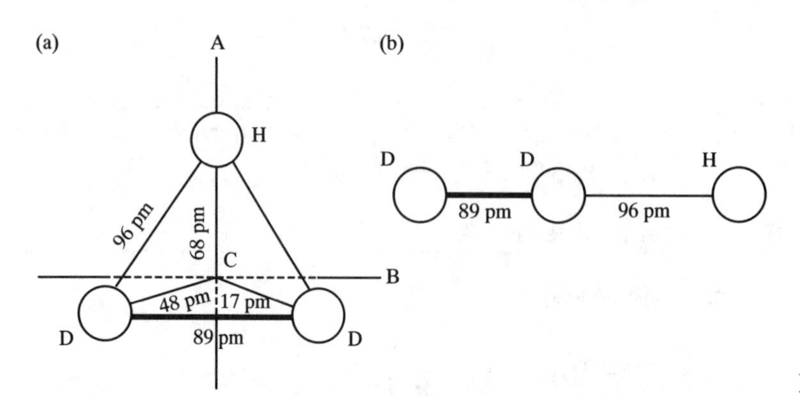

Figure 27.4

$$I_A = 2m_D \times (44\,\text{pm})^2 = 1.3 \times 10^{-47}\,\text{kg m}^2$$

$$I_B = m_H \times (68\,\text{pm})^2 + 2m_D \times (17\,\text{pm})^2 = 9.6 \times 10^{-48}\,\text{kg m}^2$$

$$I_C = m_H \times (68\,\text{pm})^2 + 2m_D \times (48\,\text{pm})^2 = 2.3 \times 10^{-47}\,\text{kg m}^2$$

The rotational constants are therefore

$$A = \frac{\hbar}{4\pi c I_A} = \frac{1.054 \times 10^{-34}\,\text{J s}}{(4\pi) \times (2.998 \times 10^{10}\,\text{cm s}^{-1}) \times I_A} = \frac{2.8 \times 10^{-46}\,\text{cm}^{-1}}{(I_A/\text{kg m}^2)} = 22\,\text{cm}^{-1}$$

$$B = \frac{2.8 \times 10^{-46}\,\text{cm}^{-1}}{9.6 \times 10^{-48}} = 29\,\text{cm}^{-1}$$

$$C = \frac{2.8 \times 10^{-46}\,\text{cm}^{-1}}{2.3 \times 10^{-47}} = 12\,\text{cm}^{-1}$$

Since $I(D_2) = 2m_D \times (37\,\text{pm})^2 = 9.1 \times 10^{-48}\,\text{kg m}^2$, we also have $B(D_2) = 31\,\text{cm}^{-1}$. Then from Table 20.4

$$q^{\ddagger R} = 1.027 \times \tfrac{1}{2} \times \frac{400^{3/2}}{(22 \times 29 \times 12)^{1/2}} = 47$$

$$q^{R}(D_2) = 0.695 \times \tfrac{1}{2} \times \frac{400}{31} = 4.5$$

The vibrational partition functions are

$$q^{V} = \frac{1}{1 - e^{-hc\tilde{\nu}/kT}} \text{ for each mode}$$

$$\approx \frac{1}{1 - e^{-\tilde{\nu}/280\,\text{cm}^{-1}}} \approx 1.03 \quad [\text{for } \tilde{\nu} = 1000\,\text{cm}^{-1}]$$

The complex has $2N - 6 = 3$ modes, but one is the reaction coordinate and is discarded. Hence, $q^{\ddagger V} \approx (1.03)^2 = 1.06$. For D_2 itself, $q^{V} \approx 1$. The translational partition functions are

$$\text{H:}\quad \frac{q_m^{\ominus T}}{N_A} = (2.561 \times 10^{-2}) \times (400^{5/2}) \times (1.01^{3/2}) = 8.3 \times 10^4$$

$$\text{D}_2\text{:}\quad \frac{q_m^{\ominus T}}{N_A} = 6.6 \times 10^5$$

$$\text{Complex:}\quad \frac{q_m^{\ominus T}}{N_A} = 9.2 \times 10^5$$

The electronic partition functions are

$$q^{E}(\text{H}) = 2 \text{ [doublet ground state]}$$

$$q^{E}(D_2) = 1$$

$$q^{\ddagger E}\text{ (Complex)} = 2 \text{ [odd number of electrons, presumably a doublet]}$$

Therefore, bringing all these fragments together with

$$\frac{kT}{h} = \frac{(1.381 \times 10^{-23}\,\mathrm{J\,K^{-1}}) \times (400\,\mathrm{K})}{6.626 \times 10^{-34}\,\mathrm{J\,s}} = 8.34 \times 10^{12}\,\mathrm{s^{-1}}$$

$$\frac{RT}{p^{\ominus}} = 3.33 \times 10^{-2}\,\mathrm{m^3\,mol^{-1}}$$

$$\text{gives } A = \frac{(8.34 \times 10^{12}\,\mathrm{s^{-1}}) \times (3.33 \times 10^{-2}\,\mathrm{m^3\,mol^{-1}}) \times (9.2 \times 10^5) \times (47) \times (1.06) \times (2)}{(8.3 \times 10^4) \times (6.6 \times 10^5) \times (4.5) \times (1.00) \times (2)}$$

$$= 5.2 \times 10^{10}\,\mathrm{L\,mol^{-1}\,s^{-1}}$$

$$k \approx A\,\mathrm{e}^{-E_a/RT} = 5.2 \times 10^{10}\,\mathrm{L\,mol^{-1}\,s^{-1}} \times \mathrm{e}^{-10.52} \approx \boxed{1.4 \times 10^6\,\mathrm{L\,mol^{-1}\,s^{-1}}}$$

Comment. The experimental value is about $4 \times 10^5\,\mathrm{L\,mol^{-1}\,s^{-1}}$.

P27.13 The structure of the activated complex is shown in Fig. 27.4(b). The (one) moment of inertia is [Table 16.1]

$$I = m_{\mathrm{O}}(89\,\mathrm{pm})^2 + m_{\mathrm{H}}(96\,\mathrm{pm})^2 - \frac{\{m_{\mathrm{O}}(89\,\mathrm{pm}) - m_{\mathrm{H}}(96\,\mathrm{pm})\}^2}{m}$$

$$= 4.0 \times 10^{-47}\,\mathrm{kg\,m^2}$$

$$B = \frac{2.8 \times 10^{-46}\,\mathrm{cm^{-1}}}{4.0 \times 10^{-47}} = 7.1\,\mathrm{cm^{-1}}\ [B \text{ from Problem 27.12}]$$

$$q^{\mathrm{R}} = 0.6950 \times \frac{400}{7.1}\ (\sigma = 1) = 39$$

Since $3N - 5 = 4$, there are four vibrational modes of the complex, and counting one as a reaction coordinate gives $q^{\mathrm{V}} \approx (1.03)^3 = 1.09$. All other contributions are as in Problem 27.12, which gave $5.2 \times 10^{10}\,\mathrm{L\,mol^{-1}\,s^{-1}}$. Therefore

$$A \approx (5.2 \times 10^{10}\,\mathrm{L\,mol^{-1}\,s^{-1}}) \times \left(\frac{39}{47}\right) \times \left(\frac{1.09}{1.06}\right) = 4.4 \times 10^{10}\,\mathrm{L\,mol^{-1}\,s^{-1}}$$

and hence k should be modified by the same factor (0.85), to give

$$k = \boxed{1.2 \times 10^6\,\mathrm{L\,mol^{-1}\,s^{-1}}}$$

P27.15 We consider the y-direction to be the direction of diffusion. Hence, for the activated atom the vibrational mode in this direction is lost. Therefore,

$$q^{\ddagger} = q_{\mathrm{z}}^{\ddagger\mathrm{V}} q_{\mathrm{x}}^{\ddagger\mathrm{V}} \quad \text{for the activated atom, and}$$

$$q = q_{\mathrm{x}}^{\mathrm{V}} q_{\mathrm{y}}^{\mathrm{V}} q_{\mathrm{z}}^{\mathrm{V}} \quad \text{for an atom at the bottom of a well}$$

For classical vibration, $q^{\mathrm{V}} \approx \dfrac{kT}{h\nu}$ [Section 27.5]

The diffusion process described is unimolecular, hence first-order, and therefore analogous to the second-order case of Section 27.5 [also see Problem 27.4] we may write

$$-\frac{\mathrm{d}[x]}{\mathrm{d}t} = k^{\ddagger}[x]^{\ddagger} = \nu[x]^{\ddagger} = \nu K^{\ddagger}[x] = k_1[x] \quad \left[K^{\ddagger} = \frac{[x]^{\ddagger}}{[x]}\right]$$

Thus

$$k_1 = \nu K^{\ddagger} = \nu \left(\frac{kT}{h\nu}\right) \times \left(\frac{q^{\ddagger}}{q}\right) e^{-\beta \Delta E_0} \quad \left[\beta = \frac{1}{RT} \text{here}\right]$$

where $q^{\ddagger}$ and q are the (vibrational) partition functions at the top and foot of the well respectively. Therefore

$$k_1 = \frac{kT}{h}\left(\frac{(kT/h\nu^{\ddagger})^2}{(kT/h\nu)^3}\right) e^{-\beta \Delta E_0} = \boxed{\frac{\nu^3}{\nu^{\ddagger 2}} e^{-\beta \Delta E_0}}$$

(a) $\nu^{\ddagger} = \nu; \quad k_1 = \nu e^{-\beta \Delta E_0}$. Assume $\Delta E_0 \approx E_a$; hence

$$k_1 \approx 10^{11}\,\text{Hz}\, e^{-60\times10^3/(8.314\times500)} = 5.4 \times 10^4\,\text{s}^{-1}$$

$$\text{But } D = \frac{\lambda^2}{2\tau} \approx \frac{1}{2}\lambda^2 k_1 \left[24.84;\ \tau = \frac{1}{k_1}, \text{Problem 25.8}\right]$$

$$= \tfrac{1}{2} \times (316\,\text{pm})^2 \times 5.4 \times 10^4\,\text{s}^{-1} = \boxed{2.7 \times 10^{-15}\,\text{m}^2\,\text{s}^{-1}}$$

(b) $\nu^{\ddagger} = \tfrac{1}{2}\nu; \qquad k_1 = 4\nu\, e^{-\beta \Delta E_0} = 2.2 \times 10^5\,\text{s}^{-1}$

$$D = (4) \times (2.7 \times 10^{-15}\,\text{m}^2\,\text{s}^{-1}) = \boxed{1.1 \times 10^{-14}\,\text{m}^2\,\text{s}^{-1}}$$

P27.17 The change in intensity of the beam, dI, is proportional to the number of scatterers per unit volume, $\mathcal{N}_s$, the intensity of the beam, I, and the path length dl. The constant of proportionality is defined as the collision cross-section σ. Therefore,

$$dI = -\sigma \mathcal{N}_s I\, dl \quad \text{or} \quad d \ln I = -\sigma \mathcal{N}_s\, dl$$

If the incident intensity (at $l = 0$) is I_0 and the emergent intensity is I, we can write

$$\ln \frac{I}{I_0} = -\sigma \mathcal{N}_s l \quad \text{or} \quad \boxed{I = I_0\, e^{-\sigma \mathcal{N}_s l}}$$

Solutions to additional problems

P27.19 The rate constants are:

$$k = A \exp\left(\frac{-E_a}{RT}\right) \quad [25.26]$$

$$k_1 = (1.13 \times 10^9\,\text{L}\,\text{mol}^{-1}\,\text{s}^{-1}) \exp\left(\frac{-14.1 \times 10^3\,\text{J}\,\text{mol}^{-1}}{(8.3145\,\text{J}\,\text{mol}^{-1}\,\text{K}^{-1}) \times (298\,\text{K})}\right)$$

$$= \boxed{3.82 \times 10^6\,\text{L}\,\text{mol}^{-1}\,\text{s}^{-1}}$$

$$k_2 = (6.0 \times 10^8\,\text{L}\,\text{mol}^{-1}\,\text{s}^{-1}) \exp\left(\frac{-17.5 \times 10^3\,\text{J}\,\text{mol}^{-1}}{(8.3145\,\text{J}\,\text{mol}^{-1}\,\text{K}^{-1}) \times (298\,\text{K})}\right)$$

$$= \boxed{5.1 \times 10^5\,\text{L}\,\text{mol}^{-1}\,\text{s}^{-1}}$$

$$k_3 = (1.01 \times 10^9\,\text{L}\,\text{mol}^{-1}\,\text{s}^{-1}) \exp\left(\frac{-13.6 \times 10^3\,\text{J}\,\text{mol}^{-1}}{(8.3145\,\text{J}\,\text{mol}^{-1}\,\text{K}^{-1}) \times (298\,\text{K})}\right)$$

$$= \boxed{4.17 \times 10^6\,\text{L}\,\text{mol}^{-1}\,\text{s}^{-1}}$$

Compared to reaction 1, reaction 2 shows a significant kinetic isotope effect whereas reaction 3 shows practically none. This difference should not be surprising: in reaction 2 a C−D bond is broken, whereas in reaction 3 the D atom is simply along for the ride already attached to the O atom. Compare the measured isotope effect of 0.13 to that expected in reaction 2.

$$\frac{k_2}{k_1} = \exp\left(\frac{-\hbar k_f{}^{1/2}}{2kT}\left(\frac{1}{\mu_{CH}^{1/2}} - \frac{1}{\mu_{CD}^{1/2}}\right)\right).$$

We take $\mu_{CH} \approx m_H$ and $\mu_{CD} \approx m_D \approx 2m_H$, so

$$\frac{k_2}{k_1} = \exp\left(\left(\frac{-(1.0546 \times 10^{-34}\,\mathrm{J\,s}) \times (500\,\mathrm{kg\,s^{-2}})^{1/2}}{2(1.381 \times 10^{-23}\,\mathrm{J\,K^{-1}}) \times (298\,\mathrm{K})}\right) \times \left(1 - \frac{1}{2^{1/2}}\right) \times \left(\frac{6.022 \times 10^{23}\,\mathrm{mol^{-1}}}{1 \times 10^{-3}\,\mathrm{kg\,mol^{-1}}}\right)^{1/2}\right)$$

$$= \boxed{0.13}$$

in agreement with the experimental value.

P27.21 Both approaches involve plots of $\log k$ versus $\log \gamma$, where γ is the activity coefficient. The limiting law has $\log \gamma$ proportional to $I^{1/2}$ (where I is ionic strength), so a plot of $\log k$ versus $I^{1/2}$ should give a straight line whose y-intercept is $\log k^\circ$ and whose slope is $2Az_Az_B$, where z_A and z_B are charges involved in the activated complex. The extended Debye–Hückel law has $\log \gamma$ proportional to $\left(\frac{I^{1/2}}{1+BI^{1/2}}\right)$, so it requires plotting $\log k$ versus $\left(\frac{I^{1/2}}{1+BI^{1/2}}\right)$, and it also has a slope of $2Az_Az_B$ and a y-intercept of $\log k^\circ$. The ionic strength in a 2:1 electrolyte solution is three times the molar concentration. The transformed data and plot (Fig. 27.5) follow

$[Na_2SO_4]/(\mathrm{mol\,kg^{-1}})$	0.2	0.15	0.1	0.05	0.025	0.0125	0.005
$k/(\mathrm{L^{1/2}\,mol^{-1/2}\,s^{-1}})$	0.462	0.430	0.390	0.321	0.283	0.252	0.224
$I^{1/2}$	0.775	0.671	0.548	0.387	0.274	0.194	0.122
$I^{1/2}/(1+BI^{1/2})$	0.436	0.401	0.354	0.279	0.215	0.162	0.109
$\log k$	−0.335	−0.367	−0.409	−0.493	−0.548	−0.599	−0.650

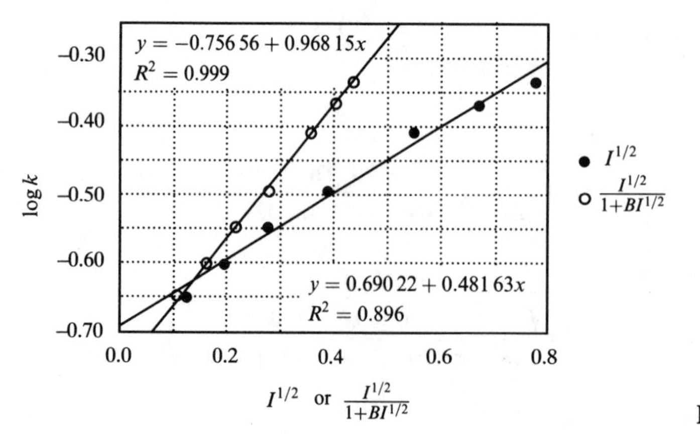

Figure 27.5

The line based on the limiting law appears curved. The zero-ionic-strength rate constant based on it is

$$k^\circ = 10^{-0.690}\,\mathrm{L^{1/2}\,mol^{-1/2}\,s^{-1}} = 0.204\,\mathrm{L^{1/2}\,mol^{-1/2}\,s^{-1}}$$

The slope is positive, so the complex must overcome repulsive interactions. The product of charges, however, works out to be 0.5, not easily interpretable in terms of charge numbers. The line based on the extended law appears straighter and has a better correlation coefficient. The zero-ionic-strength rate constant based on it is

$$k^\circ = 10^{-0.757}\,\mathrm{L^{1/2}\,mol^{-1/2}\,s^{-1}} = 0.175\,\mathrm{L^{1/2}\,mol^{-1/2}\,s^{-1}}$$

The product of charges works out to be 0.9, nearly 1, interpretable in terms of

$$\boxed{\text{a complex of two univalent ions of the same sign}}$$

P27.23 (a)

$$\frac{d[F_2O]}{dt} = -k_1[F_2O]^2 - k_2[F][F_2O] \qquad (1)$$

$$\frac{d[F]}{dt} = k_1[F_2O]^2 - k_2[F][F_2O] + 2k_3[OF]^2 - 2k_4[F]^2[F_2O] \qquad (2)$$

$$\frac{d[OF]}{dt} = k_1[F_2O]^2 + k_2[F][F_2O] - 2k_3[OF]^2 \qquad (3)$$

applying the steady-state approximation to both [F] and [OF] and adding the resulting equations gives

$$\begin{array}{ll} k_1[F_2O]^2 - k_2[F]_{SS}[F_2O] + 2k_3[OF]^2_{SS} & - 2k_4[F]^2_{SS}[F_2O] = 0 \\ k_1[F_2O]^2 + k_2[F]_{SS}[F_2O] - 2k_3[OF]^2_{SS} & = 0 \\ \hline 2k_1[F_2O]^2 & - 2k_4[F]^2_{SS}[F_2O] = 0 \end{array}$$

solving for $[F]_{SS}$ gives

$$[F]_{SS} = \left(\frac{k_1}{k_4}[F_2O]\right)^{1/2} \qquad (4)$$

substituting (4) into (1)

$$\frac{d[F_2O]}{dt} = -k_1[F_2O]^2 - k_2\left(\frac{k_1}{k_4}\right)^{1/2}[F_2O]^{3/2}$$

or

$$\boxed{-\frac{d[F_2O]}{dt} = k_1[F_2O]^2 + k_2\left(\frac{k_1}{k_4}\right)^{1/2}[F_2O]^{3/2}} \qquad (5)$$

Comparison with the experimental rate law reveals that they are consistent when we make the following identifications.

$$k = k_1 = 7.8 \times 10^{13}\mathrm{e}^{-E_1/RT}\,\mathrm{L\,mol^{-1}\,s^{-1}}$$

$$E_1 = (19350\,\mathrm{K})R = 160.9\,\mathrm{kJ\,mol^{-1}}$$

$$k' = k_2\left(\frac{k_1}{k_4}\right)^{1/2} = 2.3 \times 10^{10}\mathrm{e}^{-E'/RT}\,\mathrm{L\,mol^{-1}\,s^{-1}}$$

$$E' = (16910\,\mathrm{K})R = 140.6\,\mathrm{kJ\,mol^{-1}}$$

(b)

$$\begin{array}{lll} \frac{1}{2}O_2 + F_2 \to F_2O & \Delta_f H(F_2O) = 24.41\,\mathrm{kJ\,mol^{-1}} \\ 2F \to F_2 & \Delta H = -D(F\text{–}F) = -160.6\,\mathrm{kJ\,mol^{-1}} \\ O \to \frac{1}{2}O_2 & \Delta H = -\frac{1}{2}D(O\text{–}O) = -249.1\,\mathrm{kJ\,mol^{-1}} \\ \hline 2F + O \to F_2O & \end{array}$$

$$\begin{aligned} \Delta H(\text{FO–F}) + \Delta H(\text{O–F}) &= -\left[\Delta_f H(F_2O) - D(F\text{–}F) - \tfrac{1}{2}D(O\text{–}O)\right] \\ &= -(24.41 - 160.6 - 249.1)\,\mathrm{kJ\,mol^{-1}} \\ &= 385.3\,\mathrm{kJ\,mol^{-1}} \end{aligned}$$

We estimate that $\boxed{\Delta H(\text{FO–F}) \approx E_1 = 160.9\,\mathrm{kJ\,mol^{-1}}}$

Then

$$\begin{aligned} \Delta H(\text{O–F}) &= 385.3\,\mathrm{kJ\,mol^{-1}} - \Delta H(\text{FO–F}) \\ &\approx (385.3 - 160.9)\,\mathrm{kJ\,mol^{-1}} \end{aligned}$$

$$\boxed{\Delta H(\text{O–F}) \approx 224.4\,\mathrm{kJ\,mol^{-1}}}$$

In order to determine the activation energy of reaction (2) we assume that each rate constant can be expressed in arrhenius form, then

$$\ln k' = \ln k_2 + \tfrac{1}{2}\ln k_1 - \tfrac{1}{2}\ln k_4$$

or

$$\ln A' - \frac{E'}{RT} = \ln A_2 - \frac{E_2}{RT} + \frac{1}{2}\ln A_1 - \frac{1}{2}\frac{E_1}{RT} - \frac{1}{2}\ln A_4 + \frac{1}{2}\frac{E_4}{RT}$$

Differentiating with respect to T we obtain

$$E' = E_2 + \tfrac{1}{2}E_1 - \tfrac{1}{2}E_4 = 140.6\,\mathrm{kJ\,mol^{-1}}$$

or

$$\begin{aligned} E_2 - \tfrac{1}{2}E_4 = E' - \tfrac{1}{2}E_1 &= (140.6 - 80.4)\,\mathrm{kJ\,mol^{-1}} \\ &= 60.2\,\mathrm{kJ\,mol^{-1}} \end{aligned}$$

E_4 is expected to be small since reaction (4) is termolecular, so we set $E_4 \approx 0$, then

$$E_2 \approx \boxed{60\,\mathrm{kJ\,mol^{-1}}}$$

P27.26 Linear regression analysis of ln(rate constant) against $1/T$ yields the following results

$$\ln(k/22.4\,\mathrm{L\,mol^{-1}\,min^{-1}}) = C + B/T$$

where $C = 34.36$, standard deviation $= 0.36$
$B = -23\,227\,\mathrm{K}$, standard deviation $= 252\,\mathrm{K}$
$R = \boxed{0.999\,76}$ [good fit]

$$\ln(k'/22.4\,\mathrm{L\,mol^{-1}\,min^{-1}}) = C_2 + B_2/T$$

where $C' = 28.30$, standard deviation $= 0.84$
$B' = -21065\,\mathrm{K}$, standard deviation $= 582\,\mathrm{K}$
$R = \boxed{0.998\,48}$ [good fit]

The regression parameters can be used in the calculation of the pre-experimental factor (A) and the activation energy (E_a) using $\ln k = \ln A - E_a/RT$.

$$\ln A = C + \ln(22.4) = 37.47$$

$$A = 1.87 \times 10^{16}\,\mathrm{L\,mol^{-1}\,min^{-1}} = \boxed{3.12 \times 10^{14}\,\mathrm{L\,mol^{-1}\,s^{-1}}}$$

$$E_a = -RB = -(8.3145\,\mathrm{J\,K^{-1}\,mol^{-1}}) \times (-23\,227\,\mathrm{K}) \times \left(\frac{10^{-3}\,\mathrm{kJ}}{\mathrm{J}}\right)$$

$$= \boxed{193\,\mathrm{kJ\,mol^{-1}}}$$

$$\ln A' = C' + \ln(22.4) = 31.41$$

$$A' = 4.37 \times 10^{13}\,\mathrm{L\,mol^{-1}\,min^{-1}} = \boxed{7.29 \times 10^{11}\,\mathrm{L\,mol^{-1}\,s^{-1}}}$$

$$E_a' = -RB' = -(8.3145\,\mathrm{J\,K^{-1}\,mol^{-1}}) \times (-21065\,\mathrm{K}) \times \left(\frac{10^{-3}\,\mathrm{kJ}}{\mathrm{J}}\right)$$

$$= \boxed{175\,\mathrm{kJ\,mol^{-1}}}$$

To summarize

	$A/(\mathrm{L\,mol^{-1}\,s^{-1}})$	$E_a/(\mathrm{kJ\,mol^{-1}})$
k	$3.12 \times 10^{14} (=A)$	193
k'	$7.29 \times 10^{11} (=A')$	175

Both sets of data, k and k', fit the Arrhenius equation very well and hence are consistent with the collision theory of bimolecular gas-phase reactions which provides an equation [18] compatible with the Arrhenius equation. The numerical values for k' and A may be compared to the results of Exercise 27.4(a) and are in rough agreement at 647 K, as is the value of E_a.

28 Processes at solid surfaces

Solutions to exercises

E28.1 $$Z_W = (2.63 \times 10^{24}\,\text{m}^{-2}\,\text{s}^{-1}) \times \left(\frac{p/\text{Pa}}{\{(T/\text{K}) \times (M/\text{g mol}^{-1})\}^{1/2}}\right) \quad [1b]$$

$$= \left(\frac{(1.52 \times 10^{19}\,\text{cm}^{-2}\,\text{s}^{-1}) \times (p/\text{Pa})}{(M/\text{g mol}^{-1})^{1/2}}\right) \quad [T = 298\,\text{K}]$$

Another practical form of this equation at 298 K is

$$Z_W = \frac{(2.03 \times 10^{21}\,\text{cm}^{-2}\,\text{s}^{-1}) \times (p/\text{Torr})}{(M/\text{g mol}^{-1})^{1/2}} \quad [100\,\text{Pa} = 0.750\,\text{Torr}]$$

or

$$Z_W = \frac{(2.03 \times 10^{25}\,\text{m}^{-2}\,\text{s}^{-1}) \times (p/\text{Torr})}{(M/\text{g mol}^{-1})^{1/2}}$$

Hence, we can draw up the following table

	H_2	C_3H_8
$M/(\text{g mol}^{-1})$	2.02	44.09
$Z_W(\text{m}^{-2}\,\text{s}^{-1})$		
(i) 100 Pa	1.07×10^{25}	2.35×10^{24}
(ii) 10^{-7} Torr	1.4×10^{18}	3.1×10^{17}

E28.2 $$p/\text{Pa} = \frac{\{Z_W/(\text{m}^{-2}\,\text{s}^{-1})\} \times \{(T/\text{K}) \times (M/\text{g mol}^{-1})\}^{1/2}}{2.63 \times 10^{24}} \quad [1b]$$

$$= \frac{\{Z_W/(\text{m}^{-2}\,\text{s}^{-1})\} \times (425 \times 39.95)^{1/2}}{2.63 \times 10^{24}}$$

$$= 4.95 \times 10^{-23} \times Z_W/(\text{m}^{-2}\,\text{s}^{-1})$$

The collision rate required is

$$Z_W = \frac{4.5 \times 10^{20}\,\text{s}^{-1}}{\pi \times (0.075\,\text{cm})^2} = 2.5\bar{5} \times 10^{22}\,\text{cm}^{-2}\,\text{s}^{-1} = 2.5\bar{5} \times 10^{26}\,\text{m}^{-2}\,\text{s}^{-1}$$

Hence $p = (4.95 \times 10^{-23}\,\text{Pa}) \times (2.5\bar{5} \times 10^{26}) = \boxed{1.3 \times 10^4\,\text{Pa}}$

E28.3 $$Z_W = (2.63 \times 10^{24}\,\text{m}^{-2}\,\text{s}^{-1}) \times \left(\frac{p/\text{Pa}}{\{(T/\text{K}) \times (M/\text{g mol}^{-1})\}^{1/2}}\right) \quad [1b]$$

$$= (2.63 \times 10^{24}\,\text{m}^{-2}\,\text{s}^{-1}) \times \left(\frac{35}{(80 \times 4.00)^{1/2}}\right) = 5.1 \times 10^{24}\,\text{m}^{-2}\,\text{s}^{-1}$$

The area occupied by a Cu atom is $\left(\frac{1}{2}\right) \times (3.61 \times 10^{-10}\,\text{m})^2 = 6.52 \times 10^{-20}\,\text{m}^2$ (in an fcc unit cell, there is the equivalent of two Cu atoms per face). Therefore

$$\text{rate per Cu atom} = (5.2 \times 10^{24}\,\text{m}^{-2}\,\text{s}^{-1}) \times (6.52 \times 10^{-20}\,\text{m}^2) = \boxed{3.4 \times 10^5\,\text{s}^{-1}}$$

E28.4 $V_{\text{mon}} = 2.86\,\text{cm}^3$

$$n = \frac{pV}{RT} = \frac{(1.00\,\text{atm}) \times (2.86 \times 10^{-3}\,\text{L})}{(0.0821\,\text{L atm K}^{-1}\,\text{mol}^{-1}) \times (273\,\text{K})} = 1.28 \times 10^{-4}\,\text{mol}$$

$$N = nN_{\text{A}} = 7.69 \times 10^{19}$$

$$A = (7.69 \times 10^{19}) \times (0.165 \times 10^{-18}\,\text{m}^2) = \boxed{12.7\,\text{m}^2}$$

Comment. There is more than one method of estimating the effective cross-sectional area of an adsorbed molecule. One very simple method which is appropriate here is to obtain it from the density of the liquid.

Question. Given that the density of liquid nitrogen is $0.808\,\text{g cm}^{-3}$, what is the effective cross-sectional area of a nitrogen molecule? How does this estimate compare with the value used above?

E28.5 $\theta = \frac{V}{V_\infty}[2] = \frac{V}{V_{\text{mon}}} = \frac{Kp}{1+Kp}\,[5]$

which rearranges to [Example 28.1]

$$\frac{p}{V} = \frac{p}{V_{\text{mon}}} + \frac{1}{KV_{\text{mon}}}$$

Hence, $\frac{p_2}{V_2} - \frac{p_1}{V_1} = \frac{p_2}{V_{\text{mon}}} - \frac{p_1}{V_{\text{mon}}}$

Solving for V_{mon}

$$V_{\text{mon}} = \frac{p_2 - p_1}{\left(\frac{p_2}{V_2} - \frac{p_1}{V_1}\right)} = \frac{(760 - 142.4)\,\text{Torr}}{\left(\frac{760}{1.430} - \frac{142.4}{0.284}\right)\,\text{Torr cm}^{-3}} = \boxed{20.5\,\text{cm}^3}$$

E28.6 The enthalpy of adsorption is typical of $\boxed{\text{chemisorption}}$ (Table 28.2). The residence lifetime is

$$t_{1/2} = \tau_0 e^{E_{\text{d}}/RT}[18] \approx (1 \times 10^{-14}\,\text{s}) \times (e^{120\times10^3/(8.314\times400)})[E_{\text{d}} \approx -\Delta_{\text{ad}}H] \approx \boxed{50\,\text{s}}$$

E28.7 The average residence time for particle adsorbed on the surface is

$$t_{1/2} = \tau_0 e^{E_{\text{d}}/RT}\ [18]$$

$$E_{\text{d}} = \frac{R\ln\left(\frac{t'_{1/2}}{t_{1/2}}\right)}{\left(\frac{1}{T'} - \frac{1}{T}\right)} = \frac{(8.314\,\text{J K}^{-1}\,\text{mol}^{-1}) \times \ln\left(\frac{0.36}{3.49}\right)}{\left(\frac{1}{2548\,\text{K}} - \frac{1}{2362\,\text{K}}\right)} = \boxed{610\,\text{kJ mol}^{-1}}$$

$$\tau_0 = t_{1/2}e^{-E_{\text{d}}/RT} = (3.49\,\text{s}) \times e^{-610\times10^3/(8.314\times2362)} = \boxed{0.113 \times 10^{-12}\,\text{s}}$$

$$A = \ln 2/\tau_0 = 0.693/(0.113 \times 10^{-12}\,\text{s}) = \boxed{6.15 \times 10^{12}\,\text{s}^{-1}}$$

E28.8 $\theta = \frac{Kp}{1+Kp}$ [5], which implies that $p = \left(\frac{\theta}{1-\theta}\right)\frac{1}{K}$

(a) $p = \left(\frac{0.15}{0.85}\right) \times \left(\frac{1}{0.85\,\text{kPa}^{-1}}\right) = \boxed{0.21\,\text{kPa}}$ **(b)** $p = \left(\frac{0.95}{0.05}\right) \times \left(\frac{1}{0.85\,\text{kPa}^{-1}}\right) = \boxed{22\,\text{kPa}}$

E28.9 $$\frac{m_1}{m_2}=\frac{\theta_1}{\theta_2}=\frac{p_1}{p_2}\times\frac{1+Kp_2}{1+Kp_1}$$

which solves to

$$K=\frac{\left(\frac{m_1p_2}{m_2p_1}\right)-1}{p_2-\left(\frac{m_1p_2}{m_2}\right)}=\frac{\left(\frac{m_1}{m_2}\right)\times\left(\frac{p_2}{p_1}\right)-1}{1-\left(\frac{m_1}{m_2}\right)}\times\frac{1}{p_2}$$

$$=\frac{\left(\frac{0.44}{0.19}\right)\times\left(\frac{3.0}{26.0}\right)-1}{1-\left(\frac{0.44}{0.19}\right)}\times\frac{1}{3.0\,\text{kPa}}=0.19\,\text{kPa}^{-1}$$

Therefore,

$$\theta_1=\frac{(0.19\,\text{kPa})\times(26.0\,\text{kPa})}{(1)+(0.19\,\text{kPa}^{-1})\times(26.0\,\text{kPa})}=\boxed{0.83}\qquad\theta_2=\frac{(0.19)\times(3.0)}{(1)+(0.19)\times(3.0)}=\boxed{0.36}$$

E28.10 $t_{1/2}\approx\tau_0\mathrm{e}^{E_d/RT}\,[18]=(10^{-13}\,\text{s})\times(\mathrm{e}^{E_d/(2.48\,\text{kJ mol}^{-1})})$ [at 298 K]

(a) $E_d=15\,\text{kJ mol}^{-1}$, $t_{1/2}=(10^{-13}\,\text{s})\times(\mathrm{e}^{6.05})=\boxed{4\times10^{-11}\,\text{s}}$

(b) $E_d=150\,\text{kJ mol}^{-1}$, $t_{1/2}=(10^{-13}\,\text{s})\times(\mathrm{e}^{6.05})=\boxed{2\times10^{13}\,\text{s}}$

The latter corresponds to about 600 000 y. At 1000 K, $t_{1/2}=(10^{-13}\,\text{s})\times(\mathrm{e}^{E_d/8.314\,\text{kJ mol}^{-1}})$

(a) $t_{1/2}=\boxed{6\times10^{-13}\,\text{s}}$ **(b)** $t_{1/2}=\boxed{7\times10^{-6}\,\text{s}}$

E28.11 $\theta=\dfrac{Kp}{1+Kp}$ [5], which implies that $K=\left(\dfrac{\theta}{1-\theta}\right)\times\left(\dfrac{1}{p}\right)$

But $\ln\dfrac{K'}{K}=\dfrac{\Delta_rH}{R}\left(\dfrac{1}{T}-\dfrac{1}{T'}\right)$ [9.26]

Since θ at the new temperature is the same, $K\propto\dfrac{1}{p}$ and

$$\ln\frac{p}{p'}=\frac{\Delta_{ad}H}{R}\left(\frac{1}{T}-\frac{1}{T'}\right)=\left(\frac{-10.2\,\text{kJ mol}^{-1}}{8.314\,\text{J K}^{-1}\,\text{mol}^{-1}}\right)\times\left(\frac{1}{298\,\text{K}}-\frac{1}{313\,\text{K}}\right)=-0.197$$

which implies that $p'=(12\,\text{kPa})\times(\mathrm{e}^{0.197})=\boxed{15\,\text{kPa}}$

E28.12 $v=k\theta=\dfrac{kKp}{1+Kp}$ [Example 28.4]

(a) On gold, $\theta\approx1$, and $v=k\theta\approx$ constant, a $\boxed{\text{zeroth-order}}$ reaction.

(b) On platinum, $\theta\approx Kp$ (as $Kp\ll1$), so $v=kKp$, and the reaction is $\boxed{\text{first-order}}$

E28.13 $\theta=\dfrac{Kp}{1+Kp}$ and $\theta'=\dfrac{K'p'}{1+K'p}$

but $\theta=\theta'$, so

$$\frac{Kp}{1+Kp}=\frac{K'p'}{1+K'p'}$$

which requires $Kp = K'p'$. We also know that

$$\Delta_{\rm ad}H^{\ominus} = RT^2\left(\frac{\partial \ln K}{\partial T}\right)_\theta \quad [9]$$

and can therefore write

$$\Delta_{\rm ad}H^{\ominus} \approx RT^2\left(\frac{\ln K' - \ln K}{T'-T}\right) = \frac{RT^2\ln\left(\frac{K'}{K}\right)}{T'-T} \approx \frac{RT^2\ln\left(\frac{p}{p'}\right)}{T'-T}$$

$$\approx \frac{(8.314\,{\rm J\,K^{-1}\,mol^{-1}}) \times (220\,{\rm K})^2 \times \ln\left(\frac{4.9}{32}\right)}{60\,{\rm K}} = \boxed{-13\,{\rm kJ\,mol^{-1}}}$$

E28.14 The desorption time for a given volume is proportional to the half-life of the adsorbed species, and as

$$t_{1/2} = \tau_0 {\rm e}^{E_{\rm d}/RT} \quad [18]$$

we can write

$$E_{\rm d} = \frac{R\ln\left(\frac{t_{1/2}}{t'_{1/2}}\right)}{\left(\frac{1}{T}-\frac{1}{T'}\right)} = \frac{R\ln\left(\frac{t}{t'}\right)}{\frac{1}{T}-\frac{1}{T'}}$$

where t and t' are the two desorption times. We evaluate $E_{\rm d}$ from the data for the two temperatures

$$E_{\rm d} = \frac{8.314\,{\rm J\,K^{-1}\,mol^{-1}}}{\left(\frac{1}{1856\,{\rm K}} - \frac{1}{1978\,{\rm K}}\right)} \times \ln\frac{27}{2.0} = \boxed{65\bar{0}\,{\rm kJ\,mol^{-1}}}$$

We write

$$t = t_0{\rm e}^{65\bar{0}\times10^3/(8.314\times1856)} = t_0 \times (1.9\bar{6} \times 10^{18})$$

Therefore, since $t = 27$ min, $t_0 = 1.3\bar{8} \times 10^{-17}$ min. Consequently,

(a) At 298 K,

$$t = (1.3\bar{8} \times 10^{-17}\,{\rm min}) \times {\rm e}^{65\bar{0}\times10^3/(8.314\times298)} = \boxed{1.1 \times 10^{97}\,{\rm min}}$$

which is just about forever.

(b) At 3000 K,

$$t = (1.3\bar{8} \times 10^{-17}\,{\rm min}) \times {\rm e}^{65\bar{0}\times10^3/(8.314\times3000)} = \boxed{2.9 \times 10^{-6}\,{\rm min}}$$

Solutions to problems

Solutions to numerical problems

P28.2 We follow Example 28.1 and draw up the following table

p/Torr	0.19	0.97	1.90	4.05	7.50	11.95
$\frac{p}{V}\Big/({\rm Torr\,cm^{-3}})$	4.52	5.95	8.60	12.6	18.3	25.4

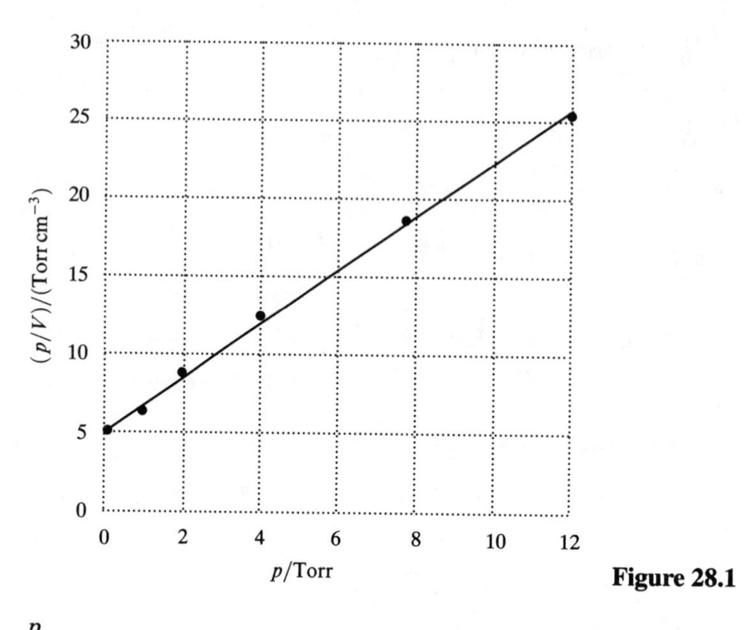

Figure 28.1

$\frac{p}{V}$ is plotted against p in Fig. 28.1.

The low-pressure points fall on a straight line with intercept 4.7 and slope 1.8. It follows that $\frac{1}{V_\infty} = 1.8\,\text{Torr cm}^{-3}/\text{Torr} = 1.8\,\text{cm}^{-3}$, or $V_\infty = 0.57\,\text{cm}^3$ and $\frac{1}{KV_\infty} = 4.7\,\text{Torr cm}^{-3}$. Therefore,

$$K = \frac{1}{(4.7\,\text{Torr cm}^{-3}) \times (0.57\,\text{cm}^3)} = \boxed{0.37\,\text{Torr}^{-1}}$$

Comment. It is unlikely that low-pressure data can be used to obtain an accurate value of the volume corresponding to complete coverage. See Problem 28.4 for adsorption data at higher pressures.

P28.4 We assume that the data fit the Langmuir isotherm; to confirm this we plot $\frac{p}{V}$ against p and expect a straight line [Example 28.1]. We draw up the following table

p/atm	0.050	0.100	0.150	0.200	0.250
$\frac{p}{V}\Big/(10^{-2}\,\text{atm mL}^{-1})$	4.1	7.52	11.5	14.7	17.9

The data are plotted in Fig. 28.2.

They fit closely to a straight line with slope $0.720\,\text{mL}^{-1}$. Hence,

$$V_\infty = \boxed{1.3\bar{9}\,\text{mL}} = 1.3\bar{9} \times 10^{-3}\,\text{L} \approx V_{\text{mon}}$$

The number of H_2 molecules corresponding to this volume is

$$N_{H_2} = \frac{pVN_A}{RT} = \frac{(1.00\,\text{atm}) \times (1.3\bar{9} \times 10^{-3}\,\text{L}) \times (6.02 \times 10^{23}\,\text{mol}^{-1})}{(0.0821\,\text{L atm K}^{-1}\,\text{mol}^{-1}) \times (273\,\text{K})} = 3.73 \times 10^{19}$$

The area occupied is the number of molecules times the area per molecule. The area per molecule can be estimated from the density of the liquid

$$A = \pi \left(\frac{3V}{4\pi}\right)^{2/3} \quad \left[V = \text{volume of molecule} = \frac{M}{\rho N_A}\right]$$

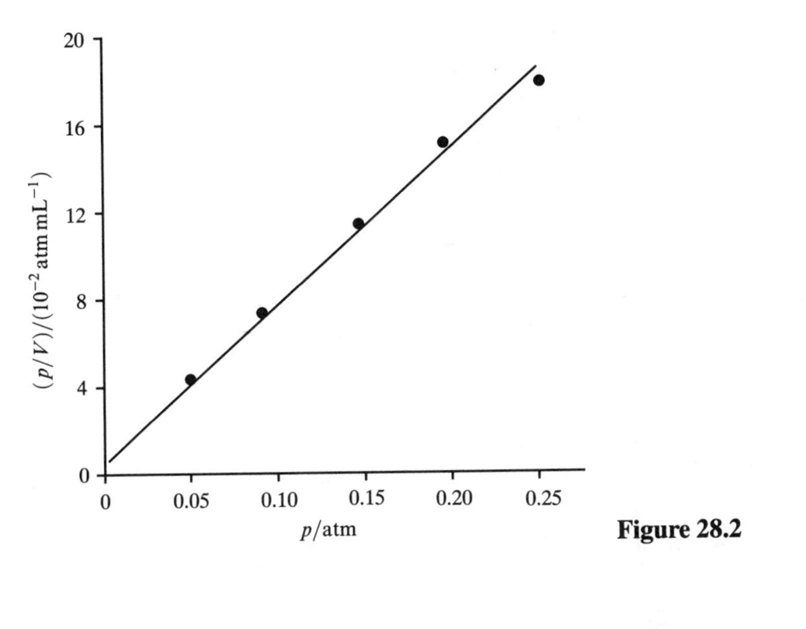

Figure 28.2

$$= \pi\left(\frac{3M}{4\pi p N_A}\right)^{2/3} = \pi\left(\frac{3\times(2.02\,\mathrm{g\,mol^{-1}})}{4\pi\times(0.0708\,\mathrm{g\,cm^{-3}})\times(6.02\times10^{23}\,\mathrm{mol^{-1}})}\right)^{2/3}$$

$$= 1.58\times10^{-15}\,\mathrm{cm^2}$$

Area occupied $= (3.73\times10^{19})\times(1.58\times10^{-15}\,\mathrm{cm^2}) = (5.9\times10^{4}\,\mathrm{cm^2}) = \boxed{5.9\,\mathrm{m^2}}$

Comment. The value for V_∞ calculated here may be compared to the value obtained in Problem 28.2. The agreement is not good and illustrates the point that these kinds of calculations provide only rough values of surface areas.

P28.6 $\theta = c_1 p^{1/c_2}$

We adapt this isotherm to a liquid by noting that $w_a \propto \theta$ and replacing p by [A], the concentration of the acid. Then $w_a = c_1[\mathrm{A}]^{1/c_2}$ (with c_1, c_2 modified constants), and hence

$$\log w_a = \log c_1 + \frac{1}{c_2}\times\log[\mathrm{A}]$$

We draw up the following table

$[\mathrm{A}]/(\mathrm{mol\,L^{-1}})$	0.05	0.10	0.50	1.0	1.5
$\log([\mathrm{A}]/(\mathrm{mol\,L^{-1}}))$	−1.30	−1.00	−0.30	−0.00	0.18
$\log(w_a/\mathrm{g})$	−1.40	−1.22	−0.92	−0.80	−0.72

These points are plotted in Fig. 28.3.

They fall on a reasonably straight line with slope 0.42 and intercept -0.80. Therefore, $c_2 = \frac{1}{0.42} = \boxed{2.4}$ and $c_1 = \boxed{0.16}$. (The units of c_1 are bizarre: $c_1 = 0.16\,\mathrm{g\,mol^{-0.42}\,dm^{1.26}}$.)

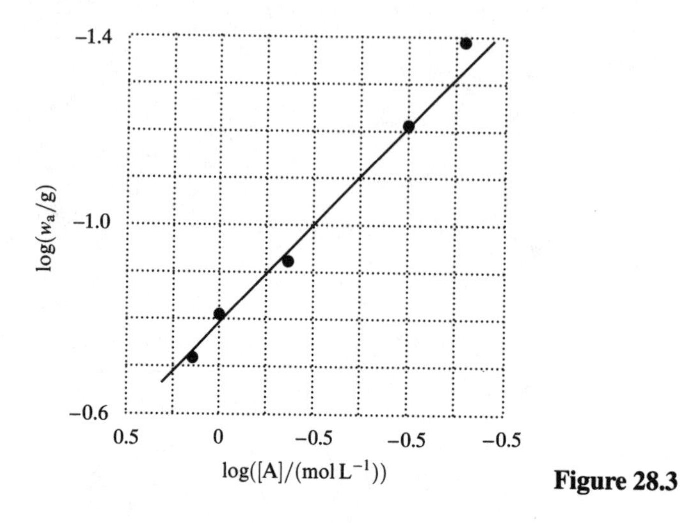

Figure 28.3

Solutions to theoretical problems

P28.9 Refer to Fig. 28.4.

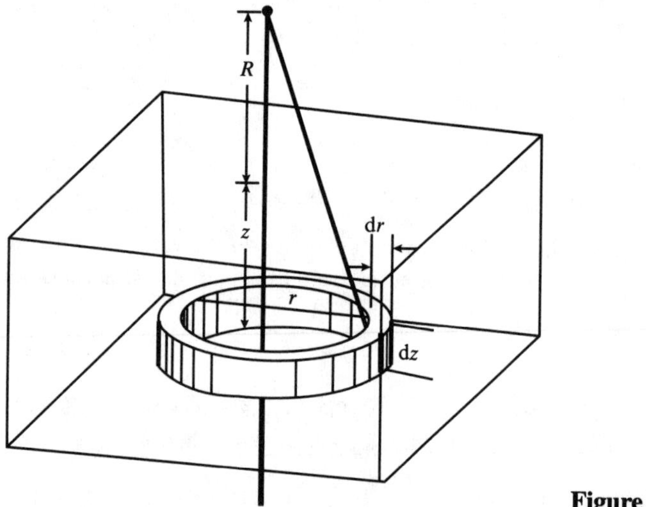

Figure 28.4

Let the number density of atoms in the solid be $\mathcal{N}$. Then the number in the annulus between r and $r + \mathrm{d}r$ and thickness $\mathrm{d}z$ at a depth z below the surface is $2\pi\mathcal{N}r\,\mathrm{d}r\,\mathrm{d}z$. The interaction energy of these atoms and the single adsorbate atom at a height R above the surface is

$$\mathrm{d}U = \frac{-2\pi\mathcal{N}r\,\mathrm{d}r\,\mathrm{d}z\,C_6}{\{(R+z)^2 + r^2\}^3}$$

if the individual atoms interact as $\dfrac{-C_6}{d^6}$, with $d^2 = (R+z)^2 + r^2$. The total interaction energy of the atom with the semi-infinite slab of uniform density is therefore

$$U = -2\pi\mathcal{N}C_6 \int_0^\infty \mathrm{d}r \int_0^\infty \mathrm{d}z \frac{r}{\{(R+z)^2 + r^2\}^3}$$

We then use

$$\int_0^\infty \frac{r\,dr}{(a^2+r^2)^3} = \frac{1}{2}\int_0^\infty \frac{d(r^2)}{(a^2+r^2)^3} = \frac{1}{2}\int_0^\infty \frac{dx}{(a^2+x)^3} = \frac{1}{4a^4}$$

and obtain

$$U = -\frac{1}{2}\pi\mathcal{N}C_6\int_0^\infty \frac{dz}{(R+z)^4} = \boxed{\frac{-\pi\mathcal{N}C_6}{6R^3}}$$

This result confirms that $U \propto \frac{1}{R^3}$. (A shorter procedure is to use a dimensional argument, but we need the explicit expression in the following.) When

$$V = 4\varepsilon\left[\left(\frac{\sigma}{R}\right)^{12} - \left(\frac{\sigma}{R}\right)^6\right] = \frac{C_{12}}{R^{12}} - \frac{C_6}{R^6}$$

we also need the contribution from C_{12}

$$U' = 2\pi\mathcal{N}C_{12}\int_0^\infty dr\int_0^\infty dz\frac{r}{\{(R+z)^2+r^2\}^6} = 2\pi\mathcal{N}C_{12}\times\frac{1}{10}\int_0^\infty \frac{dz}{(R+z)^{10}} = \frac{2\pi\mathcal{N}C_{12}}{90R^9}$$

and therefore the total interaction energy is

$$U = \frac{2\pi\mathcal{N}C_{12}}{90R^9} - \frac{\pi\mathcal{N}C_6}{6R^3}$$

We can express this result in terms of ε and σ by noting that $C_{12} = 4\varepsilon\sigma^{12}$ and $C_6 = 4\varepsilon\sigma^6$, for then

$$U = 8\pi\varepsilon\sigma^3\mathcal{N}\left[\frac{1}{90}\left(\frac{\sigma}{R}\right)^9 - \frac{1}{12}\left(\frac{\sigma}{R}\right)^3\right]$$

For the position of equilibrium, we look for the value of R for which $\frac{dU}{dR} = 0$

$$\frac{dU}{dR} = 8\pi\varepsilon\sigma^3\mathcal{N}\left[-\frac{1}{10}\left(\frac{\sigma^9}{R^{10}}\right) + \frac{1}{4}\left(\frac{\sigma^3}{R^4}\right)\right] = 0$$

Therefore, $\frac{\sigma^9}{10R^{10}} = \frac{\sigma^3}{4R^4}$ which implies that $R = \left(\frac{2}{5}\right)^{1/6}\sigma = \boxed{0.858\sigma}$. For $\sigma = 342\,\text{pm}$, $R \approx \boxed{294\,\text{pm}}$

P28.10 A general change in the Gibbs function of a one-component system with a surface is

$$dG = -S\,dT + V\,dp + \gamma\,d\sigma + \mu\,dn$$

Let $G = G(g) + G(\sigma)$ and $n = n(g) + n(\sigma)$; then

$$dG(g) = -S(g)\,dT + V(g)\,dp + \mu(g)\,dn(g)$$
$$dG(\sigma) = -S(\sigma)\,dT + \gamma\,d\sigma + \mu(\sigma)\,dn(\sigma)$$

At equilibrium, $\mu(\sigma) = \mu(g) = \mu$. At constant temperature, $dG(\sigma) = \gamma\,d\sigma + \mu\,dn(\sigma)$. Since dG is an exact differential, this expression integrates to

$$G(\sigma) = \gamma\sigma + \mu n(\sigma)$$

Therefore, $\mathrm{d}G(\sigma) = \sigma\,\mathrm{d}\gamma + \gamma\,\mathrm{d}\sigma + \mu\,\mathrm{d}n(\sigma) + n(\sigma)\,\mathrm{d}\mu$

But since $\mathrm{d}G(\sigma) = \gamma\,\mathrm{d}(\sigma) + \mu\,\mathrm{d}n(\sigma)$

we conclude that $\sigma\,\mathrm{d}\gamma + n(\sigma)\,\mathrm{d}\mu = 0$

Since $\mathrm{d}\mu = RT\,\mathrm{d}\ln p$, this relation is equivalent to

$$n(\sigma) = -\frac{\sigma\,\mathrm{d}\gamma}{\mathrm{d}\mu} = -\left(\frac{\sigma}{RT}\right) \times \left(\frac{\mathrm{d}\gamma}{\mathrm{d}\ln p}\right)$$

Now express $n(\sigma)$ as an adsorbed volume using

$$n(\sigma) = \frac{p^{\ominus} V_{\mathrm{a}}}{RT^{\ominus}}$$

and express $\mathrm{d}\gamma$ as a kind of chemical potential through

$$\mathrm{d}\mu' = \frac{RT^{\ominus}}{p^{\ominus}}\,\mathrm{d}\gamma$$

evaluated at a standard temperature and pressure ($T^{\ominus}$ and $p^{\ominus}$), then

$$\boxed{-\left(\frac{\sigma}{RT}\right) \times \left(\frac{\mathrm{d}\mu'}{\mathrm{d}\ln p}\right) = V_{\mathrm{a}}}$$

P28.11 $\mathrm{d}\mu' = -c_2\left(\frac{RT}{\sigma}\right)\mathrm{d}V_{\mathrm{a}}$

which implies that

$$\frac{\mathrm{d}\mu'}{\mathrm{d}\ln p} = \left(\frac{-c_2 RT}{\sigma}\right) \times \left(\frac{\mathrm{d}V_{\mathrm{a}}}{\mathrm{d}\ln p}\right)$$

However, we established in Problem 28.10 that

$$\frac{\mathrm{d}\mu'}{\mathrm{d}\ln p} = \frac{-RTV_{\mathrm{a}}}{\sigma}$$

Therefore,

$$-c_2\left(\frac{RT}{\sigma}\right) \times \left(\frac{\mathrm{d}V_{\mathrm{a}}}{\mathrm{d}\ln p}\right) = \frac{-RTV_{\mathrm{a}}}{\sigma}, \quad \text{or} \quad c_2\mathrm{d}\ln V_{\mathrm{a}} = \mathrm{d}\ln p$$

Hence, $\mathrm{d}\ln V_{\mathrm{a}}^{c_2} = \mathrm{d}\ln p$, and therefore $\boxed{V_{\mathrm{a}} = c_1 p^{1/c_2}}$

Solutions to additional problems

P28.13 (a) The volume of a spherical bubble is $V = \frac{4\pi r^3}{3}$. Using this expression in the perfect gas equation of state allows the evaluation of $\mathrm{d}N/\mathrm{d}t$

$$N = \frac{pV}{kT} = \frac{4\pi p r^3}{3kT}$$

$$\frac{\mathrm{d}N}{\mathrm{d}t} = \frac{4\pi p r^2}{kT}\frac{\mathrm{d}r}{\mathrm{d}t} = g(4\pi r^2)$$

$$\mathrm{d}r = \frac{gkT}{p}\,\mathrm{d}t$$

$$\int_{r_0}^{r}\mathrm{d}r = \frac{gkT}{p}\int_0^t \mathrm{d}t$$

assuming that the pressure of the bubble is constant

$$\boxed{r = r_0 + \left(\frac{gkT}{p}\right)t}$$

Thus, we find that $r = r_0 + vt$ where $\boxed{v = \frac{gkT}{p}}$.

(b) Linear regression analysis of the $r(t)$ data gives

$r_0 = 0.018\,\text{cm}$, standard deviation $= 0.0013\,\text{cm}$;

$\boxed{v = 0.0041\,\text{cm}\,\text{s}^{-1}}$, standard deviation $= 0.0005\,\text{cm}\,\text{s}^{-1}$

with $\boxed{R = 0.959}$. About 96 per cent of the variation is explained by the linear regression.

$$g = \frac{vp}{kT} = \frac{(0.0041\,\text{cm}\,\text{s}^{-1}) \times (1 \times 10^5\,\text{Pa}) \times \left(\frac{10^{-2}\,\text{m}}{\text{cm}}\right)}{(1.381 \times 10^{-23}\,\text{J}\,\text{K}^{-1}) \times (298.15\,\text{K})}$$

$$\boxed{g = 1.0 \times 10^{21}\,\text{m}^{-3}\,\text{s}^{-1}}$$

The plot of $z(t)$ (Fig. 28.5) is seen to be nonlinear. To develop an empirical description of the data we will fit the data with two regression equations and compare goodness-of-fit for them. The two forms are

$$z(t)(\text{fit 1}) = at^b \quad \text{and} \quad z(t)(\text{fit 2}) = \boxed{a(\text{e}^{bt} - 1)}$$

where a and b are the regression parameters.

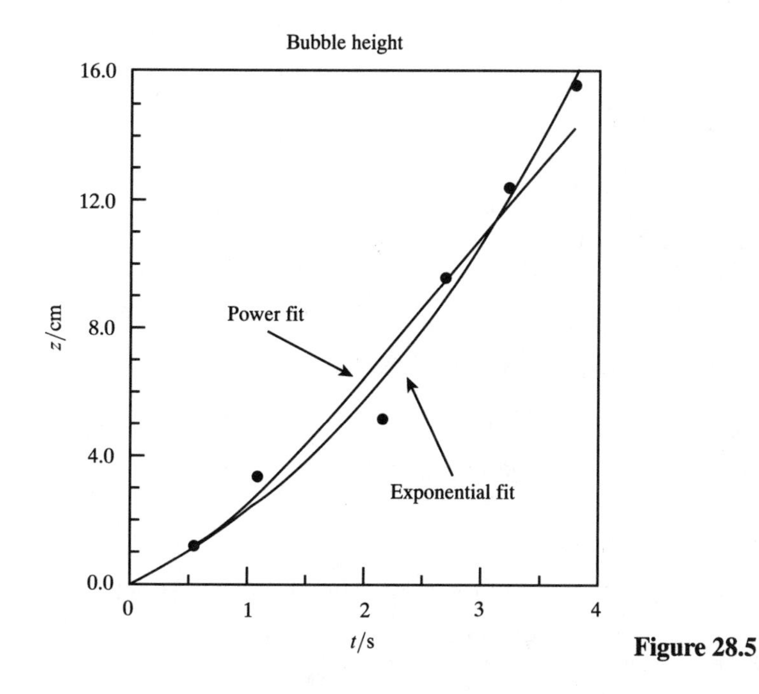

Figure 28.5

Performing the regression analysis, we find the following results,

Fit 1 (the power fit)

$a = 2.66\,\text{cm s}^{-b}$, standard deviation $= 0.098\,\text{cm s}^{-b}$
$b = 1.26$, standard deviation $= 0.11$
$R = 0.986$

Fit 2 (the exponential fit)

$\boxed{a = 5.71\,\text{cm}}$, standard deviation $= 2.24\,\text{cm}$

$\boxed{b = 0.35\,\text{s}^{-1}}$, standard deviation $= 0.08\,\text{s}^{-1}$

$\boxed{R = 0.994}$

The correlation coefficients suggest that the exponential fit is the better description of the data. However, standard deviations are relatively large and, although the exponential fit is to be preferred, we cannot totally reject the power fit possibility.

P28.15 Application of the van't Hoff equation [9] to adsorption equilibria yields

$$\left(\frac{\partial \ln K}{\partial T}\right)_\theta = \frac{\Delta_{\text{ad}} H^\ominus}{RT^2} \quad \text{or} \quad \left(\frac{\partial \ln K}{\partial(1/T)}\right)_\theta = \frac{-\Delta_{\text{ad}} H^\ominus}{R}$$

Hence, a plot (Fig. 28.6) of $\ln K$ against $1/T$ should be a straight line with slope $-\Delta_{\text{ad}} H^\ominus/R$. The transformed data and plot follow

T/K	283	298	308	318
$10^{-11}\,K$	2.642	2.078	1.286	1.085
$1000\,\text{K}/T$	3.53	3.36	3.25	3.14
$\ln K$	26.30	26.06	25.58	25.41

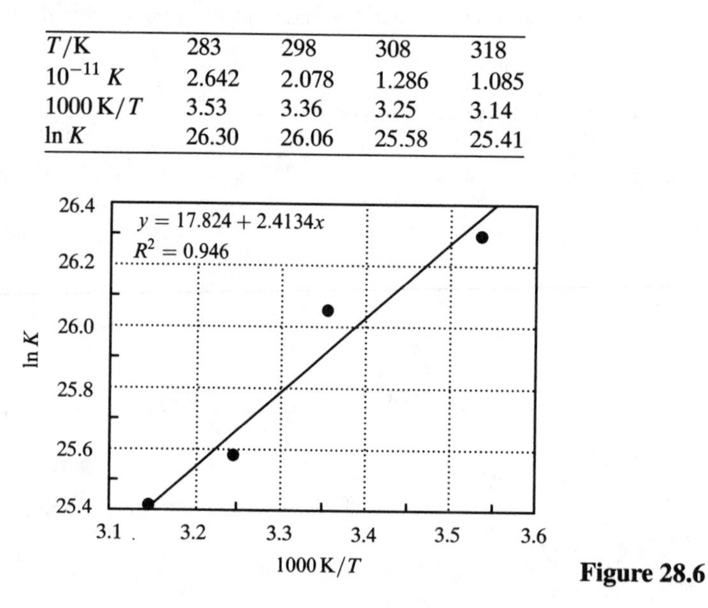

Figure 28.6

The plot is not the straightest of lines. Still, we can extract

$$\begin{aligned}\Delta_{\text{ad}} H^\ominus &= -(8.3145\,\text{J mol}^{-1}\,\text{K}^{-1}) \times (2.41 \times 10^3\,\text{K}) \\ &= -20.0 \times 10^3\,\text{J mol}^{-1} = \boxed{-20.1\,\text{kJ mol}^{-1}}\end{aligned}$$

The Gibbs energy for absorption is

$$\begin{aligned}\Delta_{\text{ad}} G^\ominus = \Delta_{\text{ad}} H^\ominus - T\Delta_{\text{ad}} S^\ominus &= -20.1\,\text{kJ mol}^{-1} - (298\,\text{K}) \times (0.146\,\text{kJ mol}^{-1}\,\text{K}^{-1}) \\ &= \boxed{-63.6\,\text{kJ mol}^{-1}}\end{aligned}$$

P28.17 Equilibrium constants vary with temperature according to the van't Hoff equation [9.26] which can be written in the form

$$\frac{K_1}{K_2} = e^{-\left[\frac{\Delta_{ad}H^{\ominus}}{R}\left(\frac{1}{T_1}-\frac{1}{T_2}\right)\right]}$$

or

$$\frac{K_1}{K_2} = \exp\left[\frac{160 \times 10^3\,\mathrm{J\,mol^{-1}}}{8.3145\,\mathrm{J\,K^{-1}\,mol^{-1}}}\left(\frac{1}{673\,\mathrm{K}} - \frac{1}{773\,\mathrm{K}}\right)\right] = \boxed{40.4}$$

As measured by the equilibrium constant of adsorption, NO is about 40 times more strongly adsorbed at 500°C than at 400°C.

P28.19 **(a)** $$\frac{1}{q_{\mathrm{VOC,RH=0}}} = \frac{1+bp_{\mathrm{VOC}}}{abc_{\mathrm{VOC}}} = \frac{1}{abc_{\mathrm{VOC}}} + \frac{1}{a}$$

Parameters of regression fit

$\theta/°\mathrm{C}$	$1/a$	$1/ab$	R	a	b/ppm^{-1}
33.6	9.07	709.8	0.9836	0.110	0.0128
41.5	10.14	890.4	0.9746	0.0986	0.0114
57.4	11.14	1599	0.9943	0.0898	0.00697
76.4	13.58	2063	0.9981	0.0736	0.00658
99	16.82	4012	0.9916	0.0595	0.00419

The linear regression fit is generally good at all temperatures with

$\boxed{R \text{ values in the range 0.975 to 0.991}}$

(b) $$\ln a = \ln k_a - \frac{\Delta_{ad}H}{R}\frac{1}{T}$$

and $$\ln b = \ln k_b - \frac{\Delta_b H}{R}\frac{1}{T}$$

Linear regression analysis of $\ln a$ versus $1/T$ gives the intercept $\ln k_a$ and slope $-\Delta_{ad}H/R$ while a similar statement can be made for a $\ln b$ versus $1/T$ plot. The temperature must be in Kelvin.

For $\ln a$ versus $1/T$

$\ln k_a = -5.605$, standard deviation $= 0.197$

$-\Delta_{ad}H/R = 1043.2\,\mathrm{K}$, standard deviation $= 65.4\,\mathrm{K}$

$R = 0.9942$ [good fit]

$$k_a = e^{-5.605} = \boxed{3.68 \times 10^{-3}}$$

$$\Delta_{ad}H = -(8.31451\,\mathrm{J\,K^{-1}\,mol^{-1}}) \times (1043.2\,\mathrm{K})$$

$$= \boxed{-8.67\,\mathrm{kJ\,mol^{-1}}}$$

For $\ln b$ versus $1/T$

$\ln(k_b/(\mathrm{ppm}^{-1})) = -10.550$, standard deviation $= 0.713$

$-\Delta_b H/R = 1895.4\,\mathrm{K}$, standard deviation $= 236.8$

$R = 0.9774$ [good fit]

$$k_b = e^{-10.550}\,\mathrm{ppm}^{-1} = \boxed{2.62 \times 10^{-5}\,\mathrm{ppm}^{-1}}$$

$$\Delta_b H = -(8.314\,51\,\mathrm{J\,K^{-1}\,mol^{-1}}) \times (1895.4\,\mathrm{K})$$

$$\boxed{\Delta_b H = -15.7\,\mathrm{kJ\,mol^{-1}}}$$

(c) k_a may be interpreted to be the maximum adsorption capacity at an adsorption enthalpy of zero, while k_b is the maximum affinity in the case for which the adsorbant–surface bonding enthalpy is zero.

P28.20 (a) $q_{\text{water}} = k(\text{RH})^{1/n}$

With a power law regression analysis we find

$\boxed{k = 0.2289}$, standard deviation $= 0.0068$

$1/n = 1.6182$, standard deviation $= 0.0093$; $\boxed{n = 0.6180}$

$R = 0.999\,508$

A linear regression analysis may be performed by transforming the equation to the following form by taking the logarithm of the Freundlich type equation

$$\ln q_{\text{water}} = \ln k + \frac{1}{n}\ln(\text{RH})$$

$\ln k = -1.4746$, standard deviation $= 0.0068$; $\boxed{k = 0.2289}$

$\frac{1}{n} = 1.6183$, standard deviation $= 0.0093$; $\boxed{n = 0.6180}$

$R = 0.999\,508$

The two methods give exactly the same result because the software package for performing the power law regression performs the transformation to linear form for you. Both methods are actually performing a linear regression.

The correlation coefficient indicates that 99.95 per cent of the data variation is explained with the Freundlich type isotherm. The Freundlich fit hypothesis looks very good.

(b) The Langmuir isotherm model describes adsorption sites that are independent and equivalent. This assumption seems to be valid for the VOC case in which molecules interact very weakly . However, water molecules interact much more strongly through forces such as hydrogen bonding and multilayers may readily form at the lower temperatures. The intermolecular forces of water apparently cause adsorption sites to become non-equivalent and dependent. In this particular case the Freundlich type isotherm becomes the better description.

(c) $r_{\text{VOC}} = 1 - q_{\text{water}}$ where $r_{\text{VOC}} \equiv q_{\text{VOC}}/q_{\text{VOC,RH=0}}$

$$r_{\text{VOC}} = 1 - k(\text{RH})^{1/n}$$
$$1 - r_{\text{VOC}} = k(\text{RH})^{1/n}$$

To determine the goodness of fit, k, and n we perform a power law regression fit of $1 - r_{\text{VOC}}$ against RH. Results are

$\boxed{k = 0.5227}$, standard deviation $= 0.0719$

$\frac{1}{n} = 1.3749$, standard deviation $= 0.0601$; $\boxed{n = 0.7273}$

$R = 0.996\,20$

Since 99.62 per cent of the variation is explained by the regression, we conclude that the hypothesis that $r_{\text{VOC}} = 1 - q_{\text{water}}$ may be very useful. The values of R and n differ significantly from those of part (a). It may be that water is adsorbing to some portions of the surface and VOC to others.

29 Dynamic electrochemistry

Solutions to exercises

E29.1 $\mathcal{E} = \dfrac{\Delta\phi}{l}$ [24.40]

$$= \frac{\sigma}{\varepsilon} = \frac{\sigma}{\varepsilon_r\varepsilon_0} = \frac{0.10\,\mathrm{C\,m^{-2}}}{(48)\times(8.854\times10^{-12}\,\mathrm{J^{-1}\,C^2\,m^{-1}})} = \boxed{2.4\times10^{8}\,\mathrm{V\,m^{-1}}}$$

Comment. Surface electric fields are very large. Relative permittivities of solutions vary with concentration and temperature. The value for pure water at 20°C is 80.4.

E29.2 $\ln j = \ln j_0 + (1-\alpha)f\eta \quad \left[25,\ f = \dfrac{F}{RT}\right]$

$$\ln\frac{j'}{j} = (1-\alpha)f(\eta'-\eta)$$

which implies that for a current density j' we require an overpotential

$$\eta' = \eta + \frac{\ln\frac{j'}{j}}{(1-\alpha)f} = (125\,\mathrm{mV}) + \frac{\ln\left(\frac{75}{55}\right)}{(1-0.39)\times(25.69\,\mathrm{mV})^{-1}} = \boxed{138\,\mathrm{mV}}$$

E29.3 Take antilogarithms of eqn 25; then

$$j_0 = j\,\mathrm{e}^{-(1-\alpha)\eta f} = (55.0\,\mathrm{mA\,cm^{-2}})\times\mathrm{e}^{-0.61\times125\,\mathrm{mV}/25.69\,\mathrm{mV}} = \boxed{2.8\,\mathrm{mA\,cm^{-2}}}$$

E29.4 $O_2(g)$ is produced at the anode in this electrolysis and $H_2(g)$ at the cathode. The net reaction is

$$2H_2O(l) \rightarrow 2H_2(g) + O_2(g)$$

For a large positive overpotential we use

$$\ln j = \ln j_0 + (1-\alpha)f\eta\ [25]$$

$$\ln\frac{j'}{j} = (1-\alpha)f(\eta'-\eta) = (0.5)\times\left(\frac{1}{0.02569\,\mathrm{V}}\right)\times(0.6\,\mathrm{V}-0.4\,\mathrm{V}) = 3.8\overline{9}$$

$$j' = j\,\mathrm{e}^{3.8\overline{9}} = (1.0\,\mathrm{mA\,cm^2})\times(4\overline{9}) = 4\overline{9}\,\mathrm{mA\,cm^2}$$

Hence, the anodic current density $\boxed{\text{increases}}$ roughly by a $\boxed{\text{factor of 50}}$ with a corresponding increase in O_2 evolution.

E29.5 $j_0 = 6.3\times10^{-6}\,\mathrm{A\,cm^{-2}}, \qquad \alpha = 0.58$ [Table 29.1]

(a) $j = j_0\{\mathrm{e}^{(1-\alpha)f\eta} - \mathrm{e}^{-\alpha f\eta}\}$ [21]

$$\frac{j}{j_0} = \mathrm{e}^{\{(1-0.58)\times(1/0.02569)\times0.20\}} - \mathrm{e}^{\{-0.58\times(1/0.02569)\times0.2\}} = (26.\overline{3}-0.011)\approx 26$$

$$j = (26)\times(6.3\times10^{-6}\,\mathrm{A\,cm^{-2}}) = \boxed{1.7\times10^{-4}\,\mathrm{A\,cm^{-2}}}$$

(b) The Tafel equation corresponds to the neglect of the second exponential above, which is very small for an overpotential of 0.2 V. Hence

$$j = \boxed{1.7\times10^{-4}\,\mathrm{A\,cm^{-2}}}$$

The validity of the Tafel equation increases with higher overpotentials, but decreases at lower overpotentials. A plot of j against η becomes linear (non-exponential) as $\eta \rightarrow 0$.

E29.6 $$j_{\text{lim}} = \frac{cRT\lambda}{zf\delta} \text{ [Example 29.3]}$$

$$= \frac{(2.5\times10^{-3}\,\text{mol L}^{-1})\times(25.69\times10^{-3}\,\text{V})\times(61.9\,\text{S cm}^2\,\text{mol}^{-1})}{0.40\times10^{-3}\,\text{m}}$$

$$= 9.9\,\text{mol L}^{-1}\,\text{V S cm}^2\,\text{mol}^{-1}\,\text{m}^{-1}$$

$$= (9.9\,\text{mol m}^{-3})\times(10^3)\times(\text{V}\,\Omega^{-1})\times(10^{-4}\,\text{m}^2\,\text{mol}^{-1}\,\text{m}^{-1})$$

$$= \boxed{0.99\,\text{A m}^{-2}} \quad [1\,\text{V}\,\Omega^{-1} = 1\,\text{A}]$$

E29.7 For the cadmium electrode $E^{\ominus} = -0.40\,\text{V}$ [Table 10.7] and the Nernst equation for this electrode [Section 10.4(f)] is

$$E = E^{\ominus} - \frac{RT}{\nu F}\ln\left(\frac{1}{[\text{Cd}^{2+}]}\right) \qquad \nu = 2$$

Since the hydrogen overpotential is 0.60 V evolution of H_2 will begin when the potential of the Cd electrode reaches -0.60 V. Thus

$$-0.60\,\text{V} = -0.40\,\text{V} + \frac{0.02569\,\text{V}}{2}\ln[\text{Cd}^{2+}]$$

$$\ln[\text{Cd}^{2+}] = \frac{-0.20\,\text{V}}{0.0128\,\text{V}} = -15.\bar{6}$$

$$[\text{Cd}^{2+}] = \boxed{2\times10^{-7}\,\text{mol L}^{-1}}$$

Comment. Essentially all Cd^{2+} has been removed by deposition before evolution of H_2 begins.

E29.8 $$\frac{j}{j_0} = e^{(1-\alpha)f\eta} - e^{-\alpha f\eta}[21] = e^{(1/2)f\eta} - e^{-(1/2)f\eta} \quad [\alpha = 0.5]$$

$$= 2\sinh\left(\frac{1}{2}f\eta\right) \quad \left[\sinh x = \frac{e^x - e^{-x}}{2}\right]$$

and we use $\frac{1}{2}f\eta = \frac{1}{2}\times\frac{\eta}{25.69\,\text{mV}} = 0.01946(\eta/\text{mV})$

Or

$$j = 2j_0\sinh\left(\tfrac{1}{2}f\eta\right) = (1.58\,\text{mA cm}^{-2})\times\sinh\left(\frac{0.01946\,\eta}{\text{mV}}\right)$$

(a) $\eta = 10\,\text{mV}$

$$j = (1.58\,\text{mA cm}^{-2})\times(\sinh 0.1946) = \boxed{0.31\,\text{mA cm}^{-2}}$$

(b) $\eta = 100\,\text{mV}$

$$j = (1.58\,\text{mA cm}^{-2})\times(\sinh 1.946) = \boxed{5.41\,\text{mA cm}^{-2}}$$

(c) $\eta = -0.5\,\text{V}$

$$j = (1.58\,\text{mA cm}^{-2})\times(\sinh -0.973) \approx \boxed{-2.19\,\text{A cm}^{-2}}$$

E29.9 $$E = E^{\ominus} + \frac{RT}{F}\ln\frac{a(\text{Fe}^{3+})}{a(\text{Fe}^{2+})} \text{ [Nernst equation]}$$

$$E/\text{mV} = 770 + 25.7\ln\frac{a(\text{Fe}^{3+})}{a(\text{Fe}^{2+})}$$

$$\eta/\text{mV} = 1000 - E/\text{mV} = 229 - 25.7 \ln \frac{a(\text{Fe}^{3+})}{a(\text{Fe}^{2+})}$$

and hence

$$j = 2j_0 \sinh\left(\frac{0.01946\,\eta}{\text{mV}}\right) \quad \text{[Exercise 29.8(a)]}$$

$$= (5.0\,\text{mA cm}^{-2}) \times \sinh\left(4.46 - 0.50 \ln \frac{a(\text{Fe}^{3+})}{a(\text{Fe}^{2+})}\right)$$

We can therefore draw up the following table

$\frac{a(\text{Fe}^{3+})}{a(\text{Fe}^{2+})}$	0.1	0.3	0.6	1.0	3.0	6.0	10.0
$j/(\text{mA cm}^{-2})$	684	395	278	215	124	88	68.0

The current density falls to zero when

$$4.46 = 0.50 \ln \frac{a(\text{Fe}^{3+})}{a(\text{Fe}^{2+})}$$

which occurs when $a(\text{Fe}^{3+}) = 7480 \times a(\text{Fe}^{2+})$.

E29.10

$$I = 2j_0 S \sinh\left(\frac{0.01946\,\eta}{\text{mV}}\right) \quad \text{[Exercise 29.8(a)]}$$

$$\eta = (51.39\,\text{mV}) \times \sinh^{-1}\left(\frac{I}{2j_0 S}\right)$$

$$= (51.39\,\text{mV}) \times \sinh^{-1}\left(\frac{20\,\text{mA}}{(2) \times (2.5\,\text{mA cm}^{-2}) \times (1.0\,\text{cm}^2)}\right)$$

$$= (51.39\,\text{mV}) \times (\sinh^{-1} 4.0) = \boxed{108\,\text{mV}}$$

E29.11 The current density of electrons is $\frac{j_0}{e}$ because each one carries a charge of magnitude e. Therefore,

(a) $\text{Pt}|\text{H}_2|\text{H}^+$; $j_0 = 0.79\,\text{mA cm}^{-2}$ [Table 29.1]

$$\frac{j_0}{e} = \frac{0.79\,\text{mA cm}^{-2}}{1.602 \times 10^{-19}\,\text{C}} = \boxed{4.9 \times 10^{15}\,\text{cm}^{-2}\,\text{s}^{-1}}$$

(b) $\text{Pt}|\text{Fe}^{3+}, \text{Fe}^{2+}$; $j_0 = 2.5\,\text{mA cm}^{-2}$

$$\frac{j_0}{e} = \frac{2.5\,\text{mA cm}^{-2}}{1.602 \times 10^{-19}\,\text{C}} = \boxed{1.6 \times 10^{16}\,\text{cm}^{-2}\,\text{s}^{-1}}$$

(c) $\text{Pb}|\text{H}_2|\text{H}^+$; $j_0 = 5.0 \times 10^{-12}\,\text{A cm}^{-2}$

$$\frac{j_0}{e} = \frac{5.0 \times 10^{-12}\,\text{A cm}^{-2}}{1.602 \times 10^{-19}\,\text{C}} = \boxed{3.1 \times 10^{7}\,\text{cm}^{-2}\,\text{s}^{-1}}$$

There are approximately $\frac{1.0\,\text{cm}^2}{(280\,\text{pm})^2} = 1.3 \times 10^{15}$ atoms in each square centimetre of surface. The numbers of electrons per atom are therefore $\boxed{3.8\,\text{s}^{-1}}$, $\boxed{12\,\text{s}^{-1}}$, and $\boxed{2.4 \times 10^{-8}\,\text{s}^{-1}}$ respectively. The last corresponds to less than one event per year.

E29.12 $\eta = \dfrac{RTj}{Fj_0}$ [23]

which implies that

$$I = Sj = \left(\frac{Sj_0F}{RT}\right)\eta$$

An ohmic conductor of resistance r obeys $\eta = Ir$, and so we can identify the resistance as

$$r = \frac{RT}{Sj_0F} = \frac{25.69 \times 10^{-3}\,\mathrm{V}}{1.0\,\mathrm{cm}^2 \times j_0} = \frac{25.69 \times 10^{-3}\,\Omega}{(j_0/\mathrm{A\,cm^{-2}})} \quad [1\,\mathrm{V} = 1\,\mathrm{A}\,\Omega]$$

(a) $\mathrm{Pt|H_2|H^+}$; $j_0 = 7.9 \times 10^{-4}\,\mathrm{A\,cm^{-2}}$

$$r = \frac{25.60 \times 10^{-3}\,\Omega}{7.9 \times 10^{-4}} = \boxed{33\,\Omega}$$

(b) $\mathrm{Hg|H_2|H^+}$; $j_0 = 0.79 \times 10^{-12}\,\mathrm{A\,cm^{-2}}$

$$r = \frac{25.69 \times 10^{-3}\,\Omega}{0.79 \times 10^{-12}} = 3.3 \times 10^{10}, \quad \text{or} \quad \boxed{33\,\mathrm{G}\Omega}$$

E29.13 For deposition of cations, a significant net current towards the electrodes is necessary. For copper and zinc, we have $E^\ominus \approx 0.34\,\mathrm{V}$ and $-0.76\,\mathrm{V}$, respectively. Therefore, deposition of copper occurs when the potential falls below 0.34 V and continues until the copper ions are exhausted to the point that the limiting current density is reached. Then a further reduction in potential to below $-0.76\,\mathrm{V}$ brings about the deposition of zinc.

Comment. The depositions will be very slow until E drops substantially below $E^\ominus$.

E29.14 See Exercise 29.7(a) for a related situation.

Hydrogen evolution occurs significantly (in the sense of having a current density of $1\,\mathrm{mA\,cm^{-2}}$, which is 6.2×10^{15} electrons $\mathrm{cm^{-2}\,s^{-1}}$, or $1.0 \times 10^{-8}\,\mathrm{mol\,cm^{-2}\,s^{-1}}$, corresponding to about $1\,\mathrm{cm^3}$ of gas per hour) when the overpotential is $\approx 0.60\,\mathrm{V}$. Since $E = E^\ominus + \left(\dfrac{RT}{F}\right)\ln a(\mathrm{H^+}) = -59\,\mathrm{mV} \times \mathrm{pH}$, this rate of evolution occurs when the potential at the electrode is about $-0.66\,\mathrm{V}$ (when pH ≈ 1). $\mathrm{Ag^+}$ ($E^\ominus = 0.80\,\mathrm{V}$) has a more positive deposition potential and so deposits first.

E29.15 We assume $\alpha \approx 0.5$; $E^\ominus(\mathrm{Zn^{2+}, Zn}) = -0.76\,\mathrm{V}$.

Zinc will deposit from a solution of unit activity when the potential is below $-0.76\,\mathrm{V}$. The hydrogen ion current toward the zinc electrode is then

$$j(\mathrm{H^+}) = j_0\mathrm{e}^{-\alpha f\eta}\ [20]$$

$$j(\mathrm{H^+}) = (5 \times 10^{-11}\,\mathrm{A\,cm^{-2}}) \times (\mathrm{e}^{760/51.4}) \quad \left[\eta = -760\,\mathrm{mV},\ f = \frac{1}{25.7\,\mathrm{mV}}\right]$$

$$= 1.3 \times 10^{-4}\,\mathrm{A\,cm^{-2}}, \quad \text{or} \quad 0.13\,\mathrm{mA\,cm^{-2}}$$

Using the criterion that $j > 1\,\mathrm{mA\,cm^{-2}}$ [Exercise 29.14(a)] for significant evolution of hydrogen, this value of j corresponds to a negligible rate of evolution of hydrogen, and so $\boxed{\text{zinc may be deposited}}$ from the solution.

E29.16 Since $E^\ominus(\mathrm{Mg, Mg^{2+}}) = -2.37\,\mathrm{V}$, magnesium deposition will occur when the potential is reduced to below this value. The hydrogen ion current density is then (Exercise 29.15(a))

$$j(\mathrm{H^+}) = (5\times10^{-11}\,\mathrm{A\,cm^{-2}})\times(\mathrm{e}^{2370/51.4}) = 5.3\times10^{9}\,\mathrm{A\,cm^{-2}}$$

which is a lot of hydrogen ($10^6\,\mathrm{L\,cm^{-2}\,s^{-1}}$), and so magnesium **will not be plated out**

E29.17 The cell half-reactions are

$$\mathrm{Cd(OH)_2 + 2e^- \rightarrow Cd + 2OH^-} \quad E^\ominus = -0.81\,\mathrm{V}$$
$$\mathrm{NiO(OH) + H_2O + e^- \rightarrow Ni(OH)_2 + OH^-} \quad E^\ominus = +0.49\,\mathrm{V}$$

Therefore, the standard cell potential is $\boxed{+1.30\,\mathrm{V}}$. If the cell is working reversibly yet producing 100 mA, the power it produces is

$$P = IE = (100\times10^{-3}\,\mathrm{A})\times(1.3\,\mathrm{V}) = \boxed{0.13\,\mathrm{W}}$$

E29.18 $E^\ominus = \dfrac{-\Delta_r G^\ominus}{\nu F}$

(a) $\mathrm{H_2 + \frac{1}{2}O_2 \rightarrow H_2O}$; $\Delta_r G^\ominus = -237\,\mathrm{kJ\,mol^{-1}}$
Since $\nu = 2$,

$$E^\ominus = \frac{-(-237\,\mathrm{kJ\,mol^{-1}})}{(2)\times(96.48\,\mathrm{kC\,mol^{-1}})} = \boxed{+1.23\,\mathrm{V}}$$

(b) $\mathrm{CH_4 + 2O_2 \rightarrow CO_2 + 2H_2O}$
$\Delta_r G^\ominus = 2\Delta_f G^\ominus(\mathrm{H_2O}) + \Delta_f G^\ominus(\mathrm{CO_2}) - \Delta_f G^\ominus(\mathrm{CH_4})$
$= [(2)\times(-237.1) + (-394.4) - (-50.7)]\,\mathrm{kJ\,mol^{-1}} = -817.9\,\mathrm{kJ\,mol^{-1}}$
As written, the reaction corresponds to the transfer of eight electrons. It follows that, for the species in their standard states,

$$E^\ominus = \frac{-(-817.9\,\mathrm{kJ\,mol^{-1}})}{(8)\times(96.48\,\mathrm{kC\,mol^{-1}})} = \boxed{+1.06\,\mathrm{V}}$$

E29.19 The electrode potentials of half-reactions (a), (b), and (c) are (Section 29.7)

(a) $E(\mathrm{H_2, H^+}) = -0.059\,\mathrm{V\,pH} = (-7)\times(0.059\,\mathrm{V}) = -0.41\,\mathrm{V}$
(b) $E(\mathrm{O_2, H^+}) = (1.23\,\mathrm{V}) - (0.059\,\mathrm{V})\mathrm{pH} = +0.82\,\mathrm{V}$
(c) $E(\mathrm{O_2, OH^-}) = (0.40\,\mathrm{V}) + (0.059\,\mathrm{V})\mathrm{pOH} = 0.81\,\mathrm{V}$

$$E(\mathrm{M, M^+}) = E^\ominus(\mathrm{M, M^+}) + \left(\frac{0.059\,\mathrm{V}}{z_+}\right)\log 10^{-6} = E^\ominus(\mathrm{M, M^+}) - \frac{0.35\,\mathrm{V}}{z_+}$$

Corrosion will occur if $E(\mathrm{a})$, $E(\mathrm{b})$, or $E(\mathrm{c}) > E(\mathrm{M, M^+})$

(i) $E^\ominus(\mathrm{Fe, Fe^{2+}}) = -0.44\,\mathrm{V}$, $z_+ = 2$
$E(\mathrm{Fe, Fe^{2+}}) = (-0.44 - 0.18)\,\mathrm{V} = -0.62\,\mathrm{V} < E(\mathrm{a, b, and\ c})$

(ii) $E(\mathrm{Cu, Cu^+}) = (0.52 - 0.35)\,\mathrm{V} = 0.17\,\mathrm{V}\begin{cases} > E(\mathrm{a}) \\ < E(\mathrm{b\ and\ c}) \end{cases}$

$E(\mathrm{Cu, Cu^{2+}}) = (0.34 - 0.18)\,\mathrm{V} = 0.16\,\mathrm{V}\begin{cases} > E(\mathrm{a}) \\ < E(\mathrm{b\ and\ c}) \end{cases}$

(iii) $E(\text{Pb}, \text{Pb}^{2+}) = (-0.13 - 0.18)\,\text{V} = -0.31\,\text{V} \begin{cases} > E(\text{a}) \\ < E(\text{b and c}) \end{cases}$

(iv) $E(\text{Al}, \text{Al}^{3+}) = (-1.66 - 0.12)\,\text{V} = -1.78\,\text{V} < E(\text{a, b, and c})$

(v) $E(\text{Ag}, \text{Ag}^{+}) = (0.80 - 0.35)\,\text{V} = 0.45\,\text{V} \begin{cases} > E(\text{a}) \\ < E(\text{b and c}) \end{cases}$

(vi) $E(\text{Cr}, \text{Cr}^{3+}) = (-0.74 - 0.12)\,\text{V} = -0.86\,\text{V} < E(\text{a, b, and c})$

(vii) $E(\text{Co}, \text{Co}^{2+}) = (-0.28 - 0.15)\,\text{V} = -0.43\,\text{V} < E(\text{a, b, and c})$

Therefore, the metals with a thermodynamic tendency to corrode in moist conditions at pH = 7 are $\boxed{\text{Fe, Al, Co, Cr}}$ if oxygen is absent, but, if oxygen is present, all seven elements have a tendency to corrode.

E29.20

$$\frac{(1.0\,\text{A}\,\text{m}^{-2}) \times (3.16 \times 10^{7}\,\text{s}\,\text{yr}^{-1})}{9.65 \times 10^{4}\,\text{C}\,\text{mol}^{-1}} = 32\bar{7}\,\text{mol}\,\text{e}^{-}\,\text{m}^{-2}\,\text{yr}^{-1} = 16\bar{4}\,\text{mol Fe}\,\text{m}^{-2}\,\text{yr}^{-1}$$

$$\frac{(16\bar{4}\,\text{mol}\,\text{m}^{-2}\,\text{yr}^{-1}) \times (55.85\,\text{g}\,\text{mol}^{-1})}{7.87 \times 10^{6}\,\text{g}\,\text{m}^{-3}} = 1.2 \times 10^{-3}\,\text{m}\,\text{yr}^{-1} = \boxed{1.2\,\text{mm}\,\text{yr}^{-1}}$$

Solutions to problems

Solutions to numerical problems

P29.1 $\ln j = \ln j_0 + (1 - \alpha) f \eta$ [25]

Draw up the following table

η/mV	50	100	150	200	250
$\ln(j/\text{mA}\,\text{cm}^{-2})$	0.98	2.19	3.40	4.61	5.81

The points are plotted in Fig. 29.1.

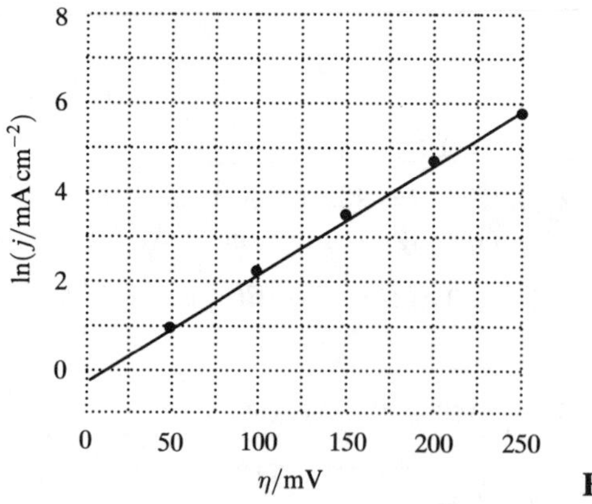

Figure 29.1

The intercept is at -0.25, and so $j_0/(\text{mA}\,\text{cm}^{-2}) = \text{e}^{-0.25} = \boxed{0.78}$. The slope is 0.0243, and so $\frac{(1-\alpha)F}{RT} = 0.0243\,\text{mV}^{-1}$. It follows that $1 - \alpha = 0.62$, and so $\boxed{\alpha = 0.38}$. If η were large but negative.

$$|j| \approx j_0 \text{e}^{-\alpha f \eta}[26] = (0.78\,\text{mA}\,\text{cm}^{-2}) \times (\text{e}^{-0.38\,\eta/25.7\,\text{mV}})$$
$$= (0.78\,\text{mA}\,\text{cm}^{-2}) \times (\text{e}^{-0.015(\eta/\text{mV})})$$

and we draw up the following table

η/mV	−50	−100	−150	−200	−250
$j/(\mathrm{mA\,cm^{-2}})$	1.65	3.50	7.40	15.7	33.2

P29.3 $j_{\mathrm{lim}} = \dfrac{zFDc}{\delta}$ [37], and so $\delta = \dfrac{FDc}{j_{\mathrm{lim}}}$ $[z=1]$

Therefore,

$$\delta = \frac{(9.65\times10^4\,\mathrm{C\,mol^{-1}})\times(1.14\times10^{-9}\,\mathrm{m^2\,s^{-1}})\times(0.66\,\mathrm{mol\,m^{-3}})}{28.9\times10^{-2}\,\mathrm{A\,m^{-2}}}$$

$$= 2.5\times10^{-4}\,\mathrm{m}, \quad \text{or} \quad \boxed{0.25\,\mathrm{mm}}$$

P29.5 $E' = E - IR_{\mathrm{s}} - \dfrac{2RT}{zF}\ln g(I)$ [46]

$$g = \frac{\left(\frac{I}{Aj}\right)^{2z}}{\left[\left(1-\frac{I}{Aj_{\mathrm{lim,L}}}\right)\times\left(1-\frac{I}{Aj_{\mathrm{lim,R}}}\right)\right]^{1/2}}$$

with $j_{\mathrm{lim}} = \dfrac{cRT\lambda}{zF\delta}$ [Example 29.3] $= a\lambda$

$$R_{\mathrm{s}} = \frac{l}{\kappa A} = \frac{1}{cA\Lambda_{\mathrm{m}}} \quad \text{with } \Lambda_{\mathrm{m}} = \lambda_+ + \lambda_-$$

Therefore, $E' = E - \dfrac{Il}{cA\Lambda_{\mathrm{m}}} - \dfrac{2RT}{zF}\ln g(I)$

$$\text{with } g(I) = \frac{\left(\frac{I^2}{A^2 j_{\mathrm{LO}} j_{\mathrm{RO}}}\right)^{z}}{\left[1-\left(\frac{I}{Aa_{\mathrm{L}}\lambda_{\mathrm{L}+}}\right)\right]^{1/2}\left[1-\left(\frac{I}{Aa_{\mathrm{R}}\lambda_{\mathrm{R}+}}\right)\right]^{1/2}}$$

with $a_{\mathrm{L}} = \dfrac{RTc_{\mathrm{L}}}{z_{\mathrm{L}}F\delta_{\mathrm{L}}}$ and $a_{\mathrm{R}} = \dfrac{RTc_{\mathrm{R}}}{z_{\mathrm{R}}F\delta_{\mathrm{R}}}$

For the cell $\mathrm{Zn|ZnSO_4(aq)||CuSO_4(aq)|Cu}$, $l = 5\,\mathrm{cm}$, $A = 5\,\mathrm{cm^2}$, $c(\mathrm{M_L^+}) = c(\mathrm{M_R^+}) = 1\,\mathrm{mol\,L^{-1}}$, $z_{\mathrm{L}} = z_{\mathrm{R}} = 2$, $\lambda_{\mathrm{L}+} = 107\,\mathrm{S\,cm^2\,mol^{-1}}$, $\lambda_{\mathrm{R}+} = 106\,\mathrm{S\,cm^2\,mol^{-1}} \approx \lambda_{\mathrm{L}+}$, $\lambda_- = \lambda_{\mathrm{SO_4^{2-}}} = 160\,\mathrm{S\,cm^2\,mol^{-1}}$. $\Lambda_{\mathrm{m}} \approx (107+160)\,\mathrm{S\,cm^2\,mol^{-1}} = 267\,\mathrm{S\,cm^2\,mol^{-1}}$ for both electrolyte solutions. We take $\delta \approx 0.25\,\mathrm{mm}$ [Example 29.3] and $j_{\mathrm{LO}} \approx j_{\mathrm{RO}} \approx 1\,\mathrm{mA\,cm^{-2}}$. We can also take

$$E^{\ominus}(a\approx1) = E^{\ominus}(\mathrm{Cu,Cu^{2+}}) - E^{\ominus}(\mathrm{Zn,Zn^{2+}}) = [0.34-(-0.76)]\,\mathrm{V} = 1.10\,\mathrm{V}$$

$$R_{\mathrm{s}} = \frac{5\,\mathrm{cm}}{(1\,\mathrm{M})\times(267\,\mathrm{S\,cm^2\,mol^{-1}})\times(5\,\mathrm{cm^2})} = 3.\overline{8}\,\Omega$$

$$j_{\mathrm{lim}} = j_{\mathrm{lim}}^{+} = \frac{1}{2}\times\left(\frac{(0.0257\,\mathrm{V})\times(107\,\mathrm{S\,cm^2\,mol^{-1}})\times(1\,\mathrm{M})}{0.25\times10^{-3}\,\mathrm{m}}\right) \approx 5.5\times10^{-2}\,\mathrm{S\,V\,cm^{-2}}$$

$$= 5.5\times10^{-2}\,\mathrm{A\,cm^{-2}}$$

It follows that

$$E'/\mathrm{V} = (1.10) - 3.7\overline{5}(I/\mathrm{A}) - (0.0257)\ln\left(\frac{(I/5\times10^{-3}\,\mathrm{A})^4}{1-3.6(I/\mathrm{A})}\right)$$

$$= (1.10) - 3.7\overline{5}(I/\mathrm{A}) - (0.0257)\ln\left(\frac{1.6\times10^{9}(I/\mathrm{A})^4}{1-3.6(I/\mathrm{A})}\right)$$

This function is plotted in Fig. 29.2.

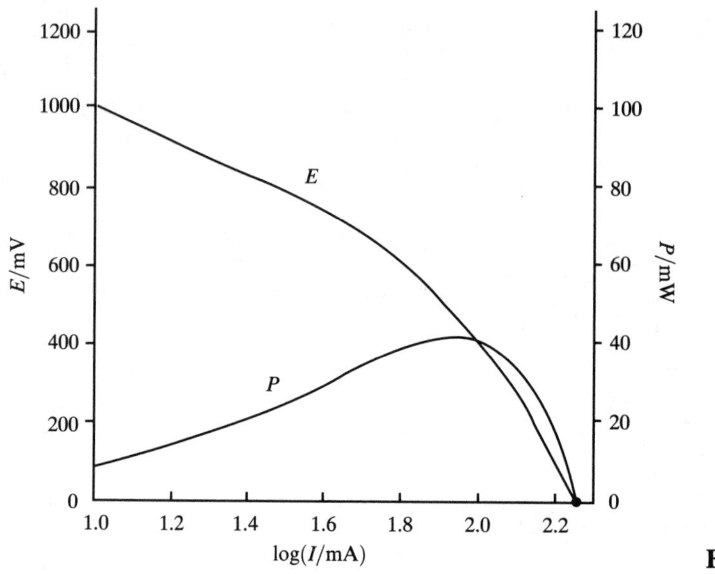

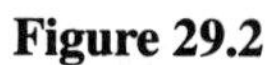
Figure 29.2

The power is

$$P = IE'$$

and so $P/\mathrm{W} = 1.10(I/\mathrm{A}) - 3.7\overline{5}(I/\mathrm{A})^2 - 0.0257(I/\mathrm{A})\ln\left(\dfrac{1.6\times10^9(I/\mathrm{A})^4}{1-3.6(I/\mathrm{A})}\right)$

This function is also plotted in Fig. 29.2. Maximum power is delivered at about $\boxed{87\,\mathrm{mA}}$ and 0.46 V, and is about 40 mW.

P29.6 $E' = E - \left(\dfrac{4RT}{F}\right)\ln\left(\dfrac{I}{A\bar{j}}\right) - IR_\mathrm{s}$ [44]

$$P = IE' = IE - aI\ln\left(\frac{I}{I_0}\right) - I^2R_\mathrm{s}$$

where $a = \dfrac{4RT}{F}$ and $I_0 = A\bar{j}$. For maximum power,

$$\frac{\mathrm{d}P}{\mathrm{d}I} = E - a\ln\left(\frac{I}{I_0}\right) - a - 2IR_\mathrm{s} = 0$$

which requires

$$\ln\left(\frac{I}{I_0}\right) = \left(\frac{E}{a} - 1\right) - \frac{2IR_\mathrm{s}}{a}$$

This expression may be written

$$\ln\left(\frac{I}{I_0}\right) = c_1 - c_2I; \qquad c_1 = \frac{E}{a} - 1, \quad c_2 = \frac{2R_\mathrm{s}}{a} = \frac{FR_\mathrm{s}}{2RT}$$

For the present calculation, use the data in Problem 29.5. Then

$$I_0 = A\bar{j} = (5\,\mathrm{cm}^2)\times(1\,\mathrm{mA\,cm^{-2}}) = 5\,\mathrm{mA}$$

$$c_1 = \frac{(1.10\,\text{V})}{(4) \times (0.0257\,\text{V})} - 1 = 10.7$$

$$c_2 = \frac{(3.7\bar{5}\,\Omega)}{(2) \times (0.0257\,\text{V})} = 73\,\Omega\,\text{V}^{-1} = 73\,\text{A}^{-1}$$

That is, $\ln(0.20I/\text{mA}) = 10.7 - 0.073(I/\text{mA})$

We then draw up the following table

I/mA	103	104	105	106	107
$\ln(0.20I/\text{mA})$	3.025	3.034	3.044	3.054	3.063
$10.7 - 0.073(I/\text{mA})$	3.181	3.108	3.035	2.962	2.889

The two sets of points are plotted in Fig. 29.3. The lines intersect at $I = 105\,\text{mA}$, which therefore

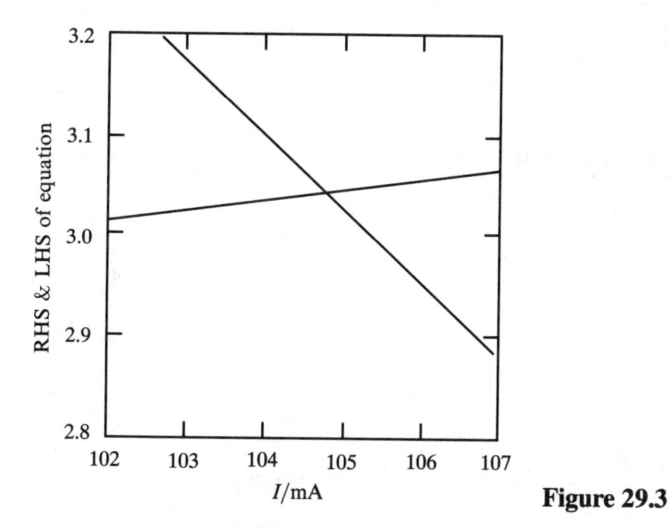

Figure 29.3

corresponds to the current at which maximum power is delivered. The power at this current is

$$P = (105\,\text{mA}) \times (1.10\,\text{V}) - (0.103\,\text{V}) \times (105\,\text{mA}) \times \ln\left(\frac{105}{5}\right) - (105\,\text{mA})^2 \times (3.7\bar{5}\,\Omega)$$

$$= 41\,\text{mW}$$

P29.8 Corrosion occurs by way of the reaction

$$Fe + 2H^+ \rightarrow Fe^{2+} + H_2$$

The half-reactions at the anode and cathode are

Anode: $Fe \rightarrow Fe^{2+} + 2e^-$

Cathode: $2H^+ + 2e^- \rightarrow H_2$

$\Delta\phi_{\text{corr}} = (-0.720\,\text{V}) + (0.2802\,\text{V}) = -0.440\,\text{V}$

$\Delta\phi_{\text{corr}} = \eta(\text{H}) + \Delta\phi_{\text{e}}(\text{H})$ [*Justification* 29.2]

$\Delta\phi_{\text{e}}(\text{H}) = (-0.0592\,\text{V}) \times \text{pH} = (-0.0592\,\text{V}) \times 3 = -0.177\bar{6}\,\text{V}$

$$\eta(\text{H}) = -\frac{1}{\alpha f}\ln\frac{j_{\text{corr}}}{j_0(\text{H})} \quad [27]$$

Then, $\Delta\phi_{\text{corr}} = -0.440\text{ V} = -\dfrac{1}{\alpha f}\ln\dfrac{j_{\text{corr}}}{j_0(\text{H})} - 0.177\overline{6}\text{ V}$

and $\ln\dfrac{j_{\text{corr}}}{j_0(\text{H})} = (0.262\text{ V}) \times \alpha f = (0.262\text{ V}) \times (18\text{ V}^{-1}) = 4.7\overline{1}\overline{6}$

$$j_{\text{corr}} = j_0(\text{H}) \times e^{4.71\overline{6}} = (1.0 \times 10^{-7}\text{ A cm}^{-2}) \times (112) = 1.1\overline{2} \times 10^{-5}\text{ A cm}^{-2}$$

Faraday's laws give the amount of iron corroded

$$n = \frac{I_{\text{corr}}t}{zF} = \frac{(1.1\overline{2} \times 10^{-5}\text{ A cm}^{-2}) \times (8.64 \times 10^4\text{ s d}^{-1})}{(2) \times (9.65 \times 10^4\text{ C mol}^{-1})} = 5.0 \times 10^{-6}\text{ mol cm}^{-2}\text{ d}^{-1}$$

$$m = n \times (55.85\text{ g mol}^{-1}) = (5.0 \times 10^{-6}\text{ mol cm}^{-2}\text{ d}^{-1}) \times (55.85 \times 10^3\text{ mg mol}^{-1})$$

$$= \boxed{0.28\text{ mg cm}^{-2}\text{ d}^{-1}}$$

Solutions to theoretical problems

P29.9

$$j = j_0\left\{e^{(1-\alpha)f\eta} - e^{-\alpha f\eta}\right\}\ [21]$$

$$= j_0\left\{1 + (1-\alpha)\eta f + \tfrac{1}{2}(1-\alpha)^2\eta^2 f^2 + \cdots - 1 + \alpha f\eta - \tfrac{1}{2}\alpha^2\eta^2 f^2 + \cdots\right\}$$

$$= j_0\left\{\eta f + \tfrac{1}{2}(\eta f)^2(1-2\alpha) + \cdots\right\}$$

$$\langle j\rangle = j_0\left\{\langle\eta\rangle f + \tfrac{1}{2}(1-2\alpha)f^2\langle n^2\rangle + \cdots\right\}$$

$$\langle\eta\rangle = 0 \quad \text{because} \quad \frac{\omega}{2\pi}\int_0^{2\pi/\omega}\cos\omega t\,dt = 0 \quad \left[\frac{2\pi}{\omega}\text{ is the period}\right]$$

$$\langle\eta^2\rangle = \frac{1}{2}\eta_0^2 \quad \text{because} \quad \frac{\omega}{2\pi}\int_0^{2\pi/\omega}\cos^2\omega t\,dt = \frac{1}{2}$$

Therefore, $\boxed{\langle j\rangle = \tfrac{1}{4}(1-2\alpha)f^2 j_0\eta_0^2}$

and $\langle j\rangle = 0$ when $\alpha = \tfrac{1}{2}$. For the mean current,

$$\langle I\rangle = \tfrac{1}{4}(1-2\alpha)f^2 j_0 S\eta_0{}^2$$

$$= \frac{1}{4} \times (1-0.76) \times \left(\frac{(7.90 \times 10^{-4}\text{ A cm}^{-2}) \times (1.0\text{ cm}^2)}{(0.0257\text{ V})^2}\right) \times (10\text{ mV})^2$$

$$= \boxed{7.2\ \mu\text{A}}$$

P29.11

$$j = \left(\frac{cFD}{\delta}\right) \times (1 - e^{f\eta^c})\ [39;\ z=1] = \boxed{j_L(1 - e^{F\eta^c/RT})}$$

The form of this expression is illustrated in Fig. 29.4.

For the anion current, the sign of η^c is changed, and the current of anions approaches its limiting value as η^c becomes more positive (Fig. 29.4).

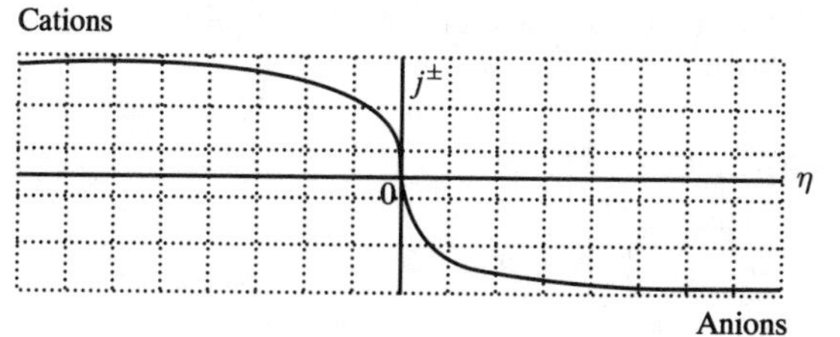

Figure 29.4

Solutions to additional problems

P29.12 $Fe^{2+} + 2e^- \rightarrow Fe \quad \nu = 2; \quad E^{\ominus} = -0.447\,V$

(a)
$$E_0 = E^{\ominus} - \frac{RT}{\nu F}\ln Q \ [10.45]$$
$$= E^{\ominus} - \frac{RT}{\nu F}\ln\frac{1}{[Fe^{2+}]} \text{ assuming } \gamma_{Fe^{2+}} = 1$$
$$= -0.447\,V - \frac{25.693\times10^{-3}\,V}{2}\ln\left(\frac{mol\,L^{-1}}{1.70\times10^{-6}\,mol\,L^{-1}}\right)$$

$$\boxed{E_0 = -0.618\,V}$$

$\eta = E' - E_0$ [18]

η values are reported in the following table.

(b)
$$j = \frac{\nu F}{A}\frac{dn_{Fe}}{dt} = \frac{2(96485\,C\,mol^{-1})}{9.1\,cm^2}\frac{dn_{Fe}}{dt}$$

j values are reported in the following table.

$$j = j_0\left(e^{(1-\alpha)f\eta} - e^{-\alpha f\eta}\right) = j_0 e^{-\alpha f\eta}\{e^{f\eta} - 1\}$$
$$= -j_c\{e^{f\eta} - 1\} \ [20, 21]$$
$$j_c = \frac{-j}{e^{f\eta} - 1}$$

j_c values are reported in the following table

$\frac{dn_{Fe}}{dt}\Big/(10^{-12}\,mol\,s^{-1})$	$-E'/mV$	$-\eta/mV$	$j/(\mu A\,cm^{-2})$	$j_c/(\mu A\,cm^{-2})$
1.47	702	84	0.0312	0.0324
2.18	727	109	0.0462	0.0469
3.11	752	134	0.0659	0.0663
7.26	812	194	0.154	0.154

(c) $j_c = j_0\,e^{-\alpha f\eta}$ [20]

$\ln j_c = \ln j_0 - \alpha f\eta$

Performing a linear regression analysis of the $\ln j_c$ versus η data, we find

$\ln j_0 = -4.608$, standard deviation $= 0.015$
$\alpha f = 0.01413\,mV$, standard deviation $= 0.00011$

$$\boxed{R = 0.999\,94}$$

The correlation coefficient and the standard deviation indicate that the plot provides an excellent description of the data.

$$j_0 = e^{-4.608} \quad \text{or} \quad \boxed{j_0 = 0.009\,97\ \mu\text{A cm}^{-2}}$$

$$\alpha = \frac{0.01413}{f} = (0.01413\,\text{mV}^{-1}) \times (25.693\,\text{mV})$$

$$\boxed{\alpha = 0.363}$$

P29.15 The accompanying Tafel plot (Fig. 29.5) of $\ln j$ against E shows no region of linearity so the Tafel equation cannot be used to determine j_0 and α.

$$\text{M}_{\text{sol}} \overset{K_1}{\rightleftharpoons} \text{M}_{\text{ads}}$$

$$\text{M}_{\text{ads}} + \text{H}^+ + \text{e}^- \overset{K_2}{\rightleftharpoons} \text{MH}_{\text{ads}}$$

$$2\text{MH}_{\text{ads}} \xrightarrow[\text{rate-determining}]{k_3} \text{HMMH}$$

Assuming that the dimerization is rate-determining, two electrons are transferred per molecule of HMMH and $z = 2$. It is also reasonable to suppose that the first two reactions are at quasi-equilibrium. According to reaction 3, the current density is proportional to the square of the functional surface coverage by MH_{ads}, θ_{MH}.

$$j = zFk_3\,\theta_{\text{MH}}^2$$

$$\ln j = \ln(zFk_3) + 2\ln\theta_{\text{MH}} \qquad (1)$$

The characteristics of this equation differ from those of the Tafel equation at high negative overpotentials

$$\ln j = \ln j_0 - \alpha f \eta \ [27]$$

at low concentrations of M the value of θ_{MH} changes with the overpotential in a non-exponential manner. This makes $\ln j$ non-linear throughout the potential range.

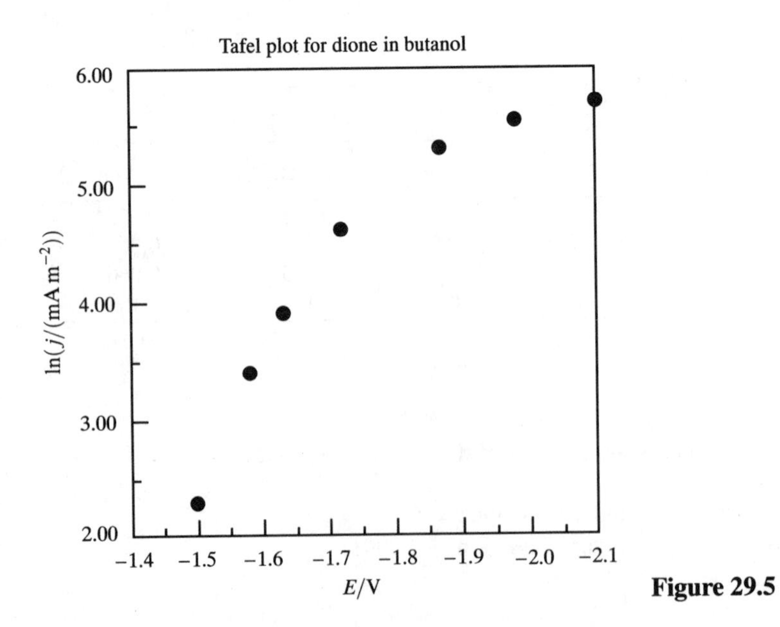

Figure 29.5

P29.16 At large positive values of the overpotential the current density is anodic.

$$j = j_0[e^{(1-\alpha)f\eta} - e^{-\alpha f\eta}]\ [21]$$
$$\approx j_0\, e^{(1-\alpha)f\eta} = j_a\ [20]$$
$$\ln j = \ln j_0 + (1-\alpha)f\eta$$

Performing a linear regression analysis of $\ln j$ against η, we find

$\ln(j_0/(\text{mA m}^{-2})) = -10.826$, standard deviation $= 0.287$
$(1-\alpha)f = 19.550\ \text{V}^{-1}$, standard deviation $= 0.355$

$\boxed{R = 0.999\,01}$

$$j_0 = e^{-10.826}\ \text{mA m}^{-2} = \boxed{2.00 \times 10^{-5}\ \text{mA m}^{-2}}$$

$$\alpha = 1 - \frac{19.550\ \text{V}^{-1}}{f} = 1 - \frac{19.550\ \text{V}^{-1}}{(0.025\,693\ \text{V})^{-1}}$$

$\boxed{\alpha = 0.498}$

The linear regression explains 99.90 per cent of the variation in a $\ln j$ against η plot and standard deviations are low. There are $\boxed{\text{no}}$ deviations from the Tafel equation/plot.